Collected Papers of Hans Rademacher

Volume II

Mathematicians of Our Time

Gian-Carlo Rota, series editor

1. Charles Loewner: Theory of Continuous Groups
   notes by Harley Flanders and Murray H. Protter

2. Oscar Zariski: Collected Papers
   Volume I
   Foundations of Algebraic Geometry and Resolution of Singularities
   edited by H. Hironaka and D. Mumford

3. Collected Papers of Hans Rademacher
   Volume I
   edited by Emil Grosswald

4. Collected Papers of Hans Rademacher
   Volume II
   edited by Emil Grosswald

The MIT Press
Cambridge, Massachusetts, and London, England

Collected Papers of Hans Rademacher

Volume II

Edited by Emil Grosswald

This book was set in IBM Baskerville
by Williams Graphic Service
printed and bound by Semline Inc.
in the United States of America.

Library of Congress Cataloging in Publication Data

Rademacher, Hans 1892-1969.
Collected papers of Hans Rademacher.

(Mathematicians of our time, v. 4)
Bibliography: p.
1. Mathematics–Collected works. I. Grosswald, Emil, ed. II. Series.
QA3.R27 510′.8 72-1747
ISBN 0-262-07055-3

Contents and Bibliography

Papers (continued)

Abstracts

Problems and Solutions

Preface

Professor Gian-Carlo Rota took the initiative to assemble the papers of Hans Rademacher for publication in a reprint collection. He asked me to act as editor, I accepted, and here is the result.

In the two volumes of these *Collected Papers* one will find all the papers published by Rademacher, either alone or as joint author, essentially in chronological order. The published abstracts follow under a separate (still chronological) numbering. For completeness, several problems proposed and/or solved by Rademacher are also reprinted here.

The books written by Rademacher are not reproduced in this collection, nor is his single paper in Hebrew (On the necessity of a compass in elementary geometric constructions. *Riveon Lematematika,* vol. 1 (1964), pp. 14-19), which is, essentially, a translation of chapter 26 of his book with O. Toeplitz, *The Enjoyment of Mathematics.* These works are listed, however, in appendix 1 of volume II.

Comments and corrections are assembled in the "notes" at the end of each paper. For each paper, the first note gives the reviews of that paper in the *Jahrbuch über die Fortschritte der Mathematik* (JF), the *Zentralblatt* (Z), and the *Mathematical Reviews* (MR). Notes after the first are preceded by the original page and line numbers of the text to which they refer. The reader will note that the page numbers given in the notes refer to the original pagination of the articles, not to the pagination of the present work. Also, the line counts refer to lines of text and internal subheads (section heads, theorem heads, etc.) only; displayed equations are counted separately.

Appendix 2 of this collection is a list of the mathematicians whose dissertations were directed by Rademacher. If the list is complete (and I hope it is), this is due to a large extent to the diligence of Professor P. T. Bateman of the University of Illinois. His help in identifying Rademacher's pre-1934 students in Germany was invaluable.

Without the continuous support of Gian-Carlo Rota it would have been impossible for me to complete, or even start, this venture.

Each paper had to be read very carefully. In this task I have been helped substantially by Dr. Günter Köhler of the University of Freiburg, to whom I herewith express my thanks. During the preparation of this collection I had the support of the secretarial staff of Temple University, in particular that of Mrs. G. Ballard, who did most of the necessary typing. I avail myself of this opportunity to express my gratitude to her, to the secretarial staff, and to Temple University.

Finally, I have had the wholehearted assistance of several members of the Rademacher family. In particular, Mrs. Irma Wolpe-Rademacher, Mrs. Karin Loewy, and Mrs. Suzanne Gaspari-Rademacher contributed valuable information and personal recollections about the great man and scholar who was Hans Rademacher.

Emil Grosswald
Philadelphia, February 1972

Hans Rademacher

Biographical Sketch

Hans Rademacher was born in 1892 and he died in 1969. Two world wars occurred during his lifespan and he did not remain unaffected by the upheavals of the times. His personal life had also its own elements of drama. And yet it seems that war and emigration, marriage and divorce, and even professorships and honors were not much more than accidents in his life and that his true biography consists in the evolution of his creative spirit. In other words, the real history of his life is the history of his work. As no attempt could be made to present a complete critical analysis and appraisal of Professor Rademacher's work in this short sketch, we must hope that this highly desirable task will be undertaken soon by more competent hands.

Rademacher was born on April 3, 1892, in Wandsbek, near Hamburg (Germany). His father, Henry, had a store; his mother, Emma, kept house and helped in the store. There were two other children in the house, a brother Martin and a much younger sister Erna.

His childhood seems to have passed rather uneventfully. He went to elementary and secondary schools near his home and in Hamburg.

In 1910, at the age of 18, he matriculated at the University of Göttingen. At that time, as was to be the case throughout his whole life, his interests were broad, encompassing not only mathematics, but also foreign languages and cultures, the natural sciences and philosophy. It is, therefore, not too surprising that, being somewhat disappointed by the lectures of Felix Klein and having come under the influence of Leonhard Nelson, he felt strongly inclined to dedicate himself entirely to the study of philosophy. He actually seems to have abandoned mathematics completely for one or two semesters. In later years, he attributed his return to mathematics to the influence of Richard Courant.

He now continued his studies under Carathéodory and started work on a dissertation. However, the year was 1914 and the world exploded into World War I. Hans Rademacher became a soldier. The interruption of his work—if there ever was any—did not last long and he finished the writing of his dissertation in 1916. The dissertation was accepted and Rademacher obtained his Ph.D. from Göttingen in 1917.

The dissertation (paper 1 of the present collection), entitled "Eineindeutige Abbildungen und Messbarkeit," discusses some delicate points of differentiation and integration of real-valued functions of real variables. It should be mentioned that similar problems were being considered at about the same time in France (A. Denjoy) and England (G. C. Young). Rapid communication among the mathematicians belonging to the belligerent countries was not possible, and several discoveries seem to have been duplicated. It does not appear possible at this time (and is, anyhow, of no particular importance) to establish with certainty the priority of some of the results of that dissertation.

This imposing work of over 100 printed pages was soon followed by several others (one, paper 2, written jointly with Carathéodory) on real variables, measurability, differentiability, convergence factors and Euler summability of series. It seems that the term "total differentiability" now in common use appeared for the first time in a long paper in two parts (see papers 4, 7) published in 1919 and 1920. This phase of Rademacher's work culminated with paper 13, "Einige Sätze über Reihen von allgemeinen Orthogonalfunktionen," published in 1922. Here he introduced the systems of orthogonal functions now generally known as the Rademacher functions. Fortuitous circumstances delayed the publication of a second paper, containing the completion of the system of Rademacher functions. In the meantime, Walsh also obtained and published the completion of the system of Rademacher functions, whereupon Rademacher decided not to publish his own manuscript. The few persons who had the opportunity to read it considered it to be extremely beautiful, so that it is most unfortunate that it appears to have been lost.

After he left the army, Rademacher had been first a teacher in Wickersdorf, then a Privatdozent (assistant professor) in Berlin under Carathéodory (1919-1922). In the spring of 1922 he was called to the recently created University of Hamburg, as Ausserordentlicher Professor (associate professor) and, in April, 1925, to Breslau as Ordentlicher (full) Professor.

During his stay in Berlin (April, 1921) Rademacher had married Suzanne Gaspary, and in 1925 his daughter Karin was born. The marriage ended in divorce in 1929.

A new period in Rademacher's career started with the publication in 1923 of paper 14, "Beiträge zur Viggo Brunschen Methode in der Zahlentheorie." From here on his major effort was to be devoted to the theory of functions of a complex variable and number theory. Some of his most important contributions to mathematics were made in these fields. He perfected the sieve method and used it skillfully in the study of algebraic number fields (see papers 15, 19). He studied the additive prime number theory of these fields (papers 17, 18, 20). He generalized Goldbach's problem (paper 19) and started working on three topics that were to play a great role in his future work: the theory of the Riemann zeta function (paper 24), that of modular functions (papers 25, 26, 27), and the Dedekind sums (papers 26-29, 31). To this period belongs also the publication (see problems 2-4) of what has become known as the Rademacher-Brauer formula and that of his book with Toeplitz, *Von Zahlen und Figuren*. Finally, in 1934 appeared also *Vorlesungen über die Theorie der Polyeder, unter Einschluss der Elemente der Topologie,* based on posthumous notes of E. Steinitz.

During that same year 1934, shortly after Hitler's advent to power, Rademacher was forcibly removed from his professorship. Indeed, he not only had joined the International League for the Rights of Men, but had also become Chairman of the local (Breslau) chapter of the Deutsche Friedensgesellschaft (German Society for Peace). He retired with his daughter Karin to a small town on the Baltic coast. There he met and married Olga Frey.

The same year he came to the United States, being invited by the University of Pennsylvania in Philadelphia as a visiting Rockefeller fellow. During the following summer he returned briefly to Germany but then in the fall of 1935 he reentered the United States, seeking permanent residence and establishing himself in Swarthmore. In 1935 his son Peter was born.

At the expiration of his Rockefeller Fellowship, the University of Pennsylvania gave him the option to stay on. Yet, in spite of the fact that more than 10 years earlier he had been a full professor in Germany and that he had achieved worldwide recognition as a great mathematician, nothing better than an assistant professorship was offered to him. He accepted this "demotion" and never left the University of Pennsylvania until his retirement, although he had offers of much better conditions from other universities.

Among his reasons to stay on under often adverse circumstances seems to have been his loyalty to the University of Pennsylvania, the institution that had made it possible for him to emigrate from Germany in 1934. He may also have been reluctant to change once more the pattern of his life and to leave Swarthmore, which he liked. Indeed, he had found congenial friends among his colleagues at the university as well as among his neighbours in Swarthmore. He shared many mathematical interests with Professors A. Zygmund and I. J. Schoenberg of the University of Pennsylvania, and had also become very friendly with Professor A. Dresden, Chairman of the Department of Mathematics at Swarthmore College.

Rademacher and Schoenberg had met previously in Germany, where Schoenberg had spent several years with E. Landau. Now Schoenberg also lived in Swarthmore and soon a close friendship developed between these two outstanding mathematicians. This led to close personal relations as well as to professional cooperation. They often shared the responsibility for the Problems Seminar created by Rademacher at the University of Pennsylvania and that became justly famous from the 1930s to the 1950s; they also published some joint papers (49, 52).

It is during these years that Rademacher obtained some of his most important results in connection with the Fourier coefficients of modular forms of positive dimensions. His general method may be considered as a modification and improvement of the Hardy-Ramanujan-Littlewood circle method. As particular instances he obtained his famous formula for the partition number $p(n)$ (see papers 37, 38, 48), as well as the coefficients of the modular invariant $J(\tau)$ (papers 39, 41). In 1939, at the annual meeting of the American Mathematical Society, he gave a nontechnical, comprehensive view of much of his recent work in an invited address on "Fourier expansions of modular forms and problems of partitions" (paper 43). He also continued his study of Dedekind sums (paper 44, with A. Whiteman), general number theory (papers 42, with H. S. Zuckerman, 36, 45) and modular functions (paper 40, with H. S. Zuckerman). His work in general analysis during this period is represented by "Some remarks on F. John's Identity" (paper 35), "On the Bloch-Landau constant" (paper 47), "An iteration method for calculation with Laurent series" and "Helly's theorems on convex domains and Tchebycheff's approximation problem" (papers 49 and 52, respectively, both joint work with I. J. Schoenberg) and "On a theorem of Frobenius" (paper 51).

In spite of these professional successes he had to live through another disappointment at the University of Pennsylvania, when he was passed over at the time of the next promotions. In those years, the length of faithful service to the institution, and not professional excellence, was the main criterion for promotions—a fact that was forcefully explained to the somewhat surprised assistant professor by a most self-assured dean. Fortunately for the good name of the University of Pennsylvania, not everybody there was of that opinion and at the next following promotion, Rademacher became again a full professor. Unfortunately, Rademacher was less successful in his personal life than in his professional one and his second marriage also ended in divorce in 1947. About that time he also succeeded in his efforts to help his daughter Karin, who had spent the war years with her mother in Germany, to come and join him in the United States. He left Swarthmore and came to live in Philadelphia, near the campus of the University of Pennsylvania.

By now Rademacher had become a recognized authority in number theory. Several of his numerous former students had become themselves successful mathematicians. During his stay in Swarthmore, Rademacher had been introduced by Professor Dresden to the Society of Friends, whose ideas and ideals he found congenial. He joined the Society and stayed a member of the Quaker community to the end of his life. He also became one of the founding members of the Research Club of the University of Pennsylvania.

In 1949, Rademacher married Irma Wolpe, the sister of his old friend Schoenberg. This happy union brightened the last twenty years of Rademacher's life. Irma Wolpe-Rademacher was not only a very successful concert pianist and piano teacher, but also a most gracious hostess. The home of the Rademachers became a meeting point not only of music lovers and of mathematicians, but also of many other interesting personalities living in, or passing through Philadelphia. Many of the graduate students in mathematics as well as many of the piano students of Mrs. Wolpe-Rademacher had unforgettable opportunities to meet some of the most brilliant intellects in this stimulating setting. The informal but lively receptions in the Rademacher home on the afternoons of New Years' days soon became a tradition in Philadelphia.

During these years the mathematical productivity of Rademacher continued at a high level, being, perhaps, stimulated by a few long-overdue sabbatical leaves. He was a Philips Lecturer at Haverford College (1952). He spent the spring semester of 1952-1953 at the Institute for Advanced Study in Princeton (N.J.), lectured during the summer of 1959 at the University of Oregon in Eugene (Oregon) and in 1954-1955 at the Tata Institute in Bombay (India) and at his own alma mater in Göttingen. He visited the West Coast (The University of California at Berkeley and Los Angeles), was once more Philips Lecturer at Haverford College (1959-1960) and again spent one year (1960-1961) at the Institute for Advanced Study in Princeton. During these years, his main interests focused upon Dedekind sums (papers 53, 56, 57, 59, 67), modular functions (papers 58, 61, 65), general (especially harmonic) analysis (papers 55, 60, 62, 62b, 66) and zeta functions (papers 63, 64). His lectures in analytic number theory given at the Tata Institute were published in book form.

The year after Rademacher's return from Princeton to the campus of the University of Pennsylvania had to be his last at that institution; indeed, he was to reach the compulsory retirement age of seventy on April 3, 1962. During the academic year 1961-1962, a year-long Institute in the Theory of Numbers was organized in honor of Rademacher. Among the outside participants were Professors P. T. Bateman (University of Illinois), D. G. Cantor (Princeton University), S. Chowla (University of Colorado and Pennsylvania State University), J. Lehner (University of Maryland), C. Pisot (Paris, France), E. G. Straus (UCLA), and many others.

On his seventieth birthday, Radamacher was feted at a banquet attended by many university officials, collaborators, friends, and students; a few weeks later, at the commencement exercises, the institution he had served so faithfully awarded him the highest distinction it could offer, the Doctor Honoris Causa.

Although Rademacher had always been rather frail and his health was no more the most robust one, retirement did not mean idleness for him. First, Rademacher taught for two years (1962-1964) at New York University; afterwards he became affiliated with Rockefeller University in New York City, where he worked from 1964 to the end of his life.

In 1963 Rademacher was selected as Hedrick lecturer by the Mathematical Association of America. He chose as his topic the Dedekind sums, on which he had worked so often during his life. The lectures were to be delivered in August, at the joint summer meeting of the American Mathematical Society and the Mathematical Association of America, in Boulder, Colorado. Rademacher arrived in Boulder in excellent spirits. He took part at an Institute in Number Theory that preceded the summer meeting and actually presented at a seminar some recent results on generalized Dedekind sums (paper 67). However, possibly on account of the high altitude, he began to feel tired and sick and had to leave Boulder quite suddenly. The lectures were delivered by a former student, who followed closely the carefully prepared lecture notes.

During the following years, Rademacher published six more papers (68-73) on a diversity of topics. He also spent a considerable amount of his dwindling energies on the completion of a book on analytic number theory, started many years before. All but one chapter was finished. The missing chapter was to be written, based on lectures Rademacher had planned to give in 1967-1968 at Rockefeller University. These lectures never took place. During the night of September 27-28, 1967, Rademacher was stricken by a massive cerebral hemorrhage that partially paralyzed him and impaired his capability to speak.

In spite of the best available care and the devotion of his wife Irma and daughter Karin, his condition continued to deteriorate. We shall never know what Rademacher may have suffered from his inability to communicate, while his mental lucidity seems to have remained unimpaired.

On February 7, 1969, his torment of 17 months mercifully came to an end. A memorial service was held on February 23, 1969, at the Haverford Friends Meeting House.

His friends and students will never forget him. As for the world at large, his work has become a permanent part of mathematics. The purpose of the present collection of his published papers is to make it more easily accessible.

Papers (continued)

BESTIMMUNG EINER GEWISSEN EINHEITSWURZEL IN DER THEORIE DER MODULFUNKTIONEN

Hans Rademacher*.

1. In der Abhandlung von G. H. Hardy und S. Ramanujan "Asymptotic formulae in combinatory analysis"†, durch welche die additive Zahlentheorie um eine berühmt gewordene analytische Methode bereichert worden ist, wird von der Funktion

$$(1) \qquad f(x) = \frac{1}{(1-x)(1-x^2)(1-x^3)\dots}$$

und ihrer Transformationsformel

$$(2) \; f\left\{\exp\left(2\pi i\frac{iz+h}{k}\right)\right\} = \omega_{h,k}\sqrt{z}\exp\left\{\frac{\pi}{12k}\left(\frac{1}{z}-z\right)\right\} f\left\{\exp\left(2\pi i\frac{(i/z)+h'}{k}\right)\right\}$$

Gebrauch gemacht. Hierin ist $(h, k) = 1$, $hh' \equiv -1 \pmod k$ und $\omega_{h,k}$ eine gewisse Einheitswurzel, die in der Theorie der Modulfunktionen eine Rolle spielt. Die Bestimmung von $\omega_{h,k}$ in ihrer Abhängigkeit von h und k ist ziemlich umständlich und geschieht seit Hermite‡ mit Hilfe der Theorie der ϑ-Funktionen, wobei sich herausstellt, dass $\omega_{h,k}$ mit dem Legendre-Jacobi'schen Restsymbol zusammenhängt.

Anderseits gilt nun für die Funktion

$$g(x) = \log f(x)$$

die Transformationsformel

$$(3) \quad g\left\{\exp\left(2\pi i\frac{iz+h}{k}\right)\right\}$$
$$= g\left\{\exp\left(2\pi i\frac{(i/z)+h'}{k}\right)\right\} + \frac{\pi}{12k}\left(\frac{1}{z}-z\right) + \tfrac{1}{2}\log z + \pi i\, s(h,k),$$

mit

$$(4) \qquad s(h,k) = \sum_{\mu=1}^{k-1}\frac{\mu}{k}\left(\frac{h\mu}{k} - \left[\frac{h\mu}{k}\right] - \tfrac{1}{2}\right).$$

---

* Received 3 November, 1931; read 12 November, 1931.

† *Proc. London Math. Soc.* (2), 17 (1918), 75-115; auch *Collected Papers of S. Ramanujan*, 276–309.

‡ "Sur quelques formules relatives à la transformation des fonctions elliptiques", *Journal de Math.* (2), 3 (1858), 26-36; ferner H. Weber, *Elliptische Funktionen* (1891), 98–102; J. Tannery et J. Molk, *Fonctions elliptiques*, 2 (1896), 90–113; Hardy and Ramanujan, *loc. cit.*, § 4.3.

Diese Formel findet sich im wesentlichen bei Dedekind in seinen Erläuterungen zu einem Riemannschen Fragment über die elliptischen Modulfunktionen*. In einer demnächst erscheinenden Arbeit habe ich für die Transformationsformel (3) einen direkten, von der Theorie der ϑ-Funktionen völlig unabhängigen Beweis gegeben†. Selbstverständlich folgt (2) sogleich aus (3), und im besonderen ergibt sich

$$\omega_{h,k} = \exp\{\pi i\, s(h, k)\}. \tag{5}$$

Dies ist aber noch nicht die von Hardy und Ramanujan gebrauchte, letzten Endes auf Hermite zurückgehende Bestimmung von $\omega_{h,k}$. Es dürfte sich aber wegen der Wichtigkeit von $\omega_{h,k}$ lohnen, aus (5) jene Bestimmung von $\omega_{h,k}$ herzuleiten, zu der man auf diese Weise einen neuen Zugang gewinnt.

2. Bei der Ausführung der Summe $s(h, k)$ in (4) stöszt man auf triviale Summen und auf die Summe

$$t(h, k) = \sum_{\mu=1}^{k-1} \mu \left[\frac{h\mu}{k}\right], \tag{6}$$

die vor allem zu untersuchen ist. Zunächst sei k *ungerade*. Dann ist

$$t(h, k) = \sum_{\mu=1}^{k-1} \mu \left[\frac{h\mu}{k}\right] \equiv \sum_{\lambda=1}^{\frac{1}{2}(k-1)} \left[\frac{(2\lambda-1)h}{k}\right] \pmod 2.$$

Nun ist

$$\sum_{\lambda=1}^{\frac{1}{2}(k-1)} \left[\frac{(2\lambda-1)h}{k}\right] + \sum_{\lambda=1}^{\frac{1}{2}(k-1)} \left[\frac{2\lambda h}{k}\right] = \sum_{\mu=1}^{k-1} \left[\frac{\mu h}{k}\right] = \frac{(h-1)(k-1)}{2},$$

wo die letzte Gleichung aus der Abzählung der Gitterpunkte *im Innern* desjenigen rechtwinkligen Dreiecks folgt, dessen Eckpunkte die kartesischen Koordinaten (0, 0), (0, k) und (h, k) haben. Daher ergibt sich zunächst

$$t(h, k) \equiv \frac{(h-1)(k-1)}{2} - \sum_{\lambda=1}^{\frac{1}{2}(k-1)} \left[\frac{2\lambda h}{k}\right] \pmod 2. \tag{7}$$

Anderseits habe ich bewiesen‡

$$ht(h, k) + kt(k, h) = \tfrac{1}{12}(h-1)(k-1)(8hk-h-k-1). \tag{8}$$

---

* Riemanns *Werke* (1876), 438–447.

† H. Rademacher, "Zur Theorie der Modulfunktionen". *Journal für Math.*, 167 (1932), 312–336, § 1.

‡ *Loc. cit.*, Gleichung (2.22), ferner H. Rademacher, "Eine arithmetische Summenformel", *Monatshefte für Math. und Phys.*, 39, erscheint demnächst.

Führen wir noch h' als Lösung der Kongruenz

(9) $$hh' \equiv -1 \pmod{k}$$

ein, so folgt aus (8)

$$t(h, k) \equiv -\tfrac{1}{12}h'(h-1)(k-1)(8hk-h-k-1) \pmod{k}$$

und, wegen $k+1 \equiv 1 \pmod{k}$

$$t(h, k) \equiv -\tfrac{1}{12}h'(h-1)(k^2-1)(8hk-h-k-1),$$

(10) $$t(h, k) \equiv -\tfrac{1}{12}h'k(h-1)(k^2-1)(8h-1)+\tfrac{1}{12}h'(h^2-1)(k^2-1) \pmod{k}.$$

Nun ist

(11) $$\tfrac{1}{12}(h-1)(8h-1)(k^2-1) \equiv 0 \pmod{2},$$

denn erstens ist wegen $k \equiv 1 \pmod{2}$

$$k^2-1 \equiv 0 \pmod{8},$$

und zweitens ist

$$(h-1)(8h-1)(k^2-1) \equiv 0 \pmod{3}.$$

Dies kommt so zustande: entweder ist $3 \nmid k$, dann also $3 \mid (k^2-1)$; oder es ist $3 \mid k$, also $3 \nmid h$, da $(h, k) = 1$. Dann ist aber

$$(h-1)(8h-1) \equiv -(h-1)(h+1) \equiv 0 \pmod{3}.$$

Damit ist (11) bewiesen, und statt (10) kann man schreiben

(12) $$t(h, k) \equiv \tfrac{1}{12}h'(h^2-1)(k^2-1) \pmod{k}.$$

Aus (7) und (12) zusammen ergibt sich

(13) $$t(h, k) \equiv -k \sum_{\lambda=1}^{\frac{1}{2}(k-1)} \left[\frac{2\lambda h}{k}\right] + \tfrac{1}{2}k(h-1)(k-1) + \tfrac{1}{12}h'(h^2-1)(k^2-1) \pmod{2k},$$

wobei $(h, k) = 1$ und

$$\tfrac{1}{12}(h^2-1)(k^2-1) \equiv 0 \pmod{2}$$

benutzt worden ist; letzteres folgt ähnlich wie (11). Für (13) schreiben wir

$$\frac{1}{k}\, t(h, k) \equiv -\sum_{\lambda=1}^{\frac{1}{2}(k-1)} \left[\frac{2\lambda h}{k}\right] + \tfrac{1}{2}(h-1)(k-1) + \tfrac{1}{12}h'(h^2-1)\left(k-\frac{1}{k}\right) \pmod{2}.$$

Nun ist nach (4)

$$s(h,k)=\sum_{\mu=1}^{k-1}\frac{h\mu^2}{k^2}-\sum_{\mu=1}^{k-1}\frac{\mu}{k}\left[\frac{h\mu}{k}\right]-\tfrac{1}{2}\sum_{\mu=1}^{k-1}\frac{\mu}{k}$$

$$=\tfrac{1}{6}h(k-1)\left(2-\frac{1}{k}\right)-\frac{1}{k}\,t(h,k)-\tfrac{1}{4}(k-1),$$

also

$$s(h,k)\equiv\sum_{\lambda=1}^{\frac{1}{2}(k-1)}\left[\frac{2\lambda h}{k}\right]-\tfrac{1}{2}(h-1)(k-1)-\tfrac{1}{12}h'(h^2-1)\left(k-\frac{1}{k}\right)$$

$$+\tfrac{1}{2}h(k-1)-\tfrac{1}{6}h(k-1)\left(1+\frac{1}{k}\right)-\tfrac{1}{4}(k-1)\pmod 2,$$

(14) $$s(h,k)\equiv\sum_{\lambda=1}^{\frac{1}{2}(k-1)}\left[\frac{2\lambda h}{k}\right]+\tfrac{1}{4}(k-1)-\tfrac{1}{12}\left(k-\frac{1}{k}\right)(2h+h'h^2-h')\pmod 2.$$

Bei *ungeradem* $k>0$ und *beliebigem* zu k teilerfremden h gilt nun

(15) $$(-1)^{\sum_{\lambda=1}^{\frac{1}{2}(k-1)}\left[\frac{2\lambda h}{k}\right]}=\left(\frac{h}{k}\right)$$

für das Legendre-Jacobi'sche Symbol (h/k)*. Aus (5), (14) und (15) ergibt sich, wenn man noch die sofort aus (15) folgende Gleichung

$$\left(\frac{-1}{k}\right)=(-1)^{\frac{1}{2}(k-1)}$$

berücksichtigt,

(16) $$\omega_{h,k}=\left(\frac{-h}{k}\right)\exp\left[-\pi i\left\{\tfrac{1}{4}(k-1)+\tfrac{1}{12}\left(k-\frac{1}{k}\right)(2h+h'h^2-h')\right\}\right],$$

die zu beweisende Formel für ungerades k†.

3. Ist h ungerade, so folgt aus (14) durch Vertauschung von h und k

(17) $$s(k,h)\equiv\sum_{\lambda=1}^{\frac{1}{2}(h-1)}\left[\frac{2\lambda k}{h}\right]+\tfrac{1}{4}(h-1)-\tfrac{1}{12}\left(h-\frac{1}{h}\right)(2k+k'k^2-k')\pmod 2,$$

* Diese einfache Formel für das Jacobi'sche Restsymbol habe ich nirgends explizit gefunden. Sie folgt aber aus der Scheringschen Verallgemeinerung des Gauszschen Lemmas [E. Schering, "Zur Theorie der quadratischen Reste", *Acta Math.*, 1 (1882), 153–170, insbesondere 166, oder P. Bachmann, *Die Elemente der Zahlentheorie* (Leipzig, 1892), 148] ebenso, wie nach einem Schlusse von Kraitchik [*Théorie des nombres* (Paris, 1922), 66, 67] die Formel (15) für das Legendresche Symbol, d.h. für primzahliges k aus dem Gauszschen Lemma selbst folgt.

† Hardy and Ramanujan, *loc. cit.*, Gleichung (1.722).

wo k' die Bedingung

$$(18) \qquad kk' \equiv -1 \pmod{h}$$

erfüllt. Für die Summen $s(h, k)$ gilt nun die Reziprozitätsformel

$$s(h, k)+s(k, h) = -\tfrac{1}{4}+\tfrac{1}{12}\left(\frac{k}{h}+\frac{h}{k}+\frac{1}{hk}\right),$$

wie ich im Zusammenhang mit (8) bewiesen habe*. Dadurch wird aus (17)

$$(19) \quad s(h, k) \equiv -\sum_{\lambda=1}^{\frac{1}{2}(h-1)} \left[\frac{2\lambda k}{h}\right] - \tfrac{1}{4}(h-1) + \tfrac{1}{6}\left(h-\frac{1}{h}\right)k + \tfrac{1}{12}\left(h-\frac{1}{h}\right)k'(k^2-1)$$
$$-\tfrac{1}{4}+\tfrac{1}{12}\left(\frac{k}{h}+\frac{h}{k}+\frac{1}{hk}\right) \pmod{2}.$$

Wegen (9) und (18) gilt

$$\tfrac{1}{12}(h^2-1)(k^2-1)\,kk' + \tfrac{1}{12}(h^2-1)(k^2-1)\,hh' \equiv -\tfrac{1}{12}(h^2-1)(k^2-1)$$
$$\pmod{h} \text{ und } \pmod{2k},$$

da $\frac{1}{12}(h^2-1)(k^2-1)$ ganz und sogar *gerade* ist. Da aber $(h, 2k) = 1$, so haben wir

$$\tfrac{1}{12}(h^2-1)(k^2-1)\,kk' + \tfrac{1}{12}(h^2-1)(k^2-1)\,hh' \equiv -\tfrac{1}{12}(h^2-1)(k^2-1)$$
$$\pmod{2hk}$$

oder

$$\tfrac{1}{12}\left(h-\frac{1}{h}\right)(k^2-1)\,k' + \tfrac{1}{12}(h^2-1)\left(k-\frac{1}{k}\right)h' \equiv -\tfrac{1}{12}\left(h-\frac{1}{h}\right)\left(k-\frac{1}{k}\right)$$
$$\pmod{2}.$$

Hiermit lässt sich k' aus (19) eliminieren:

$$s(h, k) \equiv -\sum_{\lambda=1}^{\frac{1}{2}(h-1)} \left[\frac{2\lambda k}{h}\right] - \tfrac{1}{4}h + \tfrac{1}{6}\left(h-\frac{1}{h}\right)k - \tfrac{1}{12}(h^2-1)\left(k-\frac{1}{k}\right)h'$$
$$-\tfrac{1}{12}\left(h-\frac{1}{h}\right)\left(k-\frac{1}{k}\right) + \tfrac{1}{12}\left(\frac{k}{h}+\frac{h}{k}+\frac{1}{hk}\right) \pmod{2}.$$

---

* H. Rademacher, *loc. cit.*, Gleichung (2.21).

Nach einigen einfachen Rechnungen findet man

$$s(h,k) \equiv -\sum_{\lambda=1}^{\frac{1}{2}(h-1)} \left[\frac{2\lambda k}{h}\right] - \tfrac{1}{2}(h-1) + \tfrac{1}{4}(h+hk-2) - \tfrac{1}{12}\left(k-\frac{1}{k}\right)(2h+h^2h'-h') \pmod 2,$$

und da

$$(-1)^{\sum_{\lambda=1}^{\frac{1}{2}(h-1)}[2\lambda k/h]} = \left(\frac{k}{h}\right), \quad (-1)^{\frac{1}{2}(h-1)} = \left(\frac{-1}{h}\right)$$

ist, erhält man schliesslich nach (5)

$$(20) \quad \omega_{h,k} = \left(\frac{-k}{h}\right) \exp\left[-\pi i \left\{\tfrac{1}{4}(2-h-hk) + \tfrac{1}{12}\left(k-\frac{1}{k}\right)(2h+h^2h'-h')\right\}\right],$$

was für ungerades h bewiesen werden sollte*.

Sind h und k *beide* ungerade, so zeigt unser Beweis, dass (16) und (20) zugleich gelten.

* Hardy and Ramanujan, *loc cit.*, Gleichung (1.721).

Notes to Paper 29

This paper has been reviewed in the JF, vol. 58 (1932), p. 397, and the Z, vol. 3 (1932), p. 357.

Page 15, footnote †; page 18, footnote.
See paper 27 of the present collection.

Page 15, footnote ‡.
See paper 28.

Schlesier
des
17. bis 19. Jahrhunderts

55 Lebensbeschreibungen
hervorragender Schlesier aller Berufe und Stände
Preis in Ganzleinen gebunden 9 Mark

Wilhelm Foerster

Wilhelm Julius Foerster wurde am 16. Dezember 1832 zu Grünberg in Schlesien geboren. Seine Eltern, der Tuchfabrikant Wilhelm Foerster und dessen Frau, geborene Seydel, die aus einer alteingesessenen Weinbauerfamilie stammte, lebten in wohlhabenden Verhältnissen. Die

343

geistigen Interessen des Vaters, die sich vor allem auf Schulreform und freireligiöse Bewegung richteten, bürgten für eine liberale Atmosphäre im Foersterschen Hause, die für die Entwicklung des jungen Foerster bestimmend wurde.

Er besuchte zunächst die Elementarschule in seinem Heimatort und siedelte 1847 nach Breslau über, wo er in das Magdalenengymnasium eintrat. In dieser Schule herrschte zwischen Lehrern und Schülern der Geist arbeitsfreudiger Gemeinschaft, und Foerster gedenkt in seinen Lebenserinnerungen dankbar des menschlichen und sachlichen Gewinns, den er aus dem Umgang mit seinen Lehrern gezogen hat. Er erwähnt vor allem den Mathematiklehrer Sadebeck, dem die Anwendungen der Mathematik sehr am Herzen lagen, und den literarisch feingebildeten und menschlich weitblickenden Direktor Schönborn. Die Revolution im Jahre 1848 beeinflußte das Schulleben naturgemäß sehr stark. Auch in der Klasse des jungen Foerster herrschte revolutionärer Geist, und in erregten Schülerversammlungen wurde über eine Reform der Reifeprüfungsordnung diskutiert. Sehr lebendig erzählt er ein gefährliches Erlebnis dieser Zeit: er ging über eine Straße und mußte über eine verlassene Barrikade turnen, worauf aus einer Nebenstraße eine Gewehrsalve auf ihn abgefeuert wurde. Er entging nur knapp dem versehentlichen Tode als Barrikadenheld.

Schon in der Schulzeit fesselte ihn die Astronomie. Angeregt durch Alexander von Humboldts Werke, begann er das Studium astronomischer Bücher und stellte mit einem Fernrohr Himmelsbeobachtungen an. Die endgültige Entscheidung über den zu wählenden Beruf konnte er jedoch erst nach einigem Schwanken fällen, da seine Vorliebe zur Musik eine ernstliche Konkurrenz für seine Neigung zur Wissenschaft bedeutete. Er wählte die Wissenschaft und begann im Jahre 1850 an der Universität Berlin unter Encke das Studium der Astronomie und Mathematik. 1852 bis 1854 wurde er in Bonn unter Argelanders Leitung in der praktischen Astronomie ausgebildet und wandte sich der Beobachtung der kleinen Planeten zu. Er promovierte im Jahre 1854 an der Bonner Sternwarte mit einer Arbeit über die Bonner Polhöhe.

Als 1855 die Stelle des zweiten Assistenten bei Encke frei wurde, trat er dort ein und arbeitete mit dem Instrument, mit dem Galle den Planeten Neptun entdeckt hat und mit dem er selbst später den kleinen Planeten Erato fand. Inzwischen erweiterte er seinen Gesichtskreis durch große Reisen nach England, Schottland, Italien, Frankreich, Österreich und der Schweiz. Im Alter von 26 Jahren habilitierte er sich für das Fach der Astronomie in Berlin und hielt zunächst Vorlesungen über die Geschichte der Astronomie. Durch die Beschäftigung mit diesem

Gebiet kam er in enge Fühlung mit August Böckh, dem tiefen Kenner der Geschichte der Naturwissenschaften im Altertum, und dem neunzigjährigen Alexander von Humboldt. 1860 rückte er an die erste Assistentenstelle an der Sternwarte.

Er führte in diesen Jahren ein äußerst bewegtes und geselliges Leben und verkehrte vor allem in Musiker- und Künstlerkreisen. In dieser Zeit begann er auch seine populären Vorträge über Astronomie in der Singakademie, die in den höchsten Gesellschaftskreisen großen Anklang fanden. Seine glänzende Redegabe riß seine Zuhörer, unter denen sich auch Moltke befand, stets aufs neue zur Begeisterung hin.

1863 übernahm er als außerordentlicher Professor die interimistische Leitung der Sternwarte Berlin, da Encke schwer erkrankte. Schon nach einem Jahre wurde er zum definitiven Direktor ernannt, allerdings erst nach der Überwindung heftiger Widerstände von seiten der Akademie der Wissenschaften. Mit dem Astronomen der Akademie, dem später auf seinen Vorschlag gewählten Auwers, verband ihn innige Freundschaft.

Das erste Berliner Jahrzehnt war stark durch wissenschaftliche Arbeit ausgefüllt. Zunächst arbeitete er auf dem Gebiete der Sternschnuppen und organisierte Beobachtungen für den erwarteten großen Sternschnuppenfall vom 13. November 1866, das sogenannte Leonidenphänomen. Mit einem nach seinen Plänen gebauten Universaldurchgangsinstrument entdeckte sein Mitarbeiter Küstner die Polhöhenschwankungen. Diese Entdeckung gab Foerster Anlaß zu organisatorischer Ausgestaltung der Beobachtungstätigkeit, und er erreichte die internationale Überwachung der Lage der Erdachse. Durch gleichzeitige Beobachtungen in Potsdam und Honolulu wurden die Polhöhenschwankungen als reell festgestellt. Die kleinen Planeten, für die er eine alte Vorliebe hegte, beschäftigten ihn auch in dieser Zeit. Er führte mit seinen Mitarbeitern Rechnungen in großem Maßstabe aus, und aus dieser Zusammenarbeit entwickelte sich später das astronomische Recheninstitut. In allen Beobachtungen lag ihm höchste Genauigkeit am Herzen, die Kontrolle, Verbesserung und Umgestaltung der Instrumente wurden unermüdlich betrieben. Er gab auf diesem Gebiet der Technik wesentliche Anregungen, vor allem sind seine Verbindung mit Abbe und seine tätige Mitwirkung bei der Gründung der Schottschen Glaswerke zu erwähnen. Seine Bemühungen, der Technik präzise Grundlagen für ihre Arbeit zu geben, führten ihn gemeinsam mit Werner von Siemens zu dem Plan, die Physikalisch-Technische Reichsanstalt zu gründen. Merkwürdigerweise erhoben sich in der öffentlichen Meinung starke Widerstände dagegen, denn man glaubte, in dieser staatlichen Hilfe für die Technik einen ersten Schritt zum Sozialismus zu sehen. Der vorurteilsfreie Kronprinz Friedrich Wilhelm jedoch unterstützte auch hier Foersters Bestrebungen, und im Jahre 1888 wurde die Physikalisch-Technische Reichsanstalt tatsächlich gegründet.

Durch die Übernahme immer neuer organisatorischer Tätigkeiten wuchs Foersters Arbeitslast ungeheuer. Vor allem waren es die Konferenzen für Gradmessung, die er fortschreitend auf eine breitere Basis zu stellen verstand und die er zu europäischen und später zu internationalen Konferenzen aufgestaltete. Daneben beschäftigten ihn die Konferenzen für Maß und Gewicht und die Einführung des Dezimalsystems im Norddeutschen Bunde, mit deren oberster Leitung Foerster im Jahre 1869 betraut wurde. Im Verlaufe dieser Arbeiten wurde er zum Direktor der Normaleichungskommission ernannt, ein Amt, das er neben vielen andern bis 1885 verwaltete. 1891 wurde er zum Rektor der Berliner Universität gewählt. In seinem Amtsjahr bemühte er sich um die Einführung einer Selbstverwaltung der Studentenschaft, allerdings ohne Erfolg.

Sehr wichtig war ihm stets in seiner gesamten Tätigkeit die populäre Verständlichmachung seiner Wissenschaft. So gründete er das populäre Institut für Astronomie „Urania" und schrieb zahlreiche Aufsätze und Abhandlungen für Freunde der Astronomie. Sein ganzes Wirken war von liberal-humanitären Tendenzen beherrscht, die ja schon durch die liberale Atmosphäre in seinem Vaterhause in ihm geweckt und gefördert worden waren. Die Astronomie in ihrer klaren Objektivität und ihrer Bedeutung für die menschliche Kultur erschien ihm besonders geeignet und berufen, Verbindungen zwischen Menschen der verschiedensten Völker zu schaffen. Sein Optimismus auf diesem Gebiet verhalf ihm zu schönen Erfolgen bei der Organisation und Leitung internationaler wissenschaftlicher Kongresse und der Internationalisierung astronomischer, geodätischer und metrischer Arbeiten. Seit 1891 war er Vorsitzender der Internationalen Maß- und Gewichtskommission. Die Erinnerung an die ungetrübte Herzlichkeit, mit der er gleich nach dem Kriege 1870/71 bei Gelegenheit einer internationalen Gradmessungskonferenz in Paris aufgenommen war, veranlaßte ihn, noch kurz vor seinem Tode im Jahre 1920 zur Tagung der internationalen Maß- und Gewichtskommission deren Vorsitzender er auch damals noch war, zu fahren. Diesmal aber wurde er schwer enttäuscht. Die politischen Gegensätze beherrschten auch die wissenschaftliche Konferenz in einem Maße, daß er sich gezwungen sah, den Vorsitz niederzulegen. Trotz dieser bitteren Enttäuschung am Ende seines Lebens kam in ihm niemals ein Zweifel an der Durchführbarkeit seiner Ideale auf.

Um seine humanitären Bestrebungen besser zur Geltung zu bringen, als es auf wissenschaftlichem Gebiete möglich war, organisierte er als Sechzigjähriger die Gesellschaft für Ethische Kultur. Diese Bewegung wurde durch Felix Adler aus Amerika herübergebracht. Der Anlaß zu der Foersterschen Gründung war ein reaktionärer Schulgesetzentwurf im Oktober 1892, gegen den sich eine Schar von Gegnern erhob, die dann als Gesellschaft für Ethische Kultur zusammenblieb. Unter diesen Gegnern befanden sich Georg von Gizyzki und Moritz von Egidy. Egidy trat jedoch der Gesellschaft nicht bei, da er selbst eine religiöse Bewegung „Einiges Christentum" ins Leben gerufen hatte. Er stand der Gesellschaft jedoch nahe und dies um so mehr, als Foerster stets die Unabhängigkeit seiner Vereinigung von religiösen Weltanschauungen betonte und auch allen Versuchen entgegentrat, die Gesellschaft auf atheistische Prinzipien festzulegen. Überhaupt zeigte er sich stets versöhnlich und ausgleichend allen radikalen Überspannungen gegenüber.

Die Ethische Bewegung sollte erzieherisch dazu beitragen, die Menschheit durch stufenweise Hebung ihrer Gesinnung zum Aufbau einer weltumfassenden, gerechten und freien Gesellschaftsordnung zu befähigen. Den terroristischen Ausartungen sozialer Bewegungen stellte er die Forderungen geistiger Vertiefung, genauer und gründlicher Denkweise und Stärkung zu innerer Freiheit, Güte und Selbstbescheidung gegenüber. Seine Idee war, mit diesen Forderungen dem Kampf ums Dasein seine Härte zu nehmen und so die soziale Frage zu lösen. Das einzig durchgreifende Heilmittel gegen alle Übel innerhalb der menschlichen Gesellschaft schien ihm die rein sittliche Bildung der Jugend zu sein, die er von allen metaphysischen, religiösen und nationalistischen Bindungen frei wissen wollte. Wenn auch seine Bestrebungen, die in seiner bürgerlich-liberalen Grundeinstellung wurzelten, ihm von vielen Seiten, so z. B. auch von seinem alten Freunde Adolph Menzel, Gegnerschaft eintrugen, so ließ er sich dadurch jedoch keineswegs beirren. Anderseits wurde ihm aber von hohen Stellen erfreuliche Anerkennung zuteil. Die Kaiserin Friedrich bekundete lebhaftes Interesse an der ethischen Bewegung. Sie wollte sogar der Gesellschaft beitreten und unterließ es nur auf Foersters dringendes Abraten, der Konflikte für sie befürchtete.

Aus der glücklichen Ehe, die Foerster 1867 mit Ina Paschen, der Tochter eines Fachkollegen, schloß, gingen fünf Kinder hervor. Die drei Söhne sind sämtlich der Öffentlichkeit bekannt: der Älteste, der Philosoph und Pädagoge Friedrich Wilhelm Foerster, setzt in der Gesellschaft für Ethische Kultur die Bestrebungen des Vaters fort. Der zweite, Karl, hat sich durch seine Züchtungen von Staudenpflanzen in seinen großen Gärtnereien in Bornim bei Potsdam einen Namen gemacht, und der Jüngste, Ernst, hat als Schiffsbauingenieur den Bau des Ozeandampfers „Vaterland", jetzt „Leviathan", geleitet.

Wilhelm Foerster wohnte in seiner letzten Lebenszeit in dem Landhaus seines Sohnes Karl und starb dort am 18. Januar 1921. Bis zuletzt war er von hoher geistiger Frische. Er genoß Verehrung von allen Seiten und wurde wegen seiner Güte und reinen Gesinnung auch von seinen Gegnern stets anerkannt.

Schriften von W. Foerster: Lebenserinnerungen und Lebenshoffnungen, Berlin 1911. — Lebensfragen und Lebensbilder, Berlin 1904. — Sammlung wissenschaftlicher Vorträge. — Wahrheit und Wahrscheinlichkeit, Berlin 1875. — Die Begründung einer Gesellschaft für Ethische Kultur, Einleitungsrede, gehalten am 18. Oktober 1892 zu Berlin. — Nekrologe: P. Guthnick in Vierteljahrsschrift der astronomischen Gesellschaft LIX (1924), S. 5—13. — J. Bauschinger in Astronomische Nachrichten 212, Nr. 5088. — P. Spieß in Sirius, Bd. LIV (1921) S. 81 bis 85. — R. Penzig, ebenda, S. 154—158.

Hans Rademacher

Note to Paper 30

This paper has not been reviewed in mathematical review journals.

Mathematische Theorie der Genkoppelung unter Berücksichtigung der Interferenz.

Von Professor Rademacher.

Nach Mendel werden die Vererbungserscheinungen bei Bastardierungen dadurch verständlich, daß man die Erbeigenschaften paarweise betrachtet, indem man je eine dominante Eigenschaft A und eine dazugehörige rezessive Eigenschaft a einander gegenüberstellt. In den Bastarden Aa ist phänotypisch nur A erkennbar; a ist jedoch nicht verschwunden, da es in den Nachkommen wieder auftreten kann. Am übersichtlichsten wird die Verteilung bei der Rückkreuzung der Bastarde Aa mit dem Elter, der phänotypisch die rezessive Eigenschaft a trägt, also in a homozygot ist. Unter den von dem Bastard Aa erzeugten Keimzellen finden sich (infolge der Reduktionsteilung) Zellen nur mit A und nur mit a. Bei Vereinigung mit den Keimzellen a des Elters von dem Phänotypus a entstehen Organismen mit den genotypischen Anlagen Aa und aa. Die Beobachtung zeigte nun bei Mendels Versuchen, daß die Anzahlen von Aa und aa, die phänotypisch als A und a erkennbar sind, sich wie 1:1 verhalten, woraus man schließt, daß die Keimzellen A und a des Bastardes Aa schon in dem Zahlenverhältnis 1:1 gestanden haben.

Verfolgt man außer dem Merkmalspaar Aa noch ein zweites Bb bei der Vererbung, so ist es zunächst denkbar, daß ein Bastard $\binom{Aa}{Bb}$ Keimzellen von 4 Arten, nämlich mit den Anlagen $\frac{A}{B}, \frac{A}{b}, \frac{a}{B}, \frac{a}{b}$ erzeugt. Bei Rückkreuzung mit dem Elter $\frac{aa}{bb}$, der nur Keimzellen $\frac{a}{b}$ bildet, ergeben sich dann die Kombinationen

$$\binom{Aa}{Bb}, \binom{Aa}{bb}, \binom{aa}{Bb}, \binom{aa}{bb},$$

die phänotypisch die Beschaffenheiten

$$\binom{A}{B}, \binom{A}{b}, \binom{a}{B}, \binom{a}{b}$$

aufweisen. Bei solchen Versuchen an Erbsen hat Mendel festgestellt, daß die 4 Sorten in den Zahlenverhältnissen 1:1:1:1 auftreten, d. h. B und b verteilen sich völlig unabhängig von A und a, und zwar treten sie jede ebenso wahrscheinlich mit A wie mit a zusammen. Mendel glaubte hier ein besonderes, zweites, Gesetz formulieren zu können, daß Merkmale verschiedener Eigenschaftspaare sich voneinander unabhängig vererben.

Diese Behauptung hielt aber weiteren Nachprüfungen nicht stand. Während bei frei mendelnden Merkmalspaaren 50% der F_1-Generation (wie im obigen Beispiel) Merkmalskombinationen aufweisen müssen, die in der P-Generation nicht kombiniert auftreten, so gibt es Fälle, wo die in der P-Generation vereinten Merkmale vorzugsweise zusammenbleiben und in weniger als 50% der Fälle ausgetauscht erscheinen. Solche Merkmale heißen gekoppelt, und die Quote der Individuen mit ausgetauschten Merkmalskombinationen heißt die Austauschwahrscheinlichkeit zwischen diesen Erbfaktoren oder Genen.

Bei der genetisch am besten untersuchten Drosophila melanogaster findet man vier Gruppen von je untereinander gekoppelten Erbfaktoren (von denen man über 400 kennt); beim Mais gibt es zehn Koppelungsgruppen.

Innerhalb einer solchen lassen sich nun die Gene eindimensional („linear“) in eine Reihe ordnen, indem man zwei Gene als um so enger gekoppelt (näher benachbart) ansieht, je geringer ihre Austauschwahrscheinlichkeit ist. Die folgende Tabelle der Austauschwahrscheinlichkeiten zwischen sieben wichtigen Genen der ersten Koppelungsgruppe von Drosophila melanogaster läßt den Sachverhalt ohne weiteres erkennen (am Kopf der Spalten stehen Abkürzungen der an den Zeilenanfängen genannten Gene):

	sc	*ec*	*cv*	*ct*	*v*	*g*	*f*
scute		.068	.163	.245	.370	.430	.465
echinus	.068		.096	.180	.314	.387	.435
crossveinless	.163	.096		.084	.228	.319	.389
cut	.245	.180	.084		.147	.249	.334
vermillion	.370	.314	.228	.147		.110	.219
garnet	.430	.387	.319	.249	.110		.114
forked	.465	.435	.389	.334	.219	.114	

Tabelle I (beobachtet).

Man kann auch unmittelbar eine Relation „zwischen" bei je drei Genen A, B, C definieren, indem man festsetzt, dasjenige liegt zwischen den beiden anderen, das nicht zu der größten der drei Austauschwahrscheinlichkeiten f_{AB}, f_{AC}, f_{BC} gehört. Dieser „Zwischen"-begriff erfüllt, wie die Anordnung der Zahlen in der obigen Tabelle evident werden läßt, alle Postulate über den mathematischen Begriff „zwischen" (vgl. etwa Hilbert, Grundlagen der Geometrie, Axiomgruppe II). Es muß betont werden, daß diese Möglichkeit der widerspruchslosen Reihenanordnung der Gene einer Kopplungsgruppe eine besondere Erfahrungstatsache ist, die keineswegs mit dem Begriff der Koppelung schon logisch notwendig verbunden ist.

Wenn man die Anordnung der Gene in der Koppelungsgruppe nicht nur durch ihre Anordnung, sondern auch in ihren gegenseitigen Abständen metrisch erfassen will, so könnte man versuchen, als Abstandsmaß zweier Gene A, B ihre Austauschwahrscheinlichkeit f_{AB} einzuführen. Die Zahlen der Tabelle zeigen aber, daß $f_{AB} + f_{BC} > f_{AC}$ ist, z. B.

$$f_{sc,\,cv} + f_{cv,\,g} = 0.163 + 0.319 = 0.482 > 0.430 = f_{sc,\,g},$$

daß also die f nicht additiv sind und daher nicht als Längenmaß benutzt werden können. Es entsteht die Aufgabe, eine additive Funktion der Austauschwahrscheinlichkeiten zu finden, oder umgekehrt, ein Entfernungsmaß der Gene anzugeben, so daß die f als eine gleichfalls anzugebende Funktion der Entfernungen sich ergeben.

Obgleich es für die folgende Überlegung nicht unerläßlich ist, erleichtert man sich die Vorstellung, wenn man sich die Gene, von denen wir ja schon festgestellt haben, daß sie sich in eine Reihe ordnen lassen, in eben dieser Reihenfolge auf den Chromosomen lokalisiert denkt. Der Austausch zwischen zwei Erbfaktoren geschieht dann dadurch, daß im Laufe der Reduktionsteilung der Keimzellen der F_1-Generation die nebeneinander liegenden Chromosomen eines Paares an homologen Stellen „brechen" und vertauscht zusammenwachsen, was als „Crossingover" bezeichnet wird. Auf einem Chromosom können auch mehrere Brüche erfolgen; bei einer geraden Anzahl von Brüchen zwischen zwei Genen tritt kein Austausch der Partner ein, wohl aber bei einer ungeraden Anzahl von Brüchen.

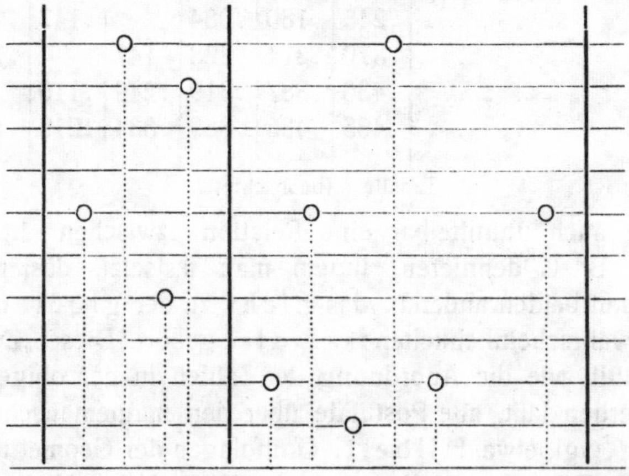

Um sogleich von einem additiven Abstandsmaß ausgehen zu können, denken wir uns sehr viele Chromosomen parallel zu einer x-Axe und mit den homologen Stellen übereinander ausgebreitet. Die Brüche seien auf den Chromosomen markiert und sämtlich auf die x-Axe projiziert. Die relative Häufigkeit von Brüchen (d. h. Zahl der Brüche dividiert durch Anzahl der Chromosomen) zwischen A und B gelte als Maß des Abstandes x von A bis B. Ist C ein Punkt zwischen A B, so setzt sich die relative Häufigkeit der Brüche zwischen A und B additiv aus den Häufigkeiten der Brüche zwischen A und C und denen zwischen C und B zusammen.

Unter den betrachteten Chromosomen sei eine Quote $w_0(x)$ ohne Bruch zwischen A und B, $w_1(x)$ mit genau e i n e m Bruch, $w_2(x)$ mit z w e i Brüchen usw. Dann setzt sich x zusammen aus

(1) $$x = w_1(x) + 2\,w_2(x) + 3\,w_3(x) + 4\,w_4(x) + \ldots$$

Es kommt nun darauf an, die Größen $w_n(x)$ einzeln zu berechnen. Hier muß nun noch eine weitere Erfahrungstatsache aus den Arbeiten der M o r g a n 'schen Schule herangezogen werden, nämlich die sog. „Interferenz". Es hat sich nämlich herausgestellt, das zwei Chromosomenbrüche nicht beliebig nahe beieinanderliegen können, sondern eine

Mindestentfernung d, die Interferenzstrecke, aufweisen müssen, wobei d für unsere Zwecke nach demselben Maßstab wie die Abstände x gemessen werden soll. Dagegen kann man annehmen, daß zwei Brüche in einem größeren Abstand als d sich gegenseitig nicht stören.

Wir fragen nach der Anzahl (d. h. nach der Quote) derjenigen Chromosomen, die sowohl zwischen ξ und $\xi + d\xi$ als auch zwischen η und $\eta + d\eta$ je einen Bruch haben. Ist der Abstand $|\xi - \eta|$ der beiden Stellen kleiner als d, so kommen auf Grund der Interferenz Chromosomen solcher Beschaffenheit nicht vor. Ist dagegen $|\xi - \eta| > d$, so stören sich beide Brüche gegenseitig nicht. Da die Zahl der Brüche proportional der Strecke ist, so gibt es $d\xi$ Brüche zwischen ξ und $\xi + d\xi$, also auch soviele Chromosomen mit Brüchen auf dieser Strecke. Und von diesen $d\xi$ Chromosomen ist der Bruchteil $d\eta$ auch mit einem Bruch zwischen η und $\eta + d\eta$ behaftet, so daß man $d\xi \cdot d\eta$ Chromosomen mit Brüchen an den angegebenen Stellen hat. Wenn man ξ und η alle Stellen zwischen A und B, also zwischen 0 und x einnehmen läßt, so erhält man danach im ganzen

$$\iint\limits_{\substack{0 \\ |\xi - \eta| > d}}^{x} d\xi\, d\eta = (x - d)_+^2 \tag{2}$$

Chromosomen. Dabei soll, wie die elementare Berechnung des Doppelintegrals leicht ergibt, $(x-d)_+^2$ folgendes bedeuten: von $x - d$ sollen nur die positiven Werte benutzt werden, und es sei einfach

$$(x-d)_+ = 0 \text{ für } x-d < 0$$
$$(x-d)_+ = x-d \text{ für } x-d > 0,$$

woraus die Bedeutung von $(x-d)_+^2$ und entsprechend die von den weiterhin auftretenden Ausdrücken $(x-2d)_+^3$, $(x-3d)_+^4$ sich ergibt.

Durch Zählung der Chromosomen mit Brüchen auf den drei Strecken ξ bis $\xi + d\xi$, η bis $\eta + d\eta$, ζ bis $\zeta + d\zeta$ und durch Summation über alle Lagen von ξ, η, ζ zwischen 0 und x erhält man analog

$$\iiint\limits_{\substack{0 \\ |\xi-\eta| > d \\ |\xi-\zeta| > d \\ |\eta-\zeta| > d}}^{x} d\xi\, d\eta\, d\zeta = (x-2d)_+^3, \tag{3}$$

wo die elementare Berechnung auch dieses Integrals als $(x-2d)_+^3$ hier der Kürze halber wieder unterdrückt sei.

Es bleibt noch übrig, die Integrale (2), (3), und die ähnlich gebauten 4fachen, 5fachen, usw. durch die Funktionen $w_1(x)$, $w_2(x)$, ... auszudrücken. In dem Integral (2) sind alle Chromosomen mit zwei

Brüchen gezählt, und zwar jedes 2 mal, da ein Bruch sowohl für die Stelle ξ als auch für η gelten kann, so daß also $2w_2(x)$ Chromosomen hierfür in Ansatz zu bringen sind. Durch einige kombinatorische Überlegungen findet man, daß in (2) jedes Chromosom mit 3 Brüchen $3\cdot 2$ mal, jedes mit 4 Brüchen $4\cdot 3$ mal gezählt wird, so daß wir haben

(2a) $(x-d)_+^2 = 2\cdot 1 w_2(x) + 3\cdot 2 w_3(x) + 4\cdot 3 w_4(x) + \dots$

Analog ergibt sich für das Integral (3)

(3a) $(x-2d)_+^3 = 3\cdot 2\cdot 1\ w_3(x) + 4\cdot 3\cdot 2\ w_4(x) + \dots$

und für das analog gebaute 4fache Integral:

(4a) $(x-3d)_+^4 = 4\cdot 3\cdot 2\cdot 1\ w_4(x) + 5\cdot 4\cdot 3\cdot 2\ w_5(x) + \dots$

In diesen Gleichungen wird die linke Seite $(x-(n-1)d)_+^n$ von gewissem n an gleich Null. Da die Summanden rechts das negative Zeichen nicht haben können, müssen sie von diesem n an gleichfalls alle 0 sein. Alle Reihen (1), (2a), (3a), (4a), ... haben also nur e n d l i c h viele Glieder. Man kann sie auffassen als lineare Gleichungen für die Unbekannten $w_1(x)$, $w_2(x)$, $w_3(x)$, ..., die man aus ihnen sukzessive eliminieren kann. Die Ausrechnung ergibt:

$$
\begin{aligned}
w_1(x) &= x - \frac{(x-d)_+^2}{1} + \frac{(x-2d)_+^3}{1\cdot 2} - \frac{(x-3d)_+^4}{1\cdot 2\cdot 3} + - \dots \\
(5)\qquad 1\cdot 2\, w_2(x) &= (x-d)_+^2 - \frac{(x-2d)_+^3}{1} + \frac{(x-3d)_+^4}{1\cdot 2} - \frac{(x-4d)_+^5}{1\cdot 2\cdot 3} + - \dots \\
1\cdot 2\cdot 3\cdot w_3(x) &= (x-2d)_+^3 - \frac{(x-3d)_+^4}{1} + \frac{(x-4d)_+^5}{1\cdot 2} - \frac{(x-5d)_+^6}{1\cdot 2\cdot 3} + - \dots
\end{aligned}
$$

u. s. f.

Die Austauschhäufigkeit f(x) zwischen A und B setzt sich nun, wie schon oben erwähnt, zusammen aus der Häufigkeit des Vorkommens von Chromosomen mit einem Bruch, mit drei Brüchen, mit fünf Brüchen, ... zwischen A und B, d. h.

$$f(x) = w_1(x) + w_3(x) + w_5(x) + \dots$$

Setzt man hierin die Werte von $w_1(x)$, $w_3(x)$, $w_5(x)$ .. aus (5) ein, so findet man schließlich, nach einigen mathematischen Umformungen:

$$(6)\qquad f(x) = \frac{x}{1} - \frac{2(x-d)_+^2}{1\cdot 2} + \frac{2^2(x-2d)_+^3}{1\cdot 2\cdot 3} - \frac{2^3(x-3d)_+^4}{1\cdot 2\cdot 3\cdot 4} + \dots$$

Hierin tritt noch d als Parameter auf, d. h. der Verlauf der Funktion f(x) ist erst nach Angabe des Wertes von d bestimmt. Man erhält also für f(x) ein ganzes Kurvenbüschel. Für d = 0 geht speziell aus (6) hervor:

$$f(x) = \frac{x}{1} - \frac{2x^2}{1\cdot 2} + \frac{2^2 x^3}{1\cdot 2\cdot 3} - \frac{2^3 x^4}{1\cdot 2\cdot 3\cdot 4} + - \ldots$$

Durch Vergleichung mit der bekannten Reihe für die Exponentialfunktion erhält man hierfür

$$f(x) = \tfrac{1}{2}(1 - e^{-2x}), \quad d = 0.$$

Diese Formel, die also für fehlende Interferenz gilt, ist schon von J. B. S. Haldane 1919 aufgestellt worden.

Um nun (6) mit der Erfahrung vergleichen zu können, muß man bedenken, daß das Vererbungsexperiment direkt die Werte von f(x), nicht die von x und d liefert. Durch Probieren mit verschiedenem d findet man zuerst die den Experimenten am besten angepaßte Kurve und dann die x-Werte, die Distanzen der verschiedenen Gene. Für die sieben Gene des Chromosoms I, die in Tabelle I, Seite 3, verzeichnet sind, hat man also außer d noch sechs Distanzen zu bestimmen, also sieben unbekannte, für die 21 beobachtete f-Werte der Tabelle vorliegen. Durch eine graphische Methode findet man als am besten den Beobachtungen angepaßt:

$$d = 0.22$$

und die Distanzen, von „*scute*" aus gemessen:

sc—ec	*sc—cv*	*sc—ct*	*sc—v*	*sc—g*	*sc—f*
0.070	0.167	0.249	0.395	0.503	0.611

Entnimmt man hieraus die Distanzen x je zweier Gene und trägt x in (6) ein mit d = 0.22, so erhält man folgende Tabelle berechneter f-Werte:

	sc	*ec*	*cv*	*ct*	*v*	*g*	*f*
scute		.070	.167	.248	.364	.423	.461
echinus	.070		.097	.179	.314	.388	.439
crossveinless	.167	.097		.082	.228	.323	.394
cut	.248	.179	.082		.146	.253	.342
vermillion	.364	.314	.228	.146		.108	.216
garnet	.423	.388	.323	.253	.108		.108
forked	.461	.439	.394	.342	.216	.108	

Tabelle II (berechnet).

Man bemerkt die gute Übereinstimmung dieser theoretischen Werte mit den beobachteten von Tabelle I. Die Abweichungen erreichen nur an einer Stelle den Betrag von 0.008.

Auch für die Chromosomen II und III von Drosophila melanogaster liegen Beobachtungen der Austauschhäufigkeiten f vor (Bridges und Morgan 1919, 1923). Die hier entwickelte Theorie gibt auch in diesen Fällen den Verlauf der f-Werte im wesentlichen wieder, allerdings mit erheblich größeren Differenzen zwischen Beobachtung und Theorie. Es darf aber bemerkt werden, daß das Chromosom I viel genauer erforscht ist als die Chromosomen II und III, und daß insbesondere die in Tabelle I angeführten Beobachtungsresultate über Chromosom I auch vom rein biologischen Standpunkt beurteilt das beste gegenwärtig verfügbare Material darstellen. Sie stammen aus den Beobachtungen von Bridges und Olbrycht (1926) an mehr als 20000 Fliegen, und zwar sind diese Experimente mit ganz besonderen Vorsichtsmaßregeln (Aufrechterhaltung einer „balanced viability") an einem Fliegenstamm unter einheitlichen Bedingungen durchgeführt. Die Tabelle I ist übrigens nicht explizit bei Bridges und Olbrycht publiziert, sondern zuerst von K. v. Körösy (1930) zusammengestellt worden.

Literatur:

C. B. Bridges and T. H. Morgan, Carnegie Institution Publication (Washington) No. 278, p. 123—304 (1919), No. 327 (1923).

C. B. Bridges and T. M. Olbrycht, The multiple stock Xple and its use. Genetics 11 (1926).

J. B. S. Haldane, The combination of linkage values, and the calculation of distances between the loci of linked factors. Journ. of Genetics 8 (1919).

David Hilbert, Grundlagen der Geometrie, 7. Aufl., Leipzig 1931.

K. v. Körösy, Versuch einer Theorie der Genkoppelung, Bibliotheca genetica XV, Berlin 1930.

Note to Paper 31

Apparently, this paper has not been reviewed in mathematical review journals.

EGY RECIPROCITÁSKÉPLETRŐL A MODULFÜGGVÉNYEK ELMÉLETÉBŐL.*

1. A

$$\log \eta(\tau) = \frac{\pi i \tau}{12} + \log \prod_{m=1}^{\infty} (1 - e^{2\pi i m \tau})$$

függvény $\Im(\tau) > 0$ mellett reguláris és, a logarithmus egyik meghatározott függvényágának megtartása után, a felső τ-félsíkban egyértékű. DEDEKIND, aki az $\eta(\tau)$ függvényt a modulfüggvények elméletébe bevezette, a $\log \eta(\tau)$ vizsgálatánál ilyenfajta összegekhez jutott:[1]

$$s(h, k) = \sum_{\mu=1}^{k} \frac{\mu}{k} \left(\left(\frac{h\mu}{k}\right)\right), \tag{1}$$

ahol h, k relatív prím pozitív egész számok. Itt s a következőkben is, valós x-re

$$((x)) = x - [x] - \frac{1}{2}.^{2} \tag{2}$$

Egyébként a $k = 1$ esetben az (1)-ben fellépő összeg üres; ekkor az a képlet az

$$s(h, 1) = 0$$

értelemben veendő.

---

* Boroszlói egyetemi tanárok budapesti látogatása alkalmából 1929 októberében a Pázmány Péter Tudomány-Egyetemen tartott előadás.

[1] *Dedekinds Erläuterungen zu Riemanns Fragmenten über die Grenzfälle der elliptischen Modulfunktionen,* RIEMANNS Werke (1876), S. 438—447.

[2] DEDEKIND az idézett helyen más jelölést használ. Az ő (m, n) jele a mi írásmódunkban $6n \cdot s(m, n)$. A mi $((x))$ jelünk DEDEKIND-nél $\left((x - \frac{1}{2})\right)$.

Az $s(h, k)$ összegekre DEDEKIND a ϑ-függvények elméletéből a figyelemreméltó

$$12s(h, k) + 12s(k, h) = -3 + \frac{h}{k} + \frac{k}{h} + \frac{1}{hk} \tag{3}$$

reciprocitásképletet nyerte, amelyben h, k relatív prím pozitív egész számok. Ez az egyenlet az (1) definíció következtében tisztán arithmetikai természetű; egy másik dolgozatomban[3] a (3)-at közvetlenül arithmetikai úton bebizonyítottam. A következőkben még egy bizonyítást akarok adni, amely ugyancsak független a modulfüggvények elméletétől, de analitikai segédeszközöket használ. Mellékeredményként a (3)-nak egy trigonometriai átalakítását fogom nyerni.

2. A $k = 1$ esetben (3) elemi számításokkal azonnal igazolható. Ezért ettől kezdve legyen $k \geqq 2$.

Közelfekvő az $((x))$ függvényre az ismert FOURIER-sort alkalmazni:

$$((x)) = -\frac{1}{\pi} \sum_{n=1}^{\infty} \frac{\sin 2\pi n x}{n}.\text{[4]}$$

Ezáltal nyerjük, hogy

$$s(h, k) = -\frac{1}{\pi k} \sum_{n=1}^{\infty} \frac{1}{n} \sum_{\mu=1}^{k} \mu \sin \frac{2\pi n h \mu}{k}. \tag{4}$$

A belső összeg elemi úton kiszámítható. Legyen

$$S_m = \sum_{\mu=1}^{k} \mu \sin \frac{2\pi m \mu}{k}. \tag{5}$$

Ha k osztója m-nek, akkor

$$S_m = 0. \tag{6a}$$

A többi esetben

$$S_m = \mathfrak{J}\Big(\sum_{\mu=1}^{k} \mu e^{\frac{2\pi i m \mu}{k}}\Big) = \mathfrak{J}(U).$$

---

[3] *Zur Theorie der Modulfunktionen*, Journ. f. d. Math., 167 (1932) S. 312—336.

[4] Egész x mellett nem érvényes a fenti összefüggés, de ezt ilyen esetben nem fogjuk használni.

Itt
$$U=\sum_{\mu=1}^{k}\mu e^{\frac{2\pi i m u}{k}}=\sum_{\mu=1}^{k}e^{\frac{2\pi i m\mu}{k}}+\sum_{\mu=1}^{k}(\mu-1)e^{\frac{2\pi i m\mu}{k}}=$$

$$=0+e^{\frac{2\pi i m}{k}}\sum_{\nu=0}^{k-1}\nu e^{\frac{2\pi i m\nu}{k}}=e^{\frac{2\pi i m}{k}}(U-ke^{2\pi i m})=e^{\frac{2\pi i m}{k}}(U-k),$$

következőleg
$$U=\frac{k}{1-e^{-\frac{2\pi i m}{k}}}\overset{4a}{=}k\frac{1-e^{\frac{2\pi i m}{k}}}{\left|1-e^{\frac{2\pi i m}{k}}\right|^2},$$

és igy
$$S_m=-k\frac{\sin\frac{2\pi m}{k}}{4\sin^2\frac{\pi m}{k}}=-\frac{k}{2}\operatorname{ctg}\frac{\pi m}{k}. \tag{6b}$$

A (4), (5), (6) szerint:
$$s(h,k)=\frac{1}{2\pi}\sum_{n=1}^{\infty}{}'\frac{1}{n}\operatorname{ctg}\frac{\pi n h}{k},$$

ahol n kihagyja a k többeseit. E szerint:
$$s(h,k)=\frac{1}{2\pi}\sum_{m=0}^{\infty}\sum_{l=1}^{k-1}\frac{1}{mk+l}\operatorname{ctg}\frac{\pi l h}{k}. \tag{7}$$

Minthogy
$$\sum_{l=1}^{k-1}\operatorname{ctg}\frac{\pi l h}{k}=0, \tag{8}$$

azért (7)-re egy DIRICHLET-féle összegezési eljárás[5] alkalmazható. Ugyanis
$$\sum_{m=0}^{M}\sum_{l=1}^{k-1}\frac{1}{mk+l}\operatorname{ctg}\frac{\pi l h}{k}=\sum_{m=0}^{M}\sum_{l=1}^{k-1}\int_0^1 x^{mk+l-1}\operatorname{ctg}\frac{\pi l h}{k}\,dx=$$
$$=\int_0^1\sum_{m=0}^{M}x^{mk}\sum_{l=1}^{k-1}x^{l-1}\operatorname{ctg}\frac{\pi l h}{k}\,dx.$$

[4a] Az U-nak ez az értéke azonnal nyerhető a
$$\sum_{\mu=1}^{k}\mu x^{u}=\frac{kx^{k+1}}{x-1}-\frac{x(x^k-1)}{(x-1)^2}$$
azonosságból. (Fordító megjegyzése.)

[5] LEJEUNE DIRICHLETS Werke, Bd. I., S. 420.

A belső összegben álló $(k-2)$-edfokú polynomnak (8) szerint zérushelye $x=1$, s ezért írható

$$\sum_{l=1}^{k-1} x^{l-1} \operatorname{ctg} \frac{\pi l h}{k} = (1-x)\, Q(x), \tag{9}$$

ahol $Q(x)$ egy $(k-3)$-adfokú polynom.[6] Ezekután:

$$\begin{aligned}\sum_{m=0}^{M} \sum_{l=1}^{k-1} \frac{1}{mk+l} \operatorname{ctg} \frac{\pi l h}{k} &= \int_0^1 \sum_{m=0}^{M} x^{mk} (1-x)\, Q(x)\, dx = \\ &= \int_0^1 \frac{1-x}{1-x^k} (1-x^{(M+1)k})\, Q(x)\, dx.\end{aligned} \tag{10}$$

Itt $0 \leqq x \leqq 1$ mellett

$$\left| \frac{1-x}{1-x^k} Q(x) \right| \leqq K,$$

$$\left| \int_0^1 \frac{1-x}{1-x^k} Q(x) . x^{(M+1)k} dx \right| \leqq K \int_0^1 x^{(M+1)k} dx = \frac{K}{(M+1)k+1},$$

amely utóbbi zérushoz konvergál $M \to \infty$ mellett. Eszerint (10)-ben végrehajtható az $M \to \infty$ határátmenet:

$$\sum_{m=0}^{\infty} \sum_{l=1}^{k-1} \frac{1}{mk+l} \operatorname{ctg} \frac{\pi l h}{k} = \int_0^1 \frac{1-x}{1-x^k} Q(x)\, dx,$$

amely helyett (7) és (9) szerint írható:

$$s(h,k) = \frac{1}{2\pi} \int_0^1 \frac{1}{1-x^k} \sum_{l=1}^{k-1} x^{l-1} \operatorname{ctg} \frac{\pi l h}{k}\, dx.$$

Részlettörtekre bontva

$$\frac{x^{l-1}}{1-x^k} = -\frac{1}{k} \sum_{\lambda=0}^{k-1} \frac{\varrho^{\lambda l}}{x-\varrho^{\lambda}},$$

ahol

$$\varrho = e^{\frac{2\pi i}{k}}.$$

[6] Azaz $2 \leqq k \leqq 3$ mellett $Q(x)$ állandó, mégpedig $Q(x) = \operatorname{ctg} \frac{\pi h}{3}$, illetőleg 0, a $k=3$, 2 esetek szerint.

Eszerint:

$$\frac{1}{1-x^k}\sum_{l=1}^{k-1} x^{l-1}\operatorname{ctg}\frac{\pi l h}{k} = -\frac{1}{k}\sum_{\lambda=0}^{k-1}\sum_{l=1}^{k-1}\frac{\varrho^{\lambda l}\operatorname{ctg}\frac{\pi l h}{k}}{x-\varrho^{\lambda}}.$$

(8) miatt eltűnik a belső összeg $\lambda=0$ mellett. A λ-összeget tehát csupán a $\lambda=1$-től $\lambda=k-1$ értékekig kell kiterjeszteni. Ezzel nyerjük, hogy

$$s(h,k) = -\frac{1}{2k\pi}\int_0^1\sum_{\lambda=1}^{k-1}\sum_{l=1}^{k-1}\frac{\varrho^{\lambda l}\operatorname{ctg}\frac{\pi l h}{k}}{x-\varrho^{\lambda}}\,dx.$$

Legyen

$$\sigma(\lambda) = \sum_{l=1}^{k-1}\varrho^{\lambda l}\operatorname{ctg}\frac{\pi l h}{k}; \tag{11}$$

akkor

$$\begin{aligned}\sigma(\lambda) &= \sum_{l=1}^{k-1}\varrho^{\lambda(k-l)}\operatorname{ctg}\frac{\pi(k-l)h}{k} = \\ &= -\sum_{l=1}^{k-1}\varrho^{-l\lambda}\operatorname{ctg}\frac{\pi l h}{k} = -\overline{\sigma(\lambda)}.\end{aligned} \tag{12}$$

Tehát $\sigma(\lambda)$ tisztán képzetes, és igy

$$\sigma(\lambda) = i\sum_{l=1}^{k-1}\sin\frac{2\pi\lambda l}{k}\operatorname{ctg}\frac{\pi l h}{k}.^{7} \tag{13}$$

Az $s(h,k)$ így alakul:

$$\begin{aligned}s(h,k) &= -\frac{1}{2k\pi}\sum_{\lambda=1}^{k-1}\sigma(\lambda)\int_0^1\frac{dx}{x-\varrho^{\lambda}} = \\ &= -\frac{1}{2k\pi}\sum_{\lambda=1}^{k-1}\sigma(\lambda)\log\left(\frac{1-\varrho^{\lambda}}{-\varrho^{\lambda}}\right) = \\ &= -\frac{1}{2k\pi}\sum_{\lambda=1}^{k-1}\sigma(\lambda)\log(1-\varrho^{-\lambda}).\end{aligned} \tag{14}$$

Itt $(1-\varrho^{-\lambda})$ a jobbfélsíkban fekszik, s a logarithmus úgy határozandó meg, hogy

---

[7] A számitásnak kissé más elrendezésével egyébként

$$\sigma(\lambda) = -2ki\left(\left(\frac{\lambda h^{\star}}{k}\right)\right)$$

adódnék, ahol $h^{\star}$ egy olyan egész szám, amelyre $h.h^{\star}\equiv 1 \pmod{k}$.

$$|\mathfrak{J}(\log(1-\varrho^{-\lambda}))| < \frac{\pi}{2} \tag{14a}$$

legyen. Ha (14) mindkét oldalán konjugált komplex értékekre térünk át, akkor (12) figyelembevételével:

$$s(h, k) = \frac{1}{2k\pi} \sum_{\lambda=1}^{k-1} \sigma(\lambda) . \log(1-\sigma^{\lambda}). \tag{15}$$

A (14)-ből és (15)-ből következik

$$s(h, k) = \frac{1}{4k\pi} \sum_{\lambda=1}^{k-1} \sigma(\lambda) \log \frac{1-\varrho^{\lambda}}{1-\varrho^{-\lambda}}.$$

A logarithmus jele alatt álló szám abszolút értéke 1, ugyanis

$$\frac{1-\varrho^{\lambda}}{1-\varrho^{-\lambda}} = -\varrho^{\lambda} = -e^{\frac{2\pi i\lambda}{k}},$$

tehát a logarithmus tisztán képzetes, mégpedig (14a) miatt

$$-\pi < \mathfrak{J}\left(\log \frac{1-\varrho^{\lambda}}{1-\varrho^{-\lambda}}\right) < \pi.$$

E szerint

$$\log \frac{1-\varrho^{\lambda}}{1-\varrho^{-\lambda}} = \frac{2\pi i\lambda}{k} - \pi i,$$

$$s(h, k) = \frac{i}{4k} \sum_{\lambda=1}^{k-1} \sigma(\lambda) \left(\frac{2\lambda}{k} - 1\right).$$

Minthogy (11) és (8) szerint $\sigma(0) = 0$, azért

$$\sum_{\lambda=1}^{k-1} \sigma(\lambda) = \sum_{\lambda=0}^{k-1} \sigma(\lambda) = \sum_{l=1}^{k-1} \operatorname{ctg} \frac{\pi l h}{k} \sum_{=}^{-} \varrho^{\lambda l} = 0.$$

Következőleg

$$s(h, k) = \frac{2i}{4k^2} \sum_{\lambda=1}^{k-1} \lambda \sigma(\lambda)$$

és (13) szerint

$$s(h, k) = -\frac{1}{2k^2} \sum_{l=1}^{k-1} \operatorname{ctg} \frac{\pi l h}{k} \sum_{\lambda=1}^{k-1} \lambda \sin \frac{2\pi\lambda l}{k}.$$

Az (5) és (6b) szerint azonban a belső összeget már kiszámítottuk, tehát végül

$$s(h, k) = \frac{1}{4k} \sum_{l=1}^{k-1} \operatorname{ctg} \frac{\pi l h}{k} \operatorname{ctg} \frac{\pi l}{k}. \tag{16}$$

3. Most már bebizonyítjuk (3)-at azáltal, hogy levezetjük a reciprocitásképletet a (16)-beli trigonometriai összegre. Ebből a célból tekintjük az

$$\frac{1}{2\pi i}\int \pi \operatorname{ctg} \pi z \operatorname{ctg} \frac{\pi z}{k} \operatorname{ctg} \frac{\pi h z}{k}\, dz$$

integrált, kiterjesztve annak az R derékszögű négyszögnek oldalaira, amelynek csúcsai

$$k-\varepsilon+Ni,\ -\varepsilon+Ni,\ -\varepsilon-Ni,\ k-\varepsilon-Ni \quad \left(0<\varepsilon<\frac{1}{h}\right).$$

Az integrandus mindhárom tényezőjének közös pólusa $z=0$ az R belsejében. A második tényezőnek további pólusa nincs az R-ben. A $z=0$ pólustól eltekintve, a $\pi \operatorname{ctg} \pi z$ tényezőnek pólusai az R négyszögben

$$z=1, 2, \ldots k-1,$$

az 1 residuumokkal, a harmadik tényezőnek pólusai

$$z=\frac{k}{h}, \frac{2k}{h}, \ldots, \frac{(h-1)k}{h},$$

a $\frac{k}{\pi h}$ residuumokkal. E szerint

$$\frac{1}{2\pi i}\int \pi \operatorname{ctg} \pi z \operatorname{ctg} \frac{\pi z}{k} \operatorname{ctg} \frac{\pi h z}{k} = \tag{17}$$
$$= \sum_{l=1}^{k-1} \operatorname{ctg} \frac{\pi l}{k} \operatorname{ctg} \frac{\pi l h}{k} + \frac{k}{h}\sum_{m=1}^{k-1} \operatorname{ctg} \frac{\pi m k}{h} \operatorname{ctg} \frac{\pi m}{h} + R_0,$$

ahol R_0 a residuum a $z=0$ pontban. A $\operatorname{ctg} z$ LAURENT-kifejtése igy kezdődik:

$$\operatorname{ctg} z = \frac{1}{z} - \frac{z}{3} - \cdots;$$

ezért

$$\pi \operatorname{ctg} \pi z \operatorname{ctg} \frac{\pi z}{k} \operatorname{ctg} \frac{\pi h z}{k} =$$

$$= \pi\left(\frac{1}{\pi z} - \frac{\pi z}{3} - \cdots\right)\left(\frac{k}{\pi z} - \frac{\pi z}{3k} - \cdots\right)\left(\frac{k}{\pi h z} - \frac{\pi h z}{3k}\right)$$

alapján, mint az $\frac{1}{z}$ együtthatója:

$$R_0 = -\frac{k^2}{3h} - \frac{1}{3h} - \frac{h}{3}. \tag{18}$$

Mivel k periodusa a (17) integrandusának, azért az R négyszögnek a képzetes tengellyel párhuzamos oldalaira kiterjesztett integrálrészek elhagyhatók. Marad tehát

$$\int_R = \int_{k-\varepsilon+Ni}^{-\varepsilon+Ni} + \int_{-\varepsilon-Ni}^{k-\varepsilon-Ni}. \tag{19}$$

Az $N \to \infty$ határátmenetnél mindkét jobboldali integrál integrandusa egyenletesen konvergál egy-egy határértékhez. Minthogy

$$\operatorname{ctg} z = i\,\frac{e^{iz} + e^{-iz}}{e^{iz} - e^{-iz}},$$

azért az első jobboldali integrálban $\operatorname{ctg} z \to -i$, a másodikban $\operatorname{ctg} z \to i$. E szerint (17), ha még (18)-at tekintetbe vesszük, $N \to \infty$ mellett a következőbe megy át:

$$-\frac{k}{2i}(-i)^3 + \frac{k}{2i}i^3 = \sum_{l=1}^{k-1} \operatorname{ctg}\frac{\pi l}{k} \operatorname{ctg}\frac{\pi l h}{k} +$$

$$+ \frac{k}{h}\sum_{m-1}^{h-1} \operatorname{ctg}\frac{\pi m}{h} \operatorname{ctg}\frac{\pi m k}{h} - \frac{k^2}{3h} - \frac{1}{3h} - \frac{h}{3},$$

azaz

$$\frac{1}{k}\sum_{l=1}^{k-1} \operatorname{ctg}\frac{\pi l}{k} \operatorname{ctg}\frac{\pi l h}{k} + \frac{1}{h}\sum_{m=1}^{h-1} \operatorname{ctg}\frac{\pi m}{h} \operatorname{ctg}\frac{\pi m k}{h} =$$

$$= -1 + \frac{h}{3k} + \frac{k}{3h} + \frac{1}{3hk}.$$

Ez a (3) reciprocitásképlet trigonometriai alakban; maga (3) rögtön következik (20)-ból, (16) segélyével.

Hans Rademacher

(németből fordította *Rédei László.*)

UBER EINE REZIPROZITÄTSFORMEL AUS DER THEORIE DER MODULFUNKTIONEN.*

In der Theorie der Modulfunktionen treten folgende, zuerst von DEDEKIND [1] betrachteten Summen auf:

$$s(h,k)=\sum_{\mu=1}^{k-1}\frac{\mu}{k}\left(\left(\frac{h\mu}{k}\right)\right) \tag{1}$$

mit ganzen positiven $h, k,$ $(h, k)=1$. Hier ist zur Abkürzung für reelles, nicht-ganzes x

$$((x))=x-[x]-\frac{1}{2} \tag{2}$$

gesetzt. Für die Summen (1) gilt, wie sich aus der Theorie der Modulfunktionen ergibt, die folgende merkwürdige Reziprozitätsformel

$$12\,s(h,k)+12\,s(k,h)=-3+\frac{h}{k}+\frac{k}{h}+\frac{1}{hk}, \tag{3}$$

für die ich kürzlich einen direkten zahlentheoretischen Beweis gegeben habe,[3] und die ich hier auf andere Weise, abermals unabhängig von der Theorie der Modulfunktionen, beweise.

Aus der bekannten Formel

$$((x))=-\frac{1}{\pi}\sum_{n=1}^{\infty}\frac{\sin 2\pi nx}{n}$$

folgt nach (1)

$$s(h,k)=-\frac{1}{\pi k}\sum_{n=1}^{\infty}\frac{1}{n}\sum_{\mu=1}^{k}\mu\sin\frac{2\pi nh\mu}{k}. \tag{4}$$

Für die Summe

$$S_m=\sum_{\mu=1}^{k}\mu\sin\frac{2\pi m\mu}{k} \tag{5}$$

ergibt sich leicht

$$\begin{aligned} S_m &= 0 & m &\equiv 0 \pmod k \\ S_m &= -\frac{k}{2}\operatorname{ctg}\frac{\pi m}{k} & m &\not\equiv 0 \pmod k \end{aligned} \tag{6}$$

---

* Auszug eines an der Budapester Pázmány Péter Universität in Okt. 1929 gehaltenen Vortrages, bei Gelegenheit des Budapester Besuches von Professoren der Breslauer Universität.

Dadurch erhält man

$$s(h, k) = \frac{1}{2\pi} \sum_{m=0}^{\infty} \sum_{l=1}^{k-1} \frac{1}{mk+l} \operatorname{ctg} \frac{\pi l h}{k}. \tag{7}$$

Wegen

$$\sum_{l=1}^{k-1} \operatorname{ctg} \frac{\pi l h}{k} = 0 \tag{8}$$

kann man auf (7) eine DIRICHLETsche Summationsmethode [5] anwenden:

$$s(h, k) = \frac{1}{2\pi} \sum_{m=0}^{\infty} \sum_{l=1}^{k-1} \int_0^1 x^{mk+l-1} \operatorname{ctg} \frac{\pi l h}{k}\, dx =$$

$$= \frac{1}{2\pi} \int_0^1 \frac{1}{1-x^k} \sum_{l=1}^{k-1} x^{l-1} \operatorname{ctg} \frac{\pi l h}{k}\, dx.$$

Partialbruchzerlegung und Berücksichtigung von (8) ergibt

$$s(h, k) = -\frac{1}{2k\pi} \int_0^1 \sum_{\lambda}^{k-1} \sum_{l=1}^{k-1} \frac{\varrho^{\lambda l} \operatorname{ctg} \frac{\pi l h}{k}}{x - \varrho^{\lambda}}\, dx$$

mit

$$\varrho = e^{\frac{2\pi i}{k}}.$$

Die Summe

$$\sigma(\lambda) = \sum_{l=1}^{k-1} \varrho^{\lambda l} \operatorname{ctg} \frac{\pi l h}{k} \tag{11}$$

ist rein imaginär, d. h.

$$\sigma(\lambda) = i \sum_{l-1}^{k-1} \sin \frac{2\pi\lambda l}{k} \operatorname{ctg} \frac{\pi l h}{k}. \tag{12}$$

Die Ausrechnung von $s(h, k)$ ergibt nun

$$s(h, k) = -\frac{1}{2k\pi} \sum_{\lambda=1}^{k-1} \sigma(\lambda) \log (1 - \varrho^{-\lambda}) \tag{15}$$

mit

$$|\Im(\log (1 - \varrho^{-\lambda}))| < \frac{\pi}{2}. \tag{14a}$$

Nimmt man in (15) auf beiden Seiten den Realteil, so erhält man

$$s(h, k) = \frac{1}{2k\pi} \sum_{\lambda=1}^{k-1} \sigma(\lambda) \frac{i\pi}{2} \left(\frac{2\lambda}{k} - 1 \right) = \frac{i}{2k^2} \sum_{\lambda=1}^{k-1} \lambda \sigma(\lambda).$$

Durch Eintragung von (11) und Benützung von (6) folgt schliesslich

$$s(h,k)=\frac{1}{4k}\sum_{l=1}^{k-1}\operatorname{ctg}\frac{\pi lh}{k}\operatorname{ctg}\frac{\pi l}{k}. \tag{16}$$

Nun werde das Integral betrachtet

$$J=\int_R \pi \operatorname{ctg}\pi z \operatorname{ctg}\frac{\pi z}{k}\operatorname{ctg}\frac{\pi hz}{k}\,dz,$$

das um das Rechteck R mit den Ecken $k-\varepsilon+Ni$, $-\varepsilon+Ni$, $-\varepsilon-Ni$, $k-\varepsilon-Ni$ im positiven Sinne erstreckt; es sei $0<\varepsilon<\frac{1}{h}$. Das Residuum des Integranden für $z=0$ ist

$$R_0=-\frac{k^2}{3h}-\frac{1}{3h}-\frac{h}{3}. \tag{18}$$

Sonst hat der Integrand noch Pole erster Ordnung in

$$z=1, 2, \ldots, k-1$$

und

$$z=\frac{k}{h}, \quad \frac{2k}{h}, \ldots, \quad \frac{(h-1)k}{h}.$$

Daher ergibt die Residuenrechnung

$$J=\sum_{l=1}^{k-1}\operatorname{ctg}\frac{\pi l}{k}\operatorname{ctg}\frac{\pi lh}{k}+\frac{k}{h}\sum_{m=1}^{k-1}\operatorname{ctg}\frac{\pi mk}{h}\operatorname{ctg}\frac{\pi m}{h}+R_0. \tag{17}$$

Bei der direkten Berechnung von J heben sich die Integrale längs den Parallelen zur imagineren Axe weg. Auf den beiden horizontalen Strecken konvergiert der Integrand gleichmässig je gegen einen Limes bei $N\to\infty$, sodass sich

$$J=-\frac{k}{2i}(-i)^3+\frac{k}{2i}i^3=-k$$

ergibt. Aus dieser Gleichung zusammen mit (17) und (18) erhält man schliesslich

$$\frac{1}{k}\sum_{l=1}^{k-1}\operatorname{ctg}\frac{\pi l}{k}\operatorname{ctg}\frac{\pi lh}{k}+\frac{1}{h}\sum_{m=1}^{h-1}\operatorname{ctg}\frac{\pi m}{h}\operatorname{ctg}\frac{\pi mk}{h}=$$
$$=-1+\frac{h}{3k}+\frac{k}{3h}+\frac{1}{3hk}. \tag{20}$$

Dies ist die zu beweisende Reziprozitätsformel in trigonometrischer Form. Aus ihr folgt (3) mit Hife von (16).

Hans Rademacher.

Notes to Paper 32

This paper has been reviewed in the JF, vol. 59 (1933), p. 1060, and in the Z, vol. 8 (1934), p. 75.

Page 24, equation (1).
Read $\Sigma_{\mu=1}^{k-1}$ instead of $\Sigma_{\mu=1}^{k}$.

Page 24, equation (2).
It may be worthwhile to observe that this is the first time that the author restricts the definition (2) of $((x))$ to *nonintegral* real x. In all earlier papers (2) was defined to hold for all real x, in particular, $((n)) = -1/2$ for integral n (see e.g. paper 28, p. 224, equation (7)). In all subsequent papers, $((x))$ was redefined to represent zero for integral x (see e.g. paper 44 of the present collection, equation (1.1)).

Page 25, footnote 3.
See paper 27.

Page 28, first display line.
Read $\Sigma_{l=1}^{k-1}$ instead of $\Sigma_{=1}^{k-1}$.

Page 32, equation (2).
See note to page 24, equation (2).

Page 33, fifth display line.
Read $\Sigma_{\lambda=1}^{k-1}$ instead of Σ_{λ}^{k-1}.

Page 33, equation (12).
Read $\Sigma_{l=1}^{k-1}$ instead of Σ_{l-1}^{k-1}.

Primzahlen reell-quadratischer Zahlkörper in Winkelräumen.

Von

Hans Rademacher in Philadelphia, Pa. (U.S.A.)

Mit Hilfe der von ihm eingeführten Zetafunktionen mit Größencharakteren hat Herr Hecke für beliebige algebraische Zahlkörper einen Satz bewiesen[1]), den ich für reell-quadratische Zahlkörper spezialisiert folgendermaßen aussprechen will:

Es sei $\mathfrak{a}$ ein Ideal des Körpers, $\eta_\mathfrak{a} > 1$ die totalpositive Grundeinheit mod $\mathfrak{a}$, μ sei eine Körperzahl, μ' ihre Konjugierte und es werde

$$(1) \qquad w_\mathfrak{a}(\mu) = \frac{1}{2 \log \eta_\mathfrak{a}} \log \left| \frac{\mu}{\mu'} \right|$$

gesetzt. Ist ferner ϱ eine feste ganze Körperzahl, so gilt für die Anzahl $\pi_\mathfrak{a}(x; v)$ der mod $\mathfrak{a}$ nicht-assoziierten Primzahlen ϖ, die den Bedingungen

$$(2) \qquad \varpi > 0, \quad \varpi \equiv \varrho \ (\mathrm{mod}\ \mathfrak{a}), \quad N(\varpi) \leqq x, \quad w(\varpi) - [w(\varpi)] < v \leqq 1$$

genügen,

$$(3) \qquad \pi_\mathfrak{a}(x; v) \sim \frac{v}{h_0(\mathfrak{a})} \frac{x}{\log x},$$

wo $h_0(\mathfrak{a})$ die Anzahl der Idealklassen mod $\mathfrak{a}$ im engsten Sinne ist.

Für den Beweis dieses Satzes benutzt Herr Hecke das Weylsche Gleichverteilungskriterium, das aber seiner Natur nach nichts über den Fehler in der asymptotischen Gleichung (3) aussagt. Demgegenüber ist das Ziel der vorliegenden Arbeit, eine Abschätzung jenes Fehlers zu geben, die wir in dem folgenden Hauptsatz aussprechen:

Hauptsatz: Es gibt zwei nur von dem Körper und von dem Modul $\mathfrak{a}$ abhängende Konstanten C und c, so daß für $x \geqq 2$ unter den soeben eingeführten Bezeichnungen

$$(4) \qquad \left| \pi_\mathfrak{a}(x; v) - \frac{v}{h_0(\mathfrak{a})} \operatorname{Li} x \right| \leqq C\, x e^{-c\sqrt{\log x}}$$

gilt.

In § 1 bestimmen wir einen nullstellenfreien Streifen für $\zeta(s, \chi\lambda^m)$ und $\sigma < 1$, dessen Breite aber von dem Größencharakter λ^m abhängen

[1]) E. Hecke, Eine neue Art von Zetafunktionen und ihre Beziehungen zur Verteilung der Primzahlen II, Math. Zeitschr. 6 (1920), S. 11—51, insbesondere S. 38, Formel (52).

wird, und eine Abschätzung von $\frac{\zeta'}{\zeta}(s, \chi\lambda^m)$ in diesem Streifen, § 2 führt zu einer auch an sich bemerkenswerten Abschätzung von Summen von Größencharakteren über Primzahlen, in § 3 wird von einer Fourierreihe Gebrauch gemacht, § 4 endlich enthält den Beweis des Hauptsatzes.

Was die Bezeichnungsweise angeht, so schien mir wegen der vielen auftretenden Variablen der Gebrauch des O-Symbols nicht immer zweckmäßig. Es seien vielmehr im folgenden unter B Funktionen verstanden (nicht immer dieselben), die *beschränkt sind und deren Schranken nur vom Körper und vom Ideal* $\mathfrak{a}$ *abhängen.* C und c sind positive Konstanten (nicht immer dieselben), die auch nur vom Körper und vom Ideal $\mathfrak{a}$ abhängen. Wir können nach dieser Festsetzung also schreiben

$$|B| \leqq C.$$

§ 1.

Ein nullstellenfreier Streifen für die Heckeschen Funktionen.

Es sei

$$\lambda^m(\hat{\mu}) = e^{2\pi i w_{\mathfrak{a}}(\hat{\mu}) m} \tag{5}$$

ein Größencharakter für die idealen Zahlen $\hat{\mu}$ des Bereiches $\mathfrak{Z}$, zu dem nach Heckes Vorgang der Körper erweitert sei[2]). Mit $\chi(\hat{\mu})$ werden im folgenden stets Gruppencharaktere für die Gruppe der Idealklassen mod $\mathfrak{a}$ im engsten Sinne bezeichnet; diese Gruppe habe die Ordnung $h_0(\mathfrak{a})$. Ist h die Idealklassenzahl im gewöhnlichen Sinne und $\eta > 1$ die Grundeinheit des Körpers, so sei

$$\eta_{\mathfrak{a}} = \eta^{q_{\mathfrak{a}}}.$$

Jede Einheit ε des Körpers läßt sich dann darstellen als

$$\varepsilon = \pm \eta^l \eta_{\mathfrak{a}}^n$$

mit $0 \leqq l < q_{\mathfrak{a}}$, n beliebig ganz rational. Es gibt also $2q_{\mathfrak{a}}$ mod $\mathfrak{a}$ nichtassoziierter Einheiten, infolgedessen ist

$$h_0(\mathfrak{a}) = \frac{2^2 h\, \varphi(\mathfrak{a})}{2 q_{\mathfrak{a}}} = \frac{2h\, \varphi(\mathfrak{a})}{q_{\mathfrak{a}}}. \tag{6}$$

Ist $\chi(\hat{\mu})$ ein solcher Gruppencharakter, daß

$$\chi(\varepsilon)\lambda^m(\varepsilon) = 1 \tag{7}$$

für *sämtliche* Einheiten ε, so heiße $\chi\lambda^m$ ein *Größencharakter für Ideale.* Nicht jedes χ ergibt mit λ^m einen solchen. Für solche χ, für die (7) nicht gilt, ergibt sich jedoch

$$\sum_{(\varepsilon)_{\mathfrak{a}}} \chi(\varepsilon)\lambda^m(\varepsilon) = 0, \tag{8}$$

---

2) l. c. 1) § 2.

wo über alle mod $\mathfrak{a}$ nicht-assoziierten Einheiten zu summieren ist[3]). Durch die Zusammenziehung in der Schreibweise $\chi\lambda^m(\hat{\mu})$ werde stets ein Größencharakter für Ideale angedeutet, während wir $\chi(\hat{\mu})\lambda^m(\hat{\mu})$ schreiben wollen, wenn $\chi(\hat{\mu})$ ein beliebiger Charakter der Idealklassengruppe mod $\mathfrak{a}$ ist.

Mit dem Größencharakter für Ideale $\chi\lambda^m(\hat{\mu})$ wird nun die Heckesche Zetafunktion

$$\zeta(s, \chi\lambda^m) = \sum_{(\hat{\mu})} \frac{\chi\lambda^m(\hat{\mu})}{|N(\hat{\mu})|^s}$$

$$= \prod_{(\hat{\varpi})} \frac{1}{1 - \chi\lambda^m(\hat{\varpi})\,|N(\hat{\varpi})|^{-s}}$$

gebildet, wo $\hat{\mu}$ alle nicht-assoziierten idealen ganzen Zahlen von $\mathfrak{Z}$, $\hat{\varpi}$ alle nicht-assoziierten idealen Primzahlen von $\mathfrak{Z}$ durchläuft. Es sei noch $\chi_0(\hat{\mu})$ der Hauptcharakter mod $\mathfrak{a}$. Dann gilt

Satz 1. Für $\sigma > 1$ ist

$$(9) \qquad -3\frac{\zeta'}{\zeta}(\sigma, \chi_0) - 4\,\Re\frac{\zeta'}{\zeta}(\sigma + it, \chi\lambda^m) - \Re\frac{\zeta'}{\zeta}(\sigma + 2it, \chi^2\lambda^{2m}) \geqq 0.$$

Zum Beweise bilde man

$$\frac{\zeta'}{\zeta}(s, \chi_0) = -\sum_{(\hat{\varpi})} \log|N(\hat{\varpi})| \sum_{l=1}^{\infty} \left(\frac{\chi_0(\hat{\varpi})}{|N(\hat{\varpi})|^s}\right)^l$$

$$\frac{\zeta'}{\zeta}(s, \chi\lambda^m) = -\sum_{(\hat{\varpi})} \log|N(\hat{\varpi})| \sum_{l=1}^{\infty} \left(\frac{\chi\lambda^m(\hat{\varpi})}{|N(\hat{\varpi})|^s}\right)^l$$

und entsprechend $\frac{\zeta'}{\zeta}(s, \chi^2\lambda^{2m})$ und wende darauf die bekannte Schlußweise von de la Vallée-Poussin an[4]).

Satz 2. Für $m \neq 0$ und $-\frac{1}{2} \leqq \Re(s) \leqq 4$ gilt mit geeignetem C

$$(10) \qquad |\zeta(s, \chi\lambda^m)| < C\,\tau(t, m),$$

wo

$$(11) \qquad \tau(t, m) = (1 + |t|)^2 (1 + |m|)^2$$

ist.

Beweis: Mit einer geringen Abänderung eines früher von mir bewiesenen Satzes[5]) gilt unter den obigen Bedingungen zunächst

$$|\zeta(s, \chi\lambda^m)| \leqq C\left(1 + \left|t + \frac{\pi m}{\log \eta_{\mathfrak{a}}}\right|\right)\left(1 + \left|t - \frac{\pi m}{\log \eta_{\mathfrak{a}}}\right|\right).$$

[3]) H. Rademacher, Zur additiven Primzahltheorie algebraischer Zahlkörper I, Hamburger Abhandlungen 3 (1924), S. 120, Hilfssatz 4.

[4]) Siehe z. B. E. Landau, Vorlesungen über Zahlentheorie 2 (1927), S. 14, Satz 372.

[5]) Zur additiven Primzahltheorie algebraischer Zahlkörper III, Math. Zeitschr. 27 (1927), S. 362, Hilfssatz 15.

Die Abänderung besteht erstens in der Einbeziehung eines dortigen vom Modul $\mathfrak{a}$ abhängenden Faktors in die Konstante C, zweitens in der Unterdrückung von Exponenten auf der rechten Seite, die, wie sich aus dem dortigen Beweise ersehen läßt, nur für den Hauptcharakter $\chi\lambda^m = \chi_0$, also nur für $m = 0$ nötig sind. Da ferner

$$\left(1 + \left|t + \frac{\pi m}{\log \eta_{\mathfrak{a}}}\right|\right)\left(1 + \left|t - \frac{\pi m}{\log \eta_{\mathfrak{a}}}\right|\right) \leqq \left(1 + |t| + \frac{\pi |m|}{\log \eta_{\mathfrak{a}}}\right)^2$$
$$< C(1 + |t|)^2 (1 + |m|)^2$$

ist, so ist Satz 2 bewiesen.

Satz 3. Für $1 < \Re(s) = \sigma \leqq 4$, $m \neq 0$ ist

$$\left|\frac{1}{\zeta(s, \chi\lambda^m)}\right| < \frac{C}{\sigma - 1}.$$

Beweis: Es ist

$$\frac{1}{\zeta(s, \chi\lambda^m)} = \prod_{(\hat{\varpi})} (1 - \chi\lambda^m(\hat{\varpi})\,|N(\hat{\varpi})|^{-s}) = \sum_{(\hat{\nu})} \mu(\hat{\nu}) \frac{\chi\lambda^m(\hat{\nu})}{|N(\hat{\nu})|^s},$$

wo $\mu(\hat{\nu})$ die Möbiussche Funktion ist. Also

$$\left|\frac{1}{\zeta(s, \chi\lambda^m)}\right| < \sum_{(\hat{\nu})} \frac{1}{|N(\hat{\nu})|^s} < \frac{C}{\sigma - 1},$$

welch letzteres man z. B. gleichfalls meinem soeben zitierten Hilfssatz entnehmen kann.

Satz 4. Für $1 < \sigma_0 < 2$ und $m \neq 0$ gilt

$$(12) \qquad -\Re \frac{\zeta'}{\zeta}(\sigma_0 + i\gamma, \chi\lambda^m) < \log\left(\frac{C}{\sigma - 1}\tau(\gamma, m)\right).$$

Gibt es aber zu γ ein β, $\sigma_0 - 1 < \beta < \sigma_0$, so daß $\beta + i\gamma$ Nullstelle von $\zeta(s, \chi\lambda^m)$ ist, so gilt sogar

$$(13) \quad -\Re \frac{\zeta'}{\zeta}(\sigma_0 + i\gamma, \chi\lambda^m) < \log\left(\frac{C}{\sigma_0 - 1}\tau(\gamma, m)\right) + \left(1 - \frac{(\beta - \sigma_0)^2}{4}\right)\frac{1}{\beta - \sigma_0}.$$

Zum Beweise berufen wir uns auf einen Satz von Landau[6]). Für seine Anwendung brauchen wir nur zu bemerken:

---

[6]) Vorlesungen über Zahlentheorie 2, S. 15, Satz 374.

Zusatz bei der Korrektur (12. 1. 1935): Soeben bemerke ich, daß ich versehentlich unkorrekt zitiert habe. Der Wortlaut des Landauschen Satzes liefert nicht genau die im Text auftretenden (übrigens unwesentlichen) numerischen Konstanten, sondern in (12) eine doppelt so große rechte Seite. Man kann aber unter den Landauschen Voraussetzungen l. c. auch leicht die folgenden für unseren Satz benötigten Ungleichungen beweisen:

Es ist

$$-\Re \frac{f'}{f}(s_0) < \frac{2M}{r},$$

1. $\zeta(s, \chi\lambda^m)$ ist regulär in $|s - \sigma_0 - i\gamma| \leqq 2$,

2. $\zeta(\sigma_0 + i\gamma, \chi\lambda^m) \neq 0$,

3. $\left|\frac{\zeta(s, \chi\lambda^m)}{\zeta(\sigma_0 + i\gamma, \chi\lambda^m)}\right| < C(3 + |\gamma|)^2 (1 + |m|)^2 \frac{1}{\sigma_0 - 1}$

für $|s - \sigma_0 - i\gamma| \leqq 2$, was aus Satz 2 und Satz 3 folgt,

4. $\zeta(s, \chi\lambda^m) \neq 0$ für $\Re(s) > \sigma_0 > 1$.

Dann ergibt der erwähnte Landausche Satz die Ungleichungen (12) und (13).

Satz 5. Es gibt eine nur vom Modul $\mathfrak{a}$ (dagegen nicht von m) abhängende positive Zahl E, so daß $\zeta(s, \chi\lambda^m)$ nullstellenfrei ist für

$$\sigma \geqq 1 - \frac{1}{1000 \log \tau(t, m)}, \qquad |t| \geqq E$$

und

$$\sigma \geqq 1 - \frac{1}{1000 \log \tau(E, m)}, \qquad |t| \geqq E.$$

E darf offenbar > 1 angenommen werden.

Beweis: Nach Satz 4 ist, wenn $\beta + i\gamma$ eine Nullstelle von $\zeta(s, \chi\lambda^m)$ ist,

$$-\Re \frac{\zeta'}{\zeta}(\sigma_0 + i\gamma, \chi\lambda^m) < \log \frac{\tau(\gamma, m)}{\sigma_0 - 1} + C + \left(1 - \frac{(\beta - \sigma_0)^2}{4}\right) \frac{1}{\beta - \sigma_0},$$

und da $\tau(2\gamma, 2m) \leqq 16\,\tau(\gamma, m)$, so ist

$$-\Re \frac{\zeta'}{\zeta}(\sigma_0 + 2i\gamma, \chi^2\lambda^{2m}) < \log \frac{\tau(\gamma, m)}{\sigma_0 - 1} + C.$$

Ferner ist für hinreichend kleines $\sigma_0 - 1 > 0$

$$-\frac{\zeta'}{\zeta}(\sigma_0, \chi_0) < \frac{1{,}1}{\sigma_0 - 1}, \tag{14}$$

da $\zeta(s, \chi_0)$ einen Pol erster Ordnung für $s = 1$ besitzt. Dies in (9), Satz 1, eingesetzt, ergibt

$$0 < \frac{3{,}3}{\sigma_0 - 1} + 5 \log \frac{\tau(\gamma, m)}{\sigma_0 - 1} + C_1 + 4\left(1 - \frac{(\beta - \sigma_0)^2}{4}\right) \frac{1}{\beta - \sigma_0}.$$

---

und falls es eine Wurzel ϱ auf dem Radius zwischen $s_0 - r$ (exkl.) und s_0 (exkl.) gibt, ist sogar

$$-\Re \frac{f'}{f}(s_0) < \frac{2M}{r} - \left(1 - \left(\frac{s_0 - \varrho}{r}\right)^2\right) \frac{1}{s_0 - \varrho}.$$

Die einzige Abänderung im Landauschen Beweis besteht darin, statt der von Landau dort eingeführten Hilfsfunktion $g(s)$ die folgende zu betrachten:

$$g_1(s) = f(s) \prod_{\varrho} \frac{r - \frac{s\varrho}{r}}{\varrho - s},$$

wo ϱ alle etwaigen Wurzeln der Funktion $f(s)$ im Kreise $|s| < r$ durchläuft. Dann ist $|g_1(s)| = |f(s)|$ für $|s| = r$, und $\log g(s)$ ist regulär in $|s| \leqq r_1$ für jedes $r_1 < r$ usw.

Wir ziehen nur Nullstellen $\beta + i\gamma$ in Betracht mit

(15) $$\sigma_0 - \beta \leqq \tfrac{1}{5}$$

und haben dann

(16) $$\frac{3,96}{\sigma_0 - \beta} < \frac{3,3}{\sigma_0 - 1} + 5 \log \frac{\tau(\gamma, m)}{\sigma_0 - 1} + C_1.$$

Es werde nun

$$\sigma_0 = 1 + \frac{1}{100 \log \tau(\gamma, m)}$$

gesetzt, dann ist

(17) $$\sigma_0 - 1 = \frac{1}{100 \log \tau(\gamma, m)} > \frac{1}{100\ \tau(\gamma, m)},$$

also nach (16)

$$\frac{3,96}{\sigma_0 - \beta} < 330 \log \tau(\gamma, m) + 10 \log \tau(\gamma, m) + 5 \log 100 + C_1,$$

$$\sigma_0 - \beta = 1 - \beta - (1 - \sigma_0) > \frac{3,96}{340 \log \tau(\gamma, m) + 5 \log 100 + C_1}$$

und wegen (17)

$$1 - \beta > -\frac{3,96}{396 \log \tau(\gamma, m)} + \frac{3,96}{340 \log \tau(\gamma, m) + C_1 + 5 \log 100}.$$

Wegen

(18) $$\tau(\gamma, m) > \gamma^2$$

gibt es ein positives D, das von m unabhängig ist, so daß für $|\gamma| \geqq D$

(19) $$340 \log \tau(\gamma, m) + C_1 + 5 \log 100 \leqq 360 \log \tau(\gamma, m),$$

also für $|\gamma| \geqq D$

(20) $$1 - \beta > \frac{3,96}{\log \tau(\gamma, m)}\left(-\frac{1}{396} + \frac{1}{360}\right) = \frac{1}{1000 \log \tau(\gamma, m)}.$$

Für $|\gamma| < D$ schließen wir aus (16)

$$\sigma_0 - \beta > \frac{3,96}{\frac{3,3}{\sigma_0 - 1} + 5 \log \tau(\gamma, m) - 5 \log(\sigma_0 - 1) + C_1}$$

$$= \frac{3,96(\sigma_0 - 1)}{3,3 + 5(\sigma_0 - 1) \log \tau(\gamma, m) - 5(\sigma_0 - 1) \log(\sigma_0 - 1) + C_1(\sigma_0 - 1)},$$

und da wegen (19)

$$C_1 < 20 \log \tau(D, m)$$

ist, haben wir

$$\sigma_0 - \beta > \frac{3,96(\sigma_0 - 1)}{3,3 + 25(\sigma_0 - 1) \log \tau(D, m) - 5(\sigma_0 - 1) \log(\sigma_0 - 1)}.$$

Für hinreichend kleines $\delta = \sigma_0 - 1 > 0$ haben wir

$$-\delta \log \delta = -(\sigma_0 - 1) \log(\sigma_0 - 1) < 0,01,$$

also

$$1 - \beta > -(\sigma_0 - 1) + \frac{3,96(\sigma_0 - 1)}{3,35 + 25(\sigma_0 - 1) \log \tau(D, m)}$$

und für

(21) $$\sigma_0 - 1 = \operatorname{Min}\left(\delta, \frac{1}{100 \log \tau(D, m)}\right)$$

$$1 - \beta > -(\sigma_0 - 1) + \frac{3{,}96(\sigma_0 - 1)}{3{,}35 + 0{,}25} = \frac{\sigma_0 - 1}{10}$$

und somit

$$1 - \beta > \operatorname{Min}\left(\frac{\delta}{10}, \frac{1}{1000 \log \tau(D, m)}\right).$$

Es werde nun die Zahl $E \geqq D$ so groß genommen, daß für

(22) $$\sigma_0 - 1 \leqq \frac{1}{100 \log \tau(E, 1)}$$

stets (14) gilt. Außerdem erfülle E die Bedingung

$$\delta > \frac{1}{100 \log \tau(E, 1)},$$

so daß wegen (21)

$$\sigma_0 - 1 > \frac{1}{100 \log \tau(E, m)}$$

ist. Dann haben wir also

(23) $$\begin{aligned} 1 - \beta &> \frac{1}{1000 \log \tau(\gamma, m)} && \text{für } |\gamma| \geqq E, \\ 1 - \beta &> \frac{1}{1000 \log \tau(E, m)} && \text{für } |\gamma| \leqq E, \end{aligned}$$

vorausgesetzt, daß (15) gilt. Die andere Möglichkeit $\sigma_0 - \beta > \frac{1}{5}$ oder $1 - \beta > \frac{1}{5} - (\sigma_0 - 1)$ hat wegen (22) erst recht (23) zur Folge. Wir schließen also aus (23): Ist $\beta + i\gamma$ eine Nullstelle von $\zeta(s, \chi\lambda^m)$, so ist

$$\begin{aligned} \beta &< 1 - \frac{1}{1000 \log \tau(\gamma, m)} && \text{für } |\gamma| \geqq E, \\ \beta &< 1 - \frac{1}{1000 \log \tau(E, m)} && \text{für } |\gamma| \leqq E. \end{aligned}$$

Dies ist aber Satz 5.

Satz 6. In dem Streifen

$$\begin{aligned} 1 - \frac{1}{3000 \log \tau(t, m)} &\leqq \sigma \leqq 3 && \text{für } |t| \geqq E, \\ 1 - \frac{1}{3000 \log \tau(E, m)} &\leqq \sigma \leqq 3 && \text{für } |t| \leqq E \end{aligned}$$

gilt

$$\begin{aligned} \left|\frac{\zeta'}{\zeta}(s, \chi\lambda^m)\right| &\leqq C \log^3 \tau(t, m) && \text{für } |t| \geqq E \\ &\leqq C \log^3 \tau(E, m) && \text{für } |t| \leqq E. \end{aligned}$$

Beweis: Es werde $s_0 = 2 + it_0$ gesetzt. Der Kreis $\mathfrak{K}$ mit

$$|s - s_0| \leqq R$$

mit

$$R = 1 + \frac{1}{2000 \log \tau(t_0, m)} \qquad |t| \geqq E$$

bzw.
$$R = 1 + \frac{1}{2000 \log \tau(E, m)} \qquad |t| \leqq E$$

liegt ganz in dem nullstellenfreien Streifen des Satzes 5. Denn in $\mathfrak{K}$ ist $|t - t_0| \leqq R < 2$, also

$$\begin{aligned} \log \tau(t, m) &\leqq \log((3 + |t_0|)^2 (1 + |m|)^2) \\ &\leqq \log(9(1 + |t_0|)^2 (1 + |m|)^2) \\ &= \log \tau(t_0, m) + \log 9 \\ &< 2 \log \tau(t_0, m), \end{aligned}$$

wegen $\tau(t_0, m) \geqq \tau(E, m) \geqq \tau(1,1) = 16$.

Um einen Hilfssatz von Carathéodory und Landau [7]) auf die Funktion $\log \zeta(s, \chi \lambda^m)$ anwenden zu können, stellen wir fest:

$$\begin{aligned} |\log \zeta(s_0, \chi \lambda^m)| &= \left| \sum_{(\hat{\mathfrak{w}}, l)} \frac{1}{l} \left(\frac{\chi \lambda^m (\hat{\mathfrak{w}})}{|N(\hat{\mathfrak{w}})|^{s_0}} \right)^l \right| \\ &\leqq \sum_{(\hat{\mathfrak{w}}, l)} \frac{1}{l} \frac{1}{|N(\hat{\mathfrak{w}})|^{2l}} = C_2. \end{aligned}$$

Ferner ist in $\mathfrak{K}$ nach (10)

$$\begin{aligned} \Re \log \zeta(s, \chi \lambda^m) = \log |\zeta(s, \chi \lambda^m)| &< C + \log \tau(t, m) \\ &< C_3 + \log \tau(t_0, m). \end{aligned}$$

Dann gilt in dem Kreise $|s - s_0| \leqq r$ mit

$$r = 1 + \frac{1}{3000 \log \tau(t_0, m)} \qquad \text{für } |t_0| \geqq E,$$

bzw.

$$r = 1 + \frac{1}{3000 \log \tau(E, m)} \qquad \text{für } |t_0| \leqq E:$$

$$\left| \frac{\zeta'}{\zeta}(s, \chi \lambda^m) \right| \leqq \frac{3 \log^2 \tau(t_0, m)}{\left(\frac{1}{2000} - \frac{1}{3000} \right)^2} (C_2 + C_3 + \log \tau(t_0, m)), \; |t_0| \geqq E,$$

bzw

$$\leqq \frac{3 \log^2 \tau(E, m)}{\left(\frac{1}{2000} - \frac{1}{3000} \right)^2} (C_2 + C_3 + \log \tau(E, m)), \; |t_0| \leqq E,$$

woraus Satz 6 sofort folgt.

§ 2.

Über Primzahlen erstreckte Summen von Charakteren.

Satz 7. Für $m \neq 0$ und $x > 0$ ist

$$\left| \int_{2-i\infty}^{2+i\infty} \frac{x^s}{s^2} \frac{\zeta'}{\zeta}(s, \chi \lambda^m) \, ds \right| \leqq C \log^3 (1 + |m|) \cdot x \cdot e^{-c \frac{\log x}{\log(1 + |m|) + \sqrt{\log x}}}.$$

[7]) Landau, Vorlesungen über Zahlentheorie 1, S. 192—194, Satz 225.

Beweis: Der Integrationsweg von $2 - i\infty$ bis $2 + i\infty$ kann nach links verschoben werden bis zu dem Weg $\mathfrak{W}$, der durch

$$\sigma = 1 - \frac{1}{3000 \log \tau(t, m)}, \qquad t \leqq -E,$$

$$\sigma = 1 - \frac{1}{3000 \log \tau(E, m)}, \qquad -E \leqq t \leqq E,$$

$$\sigma = 1 - \frac{1}{3000 \log \tau(t, m)}, \qquad t \geqq E$$

beschrieben wird, da zwischen beiden Wegen $\zeta(s, \chi \lambda^m)$ nullstellen- und polfrei ist.

Dann ist nach Satz 6

$$(24) \qquad \left| \int_{\mathfrak{W}} \frac{x^s}{s^2} \frac{\zeta'}{\zeta}(s, \chi \lambda^m)\, ds \right| \leqq C \int_0^E \frac{x^{1 - \frac{1}{3000 \log \tau(E, m)}}}{1 + t^2} \log^3 \tau(E, m)\, dt$$

$$+ C \left\{ \int_E^T + \int_T^\infty \right\} \frac{x^{1 - \frac{1}{3000 \log \tau(t, m)}}}{1 + t^2} \log^3 \tau(t, m)\, dt$$

mit einem noch zu bestimmenden $T \geqq E$. Folglich

$$\left| \int_{\mathfrak{W}} \right| \leqq C \log^3 \tau(E, m)\, x^{1 - \frac{1}{3000 \log \tau(E, m)}} \int_0^\infty \frac{dt}{1 + t^2}$$

$$+ C x^{1 - \frac{1}{3000 \log \tau(T, m)}} \int_E^\infty \frac{\log^3 \tau(t, m)}{1 + t^2} dt$$

$$+ C x \int_T^\infty \frac{\log^3 \tau(t, m)}{1 + t^2} dt.$$

Nach der Definition (12) von $\tau(t, m)$ ist

$$\log \tau(t, m) = 2 \log(1 + |t|) + 2 \log(1 + |m|)$$

und daher für $m \neq 0$

$$\left| \int_{\mathfrak{W}} \right| \leqq C \log^3(1 + |m|)\, x^{1 - \frac{c}{\log(1 + E) + \log(1 + |m|)}}$$

$$+ C x^{1 - \frac{c}{\log(1 + T) + \log(1 + |m|)}} \int_1^\infty \frac{\log^3(1 + t) + \log^3(1 + |m|)}{1 + t^2} dt$$

$$+ C x \int_T^\infty \frac{\log^3(1 + t) + \log^3(1 + |m|)}{1 + t^2} dt,$$

$$(25)\qquad \left|\int\limits_{\mathfrak{W}}\right| \leqq C \log^3 (1 + |m|)\cdot x \cdot e^{-c\frac{\log x}{\log(1+E)+\log(1+|m|)}}$$
$$+ C \log^3 (1 + |m|)\cdot x \cdot e^{-c\frac{\log x}{\log(1+T)+\log(1+|m|)}}$$
$$+ C \log^3 (1 + |m|)\cdot x \cdot (1 + T)^{-1+\varepsilon}.$$

Um die Größenordnung der letzten beiden Glieder zur Übereinstimmung zu bringen, setzen wir, sofern es sich mit der Bedingung $T \geqq E$ verträgt,

$$\frac{\log x}{\log(1+T)+\log(1+|m|)} = \log(1+T),$$

also

$$(26)\qquad \log(1+T) = -\tfrac{1}{2}\log(1+|m|) + \sqrt{\tfrac{1}{4}\log^2(1+|m|)+\log x},$$

aber nur wenn die rechte Seite dieser Gleichung $\geqq \log(1+E)$ ausfällt. Ist dies nicht der Fall, so sei $T = E$ gesetzt, d. h. das Integral von E bis T in (24) fällt weg und damit auch das zweite Glied rechts in (25). Wenn wir dieses zweite Glied nun dennoch in jedem der beiden Fälle mitnehmen und darin (26) eintragen, so bleibt die Ungleichung (25) ausnahmslos richtig, also

$$\left|\int\limits_{\mathfrak{W}}\right| \leqq C \log^3 (1 + |m|)\cdot x \cdot e^{-c\frac{\log x}{\log(1+E)+\log(1+|m|)}}$$
$$+ C \log^3 (1 + |m|)\cdot x \cdot e^{-c\frac{\log x}{\frac{1}{2}\log(1+|m|)+\sqrt{\frac{1}{4}\log^2(1+|m|)+\log x}}}$$
$$+ C \log^3 (1 + |m|)\cdot x \cdot e^{-(1-\varepsilon)\left(-\frac{1}{2}\log(1+|m|)+\sqrt{\frac{1}{4}\log^2(1+|m|)+\log x}\right)}.$$

Im Exponenten des dritten Summanden können wir schreiben:

$$-\tfrac{1}{2}\log(1+|m|) + \sqrt{\tfrac{1}{4}\log^2(1+|m|)+\log x}$$
$$= \frac{\log x}{\frac{1}{2}\log(1+|m|)+\sqrt{\frac{1}{4}\log^2(1+|m|)+\log x}}.$$

Nun ist

$$\tfrac{1}{2}\log(1+|m|) + \sqrt{\tfrac{1}{4}\log^2(1+|m|)+\log x}$$
$$\leqq \begin{cases} \dfrac{1+\sqrt{5}}{2}\sqrt{\log x} & \text{für}\quad \log(1+|m|) \leqq \sqrt{\log x} \\ \dfrac{1+\sqrt{5}}{2}\log(1+|m|) & \text{für}\quad \log(1+|m|) \geqq \sqrt{\log x}, \end{cases}$$

also

$$\tfrac{1}{2}\log(1+|m|) + \sqrt{\tfrac{1}{4}\log^2(1+|m|)+\log x} \leqq \frac{1+\sqrt{5}}{2}\left(\sqrt{\log x}+\log(1+|m|)\right).$$

Daher haben wir

$$\left|\int\limits_{\mathfrak{W}}\right| \leqq C \log^3 (1 + |m|)\cdot x \cdot e^{-c\frac{\log x}{\log(1+E)+\log(1+|m|)}}$$
$$+ C \log^3 (1 + |m|)\cdot x \cdot e^{-c\frac{\log x}{\sqrt{\log x}+\log(1+|m|)}}.$$

Indem wir noch bemerken, daß hier das erste Glied der rechten Seite von kleinerer Größenordnung als das zweite ist, so haben wir mit geeignetem C Satz 7 bewiesen,

Satz 8. Es ist für $m \neq 0$

$$\sum_{\substack{(\hat{\varpi}) \\ |N(\hat{\varpi})| \leqq x}} \chi \lambda^m (\hat{\varpi}) = B \log^3 (1 + |m|) \cdot x \cdot e^{-c \frac{\log x}{\log(1+|m|) + \sqrt{\log x}}},$$

worin $(\hat{\varpi})$ unter dem Summenzeichen andeuten soll, daß nur nichtassoziierte ideale Primzahlen durchlaufen werden sollen.

Beweis: Es ist

$$-\frac{1}{2\pi i} \int_{2-i\infty}^{2+i\infty} \frac{x^s}{s^2} \frac{\zeta'}{\zeta} (s, \chi \lambda^m) \, ds$$

$$= \sum_{(\hat{\varpi})} \log |N(\hat{\varpi})| \sum_{l=1}^{\infty} (\chi \lambda^m (\hat{\varpi}))^l \frac{1}{2\pi i} \int_{2-i\infty}^{2+i\infty} \frac{\left(\frac{x}{|N(\hat{\varpi})|^l}\right)^s}{s^2} ds.$$

Dies ist nach einem bekannten Hilfssatz [8]) gleich

$$\sum_{\substack{(\hat{\varpi}),\, l \\ |N(\hat{\varpi})|^l \leqq x}} \log |N(\hat{\varpi})| (\chi \lambda^m (\hat{\varpi}))^l \log \frac{x}{|N(\hat{\varpi})|^l}.$$

Die Summe zerspalten wir in die für $l = 1$ und für $l \geqq 2$. Für diese gilt

$$\left| \sum_{\substack{(\hat{\varpi}) \\ l \geqq 2 \\ N(\hat{\varpi})^l \leqq x}} \right| \leqq \sum_{\substack{|N(\hat{\varpi})| \leqq \sqrt{x} \\ 2^l \leqq x}} \log |N(\hat{\varpi})| \log x$$

$$< C \log^2 x \sum_{|N(\hat{\varpi})| \leqq \sqrt{x}} \log |N(\hat{\varpi})| < C \sqrt{x} \log^2 x.$$

Hieraus und aus Satz 7 schließen wir

$$(27) \qquad \sum_{\substack{(\hat{\varpi}) \\ |N(\hat{\varpi})| \leqq x}} \log |N(\hat{\varpi})| \, \chi \lambda^m (\hat{\varpi}) \log \frac{x}{|N(\hat{\varpi})|}$$

$$= B \log^3 (1 + |m|) \cdot x \cdot e^{-c \frac{\log x}{\log(1+|m|) + \sqrt{\log x}}}.$$

---

[8]) Landau, Vorlesungen 2, Satz 377.

Von dieser Formel gelangt man zu der des Satzes ın der üblichen Weise: Es sei $0 < \delta = \delta(x) < 1$. Dann ist einerseits wegen (27)

$$\sum_{\substack{(\hat{\varpi}) \\ |N(\hat{\varpi})| \leqq (1+\delta)x}} \log|N(\hat{\varpi})|\,\chi\lambda^m(\hat{\varpi}) \log\frac{x(1+\delta)}{|N(\hat{\varpi})|} - \sum_{\substack{(\hat{\varpi}) \\ |N(\hat{\varpi})| \leqq x}} \log|N(\hat{\varpi})|\,\chi\lambda^m(\hat{\varpi}) \log\frac{x}{|N(\hat{\varpi})|} = B\log^3(1+|m|)\cdot x\cdot e^{-c\frac{\log x}{\log(1+|m|)+\sqrt{\log x}}}, \tag{28}$$

andererseits ist die linke Seite gleich

$$\sum_{\substack{(\hat{\varpi}) \\ x < |N(\hat{\varpi})| \leqq (1+\delta)x}} \log|N(\hat{\varpi})|\,\chi\lambda^m(\hat{\varpi}) \log\frac{x}{|N(\hat{\varpi})|} + \log(1+\delta) \sum_{|N(\hat{\varpi})| \leqq (1+\delta)x} \log|N(\hat{\varpi})|\,\chi\lambda^m(\hat{\varpi}). \tag{29}$$

Hierin ist

$$\left|\sum_{x<|N(\hat{\varpi})| \leqq (1+\delta)x} \log|N(\hat{\varpi})|\,\chi\lambda^m(\hat{\varpi}) \log\frac{x}{|N(\hat{\varpi})|}\right| \leqq \log(1+\delta) \sum_{x<|N(\hat{\varpi})| \leqq (1+\delta)x} \log|N(\hat{\varpi})| \leqq C\log(1+\delta)(\delta x + xe^{-c\sqrt{\log x}})$$

und ebenso

$$\left|\log(1+\delta) \sum_{x<|N(\hat{\varpi})| \leqq (1+\delta)x} \log|N(\hat{\varpi})|\,\chi\lambda^m(\hat{\varpi})\right| \leqq C\log(1+\delta)(\delta x + xe^{-c\sqrt{\log x}}).$$

Daher also wegen (28) und (29)

$$\sum_{|N(\hat{\varpi})| \leqq x} \log|N(\hat{\varpi})|\,\chi\lambda^m(\hat{\varpi}) = B\log^3(1+|m|)\frac{x}{\log(1+\delta)}\cdot e^{-c\frac{\log x}{\log(1+|m|)+\sqrt{\log x}}} + B\cdot\delta x + B\cdot xe^{-c\sqrt{\log x}}.$$

Setzen wir

$$g(x) = \sum_{\substack{(\hat{\varpi}) \\ |N(\hat{\varpi})| \leqq x}} \log|N(\hat{\varpi})|\,\chi\lambda^m(\hat{\varpi}) \tag{30}$$

und

$$\delta(x) = e^{-\frac{c}{2}\frac{\log x}{\log(1+|m|)+\sqrt{\log x}}},$$

so haben wir

$$g(x) = B\cdot\log^3(1+|m|)\cdot x\cdot e^{-\frac{c}{2}\frac{\log x}{\log(1+|m|)+\sqrt{\log x}}}. \tag{31}$$

Nun ist nach (30) für $x \geqq 2$

$$\Big|\sum_{\substack{(\hat{\varpi})\\ |N(\hat{\varpi})| \leqq x}} \chi\lambda^m(\hat{\varpi})\Big| = \Big|\sum_{j=2}^{[x]} \frac{g(j)-g(j-1)}{\log j}\Big|$$

$$\leqq \Big|\sum_{j=2}^{[x]} g(j)\Big(\frac{1}{\log j} - \frac{1}{\log(j+1)}\Big)\Big| + \Big|\frac{g([x])}{\log([x]+1)}\Big|$$

$$< \operatorname*{Max}_{2\leqq j\leqq x} |g(j)| \sum_{j=2}^{\infty}\Big(\frac{1}{\log j} - \frac{1}{\log(j+1)}\Big) + |g(x)|,$$

woraus nach (31) der Satz 8 mit c in neuer Bedeutung folgt.

Während es sich in dem soeben bewiesenen Satz um eine Summe über solche idealen Primzahlen handelt, die nicht-assoziiert schlechthin sind, soll im nächsten Satze über Primzahlen, die mod $\mathfrak{a}$ nicht-assoziiert sind, summiert werden. Es darf auch χ ein beliebiger Charakter der Idealklassengruppe mod $\mathfrak{a}$ im engsten Sinne sein, $\chi(\hat{\mu})\,\lambda^m(\hat{\mu})$ braucht also kein Größencharakter für Ideale (vgl. § 1) zu sein. In diesem Sinne gilt:

S a t z 9. Für $m \neq 0$ ist

$$\sum_{\substack{(\hat{\varpi})_{\mathfrak{a}}\\ |N(\hat{\varpi})| \leqq x}} \chi(\hat{\varpi})\,\lambda^m(\varpi) = B\log^3(1+|m|)\cdot x\cdot e^{-c\frac{\log x}{\log(1+|m|)+\sqrt{\log x}}}.$$

B e w e i s: ε durchlaufe ein System von mod $\mathfrak{a}$ nicht-assoziierten Einheiten. Dann ist

$$\sum_{\substack{(\hat{\varpi})_{\mathfrak{a}}\\ |N(\hat{\varpi})| \leqq x}} \chi(\hat{\varpi})\,\lambda^m(\hat{\varpi}) = \sum_{\substack{(\hat{\varpi})\\ |N(\hat{\varpi})| \leqq x}} \chi(\hat{\varpi})\,\lambda^m(\hat{\varpi}) \sum_{(\varepsilon)_{\mathfrak{a}}} \chi(\varepsilon)\,\lambda^m(\varepsilon).$$

Ist nun $\chi(\hat{\mu})\,\lambda^m(\hat{\mu})$ ein Größencharakter für Ideale, so ist nach (7) die Summe des zweiten Faktors gleich $2\,q_{\mathfrak{a}}$, und auf den ersten Faktor wenden wir Satz 8 an. Ist jedoch $\chi(\hat{\mu})\,\lambda^m(\hat{\mu})$ kein Größencharakter für Ideale, so ist nach (8) die Summe des zweiten Faktors gleich null, und damit verschwindet die ganze Summe. In beiden Fällen ist Satz 9 bewiesen.

S a t z 10. Es sei ϱ eine feste ganze totalpositive Körperzahl. In der folgenden Summe soll über solche mod $\mathfrak{a}$ nicht-assoziierten Primzahlen ϖ summiert werden, die die Eigenschaften haben:

$$\varpi \succ 0, \qquad \varpi \equiv \varrho \pmod{\mathfrak{a}}, \qquad N(\hat{\varpi}) \leqq x.$$

Dann ist für $m \neq 0$

$$\sum_{\varpi} \lambda^m(\varpi) = B\cdot\log^3(1+|m|)\cdot x\cdot e^{-c\frac{\log x}{\log(1+|m|)+\sqrt{\log x}}}.$$

Beweis: Satz 9 in Verbindung mit

$$\sum_{\chi} \overline{\chi}(\varrho) \sum_{\substack{(\hat{\varpi})_{\mathfrak{a}} \\ |N(\hat{\varpi})| \leqq x}} \chi(\hat{\varpi}) \lambda^m(\hat{\varpi}) = \sum_{\substack{(\hat{\varpi})_{\mathfrak{a}} \\ |N(\hat{\varpi})| \leqq x}} \lambda^m(\hat{\varpi}) \sum_{\chi} \overline{\chi}(\varrho) \chi(\hat{\varpi})$$

$$= h_0(\mathfrak{a}) \sum_{\substack{(\varpi)_{\mathfrak{a}} \\ \varpi \succ 0,\ \varpi \equiv \varrho \pmod{\mathfrak{a}} \\ N(\varpi) \leqq x}} \lambda^m(\varpi).$$

§ 3.

Anwendung einer Fourierschen Reihe.

Es sei $0 \leqq a < b \leqq 1$. Unter $f(y; a, b)$ werde folgende Funktion verstanden:

$$f(y; a, b) = \begin{cases} 0 & \text{für } 0 \leqq y \leqq a, \\ 2\dfrac{y-a}{b-a} & \text{für } a \leqq y \leqq \dfrac{a+b}{2}, \\ 2\dfrac{b-y}{b-a} & \text{für } \dfrac{a+b}{2} \leqq y \leqq b, \\ 0 & \text{für } b \leqq y \leqq 1; \end{cases} \tag{32}$$

ferner sei $f(y; a, b)$ über das Intervall (0, 1) hinaus periodisch mit der Periode 1 fortgesetzt.

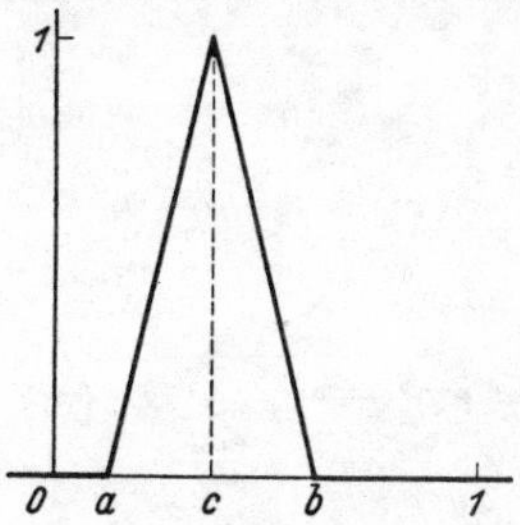

Diese Funktion hat, wie man leicht nachrechnet, folgende Fourierentwicklung:

$$f(y; a, b) = \frac{d}{2} + \frac{2}{d\pi^2} \sum_{m=1}^{\infty} \frac{\cos 2\pi m (y-c)}{m^2}$$

$$- \frac{1}{d\pi^2} \sum_{m=1}^{\infty} \frac{\cos 2\pi m (y-a)}{m^2} - \frac{1}{d\pi^2} \sum_{m=1}^{\infty} \frac{\cos 2\pi m (y-b)}{m^2}$$

mit

$$d = b - a, \qquad c = \frac{a+b}{2}.$$

Man kann sie auch in die Form

$$f(y; a, b) = \frac{d}{2} + \frac{1}{d\pi^2} \sum_{m=-\infty}^{+\infty}{}' \frac{1}{m^2} e^{2\pi i m (y-c)}$$

$$- \frac{1}{2d\pi^2} \sum_{m=-\infty}^{+\infty}{}' \frac{1}{m^2} e^{2\pi i m (y-a)} - \frac{1}{2d\pi^2} \sum_{m=-\infty}^{+\infty}{}' \frac{1}{m^2} e^{2\pi i m (y-b)}$$

setzen, da alle diese Reihen absolut konvergieren.

Wir setzen jetzt

$$y = w_{\mathfrak{a}}(\varpi) = \frac{1}{2\log \eta_{\mathfrak{a}}} \log \left| \frac{\varpi}{\varpi'} \right|$$

(siehe (1)) und summieren über ϖ:

$$(33)\qquad \sum_{\substack{(\varpi)_{\mathfrak{a}} \\ \varpi \succ 0,\ \varpi \equiv \varrho \ (\mathrm{mod}\ \mathfrak{a}) \\ N(\varpi) \leqq x}} f(w_{\mathfrak{a}}(\varpi); a, b)$$

$$= \frac{d}{2} \sum_{\varpi} 1 + \frac{1}{d\pi^2} \sum_{\varpi} \sum_{m=-\infty}^{+\infty}{}' \frac{1}{m^2} e^{2\pi i m (w_{\mathfrak{a}}(\varpi) - c)}$$

$$- \frac{1}{2d\pi^2} \sum_{\varpi} \sum_{m=-\infty}^{+\infty}{}' \frac{1}{m^2} e^{2\pi i m (w_{\mathfrak{a}}(\varpi) - a)}$$

$$- \frac{1}{2d\pi^2} \sum_{\varpi} \sum_{m=-\infty}^{+\infty}{}' \frac{1}{m^2} e^{2\pi i m (w_{\mathfrak{a}}(\varpi) - b)},$$

wobei die Summationsbedingungen der ϖ, nämlich

$$(\varpi)_{\mathfrak{a}}, \quad \varpi \succ 0, \quad \varpi \equiv \varrho \ (\mathrm{mod}\ \mathfrak{a}), \quad N(\varpi) \leqq x$$

bei allen Summen über ϖ hier und in diesem ganzen Paragraphen dieselben sind.

Nun ist nach Hecke [9])

$$\sum_{\varpi} 1 = \frac{1}{h_0(\mathfrak{a})} \int_2^x \frac{du}{\log u} + B \cdot x e^{-c\sqrt{\log x}}.$$

Ferner tragen wir (5) in (33) ein und haben dann

$$(34)\qquad \sum_{\varpi} f(w_{\mathfrak{a}}(\varpi); a, b)$$

$$= \frac{d}{2h_0(\mathfrak{a})} \int_2^x \frac{du}{\log u} + B \cdot x e^{-c\sqrt{\log x}}$$

$$+ \frac{1}{2d\pi^2} \sum_{m=-\infty}^{+\infty}{}' \frac{1}{m^2} (2e^{-2\pi i m c} - e^{-2\pi i m a} - e^{-2\pi i m b}) \sum_{(\varpi)} \lambda^m(\varpi).$$

Ziehen wir nun Satz 10 heran, so wird der dritte Ausdruck auf der rechten Seite von (34) gleich

$$B \cdot \frac{1}{d} \sum_{m=1}^{\infty} \frac{\log^3(1+m)}{m^2} x e^{-c \frac{\log x}{\log(1+m) + \sqrt{\log x}}}$$

$$= B \cdot \frac{x}{d} \sum_{m=1}^{[e^{\sqrt{\log x}} - 1]} \frac{1}{m^{3/2}} e^{-c \frac{\log x}{2\sqrt{\log x}}} + B \cdot \frac{x}{d} \sum_{[e^{\sqrt{\log x}}]}^{\infty} \frac{1}{m^{3/2}}$$

$$= B \cdot \frac{x}{d} e^{-\frac{c}{2}\sqrt{\log x}} + B \cdot \frac{x}{d} e^{-\frac{1}{2}\sqrt{\log x}}$$

$$= B \cdot \frac{x}{d} e^{-c'\sqrt{\log x}}.$$

---

[9]) l. c. [1]) S. 36, Formel (50).

Tragen wir dies in (34) ein, so haben wir, mit c in neuer Bedeutung, den

Satz 11. Ist $f(y; a, b)$ die in (32) definierte Funktion, $d = b - a$, und $h_0(\mathfrak{a})$ die Idealklassenzahl mod $\mathfrak{a}$ im engsten Sinne, so gilt

$$\sum_{\substack{(\varpi)_{\mathfrak{a}} \\ \varpi \succ 0,\ \varpi \equiv \varrho \pmod{\mathfrak{a}} \\ N(\varpi) \leqq x}} f(w_{\mathfrak{a}}(\varpi); a, b) = \frac{d}{2\, h_0(\mathfrak{a})} \int_{2}^{x} \frac{du}{\log u} + B \cdot \frac{x}{d}\, e^{-c\sqrt{\log x}}.$$

§ 4.

Beweis des Hauptsatzes.

Satz 12. Die Anzahl $N(x; a, b)$ der ganzen Körperzahlen μ mit

$$\mu \succ 0, \qquad 0 \leqq a \leqq w_{\mathfrak{a}}(\mu) \leqq b \leqq 1, \qquad N(\mu) \leqq x$$

ist

$$N(x; a, b) = B \cdot \sqrt{x} + B \cdot (b - a)\, x. \tag{35}$$

Beweis: Es handelt sich um eine Gitterpunktsabzählung. Die ganzen Körperzahlen μ sollen als Punkte mit den Koordinaten μ, μ' in ein rechtwinkliges zz'-System eingetragen werden. Sie bilden dann ein gewisses Punktgitter. Die zu zählenden Gitterpunkte liegen wegen (1) in dem Hyperbelsektor $\mathfrak{H}$

$$\eta_{\mathfrak{a}}^{2a} \leqq \frac{z}{z'} \leqq \eta_{\mathfrak{a}}^{2b}, \qquad zz' \leqq x, \qquad 0 \leqq z, \qquad 0 \leqq z'.$$

Die beiden, diesen Sektor begrenzenden Strahlen gehen vom Nullpunkt zu den Punkten

$$z_1 = \sqrt{x}\,\eta_{\mathfrak{a}}^{a}, \quad z_1' = \sqrt{x}\,\eta_{\mathfrak{a}}^{-a} \quad \text{und} \quad z_2 = \sqrt{x}\,\eta_{\mathfrak{a}}^{b}, \quad z_2' = \sqrt{x}\,\eta_{\mathfrak{a}}^{-b}, \tag{36}$$

die beide auf der begrenzenden Hyperbel $zz' = x$ liegen. Der Inhalt $J(\mathfrak{H})$ dieses Hyperbelsektors $\mathfrak{H}$ berechnet sich zu

$$J(\mathfrak{H}) = (b - a)\, x \log \eta_{\mathfrak{a}}. \tag{37}$$

Nun werde jedem Gitterpunkt μ in $\mathfrak{H}$ ein Fundamentalparallelogramm des Gitters zugeordnet, dessen eine Ecke μ bilde. Diese zu den Gitterpunkten homolog gelegenen Parallelogramme bedecken vielleicht nicht ganz $\mathfrak{H}$. Sie ragen über $\mathfrak{H}$ hinaus und mögen außerhalb $\mathfrak{H}$ noch den Streifen $\mathfrak{S}$ bedecken. Wird $\mathfrak{S}$ auf die z-Achse projiziert, so reicht die Projektion über die Strecke 0 bis z_2 höchstens an jeder Seite um eine Länge C hinaus, die nur vom gewählten Fundamentalparallelogramm, nicht aber von x und a und b abhängt. Die Ordinaten haben in der

z'-Richtung mit $\mathfrak{S}$ auch höchstens eine Strecke C gemeinsam, so daß der Inhalt $J(\mathfrak{S})$ von $\mathfrak{S}$ höchstens

(38) $$C(z_2 + 2C) = B(\sqrt{x} + 1)$$

wird. Die Zahl der Gitterpunkte ist proportional dem Flächeninhalt des von den Parallelogrammen überdeckten Gebietes, also wegen (37) und (38)

$$C(b-a)\, x \log \eta_{\mathfrak{a}} + B(\sqrt{x} + 1).$$

Da in (36) $x \geqq 1$ angenommen werden kann, ist damit Satz 12 bewiesen.

Wir kommen nunmehr auf die Funktion $f(y; a, b)$ des § 3 zurück und bilden mittels ihrer die neue Funktion

(39) $$F\left(y; \frac{d}{2}, \frac{kd}{2}\right) = \sum_{j=1}^{k} f\left(y; (j-1)\frac{d}{2}, (j+1)\frac{d}{2}\right),$$

wobei wir $\frac{kd}{2} \leqq 1$ annehmen. F hat gleichfalls die Periode 1. Da jedes f nur auf einer Strecke der Länge d (Endpunkte ausgeschlossen) von Null verschieden ist und die Funktionen f in der Summe rechts in (39) um je $\frac{d}{2}$ gegeneinander verschoben sind, so sind für jeden bestimmten Wert von y höchstens zwei Summanden und für $\frac{d}{2} < y < \frac{kd}{2}$ auch genau zwei Summanden in (39) von Null verschieden. Es sei etwa $j\frac{d}{2} < y \leqq (j+1)\frac{d}{2}$. Dann ist nach (32)

$$f\left(y; (j-1)\frac{d}{2}, (j+1)\frac{d}{2}\right) = \frac{2}{d}\left((j+1)\frac{d}{2} - y\right),$$

$$f\left(y; j\frac{d}{2}, (j+2)\frac{d}{2}\right) = \frac{2}{d}\left(y - j\frac{d}{2}\right),$$

also

$$f\left(y; (j-1)\frac{d}{2}, (j+1)\frac{d}{2}\right) + f\left(y; j\frac{d}{2}, (j+2)\frac{d}{2}\right) = 1$$

und daher

(40) $$F\left(y; \frac{d}{2}, \frac{kd}{2}\right) = 1 \quad \text{für} \quad \frac{d}{2} \leqq y \leqq \frac{kd}{2}.$$

Ferner ist

(41) $$0 \leqq F\left(y; \frac{d}{2}, \frac{kd}{2}\right) \leqq 1$$

außerhalb des Intervalls $\left(\frac{d}{2}, \frac{kd}{2}\right)$ und speziell

(42) $$F\left(y; \frac{d}{2}, \frac{kd}{2}\right) = 0 \quad \text{für} \quad (k+1)\frac{d}{2} \leqq y \leqq 1,$$

wofern überhaupt $(k+1)\frac{d}{2} \leqq 1$ ist. (Wir haben nur $k\frac{d}{2} \leqq 1$ verlangt und lassen somit ein Übergreifen des ersten und letzten Summanden in (39) zu, bestimmen daher auch (41) nicht näher.)

Für $k = 0$ ist die Summe in (39) leer, daher sei

$$F\left(y;\ \frac{d}{2},\ 0\right) = 0$$

gesetzt. Die Formel (40) verliert dann ihren Sinn, (41) gilt sinngemäß im ganzen Intervall von 0 bis 1, und (42) behält seine Gültigkeit ohne weiteres.

Es sei nun v eine positive, reelle Zahl $0 < v \leqq 1$. Wir setzen

$$k = \left[\frac{2\,v}{d}\right] \tag{43}$$

und vergleichen mit $F\left(y;\ \frac{d}{2},\ \frac{k\,d}{2}\right)$ für dieses k die Funktion

$$G(y;\ 0,\ v) = \begin{cases} 1 & \text{für} \quad 0 \leqq y < v \\ 0 & \text{für} \quad v \leqq y < 1. \end{cases} \tag{44}$$

Auch $G(y;\ 0,\ v)$ sei periodisch in y mit der Periode 1 fortgesetzt.

Da

$$k\frac{d}{2} = \left[\frac{2\,v}{d}\right]\frac{d}{2} \leqq \frac{2\,v}{d}\,\frac{d}{2} = v < \left[\frac{2\,v}{d} + 1\right]\frac{d}{2} = (k+1)\frac{d}{2}$$

ist, so haben wir

$$G(y;\ 0,\ v) - F\left(y;\ \frac{d}{2},\ \frac{k\,d}{2}\right) = 0 \tag{45}$$

für

$$\frac{d}{2} \leqq y \leqq \frac{k\,d}{2} \quad \text{und} \quad (k+1)\frac{d}{2} \leqq y \leqq 1, \tag{45a}$$

und

$$\left|G(y;\ 0,\ v) - F\left(y;\ \frac{d}{2},\ \frac{k\,d}{2}\right)\right| \leqq 1 \tag{46}$$

für

$$0 \leqq y \leqq \frac{d}{2} \quad \text{und} \quad k\frac{d}{2} \leqq y \leqq (k+1)\frac{d}{2} \tag{46a}$$

(wobei diese beiden Intervalle möglicherweise mod 1 ein gemeinsames Stück haben können).

Wir setzen nun überall $y = w_{\mathfrak{a}}(\varpi)$ und bilden

$$\sum_{\substack{(\varpi)_{\mathfrak{a}} \\ \varpi \succ 0,\ \varpi \equiv \varrho \pmod{\mathfrak{a}} \\ N(\varpi) \leqq x}} G(w_{\mathfrak{a}}(\varpi);\ 0,\ v) = \sum_{\varpi} F + \sum_{\varpi} (G - F). \tag{47}$$

Hierin ist

$$\left|\sum_{\varpi} (G - F)\right| \leqq \sum_{\substack{(\mu)_{\mathfrak{a}} \\ \mu \succ 0 \\ N(\mu) \leqq x}} |G - F|. \tag{48}$$

Man kann nun, wie aus der Definition (1) sofort hervorgeht, jede Körperzahl μ durch Multiplikation mit einer geeigneten Potenz von $\eta_{\mathfrak{a}}$ in eine solche mod $\mathfrak{a}$ assoziierte Zahl überführen, die die Eigenschaft $0 \leqq w_{\mathfrak{a}}(\mu) < 1$ besitzt, und umgekehrt können nicht zwei mod $\mathfrak{a}$ assoziierte Zahlen beide diese Ungleichung erfüllen. In (48) können wir daher rechts statt $(\mu)_{\mathfrak{a}}$ auch $0 \leqq w_{\mathfrak{a}}(\mu) < 1$ in der Summationsbedingung schreiben und schließen dann weiter mit Hilfe von (45) und (46)

$$\Big|\sum_{\varpi}(G-F)\Big| \leqq \sum_{\substack{\mu > 0 \\ 0 \leqq w_{\mathfrak{a}}(\mu) < 1 \\ N(\mu) \leqq x}} |G-F|$$

$$\leqq \sum_{\substack{\mu > 0 \\ 0 \leqq w_{\mathfrak{a}}(\mu) \leqq \frac{d}{2} \\ N(\mu) \leqq x}} 1 + \sum_{\substack{\mu > 0 \\ k\frac{d}{2} \leqq w_{\mathfrak{a}}(\mu) \leqq (k+1)\frac{d}{2} \\ N(\mu) \leqq x}} 1$$

$$= N\left(x;\ 0, \frac{d}{2}\right) + N\left(x;\ k\frac{d}{2}, (k+1)\frac{d}{2}\right)$$

Nach Satz 12 ist also

$$\sum_{\varpi}(G-F) = B \cdot \sqrt{x} + B \cdot d \cdot x. \tag{49}$$

Wir haben daher nach der Definition (39) von F und nach (47) und (49)

$$\sum_{\substack{(\varpi)_{\mathfrak{a}} \\ \varpi > 0,\ \varpi \equiv \varrho \pmod{\mathfrak{a}} \\ N(\varpi) \leqq x}} G\left(w_{\mathfrak{a}}(\varpi);\ 0, v\right)$$

$$= \sum_{j=1}^{k} \sum_{\varpi} f\left(w_{\mathfrak{a}}(\varpi);\ (j-1)\frac{d}{2},\ (j+1)\frac{d}{2}\right) + B \cdot \sqrt{x} + B \cdot d \cdot x,$$

und dies ist nach Satz 11

$$= \frac{k\,d}{2\,h_0(\mathfrak{a})} \int_2^x \frac{du}{\log u} + B \cdot k \frac{x}{d} e^{-c\sqrt{\log x}} + B \cdot \sqrt{x} + B \cdot d \cdot x$$

und wegen (43)

$$= \frac{v}{h_0(\mathfrak{a})} \int_2^x \frac{du}{\log u} + B \cdot d \cdot \int_2^x \frac{du}{\log u} + B \cdot \frac{v\,x}{d^2} e^{-c\sqrt{\log x}} + B \cdot \sqrt{x} + B \cdot d \cdot x,$$

bei der zweite Summand in den letzten aufgenommen werden kann.

Setzen wir

$$d = v^{\frac{1}{3}} e^{-\frac{c}{3}\sqrt{\log x}}$$

so haben wir

$$(50)\qquad \sum_{\substack{(\varpi)_{\mathfrak{a}} \\ \varpi \succ 0,\ \varpi \equiv \varrho \pmod{\mathfrak{a}} \\ N(\varpi) \leqq x}} G(w_{\mathfrak{a}}(\varpi);\ 0, v)$$

$$= \frac{v}{h_0(\mathfrak{a})} \int\limits_2^x \frac{du}{\log u} + B \cdot v^{\frac{1}{3}} \cdot x e^{-\frac{c}{3}\sqrt{\log x}} + B \cdot \sqrt{x}.$$

Nach der Definition (44) von $G(y;\ 0, v)$ und wegen $0 < v \leqq 1$ ist durch (50) der Hauptsatz bewiesen. Die Formel (50) besagt sogar noch etwas mehr, denn für kleines v gibt sie ein wie $v^{\frac{1}{3}}$ gegen 0 gehendes, aber in x stärkeres Restglied und ein in x schwächeres, das dafür aber nicht mit v klein wird.

Niehagen (Mecklenburg), den 27. Juli 1934.

(Eingegangen am 4. 8. 1934.)

Notes to Paper 33

This paper has been reviewed in the JF, vol. 61 (1935), p. 172, and in the Z, vol. 11 (1935), p. 150. This is the first paper of Rademacher with a Philadelphia (U.S.A.) dateline (The paper had been written in Germany).

Page 209, equation (2).
Read w_a instead of w.

Page 212, equation (12).
Read σ_0 instead of σ.

Page 214, equation (15).
Read 1/5 instead of 1/0.

Page 219, line 3.
Replace the comma by a dot.

Page 221, fourth display line.
Read $\hat{\varpi}_a$ instead of ϖ.

Über die Anzahl der Primzahlen eines reell-quadratischen Zahlkörpers, deren Konjugierte unterhalb gegebener Grenzen liegen.

Von

Hans Rademacher (Philadelphia, Pa.).

1. Herr A. Walfisz hat mir neulich brieflich die Frage vorgelegt, auf wieviele Weisen sich in einem reell-quadratischen Zahlkörper eine totalpositive ganze Zahl ν als Summe einer totalpositiven Primzahl und einer totalpositiven ganzen Zahl darstellen liesse. Sind $\nu > 0$, $\nu' > 0$ die beiden Konjugierten von ν, so kommt die Frage offenbar darauf hinaus, wieviele totalpositive Primzahlen ω es gibt mit

$$0 < \omega < \nu, \quad 0 < \omega' < \nu'.$$

Mit Hilfe eines neuerdings von mir bewiesenen Satzes[1]) können wir sogar allgemeiner die Frage nach der asymptotischen Abschätzung der totalpositiven Primzahlen ω mit

$$(1.1) \qquad 0 < \omega \leqq Y, \quad 0 < \omega' \leqq Y', \quad \omega \equiv \rho \pmod{\mathfrak{a}}$$

beantworten, wo Y, Y' beliebig reell positiv sind (und nicht die Konjugierten einer Körperzahl zu sein brauchen) und $\mathfrak{a}$ ein festes Ideal, ρ eine feste ganze zu $\mathfrak{a}$ prime Körperzahl sein sollen. Stellen wir die ganzen Zahlen μ des Körpers als Punkte mit den Koordinaten μ, μ' in einem rechtwinkligen System dar, so handelt es sich um die Anzahl der Primzahlen in dem Rechteck $0 < \mu \leqq Y$, $0 < \mu' \leqq Y'$.

[1]) Primzahlen reell-quadratischer Zahlkörper in Winkelräumen, erschein demnächst in den Mathem. Annalen.

Es genügt, die Primzahlen in dem einen der beiden Dreiecke, in die das Rechteck durch die Diagonale vom Nullpunkt nach (Y, Y') zerlegt wird, abzuschätzen, da das andere Dreieck sich durch Vertauschung der beiden Konjugierten ergibt. Wir fragen also nach der Anzahl $P_\mathfrak{a}(Y, Y')$ der Primzahlen ω mit

$$(1.2)\qquad 0 < \omega \leqq Y, \qquad \frac{\omega'}{\omega} \leqq \frac{Y'}{Y},$$

$$(1.3)\qquad \omega \equiv \rho \pmod{\mathfrak{a}}.$$

Zunächst wollen wir noch

$$(1.4)\qquad \eta_\mathfrak{a}^{-1} \leqq \frac{Y'}{Y} \leqq \eta_\mathfrak{a}$$

voraussetzen, eine Einschränkung, die man nachträglich leicht aufheben kann. Dabei sei $\eta_\mathfrak{a} > 1$ die totalpositive Grundeinheit mod $\mathfrak{a}$,

2. Es seien nun q_1, q_2 zwei positive Zahlen, $q_1 < q_2$, $\frac{q_2}{q_1} < \eta_\mathfrak{a}^2$. Dann können unter den totalpositiven Körperzahlen μ mit

$$(2.1)\qquad q_1 < \frac{\mu'}{\mu} \leqq q_2$$

nicht zwei mod $\mathfrak{a}$ assoziierte vorkommen. Führen wir die Bezeichnung

$$w_\mathfrak{a}(\mu) = \frac{1}{2 \log \eta_\mathfrak{a}} \log \left| \frac{\mu}{\mu'} \right|$$

ein, so ist für die Zahlen mit der Eigenschaft (2.1)

$$-\frac{\log q_2}{2 \log \eta_\mathfrak{a}} \leqq w_\mathfrak{a}(\mu) < -\frac{\log q_1}{2 \log \eta_\mathfrak{a}}.$$

Da wegen

$$0 < \frac{\log q_2 - \log q_1}{2 \log \eta_\mathfrak{a}} < 1$$

nur die Fälle

$$(\text{A})\qquad \left[-\frac{\log q_2}{2 \log \eta_\mathfrak{a}}\right] = \left[-\frac{\log q_1}{2 \log \eta_\mathfrak{a}}\right]$$

und

$$(\text{B})\qquad \left[-\frac{\log q_2}{2 \log \eta_\mathfrak{a}}\right] + 1 = \left[-\frac{\log q_1}{2 \log \eta_\mathfrak{a}}\right]$$

vorkommen können, so ist im Falle (A) $w_\mathfrak{a}(\mu) - [w_\mathfrak{a}(\mu)]$ enthalten in dem Intervall

$$-\frac{\log q_2}{2\log \eta_{\mathfrak{a}}}-\left[-\frac{\log q_2}{2\log \eta_{\mathfrak{a}}}\right]\leqq w_{\mathfrak{a}}(\mu)-[w_{\mathfrak{a}}(\mu)]<-\frac{\log q_1}{2\log \eta_{\mathfrak{a}}}-\left[-\frac{\log q_1}{2\log \eta_{\mathfrak{a}}}\right]$$

und im Falle (B) in den beiden Intervallen

$$-\frac{\log q_2}{2\log \eta_{\mathfrak{a}}}-\left[-\frac{\log q_2}{2\log \eta_{\mathfrak{a}}}\right]\leqq w_{\mathfrak{a}}(\mu)-[w_{\mathfrak{a}}(\mu)]<1$$

oder

$$0\leqq w_{\mathfrak{a}}(\mu)-[w_{\mathfrak{a}}(\mu)]<-\frac{\log q_1}{2\log \eta_{\mathfrak{a}}}-\left[-\frac{\log q_1}{2\log \eta_{\mathfrak{a}}}\right].$$

In beiden Fällen ist die Gesamtlänge der Intervalle für $w_{\mathfrak{a}}(\mu)-[w_{\mathfrak{a}}(\mu)]$ gleich

$$\frac{\log q_2-\log q_1}{2\log \eta_{\mathfrak{a}}}.$$

Die Anwendung des „Hauptsatzes" meiner oben zitierten Arbeit ergibt also das

Lemma: *Ist*

(2.2) $$1<\frac{q_2}{q_1}<\eta_{\mathfrak{a}}^2,$$

so ist die Anzahl der totalpositiven Primzahlen ω, *die den Bedingungen*

(2.3) $$\omega\equiv\rho \pmod{\mathfrak{a}} \qquad N(\omega)\leqq x,$$

(2.4) $$q_1<\frac{\omega'}{\omega}\leqq q_2$$

genügen, gleich

(2.5) $$\frac{\log q_2-\log q_1}{4\varphi(\mathfrak{a})h\log\eta}\int_2^x\frac{du}{\log u}+B.\,x e^{-c\sqrt{\log x}}.$$

Bei der Aufstellung dieser Formel ist noch von der Beziehung

$$h_0(\mathfrak{a})\log\eta_{\mathfrak{a}}=2\varphi(\mathfrak{a})h\log\eta$$

Gebrauch gemacht worden[2]); h ist die gewöhnliche Idealklassenzahl und η die Grundeinheit $\eta>1$; B soll in (2.5) und im folgen-

[2]) l. c., Formel (6).

den Funktionen bezeichnen, die beschränkt bleiben und deren Schranken nur vom Körper und von $\mathfrak{a}$ abhängen; ferner ist auch c eine nur vom Körper und von $\mathfrak{a}$ bhängende positive Zahl.

3. Im folgenden sollen, auch wo es nicht ausdrücklich hervorgehoben wird, nur Primzahlen des Körpers betrachtet werden, die der Kongruenz (1 · 3) genügen. Es sei nun $Y' > 2$, was nach (1.4) gewiss dann der Fall sein wird, wenn

$$YY' > 4\,\eta_{\mathfrak{a}} \tag{3.1}$$

vorausgesetzt wird. Wir wählen nun positive Zahlen $y_0', y_1', \ldots, y_l'$ mit folgender Beschaffenheit:

$$2 = y_0' < y_1' < \ldots < y_l' = Y' \tag{3.2}$$

und

$$1 < \frac{y_j'}{y'_{j-1}} < \eta_{\mathfrak{a}}^2 \qquad (j = 1, 2, \ldots, l). \tag{3.3}$$

Die totalpositiven Primzahlen mit den Eigenschaften (1.2) teilen wir ein in die Klassen $C_0, C_1, \ldots, C_l$:

$$0 < \frac{\omega'}{\omega} \leqq \frac{2}{Y}, \quad \frac{2}{Y} < \frac{\omega'}{\omega} \leqq \frac{y_1'}{Y}, \ldots, \quad \frac{y'_{l-1}}{Y} < \frac{\omega'}{\omega} \leqq \frac{Y'}{Y}. \tag{3.4}$$

Nach (1.1) gilt ausserdem für $C_0, C_1, \ldots, C_l$

$$\omega \leqq Y. \tag{3.5}$$

Für $j = 1, 2, \ldots, l$ bilden wir nun Klassen $\underline{C}_j$ und C_j von Primzahlen, sodass

$$\underline{C}_j \subset C_j \subset \overline{C}_j \tag{3.6}$$

ist. Wegen (3.4) und (3.5) ist in C_j

$$N(\omega) = \omega\,\omega' \leqq \frac{\omega^2 y_j'}{Y} \leqq Y y_j'.$$

Die Klasse $\overline{C}_j$, die durch

$$\frac{y'_{j-1}}{Y} < \frac{\omega'}{\omega} \leqq \frac{y_j'}{Y} \quad \text{und} \quad N(\omega) \leqq Y y_j'$$

definiert sei, enthält also, wie (3.6) fordert, die Klasse C_j.
Die Klasse $\underline{C}_j$ sei durch

$$\frac{y'_{j-1}}{Y} < \frac{\omega'}{\omega} \leqq \frac{y_j'}{Y} \quad \text{und} \quad N(\omega) \leqq Y y'_{j-1}$$

definiert. In ihr ist

$$\omega^2 = N(\omega) \cdot \frac{\omega}{\omega'} < Y y'_{j-1} \cdot \frac{Y}{y'_{j-1}} = Y^2,$$

also

$$\omega < Y,$$

sie fällt somit ganz in C_j hinein.

4. Auf die Klassen C_j und $\overline{C}_j$ lässt sich das Lemma § 2 anwenden. Die Anzahl der Primzahlen in C_j ist danach

$$(4.1) \qquad \frac{\log y_j' - \log y'_{j-1}}{4\varphi(\mathfrak{a})\, h \log \eta} Li\,(Y y'_{j-1}) + B\,.\, YY' e^{-c\sqrt{\log YY'}}$$

die in $\overline{C}_j$ ist

$$(4.5) \qquad \frac{\log y_j' - \log y'_{j-1}}{4\varphi(\mathfrak{a})\, h \log \eta} Li\,(Y y_j') + B\,.\, YY' e^{-c\sqrt{\log YY'}}$$

Hierbei ist noch benutzt worden, dass $x e^{-c\sqrt{\log x}}$ von gewissem positiven x an monoton wächst. Da nun nach (1.4) und (3.1) $Y > 2$ und nach (3.2) $Y y_j' > 4$ ist, so durfte in den Restgliedern $Y y'_{j-1}$ bezw. $Y y_j'$ durch YY' ersetzt werden.

Die Anzahl der Primzahlen von C_0 endlich ist höchstens so gross wie die Anzahl aller totalpositiven ganzen Körperzahlen μ mit

$$0 < \frac{\mu'}{\mu} \leqq \frac{2}{Y}, \qquad 0 < \mu \leqq Y,$$

und diese wieder sind enthalten unter allen ganzen Körperzahlen

$$0 < \mu' \leqq 2, \qquad 0 \leqq \mu \leqq Y.$$

Dies sind die in einem Rechteck mit den Seiten 2 und Y enthaltenen Gitterpunkte eines gewissen durch den Körper bestimmten Gitters; daher ist die Zahl der Primzahlen in C_0 gleich

$$B\,.\,Y.$$

Für die Anzahl $P_{\mathfrak{a}}(Y, Y')$ der Primzahlen aus allen Klassen $C_0, C_1, \ldots, C_l$ zusammen haben wir also

(4.3) $$\frac{1}{4\varphi(\mathfrak{a})h\log\eta}\sum_{j=1}^{l}(\log y_j'-\log y'_{j-1}).Li(Yy'_{j-1})$$

$$+B.lYY'e^{-c\sqrt{\log YY'}}$$

$$\leqq P_{\mathfrak{a}}(Y,Y')$$

$$\leqq\frac{1}{4\varphi(\mathfrak{a})h\log\eta}\sum_{j=1}^{l}(\log y_j'-\log y'_{j-1}).Li(Yy_j')$$

$$+B.lYY'e^{-c\sqrt{\log YY'}}+BY.$$

5. Die Summe links in (4.3) heisse $\underline{S}$, die rechts heisse $\overline{S}$. Wir setzen $u_j=\log y_j'$ und haben dann

(5.1) $$\log 2=u_0<u_1<\ldots<u_l=\log Y'$$

und

$$\underline{S}=\sum_{j=1}^{l}Li(Ye^{u_{j-1}})(u_j-u_{j-1}).$$

$$\overline{S}=\sum_{j=1}^{l}Li(Ye^{u_j})(u_j-u_{j-1}).$$

Offenbar ist

(5.2) $$\underline{S}\leqq\int_{\log 2}^{\log Y'}Li(Ye^{u})\,du\leqq\overline{S}.$$

Die u_j wählen wir nun in (5.1) äquidistant, also

(5.3) $$u_j=\log 2+\frac{j}{l}(\log Y'-\log 2).$$

Dann ist

$$\underline{S}=\frac{1}{l}(\log Y'-\log 2)\sum_{j=1}^{l}Li(Ye^{u_{j-1}}),$$

$$\overline{S}=\frac{1}{l}(\log Y'-\log 2)\sum_{j=1}^{l}Li(Ye^{u_j}),$$

also

(5.4) $$\overline{S}-\underline{S}=\frac{1}{l}(\log Y'-\log 2)\,(Li(YY')-Li(2Y))$$

$$=B.\frac{1}{l}\log(YY').Li(YY')=B.\frac{1}{l}YY'.$$

Aus (4.3), (5.2) und (5.4) schliessen wir dann

$$(5.5)\qquad P_{\mathfrak{a}}(Y,Y')=\frac{1}{4\,\varphi(\mathfrak{a})\,h\log\eta}\int\limits_{\log 2}^{\log Y'} Li(Ye^{u})\,du+B.\frac{1}{l}\,YY'$$

$$+B.\,l\,YY'\,e^{-c\sqrt{\log YY'}}+B.Y.$$

Wegen (1.4) ist

$$(5.6)\qquad Y=B.\sqrt{YY'}.$$

Die Zahl l muss nun folgenden Bedingungen genügen: erstens nach (3.2)

$$(5.7)\qquad l\geqq 1,$$

zweitens wegen (5.3)

$$u_j-u_{j-1}=\frac{1}{l}(\log Y'-\log 2)$$

und wegen der Bedeutung der u_j

$$l=\frac{\log\frac{Y'}{2}}{\log\frac{y'_j}{y'_{j-1}}},$$

also nach (3.3)

$$(5.8)\qquad l>\frac{\log\frac{Y'}{2}}{2\log\eta}.$$

Für hinreichend grosses (d. h. eine gewisse nur vom Körper und $\mathfrak{a}$ abhängende Schranke überschreitendes) YY' erfüllt dann

$$l=\left[e^{\frac{c}{2}\sqrt{\log YY'}}\right]$$

die Bedingungen (5.7) und (5.8). Mit diesem l folgt aus (5.5) und (5.6)

$$(5.9)\qquad P_{\mathfrak{a}}(Y,Y')=\frac{1}{4\,\varphi(\mathfrak{a})\,h\log\eta}\int\limits_{\log 2}^{\log Y'} Li(Ye^{u})\,du+B.\,YY'\,e^{-\frac{c}{2}\sqrt{\log YY'}}$$

6. Es bleibt noch eine Umformung des Integrals in (5.6) zu leisten. Es ist

$$\int_{\log 2}^{\log Y'} Li(Ye^u)\,du = \int_{\log 2}^{\log Y'} du \int_{2}^{Ye^u} \frac{dt}{\log t} = \int_{\log 2}^{\log Y'} du \int_{2Y}^{Ye^u} \frac{dt}{\log t} + B\,.\,Y\log Y'$$

$$= \iint \frac{du\,dt}{\log t} + B\,.\,Y\log Y',$$

das Doppelintegral erstreckt über das Gebiet

$$\log 2 \leqq u \leqq \log Y', \quad 2Y \leqq t \leqq Ye^u.$$

Dasselbe Gebiet lässt sich auch beschreiben durch

$$2Y \leqq t \leqq YY', \quad \log\frac{t}{Y} \leqq u \leqq \log Y'.$$

Also ist

$$\int_{\log 2}^{\log Y'} Li(Ye^u)\,du = \int_{2Y}^{YY'} \frac{dt}{\log t} \int_{\log\frac{t}{Y}}^{\log Y'} du + B\,.\,Y\log Y'$$

$$= \int_{2Y}^{YY'} \frac{dt}{\log t}(\log YY' - \log t) + B\,.\,Y\log Y' = \log YY' \int_{2Y}^{YY'} \frac{dt}{\log t}$$

$$- (YY' - 2Y) + B\,.\,Y\log Y'$$

und durch partielle Integration:

$$= \log YY' \left[\frac{t}{\log t}\right]_{2Y}^{YY'} + \log YY' \int_{2Y}^{YY'} \frac{dt}{(\log t)^2} - (YY' - 2Y) + B\,.\,Y\log Y',$$

woraus nach einigen Vereinfachungen und Zusammenfassungen und nochmaliger Benutzung von (1.4)

$$\int_{\log 2}^{\log Y'} Li(Ye^u)\,du = \log YY' \int_{2}^{YY'} \frac{dt}{(\log t)^2} + B\,.\,\sqrt{YY'}\log YY'$$

folgt. Dies tragen wir in (5.9) ein und haben

$$(6.1)\quad P_{\mathfrak{a}}(Y, Y') = \frac{1}{4\varphi(\mathfrak{a})\,h\log\eta}\log YY' \int_{2}^{YY'} \frac{dt}{(\log t)^2} + B\,.\,YY'\,e^{-c\sqrt{\log YY'}}$$

mit c in neuer Bedeutung.

7. Nun ist $P_{\mathfrak{a}}(Y, Y')$ noch nicht die gesuchte Anzahl, sondern erst $P_{\mathfrak{a}}(Y, Y') + P_{\mathfrak{a}}'(Y', Y)$, wo $P_{\mathfrak{a}}'(Y', Y)$ sich durch Vertauschung der Konjugierten ergibt, d. h. die Anzahl der totalpositiven Primzahlen $\omega \equiv \rho \pmod{\mathfrak{a}}$ mit

$$0 < \omega' \leqq Y', \qquad \frac{\omega}{\omega'} \leqq \frac{Y}{Y'}$$

ist. Die rechte Seite in (6.1) ist aber symmetrisch in Y und Y', sodass die gesuchte Anzahl durch Verdoppelung des Hauptgliedes in (6.1) herauskommt. Zu bemerken ist nur noch, dass in $P_{\mathfrak{a}}(Y, Y') + P'_{\mathfrak{a}}(Y', Y)$ die möglicherweise vorhandenen Primzahlen $\frac{\omega}{\omega'} = \frac{Y}{Y'}$ doppelt gezählt worden sind. Doch sind diese von der Anzahl $B\sqrt{YY'}$, gehen also in dem Restglied von (6.1) unter. Damit haben wir die Anzahl der Primzahlen, die den Bedingungen (1.1) genügen, berechnet.

Es bleibt nur noch die Einschränkung (1.4) aufzuheben. Seien nämlich Y, Y' zwei beliebige positive Zahlen. Dann bilden wir durch Multiplikation mit Potenzen von $\eta_{\mathfrak{a}}$ die Zahlen Y_0, Y_0', sodass

$$Y_0 = Y\eta_{\mathfrak{a}}^{-k}, \qquad Y_0' = Y'\eta_{\mathfrak{a}}^{k},$$

(7.1) $$\eta_{\mathfrak{a}}^{-1} \leqq \frac{Y_0'}{Y_0} < \eta_{\mathfrak{a}}$$

(was übrigens die Bestimmung

$$k = -\left[\frac{\log Y' - \log Y}{2\log \eta_{\mathfrak{a}}} + \frac{1}{2}\right]$$

ergibt). Es ist

(7.2) $$Y_0 Y_0' = Y Y'$$

und wegen (6.1) und (7.1) und wegen der Überlegung zu Beginn dieses § 7 ist die Anzahl der Primzahlen ω mit

(7.3) $$0 < \omega \leqq Y_0, \quad 0 < \omega' \leqq Y_0', \quad \omega \equiv \rho \pmod{\mathfrak{a}}$$

dann gleich

$$\frac{1}{2\varphi(\mathfrak{a})\, h \log \eta} \log Y_0 Y_0' \int_2^{Y_0 Y_0'} \frac{dt}{(\log t)^2} + B \,.\, Y_0 Y_0' e^{-c\sqrt{\log Y_0 Y_0'}}.$$

Alle Primzahlen, die (7.3) genügen, erfüllen aber auch

$$0 < \omega\,\eta_{\mathfrak{a}}^{k} \leqq Y, \quad 0 < \omega'\,\eta_{\mathfrak{a}}^{-k} \leqq Y', \quad \omega\,\eta_{\mathfrak{a}}^{k} \equiv \rho \pmod{\mathfrak{a}}$$

und umgekehrt. Nennen wir die Primzahlen $\omega\,\eta_{\mathfrak{a}}^{k}$ wieder ω, so können wir unter Berücksichtigung von (7.2) den Satz aussprechen:

Satz: *Ist* $\mathfrak{a}$ *ein Ideal eines reell-quadratischen Zahlkörpers,* ρ *eine ganze zu* $\mathfrak{a}$ *prime Körperzahl, sind* Y, Y' *positive Zahlen mit* $YY' \geqq 2$, *so ist die Anzahl der Primzahlen* ω, *die*

$$\omega \equiv \rho \pmod{\mathfrak{a}}, \quad 0 < \omega \leqq Y, \quad 0 < \omega' \leqq Y'$$

genügen, gleich

$$\frac{1}{2\,\varphi(\mathfrak{a})\,h \log \eta} \log YY' \int_{2}^{YY'} \frac{dt}{(\log t)^2} + B \,.\, YY' e^{-c\sqrt{\log YY'}} \;.$$

Hierin ist h *die Klassenzahl im gewöhnlichen Sinne und* $\eta > 1$ *die Grundeinheit des Körpers.*

Bemerkenswert ist, dass nicht der Integrallogarithmus, sondern

$$\log x \int_{2}^{x} \frac{dt}{(\log t)^2}$$

mit $x = YY'$ im Hauptglied der bewiesenen Formel auftritt. Offenbar ist für $x > 2$

$$\log x \int_{2}^{x} \frac{dt}{(\log t)^2} > \int_{2}^{x} \frac{dt}{\log t},$$

zugleich aber

$$\log x \int_{2}^{x} \frac{dt}{(\log t)^2} \sim \frac{x}{\log x},$$

denn

$$\log x \int_{2}^{x} \frac{dt}{(\log t)^2} = \log x \left[\frac{t}{(\log t)^2}\right]_{2}^{x} + 2 \log x \int_{2}^{x} \frac{dt}{(\log t)^3}$$

$$= \frac{x}{\log x} + O(\log x) + 2 \log x \int_{2}^{\sqrt{x}} \frac{dt}{(\log t)^3} + 2 \log x \int_{\sqrt{x}}^{x} \frac{dt}{(\log t)^3}$$

$$= \frac{x}{\log x} + O(\log x) + O(\sqrt{x \log x}) + O\left(\frac{x}{(\log x)^2}\right)$$

$$= \frac{x}{\log x} + O\left(\frac{x}{(\log x)^2}\right).$$

Niehagen, d. 18. August 1934.

(Eingegangen am 28. August 1934.)

Notes to Paper 34

This paper has been reviewed in the JF, vol. 61 (1935), p. 172, and in the Z, vol. 11 (1935), p. 55.

Page 67, footnote, line 1.
Read *erscheint* instead of *erschein;* see paper 33 of the present collection.

Unnumbered page following page 71.
The folio should be 72 and is missing.

Page 72, seventh display line.
The exponent should read u_{j-1} (not $u_j - 1$).

Page 73, equation (5.8).
Read η_a instead of η.

SOME REMARKS ON F. JOHN'S IDENTITY.

By Hans Rademacher.

Recently F. John [1] has proved the

THEOREM. *If $f(x)$ is a periodic function of bounded variation with the period 1, and if $\gamma = p/q > 1$ is a given rational number, $(p, q) = 1$, then*

$$(1) \qquad \sum_{n=1}^{\infty} \frac{a_n(\gamma)}{n} f\left(x - \frac{\log n}{\log \gamma}\right) = \log \gamma \int_0^1 f(y)\,dy,$$

where $a_n(\gamma)$ is defined by

$$a_n(\gamma) = a_n(p/q) = \sum_{l=1}^{q} \exp\left[\frac{2\pi i n l}{q}\right] - \sum_{l=1}^{p} \exp\left[\frac{2\pi i n l}{p}\right]$$

or, which is the same,

$$(2) \qquad a_n(\gamma) = \begin{cases} 0; & p \nmid n,\ q \nmid n, \\ -p; & p \mid n,\ q \nmid n, \\ q; & p \nmid n,\ q \mid n, \\ q - p; & p \mid n,\ q \mid n. \end{cases}$$

This interesting identity induces me to make the following three simple remarks, of which the first establishes a connection with the Riemann ζ-function, the second proves (1) for the wider realm of Riemann-integrable functions, the third gives a generalization of (1).

1. The most important special case of (1) is doubtless $f(x) = e^{2\pi i k x}$, k being an integer. If we put $\lambda^{-1} = \log \gamma$, we have to prove in this case

$$\sum_{n=1}^{\infty} \frac{a_n(\gamma)}{n} \exp[2\pi i k(x - \lambda \log n)] = \lambda^{-1} \int_0^1 e^{2\pi i k y}\,dy$$

or

$$(3) \qquad \sum_{n=1}^{\infty} \frac{a_n(\gamma)}{n} \exp[-2\pi i k \lambda \log n] = \begin{cases} 0, & k \neq 0 \\ \lambda^{-1}, & k = 0. \end{cases}$$

We have

$$(4) \qquad \sum_{n=1}^{\infty} \frac{a_n(\gamma)}{n} \exp[-2\pi i k \lambda \log n] = \sum_{n=1}^{\infty} \frac{a_n(\gamma)}{n^{1+2\pi i k \lambda}}.$$

But as by the definition (2) the sum $\sum_{n=1}^{N} a_n(\gamma)$ is bounded for all N, the series

$$(5) \qquad Z(s) = \sum_{n=1}^{\infty} \frac{a_n(\gamma)}{n^s}$$

[1] F. John, "Identitäten zwischen dem Integral einer willkürlichen Funktion und unendlichen Reihen," *Mathematische Annalen*, vol. 110, pp. 718-721.

is convergent for $\Re(s) > 0$ and defines there a regular analytic function of s. On the other hand it follows from (2) and (5) that for $\Re(s) > 1$

$$\mathrm{Z}(s) = \sum_{q|n} \frac{q}{n^s} - \sum_{p|n} \frac{p}{n^s} = q^{1-s} \sum_{m=1}^{\infty} \frac{1}{m^s} - p^{1-s} \sum_{m=1}^{\infty} \frac{1}{m^s}$$

or

$$\mathrm{Z}(s) = \zeta(s)\,(q^{1-s} - p^{1-s}). \tag{6}$$

The equation (6) holds for $\Re(s) > 0$. Now we have to distinguish two cases:

1) $k \neq 0$. We find from (6)

$$\mathrm{Z}(1 + 2\pi ik\lambda) = \zeta(1 + 2\pi ik\lambda)\,(q^{-2\pi ik\lambda} - p^{-2\pi ik\lambda}).$$

But

$$q^{-2\pi ik\lambda} - p^{-2\pi ik\lambda} = \exp\left[-\frac{2\pi ik \log q}{\log p - \log q}\right] - \exp\left[-\frac{2\pi ik \log p}{\log p - \log q}\right] = 0,$$

since

$$\frac{\log q}{\log p - \log q} = \frac{\log p}{\log p - \log q} - 1. \tag{7}$$

Hence

$$\mathrm{Z}(1 + 2\pi ik\lambda) = 0. \tag{8}$$

2) $k = 0$. In this case we have by (6)

$$\mathrm{Z}(1) = \lim_{\epsilon\to 0} \mathrm{Z}(1+\epsilon) = \lim_{\epsilon\to 0} \zeta(1+\epsilon)\,(q^{-\epsilon} - p^{-\epsilon}) = \lim_{\epsilon\to 0} \frac{q^{-\epsilon} - p^{-\epsilon}}{\epsilon},$$

$$\mathrm{Z}(1) = -\log q + \log p = \log \gamma = \lambda^{-1}. \tag{9}$$

The formulae (4), (5), (8), (9) prove (3).

By means of a Fourier expansion, the equation (3) could, of course, be used to prove (1) for a rather extended class of functions $f(x)$. But this reasoning would involve some complications of convergence, which can be surmounted easily only for functions with absolutely convergent Fourier-series, e. g., functions with bounded derivative. However, instead of pursuing this method, we proceed to prove (1) directly in our next remark.

2. In order to study the expression

$$\lim_{N\to\infty} \sum_{n=1}^{N} \frac{a_n(\gamma)}{n} f(x - \lambda \log n)$$

for a Riemann-integrable function $f(y)$ it is obviously sufficient to consider only such N as are divisible by pq, since the $a_n(\gamma)$ and $f(y)$ are bounded. Now we have by (2)

$$
\begin{aligned}
S_M(f) &= \sum_{n=1}^{Mpq} \frac{a_n(\gamma)}{n} f(x - \lambda \log n) \\
&= \sum_{\substack{1 \leqq n \leqq Mpq \\ q \mid n}} \frac{q}{n} f(x - \lambda \log n) - \sum_{\substack{1 \leqq n \leqq Mpq \\ p \mid n}} \frac{p}{n} f(x - \lambda \log n) \\
&= \sum_{m=1}^{Mp} \frac{1}{m} f(x - \lambda \log mq) - \sum_{m=1}^{Mq} \frac{1}{m} f(x - \lambda \log mp) \\
&= \sum_{m=Mq+1}^{Mp} \frac{1}{m} f(x - \lambda \log m - \lambda \log q),
\end{aligned}
$$

upon making use of (7) and of the periodicity of $f(y)$. Moreover, there is no loss of generality in setting $x = \lambda \log q$, for if (1) is true for any special value x_0, it is also true for any other x, as $f(y)$ and $f(y - x_0 + x)$, regarded as functions of y, are both periodic and Riemann-integrable. Hence all we have to prove is

$$
(10) \qquad \lim_{M \to \infty} S_M(f) = \lim_{M \to \infty} \sum_{m=Mq+1}^{Mp} \frac{1}{m} f(-\lambda \log m) = \lambda^{-1} \int_0^1 f(y)\, dy.
$$

Now let us first treat the special "step-function"

$$
(11) \qquad \phi_\alpha(y) = \begin{cases} 1, & 0 \leqq y < \alpha \\ 0, & \alpha \leqq y < 1, \end{cases}
$$

$\phi_\alpha(y)$ being defined in points outside the interval $0 \leqq y < 1$ by periodic repetition of (11) modulo 1. The parameter α is supposed to be such that $0 \leqq \alpha \leqq 1$; for the extreme values $\alpha = 0$ or 1 one of the two inequalities in (11) cannot be fulfilled. We have therefore $\phi_0(y) = 0$ and $\phi_1(y) = 1$ for all y.

For $f = \phi_\alpha$ (10) becomes

$$
(12) \qquad \lim_{M \to \infty} S_M(\phi_\alpha) = \lim_{M \to \infty} \sum_{m=Mq+1}^{Mp} \frac{1}{m} \phi_\alpha(-\lambda \log m) = \alpha \lambda^{-1}.
$$

As this is trivially true for $\alpha = 0$, we can assume $0 < \alpha \leqq 1$. Now we have by (11)

$$
(13) \qquad \sum_{m=Mq+1}^{Mp} \frac{1}{m} \phi_\alpha(-\lambda \log m) = \sum_{\substack{Mq < m \leqq Mp \\ 0 \leqq -\lambda \log m < \alpha \pmod 1}} \frac{1}{m},
$$

where $0 \leqq x < \alpha \pmod 1$ means, of course, $0 \leqq x - [x] < \alpha$. But the conditions of summation on the right-hand side of (13) can be written

$$
\text{(a)} \qquad \lambda(\log M + \log q) < \lambda \log m \leqq \lambda(\log M + \log p),
$$

$$
\text{(b)} \qquad 1 - \alpha < \lambda \log m \leqq 1 \pmod 1.
$$

Condition (a) assigns to $\lambda \log m$ an interval of length $\lambda(\log p - \log q) = 1$. Thus of the infinite set of intervals (b) of length α, which are periodic modulo 1, either just one falls in (a) or two parts of intervals of (b), together of length α, lie in (a), so that the summation in (13) is either of the type

$$\lambda(\log M + \log q) + u_M < \lambda \log m \leqq \lambda(\log M + \log q) + u_M + \alpha,$$

where $0 \leqq u_M \leqq 1 - \alpha$, or of the type

$$\lambda(\log M + \log q) < \lambda \log m \leqq \lambda(\log M + \log q) + \beta_1$$
$$\lambda(\log M + \log p) - \beta_2 < \lambda \log m \leqq \lambda(\log M + \log p), \qquad \beta_1 + \beta_2 = \alpha.$$

Therefore (12) will be proved if we show that

$$(14) \qquad \lim_{M\to\infty} \sum_{\log M+v < \log m \leqq \log M+v+\beta\lambda^{-1}} \frac{1}{m} = \beta\lambda^{-1},$$

where $v = v_M$ may be any number lying between assigned bounds, $c \leqq v_M < C$. Now for $m > 1$

$$\int_m^{m+1} \frac{dt}{t} < \frac{1}{m} < \int_{m-1}^{m} \frac{dt}{t}$$

and consequently

$$\int_{Me^v}^{Me^v\gamma^\beta} \frac{dt}{t} < \sum_{Me^v < m \leqq Me^v\gamma^\beta} \frac{1}{m} < \frac{1}{Me^v} + \int_{Me^v}^{Me^v\gamma^\beta} \frac{dt}{t}$$

or

$$\beta < \sum_{Me^v < m \leqq Me^v\gamma^\beta} \frac{1}{m} < \frac{1}{Me^c} + \beta,$$

which proves (14) and therefore also (12).

Now any periodic step-function of period 1 can be built up as a linear combination of a finite number of step-functions $\phi_a(y)$ of the special type (11) with different parameters α. Hence (10) is proved for arbitrary step-functions with a finite number of steps.

If, finally, $f(y)$ is a periodic Riemann-integrable function, we can, to any given $\epsilon > 0$, assign two step-functions $\phi(y)$ and $\Phi(y)$ of period 1, such that

$$(15) \qquad \phi(y) \leqq f(y) \leqq \Phi(y)$$

and

$$(16) \qquad \int_0^1 (\Phi(y) - \phi(y))\,dy < \epsilon.$$

Since (10) is valid for $\phi(y)$ and $\Phi(y)$, we have

$$(17) \qquad \begin{aligned} \lim_{M\to\infty} S_M(\phi) &= \lambda^{-1} \int_0^1 \phi(y)\,dy, \\ \lim_{M\to\infty} S_M(\Phi) &= \lambda^{-1} \int_0^1 \Phi(y)\,dy. \end{aligned}$$

But $S_M(f)$ shows in (10) only *positive* coefficients of $f(-\lambda \log m)$ and therefore we have from (15)

$$S_M(\phi) \leqq S_M(f) \leqq S_M(\Phi).$$

From this and (17) we conclude

$$\lambda^{-1} \int_0^1 \phi(y)\,dy \leqq \underline{\lim}_{M\to\infty} S_M(f) \leqq \overline{\lim}_{M\to\infty} S_M(f) \leqq \lambda^{-1} \int_0^1 \Phi(y)\,dy.$$

But according to (16) this proves (10) for any Riemann-integrable function $f(y)$, for which, therefore, John's identity (1) is true.

3. The relation between the identity (1) and the Riemann ζ-function, discussed in § 1, suggests the possibility of finding similar identities related to other ζ-functions.

Let K be a field of algebraic numbers, of degree n; let γ be a number of the field with $|N(\gamma)| > 1$ and

$$\gamma = \mathfrak{a}/\mathfrak{b}, \qquad (\mathfrak{a}, \mathfrak{b}) = 1.$$

Now for ideals $\mathfrak{n}$ of the field we introduce, in analogy with (2), the arithmetic function $a_{\mathfrak{n}}(\gamma)$ through the definition

$$a_{\mathfrak{n}}(\gamma) = \begin{cases} 0 & \mathfrak{a} \nmid \mathfrak{n} \quad \mathfrak{b} \nmid \mathfrak{n} \\ -N(\mathfrak{a}) & \mathfrak{a} \mid \mathfrak{n} \quad \mathfrak{b} \nmid \mathfrak{n} \\ N(\mathfrak{b}) & \mathfrak{a} \nmid \mathfrak{n} \quad \mathfrak{b} \mid \mathfrak{n} \\ N(\mathfrak{b}) - N(\mathfrak{a}) & \mathfrak{a} \mid \mathfrak{n} \quad \mathfrak{b} \mid \mathfrak{n}. \end{cases} \tag{18}$$

Let $\mathfrak{C}$ be a class of ideals of K. We shall then prove the

THEOREM. *If $f(x)$ is R-integrable and of period 1, the equation*

$$\sum_{\mathfrak{n} \in \mathfrak{C}} \frac{a_{\mathfrak{n}}(\gamma)}{N(\mathfrak{n})} f\left(x - \frac{\log N(\mathfrak{n})}{\log |N(\gamma)|}\right) = \kappa \log |N(\gamma)| \int_0^1 f(y)\,dy \tag{19}$$

is valid, the summands on the left-hand side being arranged according to increasing $N(\mathfrak{n})$. In (19) κ is a constant depending only on the field, namely

$$\kappa = \frac{2^{r_1+r_2}\pi^{r_2} R}{w\,|\sqrt{d}\,|}$$

(r_1 real and $2r_2$ complex fields among the conjugate fields, w number of roots of unity contained in K*, d discriminant, R regulator of* K*).*

We could repeat our argument of § 1 with one change: viz., the convergence of the series

$$Z(s) = \sum_{\mathfrak{n} \in \mathfrak{C}} \frac{a_{\mathfrak{n}}(\gamma)}{N(\mathfrak{n})^s}$$

can be proved only for $\Re(s) > 1 - [2/(n+1)]$, for which purpose we should have to use Landau's estimate [2] of the "ideal-function"

$$H(x; \mathfrak{C}) = \sum_{\substack{\mathfrak{n} \in \mathfrak{C} \\ N(\mathfrak{n}) \leqq x}} 1 = \kappa \cancel{\log} x + O(x^{1-[2/(n+1)]}). \tag{20}$$

But instead of giving further details of this reasoning we prefer to pass immediately to the generalization of § 2, which is not quite so obvious.

For our proof we start with the remark that for a fixed A

$$\sum_{\substack{\mathfrak{n} \in \mathfrak{C} \\ MA < N(\mathfrak{n}) \leqq (M+1)A}} \frac{1}{N(\mathfrak{n})} \to 0 \quad \text{as} \quad M \to \infty.$$

Indeed, we have from (20)

$$\sum_{\substack{\mathfrak{n} \in \mathfrak{C} \\ MA < N(\mathfrak{n}) \leqq (M+1)A}} \frac{1}{N(\mathfrak{n})} \leqq \frac{1}{MA} \sum_{\substack{\mathfrak{n} \in \mathfrak{C} \\ MA < N(\mathfrak{n}) \leqq (M+1)A}} 1$$

$$= \frac{1}{MA} \kappa A + O\left(\frac{1}{M} M^{1-[2/(n+1)]}\right) = O(M^{-[2/(n+1)]}).$$

Hence for the study of

$$\lim_{N \to \infty} \sum_{\substack{\mathfrak{n} \in \mathfrak{C} \\ N(\mathfrak{n}) \leqq N}} \frac{a_{\mathfrak{n}}(\gamma)}{N(\mathfrak{n})} f(x - \Lambda \log N(\mathfrak{n})),$$

where $\Lambda = (\log |N(\gamma)|)^{-1}$, it is sufficient to consider only such N as are divisible by $N(\mathfrak{ab})$, and to determine the limit of

$$S^*_M(f) = \sum_{\substack{\mathfrak{n} \in \mathfrak{C} \\ N(\mathfrak{n}) \leqq MN(\mathfrak{ab})}} \frac{a_{\mathfrak{n}}(\gamma)}{N(\mathfrak{n})} f(x - \Lambda \log N(\mathfrak{n}))$$

as $M \to \infty$. Now we have by (18)

$$S^*_M(f) = \sum_{\substack{\mathfrak{n} \in \mathfrak{C} \\ \mathfrak{b} | \mathfrak{n} \\ N(\mathfrak{n}) \leqq MN(\mathfrak{ab})}} \frac{N(\mathfrak{b})}{N(\mathfrak{n})} f(x - \Lambda \log N(\mathfrak{n})) - \sum_{\substack{\mathfrak{n} \in \mathfrak{C} \\ \mathfrak{a} | \mathfrak{n} \\ N(\mathfrak{n}) \leqq MN(\mathfrak{ab})}} \frac{N(\mathfrak{a})}{N(\mathfrak{n})} f(x - \Lambda \log N(\mathfrak{n})).$$

The conditions $\mathfrak{b} | \mathfrak{n}$ and $\mathfrak{a} | \mathfrak{n}$ may be replaced by $\mathfrak{n} = \mathfrak{b}\mathfrak{m}_1$, $\mathfrak{n} = \mathfrak{a}\mathfrak{m}_2$ respectively. As we have $\mathfrak{a} = \gamma\mathfrak{b}$, $\mathfrak{a}$ and $\mathfrak{b}$ belong to the same class, say $\mathfrak{C}_1$. Therefore the ideals $\mathfrak{m}_1$ and $\mathfrak{m}_2$ lie both in the class

[2] E. Landau, *Einführung in die elementare und analytische Theorie der algebraischen Zahlen und Ideale*, Leipzig, 1918, p. 131, Satz 210.

$$\mathfrak{C}' = \mathfrak{C}\mathfrak{C}_1^{-1}.$$

Hence we get, writing $\mathfrak{m}$ instead of $\mathfrak{m}_1$ and $\mathfrak{m}_2$,

$$S^*_M(f) = \sum_{\substack{\mathfrak{m}\in\mathfrak{C}' \\ N(\mathfrak{m})\leqq MN(\mathfrak{a})}} \frac{1}{N(\mathfrak{m})} f(x - \Lambda(\log N(\mathfrak{m}) + \log N(\mathfrak{b})))$$
$$- \sum_{\substack{\mathfrak{m}\in\mathfrak{C}' \\ N(\mathfrak{m})\leqq MN(\mathfrak{b})}} \frac{1}{N(\mathfrak{m})} f(x - \Lambda(\log N(\mathfrak{m}) + \log N(\mathfrak{a}))).$$

Because of $\Lambda^{-1} = \log|N(\gamma)| = \log N(\mathfrak{a}) - \log N(\mathfrak{b})$, we have

$$\Lambda(\log N(\mathfrak{m}) + \log N(\mathfrak{b})) - \Lambda(\log N(\mathfrak{m}) + \log N(\mathfrak{a})) = -1.$$

From this and the periodicity of $f(y)$ we conclude

$$S^*_M(f) = \sum_{\substack{\mathfrak{m}\in\mathfrak{C}' \\ MN(\mathfrak{b}) < N(\mathfrak{m})\leqq MN(\mathfrak{a})}} \frac{1}{N(\mathfrak{m})} f(x - \Lambda\log N(\mathfrak{m}) - \Lambda\log N(\mathfrak{b})).$$

As we saw in § 2, the choice of a special value for x involves no loss of generality. We put $x = \Lambda\log N(\mathfrak{b})$ and then have to prove

$$(21)\qquad \lim_{M\to\infty} S^*(f) = \lim_{M\to\infty} \sum_{\substack{\mathfrak{m}\in\mathfrak{C}' \\ MN(\mathfrak{b}) < N(\mathfrak{m})\leqq MN(\mathfrak{a})}} \frac{1}{N(\mathfrak{m})} f(-\Lambda\log N(\mathfrak{m})) = \Lambda^{-1}\int_0^1 f(y)\,dy.$$

For the required proof we need the relation

$$(22)\qquad L(x) = \sum_{\substack{\mathfrak{m}\in\mathfrak{C}' \\ N(\mathfrak{m})\leqq x}} \frac{1}{N(\mathfrak{m})} = \kappa\log x + C + O(x^{-[2/(n+1)]}),$$

which follows from (20) by the customary process of Abel's partial summation. The constant C in (22) may depend on the class $\mathfrak{C}'$.

In complete analogy with § 2, we prove (21) only for the special step-function $\phi_a(y)$, defined in (11). We have

$$S^*_M(\phi_a) = \sum_{\substack{\mathfrak{m}\in\mathfrak{C}' \\ MN(\mathfrak{b}) < N(\mathfrak{m})\leqq MN(\mathfrak{a})}} \frac{1}{N(\mathfrak{m})} \phi_a(-\Lambda\log N(\mathfrak{m})) = \sum_{\substack{\mathfrak{m}\in\mathfrak{C}' \\ MN(\mathfrak{b}) < N(\mathfrak{m})\leqq MN(\mathfrak{a}) \\ 0\leqq -\Lambda\log N(\mathfrak{m}) < a \pmod 1}} \frac{1}{N(\mathfrak{m})}.$$

The conditions of this sum may be treated like those of (13), and our problem is then reduced to the proof of

$$\lim_{M\to\infty} \sum_{\substack{\mathfrak{m}\in\mathfrak{C}' \\ \log M+v < \log N(\mathfrak{m})\leqq \log M+v+\beta\Lambda^{-1}}} \frac{1}{N(\mathfrak{m})} = \kappa\beta\Lambda^{-1}.$$

According to (22) the left-hand side is equal to

$$\begin{aligned}&\lim_{M\to\infty}\{L(Me^v \mid N(\gamma)\mid^\beta)-L(Me^v)\}\\ &=\lim_{M\to\infty}\{\kappa(\log M+v+\beta\Lambda^{-1})-\kappa(\log M+v)+O(M^{-[2/(n+1)]})\}\\ &=\kappa\beta\Lambda^{-1},\end{aligned}$$

which was to be proved.

The further arguments are quite the same as in § 2. We first consider arbitrary step-functions and can then enclose a given R-integrable function $f(y)$ between two step-functions $\phi(y)$ and $\Phi(y)$ whose integrals differ by as little as we wish. In this way the theorem of this paragraph is fully proved.

I close this article with a special example of the generalized John's identity (19). Let K be Gauss's field of complex numbers $a+bi$. We choose $\gamma=1+i$. As there is only the principal class of ideals, we can replace the ideals by integers $m+ni$ of the field. We have only to observe that each principal ideal is represented by four associated numbers. If we therefore sum over all integers (with the omission of 0), we get on both sides of (19) the four-fold amount. We notice further that $\kappa=\pi/4$ in this case and that

$$a_{\mathfrak{n}}(\gamma)=a_{m+ni}(1+i)=\begin{cases}1, & m\not\equiv n \pmod 2\\ -1, & m\equiv n \pmod 2),\end{cases}$$

or

$$a_{m+ni}(1+i)=(-1)^{m+n+1}.$$

Hence we have the equation

$$\sum_{m,n}{}' \frac{(-1)^{m+n+1}}{m^2+n^2} f\left(x-\frac{\log(m^2+n^2)}{\log 2}\right)=\pi\log 2\int_0^1 f(y)\,dy,$$

the sum being extended over all pairs (m,n) with the omission of $(0,0)$ an arranged according to increasing values of m^2+n^2.

UNIVERSITY OF PENNSYLVANIA,
PHILADELPHIA, PA.

Notes to Paper 35

This paper has been reviewed in the JF, vol. 62 (1936), p. 152, and in the Z, vol. 13 (1936), p. 107. This is the first paper of Rademacher written in English and published in an American journal.

Page 170, last display line.
Author's correction (N instead of ∞).

Page 172, sixth display line.
The term $-1/Me^{v}$ is missing on the left.

Page 172, seventh display line.
Replace β by $\beta\lambda^{-1}$, c by v, and add $-1/Me^{v}$ on the left.

Page 174, equation (20).
Author's correction (κx instead of $\kappa \log x$).

Page 175, equation (21).
Insert factor κ on the right-hand side.

ON PRIME NUMBERS OF REAL QUADRATIC FIELDS IN RECTANGLES*

BY
HANS RADEMACHER

In a recently published paper† I investigated the number $P(x, x')$ of primes ω in a real quadratic field satisfying the inequalities $0<\omega<x$, $0<\omega'<x'$. The method used there depends on certain refined estimates of the "angular" distribution of prime numbers in real quadratic fields.‡ In the present paper I propose to give another more direct proof for the estimate of $P(x, x')$, which starts from the obvious remark that for a totally positive unit η of the field we have

$$P(x, x') = P(x\eta, x'\eta'). \tag{1}$$

This fact leads to the periodicity of the function $P(x\eta^v, x'\eta'^v)$ with respect to the variable v, and subsequently to a Fourier development of $P(x\eta^v, x'\eta'^v)$ as function of v.§

Our function $P(x, x')$ is a special case of a more general type, which may be described as follows: Let $f(\mu, \mu')$ be a function defined for all integers μ of the real quadratic field k (μ' being the conjugate of μ) having the property

$$f(\mu, \mu') = f(\mu\eta, \mu'\eta') \tag{2}$$

for all totally positive units η of the field. Then we have

$$\begin{aligned} F(x, x') &= \sum_{\substack{0<\mu<x \\ 0<\mu'<x'}} f(\mu, \mu') = \sum_{\substack{0<\mu\eta<x\eta \\ 0<\mu'\eta'<x'\eta'}} f(\mu, \mu') = \sum_{\substack{0<\mu\eta<x\eta \\ 0<\mu'\eta'<x'\eta'}} f(\mu\eta, \mu'\eta') \\ &= \sum_{\substack{0<\nu<x\eta \\ 0<\nu'<x'\eta'}} f(\nu, \nu') = F(x\eta, x'\eta'), \end{aligned}$$

analogous to (1). But $F(x, x')$ is of course discontinuous and hence would not furnish an absolutely convergent Fourier series, which however we need

* Presented to the Society, October 26, 1935; received by the editors October 4, 1935.

† *Über die Anzahl der Primzahlen eines reell-quadratischen Zahlkörpers, deren Konjugierte unterhalb gegebener Grenzen liegen*, Acta Arithmetica, vol. 1 (1935), pp. 67–77, subsequently cited as "K."

‡ *Primzahlen reell-quadratischer Zahlkörper in Winkelräumen*, Mathematische Annalen, vol. 111 (1935), pp. 209–228, subsequently referred to as "W."

§ My attention was drawn to this use of the Fourier development by a remark of Siegel, who, as I have heard from him, some years ago found an identity similar to formula (13). In my former method the Fourier expansion was used at another step of the proof, viz., in connection with the angular distribution of the primes.

for the application to the estimate of the number of primes in certain rectangles (§2). We therefore prefer to investigate

$$F_1(x, x') = \sum_{0 \dashv \mu \dashv x} (x - \mu)(x' - \mu')f(\mu, \mu')$$

(the notation $0 \dashv \mu \dashv x$ being an abbreviation of both the inequalities $0<\mu<x$ and $0<\mu'<x'$ together), which yields an absolutely convergent Fourier series.

In §2 we specialize $f(\mu, \mu')$ and hence $F_1(x, x')$ for our prime-number problem and then have to make use of some results given in "W" concerning Hecke's $\zeta(s, \lambda)$-functions. In §3 we are in a position to go back from $F_1(x, x')$ to $F(x, x')$ in this special case. Our principal results are the formulas (13), (33), (43).

For the sake of simplicity we content ourselves with these formulas. Of course no fundamental changes would occur if we introduce an ideal modul $\mathfrak{a}$ and a fixed algebraic integer κ and then sum only over the integers $\mu \equiv \kappa$ (mod $\mathfrak{a}$). Instead of the totally positive fundamental unit η we should have to use the totally positive fundamental unit $\eta_{\mathfrak{a}}$ mod $\mathfrak{a}$, i.e., with $\eta_{\mathfrak{a}} \equiv 1$ (mod $\mathfrak{a}$). But the more general result having been already given as a theorem of "K," page 76, and our main interest being at present the exhibition of the other method, we confine ourselves to the case $\mathfrak{a} = (1)$.

1. A Fourier expansion

Let $\eta > 0$ be the totally positive fundamental unit, and especially $\eta > 1$; we then have

$$\eta' > 0, \qquad \eta\eta' = 1, \qquad \eta' = \eta^{-1}.$$

Let $f(\mu, \mu')$ have the property (2). Then we build up the function

$$F_1(x, x') = \sum_{0 \dashv \mu \dashv x} (x - \mu)(x' - \mu')f(\mu, \mu'). \tag{3}$$

Now we have

$$\begin{aligned} F_1(x\eta, x'\eta') &= \sum_{0 \dashv \mu \dashv x\eta} (x\eta - \mu)(x'\eta' - \mu')f(\mu, \mu') \\ &= \sum_{0 \dashv \nu\eta \dashv x\eta} (x\eta - \nu\eta)(x'\eta' - \nu'\eta')f(\nu\eta, \nu'\eta') \\ &= \sum_{0 \dashv \nu \dashv x} (x - \nu)(x' - \nu')f(\nu, \nu') = F_1(x, x'). \end{aligned}$$

Therefore $F_1(x\eta^v, x'\eta^{-v})$ as function of v is periodic with the period 1. It has obviously a bounded derivative and therefore can be developed into an absolutely convergent Fourier series, which we write down immediately for the special value $v=0$:

$$F_1(x, x') = \sum_{n=-\infty}^{+\infty} \int_0^1 e^{-2\pi i n v} F_1(x\eta^v, x'\eta^{-v})dv$$

$$= \sum_{n=-\infty}^{+\infty} \int_0^1 e^{-2\pi i n v} \sum_{\substack{0<\mu<x\eta^v \\ 0<\mu'<x'\eta^{-v}}} (x\eta^v - \mu)(x'\eta^{-v} - \mu')f(\mu, \mu')dv.$$

If we collect associated numbers, i.e., numbers differing only by factors that are powers of η,* we get

$$F_1(x, x')$$
$$= \sum_{n=-\infty}^{+\infty} \int_0^1 e^{-2\pi i n v} \sum_{\substack{(\mu) \\ \mu \succ 0 \\ N(\mu)<xx'}} f(\mu, \mu') \sum_{\substack{k \\ 0<\mu\eta^k<x\eta^v \\ 0<\mu'\eta^{-k}<x'\eta^{-v}}} (x\eta^v - \mu\eta^k)(x'\eta^{-v} - \mu'\eta^{-k})dv,$$

where the notation $(\mu)_1$ indicates that only one representative μ is taken out of each set of associated numbers (in the narrowest sense). The finite summation over $(\mu)_1$ may at once be interchanged with the integration, and in the exponential function we can replace v by $v-k$:

$$F_1(x, x')$$
$$= \sum_{n=-\infty}^{+\infty} \sum_{\substack{(\mu)_1 \\ \mu \succ 0 \\ N(\mu)<xx'}} f(\mu, \mu') \int_0^1 \sum_{\substack{k \\ 0<\mu\eta^k<x\eta^v \\ 0<\mu'\eta^{-k}<x'\eta^{-v}}} e^{-2\pi i n(v-k)}(x\eta^v - \mu\eta^k)(x'\eta^{-v} - \mu'\eta^{-k})dv.$$

If we now interchange also the integration and the summation with respect to k, we observe that v is not only bounded by $0 \leqq v \leqq 1$, but for each k also by $\mu\eta^k < x\eta^v$ and by $\mu'\eta^{-k} < x'\eta^{-v}$, which we can express as follows:

$$F_1(x, x') = \sum_{n=-\infty}^{+\infty} \sum_{\substack{(\mu)_1 \\ \mu \succ 0 \\ N(\mu)<xx'}} f(\mu, \mu') \sum_k \int e^{-2\pi i n(v-k)}(x\eta^{v-k} - \mu)(x'\eta^{-v+k} - \mu')dv,$$

the range of integration in v for each k being determined by the two conditions

$$-k \leqq v - k \leqq 1 - k, \qquad \frac{\mu}{x} < \eta^{v-k} < \frac{x'}{\mu'};$$

only those values of k are admitted for which the second condition yields a v between 0 and 1. We substitute w for $v-k$:

* More precisely these numbers should be called "associated in the narrowest sense," as only totally positive units and not all units are admitted as factors. We shall have to recall this distinction later, on p. 387.

$$F_1(x, x') = \sum_{-\infty}^{+\infty} \sum_{\substack{(\mu)_1 \\ \mu \succ 0 \\ N(\mu) < xx'}} f(\mu, \mu') \sum_k \int e^{-2\pi i n w}(x\eta^w - \mu)(x'\eta^{-w} - \mu')dw,$$

with the conditions of integration

$$-k \leqq w \leqq 1 - k, \qquad \frac{\log \dfrac{\mu}{x}}{\log \eta} < w < \frac{\log \dfrac{x'}{\mu'}}{\log \eta}.$$

But k running through all appropriate integers, the integrals for successive k unite to *one* integral, the boundaries of which are given by the second condition alone:

$$(4) \qquad \begin{aligned} F_1(x, x') &= \sum_{n=-\infty}^{+\infty} \sum_{\substack{(\mu)_1 \\ \mu \succ 0 \\ N(\mu) < xx'}} f(\mu, \mu') \int_{\log (\mu/x)/\log \eta}^{\log (x'/\mu')/\log \eta} e^{-2\pi i n w}(x\eta^w - \mu)(x'\eta^{-w} - \mu')dw \\ &= \sum_{n=-\infty}^{+\infty} \sum_{\substack{(\mu)_1 \\ \mu \succ 0 \\ N(\mu) < xx'}} f(\mu, \mu') I_n(\mu, \mu'), \end{aligned}$$

say. We have now to evaluate the integrals $I_n(\mu, \mu')$.

First, for $n \neq 0$, we get

$$\begin{aligned} I_n(\mu, \mu') = &\int_{\log (\mu/x)/\log \eta}^{\log (x'/\mu')/\log \eta} \{(xx' + \mu\mu')e^{-2\pi i n w} - \mu x' e^{-2\pi i n w - w \log \eta} \\ &- \mu' x e^{-2\pi i n w + w \log \eta}\} dw \\ = &\left(\frac{x'}{\mu'}\right)^{-2\pi i n/\log \eta} \left\{- \frac{xx' + N(\mu)}{2\pi i n} + \frac{N(\mu)}{2\pi i n + \log \eta} + \frac{xx'}{2\pi i n - \log \eta}\right\} \\ &+ \left(\frac{\mu}{x}\right)^{-2\pi i n/\log \eta} \left\{\frac{xx' + N(\mu)}{2\pi i n} - \frac{xx'}{2\pi i n + \log \eta} - \frac{N(\mu)}{2\pi i n - \log \eta}\right\} \\ = &\frac{\log \eta}{2\pi i n}\left(\frac{x}{x'}\right)^{\pi i n/\log \eta} (xx')^{-\pi i n/\log \eta} \left(\frac{\mu}{\mu'}\right)^{-\pi i n/\log \eta} N(\mu)^{\pi i n/\log \eta} \\ &\cdot \left\{\frac{xx'}{2\pi i n - \log \eta} - \frac{N(\mu)}{2\pi i n + \log \eta}\right\} \\ &+ \frac{\log \eta}{2\pi i n}\left(\frac{x}{x'}\right)^{\pi i n/\log \eta} (xx')^{\pi i n/\log \eta} \left(\frac{\mu}{\mu'}\right)^{-\pi i n/\log \eta} N(\mu)^{-\pi i n/\log \eta} \\ &\cdot \left\{\frac{xx'}{2\pi i n + \log \eta} - \frac{N(\mu)}{2\pi i n - \log \eta}\right\}, \end{aligned}$$

and finally

$$
I_n(\mu, \mu') = \frac{\log \eta}{2\pi i n}\left(\frac{x}{x'}\right)^{\pi i n/\log \eta}\left(\frac{\mu}{\mu'}\right)^{-\pi i n/\log \eta} x x'
$$

$$
\tag{5} \left[\frac{\left(\frac{N(\mu)}{xx'}\right)^{\pi i n/\log \eta} - \left(\frac{N(\mu)}{xx'}\right)^{1-\pi i n/\log \eta}}{2\pi i n - \log \eta} + \frac{\left(\frac{N(\mu)}{xx'}\right)^{-\pi i n/\log \eta} - \left(\frac{N(\mu)}{xx'}\right)^{1+\pi i n/\log \eta}}{2\pi i n + \log \eta}\right].
$$

For $n=0$ we get by an easy calculation

$$
\tag{6} \begin{aligned} I_0(\mu, \mu') &= \int_{\log (\mu/x)/\log \eta}^{\log (x'/\mu')/\log \eta} \left\{(xx' + \mu\mu') - \mu x' e^{-w \log \eta} - \mu' x e^{w \log \eta}\right\} dw \\ &= \frac{1}{\log \eta}\left\{(xx' + N(\mu)) \log \frac{xx'}{N(\mu)} - 2(xx' - N(\mu))\right\}. \end{aligned}
$$

Before introducing (5) and (6) into (4) we make use of the following

LEMMA. *Let* $0<y$, $y^s = y^{s \log y}$ *with the principal value of* $\log y$, *and* α, β *complex numbers*, c *real with* $c > \max(\Re(\alpha), \Re(\beta))$. *Then*

$$
\tag{7} \frac{1}{2\pi i}\int_{c-i\infty}^{c+i\infty} \frac{y^s}{(s+\alpha)(s+\beta)} ds = \begin{cases} 0, & 0 < y \leqq 1, \\ \dfrac{y^{-\alpha}}{\beta - \alpha} - \dfrac{y^{-\beta}}{\beta - \alpha}, & 1 \leqq y. \end{cases}
$$

To prove the lemma we consider first an integral extended from $c-i\Omega$ to $c+i\Omega$ with large positive Ω. This path of integration parallel to the imaginary axis is then to be replaced by a half-circle with radius Ω, center c; for $0<y\leqq 1$ we take the half-circle to the right-hand side (side of the positive real part in the s-plane), for $1\leqq y$ we take the half-circle to the left-hand side. An easy estimate shows that on both half-circles the integral tends to zero with infinitely increasing radius Ω. In the first case no pole is enclosed between the new and the old path of integration, in the second the poles $-\alpha$ and $-\beta$. The calculus of residues then yields the result.

An application of (7) to (5) gives for $0<N(\mu)/(xx')<1$

$$
\tag{8} I_n(\mu, \mu') = \frac{1}{2\pi i n}\left(\frac{x}{x'}\right)^{i\gamma n}\left(\frac{\mu}{\mu'}\right)^{-i\gamma n} x x'\left[-\frac{1}{2\pi i}\int_{c-i\infty}^{c+i\infty} \frac{\left(\frac{xx'}{N(\mu)}\right)^s}{(s+i\gamma n)(s+1-i\gamma n)} ds\right.
$$

$$+ \frac{1}{2\pi i}\int_{c-i\infty}^{c+i\infty} \frac{\left(\frac{xx'}{N(\mu)}\right)^s}{(s - i\gamma n)(s + 1 + i\gamma n)}\, ds \Bigg]$$

with $c>0$ and the abbreviation

$$\gamma = \frac{\pi}{\log \eta}. \tag{9}$$

The two integrals can be contracted into one:

$$I_n(\mu, \mu') = \frac{xx'}{\log \eta}\left(\frac{x}{x'}\right)^{i\gamma n}\left(\frac{\mu}{\mu'}\right)^{-i\gamma n} \cdot \frac{1}{2\pi i}\int_{c-i\infty}^{c+i\infty} \frac{\left(\frac{xx'}{N(\mu)}\right)^s}{(s + i\gamma n)(s - i\gamma n)(s + 1 + i\gamma n)(s + 1 - i\gamma n)}\, ds. \tag{10}$$

The definition (4) shows clearly that $I_n(\mu, \mu')$ depends on n continuously. Hence (10) is valid for $n=0$ also. This could of course be verified by direct reference to (6).

By our lemma the integrals in (8), and therefore also the integrals in (10), are equal to zero for $1 \leqq N(\mu)/(xx')$ or $N(\mu) \geqq xx'$. Hence it is unnecessary after the introduction of (10) into (4) to restrict the summation to the range $N(\mu) \leqq xx'$, and thus we get

$$F_1(x, x') = \frac{xx'}{\log \eta}\sum_{n=-\infty}^{+\infty}\left(\frac{x}{x'}\right)^{i\gamma n}\sum_{\substack{(\mu)_1 \\ \mu \not\equiv 0}} f(\mu, \mu')\left(\frac{\mu}{\mu'}\right)^{-i\gamma n} \cdot \frac{1}{2\pi i}\int_{c-i\infty}^{c+i\infty} \frac{\left(\frac{xx'}{N(\mu)}\right)^s}{(s + i\gamma n)(s - i\gamma n)(s + 1 + i\gamma n)(s + 1 - i\gamma n)}\, ds. \tag{11}$$

Up to this point the function $f(\mu, \mu')$ has been subject to no other conditions than (2). Let us now assume

$$f(\mu, \mu') = O(|\, N(\mu)\,|^a)$$

with a certain real a. If we then introduce

$$Z_n(s) = \sum_{\substack{(\mu)_1 \\ \mu \not\equiv 0}} \frac{f(\mu, \mu')\left(\frac{\mu}{\mu'}\right)^{-i\gamma n}}{N(\mu)^s}, \tag{12}$$

the series is absolutely convergent for $\Re(s) > a+1$. In this half-plane the function $Z_n(s)$ is certainly regular. The number $c>0$, determining the path of integration in (10), can be chosen as greater than $a+1$. Then the interchange of the summation over μ and the integration is justified. From (11) and (12) we get

$$
\begin{aligned}
F_1(x, x') &= \sum_{0 \prec \mu \prec x} (x-\mu)(x'-\mu')f(\mu, \mu') = \frac{xx'}{\log \eta} \sum_{n=-\infty}^{+\infty} \left(\frac{x}{x'}\right)^{i\gamma n} \\
&\quad \cdot \frac{1}{2\pi i} \int_{c-i\infty}^{c+i\infty} \frac{(xx')^s Z_n(s)}{(s+i\gamma n)(s-i\gamma n)(s+1+i\gamma n)(s+1-i\gamma n)}\, ds.
\end{aligned} \tag{13}
$$

If we treat the term $n=0$ separately by using its original form (6), we have

$$
\begin{aligned}
F_1(x, x') &= \frac{1}{\log \eta} \sum_{\substack{(\mu)_1 \\ \mu \succ 0 \\ N(\mu) < xx'}} f(\mu, \mu') \left\{ (xx' + N(\mu)) \log \frac{xx'}{N(\mu)} - 2(xx' - N(\mu)) \right\} \\
&\quad + \frac{xx'}{\log \eta} \sum_{n=-\infty}^{+\infty}{}' \left(\frac{x}{x'}\right)^{i\gamma n} \\
&\quad \cdot \frac{1}{2\pi i} \int_{c-i\infty}^{c+i\infty} \frac{(xx')^s Z_n(s)}{(s+i\gamma n)(s-i\gamma n)(s+1+i\gamma n)(s+1-i\gamma n)}\, ds,
\end{aligned} \tag{13a}
$$

the prime at the summation sign meaning the omission of $n=0$. In formulas (13) and (13a) we have attained the objects of this paragraph. It may here be added that by quite analogous reasoning and calculation we can derive the formula

$$
\begin{aligned}
F_0(x, x') &= \sum_{0 \prec \mu \prec x}{}^{*} f(\mu, \mu') \\
&= \frac{1}{\log \eta} \sum_{n=-\infty}^{+\infty} \left(\frac{x}{x'}\right)^{i\gamma n} \frac{1}{2\pi i} \int_{c-i\infty}^{c+i\infty} \frac{(xx')^s Z_n(s)}{(s+i\gamma n)(s-i\gamma n)}\, ds,
\end{aligned} \tag{14}
$$

where $\sum^*$ indicates a special treatment of the boundary summands: a term $f(\mu, \mu')$ with $\mu = x$, $\mu' < x'$ or with $\mu < x$, $\mu' = x'$ is only to be taken into account as $\frac{1}{2} f(\mu, \mu')$, whereas a term with $\mu = x$, $\mu' = x'$ does not count at all. But the series in (14) with summation over n is not absolutely convergent and therefore not suited to our further applications.

I am in the possession of a more general formula, of which (13) and (14) are special cases and which I hope to communicate on another occasion.

2. Estimate of prime numbers in rectangles

We shall now specialize our formula (13) for treating our prime-number problem. Let us put

$$f(\mu, \mu') = \begin{cases} 1 & \text{for } (\mu) \text{ prime ideal,} \\ 0 & \text{otherwise.} \end{cases} \tag{15}$$

For this special $f(\mu, \mu')$ the function $F(x, x')$ may be called $P_1(x, x')$. In this case we have

$$Z_n(s) = \sum_{\substack{(\omega)_1 \\ \omega \succ 0}} \frac{\left(\dfrac{\omega}{\omega'}\right)^{-i\gamma n}}{N(\omega)^s}, \tag{16}$$

ω running only through prime numbers, i.e., integers, whose principal ideals (ω) are prime ideals. We are in a position to discuss the function by reducing it to Hecke's well known $\zeta(s, \lambda)$-functions.

For this purpose we introduce following Hecke† "ideal numbers" $\hat{\mu}$, which together with the numbers of the algebraic field k constitute a certain larger realm $\mathfrak{Z}$ in which multiplication, division, and the operation of determining the greatest common divisor are possible without exception, and in which moreover all units belong to the given field k. These numbers can be separated into 2^2h classes under the stipulation that two belong to the same class when and only when their quotient is a totally positive (integral or fractional) algebraic number of the field. The number h is the ordinary class number.

Let $\chi(\hat{\mu})$ be a character of the Abelian class group of order 2^2h; the unit element of this class group being the class of all totally positive algebraic numbers of the field k, we have $\chi(\mu) = 1$ for $\mu \succ 0$, and especially $\chi(\eta) = 1$ for all characters χ. Hence we have

$$Z_n(s) = \frac{1}{4h} \sum_{\chi} \sum_{(\hat{\omega})_1} \frac{\chi(\hat{\omega})\lambda^n(\hat{\omega})}{|N(\hat{\omega})|^s},$$

where we use Hecke's notation

$$\lambda(\hat{\mu}) = \left(\frac{\hat{\mu}}{\hat{\mu}'}\right)^{-i\gamma} = \left(\frac{\hat{\mu}}{\hat{\mu}'}\right)^{-\pi i/\log \eta}$$

and $(\hat{\omega})_1$ means again that out of each set of ideal numbers associated in the narrow sense we have only to take one representative. But two numbers not associated in the narrow sense may well be associated in the ordinary sense. If we select only non-associated numbers in the ordinary sense, we must consider units θ which are not narrowly associated:

$$Z_n(s) = \frac{1}{4h} \sum_{\chi} \sum_{(\hat{\omega})} \frac{\chi(\hat{\omega})\lambda^n(\hat{\omega})}{|N(\hat{\omega})|^s} \sum_{(\theta)_1} \chi(\theta)\lambda^n(\theta). \tag{17}$$

† E. Hecke, *Eine neue Art von Zetafunktionen und ihre Beziehungen zur Verteilung der Primzahlen*, II, Mathematische Zeitschrift, vol. 6 (1920), pp. 11–51.

The units not associated in the narrow sense form a group of order $2q$, where q is defined by

$$\eta = \epsilon^q \tag{18}$$

with $\epsilon > 1$ the ordinary fundamental unit. (For $q = 1$ we have only $+1$ and -1 not associated in the narrow sense; for $q = 2$ the units $+1$, -1, $+\epsilon$, $-\epsilon$.) Obviously $\chi(\theta)\lambda^n(\theta)$ is a character of this group of units.

Now we have to distinguish two cases:

(i) $\chi(\theta)\lambda^n(\theta) = 1$ for all units θ. This occurs for certain χ, which have the property $\chi(\theta) = \lambda^{-n}(\theta)$ for the finite group of not narrowly associated units. Such χ always exist. Indeed, for given λ^n they are determined at first only for the subgroup of such classes which contain units. But according to a general property of characters of Abelian groups it is always possible to extend a character given on a subgroup to the total enclosing group. The number of such characters is equal to the index of the subgroup in the total group; in our case therefore $4h/(2q)$. For such χ we have

$$\sum_{(\theta)_1} \chi(\theta)\lambda^n(\theta) = 2q = \frac{2 \log \eta}{\log \epsilon} \tag{19}$$

because of (18). In this case we call (in a slight modification of Hecke's terminology) the product $\chi(\hat{\mu})\lambda^n(\hat{\mu})$ an "angular character for ideals" and use the abbreviated symbol $\chi\lambda^n(\hat{\mu})$. In fact because $\chi\lambda^n(\theta) = 1$ the character $\chi\lambda^n(\hat{\mu})$ has the same value for all $\hat{\mu}$ representing the same ideal.

(ii) Not for all units θ is $\chi(\theta)\lambda^n(\theta) = 1$. Then $\chi(\theta)\lambda^n(\theta)$ is not the principal character for the subgroup of the classes containing units, hence we have

$$\sum_{(\theta)_1} \chi(\theta)\lambda^n(\theta) = 0.$$

In (17) therefore we have only to consider such χ as give rise to an angular character $\chi\lambda^n(\hat{\mu})$ for ideals (such χ exist, as mentioned above, in number $2h/q$). If in a summation over χ we have to select such χ in the manner mentioned with regard to λ^n we will mark it by a subscript λ^n attached to the summation sign. From (17) and (19) we have

$$Z_n(s) = \frac{\log \eta}{2h \log \epsilon} \sum_{\chi}{}_{\lambda^n} \sum_{(\hat{\omega})} \frac{\chi\lambda^n(\hat{\omega})}{|N(\hat{\omega})|^s} = \frac{\log \eta}{2h \log \epsilon} \sum_{\chi}{}_{\lambda^n} Z(s, \chi\lambda^n), \tag{20}$$

say.

On the other hand we have the definition of Hecke's $\zeta(s, \lambda)$-functions:

$$\zeta(s, \chi\lambda^n) = \sum_{(\hat{\mu})} \frac{\chi\lambda^n(\hat{\mu})}{|N(\hat{\mu})|^s} = \prod_{(\hat{\omega})} \left(\frac{\chi\lambda^n(\hat{\omega})}{|N(\hat{\omega})|^s} \right)^{-1}$$

valid for $\Re(s) > 1$, and hence

$$\begin{aligned}\log \zeta(s, \chi\lambda^n) &= \sum_{(\hat{\omega})} \sum_{m=1}^{\infty} \frac{1}{m}\left(\frac{\chi\lambda^n(\hat{\omega})}{|N(\hat{\omega})|^s}\right)^m \\ &= \sum_{(\hat{\omega})} \frac{\chi\lambda^n(\hat{\omega})}{|N(\hat{\omega})|^s} + \sum_{(\hat{\omega})} \sum_{m=2}^{\infty} \frac{(\chi\lambda^n(\hat{\omega}))^m}{m|N(\hat{\omega})|^{ms}} \\ &= Z(s, \chi\lambda^n) + \Xi(s, \chi\lambda^n),\end{aligned}$$

say.

Now $\Xi(s, \chi\lambda^n)$ is regular for $\Re(s) > \frac{1}{2}$, since the defining series is absolutely and uniformly convergent for $\Re(s) \geqq \sigma_0 > \frac{1}{2}$:

$$\begin{aligned}\left|\sum_{(\hat{\omega})} \sum_{m=2}^{\infty} \frac{(\chi\lambda^n(\hat{\omega}))^m}{m|N(\hat{\omega})|^{ms}}\right| &< \frac{1}{2} \sum_{(\hat{\omega})} \frac{1}{|N(\hat{\omega})|^{2\sigma_0}} \frac{1}{1 - \dfrac{1}{|N(\hat{\omega})|^{\sigma_0}}} \\ &< \frac{1}{2} \sum_{(\hat{\omega})} \frac{1}{|N(\hat{\omega})|^{2\sigma_0}} \frac{1}{1 - 2^{-\sigma_0}} < \frac{1}{2 - 2^{1-\sigma_0}} \sum_{(\hat{\mu})} \frac{1}{|N(\hat{\mu})|^{2\sigma_0}} \\ &= \frac{1}{2 - 2^{1-\sigma_0}} \zeta(2\sigma_0),\end{aligned}$$

which shows moreover that for $\Re(s) \geqq \sigma_0 > \frac{1}{2}$ the function $\Xi(s, \chi\lambda^n)$ is bounded. We can write

$$Z_n(s) = \frac{\log \eta}{2h \log \epsilon} \sum_{\chi} \lambda^n \log \zeta(s, \chi\lambda^n) + B_n(s), \tag{21}$$

where $B_n(s)$ is bounded for $\Re(s) \geqq \sigma_0 > \frac{1}{2}$ and all n. If we insert (21) and (15) in (13a), we get

$$\begin{aligned}P_1(x, x') = \sum_{0 \prec \omega \prec x} (x - \omega)(x' - \omega') &= + H \frac{xx'}{2h \log \epsilon} \sum_{n=-\infty}^{+\infty} \left(\frac{x}{x'}\right)^{i\gamma n} \\ &\cdot \frac{1}{2\pi i} \sum_{\chi} \lambda^n \int_{2-i\infty}^{2+i\infty} \frac{(xx')^s \log \zeta(s, \chi\lambda^n)}{(s + i\gamma n)(s - i\gamma n)(s + 1 + i\gamma n)(s + 1 - i\gamma n)} ds \\ &+ \frac{xx'}{\log \eta} \sum_{n=-\infty}^{+\infty} \left(\frac{x}{x'}\right)^{i\gamma n} \\ &\cdot \frac{1}{2\pi i} \int_{2-i\infty}^{2+i\infty} \frac{(xx')^s B_n(s)}{(s + i\gamma n)(s - i\gamma n)(s + 1 + i\gamma n)(s + 1 - i\gamma n)} ds\end{aligned} \tag{22}$$

with

$$H = \frac{1}{\log \eta} \sum_{\substack{(\omega)_1 \\ \omega \succ 0 \\ N(\omega) < xx'}} \left\{(xx' + N(\omega)) \log \frac{xx'}{N(\omega)} - 2(xx' - N(\omega))\right\}. \tag{22a}$$

The two infinite sums over $n \neq 0$ in (22) are now to be estimated. We begin with the second of these sums, which is easier to handle. In the following the letter C is used for positive constants, not necessarily always the same.

The functions $B_n(s)$ being regular and bounded for $\Re(s) \geqq \sigma_0 > \frac{1}{2}$, we can shift the path of integration to the left up to the abscissa $\frac{3}{4}$. Then we have

$$\left| \int_{3/4-i\infty}^{3/4+i\infty} \frac{(xx')^s B_n(s)}{(s+i\gamma n)(s-i\gamma n)(s+1+i\gamma n)(s+1-i\gamma n)} ds \right|$$
$$< C(xx')^{3/4} \int_{-\infty}^{+\infty} \frac{dt}{(\frac{3}{4}+|t+\gamma n|)(\frac{3}{4}+|t-\gamma n|)(\frac{7}{4}+|t+\gamma n|)(\frac{7}{4}+|t-\gamma n|)}.$$

By reason of symmetry the latter integral from $-\infty$ to $+\infty$ can be replaced by twice the integral extended from 0 to $+\infty$; without loss of generality we can further suppose $n>0$:

$$\int_{-\infty}^{+\infty} < \frac{2}{\left(\frac{3}{4}+\gamma n\right)\left(\frac{7}{4}+\gamma n\right)} \int_0^{\gamma n} \frac{dt}{\left(\frac{3}{4}+\gamma n - t\right)\left(\frac{7}{4}+\gamma n - t\right)}$$
$$+ \frac{2}{\left(\frac{3}{4}+2\gamma n\right)\left(\frac{7}{4}+2\gamma n\right)} \int_{\gamma n}^{\infty} \frac{dt}{\left(\frac{3}{4}+t-\gamma n\right)\left(\frac{7}{4}+t-\gamma n\right)}$$
$$< \frac{5}{2(\gamma n)^2} \int_0^{\infty} \frac{dt}{\left(\frac{3}{4}+t\right)\left(\frac{7}{4}+t\right)} = \frac{C}{n^2}.$$

Hence

$$\left| \frac{xx'}{\log \eta} \sum_{n=-\infty}^{+\infty} \left(\frac{x}{x'}\right)^{i\gamma n} \frac{1}{2\pi i} \int_{2-i\infty}^{2+i\infty} \frac{(xx')^s B_n(s)}{(s+i\gamma n)(s-i\gamma n)(s+1+i\gamma n)(s+1-i\gamma n)} ds \right|$$
$$< C(xx')^{7/4} \sum_{n=1}^{\infty} \frac{1}{n^2} = C(xx')^{7/4}. \tag{23}$$

For the estimate of the first infinite sum over n in (22) we shall make use of the following

LEMMA. *There exists an absolute constant $E \geqq 1$ such that in the region (boundaries included) of the plane of the complex variable $s=\sigma+it$*

$$1 - \frac{1}{6000(\log(1+|n|)+\log(1+|t|))} \leqq \sigma \leqq 3, \qquad |t| \geqq E,$$

$$1-\frac{1}{6000(\log(1+|n|)+\log(1+E))}\leqq\sigma\leqq 3,\qquad |t|\leqq E,$$

the function $\log\zeta(s,\chi\lambda^n)$, *for* $n\neq 0$, *is regular and satisfies respectively the inequalities*

$$\begin{aligned}\log\zeta(s,\chi\lambda^n)| &\leqq C(\log^2(1+|n|)+\log^2(1+|t|)), &\quad |t|\geqq E,\\ &\leqq C(\log^2(1+|n|)+\log^2(1+E)), &\quad |t|\leqq E.\end{aligned}$$

As to the proof, this lemma is only a slight modification of Theorem 6 in my paper "W." We have only to consider $\log\zeta(s,\chi\lambda^n)$ instead of $(\zeta'/\zeta)(s,\chi\lambda^n)$ in that theorem. The Carathéodory-Landau lemma permits then all the necessary conclusions in the proof. Moreover, in the wording of the present lemma we have replaced $\tau(t,n)$ by its definition $\tau(t,n)=(1+|t|)^2(1+|n|)^2$.

By means of this lemma we are in a position to transform the path of integration from $2-i\infty$ to $2+i\infty$ into the following path P:

$$\sigma=1-\frac{1}{6000(\log(1+|n|)+\log(1-t))},\qquad t\leqq -E;$$

$$\sigma=1-\frac{1}{6000(\log(1+|n|)+\log(1+E))},\qquad -E\leqq t\leqq E;$$

$$\sigma=1-\frac{1}{6000(\log(1+|n|)+\log(1+t))},\qquad E\leqq t.$$

We thus have

$$\begin{aligned}&\left|\int_P\frac{(xx')^s\log\zeta(s,\chi\lambda^n)}{(s+i\gamma n)(s-i\gamma n)(s+1+i\gamma n)(s+1-i\gamma n)}ds\right|\\ &\leqq C\int_0^E\frac{(xx')^{1-1/(6000(\log(1+E)+\log(1+|n|)))}(\log^2(1+E)+\log^2(1+n|))|}{(\frac{1}{2}+|t+\gamma n|)(\frac{1}{2}+|t-\gamma n|)(\frac{3}{2}+|t+\gamma n|)(\frac{3}{2}+|t-\gamma n|)}dt\\ &\quad+C\int_E^\infty\frac{(xx')^{1-1/(6000(\log(1+t)+\log(1+|n|)))}(\log^2(1+t)+\log^2(1+|n|))}{(\frac{1}{2}+|t+\gamma n|)(\frac{1}{2}+|t-\gamma n|)(\frac{3}{2}+|t+\gamma n|)(\frac{3}{2}+|t-\gamma n|)}dt\\ &=I_1+I_2,\end{aligned}\tag{24}$$

say. Without loss of generality we can again assume $n>0$. We have

$$I_1\leqq C(xx')^{1-c/\log(1+n)}$$
$$\cdot\log^2(1+n)\int_0^\infty\frac{dt}{(\frac{1}{2}+|t+\gamma n|)(\frac{1}{2}+|t-\gamma n|)(\frac{3}{2}+|t+\gamma n|)(\frac{3}{2}+|t-\gamma n|)},$$

and the same conclusions which led to (23) yield here

$$I_1 \leqq C(xx')^{1-c/\log(1+n)} \frac{\log^2(1+n)}{n^2}. \tag{25}$$

As to I_2, we intersect the path of integration at the points

$$K = \max(E, \gamma n), \qquad L = \max(E, 2\gamma n e^{(\log xx')^{1/2}})$$

and have

$$\int_E^\infty = \int_E^K + \int_K^L + \int_L^\infty = J_1 + J_2 + J_3,$$

say. We assume always $xx' \geqq 1$; it follows that $(\log xx')^{1/2} \geqq 0$ and hence $K \leqq L$. It may be that $E=K$ or $E=K=L$, in which cases $J_1=0$, or $J_1=J_2=0$, respectively.

For J_1, only $E<K$ remains to be considered. We have

$$\begin{aligned} J_1 &\leqq \int_E^{\gamma n} \frac{(xx')^{1-c/(\log(1+t)+\log(1+n))}(\log^2(1+t)+\log^2(1+n))}{(\frac{1}{2}+\gamma n+t)(\frac{1}{2}+\gamma n-t)(\frac{3}{2}+\gamma n+t)(\frac{3}{2}+\gamma n-t)}dt \\ &< C\frac{(xx')^{1-c/\log(1+n)}\log^2(1+n)}{(\frac{1}{2}+\gamma n)(\frac{3}{2}+\gamma n)}\int_0^{\gamma n}\frac{dt}{(\frac{1}{2}+\gamma n-t)(\frac{3}{2}+\gamma n-t)} \\ &< C(xx')^{1-c/\log(1+n)}\frac{\log^2(1+n)}{n^2}. \end{aligned}$$

This estimate of course is also valid for $E=K$.

Secondly, for $K=L$ we have $J_2=0$. If we assume $K<L$ we have

$$\begin{aligned} J_2 &\leqq \int_{\gamma n}^{2\gamma n \exp((\log xx')^{1/2})} \frac{(xx')^{1-c/(\log(1+n)+\log(1+t))}(\log^2(1+n)+\log^2(1+t))}{(\frac{1}{2}+t+\gamma n)(\frac{1}{2}+t-\gamma n)(\frac{3}{2}+t+\gamma n)(\frac{3}{2}+t-\gamma n)}dt \\ &< C\frac{(xx')^{1-c/(\log(1+n)+(\log xx')^{1/2})}}{(\frac{1}{2}+2\gamma n)(\frac{3}{2}+2\gamma n)}\int_{\gamma n}^\infty \frac{\log^2(1+t)}{(\frac{1}{2}+t-\gamma n)(\frac{3}{2}+t-\gamma n)}dt \\ &< \frac{C}{n^2}(xx')^{1-c/(\log(1+n)+(\log xx')^{1/2})}. \end{aligned}$$

Finally we have

$$\begin{aligned} J_3 &\leqq \int_{2\gamma n\exp((\log xx')^{1/2})}^\infty < \frac{xx'}{(3\gamma n)^2}\int_{2\gamma n\exp((\log xx')^{1/2})}^\infty \frac{\log^2(1+n)+\log^2(1+t)}{(t-\gamma n)^2}dt \\ &< C\frac{xx'}{n^2}\int_{\gamma n\exp((\log xx')^{1/2})}^\infty \frac{\log^2(1+t)}{t^2}dt < C\frac{xx'}{n^2}\int_{\gamma n\exp((\log xx')^{1/2})}^\infty \frac{dt}{t^{3/2}} \\ &< C\frac{xx'}{n^2}e^{-(1/2)(\log xx')^{1/2}}. \end{aligned}$$

These estimates together furnish the inequality

$$(26)\quad I_2 = C(J_1 + J_2 + J_3) < Cxx'e^{-c \log xx'/(\log(1+|n|)+(\log xx')^{1/2})} \frac{\log^2(1+|n|)}{n^2}.$$

Collecting our results we get from (22), (23), (24), (25), (26)

$$\sum_{0 \dashv \omega \dashv x} (x-\omega)(x'-\omega')$$

$$= H + O((xx')^{7/4}) + O\left(xx'\sum_{n=1}^{\infty} \frac{\log^2(1+n)}{n^2} e^{-c \log xx'/(\log(1+n)+(\log xx')^{1/2})}\right).$$

In order to calculate the sum in the second O-term we put

$$\sum_{n=1}^{\infty} = \sum_{n=1}^{[\exp((\log xx')^{1/2})]-1} + \sum_{[\exp((\log xx')^{1/2})]}^{\infty}$$

$$< e^{-(c/2)(\log xx')^{1/2}} \sum_{n=1}^{\infty} \frac{\log^2 n}{n^2} + \sum_{n=[\exp((\log xx')^{1/2})]}^{\infty} \frac{\log^2 n}{n^2}$$

$$< Ce^{-(c/2)(\log xx')^{1/2}} + e^{-(1/2)(\log xx')^{1/2}} \sum_{n=1}^{\infty} \frac{\log^2 n}{n^{3/2}} < Ce^{-c'(\log xx')^{1/2}}.$$

Hence our result reads as follows:

$$(27)\quad P_1(x, x') = \sum_{0 \dashv \omega \dashv x} (x-\omega)(x'-\omega') = H + O((xx')^2 e^{-c(\log xx')^{1/2}})$$

with c in a new meaning.

Our next aim is to give a concise formula for H. We start with the formulas

$$(28)\quad \sum_{\substack{(\omega)_1 \\ \omega \succ 0 \\ N(\omega)<y}} 1 = \frac{q}{2h}\int_2^y \frac{dt}{\log t} + O(ye^{-c(\log y)^{1/2}})$$

and

$$(29)\quad \sum_{\substack{(\omega)_1 \\ \omega \succ 0 \\ N(\omega)<y}} \log N(\omega) = \frac{q}{2h} y + O(ye^{-c(\log y)^{1/2}})$$

where q has the same meaning of as in (18). These equations are well known.* The divisor $2h/q$ indicates the number of classes of ideals, from which in the summations only one is considered.

If we have the equation

* Edmund Landau, *Über Ideale und Primideale in Idealklassen*, Mathematische Zeitschrift, vol. 2 (1918), pp. 52–154, Theorem LXXXIII and Theorem LXXXV.

$$\sum_{\substack{(\omega)_1 \\ \omega \succ 0 \\ N(\omega)<y}} c_\omega = g(y),$$

we can conclude

$$(30)\qquad \int_2^Y g(y)dy = \int_2^Y \sum_{\substack{(\omega)_1 \\ \omega \succ 0 \\ N(\omega)<y}} c_\omega dy = \sum_{\substack{(\omega)_1 \\ \omega \succ 0 \\ N(\omega)<Y}} c_\omega \int_{N(\omega)}^Y dy = \sum_{\substack{(\omega)_1 \\ \omega \succ 0 \\ N(\omega)<Y}} c_\omega(Y - N(\omega)).$$

From (30), (28) and (29) we deduce especially for $c_\omega = 1$ and $c_\omega = \log N(\omega)$

$$\sum_{\substack{(\omega)_1 \\ \omega \succ 0 \\ N(\omega)<Y}} (Y - N(\omega)) = \frac{q}{2h}\int_2^Y dy \int_2^y \frac{dt}{\log t} + O\left(\int_2^Y ye^{-c(\log y)^{1/2}}dy\right),$$

$$\sum_{\substack{(\omega)_1 \\ \omega \succ 0 \\ N(\omega)<Y}} (Y - N(\omega)) \log N(\omega) = \frac{q}{2h}\int_2^Y y\,dy + O\left(\int_2^Y ye^{-c(\log y)^{1/2}}dy\right).$$

Now we have

$$\int_2^Y dy \int_2^y \frac{dt}{\log t} = \int_2^Y \frac{dt}{\log t}\int_t^Y dy = \int_2^Y \frac{(Y-t)dt}{\log t}$$

and

$$\int_2^Y ye^{-c(\log y)^{1/2}}dy = \int_2^{Y^{1/2}} + \int_{Y^{1/2}}^Y < \frac{Y}{2} + e^{-c((\log Y)/2)^{1/2}}\int_{Y^{1/2}}^Y y\,dy$$
$$= O(Y^2 e^{-c'(\log Y)^{1/2}}),$$

and therefore

$$\sum_{\substack{(\omega)_1 \\ \omega \succ 0 \\ N(\omega)<Y}} (Y - N(\omega)) = \frac{q}{2h}\int_2^Y \frac{(Y-t)dt}{\log t} + O(Y^2 e^{-c(\log Y)^{1/2}}),$$

$$\sum_{\substack{(\omega)_1 \\ \omega \succ 0 \\ N(\omega)<Y}} (Y - N(\omega)) \log N(\omega) = \frac{q}{2h}\,\frac{Y^2}{2} + O(Y^2 e^{-c(\log Y)^{1/2}}).$$

These equations together with (28) and (29) yield

$$(31)\qquad \sum_{\substack{(\omega)_1 \\ \omega \succ 0 \\ N(\omega)<Y}} N(\omega) = \frac{q}{2h}\int_2^Y \frac{t\,dt}{\log t} + O(Y^2 e^{-c(\log Y)^{1/2}}),$$

$$\sum_{\substack{(\omega)_1 \\ \omega \succ 0 \\ N(\omega)<Y}} N(\omega) \log N(\omega) = \frac{q}{2h \cdot 2}\frac{Y^2}{2} + O(Y^2 e^{-c(\log Y)^{1/2}}). \tag{32}$$

With the abbreviation $Y = xx'$ we get from (22a)

$$H = \frac{1}{\log \eta} \sum_{\substack{(\omega)_1 \\ \omega \succ 0 \\ N(\omega)<Y}} \{Y(\log Y - 2) + N(\omega)(\log Y + 2) - Y \log N(\omega) - N(\omega) \log N(\omega)\}$$

and hence

$$H = \frac{1}{\log \eta}\frac{q}{2h}\left\{Y(\log Y - 2)\int_2^Y \frac{dt}{\log t} + (\log Y + 2)\int_2^Y \frac{t\,dt}{\log t} - \frac{3Y^2}{2}\right\} + O(Y^2 e^{-c(\log Y)^{1/2}}),$$

and finally from (27), in view of (18),

$$\begin{aligned} P_1(x, x') &= \sum_{0 \prec \omega \prec x} (x - \omega)(x' - \omega') \\ &= \frac{1}{2h \log \epsilon}\left\{Y(\log Y - 2)\int_2^Y \frac{dt}{\log t} + (\log Y + 2)\int_2^Y \frac{t\,dt}{\log t} - \frac{3Y^2}{2}\right\} \\ &\quad + O(Y^2 e^{-c(\log Y)^{1/2}}), \qquad Y = xx'. \end{aligned} \tag{33}$$

3. Passing from $P_1(x, x')$ to $P_0(x, x')$

Let δ be a positive number, $\delta < x$, which may be at our disposal until later. We have on one hand

$$P_1(x + \delta, x') - P(x, x') = \delta \sum_{\substack{0<\omega<x \\ 0<\omega'<x'}} (x' - \omega') + \sum_{\substack{x \leqq \omega < x+\delta \\ 0<\omega'<x'}} (x + \delta - \omega)(x' - \omega') \tag{34}$$

and on the other hand from (33)

$$\begin{aligned} &2h \log \epsilon (P_1(x + \delta, x') - P_1(x, x')) \\ &= ((x + \delta)x'(\log (x + \delta)x' - 2) - xx'(\log xx' - 2))\int_2^{xx'} \frac{dt}{\log t} \\ &\quad + (x + \delta)x'(\log (x + \delta)x' - 2)\int_{xx'}^{(x+\delta)x'} \frac{dt}{\log t} \\ &\quad + (\log (x + \delta)x' - \log xx')\int_2^{xx'} \frac{t\,dt}{\log t} \\ &\quad + (\log (x + \delta)x' + 2)\int_{xx'}^{(x+\delta)x'} \frac{t\,dt}{\log t} - \frac{3}{2}\delta x'(2xx' + \delta x') \\ &\quad + O((xx')^2 e^{-c(\log xx')^{1/2}}). \end{aligned} \tag{35}$$

We give now an estimate of the second sum on the right-hand side in (34),

$$\sum_{\substack{x \leqq \omega < x+\delta \\ 0<\omega'<x'}} (x + \delta - \omega)(x' - \omega') = O\left(\delta x' \sum_{\substack{x \leqq \mu < x+\delta \\ 0<\mu'<x'}} 1\right),$$

where μ runs through all integers in the assigned rectangle. The integers forming a point lattice, their number is of the order of the rectangle's area, plus an error arising from the boundaries. But as the boundaries of the rectangle of one small side δ and one long side x' are relatively long, we have to transform the rectangle into a more suitable shape. In fact, there are as many integers in

$$x \leqq \mu < x + \delta, \qquad 0 < \mu' < x,$$

as in

$$x\eta^l \leqq \mu < (x + \delta)\eta^l, \qquad 0 < \mu' < x'\eta^{-l},$$

and we can choose the exponent l in such a way as to have $\delta\eta^l$ and $x'\eta^{-l}$ of the same order, i.e.,

$$\delta\eta^l = O((\delta x')^{1/2}), \qquad x'\eta^{-l} = O((\delta x')^{1/2}).$$

Then the length of the boundary of the rectangle is $O((\delta x')^{1/2})$, and we have

$$\sum_{\substack{x \leqq \mu < x+\delta \\ 0<\mu'<x'}} 1 = O(\delta x' + (\delta x')^{1/2} + 1)$$

and hence

$$(36) \qquad \sum_{\substack{x \leqq \omega < x+\delta \\ 0<\omega<x'}} (x + \delta - \omega)(x' - \omega') = O(\delta^2 x'^2) + O(\delta^{3/2}x'^{3/2}) + O(\delta x').$$

Moreover we have

$$(37) \qquad \log\,(x + \delta)x' - \log\,xx' = \log\left(1 + \frac{\delta}{x}\right) = \frac{\delta}{x} + O\left(\frac{\delta^2}{x^2}\right),$$

$$\begin{aligned} &(x + \delta)x' \log\,(x + \delta)x' - xx' \log\,xx' \\ &\qquad = xx'(\log\,(x + \delta)x' - \log\,xx') + \delta x' \log\,(x + \delta)x' \\ (38) \qquad &\qquad = \delta x' + O\left(\frac{\delta^2 x'}{x}\right) + \delta x' \log\,xx' + O\left(\frac{\delta^2 x'}{x}\right) \\ &\qquad = \delta x'(1 + \log\,xx') + O\left(\frac{\delta^2 x'}{x}\right), \end{aligned}$$

and

$$\tag{39} \begin{aligned} \int_{xx'}^{(x+\delta)x'} \frac{dt}{\log t} &= \frac{\delta x'}{\log xx'} + O\left(\delta x'\left(\frac{1}{\log xx'} - \frac{1}{\log (x+\delta)x'}\right)\right) \\ &= \frac{\delta x'}{\log xx'} + O\left(\frac{\delta^2 x'}{x(\log xx')^2}\right), \end{aligned}$$

and similarly

$$\tag{40} \begin{aligned} \int_{xx'}^{(x+\delta)x'} \frac{t\,dt}{\log t} &= \delta x' \frac{xx'}{\log xx'} + O\left(\delta x'\left(\frac{(x+\delta)x'}{\log (x+\delta)x'} - \frac{xx'}{\log xx'}\right)\right) \\ &= \delta x' \frac{xx'}{\log xx'} + O\left(\frac{\delta^2 x'^2}{\log xx'}\right). \end{aligned}$$

The equations (34) to (40) give, after division by δ and after due simplifications,

$$\tag{41} \begin{aligned} 2h \log \epsilon\, Q(x, x') &= 2h \log \epsilon \sum_{\substack{0<\omega<x \\ 0<\omega'<x}} (x' - \omega') \\ &= x'(\log xx' - 1) \int_2^{xx'} \frac{dt}{\log t} + \frac{1}{x} \int_2^{xx'} \frac{t\,dt}{\log t} \\ &\quad - xx'^2 + O(\delta x'^2) + O(\delta^{1/2} x'^{3/2}) + O(x') \\ &\quad + O\left(\frac{(xx')^2 e^{-c(\log xx')^{1/2}}}{\delta}\right). \end{aligned}$$

The same process that we applied to x is now to be used with respect to x'. Let δ' be positive, $\delta' < x'$. We have on one hand

$$\begin{aligned} Q(x, x'+\delta') - Q(x, x') &= \sum_{\substack{0<\omega<x \\ 0<\omega'<x'+\delta'}} (x' + \delta' - \omega') - \sum_{\substack{0<\omega<x \\ 0<\omega'<x'}} (x' - \omega') \\ &= \delta' \sum_{\substack{0<\omega<x \\ 0<\omega'<x'}} 1 + \sum_{\substack{0<\omega<x \\ x'\leqq\omega'<x'+\delta'}} (x' + \delta' - \omega'). \end{aligned}$$

Here

$$\sum_{\substack{0<\omega<x \\ x'\leqq\omega'<x'+\delta'}} (x' + \delta' - \omega') = O\left(\delta' \sum_{\substack{0<\mu<x \\ x'\leqq\mu'<x'+\delta'}} 1\right) = O(\delta'^2 x + \delta'^{3/2} x^{1/2} + \delta')$$

in analogy to our former argument. From these equations and from (41) we deduce on the other hand

$$
\begin{aligned}
&2h \log \epsilon \cdot \delta' \sum_{\substack{0<\omega<x \\ 0<\omega'<x'}} 1 \\
&\quad = ((x' + \delta')(\log x(x' + \delta') - 1) - x'(\log xx' - 1)) \int_2^{xx'} \frac{dt}{\log t} \\
&\qquad + (x + \delta')(\log x(x' + \delta') - 1) \int_{xx'}^{x(x'+\delta')} \frac{dt}{\log t} + \frac{1}{x} \int_{xx'}^{x(x'+\delta')} \frac{t\,dt}{\log t} \\
&\qquad - x\delta'(2x' + \delta') + O(\delta'^2 x) + O(\delta'^{3/2} x^{1/2}) + O(\delta') + O(\delta x'^2) \\
&\qquad + O(\delta^{1/2} x'^{3/2}) + O(x') + O(\delta^{-1}(xx')^2 e^{-c(\log xx')^{1/2}}).
\end{aligned} \tag{42}
$$

If we make use of (38), (39), (40) after replacing x, δ by x', δ' and vice versa, we get from (42) after division by δ' and some easy calculations

$$
\begin{aligned}
2h \log \epsilon \sum_{\substack{0<\omega<x \\ 0<\omega'<x'}} 1 &= \log xx' \int_2^{xx'} \frac{dt}{\log t} - xx' + O(\delta' x) + O(\delta'^{1/2} x^{1/2}) + O(1) \\
&\quad + O\left(\frac{\delta}{\delta'} x'^2\right) + O\left(\frac{\delta^{1/2} x'^{3/2}}{\delta'}\right) + O\left(\frac{x'}{\delta'}\right) + O\left(\frac{(xx')^2}{\delta\delta'} e^{-c(\log xx')^{1/2}}\right).
\end{aligned}
$$

Now we put

$$
\delta = x e^{-(c/2)(\log xx')^{1/2}}, \qquad \delta' = x' e^{-(c/4)(\log xx')^{1/2}},
$$

and have then

$$
\sum_{\substack{0<\omega<x \\ 0<\omega'<x'}} 1 = \frac{1}{2h \log \epsilon} \left(\log xx' \int_2^{xx'} \frac{dt}{\log t} - xx' \right) + O(xx' e^{-(c/4)(\log xx')^{1/2}}).
$$

As we find by partial integration

$$
\log Y \int_2^Y \frac{dt}{\log t} = Y + \log Y \int_2^Y \frac{dt}{(\log t)^2} + O(\log Y),
$$

we have finally, with c in a new meaning,

$$
P_0(x, x') = \sum_{\substack{0<\omega<x \\ 0<\omega'<x'}} 1 = \frac{1}{2h \log \epsilon} \log xx' \int_2^{xx'} \frac{dt}{(\log t)^2} + O(xx' e^{-c(\log xx')^{1/2}}), \tag{43}
$$

which is the special case $\mathfrak{a} = (1)$ of the theorem on page 76 of my paper "K."

UNIVERSITY OF PENNSYLVANIA,
PHILADELPHIA, PA.

This paper has been reviewed in the JF, vol. 62 (1936), p. 166, and in the Z, vol. 14 (1936), p. 342.

Page 380, footnote †.
See paper 34 of the present collection.

Page 380, footnote ‡.
See paper 33.

Page 382, second display line.
Read $x'\eta^{-v}$ instead of $x'^{\eta-v}$.

Page 382, fourth display line.
Read $\Sigma_{(\mu)_1}$ instead of $\Sigma_{(\mu)}$.

Page 382, last display line.
Read x'/μ' instead of x'/μ.

Page 386, fourth line of text from bottom.
It is not clear what the formula is to which the author refers. It may well be related to the subject of a communication made to the American Mathematical Society about that time (see *Bull. Amer. Math. Soc.*, vol. 42 (1936), p. 634; abstract A4 of the present collection) on Dedekind's ζ-function and Hecke's $\zeta(s,\lambda)$-functions in totally real algebraic fields. It seems that the author never returned to this topic; in any case, he did not publish any paper on this subject.

Page 388, last display line.
The symbols 1− are missing in the last bracket.

Page 389, eighth display line.
Read $H+$ instead of $+H$.

Page 389, ninth display line.
Read Σ' for Σ (twice).

Page 391, eighth display line.
Read $\log^2(1+|n|)$ instead of $\log^2(1+n|)$.

A CONVERGENT SERIES FOR THE PARTITION FUNCTION $p(n)$

By Hans Rademacher

Department of Mathematics, University of Pennsylvania

Communicated January 9, 1937

I. For a discussion of the number $p(n)$ of unrestricted partitions of n there has been used since Euler[1] the generating function

$$f(x) = 1 + \sum_{n=1}^{\infty} p(n)x^n = \frac{1}{(1-x)(1-x^2)(1-x^3)\dots}. \quad (1)$$

A purely formal treatment of (1) leads to classical formulae of recurrence. In 1917 Hardy and Ramanujan[2] applied on (1) the theory of functions of complex variables and developed a method (Cauchy integral, "Farey-dissection," approximation functions) which yields an asymptotic formula for $p(n)$. Their result is the following:

$$p(n) = \frac{1}{2\pi\sqrt{2}} \sum_{k \leqq \alpha\sqrt{n}} A_k(n)\sqrt{k}\,\frac{d}{dn}\left(\frac{e^{\frac{C\sqrt{n - 1/24}}{k}}}{\sqrt{n - 1/24}}\right) + O(n^{-1/4}) \quad (2)$$

with α as an arbitrary constant, and

$$A_k(n) = \sum_{\substack{h \bmod k \\ (h, k) = 1}} \omega_{h,k}\, e^{-\frac{2\pi i h n}{k}}, \quad C = \pi\sqrt{2/3}. \quad (3)$$

The symbol $\omega_{h,k}$ designates certain $24k$th roots of unity which have their origin in the theory of the transformation of modular functions.[3] The formula (2) is noteworthy for the fact that its error term tends to zero with increasing n. Tests for $n = 100$ and $n = 200$ showed indeed that $p(n)$ can be obtained as the nearest integer to the sum expression in (2) for a small number of summands.

The following revision and modification of the Hardy-Ramanujan method permits us to replace the asymptotic formula (2) by the equality

$$p(n) = \frac{1}{\pi\sqrt{2}} \sum_{k=1}^{\infty} A_k(n)\sqrt{k}\,\frac{d}{dn}\left(\frac{\sinh \frac{C\sqrt{n - 1/24}}{k}}{\sqrt{n - 1/24}}\right), \quad (4)$$

in which the series is absolutely convergent. The asymptotic formula (2) is then an immediate consequence of (4).

By Cauchy's theorem we deduce from (1)

$$p(n) = \frac{1}{2\pi i}\int_C \frac{f(x)}{x^{n+1}}\,dx, \quad (5)$$

where C is a certain circle enclosing the origin and interior to the unit circle. The approach to a treatment of (5) is given by the formula

$$f\left(e^{\frac{2\pi i h}{k} - \frac{2\pi z}{k}}\right) = \omega_{h,k}\sqrt{z}\exp\left(\frac{\pi}{12kz} - \frac{\pi z}{12k}\right) f\left(e^{\frac{2\pi i h'}{k} - \frac{2\pi}{kz}}\right) \quad (6)$$

with $\Re(z) > 0$, $\sqrt{z}$ denoting the principal branch of the square root and h' determined as a solution of the congruence $hh' \equiv -1 \pmod{k}$. The formula (6) itself is a consequence of a transformation formula for $\eta(\tau)$, a modular function introduced by Dedekind,[4] which is essentially the 24th root of the "discriminant."

In (5) we choose C as the circle of radius $e^{-2\pi N^{-2}}$ and divide it by the Farey-dissection of order N. The Farey arc $\xi_{h,k}$ may be described by

$$x = e^{\frac{2\pi i h}{k} - 2\pi N^{-2} + 2\pi i\varphi}, -\vartheta'_{h,k} \leqq \varphi \leqq \vartheta''_{h,k}.$$

Introducing (6) into the integrand we get

$$p(n) = \frac{1}{2\pi i} \sum_{\substack{(h,k)=1 \\ 0 \leqq h < k \leqq N}} \int_{\xi h,k} \frac{f(x)}{x^{n+1}} dx$$

$$= e^{2\pi n N^{-2}} \sum_{\substack{(h,k=1) \\ 0 \leqq h < k \leqq N}} \omega_{h,k} e^{\frac{-2\pi i h n}{k}} \int_{-\vartheta'_{h,k}}^{\vartheta''_{h,k}} \Psi(k(N^{-2} - i\varphi)) f(e^{\frac{2\pi i h'}{k} - \frac{1}{k^2(N^{-2}i\varphi)}}) e^{-2\pi i \varphi n} d\varphi,$$

where we have used the abbreviation

$$\Psi(z) = \sqrt{z} \exp\left(\frac{\pi}{12kz} - \frac{\pi z}{12k}\right). \tag{7}$$

Instead of introducing a special auxiliary function in order to approximate the integral, as Hardy and Ramanujan do, we use $\Psi(z)$ itself, since $f(x')$ approaches 1 very rapidly when x' tends to zero. If we carry out the proper estimates we obtain

$$p(n) = e^{2\pi n N^{-2}} \sum_{\substack{h,k=1 \\ 0 \leqq h < k \leqq N}} \omega_{h,k} e^{-\frac{2\pi i h n}{k}} \int_{-\vartheta'_{h,k}}^{\vartheta''_{h,k}} \Psi(k(N^{-2} - i\varphi))\, e^{-2\pi i n\varphi}\, d\varphi + \Theta_1,$$

$$|\Theta_1| < B_1 e^{2\pi n N^{-2}} N^{-1/2}. \tag{8}$$

It may be noticed that in our reasonings the order N of the Farey-dissection is independent of the represented number n.

In (8) we substitute $w = N^{-2} - i\varphi$ and consider the integral in the complex w-plane. In order to get a unique determination of $\Psi(kw)$ we cut the w-plane along the negative real axis. The integral in (8) is then replaced by a loop-integral extended over a path surrounding the cut and connecting the following points by straight segments:

$$-\infty,\ -\epsilon,\ -\epsilon - i\vartheta''_{h,k},\ N^{-2} - i\vartheta''_{h,k},\ N^{-2} + i\vartheta'_{h,k},\ -\epsilon + i\vartheta'_{h,k},\ -\epsilon,\ -\infty.$$

The estimate of the errors is based on the fact that the essential singularity of $\Psi(kw)$ in $w = 0$ arising from a factor $e^{\frac{a}{w}}$ with positive a has bounded values for $\Re(w) \leqq 0$ and that more precisely for $w = u + iv$ we have

$$\left|e^{\frac{1}{w}}\right| = e^{\frac{u}{u^2+v^2}} \leqq e^b$$

for $u \leqq b(u^2 + v^2)$, which, for $b > 0$, is fulfilled *a fortiori* if $u \leqq bv^2$. This, by the way, is the reason for choosing the radius $e^{-2\pi N^{-2}}$ together with the order N of the Farey-dissection.

The calculation leads to

$$p(n) = \sum_{k=1}^{N} A_k(n)\varphi_k(n) + \Theta_2, \tag{9}$$

$$|\Theta_2| < B_2\, e^{2\pi n N^{-2}} N^{-1/2},$$

where $A_k(n)$ is the quantity defined in (3) and

$$\varphi_k(n) = \frac{\sqrt{k}}{i}\int_{-\infty}^{(0+)} \sqrt{w}\, e^{\frac{\pi}{12k^2 w}}\, e^{2\pi(n-1/24)w}\, dw$$

$$+ 2\sqrt{k}\int_{0}^{\infty} \sqrt{u}\, e^{-\frac{\pi}{12k^2 u}}\, e^{-2\pi(n-1/24)u}\, du. \tag{10}$$

The remainder term in (9) tends to zero for $N \longrightarrow \infty$ and fixed n. Hence we have

$$p(n) = \sum_{k=1}^{\infty} A_k(n)\varphi_k(n). \tag{11}$$

The integrals in (10) are well known, and we get in this way the proof of (4).

From (4) it is possible to derive a definite expression for the O-term in (2). We obtain

$$p(n) = \frac{1}{2\pi\sqrt{2}}\sum_{k=1}^{N} A_k(n)\sqrt{k}\,\frac{d}{dn}\left(\frac{e^{\frac{C\sqrt{n-1/24}}{k}}}{\sqrt{n-1/24}}\right) \tag{12}$$

$$+ \Theta\cdot\left\{\frac{44\pi^2}{225\sqrt{3}}N^{-1/2} + \frac{\pi\sqrt{2}}{75}\left(\frac{N}{n-1}\right)^{1/2}\sinh\frac{C\sqrt{n}}{N}\right.$$

$$\left.+ \frac{1}{2\sqrt{3}}\,\frac{(N+1)^{3/2}}{n-1}\, e^{-\frac{C\sqrt{n-1}}{N}}\left(\frac{1}{3} + \frac{\sqrt{3}}{5\pi\sqrt{2}}\,\frac{N+1}{(n-1)^{1/2}}\right)\right\},\quad |\Theta| < 1,$$

a formula which proves to be useful for numerical computation of $p(n)$.

A detailed account of the outlined proofs will soon be published in the "Proceedings of the London Mathematical Society."

II. The result (4) can in turn be used for a treatment of $f(x)$, which because of its close relation to modular functions deserves some attention for itself. We get from (1), (4) and (3)

$$f(x) = 1 + \frac{1}{\pi\sqrt{2}}\sum_{\substack{(h,k)=1\\ h \bmod k}} \omega_{h,k}\sqrt{k}\,\Phi_k\left(xe^{-\frac{2\pi i h}{k}}\right),\quad |x| < 1, \tag{13}$$

with

$$\Phi_k(z) = \sum_{n=1}^{\infty} \frac{d}{dn}\left(\frac{\sinh \frac{C\sqrt{n - {}^1/_{24}}}{k}}{\sqrt{n - {}^1/_{24}}}\right)\cdot z^n. \tag{14}$$

But here

$$g_k(n) = \frac{d}{dn}\left(\frac{\sinh \frac{C\sqrt{n - {}^1/_{24}}}{k}}{\sqrt{n - {}^1/_{24}}}\right) = \sum_{\nu=1}^{\infty} \frac{\left(\frac{C}{k}\right)^{2\nu+1} \nu}{(2\nu+1)!}(n - {}^1/_{24})^{\nu-1} \tag{15}$$

is an entire transcendental function in n of order $^1/_2$. Hence by a theorem of Wigert[5] the function

$$\Phi_k(z) = \sum_{n=1}^{\infty} g_k(n) z^n,$$

originally defined only in the interior of the unit circle, can be continued over the whole z-plane, is regular at $z = \infty$ and has only one isolated essential singularity at $z = 1$.

The expansion (13) can therefore be taken as a generalized Mittag-Leffler decomposition into "partial fractions," each term giving rise to an essential singularity at a root of unity $x = e^{\frac{2\pi i h}{k}}$. The particularity of (13) lies in the fact that all singularities together build up the natural boundary $|x| = 1$ for the function $f(x)$.

The $\Phi_k(z)$ have, by Wigert's theorem, also a meaning outside the unit circle, and a simple consideration shows that the right-hand side of (13) is again convergent for $|x| > 1$. This defines there an analytic function:

$$f^*(x) = 1 + \frac{1}{\pi\sqrt{2}} \sum_{\substack{(h,k)=1 \\ h \bmod k}} \omega_{h,k}\sqrt{k}\,\Phi_k\left(xe^{-\frac{2\pi i h}{k}}\right), \quad |x| > 1. \tag{13a}$$

The functions $f(x)$ and $f^*(x)$ are determined by the same expansion (13), (13a), in spite of the fact that they are separated by the unit circle as natural boundary of $f(x)$.

We can establish a power series for $f^*(x)$ around $x = \infty$. From (14) and (15) we get

$$\Phi_k(z) = \sum_{\nu=1}^{\infty} \frac{\left(\frac{C}{k}\right)^{2\nu+1} \nu}{(2\nu+1)!} \sum_{n=1}^{\infty} (n - {}^1/_{24})^{\nu-1} z^n = \sum_{\nu=1}^{\infty} \frac{\left(\frac{C}{k}\right)^{2\nu+1} \nu}{(2\nu+1)!} \varphi_\nu(z),$$

say. We have

$$\varphi_1(z) = \sum_{n=1}^{\infty} z^n = \frac{z}{1-z} = -\frac{1}{1-z^{-1}} = -\sum_{m=0}^{\infty} z^{-m}$$

and

$$\varphi_{\nu+1}(z) = \left(z\frac{d}{dz} - {}^1/_{24}\right)\varphi_\nu(z).$$

Consequently around $z = \infty$ the expansions

$$\varphi_\nu(z) = (-1)^\nu \sum_{m=0}^{\infty} (m + {}^1/_{24})^{\nu-1} z^{-m}, \; |z| > 1$$

are valid, and we have

$$\Phi_k(z) = \sum_{\nu=1}^{\infty} \frac{\left(\frac{C}{k}\right)^{2\nu+1} \nu(-1)^\nu}{(2\nu+1)!} \sum_{m=0}^{\infty} (m + {}^1/_{24})^{\nu-1} z^{-m}$$

$$= \sum_{m=0}^{\infty} \frac{d}{dm}\left(\frac{\sin \frac{C\sqrt{m + {}^1/_{24}}}{k}}{\sqrt{m + {}^1/_{24}}}\right) \cdot z^{-m}$$

with the trigonometric sine instead of the hyperbolic sine. With this determination (13a) takes the form

$$f^*(x) = 1 + \frac{1}{\pi\sqrt{2}} \sum_{\substack{h,k=1 \\ h \bmod k}} \omega_{h,k}\sqrt{k} \sum_{m=0}^{\infty} \frac{d}{dm}\left(\frac{\sin \frac{C\sqrt{m + {}^1/_{24}}}{k}}{\sqrt{m + {}^1/_{24}}}\right) \cdot \left(xe^{-\frac{2\pi ih}{k}}\right)^{-m}$$

$$= 1 + \sum_{m=0}^{\infty} p^*(m)x^{-m}, \qquad (16)$$

where

$$p^*(m) = \frac{1}{\pi\sqrt{2}} \sum_{k=1}^{\infty} A_k(-m)\sqrt{k}\,\frac{d}{dm}\left(\frac{\sin \frac{C\sqrt{m + 1/24}}{k}}{\sqrt{m + 1/24}}\right). \qquad (17)$$

A formal comparison of (4) and (17) can be condensed in the equation

$$p^*(m) = -p(-m).$$

Whereas $p(n)$ is very rapidly increasing with increasing n, $p^*(m)$ turns out to be rather small. An estimate readily obtained is

$$p^*(m) = O(m^{-1/4}).$$

The function $f^*(x)$ requires a further investigation. It is quite possible that $f^*(x)$ is identically zero for $|x| > 1$. Not only certain function-theoretical considerations, but also some numerical computations pertaining to the first coefficients $p^*(m)$, which Dr. D. H. Lehmer was kind enough to carry out, seem to point in this direction.

The whole theory admits an application to a class of functions related to certain "modular forms" of positive dimensions, e.g., to $f^r(x)$ for $2 \leqq r \leqq 24$. Dr. Zuckerman and I are studying these generalizations in a paper now under preparation.

[1] Introductio in analysin infinitorum (1748), cap. 16, "De partitione numerorum," Opera omnia, ser. 1, **8**; "De partitione numerorum" (1753) Opera omnia, ser. 1, **2**, p. 280.

[2] "Asymptotic formulae in combinatory analysis," *Proc. Lond. Math. Soc.* (2), **17**, 75–115 (1918); also Collected Papers of S. Ramanujan, 276–309.

[3] Explicit expressions for $\omega_{h,k}$ can be found in Hardy-Ramanujan, loc. cit., formulae (1.721) and (1.722); cf. also H. Rademacher, "Bestimmung einer gewissen Einheitswurzel in der Theorie der Modulfunktionen," *Jour. Lond. Math. Soc.*, **7**, 14–19 (1932).

[4] "Erläuterungen, &c.," Riemann's Werke, 438–447 (1876); cf. also H. Weber, *Elliptische Funktionen*, 97–102 (1891), and H. Rademacher, *Jour. reine angewandte Math.*, **167**, 312–336 (1931).

[5] "Sur un théorème concernant les fonctions entières," *Arkiv Mat., Astron. Fysik*, **11**, No. 22 (1916).

Notes to Paper 37

This paper has been reviewed in the JF, vol. 63 (1937), p. 140, and in the Z, vol. 16 (1937), pp. 246-247.

Page 80, first display line.
Read $\xi_{h,k}$ instead of $\xi h,k$ (limit of integral).

Page 82, equation (15).
Read ν instead of v in the numerator.

Page 84, footnote 3, line 2.
See paper no. 29 of the present collection.

Page 84, footnote 4, line 2.
See paper 27.

ON THE PARTITION FUNCTION $p(n)$.

By Hans Rademacher.

[Received 30 November, 1936.—Read 10 December, 1936.]

1. *Introduction.* The number $p(n)$ of unrestricted partitions of n appears in the generating function

$$(1.1) \qquad f(x) = 1 + \sum_{n=1}^{\infty} p(n)\, x^n = \frac{1}{(1-x)(1-x^2)(1-x^3)\ldots},$$

which was discussed by Euler in his investigations on "Partitio numerorum"*. Several interesting formulae of recurrence were derived from (1.1), but no independent representation of $p(n)$ was known. Not even the order of magnitude of $p(n)$ for large n was ever examined, until, in 1917, G. H. Hardy and S. Ramanujan† applied to (1.1) their fundamental analytic method, which has been since used so successfully, in many variants, for numerous problems of additive number theory. Their result was an asymptotic formula for $p(n)$, which is surprising in the analytic theory of numbers in so far as it contains an error term approaching zero with increasing n. It reads as follows:

$$(1.2) \qquad p(n) = \frac{1}{2\pi\sqrt{2}} \sum_{k \leqslant \alpha\sqrt{n}} A_k(n)\, k^{\frac{1}{2}} \frac{d}{dn}\left(\frac{e^{C\lambda_n/k}}{\lambda_n}\right) + O(n^{-\frac{1}{4}}),$$

with α an arbitrary constant,

$$(1.3) \qquad C = \pi\sqrt{(\tfrac{2}{3})}, \quad \lambda_n = \sqrt{(n - \tfrac{1}{24})},$$

---

* *Introductio in analysin infinitorum* (1748), cap. 16, "De partitione numerorum", *Opera omnia* (1), 8, in particular p. 318; "De partitione numerorum (1753)", *Opera omnia* (1), 2, 280.

† "Asymptotic formulae in combinatory analysis", *Proc. London Math. Soc.* (2), 17 (1918), 75–115, also *Collected papers of S. Ramanujan*, 276–309; "Une formule asymptotique pour le nombre des partitions de n", *Comptes rendus*, 164 (1917), 35–38.

and

$$A_k(n) = \sum_{\substack{h \bmod k \\ (h,k)=1}} \omega_{h,k}\, e^{-2\pi i h n/k}. \tag{1.4}$$

The symbol $\omega_{h,k}$ means here and in the following a $24k$-th root of unity given for odd h by

$$\omega_{h,k} = \left(\frac{-k}{h}\right) \exp\left(-\left\{\tfrac{1}{4}(2-hk-h) + \tfrac{1}{12}\left(k-\frac{1}{k}\right)(2h-h'+h^2h')\right\}\pi i\right), \tag{1.51}$$

and for odd k by

$$\omega_{h,k} = \left(\frac{-h}{k}\right) \exp\left(-\left\{\tfrac{1}{4}(k-1) + \tfrac{1}{12}\left(k-\frac{1}{k}\right)(2h-h'+h^2h')\right\}\pi i\right), \tag{1.52}$$

where $\left(\frac{a}{b}\right)$ denotes the Legendre-Jacobi symbol and h' is any solution of the congruence

$$hh' \equiv -1 \pmod{k}. \tag{1.6}$$

It may be mentioned, by the way, that $\omega_{h,k}$ admits also quite a different representation, viz.

$$\omega_{h,k} = \exp\left(\pi i \sum_{\mu=1}^{k-1} \frac{\mu}{k}\left(\frac{h\mu}{k} - \left[\frac{h\mu}{k}\right] - \frac{1}{2}\right)\right), \tag{1.7}$$

whose equivalence with (1.51) and (1.52) I proved some years ago*; $[u]$ means here, as usual, the greatest integer not exceeding u.

The object of the present paper is now to replace the asymptotic formula (1.2) by the *equality*

$$p(n) = \frac{1}{\pi\sqrt{2}} \sum_{k=1}^{\infty} A_k(n)\, k^{\frac{1}{2}} \frac{d}{dn}\left(\frac{\sinh C\lambda_n/k}{\lambda_n}\right). \tag{1.8}$$

The asymptotic formula (1.2) will then appear as an immediate corollary of the equation (1.8).

2. *The Farey dissection.* By means of Cauchy's integral formula we obtain from (1.1)

$$p(n) = \frac{1}{2\pi i}\int_C \frac{f(x)}{x^{n+1}}\, dx. \tag{2.1}$$

* "Bestimmung einer gewissen Einheitswurzel in der Theorie der Modulfunktionen", *Journal London Math. Soc.*, 7 (1932), 14–19.

The path of integration may be the circle C defined as

(2.2) $$|x| = e^{-2\pi N^{-2}},$$

where N is a certain positive integer at our disposal. We take the Farey series of order N and divide the circle C in the usual manner into Farey arcs $\xi_{h,k}$. We get thus

(2.3) $$p(n) = \sum_{\substack{(h,k)=1\\0 \leqslant h < k \leqslant N}} \frac{1}{2\pi i} \int_{\xi_{h,k}} \frac{f(x)}{x^{n+1}}\, dx.$$

The theory of the transformation of modular functions provides the formula*

(2.4) $$f(e^{\frac{2\pi i h}{k} - \frac{2\pi z}{k}}) = \omega_{h,k}\, z^{\frac{1}{2}} \exp\left(\frac{\pi}{12kz} - \frac{\pi z}{12k}\right) f(e^{\frac{2\pi i h'}{k} - \frac{2\pi}{kz}}),$$

valid for $\Re(z) > 0$ and with the principal branch as the determination of $z^{\frac{1}{2}}$; $\omega_{h,k}$ and h' are explained by (1.51), (1.52), (1.6), (1.7).

We put $$z = k(N^{-2} - i\phi)$$

and on the Farey arc $\xi_{h,k}$ therefore

(2.5) $$x = e^{2\pi i h/k - 2\pi N^{-2} + 2\pi i \phi},$$

which is in accordance with (2.2). The angle ϕ on $\xi_{h,k}$ is included between bounds

(2.6) $$-\vartheta'_{h,k} \leqslant \phi \leqslant \vartheta''_{h,k},$$

where

(2.7) $$\frac{1}{2kN} \leqslant \vartheta_{h,k} < \frac{1}{kN}, \quad \frac{1}{2kN} \leqslant \vartheta''_{h,k} < \frac{1}{kN}.$$

From (2.3), (2.4), (2.5), (2.6) we deduce

$$p(n) = \sum_{\substack{(h,k=1)\\0 \leqslant h < k \leqslant N}} e^{-2\pi n N^{-2}} e^{-2\pi i h n/k} \int_{-\vartheta'_{h,k}}^{\vartheta''_{h,k}} f(e^{2\pi i h/k - 2\pi(N^{-2} - i\phi)})\, e^{-2\pi i n \phi}\, d\phi,$$

(2.81) $$p(n) = e^{-2\pi n N^{-2}} \sum_{\substack{(h,k)=1\\0 \leqslant h < k \leqslant N}} \omega_{h,k}\, e^{-2\pi i h n/k}$$
$$\times \int_{-\vartheta'_{h,k}}^{\vartheta'_{h,k}} \Psi\Big(k(N^{-2} - i\phi)\Big) f(e^{\frac{2\pi i h'}{k} - \frac{2\pi}{k^2(N^{-2} - i\phi)}})\, e^{-2\pi i n \phi}\, d\phi,$$

* Hardy-Ramanujan, *loc. cit.*, 93, Lemma 4.31.

where we use the abbreviation

$$(2.82)\qquad \Psi(z) = z^{\frac{1}{2}} \exp\left(\frac{\pi}{12kz} - \frac{\pi z}{12k}\right).$$

3. *Approximation of the integrand.* The transformation formula was applied in order to get under the sign of integration in (2.81) a value of the function $f(x)$ which tends very rapidly to 1. Hence we replace (2.81) by

$$(3.1)\qquad p(n) = e^{2\pi n N^{-2}} \sum_{\substack{(h,k)=1 \\ 0\leqslant h<k\leqslant N}} \omega_{h,k}\, e^{-2\pi i h n/k} \int_{-\vartheta'_{h,k}}^{\vartheta''_{h,k}} \Psi\left(k(N^{-2}-i\phi)\right) e^{-2\pi i n\phi}\, d\phi$$

$$+ e^{2\pi n N^{-2}} \sum_{\substack{(h,k)=1 \\ 0\leqslant h<k\leqslant N}} \omega_{h,k}\, e^{-2\pi i h n/k}$$

$$\times \int_{-\vartheta'_{h,k}}^{\vartheta''_{h,k}} \Psi\left(k(N^{-2}-i\phi)\right) \{f(e^{\frac{2\pi i h'}{k} - \frac{2\pi}{k^2(N^{-2}-i\phi)}}) - 1\}\, e^{-2\pi i n \phi}\, d\phi$$

$$= e^{2\pi n N^{-2}} \sum_{h,k} \omega_{h,k}\, e^{-2\pi i h n/k} J_{h,k} + e^{2\pi n N^{-2}} \sum_{h,k} \omega_{h,k}\, e^{-2\pi i h n/k} J'_{h,k},$$

say. We shall estimate $J'_{h,k}$.

Putting

$$(3.2)\qquad z = k(N^{-2} - i\phi),$$

we have to investigate

$$\Psi(z)\{f(e^{\frac{2\pi i h'}{k} - \frac{2\pi}{kz}}) - 1\} = z^{\frac{1}{2}}\, e^{\frac{\pi}{12k}\left(\frac{1}{z} - z\right)} \sum_{m=1}^{\infty} p(m)\, e^{\left(\frac{2\pi i h'}{k} - \frac{2\pi}{kz}\right)m},$$

from which we obtain

$$(3.3)\qquad |\Psi(z)\{f(e^{\frac{2\pi i h'}{k} - \frac{2\pi}{kz}}) - 1\}| \leqslant |z^{\frac{1}{2}}|\, e^{-\frac{\pi}{12N^2}} \sum_{m=1}^{\infty} p(m)\, e^{-\frac{2\pi}{k}(m-\frac{1}{24})\Re\left(\frac{1}{z}\right)}.$$

Now we have

$$\frac{1}{z} = \frac{1}{k(N^{-2}-i\phi)} = \frac{N^{-2}+i\phi}{k(N^{-4}+\phi^2)},$$

and for $-\vartheta'_{h,k} \leqslant \phi \leqslant \vartheta''_{h,k}$, by reason of (2.7),

$$(3.4)\qquad \frac{1}{k}\Re\left(\frac{1}{z}\right) = \frac{N^{-2}}{k^2 N^{-4} + k^2\phi^2} > \frac{N^{-2}}{k^2 N^{-4} + N^{-2}} = \frac{1}{k^2 N^{-2} + 1} \geqslant \tfrac{1}{2}.$$

Furthermore,

$$(3.5)\qquad |z^{\frac{1}{2}}| = |\sqrt{\{k(N^{-2}-i\phi)\}}| = \{k^2(N^{-4}+\phi^2)\}^{\frac{1}{4}} < (k^2 N^{-4} + N^{-2})^{\frac{1}{4}} \leqslant 2^{\frac{1}{4}} N^{-\frac{1}{2}}.$$

The formulae (3 ?), (3.4), (3.5) together yield the estimate

$$(3.6) \quad |\Psi(z)\{f(e^{\frac{2\pi ih'}{k}-\frac{2\pi}{kz}})-1\}| < 2^{\frac{1}{4}} N^{-\frac{1}{2}} \sum_{m=1}^{\infty} p(m)\,\dot{e}^{-\pi(m-\frac{1}{24})} = B_1 N^{-\frac{1}{2}}.$$

The intervals of integration in (3.1) have together the length 1. Therefore (3.6) leads to

$$\left| e^{2\pi nN^{-2}} \sum_{\substack{(h,\,k)=1 \\ 0 \leqslant h < k \leqslant N}} \omega_{h,\,k}\, e^{-2\pi ihn/k} J'_{h,\,k} \right| < B_1 e^{2\pi nN^{-2}} N^{-\frac{1}{2}},$$

and from (3.1) we get the result

$$(3.71) \qquad p(n) = e^{2\pi nN^{-2}} \sum_{\substack{(h,\,k)=1 \\ 0 \leqslant h < k \leqslant N}} \omega_{h,\,k}\, e^{-2\pi ihn/k} J_{h,\,k} + \theta_1,$$

$$(3.72) \qquad\qquad |\theta_1| < B_1 e^{2\pi nN^{-2}} N^{-\frac{1}{2}}.$$

4. *A loop integral.* We have, by (3.1) and (2.82),

$$e^{2\pi nN^{-2}} J_{h,\,k} = e^{2\pi nN^{-2}} \int_{-\vartheta'_{h,\,k}}^{\vartheta''_{h,\,k}} \sqrt{\{k(N^{-2}-i\phi)\}}\, e^{\frac{\pi}{12k}\left(\frac{1}{k(N^{-2}-i\phi)}-k(N^{-2}-i\phi)\right)} e^{-2\pi in\phi}\, d\phi.$$

This is transformed by the substitution

$$w = N^{-2} - i\phi$$

into

$$e^{2\pi nN^{-2}} J_{h,\,k} = e^{2\pi nN^{-2}} \int_{N^{-2}+i\vartheta'_{h,\,k}}^{N^{-2}-i\vartheta''_{h,\,k}} \sqrt{(kw)}\, e^{\frac{\pi}{12k}\left(\frac{1}{kw}-kw\right)} e^{2\pi n(w-N^{-2})}\, idw,$$

$$(4.1) \qquad e^{2\pi nN^{-2}} J_{h,\,k} = \frac{\sqrt{k}}{i} \int_{N^{-2}-i\vartheta''_{h,\,k}}^{N^{-2}+i\vartheta'_{h,\,k}} \sqrt{w}\, e^{\pi/12k^2\, w}\, e^{2\pi(n-\frac{1}{24})\, w}\, dw.$$

The integrand

$$(4.2) \qquad\qquad \sqrt{w}\, e^{\pi/12k^2\, w}\, e^{2\pi(n-\frac{1}{24})\, w}$$

is one-valued in the complex w-plane cut along the negative real axis from 0 to $-\infty$. In order to calculate the integral (4.1) we introduce a loop integral along the path which joins the following points by straight segments:

$$-\infty, \quad -\epsilon, \quad -\epsilon-i\vartheta''_{h,\,k}, \quad N^{-2}-i\vartheta''_{h,\,k}, \quad N^{-2}+i\vartheta'_{h,\,k}, \quad -\epsilon+i\vartheta'_{h,\,k}, \quad -\epsilon, -\infty.$$

Because of the cut we have to follow first the lower border of the cut from $-\infty$ to $-\epsilon$ and ~~and~~ on the upper border from $-\epsilon$ to $-\infty$. The figure

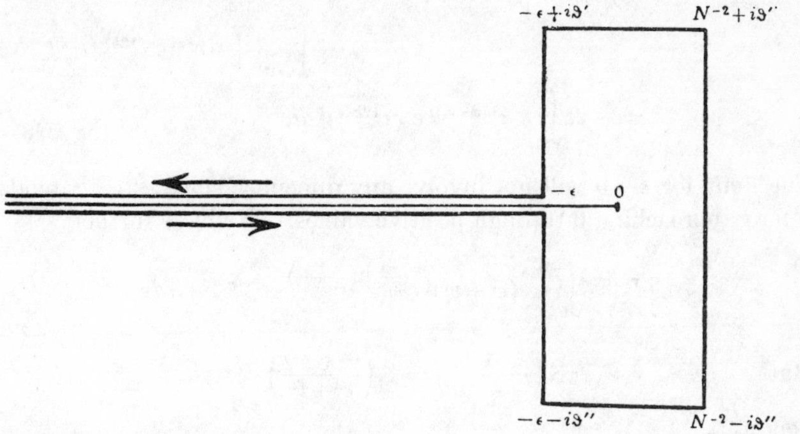

indicates the path of integration in the w-plane. The arbitrary positive number ϵ may be chosen to satisfy

$$0 < \epsilon < N^{-2}. \tag{4.3}$$

If we denote (after Watson) by $\int_{-\infty}^{(0+)}$ a loop integral beginning at $-\infty$, surrounding the point 0 in the positive sense and returning to $-\infty$, we get, from (4.1),

(4.4) $e^{2\pi n N^{-2}} J_{h,k}$

$$= \frac{k^{\frac{1}{2}}}{i}\int_{-\infty}^{(0+)} - \frac{k^{\frac{1}{2}}}{i}\left\{\int_{-\infty}^{-\epsilon} + \int_{-\epsilon}^{-\epsilon - i\vartheta''_{h,k}} + \int_{-\epsilon - i\vartheta''_{h,k}}^{N^{-2} - i\vartheta''_{h,k}} + \int_{N^{-2}+i\vartheta'_{h,k}}^{-\epsilon + i\vartheta'_{h,k}} + \int_{-\epsilon + i\vartheta'_{h,k}}^{-\epsilon} + \int_{-\epsilon}^{-\infty}\right\}$$

$$= \frac{k^{\frac{1}{2}}}{i} L_k - \frac{k^{\frac{1}{2}}}{i}\{I_1 + I_2 + I_3 + I_4 + I_5 + I_6\},$$

say. The integrals are understood all to contain the same integrand (4.2).

5. *Convergence of a series for* $p(n)$. Our next steps are to estimate the integrals I_2, I_3, I_4, I_5 and to make ϵ approach 0. The integrals I_1 and I_6 have later to be evaluated in exact form and will furnish an essential contribution to our final formula.

We have

$$(5.1)\quad I_1+I_6=\int_{-\infty}^{-\epsilon}\sqrt{|u|}\,e^{-\frac{1}{4}\pi i}\,e^{\pi/12k^2u}\,e^{2\pi(n-\frac{1}{24})u}\,du$$

$$+\int_{-\epsilon}^{-\infty}\sqrt{|u|}\,e^{\frac{1}{4}\pi i}\,e^{\pi/12k^2u}\,e^{2\pi(n-\frac{1}{24})u}\,du$$

$$=-2i\int_{\epsilon}^{\infty}e^{-2\pi(n-\frac{1}{24})t}\,e^{-\pi/12k^2t}\,t^{\frac{1}{2}}\,dt.$$

The limit for $\epsilon\to 0$ will not involve any difficulty, since $e^{-\pi/12k^2t}$ tends to 0 for t approaching 0 through positive values. We have further

$$|I_2|\leqslant\int_0^{-\vartheta''_{h,k}}(\epsilon^2+v^2)^{\frac{1}{4}}\,e^{\frac{\pi}{12k^2}\Re\left(\frac{1}{-\epsilon+iv}\right)}\,e^{-2\pi(n-\frac{1}{24})\epsilon}\,|dv|.$$

But

$$\Re\left(\frac{1}{-\epsilon+iv}\right)=\Re\left(\frac{-\epsilon-iv}{\epsilon^2+v^2}\right)<0;$$

hence

$$(5.2)\qquad |I_2|<(\epsilon^2+\vartheta''^2_{h,k})^{\frac{1}{4}}\,\vartheta''_{h,k}<\left(\epsilon^2+\frac{1}{k^2N^2}\right)^{\frac{1}{4}}\frac{1}{kN},$$

and similarly

$$(5.3)\qquad |I_5|<(\epsilon^2+\vartheta'^2_{h,k})^{\frac{1}{4}}\,\vartheta'_{h,k}<\left(\epsilon^2+\frac{1}{k^2N^2}\right)^{\frac{1}{4}}\frac{1}{kN}.$$

For the two remaining integrals we find

$$|I_3|\leqslant\int_{-\epsilon}^{N^{-2}}(u^2+\vartheta''^2_{h,k})^{\frac{1}{4}}\,e^{\frac{\pi}{12k^2}\Re\left(\frac{1}{u-i\vartheta''_{h,k}}\right)}\,e^{2\pi(n-\frac{1}{24})u}\,du.$$

Here we have

$$\frac{1}{k^2}\Re\left(\frac{1}{u-i\vartheta''_{h,k}}\right)=\frac{u}{k^2(u^2+\vartheta''^2_{h,k})}\leqslant\frac{N^{-2}}{k^2\vartheta''^2_{h,k}}\leqslant 4.$$

This together with (4.3) leads to

$$|I_3|<(\epsilon+N^{-2})(N^{-4}+\vartheta''^2_{h,k})^{\frac{1}{4}}\,e^{\frac{1}{3}\pi}\,e^{2\pi nN^{-2}}$$

$$<(\epsilon+N^{-2})\left(N^{-4}+\frac{1}{k^2N^2}\right)^{\frac{1}{4}}e^{\frac{1}{3}\pi}\,e^{2\pi nN^{-2}},$$

i.e.

$$(5.4)\qquad |I_3|<(\epsilon+N^{-2})\,2^{\frac{1}{4}}\,k^{-\frac{1}{2}}\,N^{-\frac{1}{2}}\,e^{\frac{1}{3}\pi}\,e^{2\pi nN^{-2}}.$$

The analogue is

$$(5.5)\qquad |I_4|<(\epsilon+N^{-2})\,2^{\frac{1}{4}}\,k^{-\frac{1}{2}}\,N^{-\frac{1}{2}}\,e^{\frac{1}{3}\pi}\,e^{2\pi nN^{-2}}.$$

We now make $\epsilon \to 0$ in (5.1), (5.2), (5.3), (5.4), (5.5), and insert the results in (4.4). We get

$$(5.61) \qquad e^{2\pi n N^{-2}} J_{h,k} = \frac{k^{\frac12}}{i} L_k + 2 k^{\frac12} H_k + U_{h,k} + V_{h,k}$$

with

$$(5.62) \qquad L_k = \int_{-\infty}^{(0+)} w^{\frac12} e^{\pi/12k^2 w} e^{2\pi (n-\frac{1}{24}) w} \, dw,$$

$$(5.63) \qquad H_k = \int_0^{\infty} t^{\frac12} e^{-\pi/12k^2 t} e^{-2\pi (n-\frac{1}{24}) t} \, dt,$$

$$(5.64) \qquad |U_{h,k}| \leqslant 2N^{-\frac32} k^{-1},$$

$$(5.65) \qquad |V_{h,k}| \leqslant 2^{\frac54} e^{\frac13 \pi} e^{2\pi n N^{-2}} N^{-\frac52}.$$

Putting

$$(5.7) \qquad \psi_k(n) = \frac{k^{\frac12}}{i} L_k + 2 k^{\frac12} H_k$$

and introducing (5.61) into (3.71), we obtain

$$p(n) = \sum_{\substack{(h,k)=1 \\ 0 \leqslant h < k \leqslant N}} \omega_{h,k} e^{-2\pi i h n/k} \psi_k(n) + \theta_1 + \sum_{\substack{(h,k)=1 \\ 0 \leqslant h < k \leqslant N}} U_{h,k} + \sum_{\substack{(h,k)=1 \\ 0 \leqslant h < k \leqslant N}} V_{h,k}.$$

But we have, from (5.64),

$$\left| \sum_{\substack{(h,k)=1 \\ 0 \leqslant h < k \leqslant N}} U_{h,k} \right| \leqslant 2N^{-\frac32} \sum_{0 < k \leqslant N} 1 = 2N^{-\frac12},$$

and, from (5.65),

$$\left| \sum_{\substack{(h,k)=1 \\ 0 \leqslant h < k \leqslant N}} V_{h,k} \right| \leqslant 2^{\frac54} e^{\frac13 \pi} e^{2\pi n N^{-2}} N^{-\frac52} \sum_{0 < k \leqslant N} k \leqslant 2^{\frac54} e^{\frac13 \pi} e^{2\pi n N^{-2}} N^{-\frac12},$$

and hence, observing (3.72), and with the notation (1.4), we get

$$(5.81) \qquad p(n) = \sum_{k=1}^{N} A_k(n) \psi_k(n) + R(n, N),$$

$$(5.82) \qquad |R(n, N)| < B_2 e^{2\pi n N^{-2}} N^{-\frac12}.$$

If we keep n fixed, we can infer from (5.82) that when $N \to \infty$ the remainder tends to zero. This means that the series extended over all k is convergent and has $p(n)$ as its value: *i.e.*

$$(5.9) \qquad p(n) = \sum_{k=1}^{\infty} A_k(n) \psi_k(n).$$

6. *Evaluation of an integral.* The final steps are directed towards the calculation of $\psi_k(n)$, *i.e.* of the integrals L_k and H_k. We have, from (5.62)*,

$$\frac{1}{i} L_k = \frac{1}{i}\int_{-\infty}^{(0+)} w^{\frac{1}{2}} e^{\pi/12k^2 w} e^{2\pi(n-\frac{1}{24})w} dw$$

$$= \frac{1}{i}\int_{-\infty}^{(0+)} w^{\frac{1}{2}} e^{2\pi(n-\frac{1}{24})w} \sum_{l=0}^{\infty} \frac{(\pi/12k^2 w)^l}{l!} dw$$

$$= \frac{1}{i}\sum_{l=0}^{\infty} \frac{(\pi/12k^2)^l}{l\,!}\int_{-\infty}^{(0+)} w^{\frac{1}{2}} e^{2\pi(n-\frac{1}{24})w} w^{-l} dw$$

$$= 2\pi \sum_{l=0}^{\infty} \frac{(\pi/12k^2)^l}{l\,!} \{2\pi(n-\tfrac{1}{24})\}^{l-\frac{3}{2}} \frac{1}{2\pi i}\int_{-\infty}^{(0+)} e^z z^{-l+\frac{1}{2}} dz.$$

We apply Hankel's formula

$$\frac{1}{\Gamma(s)} = \frac{1}{2\pi i}\int_{-\infty}^{(0+)} e^z z^{-s} dz$$

and obtain

$$(6.1) \qquad \frac{1}{i} L_k = \frac{1}{\sqrt{(2\pi)}} (n-\tfrac{1}{24})^{-\frac{3}{2}} \sum_{l=0}^{\infty} \frac{\{\pi^2(n-\frac{1}{24})/6k^2\}^l}{l!\,\Gamma(l-\frac{1}{2})}.$$

This infinite series may be studied separately: we have

$$\sum_{l=0}^{\infty} \frac{X^l}{l!\,\Gamma(l-\frac{1}{2})} = \frac{1}{-2\pi^{\frac{1}{2}}} + \frac{X}{1!\,\pi^{\frac{1}{2}}} + \frac{X^2}{2!\,\frac{1}{2}\pi^{\frac{1}{2}}} + \frac{X^3}{3!\,\frac{1}{2}.\frac{3}{2}\pi^{\frac{1}{2}}} + \dots$$

$$= \frac{1}{\pi^{\frac{1}{2}}}\left\{-\tfrac{1}{2} + \frac{2X}{2!} + \frac{3.2^3 X^2}{4!} + \frac{5.2^5 X^3}{6!} + \dots\right\}$$

$$= \frac{1}{2\pi^{\frac{1}{2}}}\left\{-1 + \frac{4X}{2!} + \frac{3(4X)^2}{4!} + \frac{5(4X)^3}{6!} + \dots\right\}.$$

This yields, for $4X = Y^2$,

$$\sum_{l=0}^{\infty} \frac{(\frac{1}{4}Y^2)^l}{l!\,\Gamma(l-\frac{1}{2})} = \frac{1}{2\pi^{\frac{1}{2}}}\left\{-1 + Y^2\left(\frac{1}{2!} + \frac{3Y^2}{4!} + \frac{5Y^4}{6!} + \dots\right)\right\}$$

$$= \frac{1}{2\pi^{\frac{1}{2}}}\left\{-1 + Y^2 \frac{d}{dY}\left(\frac{Y}{2!} + \frac{Y^3}{4!} + \frac{Y^5}{6!} + \dots\right)\right\}$$

$$= \frac{1}{2\pi^{\frac{1}{2}}}\left\{-1 + Y^2 \frac{d}{dY}\left(\frac{\cosh Y - 1}{Y}\right)\right\},$$

$$(6.2) \qquad \sum_{l=1}^{\infty} \frac{(\frac{1}{4}Y^2)^l}{l!\,\Gamma(l-\frac{1}{2})} = \frac{1}{2\pi^{\frac{1}{2}}} Y^2 \frac{d}{dY}\left(\frac{\cosh Y}{Y}\right).$$

---

* The integral L_k is well known and is connected with Bessel functions of purely imaginary argument: *cf.* Watson, *Bessel functions*, 181. We prefer the direct calculation.

In order to use (6.2) for (6.1) we have to put

$$\tfrac{1}{4}Y^2 = \frac{\pi^2}{6k^2}\left(n-\tfrac{1}{24}\right),$$

$$Y = \frac{\pi}{k}\sqrt{\left(\tfrac{2}{3}\right)}\sqrt{\left(n-\tfrac{1}{24}\right)} = \frac{C\lambda_n}{k} \tag{6.3}$$

in the notation (1.3). Moreover, we have

$$\frac{d}{dn}\left(\frac{\cosh Y}{Y}\right) = \frac{d}{dY}\left(\frac{\cosh Y}{Y}\right)\frac{dY}{dn} = \frac{d}{dY}\left(\frac{\cosh Y}{Y}\right)\frac{C}{k\,.\,2\lambda_n}. \tag{6.4}$$

Combining (6.1), (6.2), (6.3), (6.4), we get

$$\frac{1}{i}L_k = \frac{1}{\sqrt{(2\pi)}}\left(n-\tfrac{1}{24}\right)^{-\frac{3}{4}}\frac{1}{2\sqrt{\pi}}\frac{C^2\lambda_n^{\,2}}{k^2}\frac{2k\lambda_n}{C}\frac{d}{dn}\left(\frac{\cosh Y}{Y}\right)$$

$$= \frac{1}{\pi\sqrt{2}}\frac{C}{k}\frac{d}{dn}\left(\frac{\cosh Y}{Y}\right),$$

and finally

$$\frac{1}{i}L_k = \frac{1}{\pi\sqrt{2}}\frac{d}{dn}\left(\frac{\cosh C\lambda_n/k}{\lambda_n}\right). \tag{6.5}$$

7. *Evaluation of another integral and proof of the theorem.* There remains only the integral (5.63), namely

$$H_k = \int_0^\infty t^{\frac{1}{2}}\, e^{-\pi/12k^2 t}\, e^{-2\pi(n-\frac{1}{24})t}\, dt.$$

Now

$$\int_0^\infty e^{-c^2t-a^2/t}\,t^{-\frac{1}{2}}\,dt = 2\int_0^\infty e^{-c^2u^2-a^2/u^2}\,du = \frac{\sqrt{\pi}}{c}e^{-2ac},$$

and so

$$\int_0^\infty e^{-c^2t-a^2/t}\,t^{\frac{1}{2}}\,dt = -\frac{\sqrt{\pi}}{2c}\frac{d}{dc}\left(\frac{e^{-2ac}}{c}\right).$$

Putting $c = \sqrt{(2\pi)}\,\lambda_n$, $a = \dfrac{1}{k}\sqrt{\left(\dfrac{\pi}{12}\right)}$, we find that

$$H_k = -\frac{1}{2\sqrt{2}\,\lambda_n}\frac{d}{d\lambda_n}\left(\frac{e^{-C\lambda_n/k}}{\sqrt{(2\pi)}\,\lambda_n}\right)\frac{1}{\sqrt{(2\pi)}},$$

and hence

$$H_k = -\frac{1}{2\pi\sqrt{2}}\frac{d}{dn}\left(\frac{e^{-C\lambda_n/k}}{\lambda_n}\right). \tag{7.1}$$

From (5.7), (6.5), and (7.1) we deduce

$$\psi_k(n) = \frac{\sqrt{k}}{\pi\sqrt{2}} \frac{d}{dn}\left(\frac{\cosh(C\lambda_n/k)}{\lambda_n} - \frac{e^{-C\lambda_n/k}}{\lambda_n}\right),$$

so that, finally,

$$\psi_k(n) = \frac{\sqrt{k}}{\pi\sqrt{2}} \frac{d}{dn}\left(\frac{\sinh(C\lambda_n/k)}{\lambda_n}\right). \tag{7.2}$$

This together with (5.9) proves the

THEOREM. *The number $p(n)$ of unrestricted partitions of n is given by the convergent series*

$$p(n) = \frac{1}{\pi\sqrt{2}} \sum_{k=1}^{\infty} A_k(n)\, k^{\frac{1}{2}} \frac{d}{dn}\left(\frac{\sinh(C\lambda_n/k)}{\lambda_n}\right), \tag{7.3}$$

where C, λ_n, $A_k(n)$ are defined by (1.3) *and* (1.4).

8. *Asymptotic formulae.* We proceed to discuss asymptotic formulae as consequences of our theorem. If we put, in (5.81), (5.82),

$$N = [a\sqrt{n}]$$

we get

$$p(n) = \sum_{k=1}^{[a\sqrt{n}]} A_k(n)\,\psi_k(n) + O(n^{-\frac{1}{4}}). \tag{8.1}$$

This is not yet the Hardy-Ramanujan formula (1.2). In order to obtain the latter we have to estimate the error which arises from replacing $\sinh(C\lambda_n/k)$ in (8.1) by $\frac{1}{2}\exp(C\lambda_n/k)$. We get

$$\sum_{1 \leqslant k \leqslant a\sqrt{n}} A_k k^{\frac{1}{2}} \frac{d}{dn}\left(\frac{e^{-C\lambda_n/k}}{\lambda_n}\right) = O\left\{\sum_{1 \leqslant k \leqslant a\sqrt{n}} k^{\frac{3}{2}}\left(\frac{1}{kn} + \frac{1}{n^{\frac{3}{2}}}\right)\right\}$$

$$= O\left(n^{-1} \sum_{1 \leqslant k \leqslant a\sqrt{n}} k^{\frac{1}{2}}\right) + O\left(n^{-\frac{3}{2}} \sum_{1 \leqslant k \leqslant a\sqrt{n}} k^{\frac{3}{2}}\right)$$

$$= O(n^{-1} n^{\frac{3}{4}}) + O(n^{-\frac{3}{2}} n^{\frac{5}{4}}) = O(n^{-\frac{1}{4}}).$$

This together with (8.1) proves (1.2).

For the purpose of numerical computation, however, it is desirable to know suitable bounds for the constants involved in the O-terms. Such an estimate is better obtained from the convergent series (7.3) directly. We have

$$\left|\frac{1}{\pi\sqrt{2}}\sum_{k=N+1}^{\infty} A_k(n)\, k^{\frac{1}{2}} \frac{d}{dn}\left(\frac{\sinh(C\lambda_n/k)}{\lambda_n}\right)\right|$$

$$\leqslant \frac{1}{\pi\sqrt{2}}\sum_{k=N+1}^{\infty} k^{\frac{3}{2}} \frac{d}{dn}\sum_{\nu=0}^{\infty}\frac{(C/k)^{2\nu+1}(n-\frac{1}{24})^{\nu}}{(2\nu+1)!}$$

$$= \frac{1}{\pi\sqrt{2}}\sum_{k=N+1}^{\infty} k^{\frac{3}{2}} \sum_{\nu=1}^{\infty}\frac{\nu}{(2\nu+1)!}\left(\frac{C}{k}\right)^{2\nu+1}\lambda_n^{2\nu-2}$$

$$< \frac{C^2}{\pi\sqrt{2}\lambda_n}\sum_{\nu=1}^{\infty}\frac{\nu}{(2\nu+1)!}(C\lambda_n)^{2\nu-1}\int_N^{\infty}\frac{dk}{k^{2\nu-\frac{1}{2}}}$$

$$= \frac{C^2}{\pi\sqrt{2}\lambda_n}\sum_{\nu=1}^{\infty}\frac{\nu}{(2\nu-\frac{3}{2})(2\nu+1)!}\frac{(C\lambda_n)^{2\nu-1}}{N^{2\nu-\frac{3}{2}}}$$

$$= \frac{C^2 N^{\frac{1}{2}}}{\pi\sqrt{2}\lambda_n}\left\{\frac{1}{3}\frac{C\lambda_n}{N}+\sum_{\nu=2}^{\infty}\frac{(C\lambda_n/N)^{2\nu-1}}{(2\nu-1)!\,(2\nu+1)\,(4\nu-3)}\right\}$$

$$< \frac{C^2 N^{\frac{1}{2}}}{\pi\sqrt{2}\lambda_n}\left\{\frac{1}{3}\frac{C\lambda_n}{N}+\frac{1}{5^2}\sum_{\nu=2}^{\infty}\frac{(C\lambda_n/N)^{2\nu-1}}{(2\nu-1)!}\right\}$$

$$= \frac{C^2 N^{\frac{1}{2}}}{\pi\sqrt{2}\lambda_n}\left\{\left(\tfrac{1}{3}-\tfrac{1}{25}\right)\frac{C\lambda_n}{N}+\tfrac{1}{25}\sinh\frac{C\lambda_n}{N}\right\},$$

which with (7.3) leads to

$$(8.2)\quad p(n)=\frac{1}{\pi\sqrt{2}}\sum_{k=1}^{N} A_k(n)\, k^{\frac{1}{2}}\frac{d}{dn}\left(\frac{\sinh(C\lambda_n/k)}{\lambda_n}\right)$$

$$+\vartheta'\left\{\frac{44\pi^2}{225\sqrt{3}}N^{-\frac{1}{2}}+\frac{\pi\sqrt{2}}{75}\left(\frac{N}{n-1}\right)^{\frac{1}{2}}\sinh\frac{C\sqrt{n}}{N}\right\}\quad (|\vartheta'|<1).$$

Finally we replace $\sinh C\lambda_n/k$ in (8.2) by $\frac{1}{2}\exp(C\lambda_n/k)$, committing thereby an error which lends itself to the following estimate:

$$\left|\frac{1}{2\pi\sqrt{2}}\sum_{k=1}^{N} A_k(n)\, k^{\frac{1}{2}}\frac{d}{dn}\left(\frac{e^{-C\lambda_n/k}}{\lambda_n}\right)\right|$$

$$\leqslant \frac{1}{2\pi\sqrt{2}}\sum_{k=1}^{N} k^{\frac{3}{2}}\left(\frac{e^{-C\lambda_n/k}}{2\lambda_n{}^2}\frac{C}{k}+\frac{e^{-C\lambda_n/k}}{2\lambda_n{}^3}\right)$$

$$= \frac{C}{4\pi\sqrt{2}}\frac{1}{n-\frac{1}{24}}\sum_{k=1}^{N} k^{\frac{1}{2}}e^{-C\lambda_n/k}+\frac{1}{4\pi\sqrt{2}}\frac{1}{(n-\frac{1}{24})^{\frac{3}{2}}}\sum_{k=1}^{N} k^{\frac{3}{2}}e^{-C\lambda_n/k}$$

$$< \frac{C}{4\pi\sqrt{2}}\frac{1}{n-1}e^{-C\lambda_n/N}\int_0^{N+1}k^{\frac{1}{2}}\,dk+\frac{1}{4\pi\sqrt{2}}\frac{1}{(n-1)^{\frac{3}{2}}}e^{-C\lambda_n/N}\int_0^{N+1}k^{\frac{3}{2}}\,dk$$

$$< \frac{1}{2\sqrt{3}}\frac{(N+1)^{\frac{3}{2}}}{n-1}e^{-C\lambda_n/N}\left(\frac{1}{3}+\frac{\sqrt{3}}{5\pi\sqrt{3}}\frac{N+1}{(n-1)^{\frac{1}{2}}}\right).$$

This together with (8.2) yields the

COROLLARY.

$$(8.3)\quad p(n)=\frac{1}{2\pi\sqrt{2}}\sum_{k=1}^{N} A_k(n)\,k^{\frac{1}{2}}\,\frac{d}{dn}\left(\frac{e^{C\lambda_n/k}}{\lambda_n}\right)$$

$$+\vartheta\left\{\frac{44\pi^2}{225\sqrt{3}}N^{-\frac{1}{2}}+\frac{\pi\sqrt{2}}{75}\left(\frac{N}{n-1}\right)^{\frac{1}{2}}\sinh\frac{C\sqrt{n}}{N}\right.$$

$$\left.+\frac{1}{2\sqrt{3}}\,\frac{(N+1)^{\frac{3}{2}}}{n-1}\,e^{-C(n-1)^{\frac{1}{2}}/N}\left(\tfrac{1}{3}+\frac{\sqrt{3}}{5\pi\sqrt{2}}\,\frac{N+1}{(n-1)^{\frac{1}{2}}}\right)\right\}.\ (|\vartheta|<1).$$

9. *A numerical application.* Recently D. H. Lehmer* has used (1.2) for the computation of $p(599)$ and $p(721)$. By some special cases of Ramanujan's conjecture which are already proved we know that

$$(9.1)\qquad p(599)\equiv 0 \pmod{5^3},$$

$$(9.2)\qquad p(721)\equiv 0 \pmod{11^2}.$$

Now (1.2) asserts only that the error after $[a\sqrt{n}]$ terms tends to zero like $n^{-\frac{1}{4}}$. But since the constants in the O-term were not known, a conclusive control of the numerical result was not possible. Lehmer had to admit the possibility that his result for $p(599)$ differed from the true value by a multiple of 5^3, in accordance with (9.1), and $p(721)$ differed by a multiple of 11^2, in accordance with (9.2).

Our corollary (8.3) will now easily justify Lehmer's use of the Hardy-Ramanujan formula for the computation of $p(599)$ and $p(721)$, since the possible errors turn out to be not only less than $\frac{1}{2}5^3$ and $\frac{1}{2}11^2$ respectively, which obviously would suffice, but even less than $\frac{1}{2}$ each.

If we modify the error term in (8.3) slightly by substituting $\frac{1}{2}\exp(C\sqrt{n}/N)$ for $\sinh(C\sqrt{n}/N)$, we introduce a certain increase of the error, which in the present cases is not very noticeable. We get then:

(1) For $n=599$, $N=18$, an error of at most

$$\frac{44\pi^2}{225\sqrt{3}\sqrt{18}}+\frac{\pi\sqrt{2}}{150}\left(\frac{18}{598}\right)e^{\pi\sqrt{(\frac{2}{3})}\frac{\sqrt{(599)}}{18}}$$

$$+\frac{19^{\frac{3}{2}}}{2\sqrt{3}\,.\,598}\,e^{-\pi\sqrt{(\frac{2}{3})}\frac{\sqrt{(598)}}{18}}\left(\tfrac{1}{3}+\frac{\sqrt{3}\,.\,19}{5\pi\sqrt{2}\sqrt{598}}\right)$$

$$=0{\cdot}26265+0{\cdot}16810+0{\cdot}00048=0{\cdot}43123$$

* "On a conjecture of Ramanujan", *Journal London Math. Soc.*, 11 (1936), 114–118.

in absolute value. This renders Lehmer's hypothetical figure

$$p(599) = 4353\ 50207\ 84031\ 73482\ 70000$$

definitive.

(2) For $n = 721$, $N = 21$, at most an absolute value

$$\frac{44\pi^2}{225\sqrt{3}\cdot\sqrt{21}} + \frac{\pi\sqrt{2}}{150}\left(\frac{21}{720}\right)^{\frac{1}{4}} e^{\pi\sqrt{(\frac{2}{3})}\frac{\sqrt{(721)}}{21}} + \frac{22^{\frac{3}{2}}}{2\sqrt{3}\,.\,720} e^{-\pi\sqrt{(\frac{2}{3})}\frac{\sqrt{(720)}}{21}}\left(\frac{1}{3} + \frac{\sqrt{3}\,.\,22}{5\pi\sqrt{2}\sqrt{720}}\right)$$

$$= 0{\cdot}24316 + 0{\cdot}13441 + 0{\cdot}00062 = 0{\cdot}37819$$

for the error. This vindicates Lehmer's result of

$$p(721) = 16\ 10617\ 55750\ 27947\ 76355\ 34762.$$

The smallness of the error estimates in both cases, which made it indeed unnecessary to refer to the divisibility property (9.1), (9.2), seems to promise an applicability of (8.3) for numerical computation of $p(n)$ in a wider range.

University of Pennsylvania,
Philadelphia.

This paper has been reviewed in the JF, vol. 63 (1937), p. 140, and in the Z, vol. 17 (1937-38), pp. 55-56.

Page 242, footnote.
See paper 29 of the present collection.

Page 243, fourth display line from bottom.
Read $\vartheta'_{h,k}$ instead of $\vartheta_{h,k}$.

Page 243, last display line.
The upper limit of the integral is $\vartheta''_{h,k}$ (not $\vartheta'_{h,k}$).

Page 246, marginal note.
Author's correction (suppress "end").

Page 251, third display line from bottom.
Author's correction: read $A_k k^{1/2} \dfrac{d}{dn}(\ldots)$.

Page 252, marginal note.
Author's correction (replace x by n).

THE FOURIER COEFFICIENTS OF THE MODULAR INVARIANT $J(\tau)$.*

By Hans Rademacher.

1. Recently Dr. Zuckerman and I have developed general formulae for the Fourier coefficients of modular forms of positive dimensions.[1] We remarked at the end of our paper that the series obtained would be convergent also for forms of dimension zero, i. e. for *modular functions*, among which $J(\tau)$ can be regarded as fundamental. The question arises whether the formally constructed series for the coefficients of $J(\tau)$ actually represents them.

The solution of this problem requires a thorough revision of our method, since we had essentially made use of the positivity of the dimension of the modular forms. A method due to Kloosterman [2] and later extended by Estermann [3] gives the clue. Kloostermann's method consists of two devices: first it improves the estimate of certain sums $A_k(n)$ of roots of unity from the trivial one

$$| A_k(n) | \leqq k$$

to

$$(1.1) \qquad | A_k(n) | \leqq C k^{1-\beta+\epsilon} \cdot (k, n)^{\beta},$$

where β, according to results of Salié [4] and Davenport [5] can be taken as $\beta = \frac{1}{3}$. Secondly it collects the Farey arcs $\xi_{h,k}$ belonging to the same k and treats the resulting sum as a whole instead of estimating the summands separately. Both of these expedients will be used in the following paper.

* Received February 20, 1938.

[1] "On the Fourier coefficients of certain modular forms of positive dimension," to be published in the *Annals of Mathematics*.

[2] H. D. Kloosterman, "Asymptotische Formeln für die Fourierkoeffizienten ganzer Modulformen," *Abhandlungen Hamburg. Math. Seminar*, vol. 5 (1927), pp. 337-352.

[3] T. Estermann, "Vereinfachter Beweis eines Satzes von Kloosterman," *ibid.*, vol. 7 (1929), pp. 82-98.

[4] H. Salié, "Zur Abschätzung der Fourierkoeffizienten ganzer Modulformen," *Mathematische Zeitschrift*, vol. 36 (1933), pp. 263-278.

[5] H. Davenport, "On certain exponential sums," *Journal f. d. reine u. angew. Mathematik*, vol. 169 (1933), pp. 158-176.

2. Our problem is to investigate the coefficients of the expansion

$$12^3 J(\tau) = e^{-2\pi i\tau} + \sum_{n=0}^{\infty} c_n e^{2\pi i n\tau} = f(e^{2\pi i\tau}), \tag{2.1}$$

where $J(\tau)$ is defined by means of the modular forms $g_2(\omega_1, \omega_2)$ and $g_3(\omega_1, \omega_2)$ as

$$J(\tau) = \frac{g_2^3(1,\tau)}{g_2^3(1,\tau) - 27g_3^2(1,\tau)} = \frac{g_2^3(1,\tau)}{\Delta(1,\tau)}, \qquad \Im(\tau) > 0. \tag{2.2}$$

A consequence of (2.2) is, by the way, the formula [6]

$$12^3 J(\tau) = \frac{\left\{1 + 240 \sum_{m=1}^{\infty} \frac{m^3 x^m}{1 - x^m}\right\}^3}{x\left\{\sum_{\lambda=-\infty}^{+\infty} (-1)^\lambda x^{\frac{\lambda(3\lambda-1)}{2}}\right\}^{24}}, \qquad x = e^{2\pi i\tau}, \tag{2.3}$$

which shows that the coefficients c_n in (2.1) are integers. From (2.2) we infer the invariance of $J(\tau)$ with respect to the transformations of the full modular group:

$$J\left(\frac{a\tau + b}{c\tau + d}\right) = J(\tau). \tag{2.4}$$

We shall see that the equations (2.1) and (2.4) will completely suffice for the determination of the coefficients c_n, provided $n \geqq 1$. For

$$\tau = \frac{iz}{k} + \frac{h}{k}, \qquad \Re(z) > 0,$$

$$\begin{pmatrix} a & b \\ c & d \end{pmatrix} = \begin{pmatrix} h' & -\frac{hh'+1}{k} \\ k & -h \end{pmatrix}$$

with

$$hh' \equiv -1 \pmod{k}, \tag{2.5}$$

equation (2.4) goes over into

$$J\left(\frac{iz}{k} + \frac{h}{k}\right) = J\left(\frac{i}{kz} + \frac{h'}{k}\right)$$

or, in the notation of (2.1),

$$f\left(e^{-\frac{2\pi z}{k} + \frac{2\pi i h}{k}}\right) = f\left(e^{-\frac{2\pi}{kz} + \frac{2\pi i h'}{k}}\right). \tag{2.6}$$

3. From (2.1) we obtain

$$c_n = \frac{1}{2\pi i}\int_C \frac{f(x)}{x^{n+1}}\,dx = \sum_{\substack{h,k \\ 0 \leqq h < k \leqq N}}' \frac{1}{2\pi i}\int_{\xi_{h,k}} \frac{f(x)}{x^{n+1}}\,dx,[7]$$

[6] Klein–Fricke, *Vorlesungen über die Theorie der Modulfunktionen*, vol. I, p. 154.

[7] Σ' means here and subsequently that h runs over integers prime to k.

where the $\xi_{h,k}$ may be the Farey arcs of order N of the circle C

$$|x| = e^{-2\pi N^{-2}}.$$

If we introduce on $\xi_{h,k}$ the new variable ϕ through

$$x = \exp\left(-2\pi N^{-2} + \frac{2\pi i h}{k} + 2\pi i \phi\right)$$

we get

$$(3.1) \qquad c_n = \sum_{\substack{h,k \\ 0 \leqq h < k \leqq N}}' e^{-\frac{2\pi i n h}{k}} \int_{-\vartheta'_{h,k}}^{\vartheta''_{h,k}} f\left(e^{\frac{2\pi i h}{k} - 2\pi(N^{-2} - i\phi)}\right) e^{2\pi n(N^{-2} - i\phi)} \, d\phi.$$

For a later purpose we need here the determination of $\vartheta'_{h,k}$ and $\vartheta''_{h,k}$ in terms of h and k. In the Farey series of order N we consider the fraction h/k with its two neighbors:

$$(3.2) \qquad \frac{h_1}{k_1} < \frac{h}{k} < \frac{h_2}{k_2}, \qquad k, k_1, k_2 \leqq N.$$

We have here

$$hk_1 - h_1 k = 1, \qquad h_2 k - h k_2 = 1,$$

and therefore

$$hk_1 \equiv 1 \pmod{k}, \qquad hk_2 \equiv -1 \pmod{k}$$

or, from (2.5),

$$(3.3) \qquad k_1 \equiv -h', \qquad k_2 \equiv h' \pmod{k}.$$

The Farey segment around h/k is bounded by the mediants between the fractions (3.2)

$$\frac{h_1 + h}{k_1 + k}, \qquad \frac{h_2 + h}{k_2 + k}.$$

Since these mediants do not belong to the Farey series of order N we have

$$k_1 + k > N, \qquad k_2 + k > N,$$

which conditions, together with (3.2), enclose k_1 and k_2 in the intervals

$$(3.4) \qquad N - k < k_1 \leqq N, \qquad N - k < k_2 \leqq N.$$

The formulae (3.3) and (3.4) determine k_1 and k_2 uniquely as functions of h and k. In particular we have

$$(3.5) \qquad \vartheta'_{h,k} = \frac{1}{k(k_1 + k)}, \qquad \vartheta''_{h,k} = \frac{1}{k(k_2 + k)}.$$

4. In (3.1) we apply the transformation formula (2.6) and obtain

$$(4.1)\qquad c_n = \sum_{\substack{h,k \\ 0 \leqq h < k \leqq N}}' e^{-\frac{2\pi i n h}{k}} \int_{-\vartheta'_{h,k}}^{\vartheta''_{h,k}} f(e^{\frac{2\pi i h'}{k} - \frac{2\pi}{k^2 w}}) e^{2\pi n w} \, d\phi$$

with the abbreviation

$$(4.2)\qquad w = N^{-2} - i\phi.$$

If we now write

$$(4.3)\qquad \begin{aligned} f(x) &= x^{-1} + D(x), \\ D(x) &= \sum_{m=0}^{\infty} c_m x^m \end{aligned}$$

we can accordingly split the expression (4.1) into two parts:

$$(4.4)\qquad c_n = Q(n) + R(n)$$

with

$$(4.41)\qquad Q(n) = \sum_{\substack{h,k \\ 0 \leqq h < k \leqq N}}' e^{-\frac{2\pi i}{k}(nh+h')} \int_{-\vartheta'_{h,k}}^{\vartheta''_{h,k}} e^{\frac{2\pi}{k^2 w} + 2\pi n w} \, d\phi,$$

$$(4.42)\qquad R(n) = \sum_{\substack{h,k \\ 0 \leqq h < k \leqq N}}' e^{-\frac{2\pi i n h}{k}} \int_{-\vartheta'_{h,k}}^{\vartheta''_{h,k}} D(e^{\frac{2\pi i h'}{k} - \frac{2\pi}{k^2 w}}) e^{2\pi n w} \, d\phi.$$

In $Q(n)$, which we consider first, we divide the intervals of integration into three parts according to

$$-\vartheta'_{h,k} = -\frac{1}{k(k_1+k)} \leqq -\frac{1}{k(N+k)} < \frac{1}{k(N+k)} \leqq \frac{1}{k(k_2+k)} = \vartheta''_{h,k}$$

and get

$$(4.5)\qquad \begin{aligned} Q(n) = &\sum_{k=1}^{N} \sum_{h \bmod k}' e^{-\frac{2\pi i}{k}(nh+h')} \int_{-\frac{1}{k(N+k)}}^{\frac{1}{k(N+k)}} \\ &+ \sum_{k=1}^{N} \sum_{h \bmod k}' e^{-\frac{2\pi i}{k}(nh+h')} \int_{-\frac{1}{k(k_1+k)}}^{-\frac{1}{k(N+k)}} \\ &+ \sum_{k=1}^{N} \sum_{h \bmod k}' e^{-\frac{2\pi i}{k}(nh+h')} \int_{\frac{1}{k(N+k)}}^{\frac{1}{k(k_2+k)}} \\ = &\; Q_0(n) + Q_1(n) + Q_2(n), \end{aligned}$$

say. The integrand in all three integrals of (4.5) is the same as in (4.41).

5. In $Q_0(n)$ we can immediately perform the summation with respect to h since the integral is independent of h. Setting

$$(5.1)\qquad A_k(n) = \sum_{h \bmod k}{}' e^{-\frac{2\pi i}{k}(nh+h')}$$

we get

$$Q_0(n) = \sum_{k=1}^{N} A_k(n) \int_{-\frac{1}{k(N+k)}}^{\frac{1}{k(N+k)}} e^{\frac{2\pi}{k^2 w}+2\pi n w}\, d\phi$$

or

$$(5.2)\qquad Q_0(n) = \sum_{k=1}^{N} A_k(n) \frac{1}{i} \int_{N^{-2}-\frac{i}{k(N+k)}}^{N^{-2}+\frac{i}{k(N+k)}} e^{\frac{2\pi}{k^2 w}+2\pi n w}\, dw,$$

where we have introduced w from (4.2) as variable of integration. We remark further that $A_k(n)$ is a Kloosterman sum (cf. the references in § 1) and can therefore be estimated as

$$(5.3)\qquad |A_k(n)| < Ck^{2/3+\epsilon} \cdot (k,n)^{1/3},$$

where (k,n) is the greatest common divisor of k and n. In the complex w-plane we consider now the closed rectangular path R with the four vertices

$$\pm N^{-2} \pm \frac{i}{k(N+k)}.$$

We take R as surrounding 0 in the positive sense. Then we have

$$(5.4)\qquad Q_0(n) = 2\pi \sum_{k=1}^{N} A_k(n) \frac{1}{2\pi i} \int_R e^{\frac{2\pi}{k^2 w}+2\pi n w}\, dw$$

$$-\frac{1}{i}\sum_{k=1}^{N} A_k(n) \left\{ \int_{N^{-2}+\frac{i}{k(N+k)}}^{-N^{-2}+\frac{i}{k(N+k)}} + \int_{-N^{-2}+\frac{i}{k(N+k)}}^{-N^{-2}-\frac{i}{k(N+k)}} + \int_{-N^{-2}-\frac{i}{k(N+k)}}^{N^{-2}-\frac{i}{k(N+k)}} \right\}$$

$$= 2\pi \sum_{k=1}^{N} A_k(n) L_k(n) - \frac{1}{i}\sum_{k=1}^{N} A_k(n)\{J_1 + J_2 + J_3\},$$

say, where all four integrals have the same integrand.

For an estimation of J_1 and J_3 we observe that on their paths of integration we have

$$w = u \pm \frac{i}{k(N+k)},$$
$$-N^{-2} \leqq u \leqq N^{-2},$$
$$\Re(w) = u \leqq N^{-2},$$
$$\Re\left(\frac{1}{w}\right) = \frac{u}{u^2 + \frac{1}{k^2(N+k)^2}} < N^{-2}k^2(N+k)^2 \leqq 4k^2,$$

so that

$$\left| e^{\frac{2\pi}{k^2 w} + 2\pi n w} \right| \leqq e^{8\pi + 2\pi n N^{-2}}$$

and therefore

$$(5.5) \qquad \begin{matrix} |J_1| \\ |J_3| \end{matrix} \leqq 2N^{-2} e^{8\pi + 2\pi n N^{-2}}.$$

In J_2 we have

$$w = -N^{-2} + iv,$$
$$-\frac{1}{k(N+k)} \leqq v \leqq \frac{1}{k(N+k)},$$
$$\Re(w) = -N^{-2} < 0, \qquad \Re\left(\frac{1}{w}\right) = \frac{-N^{-2}}{N^{-4} + v^2} < 0,$$

hence

$$\left| e^{\frac{2\pi}{k^2 w} + 2\pi n w} \right| < 1$$

and therefore

$$(5.6) \qquad |J_2| < \frac{2}{k(N+k)} < 2k^{-1}N^{-1}.$$

Combining (5.3), (5.5), and (5.6) we obtain

$$\sum_{k=1}^{N} A_k(n)\{J_1 + J_2 + J_3\} = O(e^{2\pi n N^{-2}} \sum_{k=1}^{N} k^{2/3+\epsilon}(n,k)^{1/3} k^{-1} N^{-1})$$

and for $n \geqq 1$, which we assume from now on, we have $(n, k) \leqq n$ and hence

$$(5.7) \qquad \sum_{k=1}^{N} A_k(n)\{J_1 + J_2 + J_3\} = O(e^{2\pi n N^{-2}} n^{1/3} N^{-1/3+\epsilon}).$$

Furthermore we have

$$L_k(n) = \frac{1}{2\pi i}\int_R e^{\frac{2\pi}{k^2 w}+2\pi n w}\,dw$$

$$= \frac{1}{2\pi i}\int_R \sum_{\mu=0}^{\infty}\frac{\left(\frac{2\pi}{k^2 w}\right)^{\mu}}{\mu!}\sum_{\nu=0}^{\infty}\frac{(2\pi n w)^{\nu}}{\nu!}\,dw$$

$$= \frac{1}{k\sqrt{n}}\sum_{\nu=0}^{\infty}\frac{\left(\frac{2\pi\sqrt{n}}{k}\right)^{2\nu+1}}{\nu!\,(\nu+1)!}$$

or

$$L_k(n) = \frac{1}{k\sqrt{n}} I_1\left(\frac{4\pi\sqrt{n}}{k}\right), \tag{5.8}$$

where $I_1(z)$ is the Bessel function of first order with purely imaginary argument. From (5.4), (5.7), (5.8) we deduce

$$Q_0(n) = \frac{2\pi}{\sqrt{n}}\sum_{k=1}^{N}\frac{A_k(n)}{k} I_1\left(\frac{4\pi\sqrt{n}}{k}\right) + O(e^{2\pi n N^{-2}} n^{1/3} N^{-1/3+\epsilon}). \tag{5.9}$$

6. We now turn our attention to $Q_1(n)$ and $Q_2(n)$ in (4.5), of which we discuss only $Q_2(n)$ in detail since $Q_1(n)$ can be treated in quite the same manner. We have from (4.5)

$$Q_2(n) = \sum_{k=1}^{N}\ \sum_{h \bmod k}{}' e^{-\frac{2\pi i}{k}(nh+h')}\int_{\frac{1}{k(N+k)}}^{\frac{1}{k(k_2+k)}} e^{\frac{2\pi}{k^2 w}+2\pi n w}\,d\phi$$

$$= \sum_{k=1}^{N}\ \sum_{h \bmod k}{}' e^{-\frac{2\pi i}{k}(nh+h')}\sum_{l=k_2+k}^{N+k-1}\int_{\frac{1}{k(l+1)}}^{\frac{1}{kl}} e^{\frac{2\pi}{k^2 w}+2\pi n w}\,d\phi,$$

$$Q_2(n) = \sum_{k=1}^{N}\sum_{l=N+1}^{N+k-1}\int_{\frac{1}{k(l+1)}}^{\frac{1}{kl}} e^{\frac{2\pi}{k^2 w}+2\pi n w}\,d\phi \sum_{\substack{h \bmod k\\ N<k_2+k\leqq l}}{}' e^{-\frac{2\pi i}{k}(nh+h')} \tag{6.1}$$

In the inner sum of the last expression the restriction imposed on k_2 means, in consequence of (3.3), a restriction of h' to an interval modulo k, which is equivalent to one interval or to two intervals in the range $0 \leqq h' < k$. Therefore the sum in question is an incomplete Kloosterman sum, for which we have the estimate [8]

---

[8] This estimate of the incomplete sum does not seem to appear explicitly in the

$$(6.2)\qquad \sum_{\substack{h \bmod k\\ N-k<k_2\leqq l-k}}' e^{-\frac{2\pi i}{k}(nh+h')} = O(k^{2/3+\epsilon}(n,k)^{1/3}) = O(k^{2/3+\epsilon}n^{1/3}).$$

In the integral in (6.1) we have

$$\begin{aligned}\Re\left(\frac{2\pi}{k^2w}+2\pi nw\right) &= \Re\left(\frac{2\pi}{k^2(N^{-2}-i\phi)}+2\pi n(N^{-2}-i\phi)\right)\\ &= 2\pi\left(\frac{N^{-2}}{k^2(N^{-4}+\phi^2)}+nN^{-2}\right)\leqq 2\pi\left(\frac{N^{-2}}{k^2N^{-4}+\dfrac{1}{(k+N)^2}}+nN^{-2}\right)\\ &< 2\pi\left(\left(\frac{k+N}{N}\right)^2+nN^{-2}\right)\leqq 8\pi+2\pi nN^{-2},\end{aligned}$$

and from this and (6.2) we get

$$\begin{aligned}Q_2(n) &= O\left(e^{2\pi nN^{-2}}n^{1/3}\sum_{k=1}^{N}\sum_{l=N+1}^{N+k-1}\left(\frac{1}{kl}-\frac{1}{k(l+1)}\right)k^{2/3+\epsilon}\right)\\ &= O\left(e^{2\pi nN^{-2}}n^{1/3}\sum_{k=1}^{N}\frac{1}{k^{1/3-\epsilon}N}\right),\end{aligned}$$

$$(6.3)\qquad Q_2(n) = O(e^{2\pi nN^{-2}}n^{1/3}N^{-1/3+\epsilon}).$$

Since a similar result is valid for $Q_1(n)$ we derive from (4.5), (5.9), and (6.3):

$$(6.4)\qquad Q(n) = \frac{2\pi}{\sqrt{n}}\sum_{k=1}^{N}\frac{A_k(n)}{k}I_1\left(\frac{4\pi\sqrt{n}}{k}\right)+O(e^{2\pi nN^{-2}}n^{1/3}N^{-1/3+\epsilon}).$$

7. From (4.42) and (4.3) we obtain

$$R(n) = \sum_{k=1}^{N}\sum_{h \bmod k}' e^{-\frac{2\pi i nh}{k}}\int_{-\vartheta'_{h,k}}^{\vartheta''_{h,k}}\sum_{m=0}^{\infty}c_m e^{\frac{2\pi i h'm}{k}-\frac{2\pi m}{k^2w}}e^{2\pi nw}\,d\phi.$$

We decompose again the Farey segment $-\vartheta'_{h,k}\leqq\phi\leqq\vartheta''_{h,k}$ in the Kloosterman manner and have, after an interchange of the summations with respect to h and m,

---

literature. Incomplete sums for general k are given in Kloosterman's paper with a less precise estimate, and Davenport treats only the case k equal to a prime number, but with the precision (6.2). By the device, however, which Estermann uses *loc. cit.*, p. 94, or by the other one, which Davenport applies *loc. cit.*, pp. 173, 174, we can reduce the estimation of the incomplete sum to that of the complete sum. Thus we obtain (6.2) from Salié's estimate, *loc. cit.*, p. 264. We are for our purpose not particularly interested in the lowest possible value of the exponent of k in (6.2), as long as it is a constant less than 1.

$$(7.1)\quad R(n) = \sum_{k=1}^{N} \sum_{m=0}^{\infty} c_m \int_{-\frac{1}{k(N+k)}}^{\frac{1}{k(N+k)}} e^{-\frac{2\pi m}{k^2 w} + 2\pi n w} d\phi \sum_{h \bmod k}' e^{-\frac{2\pi i}{k}(nh - mh')}$$

$$+ \sum_{k=1}^{N} \sum_{m=0}^{\infty} c_m \sum_{h \bmod k}' e^{-\frac{2\pi i}{k}(nh-mh')} \sum_{l=k_1+k}^{N+k-1} \int_{-\frac{1}{kl}}^{-\frac{1}{k(l+1)}} e^{-\frac{2\pi m}{k^2 w} + 2\pi n w} d\phi$$

$$+ \sum_{k=1}^{N} \sum_{m=0}^{\infty} c_m \sum_{h \bmod k}' e^{-\frac{2\pi i}{k}(nh-mh')} \sum_{l=k_2+k}^{N+k-1} \int_{\frac{1}{k(l+1)}}^{\frac{1}{kl}} e^{-\frac{2\pi m}{k^2 w} + 2\pi n w} d\phi$$

$$= S_1 + S_2 + S_3.$$

In all integrals of (7.1) we have

$$(7.2)\quad \Re\left(\frac{2\pi m}{k^2 w}\right) = \frac{2\pi m N^{-2}}{k^2(N^{-4} + \phi^2)} \geqq \frac{2\pi m}{k^2 N^{-2} + N^2 k^2 \vartheta^2_{h,k}} \geqq \frac{2\pi m}{1+1} = \pi m.$$

The complete Kloosterman sum in S_1 admits of an estimate

$$\sum_{h \bmod k}' e^{-\frac{2\pi i}{k}(nh-mh')} = O(k^{2/3+\epsilon}(n,k)^{1/3}) = O(k^{2/3+\epsilon} n^{1/3})$$

which holds uniformly in m. We obtain therefore

$$S_1 = O\left(\sum_{k=1}^{N} \sum_{m=0}^{\infty} | c_m | \frac{2}{kN} e^{-\pi m + 2\pi n N^{-2}} k^{2/3+\epsilon} n^{1/3}\right)$$

$$= O\left(e^{2\pi n N^{-2}} n^{1/3} N^{-1} \sum_{m=0}^{\infty} | c_m | e^{-\pi m} \sum_{k=1}^{N} k^{-1/3+\epsilon}\right),$$

$$(7.3)\quad S_1 = O(e^{2\pi n N^{-2}} n^{1/3} N^{-1/3+\epsilon}).$$

The sums S_2 and S_3 are both of the same structure so that we need to treat only one of them. By interchanging the summations with respect to h and l we get

$$S_2 = \sum_{k=1}^{N} \sum_{m=0}^{\infty} c_m \sum_{l=N+1}^{N+k-1} \int_{-\frac{1}{kl}}^{-\frac{1}{k(l+1)}} e^{-\frac{2\pi m}{k^2 w} + 2\pi n w} d\phi \sum_{\substack{h \bmod k \\ N < k + k_1 \leqq l}}' e^{-\frac{2\pi i}{k}(nh - mh')}$$

The inner sum is an incomplete Kloosterman sum, for which we have, uniformly in m, the estimate

$$\sum_{\substack{h \bmod k \\ N-k<k_1\leqq l-k}}' e^{-\frac{2\pi i}{k}(nh-mh')} = O(k^{2/3+\epsilon}(n,k)^{1/3}) = O(k^{2/3+\epsilon}n^{1/3})$$

Therefore, taking note of (7. 2), we get

$$S_2 = O\left(\sum_{k=1}^{N}\sum_{m=0}^{\infty} |c_m| \frac{1}{kN} e^{-\pi m+2\pi nN^{-2}} k^{2/3+\epsilon} n^{1/3}\right),$$

$$S_2 = O(e^{2\pi nN^{-2}} n^{1/3} N^{-1/3+\epsilon}). \tag{7.4}$$

From (7. 1), (7. 3), and (7. 4) we infer

$$R(n) = O(e^{2\pi nN^{-2}} n^{1/3} N^{-1/3+\epsilon}) \tag{7.5}$$

and then from (4. 4), (6. 4), (7. 5)

$$c_n = \frac{2\pi}{\sqrt{n}} \sum_{k=1}^{N} \frac{A_k(n)}{k} I_1\left(\frac{4\pi\sqrt{n}}{k}\right) + O(e^{2\pi nN^{-2}} n^{1/3} N^{-1/3+\epsilon}). \tag{7.6}$$

Now we keep here $n > 0$ fixed and let N tend to infinity. The error term then tends to zero. Thus we obtain our main result, which we state in the following

THEOREM. *In the Fourier expansion for the modular function* $J(\tau)$

$$12^3 J(\tau) = e^{-2\pi i t} + c_0 + \sum_{n=1}^{\infty} c_n e^{2\pi i n\tau} \tag{7.71}$$

the coefficients c_n, $n \geqq 1$, *are determined by the convergent series*

$$c_n = \frac{2\pi}{\sqrt{n}} \sum_{k=1}^{\infty} \frac{A_k(n)}{k} I_1\left(\frac{4\pi\sqrt{n}}{k}\right) \tag{7.72}$$

with

$$A_k(n) = \sum_{h \bmod k}' e^{-\frac{2\pi i}{k}(nh+h')}, \qquad hh' \equiv -1 \pmod{k}. \tag{7.73}$$

8. We had to exclude $n = 0$ in our discussion. This peculiarity is not caused by $\sqrt{n}$ appearing in the denominator in (7. 6), as the computation of $L_k(n)$ in the lines preceding (5. 8) shows that $n = 0$ is not exceptional in this respect:

$$\left[\frac{1}{k\sqrt{n}} I_1\left(\frac{4\pi\sqrt{n}}{k}\right)\right]_{n=0} = L_k(0) = \frac{2\pi}{k^2}. \tag{8.1}$$

The estimates of the incomplete Kloosterman sums, however, would break down for $n = 0$. By suitable examples it is easy to see that incomplete Kloosterman sums with $n = 0$ do not admit of a better general estimate than $O(k)$, which would, of course, not suffice in our reasonings. The series (7. 72), on the other hand, will remain convergent for $n = 0$, but does not even accidentally represent the coefficient c_0, which can directly be obtained from (2. 3) as

$$c_0 = 744. \tag{8.2}$$

Indeed, we get

$$A_k(0) = \sum_{h \bmod k}' e^{-\frac{2\pi i}{k}h'} = \mu(k)$$

with the Möbius symbol $\mu(k)$, and hence from (8. 1) and (7. 72)

$$2\pi \sum_{k=1}^{\infty} A_k(0) \frac{2\pi}{k^2} = 4\pi^2 \sum_{k=1}^{\infty} \frac{\mu(k)}{k^2} = 4\pi^2 \frac{1}{\zeta(2)} = 4\pi^2 \cdot \frac{6}{\pi^2} = 24,$$

different from (8. 2). There is of course no reason to expect that c_0 might be found by our method, which makes use only of the behavior of $J(\tau)$ at $\tau = i\infty$, expressed by (2. 1), and of the invariance stated in (2. 4). Both properties remain obviously unchanged for any $J(\tau) + C$ instead of $J(\tau)$.

9. The coefficients c_n, which can be found from (2. 3) by troublesome computations, which for higher n are practically inexecutable, do not seem to have attracted much attention before. All I could discover in the literature were, besides c_0, the two coefficients

$$c_1 = 196\,884, \qquad c_2 = 21\,493\,760.$$

Our convergent series (7. 72) gives another approach to the actual computation of the c_n, which, as we know, have to be integers. Unfortunately, the convergence of (7. 72) is rather slow, so that we should need quite a number of terms in order to get an error which is safely less than 1/2. Nevertheless, it is interesting to see that the first few terms of the series already furnish the bulk of the considerable amounts of those coefficients.

We get, for $n = 1$:

$$\begin{aligned}
2\pi \frac{A_1(1)}{1} I_1(4\pi) &= 196\,550.665 \\
2\pi \frac{A_2(1)}{2} I_1\left(\frac{4\pi}{2}\right) &= 250.822 \\
2\pi \frac{A_3(1)}{3} I_1\left(\frac{4\pi}{3}\right) &= 48.535 \\
2\pi \frac{A_4(1)}{4} I_1\left(\frac{4\pi}{4}\right) &= 14.110 \\
2\pi \frac{A_5(1)}{5} I_1\left(\frac{4\pi}{5}\right) &= 8.380 \\
\Sigma &= 196\,872.512,
\end{aligned}$$

and for $n=2$:

$$\frac{2\pi}{\sqrt{2}}\ \frac{A_1(2)}{1}\ I_1(4\pi\sqrt{2}) \quad = 21\,495\,869.\,279$$
$$\frac{2\pi}{\sqrt{2}}\ \frac{A_2(2)}{2}\ I_1\left(\frac{4\pi\sqrt{2}}{2}\right) = \quad -2\,054.\,739$$
$$\frac{2\pi}{\sqrt{2}}\ \frac{A_3(2)}{3}\ I_1\left(\frac{4\pi\sqrt{2}}{3}\right) = \quad -\quad 84.\,640$$
$$\frac{2\pi}{\sqrt{2}}\ \frac{A_4(2)}{4}\ I_1\left(\frac{4\pi\sqrt{2}}{4}\right) = \quad 0.\,000$$
$$\frac{2\pi}{\sqrt{2}}\ \frac{A_5(2)}{5}\ I_1\left(\frac{4\pi\sqrt{2}}{5}\right) = \quad 7.\,012$$
$$\sum = 21\,493\,736.\,912.$$

These values are in error only by the comparatively small amounts of — 11. 488 and — 23. 088 respectively.

UNIVERSITY OF PENNSYLVANIA,
PHILADELPHIA, PA.

Notes to Paper 39

This paper has been reviewed in the JF, vol. 64 (1938), p. 122, and in the Z, vol. 18 (1938), p. 246.

Page 501, footnote 1.
See paper 40 of the present collection.

Page 511, fourth display line.
See paper 41, p. 237, footnote 2.

ON THE FOURIER COEFFICIENTS OF CERTAIN MODULAR FORMS OF POSITIVE DIMENSION

BY HANS RADEMACHER AND HERBERT S. ZUCKERMAN[1]

(Received October 13, 1937)

1. Introduction

One of us has recently obtained an exact formula for the number $p(n)$ of unrestricted partitions of n.[2] The proof, which makes use of the Hardy-Littlewood-Ramanujan method, rests essentially on the fact that the generating function

$$f(x) = 1 + \sum_{n=1}^{\infty} p(n)x^n = \frac{1}{(1-x)(1-x^2)(1-x^3)\cdots}, \qquad |x| < 1, \tag{1.1}$$

is very closely related to a certain modular form, usually called $\eta(\tau)$. We now propose to apply the new method to general modular forms of a type presently to be specified.

An analytic function $H(\omega_1, \omega_2)$ of two complex variables ω_1, ω_2 subject to the restriction $\Im(\omega_2/\omega_1) > 0$, is called a modular form if
(A) it is homogeneous:

$$H(\lambda\omega_1, \lambda\omega_2) = \lambda^r H(\omega_1, \omega_2),$$

the real number r being called its dimension;
(B) it is invariant with respect to all transformations

$$\omega_2' = a\omega_2 + b\omega_1$$
$$\omega_1' = c\omega_2 + d\omega_1$$

of the full modular group (or, a case which we do not consider in this paper, invariant under all transformations of a subgroup of the modular group):

$$H(\omega_1', \omega_2') = H(\omega_1, \omega_2).$$

Writing

$$\tau = \frac{\omega_2}{\omega_1}, \qquad \tau' = \frac{\omega_2'}{\omega_1'} = \frac{a\tau + b}{c\tau + d}$$

---

[1] National Research Fellow.

[2] *A convergent series for the partition function* $p(n)$, Proceedings of the National Academy of Sciences, vol. 23, no. 2., pp. 78–84 (1937); *On the partition function* $p(n)$, Proceedings of the London Mathematical Society, (2), vol. 43 (1937), pp. 241–154.

we get from (A) and (B)

$$H(1, \tau') = (c\tau + d)^{-r} H(1, \tau). \tag{1.21}$$

The function $H(1, \tau)$, in spite of the inhomogeneous notation, may also be called a modular form of dimension r. For non-integral values of r we should have to determine the branch of the many-valued function $(c\tau + d)^{-r}$. Instead of doing this we generalize the notion of modular forms by admitting a factor ϵ of absolute value 1 and independent of τ, such that we finally define a modular form of dimension r as a function $F(\tau)$ analytic in the upper τ-half-plane and satisfying a functional relation

$$F(\tau') = \epsilon \cdot (-i(c\tau + d))^{-r} F(\tau), \tag{1.22}$$

for every modular transformation

$$\tau' = \frac{a\tau + b}{c\tau + d} \tag{1.3}$$

of the modular group (or of one of its subgroups only). Since for $c \neq 0$ we can choose in (1.3) $c > 0$, we can assign

$$-\pi/2 < \arg(-i(c\tau + d)) < \pi/2 \tag{1.4}$$

and determine $(-i(c\tau + d))^{-r}$ as $|(c\tau + d)^{-r}| \exp(-ir \arg(-i(c\tau + d)))$. The multiplier ϵ, being independent of τ, depends only on the particular substitution (1.3) and may therefore, if necessary, also be written

$$\epsilon = \epsilon(a, b, c, d).$$

The case $c = 0$, which reduces (1.3) to

$$\tau' = \tau + b$$

is conveniently treated separately. We write

$$F(\tau + b) = \epsilon_b F(\tau) \tag{1.51}$$

and in particular

$$F(\tau + 1) = \epsilon_1 F(\tau) = e^{2\pi i\alpha} F(\tau) \tag{1.52}$$

where α can be restricted to

$$0 \leqq \alpha < 1. \tag{1.6}$$

From (1.52) we get

$$e^{-2\pi i\alpha(\tau+1)} F(\tau + 1) = e^{-2\pi i\alpha\tau} F(\tau),$$

from which we infer the validity of a Fourier expansion

$$e^{-2\pi i\alpha\tau} F(\tau) = \sum_{m=-\infty}^{+\infty} a_m e^{2\pi i m\tau}. \tag{1.7}$$

We now restrict our considerations to functions $F(\tau)$ whose developments (1.7) have at least one, but only a finite number of negative exponents m:

$$F(\tau) = e^{2\pi i\alpha\tau} \sum_{m=-\mu}^{\infty} a_m e^{2\pi i m\tau} \tag{1.71}$$

The function $e^{-2\pi i\alpha\tau}F(\tau)$ has a pole at the parabolic point $\tau = i\infty$, if we measure the singularity by the uniformizing variable $x = e^{2\pi i\tau}$. The segment

$$P(e^{2\pi i\tau}) = \sum_{m=-\mu}^{-1} a_m e^{2\pi i m\tau} \tag{1.8}$$

is the "principal part" of the function at that singularity. We assume moreover $F(\tau)$ to be regular at all finite points of the upper half-plane, so that the series (1.71) is convergent for $\Im(\tau) > 0$.

Our method requires eventually for its applicability the positivity of the dimension r (cf. end of no. 4).

It is now our aim to determine the coefficients a_m for $m \geqq 0$ as they arise from the combination of two sources, viz. the functional relation (1.22), (1.52) on one side, and the principal part (1.8) of $e^{-2\pi i\alpha\tau}F(\tau)$ at the parabolic point $\tau = i\infty$ of the fundamental region on the other side; P itself is given by the coefficients $a_{-\mu}, \cdots, a_{-1}$.

It is convenient to introduce the variable $x = e^{2\pi i\tau}$ and to put

$$e^{-2\pi i\alpha\tau}F(\tau) = f(x) = \sum_{m=-\mu}^{\infty} a_m x^m. \tag{1.9}$$

The function $f(x)$ is analytic inside the unit circle and has a pole of order μ at $x = 0$. The rational function

$$P(x) = a_{-\mu}x^{-\mu} + \cdots + a_{-1}x^{-1} \tag{1.91}$$

is the principal part of $f(x)$ at that pole.[3]

---

[3] It may be noticed that our present notation is slightly at variance with the notation used in (1.1) and that adopted in the papers cited in the preceding footnote. There we had as $F(\tau)$ the special function $\eta(\tau)^{-1}$. Writing it in the form (1.71) we have

$$\eta(\tau)^{-1} = e^{23\pi i\tau/12}\{e^{-2\pi i\tau} + a_0 + a_1 e^{2\pi i\tau} + \cdots\}$$

and hence in analogy to (1.9)

$$\begin{aligned} e^{-23\pi i\tau/12}\eta(\tau)^{-1} = f(x) &= x^{-1} + a_0 + a_1 x + \cdots \\ &= x^{-1}(1 + a_0 x + a_1 x^2 + \cdots) \\ &= x^{-1}\left(1 + \sum_{n=1}^{\infty} p(n)x^n\right). \end{aligned}$$

Our present function $f(x)$ therefore differs by a factor x^{-1} from that which is called $f(x)$ in (1.1) and the other places mentioned above.

2. Cauchy integral and Farey dissection

We have now, by Cauchy's integral formula,

$$a_m = \frac{1}{2\pi i}\int_C \frac{f(x)}{x^{m+1}}\,dx,$$

where C may be the circle

$$|x| = e^{-2\pi N^{-2}}$$

and N a positive integer. We make the usual Farey dissection of the circle C into arcs $\xi_{h,k}$, using the Farey series of order N. Thus we have

$$a_m = \sum_{\substack{h,k\\ 0\leqq h<k\leqq N\\ (h,k)=1}} \frac{1}{2\pi i}\int_{\xi_{h,k}} \frac{f(x)}{x^{m+1}}\,dx. \tag{2.1}$$

We can describe the Farey arc $\xi_{h,k}$ by

$$x = \exp(-2\pi N^{-2} + 2\pi ih/k + 2\pi i\varphi) \tag{2.21}$$

$$-\vartheta'_{h,k} \leqq \varphi \leqq \vartheta''_{h,k} \tag{2.22}$$

and get through the substitution (2.21)

$$a_m = e^{2\pi N^{-2}m} \sum_{\substack{h,k\\ 0\leqq h<k\leqq N\\ (h,k)=1}} e^{-2\pi ihm/k} \times \int_{-\vartheta'}^{\vartheta''} f[\exp(2\pi ih/k - 2\pi(N^{-2} - i\varphi))]e^{-2\pi im\varphi}\,d\varphi, \tag{2.3}$$

where we have briefly written ϑ', ϑ'' instead of $\vartheta'_{h,k}$, $\vartheta''_{h,k}$.

In order to make use of the transformation formula (1.22) we write, for $c > 0$,

$$-i(c\tau + d) = z, \qquad \tau = \frac{iz}{c} - \frac{d}{c}, \qquad \Re(z) > 0,$$

and choose the modular transformation

$$\begin{pmatrix} a & b\\ c & d\end{pmatrix} = \begin{pmatrix} h' & -\dfrac{hh'+1}{k}\\ k & -h\end{pmatrix}$$

where h' is a solution of $hh' \equiv -1 \pmod{k}$. We then have

$$\tau = \frac{iz}{k} + \frac{h}{k}, \qquad \tau' = \frac{i}{kz} + \frac{h'}{k}$$

and therefore, by (1.22),

$$F\left(\frac{i}{kz} + \frac{h'}{k}\right) = \epsilon z^{-r} F\left(\frac{iz}{k} + \frac{h}{k}\right)$$

and from (1.9)

$$f(e^{-2\pi(z-ih)/k}) = \epsilon^{-1} z^r e^{2\pi\alpha(z-ih)/k} e^{2\pi i\alpha h'/k} e^{-2\pi\alpha/kz} f(e^{2\pi ih'/k} e^{-2\pi/kz}), \tag{2.4}$$

where

$$\epsilon = \epsilon(h', -(hh'+1)/k, k, -h).$$

We could make h' unique by adding the restriction $0 \leqq h' < k$, but this is unessential, since (2.4) shows that the expression

$$\Omega_{h,k} = \epsilon^{-1} e^{2\pi i\alpha(h'-h)/k} \tag{2.5}$$

must be independent of the special choice of h'.[4]

Introducing (2.5) and

$$\Psi_k(z) = z^r e^{2\pi\alpha(z-1/z)/k} \tag{2.6}$$

we can replace (2.4) by

$$f(e^{-2\pi(z-ih)/k}) = \Omega_{h,k}\Psi_k(z) f(e^{2\pi ih'/k} e^{-2\pi/kz}).$$

This equation applied to (2.3) yields

$$\begin{aligned} a_m = e^{2\pi N^{-2}m} & \sum_{\substack{h,k \\ 0\leqq h<k\leqq N \\ (h,k)=1}} \Omega_{h,k} e^{-2\pi ihm/k} \\ & \int_{-\vartheta'}^{\vartheta''} \Psi_k(k(N^{-2}-i\varphi)) f(\exp[(2\pi/k)(ih' - k^{-1}(N^{-2}-i\varphi)^{-1})]) e^{-2\pi im\varphi}\, d\varphi. \end{aligned} \tag{2.7}$$

3. An estimation

We are now prepared to make full use of the fact that $f(x)$ in the neighborhood of $x = 0$ is dominated by the principal part $P(x)$. For that purpose we split (2.7) into two parts

$$a_m = Q(m) + R(m) \tag{3.1}$$

where

$$\begin{aligned} Q(m) = e^{2\pi N^{-2}m} & \sum_{\substack{h,k \\ 0\leqq h<k\leqq N \\ (h,k)=1}} \Omega_{h,k} e^{-2\pi ihm/k} \\ & \int_{-\vartheta'}^{\vartheta''} \Psi_k(k(N^{-2}-i\varphi)) P(\exp[(2\pi/k)(ih' - k^{-1}(N^{-2}-i\varphi)^{-1})]) e^{-2\pi im\varphi}\, d\varphi \end{aligned} \tag{3.2}$$

---

[4] This means that the ϵ's must, for consistency of (1.22) and (1.52), satisfy (among others) the relation

$$\epsilon(a, b, c, d) = \epsilon(a+c, b+d, c, d)e^{-2\pi i\alpha}, \quad c > 0.$$

and

$$R(m) = e^{2\pi N^{-2}m} \sum_{\substack{h,k \\ 0 \le h < k \le N \\ (h,k)=1}} \Omega_{h,k} e^{-2\pi i h m/k} \tag{3.3}$$
$$\int_{-\vartheta'}^{\vartheta''} \Psi_k(k(N^{-2} - i\varphi)) D\,(\exp\,[(2\pi/k)(ih' - k^{-1}(N^{-2} - i\varphi)^{-1})]) e^{-2\pi i m\varphi}\, d\varphi$$

with

$$D(x) = f(x) - P(x) = \sum_{m=0}^{\infty} a_m x^m. \tag{3.4}$$

We are now going to make an estimate of $R(m)$. From the theory of Farey fractions we have

$$1/2kN \le \vartheta' \le 1/kN, \qquad 1/2kN \le \vartheta'' \le 1/kN$$

and therefore, since $k \le N$, we find for $-\vartheta' \le \varphi \le \vartheta''$

$$\Re(k(N^{-2} - i\varphi)) = kN^{-2},$$

$$\Re\left(\frac{1}{k(N^{-2} - i\varphi)}\right) = \frac{N^{-2}}{k(N^{-4} + \varphi^2)} \ge \frac{N^{-2}}{k(N^{-4} + k^{-2}N^{-2})} = \frac{k}{k^2N^{-2} + 1} \ge \frac{k}{2},$$

$$|\,k(N^{-2} - i\varphi)\,| = k(N^{-4} + \varphi^2)^{\frac{1}{2}} \le (k^2N^{-4} + N^{-2})^{\frac{1}{2}} \le 2^{\frac{1}{2}}N^{-1}.$$

Using these results we have by (2.6), (3.4)

$$|\,\Psi_k(k(N^{-2} - i\varphi)) \cdot D\,(\exp\,[(2\pi/k)(ih' - k^{-1}(N^{-2} - i\varphi)^{-1})])\,|$$
$$\le 2^{r/2}N^{-r} \exp\,[2\pi\alpha(N^{-2} - k^{-1}\Re(k^{-1}(N^{-2} - i\varphi)^{-1}))]$$
$$\times \sum_{m=0}^{\infty} |\,a_m\,| \exp\,[-2\pi m \Re(k^{-2}(N^{-2} - i\varphi)^{-1})] \tag{3.5}$$
$$\le 2^{r/2}N^{-r} e^{2\pi\alpha N^{-2}} e^{-\pi\alpha} \sum_{m=0}^{\infty} |\,a_m\,| e^{-\pi m} = CN^{-r} e^{2\pi\alpha N^{-2}},$$

where

$$C = 2^{r/2} e^{-\pi\alpha} \sum_{m=0}^{\infty} |\,a_m\,| e^{-\pi m}$$

is finite since $|\,e^{-\pi}\,| < 1$ and the series (1.9) is convergent inside the unit circle. Combining (3.5) with (3.3) we have

$$|\,R(m)\,| \le e^{2\pi N^{-2}m} \sum_{\substack{h,k \\ 0 \le h < k \le N \\ (h,k)=1}} CN^{-r} e^{2\pi\alpha N^{-2}} \int_{-\vartheta'}^{\vartheta''} d\varphi = CN^{-r} e^{2\pi(m+\alpha)N^{-2}}. \tag{3.6}$$

4. A convergent series for a_m

We now evaluate $Q(m)$, under the condition

$$m + \alpha > 0, \tag{4.1}$$

which will be necessary for the convergence of a certain integral. The condition (4.1) admits, because of (1.6), all $m \geqq 0$, with the single exception $m = 0$ for $\alpha = 0$, a case which we shall have to treat separately.

Making the substitution

$$w = N^{-2} - i\varphi$$

in (3.2) we have

$$Q(m) = \sum_{\substack{h,k \\ 0 \leq h < k \leq N \\ (h,k)=1}} \Omega_{h,k} e^{-2\pi i h m/k} \frac{1}{i} \int_{N^{-2}-i\vartheta''}^{N^{-2}+i\vartheta'} \Psi_k(kw) P(e^{2\pi i h'/k} e^{-2\pi/k^2 w}) e^{2\pi m w}\, dw$$

which, by (2.6) and (1.91), can be written as

$$Q(m) = \sum_{\substack{h,k \\ 0 \leq h < k \leq N \\ (h,k)=1}} \Omega_{h,k} e^{-2\pi i h m/k} k^r \sum_{\nu=1}^{\mu} a_{-\nu} e^{-2\pi i h' \nu/k} I_k(m, \nu), \tag{4.21}$$

where

$$I_k(m, \nu) = \frac{1}{i} \int_{N^{-2}-i\vartheta''}^{N^{-2}+i\vartheta'} w^r \exp\,[2\pi((m + \alpha)w + (\nu - \alpha)k^{-2}w^{-1})]\, dw, \tag{4.22}$$

$$\nu = 1, \cdots, \mu.$$

We cut the complex w-plane from 0 to $-\infty$ along the negative real axis and consider a path of integration encircling the cut in the positive sense and connecting the points

$$-\infty,\ -\epsilon,\ -\epsilon - i\vartheta'',\ N^{-2} - i\vartheta'',\ N^{-2} + i\vartheta',\ -\epsilon + i\vartheta',\ -\epsilon,\ -\infty \tag{4.3}$$

by straight lines. Then we can write

$$\begin{aligned} I_k(m, \nu) &= \frac{1}{i} \int_{-\infty}^{(0+)} - \frac{1}{i} \int_{-\infty}^{-\epsilon} - \frac{1}{i} \int_{-\epsilon}^{-\epsilon-i\vartheta''} - \frac{1}{i} \int_{-\epsilon-i\vartheta''}^{N^{-2}-i\vartheta''} - \frac{1}{i} \int_{N^{-2}+i\vartheta'}^{-\epsilon+i\vartheta'} \\ &\quad - \frac{1}{i} \int_{-\epsilon+i\vartheta'}^{-\epsilon} - \frac{1}{i} \int_{-\epsilon}^{-\infty} = L_k(m, \nu) - J_1 - J_2 - J_3 - J_4 - J_5 - J_6, \end{aligned} \tag{4.41}$$

say. All these integrals have the same integrand

$$w^r \exp\,[2\pi((m + \alpha)w + (\nu - \alpha)k^{-2}w^{-1})]. \tag{4.42}$$

The integral J_1 is to be taken on the border below the cut, J_6 above the cut. We assume moreover

$$0 < \epsilon < N^{-2}.$$

In the integral J_2 we have

$$w = -\epsilon + iv, \qquad 0 \geqq v \geqq -\vartheta'',$$

$$\Re(w) = -\epsilon, \qquad \Re\left(\frac{1}{w}\right) = \frac{-\epsilon}{\epsilon^2 + v^2} < 0,$$

$$|w| = (\epsilon^2 + v^2)^{\frac{1}{2}} \leqq (N^{-4} + k^{-2}N^{-2})^{\frac{1}{2}} \leqq 2^{\frac{1}{2}}k^{-1}N^{-1},$$

and therefore

$$|J_2| \leqq \vartheta'' 2^{r/2} k^{-r} N^{-r} e^{-2\pi(m+\alpha)\epsilon} < 2^{r/2} k^{-r-1} N^{-r-1}. \tag{4.51}$$

Similarly we have

$$|J_5| < 2^{r/2} k^{-r-1} N^{-r-1}. \tag{4.52}$$

In the integral J_3 we have

$$w = u - i\vartheta'', \qquad -N^{-2} < -\epsilon \leqq u \leqq N^{-2},$$

$$\Re(w) = u \leqq N^{-2}, \qquad \Re\left(\frac{1}{w}\right) = \frac{u}{u^2 + \vartheta''^2} \leqq \frac{N^{-2}}{\vartheta''^2} \leqq 4k^2,$$

$$|w| = (u^2 + \vartheta''^2)^{\frac{1}{2}} \leqq (N^{-4} + N^{-2}k^{-2})^{\frac{1}{2}} \leqq 2^{\frac{1}{2}}k^{-1}N^{-1}$$

and therefore

$$\begin{aligned} |J_3| &\leqq (N^{-2} + \epsilon)2^{r/2}k^{-r}N^{-r} \exp\,[2\pi(m+\alpha)N^{-2} + 8\pi(\nu - \alpha)] \\ &\leqq 2^{1+r/2}k^{-r-1}N^{-r-1} \exp\,[2\pi(m+\alpha)N^{-2} + 8\pi(\nu-\alpha)], \end{aligned} \tag{4.61}$$

and similarly

$$|J_4| \leqq 2^{1+r/2}k^{-r-1}N^{-r-1} \exp\,[2\pi(m+\alpha)N^{-2} + 8\pi(\nu-\alpha)]. \tag{4.62}$$

Finally we have

$$\begin{aligned} J_1 + J_6 &= \frac{1}{i}\int_{-\infty}^{-\epsilon} |w|^r e^{-\pi i r} \exp\,[2\pi(m+\alpha)w + 2\pi(\nu-\alpha)k^{-2}w^{-1}]\,dw \\ &\quad + \frac{1}{i}\int_{-\epsilon}^{-\infty} |w|^r e^{\pi i r} \exp\,[2\pi(m+\alpha)w + 2\pi(\nu-\alpha)k^{-2}w^{-1}]\,dw \\ &= -2\sin \pi r \int_{\epsilon}^{\infty} t^r \exp\,[-2\pi(m+\alpha)t - 2\pi(\nu-\alpha)k^{-2}t^{-1}]\,dt, \end{aligned} \tag{4.7}$$

where the condition (4.1) ensures the convergence of the integrals. Combining (4.41), (4.51), (4.52), (4.61), (4.62), (4.7) and making $\epsilon \to 0$, we have

$$\begin{aligned} I_k(m,\nu) &= L_k(m,\nu) + 2\sin \pi r \int_0^{\infty} t^r \exp\,[-2\pi(m+\alpha)t - 2\pi(\nu-\alpha)k^{-2}t^{-1}]\,dt \\ &\quad + 6\Theta\, 2^{r/2} e^{8\pi(\mu-\alpha)} k^{-r-1} N^{-r-1} e^{2\pi(m+\alpha)N^{-2}}, \qquad |\Theta| < 1, \qquad \nu = 1, \cdots, \mu. \end{aligned}$$

Introducing this into (4.21) we obtain

$$
\begin{aligned}
Q(m) = &\sum_{\substack{h,k \\ 0 \leqq h < k \leqq N \\ (h,k)=1}} \Omega_{h,k} e^{-2\pi i h m/k} k^r \sum_{\nu=1}^{\mu} a_{-\nu} e^{-2\pi i h'\nu/k} \{L_k(m, \nu) + H_k(m, \nu)\} \\
&+ 6\Theta\, 2^{r/2} e^{8\pi(\mu-\alpha)} N^{-r-1} e^{2\pi(m+\alpha)N^{-2}} \sum_{\nu=1}^{\mu} |a_{-\nu}| \sum_{\substack{h,k \\ 0 \leqq h < k \leqq N \\ (h,k)=1}} k^{-1}, \qquad |\Theta| < 1,
\end{aligned} \tag{4.81}
$$

with

$$L_k(m, \nu) = \frac{1}{i} \int_{-\infty}^{(0+)} w^r \exp\,[2\pi(m + \alpha)w + 2\pi(\nu - \alpha)k^{-2}w^{-1}]\, dw, \tag{4.82}$$

$$H_k(m, \nu) = 2 \sin \pi r \int_0^{\infty} t^r \exp\,[-2\pi(m + \alpha)t - 2\pi(\nu - \alpha)k^{-2}t^{-1}]\, dt. \tag{4.83}$$

Now we have in (4.81)

$$\sum_{\substack{h,k \\ 0 \leqq h < k \leqq N \\ (h,k)=1}} k^{-1} \leqq \sum_{\substack{k \\ 0 < k \leqq N}} 1 = N,$$

and thus (4.81), (3.1) and (3.6) together give the result

$$a_m = \sum_{k=1}^{N} k^r \sum_{\substack{0 \leqq h < k \\ (h,k)=1}} \Omega_{h,k} e^{-2\pi i h m/k} \sum_{\nu=1}^{\mu} a_{-\nu} e^{-2\pi i h'\nu/k} \{L_k(m, \nu) + H_k(m, \nu)\} + O(N^{-r} e^{2\pi(m+\alpha)N^{-2}}).$$

If we keep m fixed and let N tend to infinity, the error term tends to zero, since $r > 0$. It is at this point that we must definitely make use of the positivity of the dimension r of our modular form $F(\tau)$.

The limit 0 of the error term signifies the convergence of the series thus obtained for a_m:

$$a_m = \sum_{k=1}^{\infty} k^r \sum_{\nu=1}^{\mu} a_{-\nu} A_{k,\nu}(m)\, \{L_k(m, \nu) + H_k(m, \nu)\} \tag{4.91}$$

with

$$A_{k,\nu}(m) = \sum_{\substack{0 \leqq h < k \\ (h,k)=1}} \Omega_{h,k} e^{-2\pi i (hm + h'\nu)/k}. \tag{4.92}$$

5. Evaluation of the integrals

The integrals (4.82) and (4.83) are well-known in the theory of Bessel functions. Indeed we have[5]

$$I_\rho(z) = \frac{(z/2)^\rho}{2\pi i} \int_{-\infty}^{(0+)} t^{-\rho-1} \exp\,(t + z^2/4t)\, dt \tag{5.11}$$

[5] G. N. Watson, *Theory of Bessel Functions* (Cambridge 1922), p. 181, (1), p. 183, (15).

and

$$(5.12) \qquad K_\rho(z) = \frac{(z/2)^\rho}{2} \int_0^\infty t^{-\rho-1} \exp(-t - z^2/4t)\, dt,$$

which we need here only for positive z and real ρ. The functions $I_\rho(z)$ and $K_\rho(z)$ are the Bessel functions of the first kind and the third kind respectively, with purely imaginary arguments. Comparing (4.82) and (4.83) with (5.11) and (5.12) we get

$$L_k(m, \nu) = \frac{2\pi}{k^{r+1}} \left(\frac{\nu - \alpha}{m + \alpha}\right)^{(r+1)/2} I_{-r-1}\left((4\pi/k)(\nu - \alpha)^{\frac{1}{2}}(m + \alpha)^{\frac{1}{2}}\right)$$

$$H_k(m, \nu) = \frac{4 \sin \pi r}{k^{r+1}} \left(\frac{\nu - \alpha}{m + \alpha}\right)^{(r+1)/2} K_{-r-1}\left((4\pi/k)(\nu - \alpha)^{\frac{1}{2}}(m + \alpha)^{\frac{1}{2}}\right).$$

But[6]

$$\sin \pi\rho\, K_\rho(z) = (\pi/2)(I_{-\rho}(z) - I_\rho(z)),$$

and so we have, since $\sin \pi r = \sin(-\pi(r + 1))$,

$$H_k(m, \nu) = \frac{2\pi}{k^{r+1}} \left(\frac{\nu - \alpha}{m + \alpha}\right)^{(r+1)/2} \{I_{r+1}\left((4\pi/k)(\nu - \alpha)^{\frac{1}{2}}(m + \alpha)^{\frac{1}{2}}\right) - I_{-r-1}\left((4\pi/k)(\nu - \alpha)^{\frac{1}{2}}(m + \alpha)^{\frac{1}{2}}\right)\},$$

and hence

$$L_k(m, \nu) + H_k(m, \nu) = \frac{2\pi}{k^{r+1}} \left(\frac{\nu - \alpha}{m + \alpha}\right)^{(r+1)/2} I_{r+1}\left((4\pi/k)(\nu - \alpha)^{\frac{1}{2}}(m + \alpha)^{\frac{1}{2}}\right).$$

This reduces (4.91) to

$$(5.2) \quad a_m = 2\pi \sum_{k=1}^{\infty} \frac{1}{k} \sum_{\nu=1}^{\mu} a_{-\nu} A_{k,\nu}(m) \left(\frac{\nu - \alpha}{m + \alpha}\right)^{(r+1)/2} I_{r+1}\left((4\pi/k)(\nu - \alpha)^{\frac{1}{2}}(m + \alpha)^{\frac{1}{2}}\right).$$

Each term of this infinite series consists of μ members. The summation signs in (5.2) can, however, be interchanged, so that we get the sum of μ convergent series. Indeed, a single series

$$(5.3) \qquad \sum_{k=1}^{\infty} \frac{1}{k} A_{k,\nu}(m) \left(\frac{\nu - \alpha}{m + \alpha}\right)^{(r+1)/2} I_{r+1}\left((4\pi/k)(\nu - \alpha)^{\frac{1}{2}}(m + \alpha)^{\frac{1}{2}}\right)$$

is absolutely convergent, since we have on one hand, from (4.92),

$$\left|\frac{1}{k} A_{k,\nu}(m)\right| \leq 1,$$

[6] Watson, l. c., p. 78, (6).

and on the other hand[7]

$$(5.4)\quad \left(\frac{\nu-\alpha}{m+\alpha}\right)^{(r+1)/2} I_{r+1}((4\pi/k)(\nu-\alpha)^{\frac{1}{2}}(m+\alpha)^{\frac{1}{2}}) = \frac{(2\pi(\nu-\alpha))^{r+1}}{k^{r+1}} \sum_{l=0}^{\infty} \frac{(2\pi/k)^{2l}(\nu-\alpha)^l(m+\alpha)^l}{l!\,\Gamma(r+l+2)}.$$

The series (5.3) converges therefore absolhtely for $r > 0$. It follows that (5.2) can be changed into

$$(5.5)\quad a_m = 2\pi \sum_{\nu=1}^{\mu} a_{-\nu} \sum_{k=1}^{\infty} \frac{1}{k} A_{k,\nu}(m) \left(\frac{\nu-\alpha}{m+\alpha}\right)^{(r+1)/2} I_{r+1}((4\pi/k)(\nu-\alpha)^{\frac{1}{2}}(m+\alpha)^{\frac{1}{2}}).$$

We finally notice that because of (2.5) and (4.92) the coefficients $A_{k,\nu}(m)$ can be written as

$$(5.6)\quad A_{k,\nu}(m) = \sum_{\substack{0 \leq h < k \\ (h,k)=1}} \epsilon\left(h', -\frac{hh'+1}{k}, k, -h\right)^{-1} \exp[(-2\pi i/k)((\nu-\alpha)h' + (m+\alpha)h)].$$

6. Connection between r and α

Up to the present we had assumed that $m + \alpha > 0$. We shall show, however, that (5.5) remains true for

$$(6.11)\quad m + \alpha = 0,$$

which means

$$(6.12)\quad \alpha = 0, \qquad m = 0,$$

if we interpret (5.4) in the proper way as

$$(6.2)\quad \lim_{m+\alpha\to 0} \left(\frac{\nu-\alpha}{m+\alpha}\right)^{(r+1)/2} I_{r+1}((4\pi/k)(\nu-\alpha)^{\frac{1}{2}}(m+\alpha)^{\frac{1}{2}}) = \frac{(2\pi\nu)^{r+1}}{k^{r+1}\Gamma(r+2)}.$$

For this purpose we have to study the connection between r and α.

For the two generating transformations

$$S\colon \tau' = \tau + 1,$$

$$T\colon \tau' = -\frac{1}{\tau}$$

we have respectively

$$(6.31)\quad F(\tau+1) = e^{2\pi i\alpha} F(\tau),$$

$$(6.32)\quad F\left(-\frac{1}{\tau}\right) = \epsilon_0(-i\tau)^{-r} F(\tau)$$

[7] Watson, l.c., p. 77, (2).

with

$$\epsilon_0 = \epsilon(0, -1, 1, 0).$$

Replacing in (6.32) τ by $-1/\tau$ we get

$$F(\tau) = \epsilon_0 \left(\frac{i}{\tau}\right)^{-r} F\left(-\frac{1}{\tau}\right)$$

$$(6.41) \qquad = \epsilon_0^2 \left(\frac{i}{\tau}\right)^{-r} (-i\tau)^{-r} F(\tau)$$

$$= \epsilon_0^2 \left|\frac{1}{\tau}\right|^{-r} |\tau|^{-r} \exp\left\{-ir\left(\arg\frac{i}{\tau} + \arg(-i\tau)\right)\right\} F(\tau).$$

From (1.4) we infer

$$-\pi/2 < \arg\frac{i}{\tau} < \pi/2$$

$$-\pi/2 < \arg(-i\tau) < \pi/2$$

and hence

$$(6.42) \qquad -\pi < \arg\frac{i}{\tau} + \arg(-i\tau) < \pi.$$

But we have

$$\arg\frac{i}{\tau} + \arg(-i\tau) \equiv \arg\left(\frac{i}{\tau}\cdot(-i\tau)\right) \equiv 0 \qquad (\text{mod } 2\pi),$$

which together with (6.42) leaves only the possibility

$$\arg\frac{i}{\tau} + \arg(-i\tau) = 0.$$

Thus (6.41) goes over into

$$F(\tau) = \epsilon_0^2 F(\tau),$$

and we get

$$(6.5) \qquad \epsilon_0^2 = 1.$$

From (6.31) and (6.32) we deduce

$$F\left(1 - \frac{1}{\tau}\right) = e^{2\pi i\alpha} \epsilon_0 (-i\tau)^{-r} F(\tau).$$

If we replace, in this equation, two times in succession τ by $1 - 1/\tau$ we obtain

$$F(\tau) = e^{6\pi i\alpha} \epsilon_0^3 \left(\frac{i}{\tau - 1}\right)^{-r} \left(-i\left(1 - \frac{1}{\tau}\right)\right)^{-r} (-i\tau)^{-r} F(\tau)$$

$$(6.61) \qquad = e^{6\pi i\alpha} \epsilon_0 \left|\frac{1}{\tau - 1}\right|^{-r} \left|1 - \frac{1}{\tau}\right|^{-r} |\tau|^{-r}$$

$$\times \exp\left\{-ir\left(\arg\frac{i}{\tau - 1} + \arg\left(-i\left(1 - \frac{1}{\tau}\right)\right) + \arg(-i\tau)\right)\right\} F(\tau),$$

where we have made use of (6.5). Application of (1.4) gives

$$(6.62)\qquad -\frac{3\pi}{2} < \arg\frac{i}{\tau-1} + \arg\left(-i\left(1-\frac{1}{\tau}\right)\right) + \arg(-i\tau) < \frac{3\pi}{2}.$$

On the other hand we have

$$\arg\frac{i}{\tau-1} + \arg\left(-i\left(1-\frac{1}{\tau}\right)\right) + \arg(-i\tau)$$

$$\equiv \arg\left(\frac{i}{\tau-1}\cdot\frac{-i(\tau-1)}{\tau}\cdot(-i\tau)\right) \equiv \arg(-i) \equiv -\frac{\pi}{2} \pmod{2\pi}.$$

From this and (6.62) we get

$$\arg\frac{i}{\tau-1} + \arg\left(-i\left(1-\frac{1}{\tau}\right)\right) + \arg(-i\tau) = -\frac{\pi}{2},$$

and therefore (6.61) becomes

$$(6.7)\qquad \begin{aligned} F(\tau) &= e^{6\pi i\alpha}\epsilon_0 e^{\pi i r/2}F(\tau),\\ \epsilon_0 &= e^{-6\pi i\alpha - \pi i r/2}. \end{aligned}$$

Since, by (6.5), $\epsilon_0 = \pm 1$, we conclude from (6.7) that

$$6\alpha + r/2$$

is always an integer.

This result shows in particular that $\alpha = 0$ in (6.12) is necessarily connected with an integer r (which is moreover even).

7. The exceptional case $m = \alpha = 0$; the main theorem

Now if r has an integral value the reasoning of no. 4 may be considerably simplified. In this case the integrand (4.42)

$$w^r \exp\left[2\pi(m+\alpha)w + 2\pi(\nu-\alpha)k^{-2}w^{-1}\right]$$

is unique in the whole w-plane, and we need not cut the plane along the negative real axis. Instead of using the loop path indicated by (4.3) we can integrate around $w = 0$ along a rectangle R with the vertices

$$-\epsilon - i\vartheta'', \qquad N^{-2} - i\vartheta'', \qquad N^{-2} + i\vartheta', \qquad -\epsilon + i\vartheta'$$

and get

$$\begin{aligned} I_k(m,\nu) &= \frac{1}{i}\int_R - \frac{1}{i}\int_{N^{-2}+i\vartheta'}^{-\epsilon+i\vartheta'} - \frac{1}{i}\int_{-\epsilon+i\vartheta'}^{-\epsilon-i\vartheta''} - \frac{1}{i}\int_{-\epsilon-i\vartheta''}^{N^{-2}-i\vartheta''}\\ &= L_k^*(m,\nu) - J_1^* - J_2^* - J_3^*, \end{aligned}$$

say. Here J_1^* corresponds to J_4, J_3^* to J_3, and J_2^* to $J_2 + J_5$ in (4.41). The estimations are completely analogous to those in no. 4, only that we can now disregard the convergence condition $m + \alpha > 0$, since no improper integrals appear. In this way we get instead of (4.91)

$$a_m = \sum_{k=1}^{\infty} k^r \sum_{\nu=1}^{\mu} a_{-\nu} A_{k,\nu}(m) L_k^*(m, \nu).$$

We are interested only in the case $m = 0$, $\alpha = 0$:

$$a_0 = \sum_{k=1}^{\infty} k^r \sum_{\nu=1}^{\mu} a_{-\nu} A_{k,\nu}(0) L_k^*(0, \nu)$$

with

$$L_k^*(0, \nu) = \frac{1}{i} \int^{(0+)} w^r e^{2\pi\nu/k^2 w} \, dw = \frac{2\pi}{(r+1)!} \left(\frac{2\pi\nu}{k^2}\right)^{r+1}.$$

This yields, after interchanging the summations with respect to k and ν

$$(7.1) \qquad a_0 = 2\pi \sum_{\nu=1}^{\mu} a_{-\nu} \sum_{k=1}^{\infty} \frac{1}{k} A_{k,\nu}(0) \frac{1}{(r+1)!} \left(\frac{2\pi\nu}{k}\right)^{r+1},$$

which is exactly the formula which was to be expected in view of (5.5) and (6.2).

We collect our results in the following

THEOREM 1: *Let $F(\tau)$ be a modular form of positive dimension r, regular in the upper τ-half-plane, and satisfying*

$$F\left(\frac{a\tau + b}{c\tau + d}\right) = \epsilon(a, b, c, d) \cdot (-i(c\tau + d))^{-r} F(\tau), \qquad c > 0,$$

$$F(\tau + 1) = e^{2\pi i \alpha} F(\tau), \qquad 0 \leqq \alpha < 1.$$

Also let its Fourier expansion contain only a finite number of terms with negative exponents:

$$F(\tau) = e^{2\pi i \alpha \tau} \sum_{m=-\mu}^{\infty} a_m e^{2\pi i m \tau}.$$

Then the coefficients a_m for $m \geqq 0$ are determined by those with $-\mu \leqq m \leqq -1$ through the formula

$$(7.2) \quad a_m = 2\pi \sum_{\nu=1}^{\mu} a_{-\nu} \sum_{k=1}^{\infty} \frac{1}{k} A_{k,\nu}(m) \left(\frac{\nu - \alpha}{m + \alpha}\right)^{(r+1)/2} I_{r+1}((4\pi/k)(\nu - \alpha)^{\frac{1}{2}}(m + \alpha)^{\frac{1}{2}}),$$

where $A_{k,\nu}(m)$ is given by (5.6). *For $\alpha = m = 0$ this formula is to be understood as meaning* (7.1).

We see in (7.2) that each a_m for $m \geqq 0$ depends linearly on $a_{-1}, \cdots a_{-\mu}$. Our analysis would hold true also for $a_{-1} = \cdots = a_{-\mu} = 0$, in which case consequently all a_m for $m \geqq 0$ would vanish. But $a_{-1} = \cdots = a_{-\mu} = 0$ means, as the Fourier expansion shows, that $F(\tau)$ tends to a finite value (different from 0 or not) when $\Im(\tau) \to \infty$, in which case we call $F(\tau)$ *regular at infinity*. We have therefore the

COROLLARY: *A modular form, belonging to the full modular group, of positive dimension, and regular at all finite points of the upper half-plane as well as at infinity, must necessarily vanish identically.*

8. Parametric representation of the modular forms under discussion

We can completely characterize the class of all functions $F(\tau)$ which have the properties (1.22), (1.52), and a principal part (1.8) at $\tau = i\infty$.

The function

$$\eta(\tau) = e^{\pi i\tau/12} \prod_{n=1}^{\infty} (1 - e^{2\pi i n\tau}) \tag{8.11}$$

is different from zero for all τ in the upper half-plane. Therefore for any real r the function

$$\eta(\tau)^{2r} = e^{2\pi i r\tau/12} \prod_{n=1}^{\infty} (1 - e^{2\pi i n\tau})^{2r} \tag{8.12}$$

can be uniquely defined if we take for each factor that branch which tends to 1 with $\tau \to i\infty$. We have then

$$\eta(\tau + 1)^{2r} = e^{2\pi i r/12} \eta(\tau)^{2r} \tag{8.21}$$

and

$$\eta\left(-\frac{1}{\tau}\right)^{2r} = (-i\tau)^r \eta(\tau)^{2r}, \tag{8.22}$$

where (8.21) is an immediate consequence of (8.12), and where (8.22) follows from the transformation equation for log $\eta(\tau)$, first given by Dedekind.[8] The function

$$G(\tau) = F(\tau) \cdot \eta(\tau)^{2r} \tag{8.3}$$

has, as consequence of (6.31), (6.32), (8.21), (8.22) the properties

$$G(\tau + 1) = e^{2\pi i(\alpha + r/12)} G(\tau),$$

$$G\left(-\frac{1}{\tau}\right) = \epsilon_0 G(\tau).$$

But from (6.7) we know that

$$q = 6\alpha + r/2 \tag{8.41}$$

[8] *Erläuterungen zu den Fragmenten über die Grenzfälle der elliptischen Modulfunktionen*, Riemanns Werke, 2. Aufl., pp. 466–478 especially formula (17).

is always an integer, and we can write

$$G(\tau+1) = e^{\pi iq/3}G(\tau), \tag{8.42}$$

$$G\left(-\frac{1}{\tau}\right) = e^{\pi iq}G(\tau). \tag{8.43}$$

The function $G(\tau)$ is therefore a modular form of dimension zero, i.e. a "modular function." A consequence of the last two equations is

$$G\left(1-\frac{1}{\tau}\right) = e^{4\pi iq/3}G(\tau). \tag{8.44}$$

By expanding around the point $\tau = i$ we can see from (8.43) that, for odd q, $G(\tau)$ must have a zero of odd order. For this purpose we put

$$t = \frac{\tau - i}{\tau + i}, \qquad \tau = -i\frac{t+1}{t-1},$$

$$G(\tau) = G\left(-i\frac{t+1}{t-1}\right) = G_1(t):$$

The equation (8.43) goes over into

$$G_1(-t) = (-1)^q G_1(t),$$

which proves the statement.

Similarly, if we put

$$t = \frac{\tau - \rho}{\tau + \rho^2}, \qquad \tau = -\rho\frac{t\rho+1}{t-1}, \qquad (\rho = e^{\pi i/3})$$

$$G(\tau) = G\left(-\rho\frac{t\rho+1}{t-1}\right) = G_2(t),$$

we change (8.44) into

$$G_2(-\rho t) = (-\rho)^q G_2(t),$$

which means that in the power series

$$G_2(t) = c_0 + c_1 t + c_2 t^2 + \cdots$$

only such c_n can be different from 0 for which

$$n \equiv q \pmod 3.$$

If therefore $q \not\equiv 0 \pmod 3$, $G(\tau)$ has at $\tau = \rho$ (and because of (8.42) also at $\tau = \rho^2$) a zero of an order congruent to q modulo 3.

If q is divisible by 6, the equations (8.42), (8.43) show that $G(\tau)$ is an absolute modular function. Otherwise we consider the function

$$\varphi_{\beta,\gamma}(\tau) = \sqrt{J(\tau) - 1}^{\beta} \cdot \sqrt[3]{J(\tau)}^{\gamma}, \tag{8.51}$$

$$\begin{aligned} \beta &= 0, 1, & \beta &\equiv q \pmod 2, \\ \gamma &= 0, 1, 2, & \gamma &\equiv q \pmod 3. \end{aligned} \tag{8.52}$$

Here $J(\tau)$ is the absolute modular invariant, which we normalize as

$$J(\tau) = J\left(\frac{\omega_2}{\omega_1}\right) = \frac{g_2^3}{g_2^3 - 27g_3^2} = \frac{g_2^3(\omega_1, \omega_2)}{\Delta(\omega_1, \omega_2)}, \tag{8.53}$$

where we have

$$\Delta(\omega_1, \omega_2) = g_2^3 - 27g_3^2 = \frac{\pi^{12}}{\omega_1^{12}} \eta(\tau)^{24},$$

$$g_2 = g_2(\omega_1, \omega_2) = 60 \sum{}' \frac{1}{(2m_1\omega_1 + 2m_2\omega_2)^4},$$

$$g_3 = g_3(\omega_1, \omega_2) = 140 \sum{}' \frac{1}{(2m_1\omega_1 + 2m_2\omega_2)^6}.$$

With these expressions we define the roots in (8.51) as

$$\sqrt{J(\tau) - 1} = \frac{3^{\frac{3}{2}}}{\pi^6} \cdot \frac{g_3(1, \tau)}{\eta(\tau)^{12}}, \tag{8.61}$$

$$\sqrt[3]{J(\tau)} = \frac{1}{\pi^4} \cdot \frac{g_2(1, \tau)}{\eta(\tau)^8} \tag{8.62}$$

For these functions we find easily, by means of (8.21), (8.22),

$$\sqrt{J(\tau + 1) - 1} = -\sqrt{J(\tau) - 1}, \qquad \sqrt{J\left(-\frac{1}{\tau}\right) - 1} = -\sqrt{J(\tau) - 1} \tag{8.63}$$

and

$$\sqrt[3]{J(\tau + 1)} = e^{-2\pi i/3} \sqrt[3]{J(\tau)}, \qquad \sqrt[3]{J\left(-\frac{1}{\tau}\right)} = \sqrt[3]{J(\tau)}, \tag{8.64}$$

so that we have

$$\varphi_{\beta,\gamma}(\tau + 1) = e^{\pi i q/3} \varphi_{\beta,\gamma}(\tau), \tag{8.71}$$

$$\varphi_{\beta,\gamma}\left(-\frac{1}{\tau}\right) = e^{\pi i q} \varphi_{\beta,\gamma}(\tau), \tag{8.72}$$

since β and γ satisfy the congruences (8.52). The function $\varphi_{\beta,\gamma}(\tau)$ has a zero of order β at $\tau = i$ and a zero of order γ at $\tau = \rho$. The function

(8.81) $$H(\tau) = \frac{G(\tau)}{\varphi_{\beta,\gamma}(\tau)}$$

(which now may include the trivial case $\varphi_{0,0}(\tau) = 1$ for $q \equiv 0 \pmod 6$) is therefore regular in the upper half-plane, and in consequence of (8.42), (8.43), (8.71), (8.72) it fulfills the equations

(8.82) $$H(\tau + 1) = H(\tau), \qquad H\left(-\frac{1}{\tau}\right) = H(\tau).$$

If therefore $H(\tau)$ is not a constant it must be an entire rational function of $J(\tau)$.

Combining this result with the definitions (8.3), (8.81) we find that

(8.91) $$F(\tau) = \eta(\tau)^{-2r}\varphi_{\beta,\gamma}(\tau)\{B_\kappa J(\tau)^\kappa + \cdots + B_1 J(\tau) + B_0\}$$

or, because of (8.51), (8.61), (8.62),

(8.92) $$F(\tau) = \eta(\tau)^{-2r-12\beta-8\gamma} g_3(1, \tau)^\beta g_2(1, \tau)^\gamma \{C_\kappa J(\tau)^\kappa + \cdots + C_1 J(\tau) + C_0\}$$

is the most general function satisfying our postulates (1.22), (1.52), (1.71).

The independent parameters of $F(\tau)$ are $r, \beta, \gamma, \kappa, C_0, C_1, \cdots C_\kappa$, which are subject to the restrictions that r is real, β, γ, κ are integers and

(8.93) $$r > 0; \qquad \beta = 0, 1, \qquad \gamma = 0, 1, 2; \qquad \kappa \geqq 0.$$

In order to express the numbers α and μ of (1.71) by these parameters we compute the first term with negative exponent of

$$\eta(\tau)^{-2r} \sqrt{J(\tau) - 1}^{\,\beta} \cdot \sqrt[3]{J(\tau)}^{\,\gamma} \cdot J(\tau)^\kappa$$

and compare it with (1.71). Since we have

$$J(\tau) = \frac{e^{-2\pi i\tau}}{3^3 \cdot 2^6}(1 + 744e^{2\pi i\tau} + \cdots)$$

and because of (8.11) this comparison yields

$$\frac{r}{12} + \frac{\beta}{2} + \frac{\gamma}{3} + \kappa = \mu - \alpha,$$

$$\mu = \kappa + \frac{r}{12} + \frac{\beta}{2} + \frac{\gamma}{3} + \alpha.$$

This is indeed an integer, since in consequence of (8.52) and (8.41)

$$3\beta + 2\gamma \equiv -6\alpha - r/2 \pmod 6.$$

We see moreover that under the conditions (8.93) always $\mu \geqq 1$, as it has to be. Observing that $0 \leqq \alpha < 1$, we can express α and μ in the following manner

$$\alpha = -\frac{r}{12} - \frac{\beta}{2} - \frac{\gamma}{3} - \left[-\frac{r}{12} - \frac{\beta}{2} - \frac{\gamma}{3}\right], \tag{8.94}$$

$$\mu = \kappa - \left[-\frac{r}{12} - \frac{\beta}{2} - \frac{\gamma}{3}\right], \tag{8.95}$$

where $[x]$ denotes the greatest integer not surpassing x.

The coefficients $a_{-\mu}, \cdots, a_{-1}$ cannot be obtained in this concise manner. They need, in each case, a separate calculation from the first coefficients of $\eta(\tau)^{-1}$ and $J(\tau)$ as the examples below illustrate.

9. The multiplier ϵ

The multiplier ϵ in (1.22) and subsequent equations can, however, be explicitly expressed in terms of the parameters r, β, γ. For this purpose we study the multipliers of

$$\eta(\tau)^{-2r}, \qquad \varphi_{\beta,0}(\tau) = \sqrt{J(\tau) - 1}^{\,\beta} \qquad \varphi_{0,\gamma}(\tau) = \sqrt[3]{J(\tau)}^{\,\gamma}$$

separately.

Since r can be any real positive number we have to go back to log $\eta(\tau)$, using Dedekind's formula. We have[9]

$$\eta\left(\frac{a\tau + b}{c\tau + d}\right)^{-2r} = \epsilon' \cdot (-i(c\tau + d))^{-r} \eta(\tau)^{-2r} \tag{9.11}$$

with

$$\begin{aligned} \epsilon' &= \exp\left\{-2\pi i r\left(\frac{a + d}{12c} - s(a, c)\right)\right\}, \qquad c > 0, \\ s(a, c) &= \sum_{l=1}^{c-1} \frac{l}{c}\left(\frac{al}{c} - \left[\frac{al}{c}\right] - \frac{1}{2}\right). \end{aligned} \tag{9.12}$$

For $\varphi_{1,0}(\tau) = (J(\tau) - 1)^{1/2}$ we make use of (8.63), writing it

$$\varphi_{1,0}(\tau + 1) = -\varphi_{1,0}(\tau), \qquad \varphi_{1,0}(-1/\tau) = -\varphi_{1,0}(\tau). \tag{9.21}$$

It is easily seen that $\varphi_{1,0}(\tau)$ is invariant under the transformations

$$\begin{pmatrix} 1 & 2 \\ 0 & 1 \end{pmatrix}, \quad \begin{pmatrix} 1 & 0 \\ 2 & 1 \end{pmatrix}, \tag{9.22}$$

[9] H. Rademacher, *Zur Theorie der Modulfunktionen*, Journal für die reine und angewandte Mathematik, vol. 167 (1931), 312–336, in particular p. 322, formula (3.19).

which are the generators of the principal congruence subgroup mod 2 of index 6 in the complete modular group.[10] A system of representatives of the factor group is given by

$$\begin{pmatrix}1 & 0\\0 & 1\end{pmatrix}, \quad \begin{pmatrix}1 & 1\\0 & 1\end{pmatrix}, \quad \begin{pmatrix}1 & 0\\1 & 1\end{pmatrix},$$
$$(9.23) \qquad \begin{pmatrix}0 & -1\\1 & 0\end{pmatrix}, \quad \begin{pmatrix}0 & -1\\1 & 1\end{pmatrix}, \quad \begin{pmatrix}1 & -1\\1 & 0\end{pmatrix},$$

and we find from (9.21)

$$\varphi_{1,0}(\tau) = \varphi_{1,0}(\tau), \qquad \varphi_{1,0}(\tau+1) = -\varphi_{1,0}(\tau), \qquad \varphi_{1,0}\left(\frac{\tau}{\tau+1}\right) = -\varphi_{1,0}(\tau),$$
$$(9.24) \qquad \varphi_{1,0}\left(-\frac{1}{\tau}\right) = -\varphi_{1,0}(\tau), \qquad \varphi_{1,0}\left(\frac{-1}{\tau+1}\right) = \varphi_{1,0}(\tau), \quad \varphi_{1,0}\left(\frac{\tau-1}{\tau}\right) = \varphi_{1,0}(\tau).$$

Now all cases (9.24) and hence all modular substitutions satisfy

$$(9.25) \qquad \varphi_{1,0}\left(\frac{a\tau+b}{c\tau+d}\right) = \epsilon''\cdot\varphi_{1,0}(\tau)$$

with

$$(9.26) \qquad \epsilon'' = \exp\{\pi i(b(a+d) + a(b-c) + bc)\},$$

as seen by a direct verification.

For $\varphi_{0,1}(\tau) = J(\tau)^{1/3}$ we have from (8.64)

$$(9.31) \qquad \varphi_{0,1}(\tau+1) = \xi\cdot\varphi_{0,1}(\tau), \qquad \varphi_{0,1}(-1/\tau) = \varphi_{0,1}(\tau), \qquad \xi = e^{-2\pi i/3}.$$

As a consequence of these equations the function $\varphi_{0,1}(\tau)$ turns out to be invariant with respect to the substitutions

$$(9.32) \qquad \begin{pmatrix}1 & 3\\0 & 1\end{pmatrix}, \quad \begin{pmatrix}-1 & 0\\3 & -1\end{pmatrix}, \quad \begin{pmatrix}2 & -3\\3 & -4\end{pmatrix}, \quad \begin{pmatrix}-4 & -3\\3 & 2\end{pmatrix},$$

which form a system of generators for the principal congruence subgroup modulo 3.[11] As a system of representatives for the factor group the following substitutions can be taken

$$\begin{pmatrix}1 & 0\\0 & 1\end{pmatrix}, \quad \begin{pmatrix}1 & 1\\0 & 1\end{pmatrix}, \quad \begin{pmatrix}1 & -1\\0 & 1\end{pmatrix}, \quad \begin{pmatrix}1 & 0\\1 & 1\end{pmatrix},$$
$$\begin{pmatrix}1 & 1\\1 & 2\end{pmatrix}, \quad \begin{pmatrix}1 & -1\\1 & 0\end{pmatrix}, \quad \begin{pmatrix}0 & -1\\1 & 0\end{pmatrix}, \quad \begin{pmatrix}0 & -1\\1 & 1\end{pmatrix},$$
$$\begin{pmatrix}0 & -1\\1 & -1\end{pmatrix}, \quad \begin{pmatrix}-1 & 0\\1 & -1\end{pmatrix}, \quad \begin{pmatrix}-1 & -1\\1 & 0\end{pmatrix}, \quad \begin{pmatrix}-1 & 1\\1 & -2\end{pmatrix},$$

---

[10] Klein-Fricke, *Vorlesungen über die Theorie der elliptischen Modulfunktionen*, Band 1, p. 276.

[11] Klein-Fricke, l.c., p. 354.

and we find (in the same order), applying (9.31)

$$\varphi_{0,1}(\tau) = \varphi_{0,1}(\tau), \qquad \varphi_{0,1}(\tau+1) = \xi\varphi_{0,1}(\tau), \qquad \varphi_{0,1}(\tau-1) = \xi^{-1}\varphi_{0,1}(\tau),$$

$$\varphi_{0,1}\left(\frac{\tau}{\tau+1}\right) = \xi^{-1}\varphi_{0,1}(\tau), \qquad \varphi_{0,1}\left(\frac{\tau+1}{\tau+2}\right) = \varphi_{0,1}(\tau),$$

$$\varphi_{0,1}\left(\frac{\tau-1}{\tau}\right) = \xi\varphi_{0,1}(\tau), \qquad \varphi_{0,1}\left(-\frac{1}{\tau}\right) = \varphi_{0,1}(\tau), \qquad \varphi_{0,1}\left(\frac{-1}{\tau+1}\right) = \xi\varphi_{0,1}(\tau),$$

$$\varphi_{0,1}\left(\frac{-1}{\tau-1}\right) = \xi^{-1}\varphi_{0,1}(\tau), \qquad \varphi_{0,1}\left(\frac{-\tau}{\tau-1}\right) = \xi\varphi_{0,1}(\tau),$$

$$\varphi_{0,1}\left(\frac{-\tau-1}{\tau}\right) = \xi^{-1}\varphi_{0,1}(\tau), \qquad \varphi_{0,1}\left(\frac{-\tau+1}{\tau-2}\right) = \varphi_{0,1}(\tau).$$

We can verify for these 12 cases directly that

$$\varphi_{0,1}\left(\frac{a\tau+b}{c\tau+d}\right) = \epsilon'''\cdot\varphi_{0,1}(\tau) \tag{9.33}$$

with

$$\epsilon''' = \xi^{-(a+d)(b-c)(ad+bc)} = \exp\{(2\pi i/3)(a+d)(b-c)(ad+bc)\}, \tag{9.34}$$

which, because of the invariance with respect to the transformations (9.32), is then true for all modular transformations. We remember

$$\varphi_{\beta,\gamma}(\tau) = \varphi_{1,0}(\tau)^{\beta}\cdot\varphi_{0,1}(\tau)^{\gamma},$$

and collecting the results from (8.92), (9.11), (9.12), (9.25), (9.26), (9.33), (9.34) we can state the

THEOREM 2: *The modular form of dimension r*

$$F(\tau) = \eta(\tau)^{-2r-12\beta-8\gamma} g_3(1,\tau)^{\beta} g_2(1,\tau)^{\gamma}\{C_\kappa J(\tau)^\kappa + \cdots + C_1 J(\tau) + C_0\} \tag{9.41}$$

satisfies the transformation equations

$$F\left(\frac{a\tau+b}{c\tau+d}\right) = \epsilon(a,b,c,d)\cdot(-i(c\tau+d))^{-r}F(\tau), \qquad c > 0, \tag{9.42}$$

and

$$F(\tau+1) = e^{2\pi i \alpha}F(\tau), \tag{9.43}$$

with the multiplier

$$\begin{aligned} \epsilon(a,b,c,d) = \exp\Bigg\{2\pi i\Bigg(& rs(a,c) - r\,\frac{a+d}{12c} \\ &+ \frac{\beta}{2}(b(a+d) + a(b-c) + bc) + \frac{\gamma}{3}(a+d)(b-c)(ad+bc)\Bigg)\Bigg\},[12] \end{aligned} \tag{9.44}$$

[12] It can easily be proved that the multiplier ϵ defined by (9.44) fulfills, because of (9.45) the condition mentioned in footnote 4.

and with

$$\alpha = -\frac{r}{12} - \frac{\beta}{2} - \frac{\gamma}{3} - \left[-\frac{r}{12} - \frac{\beta}{2} - \frac{\gamma}{3}\right]. \tag{9.45}$$

Conversely, all modular forms regular in the upper half-plane, with at most a polar singularity at $i\infty$, *and satisfying* (9.42), (9.43), *are contained in* (9.41).

We finally apply (9.44) to obtain an expression for $A_{k,\nu}(m)$ of (5.6). By a straightforward calculation making use of

$$s(h', k) = -s(h, k),$$[13]

we obtain for $A_{k,\nu}(m)$ the formulae (9.53), (9.54) of the following

THEOREM 3: *The modular form* (9.41) *of positive dimension* r *admits for* $\mathfrak{I}(\tau) > 0$ *the Fourier expansion*

$$F(\tau) = e^{2\pi i\alpha\tau}\left\{a_{-\mu}e^{-2\pi i\mu\tau} + \cdots + a_{-1}e^{-2\pi i\tau} + \sum_{m=0}^{\infty} a_m e^{2\pi i m\tau}\right\} \tag{9.51}$$

with the coefficients

$$a_m = 2\pi \sum_{\nu=1}^{\mu} a_{-\nu} \sum_{k=1}^{\infty} \frac{1}{k} A_{k,\nu}(m)\left(\frac{\nu-\alpha}{m+\alpha}\right)^{(r+1)/2} I_{r+1}((4\pi/k)(\nu-\alpha)^{\frac{1}{2}}(m+\alpha)^{\frac{1}{2}}), \tag{9.52}$$

for $m \geqq 0$, *where the number* μ *is determined by* (8.95) *and where*

$$\begin{aligned} A_{k,\nu}(m) = \sum_{\substack{0 \leqq h < k \\ (h,k)=1}} \omega_r(h,k)\cdot\xi_1(h,k)^{\beta}\cdot\xi_2(h,k)^{\gamma} \\ \times \exp\left[-(2\pi i/k)((\nu-\mu+\kappa)h' + (m+\mu-\kappa)h)\right] \end{aligned} \tag{9.53}$$

with

$$\begin{aligned} \omega_r(h,k) &= \exp\{2\pi i r s(h,k)\}, \\ \xi_1(h,k) &= \exp\left\{\pi i\left(-\frac{h'(h^2+1)}{k} + h'k + hh' + 1\right)\right\}, \\ \xi_2(h,k) &= \exp\left\{+\frac{2\pi i}{3}(h-h')\left(\left(\frac{hh'+1}{k} + k\right)(2hh'+1) + \frac{1}{k}\right)\right\}. \end{aligned} \tag{9.54}$$

It is of interest to notice that the real number r enters only into $\omega_r(h, k)$, whereas the three other factors of the summands in (9.53) are always roots of unity.

10. Examples

The following examples may show how in each case the coefficients $a_{-\mu}, \cdots, a_{-1}$ of the principal part at $\tau = i\infty$ can be directly computed. As far as the selection of the examples is concerned, we have restricted ourselves

[13] Rademacher, l.c. footnote 9, formula (2.13).

to functions which lead to not too complicated coefficients $A_{k,\nu}(m)$. In some examples we get the same coefficients which appeared first in the Hardy-Ramanujan formula. We shall give the formulae without further explanations, since they can be understood as simple applications of Theorems 2 and 3.

I. $$r = \tfrac{1}{2}, \quad \beta = \gamma = \kappa = 0:$$

$$F(\tau) = \eta(\tau)^{-1},$$

$$\alpha = -1/24 - [-1/24] = 23/24,$$

$$\mu = 0 - [-1/24] = 1.$$

(10.11) $$\eta(\tau)^{-1} = e^{23\pi i\tau/12} e^{-2\pi i\tau} \prod_{l=1}^{\infty} (1 - e^{2\pi i l\tau})^{-1} = e^{23\pi i\tau/12} \cdot \left\{ e^{-2\pi i\tau} + \sum_{m=0}^{\infty} a_m e^{2\pi i m\tau} \right\}.$$

Hence:

$$a_{-1} = 1,$$

and

(10.12) $$a_m = 2\pi \sum_{k=1}^{\infty} \frac{1}{k} A_{k,1}(m) \left(\frac{\frac{1}{24}}{m + \frac{23}{24}} \right)^{\frac{3}{4}} I_{\frac{3}{2}}\left(\frac{4\pi}{k} \left[\frac{1}{24}\left(m + \frac{23}{24}\right) \right]^{\frac{1}{2}} \right)$$

with

(10.13)
$$A_{k,1}(m) = \sum_{\substack{0 \leq h < k \\ (h,k)=1}} \omega_{\frac{1}{2}}(h, k)\, e^{-2\pi i (m+1)h/k},$$
$$\omega_{\frac{1}{2}}(h, k) = \exp\{\pi i\, s(h, k)\} = \omega_{h,k},$$
$$A_{k,1}(m) = A_k(m + 1)$$

where $\omega_{h,k}$ and $A_k(r)$ are the notations used by Hardy and Ramanujan.[14] Now we have

(10.14) $$\frac{d}{dm}\left(\frac{I_\rho(A(m + \alpha)^{\frac{1}{2}})}{(m + \alpha)^{\rho/2}} \right) = \frac{A}{2} \frac{I_{\rho+1}(A(m + \alpha)^{\frac{1}{2}})}{(m + \alpha)^{(\rho+1)/2}}.$$ [15]

Since moreover

(10.15) $$I_{\frac{1}{2}}(z) = (2/\pi z)^{\frac{1}{2}} \sinh z,$$ [16]

---

[14] Proc. London Math. Soc. (2) vol. 17 (1918), 75–115, in particular formulae (1.721), (1.722), (1.73); cf. Rademacher, Journal London Math. Soc. 7 (1932), 14–19, formula (5).

[15] An immediate consequence of formula (6), in Watson, l.c., p. 79.

[16] Watson, l.c., p. 80, (10).

the equation (10.12) yields, after some simplifications,

$$a_m = \frac{1}{\pi\sqrt{2}} \sum_{k=1}^{\infty} A_k(m+1) k^{\frac{1}{2}} \frac{d}{dm} \left(\frac{\sinh \left(\frac{4\pi}{k} \left[\frac{1}{24} \left(m + \frac{23}{24} \right) \right]^{\frac{1}{2}} \right)}{\left[m + \frac{23}{24} \right]^{\frac{1}{2}}} \right).$$

Taking note of the remark in footnote[3] and putting $m + 1 = n$, we have again

$$(10.16) \qquad p(n) = \frac{1}{\pi\sqrt{2}} \sum_{k=1}^{\infty} A_k(n) k^{\frac{1}{2}} \frac{d}{dn} \left(\frac{\sinh \left(\frac{\pi}{k} \left[\frac{2}{3} \left(n - \frac{1}{24} \right) \right]^{\frac{1}{2}} \right)}{\left[n - \frac{1}{24} \right]^{\frac{1}{2}}} \right),$$

as proved in a previous paper.

II. $\qquad r = 12, \qquad \beta = \gamma = \kappa = 0$:

$$F(\tau) = \eta(\tau)^{-24} = \pi^{12} \Delta(1, \tau)^{-1},$$

$$(10.21) \qquad \alpha = -1 - [-1] = 0,$$

$$\mu = 0 - [-1] = 1.$$

Now

$$\eta(\tau)^{-24} = e^{-2\pi i \tau} \prod_{l=1}^{\infty} (1 - e^{2\pi i l \tau})^{-24}$$

$$(10.22) \qquad = e^{-2\pi i \tau} \{1 + e^{2\pi i \tau} + \cdots\}^{24}$$

$$= e^{-2\pi i \tau} + \sum_{m=0}^{\infty} a_m e^{2\pi i m \tau},$$

from which we infer

$$a_{-1} = 1.$$

Therefore

$$(10.23) \qquad a_m = 2\pi \sum_{k=1}^{\infty} \frac{1}{k} A_{k,1}(m)\, m^{-13/2} I_{13}(4\pi m^{\frac{1}{2}} k^{-1}).$$

Here we have

$$A_{k,1}(m) = \sum_{\substack{h \bmod k \\ (h,k)=1}} \omega_{12}(h, k)\, e^{-2\pi i (m+1) h/k},$$

$$\omega_{12}(h, k) = \exp \{2\pi i \cdot 12 s(h, k)\}.$$

Since

$$(10.24) \qquad 12 s(h, k) \equiv \frac{h - h'}{k} \pmod 1,$$ [17]

[17] Rademacher, l.c.[9], p. 321, formula (2.51).

we have

$$\omega_{12}(h,k) \doteq e^{2\pi i(h-h')/k}$$

and therefore

$$A_{k,1}(m) = \sum_{\substack{h \bmod k \\ (h,k)=1}} e^{-2\pi i(h'+mh)/k}, \tag{10.25}$$

Thus

$$a_m = 2\pi \sum_{k=1}^{\infty} \frac{1}{k} \sum_{\substack{h \bmod k \\ (h,k)=1}} e^{-2\pi i(h'+mh)/k} m^{-13/2} I_{13}(4\pi m^{\frac{1}{2}} k^{-1}). \tag{10.26}$$

This formula permits of an explicit determination of a_0 and so leads to a certain verification, since (10.22) furnishes directly $a_0 = 24$. On the other hand we have

$$A_{k,1}(0) = \sum_{\substack{h \bmod k \\ (h,k)=1}} e^{-2\pi i h'/k} = \mu(k),$$

$\mu(k)$ designating the Möbius function, and according to (6.2) we are to take the interpretation

$$m^{-13/2} I_{13}(4\pi m^{\frac{1}{2}} k^{-1})_{m=0} = \frac{(2\pi)^{13}}{k^{13} \cdot 13!}.$$

Therefore (10.23) yields

$$a_0 = \frac{(2\pi)^{14}}{13!} \sum_{k=1}^{\infty} \frac{\mu(k)}{k^{14}} = \frac{(2\pi)^{14}}{13!} \frac{1}{\zeta(14)} = 28/B_7 = 24,$$

in accordance with the previously obtained value.

III. $$12 < r \leqq 24, \qquad \beta = \gamma = \kappa = 0.$$

$$\alpha = -r/12 - [-r/12] = 2 - r/12, \qquad \mu = 0 - [-r/12] = 2,$$

$$F(\tau) = \eta(\tau)^{-2r} = e^{2\pi i\tau(2-r/12)} \left\{ e^{-4\pi i\tau} + 2re^{-2\pi i\tau} + \sum_{m=0}^{\infty} a_m e^{2\pi i m\tau} \right\}.$$

Therefore

$$a_{-2} = 1, \qquad a_{-1} = 2r.$$

Let us consider e.g. the special case $r = 25/2$:

$$F(\tau) = \eta(\tau)^{-25} = e^{23\pi i\tau/12} \left\{ e^{-4\pi i\tau} + 25e^{-2\pi i\tau} + \sum_{m=0}^{\infty} a_m e^{2\pi i m\tau} \right\}, \tag{10.31}$$

$$a_{-2} = 1, \qquad a_{-1} = 25. \tag{10.32}$$

We find by means of (10.24) and (10.13)

$$A_{k,1}(m) = \sum_{\substack{h \bmod k \\ (h,k)=1}} \omega_{h,k} e^{-2\pi i(m+1)h/k}, \tag{10.33}$$

$$A_{k,2}(m) = \sum_{\substack{h \bmod k \\ (h,k)=1}} \omega_{h,k} e^{-2\pi i(h'+(m+1)h)/k}, \tag{10.34}$$

$$\begin{aligned} a_m = 2\pi \Bigg\{ & 25 \sum_{k=1}^{\infty} \frac{1}{k} A_{k,1}(m) \left(\frac{\frac{1}{24}}{m + \frac{23}{24}} \right)^{27/4} I_{27/2}\left(\frac{4\pi}{k} \left[\frac{1}{24}\left(m + \frac{23}{24} \right) \right]^{\frac{1}{2}} \right) \\ & + \sum_{k=1}^{\infty} \frac{1}{k} A_{k,2}(m) \left(\frac{\frac{25}{24}}{m + \frac{23}{24}} \right)^{27/4} I_{27/2}\left(\frac{4\pi}{k} \left[\frac{25}{24}\left(m + \frac{23}{24} \right) \right]^{\frac{1}{2}} \right) \Bigg\}. \end{aligned} \tag{10.35}$$

If we write, in order to exhibit the linearity in the coefficients a_{-2} and a_{-1}

$$\begin{aligned} F(\tau) &= a_{-2} e^{23\pi i\tau/12} \left\{ e^{-4\pi i\tau} + 2\pi \sum_{m=0}^{\infty} e^{2\pi i m\tau} \sum_{k=1}^{\infty} \frac{1}{k} A_{k,2}(m)\, 5^{27/2} \frac{I_{27/2}\left(\frac{5\pi}{k} \left[\frac{2}{3}\left(m + \frac{23}{24} \right) \right]^{\frac{1}{2}} \right)}{(24m+23)^{27/4}} \right\} \\ &\quad + a_{-1} e^{23\pi i\tau/12} \left\{ e^{-2\pi i\tau} + 2\pi \sum_{m=0}^{\infty} e^{2\pi i m\tau} \sum_{k=1}^{\infty} \frac{1}{k} A_{k,1}(m) \frac{I_{27/2}\left(\frac{\pi}{k} \left[\frac{2}{3}\left(m + \frac{23}{24} \right) \right]^{\frac{1}{2}} \right)}{(24m+23)^{27/4}} \right\} \\ &= a_{-2} F_2(\tau) + a_{-1} F_1(\tau), \end{aligned} \tag{10.36}$$

we see that a modular form arises only for the ratio

$$a_{-2} : a_{-1} = 1 : 25. \tag{10.37}$$

Indeed, no other linear combination of $F_2(\tau)$ and $F_1(\tau)$, nor $F_2(\tau)$ or $F_1(\tau)$ separately can be a modular form (at least not for the full modular group). A comparison with (9.52) shows that any such combination would belong to

$$\alpha = \frac{23}{24}, \qquad r = \frac{25}{2}.$$ [18]

[18] This argument might need some further explanation. The question is, whether, given a function of type (9.51), the values α, r and μ in (9.52) are (if at all) *uniquely* determined. This is, indeed, the case. The value of α with $0 \leqq \alpha < 1$ can be derived from the fact that $F(\tau) \exp(-2\pi i\alpha\tau)$ is periodic in τ mod 1. The value for r appears in the form $(r+1)/2$ as exponent in (9.52), and since

$$I_{r+1}(4\pi(\mu-\alpha)^{\frac{1}{2}}(m+\alpha)^{\frac{1}{2}}) \sim \frac{\exp[4\pi(\mu-\alpha)^{\frac{1}{2}}(m+\alpha)^{\frac{1}{2}}]}{2\pi 2^{\frac{1}{2}}(\mu-\alpha)^{\frac{1}{4}}(m+\alpha)^{\frac{1}{4}}}$$

we have

$$a_m \sim C \frac{\exp[4\pi(\mu-\alpha)^{\frac{1}{2}}(m+\alpha)^{\frac{1}{2}}]}{(m+\alpha)^{r/2+3/4}}.$$

Hence we obtain

The diophantine equation (9.45) for β and γ would become

$$\frac{23}{24} \equiv -\frac{25}{24} - \frac{\beta}{2} - \frac{\gamma}{3} \pmod 1,$$

which yields, among the admissible values of β and γ, only

$$\beta = 0, \quad \gamma = 0.$$

Moreover we should have in general $\mu = 2$, and $\mu = 1$ only for the case $a_{-2} = 0$. We obtain from (8.95)

$$2 \geqq \kappa - [-25/24] = \kappa + 2$$

which together with $\kappa \geqq 0$ leads to the determination

$$\kappa = 0.$$

Our reasoning shows that any linear combination

$$a_{-2}F_2(\tau) + a_{-1}F_1(\tau) \tag{10.38}$$

which is a modular form has the uniquely determined values for r, β, γ, r which we just have computed. But these numbers in turn determine, by (9.41) of Theorem 2, uniquely (up to a constant factor) the modular form to which they belong. Thus (10.31) is the only modular form contained in the linear set (10.38), and therefore the condition (10.37) is necessary.

This discussion shows that in general the coefficients of the principal part (1.8) are not independent of each other. This is not surprising, since it follows from (8.95) that

$$\mu = \kappa + 1$$

occurs only for

$$\frac{r}{12} + \frac{\beta}{2} + \frac{\gamma}{3} \leqq 1.$$

Otherwise we have

$$\mu > \kappa + 1,$$

which means that there are less independent parameters $C_0, C_1, \cdots C_\kappa$ than coefficients $a_{-1}, \cdots, a_{-\mu}$.

$$\mu - \alpha = \lim_{m\to\infty} \frac{(\log a_m)^2}{16\pi^2 m}$$

and then

$$\frac{r}{2} + \frac{3}{4} = \lim_{m\to\infty} \frac{4\pi(\mu-\alpha)^{\frac{1}{2}}(m+\alpha)^{\frac{1}{2}} - \log a_m}{\log m}$$

as *necessary* conditions for μ and r.

IV. All our preceding examples assumed $\kappa = 0$. We give a last example with $\kappa > 0$:

$$r = \tfrac{1}{2}, \qquad \beta = \gamma = 0, \qquad \kappa = 1:$$

$$F(\tau) = \eta(\tau)^{-1}\{3^3 \cdot 2^6 J(\tau) + C_0\} \tag{10.41}$$

with arbitrary C_0. We find

$$\alpha = 23/24, \qquad \mu = 2$$

and have

$$\begin{aligned} F(\tau) &= e^{23\pi i\tau/12}\{e^{-2\pi i\tau} + 1 + \cdots\}\{e^{-2\pi i\tau} + (744 + C_0) + \cdots\} \\ &= e^{23\pi i\tau/12}\left\{e^{-4\pi i\tau} + (745 + C_0)e^{-2\pi i\tau} + \sum_{m=0}^{\infty} a_m e^{2\pi i m\tau}\right\}, \end{aligned}$$

and hence

$$a_{-2} = 1, \qquad a_{-1} = 745 + C_0 .$$

Therefore

$$\begin{aligned} a_m &= 2\pi(745 + C_0)\sum_{k=1}^{\infty}\frac{1}{k}A_{k,1}(m)\left(\frac{\frac{1}{24}}{m + \frac{23}{24}}\right)^{3/4} I_{3/2}\left(\frac{4\pi}{k}\left[\frac{1}{24}\left(m + \frac{23}{24}\right)\right]^{\frac{1}{2}}\right) \\ &\quad + 2\pi\sum_{k=1}^{\infty}\frac{1}{k}A_{k,2}(m)\left(\frac{\frac{25}{24}}{m + \frac{23}{24}}\right)^{3/4} I_{3/2}\left(\frac{4\pi}{k}\left[\frac{25}{24}\left(m + \frac{23}{24}\right)\right]^{\frac{1}{2}}\right), \end{aligned} \tag{10.42}$$

where $A_{k,1}(m)$ and $A_{k,2}(m)$ turn out to be the numbers (10.33), (10.34) of the preceding example ($A_{k,\nu}(m)$ is, as function of r, periodic modulo 12). If we make use of example I, we obtain

$$a_m = (745 + C_0)p(m + 1) + a_m^* \tag{10.43}$$

with

$$a_m^* = \frac{1}{\pi\sqrt{2}}\sum_{k=1}^{\infty} A_{k,2}(m)\, k^{\frac{1}{2}} \frac{d}{dm}\left(\frac{\sinh\left(\frac{5\pi}{k}\left[\frac{2}{3}\left(m + \frac{23}{24}\right)\right]^{\frac{1}{2}}\right)}{\left[m + \frac{23}{24}\right]^{\frac{1}{2}}}\right). \tag{10.44}$$

The function $J(\tau)$, being of dimension zero, does not fall under the scope of our theory. Nevertheless the numbers a_m^* can be used to express rather simply the coefficients of $J(\tau)$. Indeed, we have, with $C_0 = 0$:

$$3^3 \cdot 2^6 \cdot J(\tau) = F(\tau)\eta(\tau)$$

$$= e^{23\pi i\tau/12}\left\{e^{-4\pi i\tau} + \sum_{m=0}^{\infty} a_m^* e^{2\pi i m\tau}\right\}\cdot e^{\pi i\tau/12}\prod_{l=1}^{\infty}(1 - e^{2\pi i l\tau})$$

$$+ 745\, e^{23\pi i\tau/12}\left\{e^{-2\pi i\tau} + \sum_{m=0}^{\infty}{}' p(m+1)\, e^{2\pi i m\tau}\right\}\cdot \eta(\tau)$$

$$= \left\{e^{-2\pi i\tau} + \sum_{m=0}^{\infty} a_m^* e^{2\pi i(m+1)\tau}\right\}\cdot \sum_{\lambda=-\infty}^{+\infty} (-1)^\lambda e^{2\pi i\tau[\lambda(3\lambda-1)/2]} + 745.$$

If we put

$$3^3 \cdot 2^6 \cdot J(\tau) = e^{-2\pi i\tau} + 744 + \sum_{n=1}^{\infty} c_n e^{2\pi i n\tau} \tag{10.45}$$

we obtain for the desired coefficients c_n the finite sum

$$c_n = \sum_{[\lambda(3\lambda-1)/2]\leq n-1} (-1)^\lambda a^*_{n-1-[\lambda(3\lambda-1)/2]} + \epsilon_n, \tag{10.46}$$

where $\epsilon_n = 0$, if $n + 1$ is not a pentagonal number, and $\epsilon_n = (-1)^\rho$ if $n + 1 = \rho(3\rho - 1)/2$.

11. Concluding remarks

We have discussed modular forms of positive real dimension r, generalizing in this respect the customary notion of modular forms, which admits only integers or half integers as dimensions.[19] But we can even go one step further. Nothing prevents us from introducing complex values for r in the developments of nos. 8 and 9 up to Theorem 2, inclusive. We should in (8.41) again obtain q as an integer, and α would now become a complex number.

This, however, would have the consequence that the multiplier ϵ, which appears in its simplest form in

$$F(\tau + 1) = e^{2\pi i\alpha}F(\tau)$$

loses the essential property $|\epsilon| = 1$. This is not an objection against the generalization of the dimension to complex values, but it would completely upset the estimations in nos. 3 and 4, in which $|\Omega_{h,k}| = 1$, or at least $\Omega_{h,k} = 0(1)$ is indispensable. Thus our theory is intrinsically restricted to *real* and positive dimensions of the modular forms.

It will be noticed that among our examples of no. 10 we have not mentioned certain functions which Hardy and Ramanujan discuss incompletely at the end of their paper. But those functions are not accessible by our present method since they either do not belong to the full modular group, but only to a certain congruence subgroup, or they contain functions of dimension zero.

[19] See, however, papers of H. Petersson, Mathem. Annalen *103* (1930), Acta Mathem. *58* (1932), Jahresber. Deutsche Math. Ver. *47* (1937).

The first obstacle can be overcome by an appropriate generalization of our method.[20] The second, however, leads to a problem interesting in itself. The series (7.2), in fact, would remain convergent also for $r = 0$, but a direct limit process $r \to 0$ is not obvious. However, the estimates of nos. 3 and 4 can be refined in such a way as to lead to the desired result.[21]

University of Pennsylvania,
Philadelphia.

[20] Cf. H. S. Zuckerman, *On the coefficients of certain modular forms belonging to subgroups of the modular group*, to be published in the Transactions of the Amer. Math. Soc.

[21] Cf. H. Rademacher, *The Fourier coefficients of the modular invariant* $J(\tau)$, Amer. Journal of Math., vol. 60, pp. 501–512 (1938).

Notes to Paper 40

This paper has been reviewed in the JF, vol. 64 (1938), p. 121, and in the Z, vol. 19 (1938-39), p. 22.

Page 433, footnote 2, line 1.
See paper 37 of the present collection.

Page 433, footnote 2, line 2.
See paper 38.

Page 446, equation (7.2).
Author's correction (read $\Sigma_{\nu=1}^{\mu} \ldots \Sigma_{k=1}^{\infty}$ instead of $\Sigma_{\nu=1}^{\infty} \ldots \Sigma_{k=1}^{\mu}$).

Page 451, footnote.
See paper 27.

Page 454, last display line.
Author's correction (replace the first – by a +)

Page 455, footnote 14, line 2.
See paper 29.

Page 462, footnote 21.
See paper 39.

THE FOURIER SERIES AND THE FUNCTIONAL EQUATION OF THE ABSOLUTE MODULAR INVARIANT $J(\tau)$.*

By HANS RADEMACHER.

1. The Fourier expansion for the absolute modular invariant $J(\tau)$ is

$$(1.1)\qquad 12^3 J(\tau) = e^{-2\pi i\tau} + 744 + \sum_{n=1}^{\infty} c_n e^{2\pi i n\tau}$$

with

$$(1.2)\qquad c_n = \frac{2\pi}{\sqrt{n}} \sum_{k=1}^{\infty} \frac{A_k(n)}{k} I_1\left(\frac{4\pi\sqrt{n}}{k}\right),$$

$$(1.3)\qquad A_k(n) = \sum_{\substack{h \bmod k \\ (h,k)=1}} \exp\left[-\frac{2\pi i}{k}(nh + h')\right], \qquad hh' \equiv -1 \pmod{k}.$$

This expansion was first given by H. Petersson in 1932.[1] I regret not having been aware of Petersson's priority when I recently published it again.[2] Our two proofs, however, have nothing in common, since they approach the problem from two opposite sides. Petersson discusses modular forms of *negative* dimension $r \leqq -2$. The derivative $J'(\tau)$ is clearly of dimension -2. By means of his generalized Poincaré series Petersson constructs a modular form of dimension -2, of which he shows, through a theorem of uniqueness, whose application in this case he briefly indicates, that it must be identical with $J'(\tau)$. By expanding his Poincaré series into a Fourier series he finds the series for $J'(\tau)$ and by integration that for $J(\tau)$. On the other hand, my method is a refinement of my variant of the Hardy-Ramanujan method, which originally was only applicable to modular forms of *positive* dimension. The Kloosterman method permitted its extension to modular forms of dimension zero.[3]

---

* Received August 26, 1938.

[1] Hans Petersson, "Ueber die Entwicklungskoeffizienten der automorphen Formen," *Acta Mathematica*, Bd. 58, pp. 169-215, in particular p. 202.

[2] "The Fourier coefficients of the modular invariant $J(\tau)$," *American Journal of Mathematics*, vol. 60 (1938), pp. 501-512. I avail myself of this opportunity to supplement a remark in that paper. When I mentioned there (p. 511) that only the numerical values of c_1 and c_2 seemed to appear in the literature I had overlooked the paper of W. E. H. Berwick, "An invariant modular equation of the fifth order," *Quarterly Journal of Mathematics*, vol. 47 (1916), pp. 94-103, in which the author gives the numerical values of the coefficients up to c_7.

[3] The present paper had, with another introduction, already been submitted for

If we leave aside a normalizing constant factor and an additive constant the modular function $J(\tau)$ is unambiguously characterized as the function regular in the upper τ-half-plane, having a pole of the first order at $x=0$ for the variable $x=e^{2\pi i\tau}$ and satisfying the functional relations

$$J(\tau)=J(\tau+1), \tag{1.4}$$

$$J(\tau)=J(-1/\tau). \tag{1.5}$$

The problem we are dealing with in this paper is so to speak the converse of that of determining the series (1.1) for the modular function $J(\tau)$. We ask here: given the expansion (1.1), how can we see that it represents a modular function, i.e. that the function defined by the expansion satisfies the above mentioned functional relations? Now it is obvious that (1.1) as a Fourier series has the property (1.4). We are therefore only concerned with (1.5).

Our method will, in brief words, consist in a transformation carrying (1.1) over into a certain double series. In this double series, which is not absolutely convergent, we shall have to interchange the summations with respect to the two indices of summation. The greatest part of our conclusion will be purely formal. Since only the treatment of the double series involves some analytical intricacies, we shall deal with it in a separate lemma, which we prove first, in order not to interrupt the continuity of the main reasonings.

2. Lemma. *Let τ be complex with positive imaginary part. Then*

$$\sum_{k=1}^{\infty}\lim_{N\to\infty}\sum_{\substack{m=-N\\(m,k)=1}}^{+N}\frac{e^{-(2\pi i m'/k)}}{k(k\tau-m)}=\lim_{K\to\infty}\sum_{k=1}^{K}\sum_{\substack{m=-K\\(m,k)=1}}^{+K}\frac{e^{-(2\pi i m'/k)}}{k(k\tau-m)}, \tag{2.1}$$

where m' is defined as a solution of the congruence

$$mm'\equiv-1 \pmod{k}.\text{[4]}$$

Proof. We first prove the convergence of the left-hand side of (2.1). We have

$$\sum_{\substack{m=-N\\(m,k)=1}}^{+N}\frac{e^{-(2\pi i m'/k)}}{k(k\tau-m)}=\frac{1}{k}\sum_{\substack{h \bmod k\\(h,k)=1}}e^{-(2\pi i h'/k)}\sum_{\substack{n\\|nk+h|\leq N}}\frac{1}{k\tau-h-nk};$$

hence

---

publication, when on September 14th, 1938, I received a letter from Professor Petersson, drawing my attention to his previous publication.

[4] The dash ′ is used in this meaning throughout this paper; the modulus of the defining congruence will always be clearly stated by the context.

$$\lim_{N\to\infty} \sum_{\substack{m=-N\\(m,k)=1}}^{+N} \frac{e^{-(2\pi i m'/k)}}{k(k\tau - m)} = \frac{1}{k^2} \sum_{\substack{h \bmod k\\(h,k)=1}} e^{-(2\pi i h'/k)} \lim_{N\to\infty} \sum_{\substack{n\\|nk+h|\leqq N}} \frac{1}{\tau - h/k - n}$$

$$= \frac{1}{k^2} \sum_{\substack{h \bmod k\\(h,k)=1}} e^{-(2\pi i h'/k)} \cdot 2\pi i \left(\tfrac{1}{2} - \frac{1}{1 - e^{2\pi i(\tau - h/k)}} \right)$$

$$= \frac{\pi i}{k^2} \mu(k) - \frac{2\pi i}{k^2} \sum_{\substack{h \bmod k\\(h,k)=1}} e^{-(2\pi i h'/k)} \sum_{\nu=0}^{\infty} e^{2\pi i \nu(\tau - h/k)},$$

(2.2) $$\lim_{N\to\infty} \sum_{\substack{m=-N\\(m,k)=1}}^{+N} \frac{e^{-(2\pi i m'/k)}}{k(k\tau - m)}$$

$$= \frac{\pi i \mu(k)}{k^2} - \frac{2\pi i}{k^2} \sum_{\nu=0}^{\infty} e^{2\pi i \nu\tau} \sum_{\substack{h \bmod k\\(h,k)=1}} \exp\left[-\frac{2\pi i}{k}(h' + h\nu) \right].$$

Now the inner sum of the last term is a Kloosterman sum, for which we have, after Estermann, Salié and Davenport,[5] the estimate

(2.31) $$\sum_{\substack{h \bmod k\\(h,k)=1}} \exp\left[-\frac{2\pi i}{k}(h' + h\nu) \right] = O(k^{2/3+\epsilon}).$$

We put

(2.32) $$\tau = \alpha + \beta i, \quad \beta > 0$$

and obtain

$$\sum_{\nu=0}^{\infty} e^{2\pi i \nu\tau} \sum_{\substack{h \bmod k\\(h,k)=1}} \exp\left[-\frac{2\pi i}{k}(h' + h\nu) \right] = O\left(k^{2/3+\epsilon} \frac{1}{1 - e^{-2\pi\beta}} \right).$$

Consequently from (2.2)

(2.33) $$\lim_{N\to\infty} \sum_{\substack{m=-N\\(m,k)=1}}^{+N} \frac{e^{-(2\pi i m'/k)}}{k(k\tau - m)} = O\left(k^{-(4/3)+\epsilon} \frac{1}{1 - e^{-2\pi\beta}} \right).$$

This shows the convergence of the first sum in (2.1).

We now can enunciate the lemma in the following form

(2.4) $$\lim_{K\to\infty} \sum_{k=1}^{K} \lim_{N\to\infty} \sum_{\substack{K < |m| \leqq N\\(m,k)=1}} \frac{e^{-(2\pi i m'/k)}}{k(k\tau - m)} = 0,$$

in which form we are going to prove it.

We first consider

(2.5) $$T_k^{(K)} = \lim_{N\to\infty} T_k^{(K,N)} = \lim_{N\to\infty} \sum_{\substack{K < |m| \leqq N\\(m,k)=1}} \frac{e^{-(2\pi i m'/k)}}{k(k\tau - m)}.$$

Now the function

(2.61) $$g(m) = \begin{cases} e^{-(2\pi i m'/k)} & \text{for } (m,k) = 1, \\ 0 & \text{otherwise} \end{cases}$$

[5] For references cf. my paper cited in footnote 2.

is periodic modulo k and can therefore be expressed as a "finite Fourier sum"

$$(2.62)\qquad g(m)=\sum_{j=1}^{k} B_{j,k}e^{2\pi ijm/k}.$$

The coefficients $B_{j,k}$ are found in the customary way:

$$\sum_{m=1}^{k} g(m)e^{-(2\pi ilm/k)}=\sum_{j=1}^{k} B_{j,k}\sum_{m=1}^{k} e^{2\pi i(j-l)m/k}=kB_{l,k},$$

$$B_{j,k}=\frac{1}{k}\sum_{m=1}^{k} g(m)e^{-(2\pi ijm/k)},$$

$$(2.63)\qquad B_{j,k}=\frac{1}{k}\sum_{\substack{m=1\\(m,k)=1}}^{k}\exp\left[-\frac{2\pi i}{k}(m'+jm)\right].$$

Here we have again a Kloosterman sum, so that

$$(2.64)\qquad B_{j,k}=O(k^{-1/3+\epsilon}).$$

We note separately the special case $j=k$:

$$(2.65)\qquad B_{j,k}=\frac{1}{k}\sum_{\substack{m=1\\(m,k)=1}}^{k} e^{-(2\pi im'/k)}=\frac{\mu(k)}{k}.$$

The formulae (2.61), (2.62) and (2.65) permit us to transform $T_k^{(K,N)}$ of (2.5):

$$T_k^{(K,N)}=\sum_{K<|m|\leqq N}\sum_{j=1}^{k} B_{j,k}\frac{e^{2\pi ijm/k}}{k(k\tau-m)}$$
$$=\sum_{j=1}^{k-1} B_{j,k}\sum_{K<|m|\leqq N}\frac{e^{2\pi ijm/k}}{k(k\tau-m)}+\frac{\mu(k)}{k}\sum_{K<|m|\leqq N}\frac{1}{k(k\tau-m)}.$$

This leads to

$$T_k^{(K)}=\lim_{N\to\infty}T_k^{(K,N)}$$

$$(2.7)\qquad =\frac{1}{k}\sum_{j=1}^{k-1}B_{j,k}\sum_{m=K+1}^{\infty}\frac{e^{2\pi ijm/k}}{k\tau-m}+\frac{1}{k}\sum_{j=1}^{k-1}B_{j,k}\sum_{m=K+1}^{\infty}\frac{e^{-(2\pi ijm/k)}}{k\tau+m}$$
$$+\frac{\mu(k)}{k^2}\sum_{m=K+1}^{\infty}\left(\frac{1}{k\tau-m}+\frac{1}{k\tau+m}\right)=S_1+S_2+S_3,$$

say. The convergence of the infinite sums will be shown incidentally.

With the notation (2.32) we have [6]

[6] The following estimations could be considerably shortened and simplified if we were to content ourselves with the special case of purely imaginary τ. In fact, this would suffice to prove the relation (1.5) under the same restriction, which then could be lifted by the principle of analytic continuation. However, I have not chosen this way since for the purpose of this paper it seems to me more natural to leave τ as unrestricted as is compatible with the sense of (1.1).

$$S_3 = \frac{\mu(k)}{k^2}\left\{ \sum_{m=K+1}^{\infty}\left(\frac{1}{(\alpha+i\beta)k-m} - \frac{1}{i\beta k - m}\right) + \sum_{m=K+1}^{\infty}\left(\frac{1}{(\alpha+i\beta)k+m} - \frac{1}{i\beta k + m}\right) + \sum_{m=K+1}^{\infty}\left(\frac{1}{i\beta k - m} + \frac{1}{i\beta k + m}\right)\right\},$$

(2.81) $$| S_3 | \leqq \frac{1}{k^2}\left\{ \sum_{m=K+1}^{\infty} \frac{|\alpha| k}{((\alpha k - m)^2 + \beta^2 k^2)^{\frac{1}{2}}(m^2+\beta^2k^2)^{\frac{1}{2}}} + \sum_{m=K+1}^{\infty} \frac{|\alpha| k}{((\alpha k + m)^2 + \beta^2 k^2)^{\frac{1}{2}}(m^2+\beta^2k^2)^{\frac{1}{2}}} + \sum_{m=K+1}^{\infty}\frac{2\beta k}{m^2+\beta^2k^2}\right\}$$
$$< \frac{1}{k}\left\{ \sum_{m=K+1}^{\infty} \frac{2|\alpha|}{((|\alpha| k - m)^2 + \beta^2k^2)^{\frac{1}{2}} m} + \sum_{m=K+1}^{\infty}\frac{2\beta}{m^2}\right\}$$
$$= \frac{1}{k}(S_3' + S_3'').$$

Here we have

(2.82) $$S_3'' < \int_K^{\infty} \frac{2\beta dx}{x^2} = \frac{2\beta}{K}.$$

We write

(2.83) $$S_3' = \sum_{K < m \leqq (|\alpha|+1)K} + \sum_{(|\alpha|+1)K < m} = s_1 + s_2.$$

Now (2.4) shows that we always have

$$1 \leqq k \leqq K.$$

Therefore we majorize the sum s_1 in (2.83) if we replace there the sequence

$$(|\alpha| k - m)^2, \qquad K < m \leqq (|\alpha|+1)K$$

by twice the sequence

$$m^2, \qquad 0 \leqq m \leqq (|\alpha|+1)K;$$

so that

$$s_1 \leqq 4|\alpha| \sum_{0 \leqq m \leqq (|\alpha|+1)K} \frac{1}{(m^2+\beta^2k^2)^{\frac{1}{2}} K}$$

(2.84) $$\leqq 4|\alpha|\left(\frac{1}{\beta k K} + \sum_{1 \leqq m \leqq (|\alpha|+1)K}\frac{1}{mK}\right) = O\left(\frac{\log K}{K}\right);$$

where we have suppressed under the O-symbol the irrelevant parameters α and β.

In s_2 we have

$$k \leqq K < \frac{m}{|\alpha|+1};$$

hence

$$m - |\alpha| k > m - \frac{|\alpha| m}{|\alpha| + 1} = \frac{m}{|\alpha| + 1}$$

Therefore

$$(2.85) \qquad s_2 \leqq \sum_{(|\alpha|+1)K < m} \frac{2|\alpha|(|\alpha|+1)}{m^2} = O\left(\frac{1}{K}\right).$$

The inequalities (2.81), (2.82), (2.83), (2.84), (2.85) together show that

$$(2.86) \qquad S_3 = O\left(\frac{\log K}{kK}\right), \qquad 1 \leqq k \leqq K.$$

For the estimation of S_1 we first consider a finite sum. We have

$$(2.91) \qquad \sum_{m=K+1}^{N} \frac{e^{2\pi i j m/k}}{k\tau - m} = \left\{ \int_{N+\frac{1}{2}-\tau\infty}^{N+\frac{1}{2}+\tau\infty} - \int_{K+\frac{1}{2}-\tau\infty}^{K+\frac{1}{2}+\tau\infty} \right\} \frac{e^{2\pi i j z/k}}{k\tau - z} \frac{dz}{e^{2\pi i z} - 1},$$

where the paths of integration are straight lines forming the angle

$$\delta = \arg \tau$$

with the positive real axis, where

$$(2.92) \qquad 0 < \delta < \pi.$$

The absolute convergence of the integrals is easily ascertained because of $1 \leqq j \leqq k - 1$. With

$$(2.93) \qquad z = x + (\alpha + i\beta)t$$

we have

$$|e^{2\pi i j z/k}| = e^{-(2\pi j \beta t/k)}.$$

The denominator $(e^{2\pi i z} - 1)$ is different from zero on the paths of integration. For the question of convergence we can therefore disregard an interval around $t = 0$ and need only discuss $|t| \geqq 1$ in (2.93). We obtain

$$|e^{2\pi i z} - 1| \geqq |e^{-2\pi\beta t} - 1|;$$

hence

$$\left| \frac{e^{2\pi i j z/k}}{e^{2\pi i z} - 1} \right| \leqq \frac{e^{-(2\pi j \beta t/k)}}{|e^{-2\pi\beta t} - 1|}.$$

This shows the convergence of the integrals, as well as the uniform convergence of the integrand to zero with $|t| \to \infty$, for $K + \frac{1}{2} \leqq x \leqq N + \frac{1}{2}$. If we put

$$|\tau| \int_{-\infty}^{+\infty} \frac{e^{-(2\pi j \beta t/k)}}{|e^{2\pi i(\alpha + i\beta)t} + 1|} dt = C,$$

we have in particular for the first integral in (2. 91)

$$\left| \int_{N+\frac{1}{2}-\tau\infty}^{N+\frac{1}{2}+\tau\infty} \right| \leqq C \operatorname*{Max}_{z=N+\frac{1}{2}+\tau t} \frac{1}{|k\tau - z|}$$
$$= C \operatorname*{Max}_{-\infty<t<\infty} \frac{1}{|\tau(k-t) - N - \frac{1}{2}|} = \frac{C}{(N+\frac{1}{2})\sin\delta},$$

which tends to zero with N tending to infinity. Hence

$$\sum_{m=K+1}^{\infty} \frac{e^{2\pi ijm/k}}{k\tau - m} = -\int_{K+\frac{1}{2}-\tau\infty}^{K+\frac{1}{2}+\tau\infty} \frac{e^{2\pi ijz/k}}{k\tau - z} \frac{dz}{e^{2\pi iz} - 1},$$

and so

$$\left| \sum_{m=K+1}^{\infty} \frac{e^{2\pi ijm/k}}{k\tau - m} \right| \leqq C \operatorname*{Max}_{z=K+\frac{1}{2}+\tau t} \frac{1}{|k\tau - z|} = \frac{C}{(K+\frac{1}{2})\sin\delta}.$$

If we introduce this into the definition (2. 7) of S_1 and make use of (2. 64) we obtain

(2. 94) $$S_1 = O(k^{-(1/3)+\epsilon}K^{-1}).$$

The analogous estimate holds true for S_2. Combining (2. 7), (2. 86), and (2. 94) we get

$$T_k^{(K)} = O(k^{-(1/3)+\epsilon}K^{-1}\log K), \qquad 1 \leqq k \leqq K.$$

From this we deduce further

$$\sum_{k=1}^{K} T_k^{(K)} = O(K^{-1}\log K \sum_{k=1}^{K} k^{-(1/3)+\epsilon}) = O(K^{-(1/3)+\epsilon}\log K).$$

In virtue of the definition (2. 5) of $T_k^{(K)}$ this implies the assertion (2. 4) and therefore proves the lemma.

3. We now proceed as follows. Putting

$$x = e^{2\pi i\tau},$$

we have from (1. 1)

(3. 1) $$f(x) = 12^3 J(\tau) = x^{-1} + 744 + \sum_{n=1}^{\infty} c_n x^n.$$

The function $f(x)$ is regular for $|x| < 1$ with the exception of $x = 0$. The convergence of the power series in $|x| < 1$ follows immediately from the asymptotic formula

$$c_n \sim \frac{1}{\sqrt{2}\, n^{3/4}} e^{4\pi\sqrt{n}},$$

which in turn is a corollary of (1.2).[7] The function $J(\tau)$, which in the present discussion we have to consider as *defined* by (1.1), is therefore analytic in the upper τ-half-plane.

Introducing (1.2) and (1.3) into (3.1) we obtain

$$\begin{aligned} f(x) &= x^{-1} + 744 + \sum_{n=1}^{\infty} x^n \frac{2\pi}{\sqrt{n}} \sum_{k=1}^{\infty} \frac{A_k(n)}{k} I_1\left(\frac{4\pi\sqrt{n}}{k}\right) \\ &= x^{-1} + 744 + 2\pi \sum_{k=1}^{\infty} \frac{1}{k} \sum_{n=1}^{\infty} A_k(n) \frac{x^n}{\sqrt{n}} I_1\left(\frac{4\pi\sqrt{n}}{k}\right) \\ &= x^{-1} + 744 + 2\pi \sum_{k=1}^{\infty} \frac{1}{k} \sum_{\substack{h \bmod k \\ (h,k)=1}} e^{-(2\pi i h'/k)} \sum_{n=1}^{\infty} (xe^{-(2\pi i h/k)})^n \frac{1}{\sqrt{n}} I_1\left(\frac{4\pi\sqrt{n}}{k}\right), \end{aligned}$$

$$(3.2)\quad 12^3 J(\tau) = f(x) = x^{-1} + 744 + 2\pi \sum_{k=1}^{\infty} \frac{1}{k} \sum_{\substack{h \bmod k \\ (h,k)=1}} e^{-(2\pi i h'/k)} \Phi_k(xe^{-(2\pi i h/k)})$$

with the notation

$$(3.3)\qquad \Phi_k(z) = \sum_{n=1}^{\infty} \frac{1}{\sqrt{n}} I_1\left(\frac{4\pi\sqrt{n}}{k}\right) z^n = \sum_{n=1}^{\infty} g_k(n) z^n.$$

Here

$$(3.4)\qquad g_k(w) = \frac{1}{\sqrt{w}} I_1\left(\frac{4\pi\sqrt{w}}{k}\right) = \frac{2\pi}{k} \sum_{\nu=0}^{\infty} \frac{\left(\frac{4\pi^2 w}{k^2}\right)^{\nu}}{\nu!\,(\nu+1)!}$$

is a transcendental entire function of order $\frac{1}{2}$. We remark that by a theorem of Wigert, the function $\Phi_k(z)$, which (3.3) defines only in the interior of the unit circle, can be continued analytically over the whole z-plane, with the exception of the point $z = 1$, where it has an essential singularity. The expansion (3.2) therefore shows a decomposition into "partial fractions" similar to that one which I have obtained in connection with the generating function of $p(n)$.[8] We do not pursue this remark here any further.

From (3.3) and (3.4) we infer

$$(3.5)\qquad \begin{aligned} \Phi_k(z) &= \frac{2\pi}{k} \sum_{n=1}^{\infty} z^n \sum_{\nu=0}^{\infty} \frac{\left(\frac{4\pi^2}{k^2} n\right)^{\nu}}{\nu!\,(\nu+1)!} \\ &= \frac{2\pi}{k} \sum_{\nu=0}^{\infty} \frac{\left(\frac{2\pi}{k}\right)^{2\nu}}{\nu!\,(\nu+1)!} \sum_{n=1}^{\infty} n^{\nu} z^n. \end{aligned}$$

---

[7] Cf. Watson, *Bessel Functions*, p. 203, formula (2).

[8] "A convergent series for the partition function $p(n)$," *Proceedings of the National Academy of Sciences*, vol. 23 (1937), pp. 78-84.

For the transformation of the inner sum we make use of Lipschitz's formula [9]

$$\sum_{n=1}^{\infty} n^{\nu} e^{-2\pi t n} = \frac{\Gamma(\nu+1)}{(2\pi)^{\nu+1}} \sum_{l=-\infty}^{+\infty} \frac{1}{(t+li)^{\nu+1}},$$

which holds for $\Re(t) > 0$, $\nu > 0$. We can apply it on (3.5) if we put

$$z = e^{-2\pi t}. \tag{3.51}$$

For $\nu = 0$ we have

$$\sum_{n=1}^{\infty} e^{-2\pi t n} = -1 + \frac{1}{1-e^{-2\pi t}} = -\tfrac{1}{2} + \frac{1}{2\pi} \lim_{N\to\infty} \sum_{l=-N}^{+N} \frac{1}{t+li}$$

and so

$$\Phi_k(z) = \frac{2\pi}{k}\left\{-\tfrac{1}{2} + \sum_{\nu=0}^{\infty} \frac{(2\pi/k)^{2\nu}}{\nu!\,(\nu+1)!}\,\frac{\nu!}{(2\pi)^{\nu+1}} \lim_{N\to\infty} \sum_{l=-N}^{+N} \frac{1}{(t+li)^{\nu+1}}\right\} \tag{3.61}$$

$$= -\frac{\pi}{k} + \lim_{N\to\infty} \frac{1}{k} \sum_{l=-N}^{+N} \sum_{\nu=0}^{\infty} \frac{1}{t+li}\,\frac{\left(\dfrac{2\pi}{k^2(t+li)}\right)^{\nu}}{(\nu+1)!}$$

$$= -\frac{\pi}{k} + \lim_{N\to\infty} \frac{1}{k} \sum_{l=-N}^{+N} \frac{1}{t+li} \cdot \frac{e^{2\pi/(k^2(t+li))} - 1}{2\pi/(k^2(t+li))},$$

$$\Phi_k(z) = -\frac{\pi}{k} + \lim_{N\to\infty} \frac{k}{2\pi} \sum_{l=-N}^{+N} \left(e^{2\pi/(k^2(t+li))} - 1\right). \tag{3.62}$$

We need in (3.2)

$$z = xe^{-(2\pi i h/k)} = e^{2\pi i(\tau - h/k)},$$

and therefore, because of (3.51),

$$t = -i\tau + i\frac{h}{k}.$$

Inserting this in (3.62), we obtain

$$\Phi_k(xe^{-(2\pi i h/k)}) = -\frac{\pi}{k} + \lim_{N\to\infty} \frac{k}{2\pi} \sum_{l=-N}^{+N} \left(e^{2\pi/ik(-k\tau+h+kl)} - 1\right) \tag{3.63}$$

$$= -\frac{\pi}{k} + \lim_{N\to\infty} \frac{k}{2\pi} \sum_{\substack{m\equiv h \pmod{k} \\ |m-h|\leqq kN}} \left(e^{2\pi i/k(k\tau-m)} - 1\right).$$

If we go back to the power series for the exponential function, viz. to (3.61), we see that for $\nu \geqq 1$ the limit process $N \to \infty$ may be carried out under the sign of summation with respect to l. For $\nu = 0$ the limit process

$$N \to \infty, \qquad |m-h| \leqq kN$$

is equivalent to

$$M \to \infty, \qquad |m| \leqq M,$$

[9] R. Lipschitz, "Untersuchung der Eigenschaften einer Gattung von unendlichen Reihen," *Journal für Mathematik*, Bd. 105 (1889), pp. 127-156, in particular p. 136.

since the single term

$$2\pi i/k(k\tau - m)$$

tends to zero as $|m| \to \infty$. The equation (3.63) can therefore finally be written

$$(3.64)\quad \Phi_k(xe^{-(2\pi i h/k)}) = -\frac{\pi}{k} + \lim_{M\to\infty} \frac{k}{2\pi} \sum_{\substack{m\equiv h \pmod k\\ |m|\leqq M}} (e^{2\pi i/k(k\tau-m)} - 1).$$

From this we obtain

$$\sum_{\substack{h \bmod k\\ (h,k)=1}} e^{-(2\pi i h'/k)}\Phi_k(xe^{-(2\pi i h/k)})$$

$$= -\frac{\pi}{k}\mu(k) + \lim_{M\to\infty} \frac{k}{2\pi} \sum_{\substack{|m|\leqq M\\ (m,k)=1}} e^{-(2\pi i m'/k)}(e^{2\pi i/k(k\tau-m)} - 1).$$

Returning to (3.2) we get now

$$(3.7)\quad 12^3 J(\tau) = e^{-2\pi i\tau} + 744 - 2\pi^2 \sum_{k=1}^{\infty} \frac{\mu(k)}{k^2}$$

$$+ \sum_{k=1}^{\infty} \lim_{M\to\infty} \sum_{\substack{|m|\leqq M\\ (m,k)=1}} e^{-(2\pi i m'/k)}(e^{2\pi i/k(k\tau-m)} - 1).$$

The condition $(m,k) = 1$ implies that the value $m = 0$ can be assumed only together with $k = 1$. We separate this term. Furthermore we have

$$\sum_{k=1}^{\infty} \frac{\mu(k)}{k^2} = \frac{1}{\zeta(2)} = \frac{6}{\pi^2}.$$

In consequence of these remarks, (3.7) goes over into

$$(3.8)\quad 12^3 J(\tau) = e^{-2\pi i\tau} + e^{2\pi i/\tau} + 731$$

$$+ \sum_{k=1}^{\infty} \lim_{M\to\infty} \sum_{\substack{1\leqq|m|\leqq M\\ (m,k)=1}} \left(\exp\left(2\pi i \frac{\dfrac{mm'+1}{k} - m'\tau}{k\tau - m}\right) - \exp\left(-2\pi i\frac{m'}{k}\right)\right).$$

It is from this formula that we intend to derive the functional equation (1.5).

4. For this purpose we have to show that (3.8) can be replaced by

$$(4.1)\quad 12^3 J(\tau) = e^{-2\pi i\tau} + e^{2\pi i/\tau} + 731$$

$$+ \lim_{K\to\infty} \sum_{k=1}^{K} \sum_{\substack{1\leqq|m|\leqq K\\ (m,k)=1}} \left(\exp\left(2\pi i \frac{\dfrac{mm'+1}{k} - m'\tau}{k\tau - m}\right) - \exp\left(-2\pi i\frac{m'}{k}\right)\right).$$

It is, however, simpler to consider the infinite sum in the form which it has in (3.7) instead of (3.8). Indeed we have

$$S=\sum_{k=1}^{\infty}\lim_{M\to\infty}\sum_{\substack{|m|\leqq M\\(m,k)=1}} e^{-(2\pi i m'/k)}\left(e^{2\pi i/k(k\tau-m)}-1\right)$$

$$=\sum_{k=1}^{\infty}\lim_{M\to\infty}\sum_{\substack{|m|\leqq M\\(m,k)=1}} e^{-(2\pi i m'/k)}\sum_{\lambda=1}^{\infty}\frac{1}{\lambda!}\left(\frac{2\pi i}{k(k\tau-m)}\right)^{\lambda},$$

$$S=\sum_{k=1}^{\infty}\lim_{M\to\infty}\sum_{\substack{|m|\leqq M\\(m,k)=1}} e^{-(2\pi i m'/k)}\cdot\frac{2\pi i}{k(k\tau-m)}$$
$$+\sum_{k=1}^{\infty}\sum_{\substack{m=-\infty\\(m,k)=1}}^{+\infty} e^{-(2\pi i m'/k)}\sum_{\lambda=2}^{\infty}\frac{1}{\lambda!}\left(\frac{2\pi i}{k(k\tau-m)}\right)^{\lambda}. \tag{4.2}$$

The separation into two sums is permissible since both are convergent. This is true for the first in virtue of our lemma. The second sum is absolutely convergent in all three summations together and admits therefore any rearrangement of its terms. If we apply moreover our lemma to the first member on the right-hand side of (4.2) we get

$$S=\lim_{K\to\infty}\sum_{k=1}^{K}\sum_{\substack{|m|\leqq K\\(m,k)=1}} e^{-(2\pi i m'/k)}\cdot\frac{2\pi i}{k(k\tau-m)}$$
$$+\lim_{K\to\infty}\sum_{k=1}^{K}\sum_{\substack{|m|\leqq K\\(m,k)=1}} e^{-(2\pi i m'/k)}\sum_{\lambda=2}^{\infty}\frac{1}{\lambda!}\left(\frac{2\pi i}{k(k\tau-m)}\right)^{\lambda}$$
$$=\lim_{K\to\infty}\sum_{k=1}^{K}\sum_{\substack{|m|\leqq K\\(m,k)=1}} e^{-(2\pi i m'/k)}\sum_{\lambda=1}^{\infty}\frac{1}{\lambda!}\left(\frac{2\pi i}{k(k\tau-m)}\right)^{\lambda}$$
$$=\lim_{K\to\infty}\sum_{k=1}^{K}\sum_{\substack{|m|\leqq K\\(m,k)=1}} e^{-(2\pi i m'/k)}\left(e^{2\pi i/k(k\tau-m)}-1\right).$$

This proves (4.1).

5. The rest of our reasoning is now purely formal. The sum of the first three terms of the right-hand side of (4.1) is invariant with respect to the transformation $\tau\to-1/\tau$. In order to establish (1.5) we have therefore only to prove

$$\sum_{\substack{1\leqq k\leqq K\\(m,k)=1}}\sum_{1\leqq|m|\leqq K}\left(\exp\left(2\pi i\,\frac{\frac{mm'+1}{k}-m'\tau}{k\tau-m}\right)-\exp\left(-2\pi i\frac{m'}{k}\right)\right)$$
$$=\sum_{\substack{1\leqq k\leqq K\\(m,k)=1}}\sum_{1\leqq|m|\leqq K}\left(\exp\left(2\pi i\,\frac{\frac{mm'+1}{k}\tau+m'}{-k-m\tau}\right)-\exp\left(-2\pi i\frac{m'}{k}\right)\right)$$

Since these are finite sums we can suppress on both sides the identical second terms in the parentheses. Our assertion is therefore reduced to

$$(5.1)\quad \begin{aligned}&\sum_{k=1}^{K}\sum_{\substack{m=1\\(m,k)=1}}^{K}\exp\left(2\pi i\frac{-k'-m'\tau}{k\tau-m}\right)+\sum_{k=1}^{K}\sum_{\substack{m=1\\(m,k)=1}}^{K}\exp\left(2\pi i\frac{-k'+m'\tau}{k\tau+m}\right)\\ &=\sum_{k=1}^{K}\sum_{\substack{m=1\\(m,k)=1}}^{K}\exp\left(2\pi i\frac{-k'\tau+m'}{-k-m\tau}\right)+\sum_{k=1}^{K}\sum_{\substack{m=1\\(m,k)=1}}^{K}\exp\left(2\pi i\frac{-k'\tau-m'}{-k+m\tau}\right).\end{aligned}$$

In this formula we have set

$$(5.2)\qquad -k'=\frac{mm'+1}{k}.$$

We have also used the fact that if m' belongs to m in

$$(5.22)\qquad mm'\equiv -1 \pmod{k}$$

then $-m'$ belongs to $-m$. The equation (5. 2) is symmetric with respect to m and k, since

$$(5.23)\qquad mm'+kk'+1=0;$$

so that, corresponding to (5. 22), we also have

$$(5.24)\qquad kk'\equiv -1 \pmod{m}.$$

We may write (5. 1) in the more symmetrical form

$$(5.3)\quad \begin{aligned}&\sum_{k=1}^{K}\sum_{\substack{m=1\\(m,k)=1}}^{K}\exp\left(-2\pi i\frac{m'\tau+k'}{k\tau-m}\right)+\sum_{k=1}^{K}\sum_{\substack{m=1\\(m,k)=1}}^{K}\exp\left(2\pi i\frac{m'\tau-k'}{k\tau+m}\right)\\ &=\sum_{k=1}^{K}\sum_{\substack{m=1\\(m,k)=1}}^{K}\exp\left(-2\pi i\frac{k'\tau+m'}{m\tau-k}\right)+\sum_{k=1}^{K}\sum_{\substack{m=1\\(m,k)=1}}^{K}\exp\left(2\pi i\frac{k'\tau-m'}{m\tau+k}\right).\end{aligned}$$

But this equation is obviously true, since one side goes over into the other by a mere change of notation, namely by replacing m, m' by k, k' and conversely.

We have therefore proved the functional property (1. 5) as a consequence of the Fourier expansion (1. 1) with the coefficients (1. 2), (1. 3).

UNIVERSITY OF PENNSYLVANIA,
PHILADELPHIA, PA.

In the last line of p. 242 we have

$$|\tau| \int_{-\infty}^{+\infty} \frac{e^{-(2\pi j \beta t/k)}}{|e^{2\pi i(\alpha+i\beta)t}+1|}\, dt = O\left(\frac{k}{j} + \frac{k}{k-j}\right)$$

With this we obtain on p. 243

$$\sum_{m=K+1}^{\infty} \frac{e^{2\pi i j m/k}}{k\tau - m} = O\left(kK^{-1}\left(\frac{1}{j} + \frac{1}{k-j}\right)\right).$$

and therefore

$$S_1 = O\left(k^{-\frac{1}{3}+\varepsilon} K^{-1} \sum_{j=1}^{k-1}\left(\frac{1}{j} + \frac{1}{k-j}\right)\right)$$

$$= O\left(k^{-\frac{1}{3}+\varepsilon} K^{-1} \log k\right),$$

which justifies

$$T_k^{(K)} = O\left(k^{-\frac{1}{3}+\varepsilon} K^{-1} \log K\right)$$

March 18th, 1942.

The treatment of S_1 and S_2 can be simplified, if partial summation used instead of complex integration:

$$(1) \quad \sum_{m=K+1}^{\infty} \frac{e^{\frac{2\pi i j}{k} m}}{k\tau - m} = \sum_{m=K+1}^{\infty} \frac{s_m - s_{m-1}}{k\tau - m}$$

with

$$s_m = \sum_{\mu=0}^{m} e^{\frac{2\pi i j}{k}\mu} = \frac{1 - e^{\frac{2\pi i j}{k}(m+1)}}{1 - e^{\frac{2\pi i j}{k}}}$$

$$(2) \quad |s_m| \leq \frac{2}{\left|1 - e^{\frac{2\pi i j}{k}}\right|} = \frac{1}{\sin \frac{\pi j}{k}} \leq Ck\left(\frac{1}{j} + \frac{1}{k-j}\right).$$

Continuing from (1)

$$\sum_{m=K+1}^{\infty} \frac{e^{\frac{2\pi i j}{k} m}}{k\tau - m} = \sum_{m=K+1}^{\infty} s_m \left(\frac{1}{k\tau - m} - \frac{1}{k\tau - m - 1}\right) - \frac{s_K}{k\tau - K - 1}$$

$$= - \sum_{m=K+1}^{\infty} \frac{s_m}{(k\tau - m)(k\tau - m - 1)} - \frac{s_K}{k\tau - K - 1}.$$

Because of (2) and

$$|k\tau - m| \geqq m \sin\delta$$

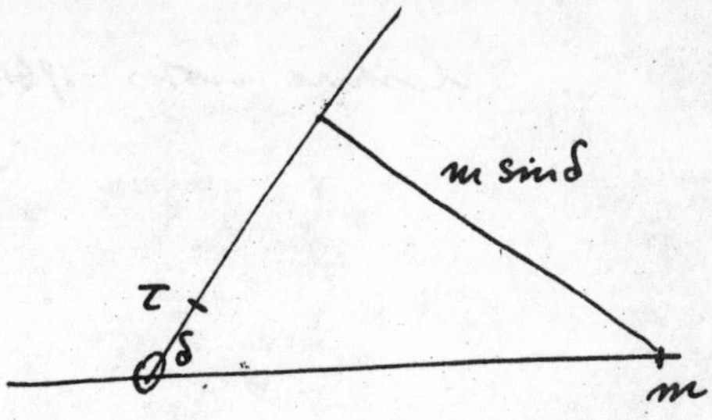

we have

$$\sum_{m=K+1}^{\infty} \frac{e^{\frac{2\pi i j}{k} m}}{k\tau - m}$$

$$= O\left(\sum_{m=K+1}^{\infty} k\left(\frac{1}{j} + \frac{1}{k-j}\right)\frac{1}{m(m+1)}\right) + O\left(k\left(\frac{1}{j}+\frac{1}{k-j}\right)K^{-1}\right)$$

$$= O\left(kK^{-1}\left(\frac{1}{j} + \frac{1}{k-j}\right)\right).$$

Therefore, ~~because of~~ in virtue of (2.7) and (2.64), we obtain

$$S_1 = O\left(K^{-1} k^{-\frac{1}{3}+\varepsilon} \sum_{j=1}^{k-1}\left(\frac{1}{j} + \frac{1}{k-j}\right)\right)$$

$$= O\left(K^{-1} k^{-\frac{1}{3}+\varepsilon} \log k\right)$$

which with (2.86) yields

$$T_k^{(K)} = O\left(k^{-\frac{1}{3}+\varepsilon} K^{-1} \log K\right), \quad 1 \leqq k \leqq K.$$

This proves (2.4) and thus the Lemma.

March 21st, 1942.

S_3 is also simplified, in my lecture notes 1961/62.

Notes to Paper 41

This paper has been reviewed in the JF, vol. 65 (1939), p. 351, and in the Z, vol. 20 (1939), p. 220.

Page 237, footnote 2, line 1.
See paper 39 of the present collection.

Page 238, equation (2.1).
Read $k = 1$ as lower limit under the third summation sign.

Page 240, equation (2.65).
Read $B_{k,k}$ instead of $B_{j,k}$.

Page 240, third display line from bottom; page 241, first display line; page 242, equation (2.86); page 243, fourth display line.
On three successive occasions (March 18, 1942, March 21, 1942, and sometime in 1962) the author reworked the estimates of S_1, S_2, S_3, and $T_k^{(K)}$. The corresponding notes, written very legibly, are reproduced in facsimile as an addendum to this paper. The lecture notes 1961-1962 with the simplified computation of S_3 were not available to the editor.

Page 242, last display line.
See paper 41a following.

Page 244, footnote 8.
See paper 37.

Page 245, ninth display line
Author's correction (read $i\,h/k$).

CORRECTION.*

By H. A. Rademacher.

Professor D. H. Lehmer has brought to my attention a mistake on pp. 242, 243 of my paper "The Fourier series and the functional equation of the absolute modular invariant $J(\tau)$," this Journal, vol. LXI (1939). It can be corrected as follows: C, the last letter on p. 242, has to be replaced by $O\{k(j^{-1}+(k-j)^{-1})\}$. With this estimate we obtain on p. 243

$$\sum_{m=K+1}^{\infty} \frac{e^{2\pi i j m/k}}{k\tau - m} = O\{kK^{-1}(j^{-1}+(k-j)^{-1})\}.$$

This leads to

$$S_1 = O\{k^{-1/3+\epsilon}K^{-1}\sum_{j=1}^{k-1}(j^{-1}+(k-j)^{-1}\}$$
$$= O(k^{-1/3+\epsilon}K^{-1}\log k),$$

which justifies the estimate given for $T_k^{(K)}$.

University of Pennsylvania.

* Received March 21, 1942.

Note to Paper 41a

This correction has been reviewed in the MR, vol. 3, p. 271.

A NEW PROOF OF TWO OF RAMANUJAN'S IDENTITIES[1]

By Hans Rademacher and Herbert S. Zuckerman[2]

(Received September 23, 1938)

1. When Ramanujan gave his first proof of the congruence properties

$$p(5n + 4) \equiv 0 \qquad (\text{mod } 5),$$
$$p(7n + 5) \equiv 0 \qquad (\text{mod } 7)$$

of the partition function $p(n)$, he observed that these congruences would be obvious consequences of the two identities

$$(1.11) \quad p(4) + p(9)x + p(14)x^2 + \cdots = 5\,\frac{\{(1 - x^5)(1 - x^{10})(1 - x^{15}) \cdots\}^5}{\{(1 - x)(1 - x^2)(1 - x^3) \cdots\}^6}$$

and

$$(1.12) \quad \begin{aligned} p(5) + p(12)x + p(19)x^2 + \cdots &= 7\,\frac{\{(1 - x^7)(1 - x^{14}) \cdots\}^3}{\{(1 - x)(1 - x^2) \cdots\}^4} \\ &\quad + 49x\,\frac{\{(1 - x^7)(1 - x^{14}) \cdots\}^7}{\{(1 - x)(1 - x^2) \cdots\}^8}. \end{aligned}$$

However he did not give a proof of these identities. Later, making use of the theory of modular functions, Darling[3] proved the first and Mordell[4] gave proofs for both identities. Recently G. N. Watson[4a] in a comprehensive study proved these identities as well as the existence of analogous ones for all powers of 5 and 7, thereby furnishing a proof for certain further conjectures of Ramanujan.

Making use of our method, an account of which we have recently published,[5] we are able to expand the right members of (1.11) and (1.12) in power series with explicitly determined coefficients. The verification of the identities which we propose to give here will then consist merely in a direct comparison of these

---

[1] Presented to the American Mathematical Society, April 15, 1938.

[2] National Research Fellow.

[3] *Proofs of certain identities and congruences enunciated by S. Ramanujan*, Proc. London Math. Soc., (2) vol. 19 (1921), pp. 350–372.

[4] *Note on certain modular relations considered by Messrs. Ramanujan, Darling, and Rogers*, Proc. London Math. Soc., (2) vol. 20 (1922), pp. 408–416.

[4a] *Ramanujans Vermutung über Zerfällungsanzahlen*, Journal für die reine und angewandte Mathematik, vol. 179 (1938), pp. 97–128.

[5] *On the Fourier coefficients of certain modular forms of positive dimension*, Annals of Mathematics, vol. 39, no. 2 (1938), pp. 433–462.

coefficients with the corresponding coefficients[6] of the expansions of the left members. Our proof still rests on the theory of modular functions but only in so far as we make use of the transformation formula of the modular form $\eta(\tau)$.

2. The right members of (1.11) and (1.12) are made up of summands of the form

$$F_p(x) = \frac{\prod_{m=1}^{\infty}(1-x^{pm})^p}{\prod_{m=1}^{\infty}(1-x^m)^{p+1}}, \tag{2.1}$$

where p is one of the prime numbers 5 or 7. Since $F_p(x)$ can be simply expressed in terms of the function

$$f(x) = \frac{1}{\prod_{m=1}^{\infty}(1-x^m)},$$

we can make use of the identity

$$f\left(\exp\left[\frac{2\pi ih}{k} - \frac{2\pi z}{k}\right]\right) = \omega_{h,k} z^{\frac{1}{2}} \exp\left[\frac{\pi}{12kz} - \frac{\pi z}{12k}\right] f\left(\exp\left[\frac{2\pi ih'}{k} - \frac{2\pi}{kz}\right]\right), \tag{2.2}$$

with $\Re(z) > 0$, where h' is any solution of the congruence

$$hh' \equiv -1 \pmod{k}$$

and where $\omega_{h,k}$ is a certain root of unity which we shall discuss later. This identity is obtained from the theory of the transformation of the function $\eta(\tau)$ and was used by Hardy and Ramanujan[7] in their discussion of $p(n)$.

In order to apply (2.2) to $f(x^p)$ we must distinguish two cases:

A) In the case $p \nmid k$ we have to replace h, in (2.2), by ph and z by pz. We then have

$$\begin{aligned} & f\left(\exp\left[\frac{2\pi iph}{k} - \frac{2\pi pz}{k}\right]\right) \\ & \quad = \omega_{ph,k}(pz)^{\frac{1}{2}} \exp\left[\frac{\pi}{12kpz} - \frac{\pi pz}{12k}\right] f\left(\exp\left[\frac{2\pi ih''}{k} - \frac{2\pi}{kpz}\right]\right) \end{aligned} \tag{2.31}$$

with $phh'' \equiv -1 \pmod{k}$ or

$$ph'' \equiv h' \pmod{k}. \tag{2.32}$$

---

[6] H. Rademacher, *A convergent series for the partition function p(n)*, Proc. Nat. Acad. Sci., vol. 23, no. 2 (1937), pp. 78–84; *On the partition function p(n)*, Proc. London Math. Soc., (2) vol. 43 (1937), pp. 241–254. Cf. also pp. 455–456 of our paper cited in footnote 5.

[7] *Asymptotic formulae in combinatory analysis*, Proc. London Math. Soc., (2) vol. 17 (1918), pp. 75–115.

B) In the case $p|k$ we replace k by k/p and have

$$f\left(\exp\left[\frac{2\pi iph}{k}-\frac{2\pi pz}{k}\right]\right) = \omega_{h,k/p} z^{\frac{1}{2}} \exp\left[\frac{\pi p}{12kz}-\frac{\pi pz}{12k}\right] f\left(\exp\left[\frac{2\pi ih'''p}{k}-\frac{2\pi p}{kz}\right]\right) \tag{2.41}$$

with $hh''' \equiv -1 \pmod{k/p}$ so that we may choose

$$h''' = h'. \tag{2.42}$$

The equations (2.2), (2.31), (2.41) enable us to write formulas for $F_\rho(x)$ with $x = \exp\left[\frac{2\pi ih}{k}-\frac{2\pi z}{k}\right]$. Again we have two cases:

A) If $p \nmid k$ we have

$$F_\rho\left(\exp\left[\frac{2\pi ih}{k}-\frac{2\pi z}{k}\right]\right) = \frac{\omega_{h,k}^{\rho+1}}{\omega_{ph,k}^{\rho}} z^{\frac{1}{2}} p^{-\rho/2} \times \exp\left[\frac{\pi}{12k}(\rho+1)\left(\frac{1}{z}-z\right)-\frac{\pi\rho}{12k}\left(\frac{1}{pz}-pz\right)\right] \frac{f\left(\exp\left[\frac{2\pi ih'}{k}-\frac{2\pi}{kz}\right]\right)^{\rho+1}}{f\left(\exp\left[\frac{2\pi ih''}{k}-\frac{2\pi}{kpz}\right]\right)^{\rho}}$$

$$= \Omega_{h,k}^{(\rho)} z^{\frac{1}{2}} p^{-\rho/2} \exp\left[\frac{\pi}{12kpz}((p-1)\rho+p)-\frac{\pi z}{12k}(\rho+1-p\rho)\right] \times \tilde{F}_\rho\left(\exp\left[\frac{2\pi ih''}{k}-\frac{2\pi}{kpz}\right]\right), \tag{2.51}$$

with the following abbreviations

$$\Omega_{h,k}^{(\rho)} = \frac{\omega_{h,k}^{\rho+1}}{\omega_{ph,k}^{\rho}} \qquad \text{for } p \nmid k, \tag{2.52}$$

$$\tilde{F}_\rho(x) = \frac{f(x^p)^{\rho+1}}{f(x)^\rho} = \frac{\prod_{m=1}^{\infty}(1-x^m)^\rho}{\prod_{m=1}^{\infty}(1-x^{pm})^{\rho+1}}. \tag{2.53}$$

It may be noted that we have eliminated h' by means of (2.32).

B) If $p|k$ we have

$$F_\rho\left(\exp\left[\frac{2\pi ih}{k}-\frac{2\pi z}{k}\right]\right) = \frac{\omega_{h,k}^{\rho+1}}{\omega_{h,k/p}^{\rho}} z^{\frac{1}{2}} \times \exp\left[\frac{\pi}{12k}(\rho+1)\left(\frac{1}{z}-z\right)-\frac{\pi\rho p}{12k}\left(\frac{1}{z}-z\right)\right] \frac{f\left(\exp\left[\frac{2\pi ih'}{k}-\frac{2\pi}{kz}\right]\right)^{\rho+1}}{f\left(\exp\left[\frac{2\pi ih'p}{k}-\frac{2\pi p}{kz}\right]\right)^{\rho}}$$

$$= \Omega_{h,k}^{(\rho)} z^{\frac{1}{2}} \exp\left[\frac{\pi}{12k}((1-p)\rho+1)\left(\frac{1}{z}-z\right)\right] F_\rho\left(\exp\left[\frac{2\pi ih'}{k}-\frac{2\pi}{kz}\right]\right) \tag{2.61}$$

with the abbreviation

$$\Omega_{h,k}^{(\rho)} = \frac{\omega_{h,k}^{p+1}}{\omega_{h,k/p}^{p}} \qquad \text{for } p \mid k. \tag{2.62}$$

3. We now introduce the power series

$$F_\rho(x) = \sum_{n=0}^{\infty} a_n^{(\rho)} x^n \tag{3.11}$$

and

$$\tilde{F}_\rho(x) = \sum_{n=0}^{\infty} \tilde{a}_n^{(\rho)} x^n \tag{3.12}$$

for the functions (2.1) and (2.53). From the definitions we have, at once,

$$a_0^{(\rho)} = 1, \qquad \tilde{a}_0^{(\rho)} = 1, \qquad \tilde{a}_1^{(\rho)} = -\rho, \qquad \tilde{a}_2^{(\rho)} = \tfrac{1}{2}\rho(\rho - 3), \tag{3.13}$$

providing $p > 2$.

Our next purpose is to determine the coefficients $a_n^{(\rho)}$ of (3.11). This is done by our method quoted above. As all the steps are quite analogous we shall omit the details and merely display the essential stages. We have

$$a_n^{(\rho)} = \frac{1}{2\pi i} \int \frac{F_\rho(x)}{x^{n+1}}\, dx, \qquad |x| = \exp\left[-\frac{2\pi}{N^2}\right],$$

$$= \sum_{h,k \le N}' \frac{1}{2\pi i} \int_{\xi_{h,k}} \frac{F_\rho(x)}{x^{n+1}}\, dx,$$

where $\xi_{h,k}$ is the Farey arc corresponding to the value h/k. We designate the end-points of this arc, measured by the angle φ from $2\pi h/k$, as $-\vartheta'_{h,k}$ and $\vartheta''_{h,k}$. The symbol $\sum'_{h,k \le N}$ designates the sum over all h and k such that $0 \le h < k \le N$ and $(h, k) = 1$.

Writing $w = N^{-2} - i\varphi$ we have

$$a_n^{(\rho)} = \sum_{h,k \le N}' e^{-\frac{2\pi i h n}{k}} \int_{-\vartheta'_{h,k}}^{\vartheta''_{h,k}} F_\rho\left(\exp\left[\frac{2\pi i h}{k} - 2\pi w\right]\right) e^{2\pi n w}\, d\varphi.$$

In the sum over k we now apply (2.51) or (2.61) according as $p \nmid k$ or $p \mid k$ and have, using the notation of (3.11) and (3.12),

$$a_n^{(\rho)} = p^{-\rho/2} \sum_{\substack{k \le N \\ p \nmid k}} k^{\frac{1}{2}} \sum_{h \bmod k}' \Omega_{h,k}^{(\rho)} e^{-\frac{2\pi i h n}{k}} \int_{-\vartheta'_{h,k}}^{\vartheta''_{h,k}} w^{\frac{1}{2}} \exp\left[\frac{2\pi}{k^2 p w} \frac{(p-1)\rho + p}{24}\right.$$

$$\left. + 2\pi w \left(\frac{\rho(p-1) - 1}{24} + n\right)\right] \sum_{m=0}^{\infty} \tilde{a}_m^{(\rho)} \exp\left[m\left(\frac{2\pi i h''}{k} - \frac{2\pi}{k^2 p w}\right)\right] d\varphi$$

$$+ \sum_{\substack{k \le N \\ p \mid k}} k^{\frac{1}{2}} \sum_{h \bmod k}' \Omega_{h,k}^{(\rho)} e^{-\frac{2\pi i h n}{k}} \int_{-\vartheta'_{h,k}}^{\vartheta''_{h,k}} w^{\frac{1}{2}} \exp\left[-\frac{2\pi}{k^2 w} \frac{\rho(p-1) - 1}{24}\right.$$

$$\left. + 2\pi w \left(\frac{\rho(p-1) - 1}{24} + n\right)\right] \sum_{m=0}^{\infty} a_m^{(\rho)} \exp\left[m\left(\frac{2\pi i h'}{k} - \frac{2\pi}{k^2 w}\right)\right] d\varphi.$$

We can now follow exactly the reasoning of sections 3, 4, and 5 of our previous paper. The factor $w^{\frac{1}{2}}$ above shows that, corresponding to the dimension r, we are to take the positive value $\frac{1}{2}$. From section 3 we see that only terms in which a factor $e^{A/w}$ with $A > 0$ appears will furnish a contribution to the final result when we let N tend to infinity. This means, in particular, that the terms corresponding to $p \mid k$ will drop out. The condition (4.1) of that paper becomes $(\rho(p - 1) - 1)/24 + n > 0$ and hence is satisfied for $n \geqq 0$. Thus, in analogy to (5.5), we find the result

$$a_n^{(\rho)} = \frac{2\pi}{p^{\rho/2}} \sum_{m=0}^{\mu} \tilde{a}_m^{(\rho)} \sum_{\substack{k=1 \\ p \nmid k}}^{\infty} \frac{1}{k} \sum_{h \bmod k}{}' \Omega_{h,k}^{(\rho)} e^{-\frac{2\pi i}{k}(hn - h''m)} \left(\frac{\dfrac{(p-1)\rho + p}{24p} - \dfrac{m}{p}}{\dfrac{\rho(p-1) - 1}{24} + n} \right)^{\frac{3}{4}} \tag{3.21}$$
$$\times I_{3/2}\left(\frac{4\pi}{k} \sqrt{\frac{(p-1)\rho + p - 24m}{24p} \left(\frac{\rho(p-1) - 1}{24} + n \right)} \right),$$

where μ is the greatest integer m for which

$$(p - 1)\rho + p - 24m > 0. \tag{3.22}$$

The values of the coefficients of the left members of (1.11) and (1.12), with which (3.21) is to be compared, are[8]

$$p(pn + \lambda) = 2\pi \sum_{k=1}^{\infty} \frac{1}{k} \sum_{h \bmod k}{}' \omega_{h,k} e^{-\frac{2\pi i h}{k}(pn+\lambda)} \left(\frac{\dfrac{1}{24}}{pn + \lambda - \dfrac{1}{24}} \right)^{\frac{3}{4}} \tag{3.3}$$
$$\times I_{3/2}\left(\frac{4\pi}{k} \sqrt{\frac{1}{24}\left(pn + \lambda - \frac{1}{24} \right)} \right).$$

4. At this point we separate the two cases $p = 5$ and $p = 7$. The first identity, (1.11), can be written as

$$\sum_{n=0}^{\infty} p(5n + 4)\, x^n = 5\, F_5(x) = 5 \sum_{n=0}^{\infty} a_n^{(5)} x^n, \tag{4.11}$$

and hence will be proved if we show

$$p(5n + 4) = 5a_n^{(5)}. \tag{4.12}$$

In this case we have

$$p = 5, \qquad \lambda = 4, \qquad \rho = 5, \tag{4.13}$$

and hence, by (3.22),

$$\mu = 1. \tag{4.14}$$

[8] We use the form in which the series for $p(n)$ is given on pp. 455–456 of our paper quoted in footnote 5.

Also from (3.13) we have

$$\tilde{a}_0^{(5)} = 1, \qquad \tilde{a}_1^{(5)} = -5. \tag{4.15}$$

Setting these values in (3.21) we have

$$\begin{aligned} 5a_n^{(5)} &= \frac{2\pi}{5^{3/2}} \sum_{\substack{k=1\\5\nmid k}}^{\infty} \frac{1}{k} \sum_{h \bmod k}{}' \Omega_{h,k}^{(5)} e^{-\frac{2\pi i h n}{k}} \left(\frac{\frac{5}{24}}{n+\frac{19}{24}}\right)^{\frac{3}{4}} I_{3/2}\left(\frac{4\pi}{k}\sqrt{\frac{5}{24}\left(n+\frac{19}{24}\right)}\right) \\ &\quad - \frac{2\pi}{5^{\frac{3}{4}}} \sum_{\substack{k=1\\5\nmid k}}^{\infty} \frac{1}{k} \sum_{h \bmod k}{}' \Omega_{h,k}^{(5)} e^{-\frac{2\pi i}{k}(hn-h'')} \left(\frac{\frac{1}{120}}{n+\frac{19}{24}}\right)^{\frac{3}{4}} I_{3/2}\left(\frac{4\pi}{5k}\sqrt{\frac{5}{24}\left(n+\frac{19}{24}\right)}\right) \end{aligned} \tag{4.21}$$

which is to be compared with

$$\begin{aligned} p(5n+4) &= 2\pi \sum_{k=1}^{\infty} \frac{1}{k} \sum_{h \bmod k}{}' \omega_{h,k} e^{-\frac{2\pi i h}{k}(5n+4)} \left(\frac{\frac{1}{24}}{5\left(n+\frac{19}{24}\right)}\right)^{\frac{3}{4}} \\ &\quad \times I_{3/2}\left(\frac{4\pi}{k}\sqrt{\frac{5}{24}\left(n+\frac{19}{24}\right)}\right), \end{aligned} \tag{4.22}$$

which we have by (3.3).

Comparing terms in (4.21) and (4.22) which have the same arguments in the Bessel functions, we see that our proof of (4.12), and hence of (1.11), will be accomplished as soon as we have shown the following three facts:

I. $$\sum_{h \bmod k}{}' \Omega_{h,k}^{(5)} e^{-\frac{2\pi i h n}{k}} = \sum_{h \bmod k} \omega_{h,k} e^{-\frac{2\pi i h}{k}(5n+4)}, \qquad 5 \nmid k, \qquad p = 5;$$

II. $$-\frac{1}{5^{\frac{3}{4}}} \sum_{h \bmod k}{}' \Omega_{h,k}^{(5)} e^{-\frac{2\pi i}{k}(hn-h'')} = \frac{1}{5} \sum_{h \bmod 5k}{}' \omega_{h,5k} e^{-\frac{2\pi i h}{5k}(5n+4)}, \qquad 5 \nmid k, \qquad p = 5;$$

III. $$\sum_{h \bmod k}{}' \omega_{h,k} e^{-\frac{2\pi i h}{k}(5n+4)} = 0 \qquad \text{for} \qquad 5^2 \mid k.$$

It is to be remembered that $\Omega_{h,k}^{(\rho)}$ depends on the value of p and that these equalities are asserted only for $p = 5$.

The second identity (1.12) is treated in a similar manner. We write it as

$$\sum_{n=0}^{\infty} p(7n+5)x^n = 7F_3(x) + 49xF_7(x) = 7a_0^{(3)} + \sum_{n=1}^{\infty} (7a_n^{(3)} + 49a_{n-1}^{(7)})x^n. \tag{4.31}$$

By (3.13) we have $a_0^{(3)} = 1$ and hence $p(5) = 7 = 7a_0^{(3)}$ and we need only show

$$p(7n+5) = 7a_n^{(3)} + 49a_{n-1}^{(7)}, \qquad n \geqq 1. \tag{4.32}$$

For $F_3(x)$ we have

$$p = 7, \quad \rho = 3, \quad \mu = 1, \quad \tilde{a}_0^{(3)} = 1, \quad \tilde{a}_1^{(3)} = -3,$$

and hence, by (3.21),

$$7a_n^{(3)} = \frac{2\pi}{7^{\frac{1}{2}}} \sum_{\substack{k=1 \\ 7\nmid k}}^{\infty} \frac{1}{k} \sum_{h \bmod k}{}' \Omega_{h,k}^{(3)} e^{-\frac{2\pi i}{k} hn} \left(\frac{\frac{25}{24}}{7\left(n + \frac{17}{24}\right)} \right)^{\frac{3}{4}} I_{3/2}\left(\frac{20\pi}{k} \sqrt{\frac{1}{7\cdot 24}\left(n + \frac{17}{24}\right)} \right)$$

(4.41)

$$- \frac{6\pi}{7^{\frac{1}{2}}} \sum_{\substack{k=1 \\ 7\nmid k}}^{\infty} \frac{1}{k} \sum_{h \bmod k}{}' \Omega_{h,k}^{(3)} e^{-\frac{2\pi i}{k}(hn - h'')} \left(\frac{\frac{1}{24}}{7\left(n + \frac{17}{24}\right)} \right)^{\frac{3}{4}} I_{3/2}\left(\frac{4\pi}{k} \sqrt{\frac{1}{7\cdot 24}\left(n + \frac{17}{24}\right)} \right).$$

For $F_7(x)$ we have

$$p = 7, \quad \rho = 7, \quad \mu = 2, \quad \tilde{a}_0^{(7)} = 1, \quad \tilde{a}_1^{(7)} = -7, \quad \tilde{a}_2^{(7)} = 14,$$

and then we find

$$49a_{n-1}^{(7)} = \frac{2\pi}{7^{3/2}} \sum_{\substack{k=1 \\ 7\nmid k}}^{\infty} \frac{1}{k} \sum_{h \bmod k}{}' \Omega_{h,k}^{(7)} e^{-\frac{2\pi i}{k} h(n-1)} \left(\frac{\frac{49}{24}}{7\left(n + \frac{17}{24}\right)} \right)^{\frac{3}{4}}$$
$$\times I_{3/2}\left(\frac{28\pi}{k} \sqrt{\frac{1}{7\cdot 24}\left(n + \frac{17}{24}\right)} \right)$$

(4.42)

$$- \frac{2\pi}{7^{\frac{1}{2}}} \sum_{\substack{k=1 \\ 7\nmid k}}^{\infty} \frac{1}{k} \sum_{h \bmod k}{}' \Omega_{h,k}^{(7)} e^{-\frac{2\pi i}{k}(h(n-1) - h'')} \left(\frac{\frac{25}{24}}{7\left(n + \frac{17}{24}\right)} \right)^{\frac{3}{4}}$$
$$\times I_{3/2}\left(\frac{20\pi}{k} \sqrt{\frac{1}{7\cdot 24}\left(n + \frac{17}{24}\right)} \right)$$

$$+ \frac{4\pi}{7^{\frac{1}{2}}} \sum_{\substack{k=1 \\ 7\nmid k}}^{\infty} \frac{1}{k} \sum_{h \bmod k}{}' \Omega_{h,k}^{(7)} e^{-\frac{2\pi i}{k}(h(n-1) - 2h'')} \left(\frac{\frac{1}{24}}{7\left(n + \frac{17}{24}\right)} \right)^{\frac{3}{4}}$$
$$\times I_{3/2}\left(\frac{4\pi}{k} \sqrt{\frac{1}{7\cdot 24}\left(n + \frac{17}{24}\right)} \right).$$

On the other side of (4.32) we have

(4.43)

$$p(7n + 5) = 2\pi \sum_{k=1}^{\infty} \frac{1}{k} \sum_{h \bmod k}{}' \omega_{h,k} e^{-\frac{2\pi i h}{k}(7n+5)} \left(\frac{\frac{1}{24}}{7\left(n + \frac{17}{24}\right)} \right)^{\frac{3}{4}}$$
$$\times I_{3/2}\left(\frac{28\pi}{k} \sqrt{\frac{1}{7\cdot 24}\left(n + \frac{17}{24}\right)} \right).$$

If we now set (4.41), (4.42) and (4.43) in (4.32) and again compare terms having the same arguments in the Bessel functions we see that our proof of (4.32), and hence of (1.12), is now reduced to show the four facts:

$$\text{I} \qquad \sum_{h \bmod k}{}' \Omega_{h,k}^{(7)} e^{-\frac{2\pi i}{k}h(n-1)} = \sum_{k \bmod k}{}' \omega_{h,k} e^{-\frac{2\pi i h}{k}(7n+5)}, \quad 7 \nmid k, p = 7;$$

$$\text{II}' \qquad -\frac{3}{7^{\frac{1}{2}}} \sum_{h \bmod k}{}' \Omega_{h,k}^{(3)} e^{-\frac{2\pi i}{k}(hn-h'')} + \frac{2}{7^{\frac{1}{2}}} \sum_{h \bmod k}{}' \Omega_{h,k}^{(7)} e^{-\frac{2\pi i}{k}(h(n-1)-2h'')}$$

$$= \frac{1}{7} \sum_{h \bmod 7k}{}' \omega_{h,7k} e^{-\frac{2\pi i h}{7k}(7n+5)}, \qquad 7 \nmid k, \qquad p = 7;$$

$$\text{III}' \qquad \sum_{h \bmod k}{}' \omega_{h,k} e^{-\frac{2\pi i h}{k}(7n+5)} = 0 \quad \text{for} \quad 7^2 | k;$$

$$\text{IV}' \qquad \sum_{h \bmod k}{}' \Omega_{h,k}^{(3)} e^{-\frac{2\pi i}{k}hn} - \sum_{h \bmod k}{}' \Omega_{h,k}^{(7)} e^{-\frac{2\pi i}{k}(h(n-1)-h'')} = 0, \quad 7 \nmid k, \quad p = 7.$$

5. For the proof of the relations stated in section 4 we shall need the two following lemmas. These lemmas are based on some recent results of D. H. Lehmer.[9] It is convenient to use the notation

$$(5.1) \qquad A_k(n) = \sum_{h \bmod k}{}' \omega_{h,k} e^{-\frac{2\pi i h}{k}n}$$

which was introduced by Hardy and Ramanujan[10] and which is in agreement with the notation of Lehmer's paper.

Lemma 1: *Let p be a prime, $p > 3$, $p \nmid k$ and let*

$$(5.21) \qquad 24\lambda \equiv 1 \pmod{p}.$$

Then

$$(5.22) \qquad A_{pk}(pn + \lambda) = A_p(\lambda) A_k(n_1)$$

where n_1 is any solution of the congruence

$$(5.23) \qquad p^2 n_1 \equiv pn + (p^2 - 1 + 24\lambda)/24 \pmod{k}.$$

It is to be noted that the coefficients in (5.23) *are integers and have the common factor p.*

Proof: It is obvious from (5.1) that $A_k(n)$ is a periodic function of n with the period k. Hence in the course of the proof we need determine the value of the argument only modulo the subscript. We have, with Lehmer, to distinguish three cases according to the divisibility of k by powers of 2.

---

[9] *On the series for the partition function*, Transactions of the American Mathematical Society, vol. 43, no. 2 (1938), pp. 271–295.

[10] Loc. cit. footnote 7, especially p. 85.

A) If k is odd we have, by Lehmer's Theorem 1,

$$A_{pk}(pn+\lambda) = A_k(n_1)A_p(n_2) \tag{5.31}$$

with

$$pn+\lambda \equiv p^2n_1 + k^2n_2 - (p^2+k^2-1)/24 \pmod{pk}. \tag{5.32}$$

Hence we have

$$24\lambda \equiv 24k^2n_2 - k^2 + 1 \pmod{p},$$

$$24\lambda - 1 \equiv k^2(24n_2 - 1) \pmod{p}$$

and then, by (5.21),

$$\begin{aligned} 24n_2 - 1 &\equiv 0 \pmod{p} \\ n_2 &\equiv \lambda \pmod{p}. \end{aligned} \tag{5.33}$$

From (5.32) we also have

$$pn+\lambda \equiv p^2n_1 - (p^2+k^2-1)/24 \pmod{k}. \tag{5.34}$$

If $3 \nmid k$ this yields

$$24(pn+\lambda) \equiv 24p^2n_1 - p^2 + 1 \pmod{k},$$

$$pn+\lambda \equiv p^2n_1 - (p^2-1)/24 \pmod{k} \tag{5.35}$$

while if $3 \mid k$ we have

$$8(pn+\lambda) \equiv 8p^2n_1 - (p^2-1)/3 \pmod{k}$$

$$pn+\lambda \equiv p^2n_1 - (p^2-1)/24 \pmod{k}. \tag{5.36}$$

The formulas (5.33), (5.35) and (5.36) complete the proof for k odd.

B) If $k = 2^\alpha k_1$, $\alpha > 1$, k_1 odd, we use Lehmer's Theorems 1 and 2 and find the following relations:

$$A_{pk}(pn+\lambda) = (-1)^{2^{\alpha-2}} A_{2^\alpha}(n_2)A_{pk_1}(n_3) \tag{5.41}$$

with

$$pn+\lambda \equiv p^2k_1^2n_2 + 2^{2\alpha}n_3 - (p^2k_1^2 - 1 + 2^{2\alpha})/24 \pmod{2^\alpha pk_1}, \tag{5.42}$$

$$A_{pk_1}(n_3) = A_p(n_4)A_{k_1}(n_5) \tag{5.43}$$

with

$$n_3 \equiv p^2n_5 + k_1^2n_4 - (p^2+k_1^2-1)/24 \pmod{pk_1}, \tag{5.44}$$

and

$$A_k(n_1) = A_{2^\alpha k_1}(n_1) = (-1)^{2^{\alpha-2}} A_{k_1}(n_5)A_{2^\alpha}(n_2) \tag{5.45}$$

with

$$n_1 \equiv k_1^2n_2 + 2^{2\alpha}n_5 - (k_1^2 - 1 + 2^{2\alpha})/24 \pmod{2^\alpha k_1}. \tag{5.46}$$

From (5.41), (5.43) and (5.45) we have

$$A_{pk}(pn + \lambda) = A_p(n_4)A_k(n_1) \tag{5.47}$$

while (5.42) and (5.44) yield

$$24\lambda - 1 \equiv 2^{2\alpha}(24n_3 - 1) \pmod{p}$$

and

$$24n_3 - 1 \equiv k_1^2(24n_4 - 1) \pmod{p}$$

and hence

$$2^{2\alpha}k_1^2(24n_4 - 1) \equiv 24\lambda - 1 \equiv 0 \pmod{p}$$

or

$$n_4 \equiv \lambda \pmod{p} \tag{5.48}$$

in virtue of (5.21). From (5.42) and (5.46) we get

$$pn + \lambda \equiv p^2k_1^2n_2 - (p^2k_1^2 - 1 + 2^{2\alpha})/24 \pmod{2^\alpha}$$

and

$$n_1 \equiv k_1^2n_2 - (k_1^2 - 1 + 2^{2\alpha})/24 \pmod{2^\alpha}$$

which, on elimination of n_2, yields

$$\begin{aligned} p^2n_1 &\equiv pn + \lambda + (1 - 2^{2\alpha})(p^2 - 1)/24 \\ &\equiv pn + \lambda + (p^2 - 1)/24 \pmod{2^\alpha}. \end{aligned} \tag{5.49}$$

Finally we use (5.42), (5.44) and (5.46) to obtain

$$pn + \lambda \equiv 2^{2\alpha}n_3 - (2^{2\alpha} - 1)/24 \pmod{k_1},$$

$$n_3 \equiv p^2n_5 - (p^2 - 1)/24 \pmod{k_1},$$

$$n_1 \equiv 2^{2\alpha}n_5 - (2^{2\alpha} - 1)/24 \pmod{k_1}.$$

Eliminating n_3 and n_5 we get

$$\begin{aligned} pn + \lambda &\equiv 2^{2\alpha}p^2n_5 - (2^{2\alpha}p^2 - 1)/24 \pmod{k_1} \\ &\equiv p^2n_1 - (p^2 - 1)/24 \pmod{k_1} \end{aligned}$$

or

$$p^2n_1 \equiv pn + \lambda + (p^2 - 1)/24 \pmod{k_1}$$

which, together with (5.49) yields (5.23). This with (5.47) and (5.48) completes the proof for this case.

C) If $k = 2k_1$, k_1 odd, we use Lehmer's Theorems 1 and 4 to obtain

$$A_{pk}(pn + \lambda) = A_{2pk_1}(pn + \lambda) = A_2(n_2)A_{pk_1}(n_3)$$

with

$$pn + \lambda \equiv 4n_3 + pk_1n_2 + (p^2k_1^2 - 1)/8$$

$$\equiv 4n_3 + p^2k_1^2n_2 + (p^2k_1^2 - 1)/8 \quad (\text{mod } 2pk_1),$$

$$A_{pk_1}(n_3) = A_p(n_4)A_{k_1}(n_5)$$

with

$$n_3 \equiv p^2n_5 + k_1^2n_4 - (p^2 + k_1^2 - 1)/24 \quad (\text{mod } pk_1),$$

and

$$A_k(n_1) = A_{2k_1}(n_1) = A_2(n_2)A_{k_1}(n_5)$$

with

$$n_1 \equiv 4n_5 + k_1n_2 + (k_1^2 - 1)/8 \equiv 4n_5 + k_1^2n_2 + (k_1^2 - 1)/8 \quad (\text{mod } 2k_1).$$

These formulas can be combined in a manner similar to that used in case B thus completing the proof of the lemma.

LEMMA 2: *Under the conditions of Lemma* 1 *and with* $\beta > 1$ *we have*

$$A_{p^\beta k}(pn + \lambda) = 0. \tag{5.51}$$

PROOF: We again treat three cases as in the proof of Lemma 1.

A) If k is odd we use Lehmer's Theorem 1 to obtain

$$A_{p^\beta k}(pn + \lambda) = A_k(n_1)A_{p^\beta}(n_2)$$

with

$$pn + \lambda \equiv p^{2\beta}n_1 + k^2n_2 - (p^{2\beta} + k^2 - 1)/24 \quad (\text{mod } p^\beta k),$$

from which we get

$$\lambda \equiv k^2n_2 - (k^2 - 1)/24 \quad (\text{mod } p),$$

$$24\lambda - 1 \equiv k^2(24n_2 - 1) \quad (\text{mod } p),$$

$$24n_2 - 1 \equiv 0 \quad (\text{mod } p). \tag{5.52}$$

Now, however, (5.52) implies

$$A_{p^\beta}(n_2) = 0$$

by Lehmer's Theorem 5 and hence our proof is complete for this case.

B) If $k = 2^\alpha k_1$, $\alpha > 1$, k_1 odd, we have, using Lehmer's Theorem 2,

$$A_{p^\beta k}(pn + \lambda) = A_{2^\alpha p^\beta k_1}(pn + \lambda) = (-1)^{2^{\alpha-2}}A_{2^\alpha}(n_2)A_{p^\beta k_1}(n_3)$$

with

$$pn + \lambda \equiv p^{2\beta}k_1^2n_2 + 2^{2\alpha}n_3 - (p^{2\beta}k_1^2 + 2^{2\alpha} - 1)/24 \; (\text{mod } 2^\alpha p^\beta k_1)$$

and hence

$$24\lambda - 1 \equiv 2^{2\alpha}(24n_3 - 1) \pmod{p},$$

$$24n_3 - 1 \equiv 0 \pmod{p},$$

$$n_3 \equiv \lambda \pmod{p}.$$

Therefore we have

$$A_{p^\beta k_1}(n_3) = 0$$

by case A which we have just proved.

C) If $k = 2k_1$, k_1 odd, Lehmer's Theorem 4 can be used to reduce this case to case A just as before.

6. We turn now to the proof of the relations asserted in section 4. We first consider those having to do with the modulus 5.

We have[11] for odd k:

$$\omega_{h,k} = \left(\frac{-h}{k}\right) \exp\left\{-\pi i\left(\frac{1}{4}(k-1) + \frac{1}{12}\left(k - \frac{1}{k}\right)(2h - h' + h^2h')\right)\right\} \tag{6.11}$$

and for odd h:

$$\omega_{h,k} = \left(\frac{-k}{h}\right) \exp\left\{-\pi i\left(\frac{1}{4}(2 - hk - h) + \frac{1}{12}\left(k - \frac{1}{k}\right)(2h - h' + h^2h')\right)\right\}, \tag{6.12}$$

and therefore by (2.52) for odd k, $5 \nmid k$, we have

$$\Omega_{h,k}^{(5)} = \frac{\omega_{h,k}^6}{\omega_{5h,k}^5} = \left(\frac{-5h}{k}\right) \exp\left\{-\pi i\left(\frac{1}{4}(k-1) + \frac{1}{12}\left(k - \frac{1}{k}\right)(12h - 6h' + 6h^2h' - 50h + 5h'' - 125h^2h'')\right)\right\}$$

with

$$5hh'' \equiv -1 \pmod{k}. \tag{6.21}$$

From this congruence we have

$$h' \equiv 5h'' \pmod{k} \tag{6.22}$$

and hence we may write

$$\Omega_{h,k}^{(5)} = \left(\frac{-5h}{k}\right) \exp\left\{-\pi i\left(\frac{1}{4}(k-1) + \frac{1}{12}\left(k - \frac{1}{k}\right)(-38h - 25h'' - 95h^2h'')\right)\right\}. \tag{6.23}$$

[11] Loc. cit. footnote 7, especially p. 85.

If we now replace h by $5h$ we see from the definition (6.21) that we must replace $5h''$ by h'' at the same time. Doing this we get, from (6.23),

$$\Omega^{(5)}_{5h,k} = \left(\frac{-h}{k}\right) \exp\left\{-\pi i\left(\frac{1}{4}(k-1) + \frac{1}{12}\left(k - \frac{1}{k}\right)(-190h - 5h'' - 475h^2h'')\right)\right\}$$

$$= \left(\frac{-h}{k}\right) \exp\left\{-\pi i\left(\frac{1}{4}(k-1) + \frac{1}{12}\left(k - \frac{1}{k}\right)(-94h - h' + h^2h' - 96h(hh'+1))\right)\right\}$$

where we have made use of (6.22). Since, by definition, $hh' + 1 \equiv 0 \pmod{k}$, this equation goes over into the form

$$\Omega^{(5)}_{5h,k} = \left(\frac{-h}{k}\right) \exp\left\{-\pi i\left(\frac{1}{4}(k-1) + \frac{1}{12}\left(k - \frac{1}{k}\right)(2h - h' + h^2h')\right)\right\} e^{-\frac{\pi i 8h}{k}}$$

$$= \omega_{h,k} e^{-\frac{\pi i 8h}{k}}.$$

Now $5h$ runs, with h, through a reduced set of residues modulo k since $5 \nmid k$ and hence we have

$$\sum_{h \bmod k}{}' \Omega^{(5)}_{h,k} e^{-\frac{2\pi i h n}{k}} = \sum_{h \bmod k}{}' \Omega^{(5)}_{5h,k} e^{-\frac{2\pi i 5hn}{k}}$$

$$= \sum_{h \bmod k}{}' \omega_{h,k} e^{-\frac{2\pi i h}{k}(5n+4)}$$

which proves I for odd k.

For the case k even we use (6.12) and find, as before,

$$\Omega^{(5)}_{h,k} = \left(\frac{-k}{5h}\right) \exp\left\{-\pi i\left(\frac{1}{4}(2 + 19hk + 19h) + \frac{1}{12}\left(k - \frac{1}{k}\right)(-38h - 25h'' - 95h^2h'')\right)\right\}, \tag{6.3}$$

$$\Omega^{(5)}_{5h,k} = \left(\frac{-k}{h}\right) \exp\left\{-\pi i\left(\frac{1}{4}(2 + 95hk + 95h) + \frac{1}{12}\left(k - \frac{1}{k}\right)(-190h - 5h'' - 475h^2h'')\right)\right\}$$

$$= \left(\frac{-k}{h}\right) \exp\left\{-\pi i\left(\frac{1}{4}(2 - hk - h) + \frac{1}{12}\left(k - \frac{1}{k}\right)(2h - h' + h^2h')\right)\right\} e^{-\frac{\pi i 8h}{k}}$$

$$= \omega_{h,k} e^{-\frac{\pi i 8h}{k}}.$$

The remainder of the proof is the same as in the previous case. Thus I is proved also for even k.

7. For the proof of II we first use Lemma 1. With $p = 5$ we have

$$\lambda = 4,$$
$$5^2 n_1 \equiv 5n + 5 \pmod{k},$$
$$n_1 \equiv 5^*(n+1) \pmod{k},$$

where 5^* is the reciprocal of 5 modulo k, that is

$$5 \cdot 5^* \equiv 1 \pmod{k}.$$

Then for $5 \nmid k$ we have

$$A_{5k}(5n+4) = A_5(4)A_k(5^*(n+1)).$$

From Lehmer's Theorem 5 we find the value

$$A_5(4) = \left(\frac{3}{5}\right) 5^{\frac{1}{2}} = -5^{\frac{1}{2}}.$$

We can therefore transform the relation II into

$$\sum_{h \bmod k}' \Omega_{h,k}^{(5)} e^{-\frac{2\pi i}{k}(hn - h'')} = \sum_{h \bmod k}' \omega_{h,k} e^{-\frac{2\pi i h}{k} 5^*(n+1)}$$

or

$$\sum_{h \bmod k}' \Omega_{h,k}^{(5)} e^{-\frac{2\pi i}{k}(hn - h'')} = \sum_{h \bmod k}' \omega_{5h,k} e^{-\frac{2\pi i h}{k}(n+1)}. \tag{7.1}$$

Now, for odd k, we take the value of $\Omega_{h,k}^{(5)}$ from (6.23) and the value of $\omega_{5h,k}$ from (6.11) and find

$$\frac{\Omega_{h,k}^{(5)} e^{\frac{2\pi i h''}{k}}}{\omega_{5h,k} e^{-\frac{2\pi i h}{k}}} = \exp\left\{-\pi i\left(\frac{1}{12}\left(k - \frac{1}{k}\right)(-48h - 24h'' - 120h^2h'') - \frac{2h'' + 2h}{k}\right)\right\}$$
$$= \exp\left\{-\frac{2\pi i}{k} h(1 + 5hh'')\right\} = 1,$$

where the last equality is a consequence of the definition $5hh'' \equiv -1 \pmod{k}$. Thus we have

$$\Omega_{h,k}^{(5)} e^{-\frac{2\pi i}{k}(hn - h'')} = \omega_{5h,k} e^{-\frac{2\pi i h}{k}(n+1)},$$

which proves (7.1) for odd k.

For even k we use (6.3) and (6.12) and find again

$$\frac{\Omega_{h,k}^{(5)} e^{\frac{2\pi i h''}{k}}}{\omega_{5h,k} e^{-\frac{2\pi i h}{k}}} = \exp\left\{-\pi i\left(\frac{1}{4}(24hk + 24h)\right.\right.$$
$$\left.\left. + \frac{1}{12}\left(k - \frac{1}{k}\right)(-48h - 24h'' - 120h^2h'') - \frac{2h'' + 2h}{k}\right)\right\} = 1$$

which completes the proof of (7.1).

8. The relation III can be written

$$A_{5^\beta k_1}(5n+4)=0$$

with $\beta>1$, $5\nmid k$ and is, as such, an immediate consequence of Lemma 2.

Thus we have proved the three relations for the modulus 5 and hence have proved Ramanujan's identity (1.11).

9. We now turn to the proof of the four relations concerning the modulus 7. The proof of I′ is analogous to that for I so we shall omit all detail and merely state the results. In this case we have $7\nmid k$ and

$$7hh''\equiv -1,\qquad 7h''\equiv h'\qquad (\text{mod } k)$$

and we obtain

$$\Omega_{h,k}^{(7)}=\left(\frac{-7h}{k}\right)\exp\left\{-\pi i\left(\frac{1}{4}(k-1)+\frac{1}{12}\left(k-\frac{1}{k}\right)(-82h-49h''-287\,h^2h'')\right)\right\}$$

for odd k, and

$$\Omega_{h,k}^{(7)}=\left(\frac{-k}{7h}\right)\exp\left\{-\pi i\left(\frac{1}{4}(2+41hk+41h)\right.\right.$$
$$\left.\left.+\frac{1}{12}\left(k-\frac{1}{h}\right)(-82h-49h''-287h^2h'')\right)\right\}$$

for even k. In both cases we get

$$\Omega_{7h,k}^{(7)}=\omega_{h,k}e^{-\frac{24\pi ih}{k}}$$

from which we have

$$\sum_{h \bmod k}{}' \Omega_{h,k}^{(7)}e^{-\frac{2\pi ih(n-1)}{k}}=\sum_{h \bmod k}{}' \Omega_{7h,k}^{(7)}e^{-\frac{14\pi ih(n-1)}{k}}$$
$$=\sum_{h \bmod k}{}' \omega_{h,k}e^{-\frac{2\pi ih}{k}(7n+5)}$$

which proves I′.

10. At this point it is convenient to prove the relation IV′. From the definition (2.52) we have, for $7\nmid k$,

$$\Omega_{h,k}^{(3)}=\frac{\omega_{h,k}^{4}}{\omega_{7h,k}^{3}},\qquad \Omega_{h,k}^{(7)}=\frac{\omega_{h,k}^{8}}{\omega_{7h,k}^{7}}$$

and hence

$$\frac{\Omega_{h,k}^{(3)}}{\Omega_{h,k}^{(7)}e^{\frac{2\pi i(h+h'')}{k}}}=\frac{\omega_{7h,k}^{4}}{\omega_{h,k}^{4}}e^{-\frac{2\pi i(h+h'')}{k}}$$

For odd k we use (6.11) to obtain

$$\frac{\omega_{7h,k}^4}{\omega_{h,k}^4} e^{-\frac{2\pi i(h+h'')}{k}}$$

$$= \exp\left\{-\pi i\left(\frac{1}{3}\left(k-\frac{1}{k}\right)(12h+6h''+42h^2h'') + \frac{2(h+h'')}{k}\right)\right\} \tag{10.11}$$

$$= \exp\left\{\frac{2\pi ih}{k}(1+7hh'')\right\} = 1.$$

For even k we use (6.12) and find

$$\frac{\omega_{7h,k}^4}{\omega_{h,k}^4} e^{-\frac{2\pi i(h+h'')}{k}} = \exp\left\{-\pi i\left(-6hk-6h + \frac{1}{3}\left(k-\frac{1}{k}\right)(12h+6h''+42h^2h'') + \frac{2(h+h'')}{k}\right)\right\} = 1 \tag{10.12}$$

as before. Hence, in both cases, we have

$$\Omega_{h,k}^{(3)} = \Omega_{h,k}^{(7)} e^{\frac{2\pi i(h+h'')}{k}} \tag{10.13}$$

which clearly proves IV′.

11. Equation (10.13) reduces II′ to

$$-\sum_{h \bmod k}{}' \Omega_{h,k}^{(3)} e^{-\frac{2\pi i}{k}(hn-h'')} = \frac{1}{\sqrt{7}} A_{7k}(7n+5) \tag{11.11}$$

where we have used the notation of (5.1). Now Lemma 1 yields the result

$$A_{7k}(7n+5) = A_7(5)A_k(7^*(n+1))$$

for $7 \nmid k$ and $7\cdot 7^* \equiv 1 \pmod{k}$. Also, by Lehmer's Theorem 5, we find

$$A_7(5) = \left(\frac{3}{7}\right)\sqrt{7} = -\sqrt{7}.$$

Therefore (11.11) can be written as

$$\sum_{h \bmod k}{}' \Omega_{h,k}^{(3)} e^{-\frac{2\pi i}{k}(hn-h'')} = A_k(7^*(n+1))$$

$$= \sum_{h \bmod k}{}' \omega_{h,k} e^{-\frac{2\pi ih}{k}7^*(n+1)}$$

or as

$$\sum_{h \bmod k}{}' \Omega_{h,k}^{(3)} e^{-\frac{2\pi i}{k}(hn-h'')} = \sum_{h \bmod k}{}' \omega_{7h,k} e^{-\frac{2\pi ih}{k}(n+1)}. \tag{11.12}$$

Now, however, by (2.52), (10.11) and (10.12) we have

$$\Omega_{h,k}^{(3)} e^{-\frac{2\pi i}{k}(hn-h'')} = \frac{\omega_{h,k}^{4}}{\omega_{7h,k}^{3}} e^{-\frac{2\pi i}{k}(hn-h'')}$$

$$= \omega_{7h,k} e^{-\frac{2\pi i h'}{k}(n+1)}$$

which proves (11.12) and hence II′.

12. The relation III′ is an immediate consequence of Lemma 2. This completes the proof of the four relations and hence proves Ramanujan's identity (1.12).

UNIVERSITY OF PENNSYLVANIA,
PHILADELPHIA, PA.

Notes to Paper 42

This paper, written jointly with H. S. Zuckerman, has been reviewed in the JF, vol. 65 (1939), p. 160, and in the Z, vol. 22 (1940), p. 8.

Page 473, footnote 5.
See paper 40 of the present collection.

Page 474, footnote 6, line 1.
See paper 37.

Page 474, footnote 6, line 2.
See paper 38.

Page 477, fourth display line.
It should cause no confusion that in an expression like $p(pn + \lambda)$, the first p stands for the partition function, while the second p stands for an arbitrary rational prime number.

Page 478, third and sixth display lines.
The missing denominator in the exponents is k.

FOURIER EXPANSIONS OF MODULAR FORMS AND PROBLEMS OF PARTITION[1]

HANS RADEMACHER

The subject which I am going to discuss in this lecture excels in the richness of its ramifications and in the diversity of its relations to other mathematical topics. I think therefore that it will serve our present purpose better not to attempt a systematic treatment, beginning with definitions and proceeding to lemmas, theorems, and proofs, but rather to look around and to envisage some outstanding marks scattered in various directions. I hope that the intrinsic relationships connecting the problems and theorems which I shall mention will nevertheless remain quite visible.

A good deal of the investigations about which I shall report can be subsumed under the heading of analytic number theory, and, more specifically, analytic additive number theory. It would, however, be a misplacement of emphasis if we were to look upon analysis, which here means function theory, only as a tool applied to the investigation of number theory. It is more the inner harmony of a system which I wish to depict, properties of functions revealing the nature of certain arithmetical facts, and properties of numbers having a bearing on the character of analytic functions.

Whereas the multiplicative number theory, which deals with questions of factorization, divisibility, prime numbers, and so on, goes back more than 2000 years to Euclid, the history of additive number theory, in any noteworthy sense, begins with Euler less than 200 years ago. In his famous treatise, *Introductio in Analysin Infinitorum* (1748), Euler devotes the sixteenth chapter, "De partitione numerorum," to problems of additive number theory. A "partition" is, after Euler, a decomposition of a natural number into summands which are natural numbers, for example, $6=1+1+4$. We can impose various restrictions on the summands; they may belong to a specified class of numbers, let us say odd numbers, or squares, or cubes, or primes; it may be required that they be all different; or their number may be preassigned. I wish to speak here only about unrestricted partitions. By the way, only the parts are essential, not their arrangement, so that we do not count two decompositions as different if they differ only in the order of the summands; we can therefore take the summands ordered according to their size.

[1] An address delivered before the Williamsburg meeting of the Society, December 29, 1938, by invitation of the Program Committee.

The first step in any characterization will be to count the partitions. Thus we have the following 11 partitions of 6:

$$6,\ \ 1+5,\ \ 2+4,\ 3+3,\ \ 1+1+4,\ \ 1+2+3,\ \ 2+2+2,$$
$$1+1+1+3,\ 1+1+2+2,\ 1+1+1+1+2,\ 1+1+1+1+1+1.$$

We write $p(6)=11$, and in general denote the number of partitions of n by $p(n)$.

As far as the method is concerned, Euler made the simple remark that, since we have $x^m \cdot x^n = x^{m+n}$, exponents of powers can easily be combined in an additive manner, and therefore products of power series can be used as "generating functions." In our case, for the unrestricted partitions, he found by simple reasoning the generating function

$$(1) \qquad f(x) = 1 + \sum_{n=1}^{\infty} p(n)x^n = \frac{1}{(1-x)(1-x^2)(1-x^3)\cdots}.$$

Indeed, in a partition of n we can first collect the equal summands and thereby express n as a sum of multiples of 1, 2, 3, $\cdots$,

$$(2) \qquad n = m_1 \cdot 1 + m_2 \cdot 2 + m_3 \cdot 3 + \cdots, \qquad m_j \geqq 0.$$

On the other side

$$\frac{1}{(1-x)(1-x^2)(1-x^3)\cdots} = \sum_{m_1=0}^{\infty} x^{m_1} \cdot \sum_{m_2=0}^{\infty} x^{m_2 \cdot 2} \cdot \sum_{m_3=0}^{\infty} x^{m_3 \cdot 3} \cdots,$$

and the power x^n therefore occurs as often in the product as n can be written in the form (2); but the number of solutions of the diophantine equation (2) is precisely $p(n)$.

Euler later investigated the infinite product

$$(3a) \qquad P(x) = (1-x)(1-x^2)(1-x^3)\cdots$$

appearing in the right member of (1). He found, first empirically[2] and later with conclusive proof,[3] that

$$(3b) \ P(x) = 1 - x - x^2 + x^5 + x^7 - - + + = \sum_{\lambda=-\infty}^{\infty} (-1)^{\lambda} x^{\lambda(3\lambda-1)/2}.$$

The discovery of the equality expressed by (3a) and (3b) marks a

[2] *Découverte d'une loi tout extraordinaire des nombres par rapport à la somme de leurs diviseurs*, *Opera Omnia*, (1), vol. 2, pp. 241–253.

[3] *Demonstratio theorematis circa ordinem in summis divisorum observatum* (1754–1755), *Opera Omnia*, (1), vol. 2, pp. 390–398).

highly important event in the history of our science. It is the first time that a ϑ-function, in a special case, appears in the literature; moreover it appears here immediately in its two aspects: as a power series with exponents formed by a *quadratic* expression of the index of summation, and secondly as an infinite product.[4] This equation leads to the identity

$$\left(1 + \sum_{n=1}^{\infty} p(n)x^n\right) \sum_{\lambda=-\infty}^{\infty} (-1)^\lambda x^{\lambda(3\lambda-1)/2} = 1$$

which furnishes a formula of recurrence for $p(n)$. Formulas of recurrence have been used, indeed, for the construction of tables of $p(n)$ for values of n up to 600 (MacMahon,[5] Gupta[6]). The function $p(n)$ is very rapidly increasing with n. Some specimens are

$$p(10) = 42, \quad p(100) = 1905\ 69292, \quad p(200) = 397\ 29990\ 29388,$$
$$p(600) = 4580\ 04788\ 00814\ 43085\ 53622.$$

Outside the range from 1 to 600 a few isolated values of $p(n)$ have been computed by D. H. Lehmer.[7] That, of course, could not be done by a formula of recurrence, but has been made possible by certain independent representations of $p(n)$, which belong to the main part of this report. Let me mention only that the largest known value of $p(n)$ is $p(14031)$, which turns out to be a figure of 127 digits.

In spite of the profound discoveries in the field of ϑ-functions and related functions due to Jacobi, Riemann, Klein, and Poincaré, the situation of our problem remained unchanged for more than one and a half centuries after Euler's investigations. It was not until 1917 that

[4] In a letter to Fuss, the first editor of Euler's unpublished works, Jacobi writes (1848), "Ich möchte mir bei dieser Gelegenheit noch erlauben, Ihnen zu sagen, warum ich mich so sehr für diese EULERsche Entdeckung interessiere. Sie ist nämlich der erste Fall gewesen, in welchem Reihen aufgetreten sind, deren Exponenten eine arithmetische Reihe *zweiter* Ordnung bilden, und auf diese Reihen ist durch mich die Theorie der elliptischen Transcendenten gegründet worden. Die EULERsche Formel ist ein spezieller Fall einer Formel, welche wohl das wichtigste und fruchtbarste ist, was ich in reiner Mathematik erfunden habe . . ." (quoted from Euler, *Opera Omnia*, (1), vol. 2, p. 192, footnote).

[5] This table was published by Hardy and Ramanujan in the paper referred to in footnote 8.

[6] *A table of partitions*, Proceedings of the London Mathematical Society, (2), vol. 39 (1935), pp. 142–149; *A table of partitions* II, Proceedings of the London Mathematical Society, (2), vol. 42 (1937), pp. 546–549.

[7] *On a conjecture of Ramanujan*, Journal of the London Mathematical Society, vol. 11 (1936), pp. 114–118. *An application of Schläfli's modular equation to a conjecture of Ramanujan*, this Bulletin, vol. 44 (1938), pp. 84–90.

G. H. Hardy and S. Ramanujan published their fundamental paper, *Asymptotic formulae in combinatory analysis.*[8] As I have already mentioned, the denominator in (1) is closely related to ϑ-functions. In a commentary on Riemann's *Collected Works*, and in another paper,[9] Dedekind had made detailed studies of the function

$$(4) \qquad \eta(\tau) = e^{\pi i \tau/12} \prod_{m=1}^{\infty} (1 - e^{2\pi i m \tau}), \qquad \Im(\tau) > 0,$$

which, under the substitution $x = e^{2\pi i \tau}$, is essentially the denominator in question.

Hardy and Ramanujan started with Cauchy's formula applied to the equation (1):

$$(5) \qquad p(n) = \frac{1}{2\pi i} \int_C \frac{f(x)}{x^{n+1}} \, dx.$$

Since $f(x)$ is regular inside the unit circle and has the unit circle as its natural boundary, C has to be a closed curve inside $|x| = 1$, surrounding the point $x = 0$.

Now the usual approach to a complex integral is to utilize the freedom in the path of integration, whether we use the calculus of residues or the method of steepest descents or similar devices. In any case we trace the path of integration in such a way that it passes through that region in which the function, by the overwhelming amount of its value, gives the heaviest contribution on a relatively short piece of the path. The function $f(x) = P(x)^{-1}$ tends rapidly to infinity if we approach radially the point $x = 1$, since each term in $P(x)$ tends to zero. Thus the neighborhood of $x = 1$ will yield the most essential contribution. The next heaviest singularity is located at $x = -1$ in whose vicinity every other factor of $P(x)$ comes close to zero. In this way the roots of unity enter, according to their denominators, but decreasing in weight with increasing denominators. The path C is taken as a circle around 0 rather close to the unit circle, and we cut it into parts, each part corresponding to a neighborhood of just one root of unity. The assembly of all proper fractions h/k with $k \leqq N$ is called the Farey series of order N, to the use of which Hardy and Ramanujan were quite naturally led by these considerations.

[8] *Asymptotic formulae in combinatory analysis*, Proceedings of the London Mathematical Society, (2), vol. 17 (1918), pp. 75–115.

[9] *Schreiben an Herrn Borchardt über die Theorie der elliptischen Modulfunktionen*, Journal für die reine und angewandte Mathematik, vol. 83 (1877), pp. 265–292; also *Gesammelte Werke*, vol. 1, pp. 175–201.

This is, of course, only a sketch of the procedure of integration. Its details have to be furnished by a study of the function $f(x)$ in the neighborhood of a root of unity $e^{2\pi ih/k}$. This is given by the formula

$$(6) \quad f(e^{2\pi ih/k-2\pi z/k}) = \omega_{h,k} z^{1/2} \exp\left(\frac{\pi}{12kz} - \frac{\pi z}{12k}\right) f(e^{2\pi ih'/k-2\pi/kz}), \qquad \Re(z) > 0.$$

Here $\omega_{h,k}$ is a certain $24k$th root of unity. This formula is of importance in our problem since in the neighborhood of the root of unity $e^{2\pi ih/k}$ we have z small, $1/z$ large, and both z and $1/z$ with positive real parts. Such a large $1/z$ involves a very small value of $e^{2\pi ih'/k-2\pi/kz}$ so that, because of (1), $f(e^{2\pi ih'/k-2\pi/kz})$ can with good approximation (which has, of course, to be appraised) be replaced by 1. This gives an elementary approximation function for $f(x)$ in the neighborhood of $e^{2\pi ih/k}$. By means of this treatment of (5), and appropriate estimations of the errors which are made when the function $f(x)$ is replaced by certain approximation functions, Hardy and Ramanujan arrived at the asymptotic formula

$$(7) \quad p(n) = \frac{1}{2\pi 2^{1/2}} \sum_{k \leqq \alpha n^{1/2}} A_k(n) k^{1/2} \frac{d}{dn}\left(\frac{\exp\left\{\frac{\pi}{k}\left[\frac{2}{3}(n-1/24)\right]^{1/2}\right\}}{(n-1/24)^{1/2}}\right) + O(n^{-1/4}),$$

with

$$(8) \quad A_k(n) = \sum_{h \bmod k, (h,k)=1} \omega_{h,k} e^{-2\pi inh/k}.$$

This formula is remarkable in analytic number theory because of its error term which tends to zero as n increases. The constant involved in the error term was not determined; actual computation of $p(100)$ and $p(200)$, however, showed that a relatively small number of terms suffice to give a value which differs from the true value by only a few thousandths of a unit. Hardy and Ramanujan raised the question whether the series (7) extended indefinitely converges or not. D. H. Lehmer[10] has shown recently, by a study of the $A_k(n)$, that the infinite series (7) is divergent.

The method applied by Hardy and Ramanujan was further developed by Hardy and Littlewood and applied to other problems, in particular to Waring's problem, where it leads also to asymptotic

[10] *On the Hardy-Ramanujan series for the partition function*, Journal of the London Mathematical Society, vol. 12 (1937), pp. 171–176.

results. In the present case, however, Hardy and Ramanujan were not aware of the full strength of their method. When it is applied with more refinement of the estimates, and when in particular the coupling of N (the order of the Farey dissection) with n (the given integer) is abolished, and N is made to go to infinity with a fixed n, then we obtain the following exact formula[11] for $p(n)$:

$$(9)\quad p(n) = \frac{1}{\pi 2^{1/2}} \sum_{k=1}^{\infty} A_k(n) k^{1/2} \frac{d}{dn}\left(\frac{\sinh \dfrac{\pi}{k}\left[\dfrac{2}{3}(n - 1/24)\right]^{1/2}}{(n - 1/24)^{1/2}} \right),$$

where the infinite series is absolutely convergent. The formula (7) not only appears as a consequence of formula (9), but also the error term in (7) can be definitely estimated. This, by the way, made it possible to compute the function $p(n)$ for the great values of n mentioned above.

The formula (9) has now been the starting point for further results. We saw that the formula (6) forms the basis for the evaluation of the integral (1). The formula (6) in turn is derived from the theory of the function $\eta(\tau)$. We have

$$(10)\qquad \eta(\tau') = \epsilon(a, b, c, d)[-i(c\tau + d)]^{1/2}\eta(\tau), \qquad c > 0,$$

$$(10\text{a})\qquad \tau' = \frac{a\tau + b}{c\tau + d}, \qquad \begin{vmatrix} a & b \\ c & d \end{vmatrix} = 1,$$

where a, b, c, d are integers and ϵ is a certain 24th root of unity. This formula is connected with (6) through the substitution

$$\tau = \frac{iz + h}{k}, \qquad \tau' = \frac{i/z + h'}{k}, \qquad hh' \equiv -1 \pmod{k}.$$

The substitutions (10a), together with the case $c=0$:

$$(10\text{b})\qquad \tau' = \tau + b,$$

for which we have, directly from (4),

$$(10\text{c})\qquad \eta(\tau') = e^{\pi ib/12}\eta(\tau),$$

are called modular substitutions. They form an infinite discontinuous

[11] Rademacher, *A convergent series for the partition function p(n)*, Proceedings of the National Academy of Sciences, vol. 23 (1937), pp. 78–84; *On the partition function p(n)*, Proceedings of the London Mathematical Society, (2), vol. 43 (1937), pp. 241–254.

group. All these considerations can be applied to modular forms in general.[12] Let us give, in brief, a definition of this concept.

A modular form is a homogeneous analytic function of two variables ω_1, ω_2 defined for $\Im(\omega_2/\omega_1)>0$:

$$H(\lambda\omega_1, \lambda\omega_2) = \lambda^r H(\omega_1, \omega_2); \tag{11}$$

the parameter r, which we take as real, is the dimension of the form. A modular form is invariant under modular transformations:

$$\omega_2' = a\omega_2 + b\omega_1, \qquad \begin{vmatrix} a & b \\ c & d \end{vmatrix} = 1,$$
$$\omega_1' = c\omega_2 + d\omega_1,$$
$$H(\omega_1', \omega_2') = H(\omega_1, \omega_2).$$

Examples are the invariants

$$g_2 = 60\sum{}' \frac{1}{(2m_1\omega_1 + 2m_2\omega_2)^4}, \qquad g_3 = 140\sum{}' \frac{1}{(2m_1\omega_1 + 2m_2\omega_2)^6}.$$

A nonhomogeneous notation is often preferable: By means of (11) we have

$$\omega_1'^{\,r} H(1, \omega_2'/\omega_1') = \omega_1^r H(1, \omega_2/\omega_1)$$

and, with $\omega_2/\omega_1=\tau$, $\omega_2'/\omega_1'=\tau'$, $\tau'=(a\tau+b)/(c\tau+d)$,

$$H(1, \tau') = (c\tau + d)^{-r} H(1, \tau). \tag{12}$$

If r is not an integer, we have to determine the branch of $(c\tau+d)^{-r}$. In order to avoid this difficulty we admit a slight generalization. We consider functions $F(\tau)$ for $\Im(\tau)>0$ with the property

$$F(\tau') = \epsilon(a, b, c, d)\cdot(-\,i(c\tau + d))^{-r} F(\tau), \tag{13a}$$

where $c>0$ and $|\epsilon|=1$ and $(-i(c\tau+d))^{-r}$ stands for the principal branch. The case $c=0$ has to be mentioned separately:

$$F(\tau + 1) = \epsilon_1 F(\tau) = e^{2\pi i\alpha} F(\tau), \qquad 0 \leqq \alpha < 1. \tag{13b}$$

A function $F(\tau)$ having the properties (13a) and (13b) may now be called a "modular form," in spite of its nonhomogeneous notation; in particular (13a) shows that it is of dimension r.

From (13b) we derive

$$e^{-2\pi i\alpha(\tau+1)} F(\tau + 1) = e^{-2\pi i\alpha\tau} F(\tau),$$

[12] Rademacher and Zuckerman, *On the Fourier coefficients of certain modular forms of positive dimension*, Annals of Mathematics, (2), vol. 39 (1938), pp. 433–462.

and note the periodicity in τ modulo 1. Therefore we have a Fourier development

(14) $$F(\tau) = e^{2\pi i\alpha\tau}\sum_{n=-\infty}^{\infty} a_n e^{2\pi i n\tau},$$

which will converge in the whole upper half-plane if $F(\tau)$ is assumed to be regular there. We then call $F(\tau)$ an "entire" modular form. One further restriction is important: we assume that (14) contains only a finite number of terms with negative exponents, or, as we can say, $F(\tau)$ has a pole at the parabolic point $\tau = i\infty$ measured in the uniformizing variable $e^{2\pi i\tau}$.

In our previous case we had

$$\eta(\tau)^{-1} = e^{2\pi i(23/24)\tau}\sum_{n=-1}^{\infty} p(n+1)e^{2\pi i n\tau}.$$

The partition function appears here therefore as the Fourier coefficients of a modular function of dimension $+1/2$ (since $\eta(\tau)$ in (10) has the dimension $-1/2$). Our method now enables us to determine the coefficients a_n for $n \geqq 0$ from the principal part of the pole at the parabolic point $\tau = i\infty$, that is, from the coefficients a_n with $n<0$.

I cannot go into details of the application of the Hardy-Ramanujan-Littlewood method. Only one essential point need be mentioned: it is important that r be positive. Indeed, this r is responsible for a term z^r corresponding to $z^{1/2}$ in (6), and, with z approaching zero, helps in a decisive way to ensure convergence.

The result is this:

If $F(\tau)$ is an entire modular form of positive dimension r,

$$F\left(\frac{a\tau+b}{c\tau+d}\right) = \epsilon(a,b,c,d)(-i(c\tau+d))^{-r}F(\tau), \qquad c>0,$$

$$F(\tau+1) = e^{2\pi i\alpha}F(\tau), \qquad 0 \leqq \alpha < 1,$$

$$F(\tau) = e^{2\pi i\alpha\tau}\sum_{m=-\mu}^{\infty} a_m e^{2\pi i m\tau},$$

then

(15) $$a_m = \sum_{\nu=1}^{\mu} a_{-\nu}\sum_{k=1}^{\infty}\frac{1}{k}A_{k,\nu}(m)\left(\frac{\nu-\alpha}{m+\alpha}\right)^{(r+1)/2} \cdot I_{r+1}((4\pi/k)[(\nu-\alpha)(m+\alpha)]^{1/2}), \qquad m \geqq 0,$$

with

$$A_{k,\nu}(m) = \sum_{0\leq h<k,(h,k)=1} \epsilon(h', -(hh'+1)/k, k, -h)^{-1} \cdot \exp\left\{-(2\pi i/k)((\nu-\alpha)h' + (m+\alpha)h)\right\}.$$

Before I come to consequences of these equations for coefficients, I wish to mention two important extensions. We have taken into consideration only modular forms invariant with respect to the full modular group

$$\tau' = \frac{a\tau + b}{c\tau + d}, \quad \begin{vmatrix} a & b \\ c & d \end{vmatrix} = 1.$$

But we can just as well consider a subgroup of the full modular group. It is only necessary that its fundamental region have a finite number of parabolic points. The most important and best known subgroups of this sort are the congruence subgroups, that is, those in which a, b, c, d are subjected to certain congruence restrictions. The principal congruence subgroup modulo p is that which has

$$\begin{pmatrix} a & b \\ c & d \end{pmatrix} \equiv \begin{pmatrix} 1 & 0 \\ 0 & 1 \end{pmatrix} \pmod{p}.$$

Zuckerman[13] has carried out this generalization.

Secondly we can overcome to a certain extent the restriction that the dimension must be positive. For this purpose we have to be more careful with our estimates. The sums $A_{k,\nu}(m)$ can immediately be estimated as $|A_{k,\nu}(m)| \leq \sum 1 = \phi(k)$. The problem of a better estimate has not been fully solved. But in certain simple cases we can get

$$(16) \qquad A_{k,\nu}(m) = O(k^{2/3+\epsilon}).$$

These improved estimates were first begun by Kloosterman[14] after whom these sums are now named, and later continued by Esterman,[15] Salié,[16] Davenport,[17] Lehmer,[18] and others. In this way we can easily

[13] *On the coefficients of certain modular forms belonging to subgroups of the modular group*, Transactions of this Society, vol. 45 (1939), pp. 298–321.

[14] *Asymptotische Formeln für die Fourierkoeffizienten ganzer Modulformen*, Abhandlungen aus dem mathematischen Seminar der Hamburgischen Universität, vol. 5 (1927), pp. 337–352.

[15] *Vereinfachter Beweis eines Satzes von Kloosterman*, Abhandlungen aus dem mathematischen Seminar der Hamburgischen Universität, vol. 7 (1939), pp. 82–98.

[16] *Zur Abschätzung der Fourierkoeffizienten ganzer Modulformen*, Mathematische Zeitschrift, vol. 36 (1933), pp. 263–278.

[17] *On certain exponential sums*, Journal für die reine und angewandte Mathematik, vol. 169 (1933), pp. 158–176.

[18] *On the series for the partition function*, Transactions of this Society, vol. 43 (1938), pp. 271–295.

include the case with dimension zero in our reasonings. A very important function belongs to that class, the modular invariant $J(\tau)$:

$$(17)\qquad J(\tau) = \frac{g_2^3}{\Delta} = \frac{g_2^3}{g_2^3 - 27g_3^2}, \qquad J\left(\frac{a\tau + b}{c\tau + d}\right) = J(\tau),$$
$$12^3 J(\tau) = e^{-2\pi i\tau} + c_0 + \sum_{n=1}^{\infty} c_n e^{2\pi i n\tau}.$$

The coefficients c_n of this fundamental function can be found by our method[19] as

$$(18)\qquad c_n = \frac{2\pi}{n^{1/2}} \sum_{k=1}^{\infty} \frac{1}{k} A_k(n) I_1((4\pi/k)n^{1/2}), \qquad n \geqq 1,$$
$$A_k(n) = \sum_{h \bmod k,\,(h,k)=1} e^{-(2\pi i/k)(nh+h')}.$$

It is of interest that a few years ago Petersson[20] found these coefficients (which are integers) by a completely different method, which operates with modular functions of negative dimensions. Indeed, $J'(\tau)$ is of dimension -2 as is readily seen,

$$J'\left(\frac{a\tau + b}{c\tau + d}\right)\frac{1}{(c\tau + d)^2} = J'(\tau),$$

from the differentiation of (17). Modular forms of negative dimension have, however, been studied as far back as Eisenstein, who found the partial fraction series for g_2 and g_3, and later by Poincaré among others. These investigations lie outside the field of our present discussion.

If I have in this way outlined definitions, methods, and direct results, I wish now to survey briefly a few consequences of our theory and some remaining problems. Let me begin with function theoretical consequences. Looking at formula (15), we see that the a_m $(m \geqq 0)$ depend linearly on $a_{-\mu}, \cdots, a_{-1}$. If all these should happen to be equal to zero, our analysis would go through just as well and would lead to $a_m = 0$ for $m \geqq 0$. Hence we have that an entire modular form of positive dimension which is regular also at the parabolic point (or points) of the fundamental region vanishes identically. This is remarkable in so far as it is not true for modular forms of negative

[19] Rademacher, *The Fourier coefficients of the modular invariant* $J(\tau)$, American Journal of Mathematics, vol. 60 (1938), pp. 501–512.

[20] *Ueber die Entwicklungskoeffizienten der automorphen Formen*, Acta Mathematica, vol. 58 (1932), pp. 169–215.

dimension, as for example $g_2(1, \tau)$, $g_3(1, \tau)$, $\vartheta_3(0|\tau)$, and $\eta(\tau)$ show.

These facts have a bearing on the generalized Kloosterman sums, which occur in the estimates. For the case of $\eta(\tau)^{-1}$ we know

$$A_k(n) = O(k^{1/2+\epsilon}), \tag{18a}$$

and for $J(\tau)$ (which leads to the original Kloosterman sums)

$$A_k(n) = O(k^{2/3+\epsilon}n^{1/3}),$$

after Salié and Davenport. It is highly probable that in the latter case a sharper estimate like (18a) will be true. However, our reasonings show that such estimates cannot be expected in all cases. Indeed, if there existed a positive constant α such that for all Kloosterman sums belonging to all modular forms we should have

$$A_{k,\nu}(n) = O(k^{1-\alpha}), \tag{18b}$$

it would be possible to repeat our investigations of the coefficients, for example, for $\eta(\tau)^{\alpha/2}$. Since this function (of dimension $-\alpha/4$) is regular at infinity, the estimate (18b) would imply the absurd consequence that all coefficients of $\eta(\tau)^{\alpha/2}$ vanish identically. An estimate better than the trivial $O(k)$ for the generalized Kloosterman sums can therefore be obtained only under certain special conditions and not uniformly for all real dimensions.

We can use the exact coefficients which we have found again in our Fourier series. We have

$$12^3 J(\tau) = e^{-2\pi i\tau} + c_0 + \sum_{n=1}^{\infty} \frac{2\pi}{n^{1/2}} e^{2\pi i n\tau} \sum_{k=1}^{\infty} \frac{1}{k} A_k(n) I_1(4\pi n^{1/2}/k). \tag{19}$$

This series, on the other hand, determines the function directly. It is clear that it satisfies the relation $J(\tau+1) = J(\tau)$. But it must also be invariant with respect to all other modular substitutions. Since all of them are generated by $\tau' = \tau+1$, $\tau' = -1/\tau$, it is of interest to show directly $J(-1/\tau) = J(\tau)$ by means of the series, or, in other words, to show that the series defines a modular function. This, indeed, can be done, as I have shown in a recent paper.[21] The proof consists of a rearrangement of terms of a conditionally convergent double series.

I hope that this proof has a prospect of further development for the problem of existence. Up to the present we have only discussed modular forms of positive dimension which are given by certain other definitions (infinite products and so on). But the problem is to con-

[21] *The Fourier series and the functional equation of the absolute modular invariant* $J(\tau)$, American Journal of Mathematics, vol. 61 (1939), pp. 237–248.

struct new ones with given principal parts at the parabolic points. For negative dimensions we have a powerful principle of generation in the Eisenstein-Poincaré series, which have no analogue here for reasons of convergence.

The exact formulas for the coefficients lead to a sort of analytic continuation which I wish to mention at least briefly. I exemplify this idea with the treatment of $f(x)=1+\sum_{n=1}^{\infty}p(n)x^n$. Introducing our value for $p(n)$ we obtain

$$f(x) = 1 + \frac{1}{\pi 2^{1/2}} \sum_{n=1}^{\infty} x^n \sum_{k=1}^{\infty} \sum_{h \bmod k,\,(h,k)=1} \omega_{h,k} e^{-2\pi i h n/k} \cdot k^{1/2} \frac{d}{dn}\left(\frac{\sinh \frac{\pi}{k}\left[\frac{2}{3}(n-1/24)\right]^{1/2}}{(n-1/24)^{1/2}} \right)$$

$$(20) \qquad = 1 + \frac{1}{\pi 2^{1/2}} \sum_{k=1}^{\infty} \sum_{h \bmod k,\,(h,k)=1} \omega_{h,k} k^{1/2} \sum_{n=1}^{\infty} (x e^{-2\pi i h/k})^n \cdot \frac{d}{dn}\left(\frac{\sinh \frac{\pi}{k}\left[\frac{2}{3}(n-1/24)\right]^{1/2}}{(n-1/24)^{1/2}} \right)$$

$$= 1 + \frac{1}{\pi 2^{1/2}} \sum_{h \bmod k,\,(h,k)=1} \omega_{h,k} k^{1/2} \Phi_k(x e^{-2\pi i h/k}),$$

with

$$(20\text{a}) \qquad \Phi_k(z) = \sum_{n=1}^{\infty} \frac{d}{dn}\left(\frac{\sinh \frac{\pi}{k}\left[\frac{2}{3}(n-1/24)\right]^{1/2}}{(n-1/24)^{1/2}} \right) z^n = \sum_{n=1}^{\infty} g_k(n) z^n.$$

Here

$$g_k(n) = \frac{d}{dn}\left(\frac{\sinh \frac{\pi}{k}\left[\frac{2}{3}(n-1/24)\right]^{1/2}}{(n-1/24)^{1/2}} \right)$$

$$(20\text{b}) \qquad = \frac{d}{dn} \sum_{\nu=0}^{\infty} \frac{[(\pi/k)(2/3)^{1/2}]^{2\nu+1}}{(2\nu+1)!} (n-1/24)^{\nu}$$

$$= \sum_{\nu=1}^{\infty} [(\pi/k)(2/3)^{1/2}]^{2\nu+1} \frac{\nu}{(2\nu+1)!} (n-1/24)^{\nu-1}$$

is a transcendental function of order 1/2 in the variable n. Hence by a theorem of Wigert $\Phi_k(z)$, which originally is defined only in $|z|<1$, can be extended over the whole z-plane and has only an isolated essential singularity at $z=1$. The formula (20) therefore represents the splitting up of $f(x)$ into functions each having just one essential singularity. All singularities together form the natural boundary $|x|=1$.

Moreover, the series (20) has now a meaning for $|x|>1$. It turns out that it converges there also. Simple considerations show that, for $|z|>1$, we get

$$\Phi_k(z) = \sum_{m=0}^{\infty} \frac{d}{dm}\left(\frac{\sin \frac{\pi}{k}\left[\frac{2}{3}(m+1/24)\right]^{1/2}}{(m+1/24)^{1/2}}\right) z^{-m}$$

and so, for $|x|>1$, if we introduce $f^*(x)$ instead of $f(x)$,

$$\begin{aligned} f^*(x) &= 1 + \frac{1}{\pi 2^{1/2}} \sum_{h \bmod k,\,(h,k)=1} \omega_{h,k} k^{1/2} \\ &\quad \cdot \sum_{m=0}^{\infty} \frac{d}{dm}\left(\frac{\sin \frac{\pi}{k}\left[\frac{2}{3}(m+1/24)\right]^{1/2}}{(m+1/24)^{1/2}}\right) \cdot (xe^{-2\pi ih/k})^{-m} \\ &= 1 + \sum_{m=0}^{\infty} p^*(m) x^{-m}, \end{aligned}$$

let us say. Here

$$\begin{aligned} p^*(m) &= \frac{1}{\pi 2^{1/2}} \sum_{k=1}^{\infty} A_k(-m) k^{1/2} \frac{d}{dm}\left(\frac{\sin \frac{\pi}{k}\left[\frac{2}{3}(m+1/24)\right]^{1/2}}{(m+1/24)^{1/2}}\right) \\ &= -p(-m), \end{aligned}$$

formally. Now $f^*(x)=0$ identically. Indeed, it turns out that we get $p(0)=1$, as is to be expected, and $p(-m)=0$ for $m \geqq 1$. This was first proved by Petersson[22] in a recent publication, making use again of modular forms of negative dimension. It can, however, also be proved by means of the Hardy-Ramanujan method.

Similar "continuations" beyond the natural boundary can be effec-

---

[22] *Die linearen Relationen zwischen den ganzen Poincaréschen Reihen von reeller Dimension zur Modulgruppe*, Abhandlungen aus dem mathematischen Seminar der Hamburgischen Universität, vol. 12 (1938), pp. 415–472.

tuated for all those power series arising from modular forms of non-negative dimension. For $J(\tau)$ we obtain, for example,

$$J(\tau) = F(e^{2\pi i\tau}) = F(x), \qquad |x| < 1,$$

with $F^*(x) = 5/12$, $(|x| > 1)$.

I come finally to some arithmetical consequences of the method connected mainly with partitions. If we admit only odd integers as parts, then the generating function for the number $q(n)$ of partitions into odd summands is

$$g(x) = 1 + \sum_{n=1}^{\infty} q(n)x^n = \frac{1}{(1-x)(1-x^3)(1-x^5)\cdots}$$

$$= \frac{(1-x^2)(1-x^4)\cdots}{(1-x)(1-x^2)\cdots} = \frac{f(x)}{f(x^2)}.$$

This can be treated by our method, but with the difficulty that we have to deal with modular forms of dimension zero and belonging to a congruence subgroup modulo 2. This has been done by Hua in an unpublished paper. The modular form here would be $\eta(2\tau)/\eta(\tau)$. As a result Hua obtains again a convergent series for $q(n)$, of which Hardy and Ramanujan had given only the first few terms.

There are a number of problems of this sort concerning restricted partitions, which can be attacked by our method. For example Niven[23] has recently determined the number of partitions of a number into summands of the form $6n \pm 1$.

A further consequence which seems more interesting to me is the following: Ramanujan discovered, first empirically, the properties

(a) $$p(5m + 4) \equiv 0 \pmod{5},$$

(b) $$p(7m + 5) \equiv 0 \pmod{7},$$

(c) $$p(11m + 6) \equiv 0 \pmod{11}.$$

He was able to prove (a) and (b) easily but remarked that they would be obvious consequences of the identities

$$\sum_{m=0}^{\infty} p(5m + 4)x^m = 5\frac{\prod(1 - x^{5\nu})^5}{\prod(1 - x^\nu)^6},$$

$$\sum_{m=0}^{\infty} p(7m + 5)x^m = 7\frac{\prod(1 - x^{7\nu})^3}{\prod(1 - x^\nu)^4} + 49x\frac{\prod(1 - x^{7\nu})^7}{\prod(1 - x^\nu)^8}.$$

He did not give a proof of these identities. Proofs were later given

[23] *On a certain partition function*, to be published.

by Darling[24] and Mordell[25] and recently again by Watson[26] in a comprehensive paper which covers much more. Now our method can be directly applied to these identities.[27] We know the coefficients of the left member and can express the right member in the same way. All we need to do is then to compare coefficients, which turns out to be not completely trivial since the expressions need a slight transformation.

Zuckerman[28] has found a new identity of this sort by our method. If we write $\prod(1-x^\nu)=\phi(x)$, then

$$\sum_{m=0}^{\infty} p(13m+6)x^m = 11\frac{\phi(x^{13})}{\phi(x)^2} + 36\cdot 13x\frac{\phi(x^{13})^3}{\phi(x)^4} + 38\cdot 13^2x^2\frac{\phi(x^{13})^5}{\phi(x)^6}$$
$$+ 20\cdot 13^3x^3\frac{\phi(x^{13})^7}{\phi(x)^8} + 6\cdot 13^4x^4\frac{\phi(x^{13})^9}{\phi(x)^{10}} + 13^5x^5\frac{\phi(x^{13})^{11}}{\phi(x)^{12}}$$
$$+ 13^5x^6\frac{\phi(x^{13})^{13}}{\phi(x)^{14}}.$$

It is regrettable that this identity does not lead to arithmetical properties of the Ramanujan sort since the factor 13 does not appear as a factor of every term of the right member.

The connection between modular functions and partitions seems to be accidental. Analogues of these concepts may be found in algebraic fields. However they are not connected as in the rational case by formulas of the type (1). Our method could probably be carried over to the discussion of these modular functions, but they would not yield any information concerning these partition functions.

UNIVERSITY OF PENNSYLVANIA

[24] *Proofs of certain identities and congruences enunciated by S. Ramanujan*, Proceedings of the London Mathematical Society, (2), vol. 19 (1921), pp. 350–372.

[25] *Note on certain modular relations considered by Messrs. Ramanujan, Darling, and Rogers*, Proceedings of the London Mathematical Society, (2), vol. 20 (1922), pp. 408–416.

[26] *Ramanujans Vermutung über Zerfällungsanzahlen*, Journal für die reine und angewandte Mathematik, vol. 179 (1938), pp. 97–128.

[27] Rademacher and Zuckerman, *A new proof of two of Ramanujan's identities*, Annals of Mathematics, (2), vol. 40 (1939), pp. 473–489.

[28] *Identities analogous to Ramanujan's identities involving the partition function*, Duke Mathematical Journal, vol. 5 (1939), pp. 88–110.

Notes to Paper 43

This paper has been reviewed in the JF, vol. 66 (1960), p. 158, in the Z, vol. 23 (1940-1941), pp. 10-11, and in the MR, vol. 1 (1940), p. 136.

Page 64, line 13 of text.
Read (5) instead of (1).

Page 64, footnote.
See papers 37 and 38 of the present collection.

Page 65, footnote.
See paper 40.

Page 68, footnote 19.
See paper 39.

Page 69, footnote.
See paper 41.

Page 73, footnote 27.
See paper 42.

THEOREMS ON DEDEKIND SUMS.*

By HANS RADEMACHER and ALBERT WHITEMAN.

1. Introduction. In his "Erläuterungen zu den Riemannschen Fragmenten über die Grenzfälle der elliptischen Modulfunktionen," Dedekind **1**[1] introduced the function

$$\eta(\tau) = e^{(\pi i\tau)/12} \prod_{m=1}^{\infty} (1 - e^{2\pi i m\tau}),$$

which has subsequently played a fundamental role in the theory of elliptic modular functions. Since $\eta(\tau)$ is regular and free of zeros in the upper τ-halfplane, $\log \eta(\tau)$ can there be defined as a regular function. The study of the modular transformations of $\log \eta(\tau)$ led Dedekind to consider certain arithmetical sums which will form the subject of this paper.

In order to give a concise definition of the "Dedekind sums," we introduce the following symbol:

$$((x)) = \begin{cases} x - [x] - \frac{1}{2} & \text{for } x \text{ not an integer,} \\ 0 & \text{for } x \text{ an integer,} \end{cases} \tag{1.1}$$

where $[x]$ denotes, as usual, the greatest integer not exceeding x. We then define a Dedekind sum[2] by

$$s(h, k) = \sum_{\mu=1}^{k} \left(\left(\frac{\mu}{k}\right)\right)\left(\left(\frac{h\mu}{k}\right)\right), \tag{1.2}$$

where the first argument h is any integer, and the second argument k is, of course, a positive integer.

The immediate object of Dedekind's memoir was to elucidate and verify eight formulae involving the logarithms of elliptic modular functions which are contained in Riemann's fragmentary notes **2**. Dedekind bases his dis-

* Received September 23, 1940; presented to the American Mathematical Society, April 26, 1940.

[1] Boldface numerals refer to the papers listed in the bibliography at the end of this paper.

[2] Dedekind's notation differs from the one adopted above, which has been used by one of the present writers in some previous papers. Dedekind's symbol (m, n) equals $6ns(m, n)$ in our notation; at the end of his memoir **1** Dedekind introduces a symbol $D(m/n)$ which is equivalent to our $6s(m, n)$. Moreover, his symbol $((x))$ is the same as our $((x + \frac{1}{2}))$. However, we do not feel induced to rectify these disagreements, since our present notation has already been adopted in other publications.

cussion on a theory of $\log \eta(\tau)$, the central result of which is the transformation formula

$$\log \eta(\tau') = \log \eta(\tau) + \tfrac{1}{2} \log(-i(c\tau + d)) + \frac{\pi i}{12c}(a + d) - \pi i s(d, c),$$

where

$$\tau' = \frac{a\tau + b}{c\tau + d}, \ c > 0,$$

is a modular substitution. From this formula Dedekind derives relations which correspond to those of Riemann, but remarks that it would take him too far afield to demonstrate actually the equivalence of both sets of formulae. Nevertheless, he lists a number of formulae concerning the sums $s(h, k)$ which would be helpful for this purpose. Some of these formulae are given with function-theoretic proofs, but most of them are stated without proof.[3]

The purpose of the present paper is to fill the gaps in Dedekind's paper and to study, in general, the directly accessible properties of the Dedekind sums $s(h, k)$. We shall first give purely arithmetic proofs of all of Dedekind's identities **1, 3** and of some analogous ones. Secondly, we shall show that Riemann's formulae indeed agree with Dedekind's results, a fact which, according to Dedekind's own statement, is by no means obvious. Thirdly, we shall derive some congruences of a new type for the $s(h, k)$ and employ them for a discussion of the sums $A_k(n)$ which appear in the series for the partition function. This will lead us to rather simple proofs of D. H. Lehmer's theorems about the multiplicativity of the $A_k(n)$.

PART I: Dedekind's Formulae and Related Formulae.

2. Fundamental properties of $s(h, k)$. We begin with a few remarks concerning the symbols introduced in (1.1) and (1.2). The symbol $((x))$ has the fundamental properties

$$((x_1)) = ((x_2)) \quad \text{for } x_1 \equiv x_2 \pmod 1, \tag{2.1}$$

and

$$((-x)) = -((x)). \tag{2.2}$$

We have therefore[4]

$$\sum_{\mu=1}^{k} \left(\left(\frac{\mu}{k}\right)\right) = \sum_{\mu \bmod k} \left(\left(\frac{\mu}{k}\right)\right) = \sum_{\mu \bmod k} \left(\left(\frac{-\mu}{k}\right)\right) = -\sum_{\mu \bmod k} \left(\left(\frac{\mu}{k}\right)\right),$$

---

[3] Dedekind, in spite of an announcement at the end of his "Erläuterungen," apparently never came back to this subject. It might be of interest to find out whether there still exist unpublished notes containing his investigations in this direction.

[4] The condition "$\mu \bmod k$" in connection with a summation sign means that the sum is extended over a complete residue system modulo k.

so that

$$\sum_{\mu \bmod k} \left(\left(\frac{\mu}{k}\right)\right) = 0. \tag{2.3}$$

In a similar manner we may show that

$$\sum_{\mu=1}^{k} \left(\left(\frac{h\mu}{k}\right)\right) = 0, \tag{2.31}$$

where h and k are not necessarily coprime.

More general than (2.3) is

$$\sum_{\mu \bmod k} \left(\left(\frac{\mu + x}{k}\right)\right) = ((x)) \qquad [33] \tag{2.4}$$ [5]

which is clear, since, first of all, both members of (2.4) are periodic modulo 1; secondly, their difference is constant for $0 \leqq x < 1$; and, thirdly, this difference, because of (2.3), is 0 for $x = 0$.

As a very useful special case of (2.4) we note

$$((x)) + ((x + \tfrac{1}{2})) = ((2x)). \tag{2.41}$$

In view of (2.1) we may also write (1.2) in the form

$$s(h, k) = \sum_{\mu \bmod k} \left(\left(\frac{\mu}{k}\right)\right)\left(\left(\frac{h\mu}{k}\right)\right). \tag{2.5}$$

An immediate consequence of this formula is

$$s(h, k) = s(h', k) \text{ with } hh' \equiv 1 \pmod{k}. \tag{2.51}$$

Indeed we have

$$\sum_{\mu \bmod k} \left(\left(\frac{\mu}{k}\right)\right)\left(\left(\frac{h\mu}{k}\right)\right) = \sum_{\mu \bmod k} \left(\left(\frac{h'\mu}{k}\right)\right)\left(\left(\frac{hh'\mu}{k}\right)\right) = \sum_{\mu \bmod k} \left(\left(\frac{h'\mu}{k}\right)\right)\left(\left(\frac{\mu}{k}\right)\right).$$

This reasoning reveals an interesting symmetry in the products over which the sum in (2.5) extends. At the risk of losing this symmetry we may simplify the definition to read

$$s(h, k) = \sum_{\mu=1}^{k} \left(\frac{\mu}{k} - \frac{1}{2}\right)\left(\left(\frac{h\mu}{k}\right)\right) = \sum_{\mu=1}^{k} \frac{\mu}{k}\left(\left(\frac{h\mu}{k}\right)\right), \tag{2.6}$$

by virtue of (2.31).

In the sequel the symbol $s(h, k)$ will usually be applied to coprime pairs h, k, although the definition (1.2), (2.5) does not exclude a common factor of h and k. It will, however, sometimes be convenient to introduce or suppress common factors of h and k by means of

---

[5] In the sequel we shall refer to formulae in Dedekind's memoir 1 by numbers in square brackets.

THEOREM 1. *For any positive integer q we have*

$$s(qh, qk) = s(h, k). \tag{2.7}$$ [23]

Proof. From the definition (1.2) it follows that

$$\begin{aligned}
s(qh, qk) &= \sum_{\mu=1}^{qk} \left(\left(\frac{\mu}{qk}\right)\right)\left(\left(\frac{qh\mu}{qk}\right)\right) \\
&= \sum_{\nu=0}^{q-1} \sum_{\rho=1}^{k} \left(\left(\frac{\nu k + \rho}{qk}\right)\right)\left(\left(\frac{h(\nu k + \rho)}{k}\right)\right) \\
&= \sum_{\rho=1}^{k} \left(\left(\frac{h\rho}{k}\right)\right) \sum_{\nu=0}^{q-1} \left(\left(\frac{\rho}{qk} + \frac{\nu}{q}\right)\right) \\
&= \sum_{\rho=1}^{k} \left(\left(\frac{h\rho}{k}\right)\right)\left(\left(\frac{\rho}{k}\right)\right) \\
&= s(h, k).
\end{aligned}$$

where we have made use of (2.1) and (2.4).

As an application of the preceding theorem we prove

THEOREM 2. *For a prime number p we have*

$$s(ph, k) + \sum_{m=0}^{p-1} s(h + mk, pk) = (p+1)s(h, k). \tag{2.8}$$ [28], [37_2]

Proof. From (2.4) we infer readily [6]

$$\sum_{m=0}^{p-1} \left(\left(y + \frac{\mu m}{p}\right)\right) = \begin{cases} ((py)) & \text{if } p \nmid \mu, \\ p((y)) & \text{if } p \mid \mu. \end{cases} \tag{2.9}$$

Now we have

$$\begin{aligned}
&s(ph, k) + \sum_{m=0}^{p-1} s(h + mk, pk) \\
&= s(ph, k) + \sum_{m=0}^{p-1} \sum_{\mu=1}^{pk} \left(\left(\frac{\mu}{pk}\right)\right)\left(\left(\frac{(h + mk)\mu}{pk}\right)\right) \\
&= s(ph, k) + \sum_{\mu=1}^{pk} \left(\left(\frac{\mu}{pk}\right)\right) \sum_{m=0}^{p-1} \left(\left(\frac{h\mu}{pk} + \frac{\mu m}{p}\right)\right) \\
&= s(ph, k) + \sum_{\mu=1}^{pk} \left(\left(\frac{\mu}{pk}\right)\right)\left(\left(\frac{ph\mu}{pk}\right)\right) + \sum_{\nu=1}^{k} \left(\left(\frac{p\nu}{pk}\right)\right) \left\{ p\left(\left(\frac{ph\nu}{pk}\right)\right) - \left(\left(\frac{p^2\nu h}{pk}\right)\right) \right\}
\end{aligned}$$

(because of (2.9))

$$\begin{aligned}
&= s(ph, k) + s(ph, pk) + p \sum_{\nu=1}^{k} \left(\left(\frac{\nu}{k}\right)\right)\left(\left(\frac{h\nu}{k}\right)\right) - \sum_{\nu=1}^{k} \left(\left(\frac{\nu}{k}\right)\right)\left(\left(\frac{ph\nu}{k}\right)\right) \\
&= s(ph, k) + s(h, k) + ps(h, k) - s(ph, k),
\end{aligned}$$

where we have made use of (2.7) in the last step. This completes the proof of the theorem.

[6] The notation $a \mid b$ means, as usual, a divides b, and $a \nmid b$ means a does not divide b.

3. The reciprocity formula for $s(h,k)$. The most important property of the Dedekind sums $s(h,k)$ is expressed by a reciprocity formula, which Dedekind derived from his study of $\log \eta(\tau)$ and for which one of us has given some different independent proofs **4**, **5**, **6**. We propose to give here a very short arithmetic proof of

THEOREM 3. *For $h > 0$, $k > 0$, $(h,k) = 1$ the equation*

$$12s(h,k) + 12s(k,h) = -3 + \frac{h}{k} + \frac{k}{h} + \frac{1}{hk} \tag{3.1}$$ [19]

holds.

Proof. We have, on the one hand,

$$\sum_{\mu=1}^{k}\left(\left(\frac{h\mu}{k}\right)\right)^2 = \sum_{\mu=1}^{k}\left(\left(\frac{\mu}{k}\right)\right)^2 = \sum_{\mu=1}^{k-1}\left(\frac{\mu}{k}-\frac{1}{2}\right)^2 = \frac{1}{k^2}\sum_{\mu=1}^{k-1}\mu^2 - \frac{1}{k}\sum_{\mu=1}^{k-1}\mu + \frac{1}{4}\sum_{\mu=1}^{k-1}1. \tag{3.2}$$

On the other hand we see that

$$\sum_{\mu=1}^{k}\left(\left(\frac{h\mu}{k}\right)\right)^2 = \sum_{\mu=1}^{k-1}\left(\frac{h\mu}{k} - \left[\frac{h\mu}{k}\right] - \frac{1}{2}\right)^2 \tag{3.3}$$

$$= 2h\sum_{\mu=1}^{k-1}\frac{\mu}{k}\left(\frac{h\mu}{k} - \left[\frac{h\mu}{k}\right] - \frac{1}{2}\right) + \sum_{\mu=1}^{k-1}\left[\frac{h\mu}{k}\right]\left(\left[\frac{h\mu}{k}\right]+1\right) - \frac{h^2}{k^2}\sum_{\mu=1}^{k-1}\mu^2 + \frac{1}{4}\sum_{\mu=1}^{k-1}1.$$

Comparing (3.2) and (3.3) and using (2.6) we get

$$2hs(h,k) + \sum_{\mu=1}^{k-1}\left[\frac{h\mu}{k}\right]\left(\left[\frac{h\mu}{k}\right]+1\right) = \frac{h^2+1}{k^2}\sum_{\mu=1}^{k-1}\mu^2 - \frac{1}{k}\sum_{\mu=1}^{k-1}\mu. \tag{3.4}$$

In the sum of the left-hand member we have clearly

$$0 \leqq \left[\frac{h\mu}{k}\right] \leqq h-1.$$

We put

$$\left[\frac{h\mu}{k}\right] = \nu - 1, \qquad \nu = 1, 2, \cdots, h, \tag{3.5}$$

and determine the number of values of μ which yield the same value of ν. Now (3.5) implies that

$$\nu - 1 < \frac{h\mu}{k} < \nu, \tag{3.6}$$

where equality is excluded since $(h,k) = 1$ and $\mu < k$; instead of (3.6) we may write

$$\frac{k(\nu-1)}{h} < \mu < \frac{k\nu}{h}.$$

Hence, if μ ranges from $\left[\frac{k(\nu-1)}{h}\right]+1$ to $\left[\frac{k\nu}{h}\right]$ for $\nu < h$, the value of $\left[\frac{h\mu}{k}\right]$ is $\nu-1$; for $\nu = h$, however, $\frac{k\nu}{h}$ is an integer and μ ranges only from $\left[\frac{k(h-1)}{h}\right]$ to $k-1$. We obtain therefore

$$(3.7)\qquad \sum_{\mu=1}^{k-1}\left[\frac{h\mu}{k}\right]\left(\left[\frac{h\mu}{k}\right]+1\right)=\sum_{\nu=1}^{h-1}(\nu-1)\nu\left\{\left[\frac{k\nu}{h}\right]-\left[\frac{k(\nu-1)}{h}\right]\right\}$$
$$+(h-1)h\left\{(k-1)-\left[\frac{k(h-1)}{h}\right]\right\}$$
$$=-2\sum_{\nu=1}^{h-1}\nu\left[\frac{k\nu}{h}\right]+(h-1)h(k-1)$$
$$=2hs(k,h)-\frac{2k}{h}\sum_{\nu=1}^{h-1}\nu^2+\sum_{\nu=1}^{h-1}\nu+(h-1)h(k-1).$$

Comparison of (3.4) with (3.7) leads directly to the reciprocity formula (3.1).

4. The auxiliary sums $r(h,k)$. Although the Dedekind sums $s(h,k)$ are the fundamental ones in the present discussion, we find it convenient to introduce the auxiliary sums

$$(4.1)\qquad r(h,k)=\sum_{\mu \bmod k}\left(\left(\frac{\mu}{k}-\frac{1}{2}\right)\right)\left(\left(\frac{h\mu}{k}\right)\right).$$

We now develop some of their properties and their relations to the $s(h,k)$. We begin with

THEOREM 4. *The sum $r(h,k)$ has the properties*

$$(4.2)\qquad r(qh,qk)=\begin{cases} s(h,k) & \textit{for } q \textit{ even},\\ r(h,k) & \textit{for } q \textit{ odd}.\end{cases}$$

The proof of this theorem is similar to the proof of Theorem 1.

THEOREM 5. *For $(h,k)=1$ we have*

$$(4.3)\qquad r(h,k)+s(h,k)=s(2h',k),$$

with $hh'\equiv 1 \pmod k$.

Proof. With the aid of (2.41) we deduce

$$r(h,k)+s(h,k)=\sum_{\mu \bmod k}\left\{\left(\left(\frac{\mu}{k}-\frac{1}{2}\right)\right)+\left(\left(\frac{\mu}{k}\right)\right)\right\}\left(\left(\frac{h\mu}{k}\right)\right)$$
$$=\sum_{\mu \bmod k}\left(\left(\frac{2\mu}{k}\right)\right)\left(\left(\frac{h\mu}{k}\right)\right)=\sum_{\mu \bmod k}\left(\left(\frac{2h'\mu}{k}\right)\right)\left(\left(\frac{hh'\mu}{k}\right)\right)$$
$$=\sum_{\mu \bmod k}\left(\left(\frac{2h'\mu}{k}\right)\right)\left(\left(\frac{\mu}{k}\right)\right)=s(2h',k),$$

so that we have proved (4.3).

COROLLARY. *For k even and $(h,k)=1$ we have*

$$r(h,k)+s(h,k)=s(2h,k). \tag{4.31}$$

This corollary follows from (4.3), (2.51), and (2.7), since

$$s(2h',k)=s\left(h',\frac{k}{2}\right)=s\left(h,\frac{k}{2}\right)=s(2h,k).$$

Theorems 4 and 5 suffice to reduce any sum $r(h,k)$ to Dedekind sums $s(h,k)$. For a later application we need also

THEOREM 6. *For odd k we have*

$$r(2h,k)+s(2h,k)=s(h,k), \tag{4.4}$$

and for h and k both odd we have

$$s(h,2k)+r(h,k)=s(h,k)+r(2h,k). \tag{4.5}$$

Proof. To derive (4.4) we use (2.41) and obtain

$$\begin{aligned} r(2h,k)+s(2h,k) &= \sum_{\mu \bmod k}\left\{\left(\left(\frac{\mu}{k}-\frac{1}{2}\right)\right)+\left(\left(\frac{\mu}{k}\right)\right)\right\}\left(\left(\frac{2h\mu}{k}\right)\right)\\ &= \sum_{\mu \bmod k}\left(\left(\frac{2\mu}{k}\right)\right)\left(\left(\frac{2h\mu}{k}\right)\right)\\ &= \sum_{\mu \bmod k}\left(\left(\frac{\mu}{k}\right)\right)\left(\left(\frac{h\mu}{k}\right)\right).\end{aligned}$$

To prove (4.5) we use the fact that k is odd in separating the even μ's from the odd μ's in the following manner:

$$\begin{aligned} s(h,2k) &= \sum_{\mu \bmod 2k}\left(\left(\frac{\mu}{2k}\right)\right)\left(\left(\frac{h\mu}{2k}\right)\right)\\ &= \sum_{\nu \bmod k}\left(\left(\frac{2\nu}{2k}\right)\right)\left(\left(\frac{2h\nu}{2k}\right)\right)+\sum_{\nu \bmod k}\left(\left(\frac{2\nu+k}{2k}\right)\right)\left(\left(\frac{h(2\nu+k)}{2k}\right)\right)\\ &= s(h,k)+\sum_{\nu \bmod k}\left(\left(\frac{\nu}{k}+\frac{1}{2}\right)\right)\left(\left(\frac{h\nu}{k}+\frac{h}{2}\right)\right).\end{aligned}$$

Now h is also odd. Hence, using (2.1) and (2.41),

$$\begin{aligned} &s(h,2k)+r(h,k)\\ &= s(h,k)+\sum_{\nu \bmod k}\left(\left(\frac{\nu}{k}+\frac{1}{2}\right)\right)\left(\left(\frac{h\nu}{k}+\frac{1}{2}\right)\right)+\sum_{\mu \bmod k}\left(\left(\frac{\mu}{k}+\frac{1}{2}\right)\right)\left(\left(\frac{h\mu}{k}\right)\right)\\ &= s(h,k)+\sum_{\mu \bmod k}\left(\left(\frac{\mu}{k}+\frac{1}{2}\right)\right)\left\{\left(\left(\frac{h\mu}{k}+\frac{1}{2}\right)\right)+\left(\left(\frac{h\mu}{k}\right)\right)\right\}\\ &= s(h,k)+\sum_{\mu \bmod k}\left(\left(\frac{\mu}{k}+\frac{1}{2}\right)\right)\left(\left(\frac{2h\mu}{k}\right)\right)\\ &= s(h,k)+r(2h,k).\end{aligned}$$

We have therefore established Theorem 6.

5. The sums $R(h,k)$ and $S(h,k)$. In order to facilitate the verification of Riemann's formulae we introduce, following Dedekind, two other functions: [7]

$$(5.1)\qquad R(h,k)=\sum_{\mu=1}^{[(k-1)/2]}\left(\left(\frac{h\mu}{k}-\frac{1}{2}\right)\right),\; S(h,k)=\sum_{\mu=1}^{[(k-1)/2]}\left(\left(\frac{h\mu}{k}\right)\right). \quad [39]$$

The following theorems and corollaries establish the essential relations between these functions.

THEOREM 7. *For any pair of integers h, k with $k>0$ we have*

$$(5.2)\qquad R(h,k)+S(h,k)=S(2h,k), \qquad [43_2]$$

$$(5.3)\qquad S(h,2k)+S(h+k,2k)=S(h,k)-R(h,k),$$

$$(5.4)\qquad R(2h,2k)=S(2h,2k)=0.$$

Proof. The formula (5.2) is very easily established. Using (2.41) we have at once

$$R(h,k)+S(h,k)=\sum_{\mu=1}^{[(k-1)/2]}\left\{\left(\left(\frac{h\mu}{k}-\frac{1}{2}\right)\right)+\left(\left(\frac{h\mu}{k}\right)\right)\right\}$$
$$=\sum_{\mu=1}^{[(k-1)/2]}\left(\left(\frac{2h\mu}{k}\right)\right)=S(2h,k).$$

In order to prove (5.3) we proceed as follows:

$$(5.5)\qquad S(h,2k)+S(h+k,2k)=\sum_{\mu=1}^{k-1}\left\{\left(\left(\frac{h\mu}{2k}\right)\right)+\left(\left(\frac{h\mu}{2k}+\frac{\mu}{2}\right)\right)\right\}$$
$$=\sum_{\mu=1}^{k-1}\left\{\left(\left(\frac{h\mu}{2k}\right)\right)+\left(\left(\frac{h\mu}{2k}+\frac{1}{2}\right)\right)\right\}-\sum_{\mu=1}^{[(k-1)/2]}\left(\left(\frac{h\mu}{k}+\frac{1}{2}\right)\right)+\sum_{\mu=1}^{[(k-1)/2]}\left(\left(\frac{h\mu}{k}\right)\right),$$

which we have obtained by separating the odd and even μ's. Now the last member of (5.5) is equal to

$$\sum_{\mu=1}^{k-1}\left(\left(\frac{h\mu}{k}\right)\right)-R(h,k)+S(h,k),$$

and this completes the proof of (5.3), in view of (2.31).

As far as (5.4) is concerned we have

$$S(2h,2k)=\sum_{\mu=1}^{k-1}\left(\left(\frac{2h\mu}{2k}\right)\right)=\sum_{\mu=1}^{k-1}\left(\left(\frac{h\mu}{k}\right)\right)=0,$$

which proves the second half of (5.4). The first half now follows immediately from (5.2).

[7] Dedekind writes $R(x)$ and $S(x)$ with $x=h/k$ instead of our $R(h,k)$ and $S(h,k)$, respectively. We have changed his notation since the sums do not depend only on the ratio h/k.

COROLLARY. *We have*

$$(5.61)\qquad R(h,k)+S(h,k)=0 \qquad \textit{for } k \textit{ even} \qquad [42_1]$$

$$(5.62)\qquad S(h,2k)+S(h+k,2k)=2S(h,k) \ \textit{for } k \textit{ even}$$

$$(5.63)\qquad R(h,2k)+R(h+k,2k)=2R(h,k) \ \textit{for } k \textit{ even} \qquad [42_2]$$

$$(5.64)\qquad S(h,2k)=S(h,k)-R(h,k) \ \textit{for } h \textit{ odd}, k \textit{ odd}$$

$$(5.65)\qquad R(h,2k)=R(h,k)-S(h,k) \ \textit{for } h \textit{ odd}, k \textit{ odd} \qquad [43_1]$$

Indeed, (5.61) follows from (5.2) and (5.4); (5.62) from (5.3) and (5.61); (5.63) from (5.62) and (5.61); (5.64) from (5.3) and (5.4) since $h+k$ is even; (5.65) from (5.64) by means of (5.61).

THEOREM 8. *For* $(h,k)=1$ *we have*

$$(5.7)\qquad R(h,k)-S(h,k)=R(h',k)-S(h',k)=\tfrac{1}{2}d, \qquad [40_3]$$

where $hh'\equiv 1 \pmod k$, *and where* d *denotes the excess of the number of positive least remainders over the number of negative least remainders in the sequence* $\mu h \bmod k$, $\mu=1,2,\cdots,\left[\frac{k-1}{2}\right]$.

Proof. We note that $((x))$ and $((x-\tfrac{1}{2}))$ vanish for the same values of x, viz., for $x\equiv 0$ or $\tfrac{1}{2} \pmod 1$, and are of opposite sign for all other values. More precisely

$$(5.8)\qquad ((x-\tfrac{1}{2}))-((x))=\begin{cases}-\tfrac{1}{2} \text{ for } ((x-\tfrac{1}{2}))<0,\\ \ \ \tfrac{1}{2} \text{ for } ((x-\tfrac{1}{2}))>0.\end{cases}$$

Therefore we obtain

$$(5.81)\qquad R(h,k)-S(h,k)=\sum_{\mu=1}^{[(k-1)/2]}\left\{\left(\left(\frac{h\mu}{k}-\frac{1}{2}\right)\right)-\left(\left(\frac{h\mu}{k}\right)\right)\right\}$$
$$=\tfrac{1}{2}\sum_{\mu=1}^{[(k-1)/2]}\operatorname{sgn}\left(\left(\frac{h\mu}{k}-\frac{1}{2}\right)\right).$$

Now it is readily seen that

$$\left(\left(\frac{h\mu}{k}-\frac{1}{2}\right)\right)=\frac{r_\mu}{k},\ \mu=1,2,\cdots,\left[\frac{k-1}{2}\right],$$

where

$$(5.82)\qquad h\mu\equiv r_\mu \pmod k, \qquad |r_\mu|\leqq\left[\frac{k-1}{2}\right].$$

Hence the last sum in (5.81) expresses the number d defined in the theorem, so that one part of the theorem is proved.

No two values of $|r_\mu|$ in (5.82) can be the same, and therefore they too cover the range from 1 to $\left[\frac{k-1}{2}\right]$. If we put

$$|r_\mu| = \nu, \qquad 1 \leqq \nu \leqq \left[\frac{k-1}{2}\right],$$

we can either have

$$h\mu \equiv \nu \pmod{k}$$

and hence

$$\mu \equiv h'\nu \pmod{k}, \qquad 1 \leqq \mu \leqq \left[\frac{k-1}{2}\right],$$

or we can have

$$h\mu \equiv -\nu \pmod{k}, \qquad 1 \leqq \nu \leqq \left[\frac{k-1}{2}\right],$$

and hence

$$-\mu \equiv h'\nu \pmod{k}, \qquad 1 \leqq \mu \leqq \left[\frac{k-1}{2}\right].$$

This shows that the positive (negative) least remainders of $\mu h \bmod k$, $\mu = 1, 2, \cdots, \left[\frac{k-1}{2}\right]$, are in one-to-one correspondence with the positive (negative) least remainders of $h'\nu \bmod k$, $\nu = 1, 2, \cdots, \left[\frac{k-1}{2}\right]$. Thus the proof of the second half of (5.8) is complete.

COROLLARY. *For k even and $(h, k) = 1$ we have*

(5.9) $$R(h,k) = R(h',k) = -S(h,k) = -S(h',k).$$

This corollary is an immediate consequence of (5.61) and (5.7).

6. Relations between $r(h,k)$, $s(h,k)$ and $R(h,k)$, $S(h,k)$. In this section we consider relations between the $R(h,k)$, $S(h,k)$ on the one hand and the $r(h,k)$, $s(h,k)$ on the other hand.

THEOREM 9. *If $(h,k) = 1$, then*

(6.1) $$S(h,k) = r(h,k) - s(h,k);$$

if, furthermore, $hh' \equiv 1 \pmod{k}$, then

(6.2) $$S(h',k) = s(2h,k) - 2s(h,k) \qquad [41]$$

and

(6.3) $$R(h',k) = r(2h,k) - 2r(h,k).$$

Proof. In order to prove (6.1) we make use of (5.8) and conclude that

$$\begin{aligned} r(h,k) - s(h,k) &= \sum_{\mu=1}^{k} \left\{ \left(\left(\frac{\mu}{k} - \frac{1}{2}\right)\right) - \left(\left(\frac{\mu}{k}\right)\right) \right\} \left(\left(\frac{h\mu}{k}\right)\right) \\ &= \tfrac{1}{2} \sum_{\mu=1}^{[(k-1)/2]} \left(\left(\frac{h\mu}{k}\right)\right) - \tfrac{1}{2} \sum_{\mu=[(k+1)/2]}^{k-1} \left(\left(\frac{h\mu}{k}\right)\right) \\ &= \tfrac{1}{2} S(h,k) - \tfrac{1}{2} \sum_{\mu=1}^{[(k-1)/2]} \left(\left(\frac{h(k-\mu)}{k}\right)\right) = S(h,k). \end{aligned}$$

The identity (6.2) now follows immediately from (6.1) and (4.3). We get thus

$$S(h,k) = s(2h',k) - 2s(h,k).$$

Interchanging h and h' and noting (2.51) we obtain (6.2).

As for the proof of (6.3), we have

$$\begin{aligned} r(2h,k) - 2r(h,k) &= \sum_{\mu \bmod k} \left(\left(\frac{\mu}{k} - \frac{1}{2}\right)\right)\left\{\left(\left(\frac{2h\mu}{k}\right)\right) - 2\left(\left(\frac{h\mu}{k}\right)\right)\right\} \\ &= \sum_{\mu=1}^{k} \left(\left(\frac{h'\mu}{k} - \frac{1}{2}\right)\right)\left\{\left(\left(\frac{2\mu}{k}\right)\right) - 2\left(\left(\frac{\mu}{k}\right)\right)\right\} \\ &= \sum_{\mu=1}^{k-1} \left(\left(\frac{h'\mu}{k} - \frac{1}{2}\right)\right)\left\{-\left[\frac{2\mu}{k}\right] + 2\left[\frac{\mu}{k}\right]\right\} + \tfrac{1}{2}\sum_{\mu=1}^{k-1}\left(\left(\frac{h'\mu}{k} - \frac{1}{2}\right)\right) \\ &= -\sum_{\mu=[(k+1)/2]}^{k-1} \left(\left(\frac{h'\mu}{k} - \frac{1}{2}\right)\right) = -\sum_{\mu=1}^{[(k-1)/2]}\left(\left(\frac{h'(k-\mu)}{k} - \frac{1}{2}\right)\right) \\ &= \sum_{\mu=1}^{[(k-1)/2]} \left(\left(\frac{h'\mu}{k} + \frac{1}{2}\right)\right) = R(h',k). \end{aligned}$$

Incidentally, a similar proof could have been given for (6.2).

COROLLARY. *For k even and $(h,k)=1$ we have*

$$S(h,k) = s(2h,k) - 2s(h,k). \tag{6.4}$$

This follows immediately from (6.2) and (5.9).

7. Further formulae. We are now in a position to derive the remaining formulae as consequences of the preceding theorems and corollaries without needing to go back to the definitions of the different functions.

THEOREM 10. *For h and k odd and $(h,k)=1$ we have*

$$R(h,k) = s(h,2k) - s(2h,k). \tag{7.1} \quad [46_1]$$

Proof. By (6.3) we know that

$$R(h,k) = r(2h',k) - 2r(h',k).$$

If we now let h' denote an odd solution of the congruence $hh' \equiv 1 \pmod{k}$, we may apply (4.5) and (4.3) and obtain

$$\begin{aligned} R(h,k) &= s(h',2k) - s(h',k) - r(h',k) \\ &= s(h',2k) - s(2h,k). \end{aligned}$$

Now since h, h' and k are all supposed to be odd it follows that the congruence $hh' \equiv 1 \pmod{k}$ implies $hh' \equiv 1 \pmod{2k}$, and therefore

$$s(h', 2k) = s(h, 2k),$$

because of (2.51). This completes the proof.

THEOREM 11. *For k odd and $(h, k) = 1$ we have*

$$S(2h, k) = s(h, k) - 2s(2h, k) \tag{7.2}$$

Proof. From (5.2), (6.2) and (6.3) we deduce

$$\begin{aligned} S(2h, k) &= R(h, k) + S(h, k) \\ &= r(2h', k) - 2r(h', k) + s(2h', k) - 2s(h', k). \end{aligned}$$

Furthermore, by (4.3) we have

$$r(2h', k) + s(2h', k) = s(2h^*, k),$$

where h^* satisfies the congruence

$$2h'h^* \equiv 1 \pmod{k}.$$

But this congruence may also be written in the form

$$2h^* \equiv h \pmod{k},$$

so that we get

$$S(2h, k) = s(h, k) - 2r(h', k) - 2s(h', k).$$

Applying (4.3) to the last two terms we obtain (7.2).

We can now finish the list of Dedekind's formulae. The remaining set of formulae is particularly interesting because these formulae express the Dedekind sums $s(h, k)$ in terms of the $R(h, k)$. This is of some importance since the summands of the latter are simpler than the summands of the former.

THEOREM 12. *For h and k odd and $(h, k) = 1$ we have*

$$s(h, k) = R(h, 2k) - \tfrac{2}{3}R(h, k) - \tfrac{2}{3}R(h', k). \tag{7.3}$$ [44₁]

Proof. By (6.2), (7.2) and (5.2)

$$\begin{aligned} 3s(h, k) &= -2S(h', k) - S(2h, k) \\ &= -2S(h', k) - R(h, k) - S(h, k). \end{aligned}$$

Let h' be odd. Then h' satisfies the congruence

$$hh' \equiv 1 \pmod{2k},$$

and we may apply (5.65) and (5.9):

$$\begin{aligned} 3s(h, k) &= 2R(h', 2k) - 2R(h', k) - R(h, k) + R(h, 2k) - R(h, k) \\ &= 3R(h, 2k) - 2R(h', k) - 2R(h, k). \end{aligned}$$

This completes our proof.

THEOREM 13. *For h and k odd and* $(h,k)=1$ *we have*

$$(7.4)\qquad s(2h,k)=R(h,2k)-\tfrac{4}{3}R(h,k)-\tfrac{1}{3}R(h',k). \qquad [44_2]$$

Proof. Solving the equations (6.2) and (7.2) for $s(2h,k)$ we get

$$\begin{aligned}3s(2h,k)&=-2S(2h,k)-S(h',k)\\&=-2R(h,k)-2S(h,k)-S(h',k).\end{aligned}$$

Let h' be odd. Then, applying (5.65) and (5.9), we derive the formula

$$\begin{aligned}3s(2h,k)&=-2R(h,k)+2R(h,2k)-2R(h,k)+R(h',2k)-R(h',k)\\&=-4R(h,k)+3R(h,2k)-R(h',k),\end{aligned}$$

which completes the proof.

THEOREM 14. *For h and k odd and* $(h,k)=1$ *we have*

$$(7.5)\qquad s\left(\frac{h+k}{2},k\right)=R(h,2k)-\tfrac{1}{3}R(h,k)-\tfrac{4}{3}R(h',k). \qquad [44_3]$$

Proof. We have obviously

$$\frac{h+k}{2}\cdot 2h\equiv 1 \pmod{k},$$

so that by (2.51)

$$s\left(\frac{h+k}{2},k\right)=s(2h',k).$$

If we again choose h' as an odd number we may apply (7.4) and (5.9) and obtain (7.5).

THEOREM 15. *For h and k odd and* $(h,k)=1$ *we have*

$$(7.6)\qquad s(h,2k)=R(h,2k)-\tfrac{1}{3}R(h,k)-\tfrac{1}{3}R(h',k). \qquad [44_4]$$

Proof. This theorem follows immediately from (7.1) and (7.4).

Dedekind remarks that the equations (7.3), (7.4), (7.5) and (7.6) together yield

$$(7.7)\qquad s(2h,k)+s\left(\frac{h+k}{2},k\right)+s(h,2k)=3s(h,k), \qquad [45]$$

which, by virtue of (2.7), agrees with (2.8) in the case $p=2$.

It will be noticed that in Theorems 12 to 15 the "denominators" are either k or $2k$ with k odd. In case the "denominator" contains higher powers of 2 the reduction to the $R(h,k)$ is not quite so short. For this purpose we need the following

THEOREM 16. *For h and k odd,* $(h,k)=1$ *and* $\lambda\geqq 1$ *we have*

$$(7.8) \qquad 2^\lambda s(h, 2^\lambda k) = R(h, 2k) + \sum_{a=1}^{\lambda} 2^{a-1} R(h, 2^a k) - \tfrac{2}{3} R(h, k) - \tfrac{2}{3} R(h', k).$$

Proof. This is obviously a generalization of (7.6). Making use of (6.4) and (2.7) we obtain the system of equations

$$2^a s(h, 2^a k) = 2^{a-1} s(h, 2^{a-1} k) - 2^{a-1} S(h, 2^a k), \qquad a = 2, 3, \cdots, \lambda.$$

Summing from $a = 2$ to $a = \lambda$ we get

$$2^\lambda s(h, 2^\lambda k) = 2s(h, 2k) - \sum_{a=2}^{\lambda} 2^{a-1} S(h, 2^a k)$$

$$= 2s(h, 2k) + \sum_{a=2}^{\lambda} 2^{a-1} R(h, 2^a k)$$

(because of (5.61))

$$= R(h, 2k) + \sum_{a=1}^{\lambda} 2^{a-1} R(h, 2^a k) - \tfrac{2}{3} R(h, k) - \tfrac{2}{3} R(h', k),$$

where we have made use of (7.6) in the last step. This completes the proof of the theorem.

Some further formulae [46] with which Dedekind closes his memoir may be deduced immediately from (7.3), (7.4), (7.5) and (7.6) by solving these linear equations for $R(h, k)$, $R(h', k)$ and $R(h, 2k)$.

PART II: Verification of Riemann's Formulae.

8. Verification of Riemann's formulae. The object of the following discussion is to complete Dedekind's verification of the formulae in Riemann's second fragment on modular functions **2**, pp. 461-465. These formulae concern the functions

$$(8.1) \qquad \log k, \qquad \log k', \qquad \log \frac{2K}{\pi}$$

which appear in the theory of elliptic functions (Legendre's notation), and which are here expressed in Jacobi's manner as series in $q = e^{\pi i \omega}$.

Riemann's problem was to study the behavior of these functions as ω tends to a rational point

$$\omega = \frac{h}{k} + iy$$

with y approaching 0 through real values. Riemann, for the sake of brevity, does not mention the infinitesimal y at all and consequently disregards the real parts of the functions (8.1), which depend on y and tend to infinity as y

tends to 0. However, the imaginary parts of the functions tend to certain finite limits depending only on the rational value $\frac{h}{k}$, the real part of ω, and it is with these limits that we are concerned here.

Riemann divides the discussion of each of the functions (8.1) into several subcases, according to the parity of h and k in $\frac{h}{k}$, whereas Dedekind's formulae [35] cover each function with one equation. Observing that Riemann writes $q = e^{xi}$ instead of Dedekind's $q = e^{\pi i\omega}$ we are led, by comparing the two sets of equations, to the following eight identities which we shall have to prove; it is always understood that $(h, k) = 1$.

$$\mathrm{I}_1 \quad \frac{h}{2k} + \sum_{\mu=1}^{k-1} (-1)^\mu \frac{h\mu}{2k} - 2\sum_{\mu=1}^{k-1} (-1)^\mu \left[\frac{h\mu}{2k} + \frac{1}{2}\right] = \frac{h}{2k} + 4s(h+k, 2k) - 4s(2h, k), \qquad h \text{ even, } k \text{ odd};$$

$$\mathrm{I}_2 \quad \frac{h-h'}{2} + \frac{h'}{2k} + 2\sum_{\mu=1}^{k-1} (-1)^\mu \left[\frac{h'\mu}{2k}\right] - 2\sum_{\mu=1}^{k-1} (-1)^\mu \left[\frac{h\mu}{2k}\right] = \frac{h}{2k} + 4s(h+k, 2k) - 4s(2h, k), \qquad h, h', k \text{ odd};$$

$$\mathrm{I}_3 \quad \frac{h}{2k} - 8\sum_{\mu=1}^{k-1} (-1)^\mu \left(\left(\frac{h\mu}{2k}\right)\right)\left(\left(\frac{\mu}{2k} - \frac{1}{2}\right)\right) = \frac{h}{2k} + 4s(h+k, 2k) - 4s(2h, k), \qquad h \text{ odd, } k \text{ even};$$

$$\mathrm{II}_1 \quad \frac{(h-2)k}{2} - 4\sum_{\mu=0}^{k/2-1} \left[\frac{h(2\mu+1)}{2k}\right] = 4s(h+k, 2k) - 4s(h, 2k), \qquad h \text{ odd, } k \text{ even};$$

$$\mathrm{II}_2 \quad (-1)^h \left\{\frac{k^2-1}{2k} h' - 4\sum_{\mu=1}^{(k-1)/2} \left[\frac{h'\mu}{k} + \frac{1}{2}\right]\right\} = 4s(h+k, 2k) - 4s(h, 2k), \qquad k \text{ odd};$$

$$\mathrm{III}_1 \quad 2\sum_{\mu=1}^{(k-1)/2} \left\{\left(\left(\frac{h\mu}{k}\right)\right) - \left(\left(\frac{h\mu}{k} - \frac{1}{2}\right)\right)\right\} = 2s(h, k) - 4s(h+k, 2k), \qquad h \text{ even, } k \text{ odd};$$

$$\mathrm{III}_2 \quad 8\sum_{\mu=1}^{k-1} (-1)^\mu \left(\left(\frac{h'\mu}{2k}\right)\right) = 2s(h, k) - 4s(h+k, 2k), \qquad h, h', k \text{ odd};$$

$$\mathrm{III}_3 \quad 2\sum_{\mu=1}^{k-1} (-1)^\mu \left(\left(\frac{h''\mu}{k}\right)\right) = 2s(h, k) - 4s(h+k, 2k), \qquad hh'' \equiv 1 \pmod{2k}, \quad h \text{ odd, } k \text{ even}.$$

The theorems of the preceding sections enable us to present the proofs in an abridged manner. In the following proofs the numbers at the end of some of the lines indicate the formulae which have been used. For the application of (7.3), $\cdots$, (7.6) the parity of h has to be observed carefully.

Proof of I_1. h even, k odd.

$$4s(h+k,2k)-4s(2h,k)=4s(h+k,2k)-4s(2(h+k),k)$$

$$=4(R(h+k,2k)-\tfrac{1}{3}R(h,k)-\tfrac{1}{3}R(h',k)) \qquad (7.6)$$

$$-4(R(h+k,2k)-\tfrac{4}{3}R(h,k)-\tfrac{1}{3}R(h',k)) \qquad (7.4)$$

$$=4R(h,k);$$

$$2\sum_{\mu=1}^{k-1}(-1)^\mu\frac{h\mu}{2k}-2\sum_{\mu=1}^{k-1}(-1)^\mu\left[\frac{h\mu}{2k}+\frac{1}{2}\right]$$

$$=2\sum_{\mu=1}^{k-1}(-1)^\mu\left(\left(\frac{h\mu}{2k}+\frac{1}{2}\right)\right) \qquad (1.1)$$

$$=-2\sum_{\mu=1}^{k-1}\left(\left(\frac{h\mu}{2k}+\frac{1}{2}\right)\right)+4\sum_{\mu=1}^{(k-1)/2}\left(\left(\frac{h\mu}{k}+\frac{1}{2}\right)\right)$$

$$=-2R(h,2k)+4R(h,k) \qquad (5.1)$$

$$=4R(h,k). \qquad (5.4)$$

Proof of I_2. h, h', k odd.

$$4s(h+k,2k)-4s(2h,k)$$

$$=4(R(h,2k)-\tfrac{1}{3}R(h,k)-\tfrac{4}{3}R(h',k)) \qquad (2.7),\ (7.5)$$

$$-4(R(h,2k)-\tfrac{4}{3}R(h,k)-\tfrac{1}{3}R(h',k)) \qquad (7.4)$$

$$=4(R(h,k)-R(h',k))$$

$$=4(S(h,k)-S(h',k)); \qquad (5.7)$$

$$2\sum_{\mu=1}^{k-1}(-1)^\mu\left\{\left[\frac{h'\mu}{2k}\right]-\left[\frac{h\mu}{2k}\right]\right\}$$

$$=-2\sum_{\mu=1}^{k-1}(-1)^\mu\left\{\left(\left(\frac{h'\mu}{2k}\right)\right)-\left(\left(\frac{h\mu}{2k}\right)\right)-\frac{h'\mu}{2k}+\frac{h\mu}{2k}\right\} \qquad (1.1)$$

$$=\frac{h'-h}{k}\sum_{\mu=1}^{k-1}(-1)^\mu\mu+2\sum_{\mu=1}^{k-1}\left\{\left(\left(\frac{h'\mu}{2k}\right)\right)-\left(\left(\frac{h\mu}{2k}\right)\right)\right\}$$

$$-4\sum_{\mu=1}^{(k-1)/2}\left\{\left(\left(\frac{h'\mu}{k}\right)\right)-\left(\left(\frac{h\mu}{k}\right)\right)\right\}$$

$$=\frac{h'-h}{k}\frac{k-1}{2}+2\{S(h',2k)-S(h,2k)\}$$

$$-4\{S(h',k)-S(h,k)\} \qquad (5.1)$$

$$=-\frac{h-h'}{2}-\frac{h'}{2k}+\frac{h}{2k}+4(S(h,k)-S(h',k)), \qquad (5.9)$$

since $hh'\equiv 1 \pmod{2k}$.

Proof of I_3. h odd, k even.

$$-8\sum_{\mu=1}^{k-1}(-1)^\mu\left(\left(\frac{h\mu}{2k}\right)\right)\left(\left(\frac{\mu}{2k}-\frac{1}{2}\right)\right)$$

$$=-4\sum_{\mu=1}^{k-1}(-1)^\mu\left(\left(\frac{h\mu}{2k}\right)\right)\left(\left(\frac{\mu}{2k}-\frac{1}{2}\right)\right)$$

$$-4\sum_{\mu=1}^{k-1}(-1)^\mu\left(\left(\frac{h(2k-\mu)}{2k}\right)\right)\left(\left(\frac{2k-\mu}{2k}-\frac{1}{2}\right)\right) \quad (2.1), (2.2)$$

$$=-4\sum_{\mu=1}^{2k}(-1)^\mu\left(\left(\frac{h\mu}{2k}\right)\right)\left(\left(\frac{\mu}{2k}-\frac{1}{2}\right)\right) \quad (1.1)$$

$$=4\sum_{\mu=1}^{2k}\left(\left(\frac{h\mu}{2k}\right)\right)\left(\left(\frac{\mu}{2k}-\frac{1}{2}\right)\right)-8\sum_{\mu=1}^{k}\left(\left(\frac{h\mu}{k}\right)\right)\left(\left(\frac{\mu}{k}-\frac{1}{2}\right)\right)$$

$$=4r(h,2k)-8r(h,k) \quad (4.1)$$

$$=4s(2h,2k)-4s(h,2k)-8s(2h,k)+8s(h,k) \quad (4.31)$$

$$=4(3s(h,k)-s(h,2k)-2s(2h,k)) \quad (2.7)$$

$$=4(s(h+k,2k)-s(2h,k)). \quad (7.7)$$

Proof of II_1. h odd, k even.

Put $k=2^\lambda k_1$, $\lambda\geqq 1$, with k_1 odd. Then

$$2^{\lambda+1}(s(h+k,2k)-s(h,2k)$$

$$=2^{\lambda+1}s(h+2^\lambda k_1,2^{\lambda+1}k_1)-2^{\lambda+1}s(h,2^{\lambda+1}k_1)$$

$$=R(h,2k_1)+\sum_{a=1}^{\lambda+1}2^{a-1}R(h+2^\lambda k_1,2^a k_1)-\tfrac{2}{3}R(h,k_1)-\tfrac{2}{3}R(h',k_1)$$

$$-R(h,2k_1)-\sum_{a=1}^{\lambda+1}2^{a-1}R(h,2^a k_1)+\tfrac{2}{3}R(h,k_1)+\tfrac{2}{3}R(h',k_1) \quad (7.8)$$

$$=2^\lambda R(h+2^\lambda k_1,2^{\lambda+1}k_1)-2^\lambda R(h,2^{\lambda+1}k_1)$$

$$=2^\lambda(R(h+k,2k)-R(h,2k))$$

or

$$(8.2) \qquad s(h+k,2k)-s(h,2k)=\tfrac{1}{2}(R(h+k,2k)-R(h,2k))$$

h, odd, k even.

On the other hand

$$-\sum_{\mu=1}^{k/2-1}\left[\frac{h(2\mu+1)}{2k}\right]=-\sum_{\mu=1}^{k-1}\left[\frac{h\mu}{2k}\right]+\sum_{\mu=1}^{k/2-1}\left[\frac{h\mu}{k}\right]$$

$$=\sum_{\mu=1}^{k-1}\left(\left(\frac{h\mu}{2k}\right)\right)-\sum_{\mu=1}^{k-1}\left(\frac{h\mu}{2k}-\frac{1}{2}\right)-\sum_{\mu=1}^{k/2-1}\left(\left(\frac{h\mu}{k}\right)\right)+\sum_{\mu=1}^{k/2-1}\left(\frac{h\mu}{k}-\frac{1}{2}\right) \quad (1.1)$$

$$=S(h,2k)-S(h,k)+\frac{k}{4}-\frac{h}{2k}\frac{(k-1)k}{2}+\frac{h}{k}\left(\frac{k}{2}-1\right)\frac{k}{4} \quad (5.1)$$

$$=-R(h,2k)+\tfrac{1}{2}(R(h,2k)+R(h+k,2k))-\frac{k}{8}(h-2)$$

(5.61), (5.63)

$$=\tfrac{1}{2}(R(h+k,2k)-R(h,2k))-\frac{k}{8}(h-2),$$

which, with (8.2), completes the proof.

Proof of II_2. k odd.

1) h odd.

$$\begin{aligned} & s(h+k,2k)-s(h,2k) \\ &\quad = R(h,2k)-\tfrac{1}{3}R(h,k)-\tfrac{4}{3}R(h',k) && (2.7),(7.5) \\ &\quad\quad -R(h,2k)+\tfrac{1}{3}R(h,k)+\tfrac{1}{3}R(h',k) && (7.6) \\ &\quad = -R(h',k). \end{aligned}$$

2) h even.

$$\begin{aligned} & s(h+k,2k)-s(h,2k) = s(h+k,2k)-s(h+k+k,2k) \\ &\quad = R(h+k,2k)-\tfrac{1}{3}R(h,k)-\tfrac{1}{3}R(h',k) && (7.6) \\ &\quad\quad -R(h+k,2k)+\tfrac{1}{3}R(h,k)+\tfrac{4}{3}R(h',k) && (2.7),(7.5) \\ &\quad = R(h',k) \end{aligned}$$

Hence, in either case, we have

$$s(h+k,2k)-s(h,2k) = (-1)^h R(h',k), \qquad k \text{ odd}. \tag{8.3}$$

On the other hand

$$\begin{aligned} -\sum_{\mu=1}^{(k-1)/2}\left[\frac{h'\mu}{2k}+\frac{1}{2}\right] &= \sum_{\mu=1}^{(k-1)/2}\left(\left(\frac{h'\mu}{2k}+\frac{1}{2}\right)\right)-\sum_{\mu=1}^{(k-1)/2}\frac{h'\mu}{2k} && (1.1) \\ &= R(h',k)-\frac{h'}{8k}(k^2-1), && (5.1) \end{aligned}$$

which, together with (8.3), completes the proof.

Proof of III_1. h even, k odd.

$$\begin{aligned} & 2s(h,k)-4s(h+k,2k) = 2s(h+k,k)-4s(h+k,2k) \\ &\quad = 2(R(h+k,2k)-\tfrac{2}{3}R(h,k)-\tfrac{2}{3}R(h',k)) && (7.3) \\ &\quad\quad -4(R(h+k,2k)-\tfrac{1}{3}R(h,k)-\tfrac{1}{3}R(h',k)) && (7.6) \\ &\quad = -2R(h+k,2k) = -2(R(h+k,k)-S(h+k,k)) && (5.65) \\ &\quad = 2(S(h,k)-R(h,k)) = 2\sum_{\mu=1}^{(k-1)/2}\left\{\left(\left(\frac{h\mu}{k}\right)\right)-\left(\left(\frac{h\mu}{k}-\frac{1}{2}\right)\right)\right\} && (5.1) \end{aligned}$$

Proof of III_2. h, h', k odd.

$$\begin{aligned} 2s(h,k)-4s(h+k,2k) &= 2(R(h,2k)-\tfrac{2}{3}R(h,k)-\tfrac{2}{3}R(h',k)) && (7.3) \\ &\quad -4(R(h,2k)-\tfrac{1}{3}R(h,k)-\tfrac{4}{3}R(h',k)) && (7.5) \\ &= -2R(h,2k)+4R(h',k); \end{aligned}$$

$$\begin{aligned} 2\sum_{\mu=1}^{k-1}(-1)^\mu\left(\left(\frac{h'\mu}{2k}\right)\right) &= -2\sum_{\mu=1}^{k-1}\left(\left(\frac{h'\mu}{2k}\right)\right)+4\sum_{\mu=1}^{(k-1)/2}\left(\left(\frac{h'\mu}{k}\right)\right) \\ &= -2S(h',2k)+4S(h',k) \\ &= -2S(h',2k)+4R(h',k)-4R(h',2k) && (5.65) \\ &= -2R(h,2k)+4R(h',k), && (5.9) \end{aligned}$$

since $hh' \equiv 1 \pmod{2k}$.

Proof of III_3. h odd, k even, $hh'' \equiv 1 \pmod{2k}$.

$$2s(h,k) - 4s(h+k, 2k) = 2(s(2(h+k), 2k) - 2s(h+k, 2k)) \qquad (2.7)$$
$$= 2S(h+k, 2k) = 4S(h,k) - 2S(h, 2k); \qquad (6.4), (5.62)$$
$$2\sum_{\mu=1}^{k-1}(-1)^{\mu}\left(\left(\frac{h''\mu}{2k}\right)\right) = -2\sum_{\mu=1}^{k-1}\left(\left(\frac{h''\mu}{2k}\right)\right) + 4\sum_{\mu=1}^{(k-1)/2}\left(\left(\frac{h''\mu}{k}\right)\right)$$
$$= -2S(h'', 2k) + 4S(h'', k) \qquad (5.1)$$
$$= -2S(h, 2k) + 4S(h, k), \qquad (5.9)$$

which completes the proof.

PART III: Congruences for the $s(h,k)$ and the Factorization of the $A_k(n)$.

9. Congruences for the $s(h,k)$. Whereas in the preceding sections we have studied equations containing the Dedekind sums $s(h,k)$, we shall now consider congruences for the $s(h,k)$. Some of these congruences will appear as direct consequences of the reciprocity formula (3.1) and others, among them one already mentioned by Dedekind, will require a return to the definition of $s(h,k)$. The main new result is contained in Theorem 20, which is supplemented by Theorems 21 and 22. We shall apply these theorems to the study of certain sums $A_k(n)$ which appear in the theory of partitions.

In this chapter h and k are always supposed to be coprime. We remark further that a congruence in which fractions appear will always be understood in the ordinary way, viz., that the difference of its members is an *integral* multiple of the modulus (that is, we do not introduce the generalization of congruences for fractions due to Gauss, *Disquisitiones arithmeticae, art.* 31).

We find it very helpful to introduce a special symbol:

Definition. The letter θ, appearing as the first factor of a product, represents 1 if the following factors are all prime to 3, and represents 3 if at least one of the following factors is divisible by 3.

A product $\theta ab \cdots$ is therefore either prime to 3 or divisible at least by 9. Using this notation we may now state

THEOREM 17. *For* $(h,k) = 1$ *we have*

$$12hks(h,k) - h^2 \equiv 1 \pmod{\theta k}. \tag{9.1}$$

Proof. From the definition (2.6) we obtain

$$12ks(h,k) = 12\sum_{\mu=1}^{k}\mu\left(\left(\frac{h\mu}{k}\right)\right) = 12\sum_{\mu=1}^{k-1}\mu\left(\frac{h\mu}{k} - \left[\frac{h\mu}{k}\right] - \frac{1}{2}\right)$$

$$(9.2)\qquad = 2h(k-1)(2k-1) - 12\sum_{\mu=1}^{k-1}\mu\left[\frac{h\mu}{k}\right] - 3k(k-1),$$

so that

$$12ks(h,k) \equiv h(k-1)(k+1) \pmod 3,$$

from which it follows immediately that

$$(9.21)\qquad 12ks(h,k) \equiv 0 \pmod 3 \text{ if and only if } 3 \nmid k.$$

On the other hand the reciprocity formula (3.1) yields

$$(9.3)\qquad 12hks(h,k) + 12khs(k,h) = -3hk + h^2 + k^2 + 1$$
$$\equiv h^2 + 1 \pmod{\theta k}.$$

Interchanging h and k in (9.2) and (9.21) we see at once that $12hs(k,h)$ is an integer and is moreover divisible by 3 if $3 \nmid h$. But if $3 \mid k$ then $3 \nmid h$ since $(h,k) = 1$.

Therefore

$$(9.4)\qquad 12khs(k,h) \equiv 0 \pmod{\theta k},$$

which, together with (9.3), completes the proof.

THEOREM 18. (Dedekind **3**, § 6). *For k odd we have*

$$(9.5)\qquad 12ks(h,k) \equiv k + 1 - 2\left(\frac{h}{k}\right) \pmod 8,$$

where $\left(\frac{h}{k}\right)$ *denotes the Jacobi symbol.*

Proof. As a consequence of (9.2) we have

$$12ks(h,k) \equiv -2h(k-1) - 4\sum_{\mu=1}^{k-1}\mu\left[\frac{h\mu}{k}\right] + k(k-1) \pmod 8$$
$$\equiv (k-1)(k-2h) - 4\sum_{\substack{\mu=1\\ \mu \text{ odd}}}^{k-1}\left[\frac{h\mu}{k}\right] \pmod 8$$
$$\equiv (k-1)(k-2h) - 4\sum_{\mu=1}^{k-1}\left[\frac{h\mu}{k}\right] + 4\sum_{\mu=1}^{(k-1)/2}\left[\frac{2h\mu}{k}\right] \pmod 8.$$

Now the first sum in the right member of the last congruence is elementary since

$$(9.6)\qquad \sum_{\mu=1}^{k-1}\left[\frac{h\mu}{k}\right] = -\sum_{\mu=1}^{k-1}\left(\left(\frac{h\mu}{k}\right)\right) + \sum_{\mu=1}^{k-1}\left(\frac{h\mu}{k} - \frac{1}{2}\right) = \tfrac{1}{2}(h-1)(k-1),$$

where we have made use of (2.31). We obtain therefore

$$(9.7)\qquad 12ks(h,k) \equiv (k-1)(k-4h+2) + 4\sum_{\mu=1}^{(k-1)/2}\left[\frac{2h\mu}{k}\right] \pmod 8.$$

The generalized Gauss's lemma in the theory of the Jacobi symbol may be stated as follows:[8]

$$(9.8)\qquad \frac{1}{2}\left\{1-\left(\frac{h}{k}\right)\right\} \equiv \sum_{\mu=1}^{(k-1)/2}\left[\frac{2h\mu}{k}\right] \pmod 2.$$

Introducing (9.8) into (9.7) we arrive, after some simplifications, at the desired result.

In order to deal also with even k we establish the following supplementary theorem:

THEOREM 19. *If k is equal to $2^\lambda k_1$, $\lambda \equiv 0$, and k_1 and h are odd integers, then*

$$(9.9)\qquad 12hks(h,k) \equiv h^2+k^2+3k+1+2k\left(\frac{k}{h}\right) \pmod{2^{\lambda+3}}.$$

Proof. From (3.1) and (9.2) we get

$$12hks(h,k) = -2k^2(h-1)(2h-1)+12k\sum_{\nu=1}^{h-1}\nu\left[\frac{2k\nu}{h}\right]$$
$$+3kh(h-1)-3hk+h^2+k^2+1.$$

Proceeding exactly as at the corresponding point in the proof of Theorem 18, we find

$$12hks(h,k) \equiv 2k^2(h-1)+4k\sum_{\nu=1}^{h-1}\left[\frac{k\nu}{h}\right]-4k\sum_{\nu=1}^{h-1/2}\left[\frac{2k\nu}{h}\right]$$
$$+3kh^2-6hk+h^2+k^2+1 \pmod{2^{\lambda+3}}.$$

Then (9.6) and (9.8) together imply

$$12hks(h,k) \equiv 2k^2(h-1)+2k(k-1)(h-1)$$
$$-2k\left\{1-\left(\frac{k}{h}\right)\right\}+3k+2hk+h^2+k^2+1 \pmod{2^{\lambda+3}},$$

where we have noted that $h^2 \equiv 1 \pmod 8$. This leads directly to (9.9).

---

[8] The generalization of Gauss's lemma is usually attributed to E. Schering, "Zur Theorie der quadratischen Reste," *Acta Mathematica*, vol. 1 (1882), pp. 153-170. Actually, it had already been obtained fifteen years earlier by Morgan Jenkins, "Proof of an Arithmetical Theorem leading, by means of Gauss's Fourth Demonstration of Legendre's Law of Reciprocity, to the extension of that Law," *Proceedings of the London Mathematical Society*, (1), vol. 2 (1867), pp. 29-32. For an exceptionally clear exposition of the proof see P. Bachmann, *Die Elemente der Zahlentheorie*, Leipzig (1892), pp. 144-148. The generalized lemma states that if k is an odd positive integer and $(h,k)=1$, then the Jacobi symbol (h/k) is equal to $(-1)^m$, where m is the number of least positive remainders exceeding $k/2$ in the sequence $\mu h \bmod k$, $\mu=1,\cdots,(k-1)/2$. In order to deduce our version of the lemma we observe that if $\mu h=\lambda k+r_\mu$, $k/2<r_\mu<k$, then $2\mu h=(2\lambda+1)k+2r_\mu-k$, and conversely. Hence, if $k/2<k$, then $[2\mu h/k]=2\lambda+1$ is an odd integer, and conversely. Consequently $m\equiv\sum_{\mu=1}^{(k-1)/2}[2\mu h/k]$ (mod 2).

10. The principal theorem. The main feature of the following theorem is that it deals with three variables a, b, c simultaneously, as distinguished from the pair of variables h and k of our previous discussions.

THEOREM 20. *Let a, b, c be positive integers such that*

$$(a, b) = (b, c) = (c, a) = 1 \tag{10.11}$$

and

$$24 \mid abc. \tag{10.12}$$

Then we have

$$\left(s(ab, c) - \frac{ab}{12c}\right) + \left(s(bc, a) - \frac{bc}{12a}\right) - \left(s(b, ac) - \frac{b}{12ac}\right) \equiv 0\ (\mathrm{mod}\ 2 \tag{10.2}$$

Proof. Multiply both members of the congruence (10.2) by $12ac$ and put

$$A = (12acs(ab, c) - a^2b) + (12acs(bc, a) - c^2b) - (12acs(b, ac) - \tag{10.21}$$

Our task is then to prove that

$$A \equiv 0 \pmod{24ac}. \tag{10.22}$$

We shall have to distinguish two principal cases according to the parity of ac on the one hand and of b on the other.

Case I. Let ac be odd. Then, because of (10.12), $8 \mid b$. We shall prove, first of all, that $A \equiv 0 \pmod{3ac}$; and, secondly, that $A \equiv 0 \pmod{8}$.

To begin with, (9.1) and (9.4) imply that the congruence

$$(12abcs(ab, c) - a^2b^2) + (12abcs(bc, a) - c^2b^2) \equiv 1$$

holds modulo θc as well as modulo θa, and is therefore also true modulo θac. Again, it follows from (9.1) that

$$12abcs(b, ac) - b^2 \equiv 1 \pmod{\theta ac}.$$

Subtracting the last two congruences and dividing by b, which is prime to ac, we obtain

$$A \equiv 0 \pmod{\theta ac}. \tag{10.23}$$

We still have to prove that $A \equiv 0 \pmod{3ac}$. If $3 \mid ac$ this follows at once from (10.23). But if $3 \nmid ac$ then, because of the hypothesis (10.12), $3 \mid b$. Furthermore, from (9.21) we infer that

$$3 \mid 12cs(ab, c), \qquad 3 \mid 12as(bc, a), \qquad 3 \mid 12acs(b, ac).$$

Hence, in this case, we see directly from (10.21) that $A \equiv 0 \pmod{3}$. Combining our results we deduce that in any event

$$A \equiv 0 \pmod{3ac}.$$

We next prove that $A \equiv 0 \pmod 8$. For this purpose we appeal to Theorem 18 and deduce, since $8 \mid b$, that

$$A \equiv a\left(c+1-2\left(\frac{ab}{c}\right)\right)+c\left(a+1-2\left(\frac{bc}{a}\right)\right)-\left(ac+1-2\left(\frac{b}{ac}\right)\right) \pmod 8$$

$$\equiv ac+a+c-1-2\left\{a\left(\frac{a}{c}\right)\left(\frac{b}{c}\right)+c\left(\frac{b}{a}\right)\left(\frac{c}{a}\right)-\left(\frac{b}{a}\right)\left(\frac{b}{c}\right)\right\} \pmod 8$$

$$\equiv -ac+a+c-1+2+2\left\{\left(\frac{c}{a}\right)-a\left(\frac{b}{c}\right)\right\}\left\{\left(\frac{a}{c}\right)-c\left(\frac{b}{a}\right)\right\}$$

$$-2(ac-1)\left\{\left(\frac{b}{c}\right)\left(\frac{b}{a}\right)-1\right\}-2\left(\frac{c}{a}\right)\left(\frac{a}{c}\right) \pmod 8.$$

In the last congruence the expression within each brace is even since a and c are odd. Moreover, the last product may be evaluated by means of the reciprocity law for Jacobi's symbol. We obtain thus

$$A \equiv 4\left(-\frac{(a-1)(c-1)}{4}+\frac{1}{2}\left(1-(-1)^{(a-1)(c-1)/4}\right)\right) \pmod 8.$$

But since for any integer n we have

(10.24) $$n \equiv \tfrac{1}{2}(1-(-1)^n) \pmod 2$$

we conclude that

$$A \equiv 0 \pmod 8.$$

This completes the proof of Theorem 20 when Case I holds.

Case II. Let ac be even. Since our theorem is symmetric in a and c we may more specifically assume that a is even. It then follows from the hypotheses (10.11) and (10.12) that indeed $8 \mid a$. We now specialize further and consider first of all

Case IIa. Let $a = 2^\lambda$, bc be odd and $3 \mid bc$. This case requires only that $\lambda \geqq 3$. However, it will be convenient, for a later purpose, to assume that $\lambda \geqq 1$ as long as possible.

Replacing a by 2^λ in (10.21) we obtain the following expression

(10.31) $$B = (12 \cdot 2^\lambda c s(2^\lambda b, c) - 2^{2\lambda} b) + (12 \cdot 2^\lambda c s(bc, 2^\lambda) - c^2 b) - (12 \cdot 2^\lambda c s(b, 2^\lambda c) - b)\,.$$

We have to prove that

(10.32) $$B \equiv 0 \pmod{24 . 2^\lambda c}.$$

We shall prove, first of all, that $B \equiv 0 \pmod 3$; and secondly, that $B \equiv 0 \pmod{2^{\lambda+3}}$.

Employing an argument similar to that used in Case I we have by (9.1) and (9.3)

$$(12\cdot 2^\lambda bcs(2^\lambda b, c) - 2^{2\lambda}b^2) + (12\cdot 2^\lambda bcs(bc, 2^\lambda) - c^2b^2) \equiv 1 \pmod{\theta c}$$

and

$$12\cdot 2^\lambda bcs(b, 2^\lambda c) - b^2 \equiv 1 \pmod{\theta c}.$$

Hence $B \equiv 0 \pmod{\theta c}$. Now if $3\,|\,c$ it follows immediately that $B \equiv 0 \pmod{3c}$. But if $3\nmid c$, then we have $3\,|\,b$, which, together with (9.21) and (9.4), implies that each summand of B is divisible by 3. In either case we conclude that

$$B \equiv 0 \pmod{3c}. \tag{10.33}$$

We next prove that $B \equiv 0 \pmod{2^{\lambda+3}}$. From Theorems 18 and 19 we infer that

$$\begin{aligned}
bB &\equiv 2^\lambda b\left(c+1-2\left(\frac{2^\lambda b}{c}\right)\right) - 2^{2\lambda}b^2 + \left(2^{2\lambda}+3\cdot 2^\lambda+1+2^{\lambda+1}\left(\frac{2^\lambda}{bc}\right)\right)\\
&\quad -\left(2^{2\lambda}c^2+3\cdot 2^\lambda c+1+2^{\lambda+1}c\left(\frac{2^\lambda c}{b}\right)\right) \pmod{2^{\lambda+3}}\\
&\equiv -2^{2\lambda}b^2+2^\lambda(3-3c+b(c+1))\\
&\quad -2^{\lambda+1}\left\{b\left(\frac{2^\lambda}{c}\right)\left(\frac{b}{c}\right)+c\left(\frac{2^\lambda}{b}\right)\left(\frac{c}{b}\right)-\left(\frac{2^\lambda}{b}\right)\left(\frac{2^\lambda}{c}\right)\right\} \pmod{2^{\lambda+3}}\\
&\equiv -2^{2\lambda}b^2+2^\lambda(c-1+bc+b)+2^{\lambda+1}\left\{b\left(\frac{2^\lambda}{c}\right)-\left(\frac{c}{b}\right)\right\}\left\{c\left(\frac{2^\lambda}{b}\right)-\left(\frac{b}{c}\right)\right\}\\
&\quad -2^{\lambda+1}\left\{\left(bc-1\right)\left(\frac{2^\lambda}{b}\right)\left(\frac{2^\lambda}{c}\right)+\left(\frac{c}{b}\right)\left(\frac{b}{c}\right)\right\} \pmod{2^{\lambda+3}}\\
&\equiv -2^{2\lambda}b^2-2^{\lambda+1}(bc-1)\left\{\left(\frac{2^\lambda}{b}\right)\left(\frac{2^\lambda}{c}\right)-1\right\}+2^\lambda(c+1-bc+b)\\
&\quad -2^{\lambda+1}(-1)^{[(b-1)(c-1)]/4} \pmod{2^{\lambda+3}}\\
&\equiv -2^{2\lambda}b^2-2^{\lambda+1}\left\{\frac{(b-1)(c-1)}{2}+\left((-1)^{[(b-1)(c-1)]/4}-1\right)\right\} \pmod{2^{\lambda+3}}\\
&\equiv -2^{2\lambda}b^2 \pmod{2^{\lambda+3}},
\end{aligned}$$

where we have again made use of the reciprocity law for Jacobi's symbol and (10.24). Dividing by b we now get

$$B \equiv -2^{2\lambda}b \pmod{2^{\lambda+3}}, \tag{10.34}$$

which for $\lambda \geqq 3$ yields

$$B \equiv 0 \pmod{2^{\lambda+3}}.$$

This completes the proof of the Theorem when Case IIa holds.

We now return to the more general

Case II. Let $a=2^{\lambda}a_1$, b, c be odd, $3|a_1bc$ and $\lambda\geqq 3$. It is easy to prove the theorem in this case by employing the results of Case I and Case IIa. For the sake of brevity we introduce the following notation:

$$t(h,k)=s(h,k)-\frac{h}{12k}.$$

Then (10. 2) becomes in the present case

$$t(2^{\lambda}a_1b,c)+t(bc,2^{\lambda}a_1)-t(b,2^{\lambda}a_1c)\equiv 0 \pmod 2. \tag{10.4}$$

But this congruence is a consequence of the three congruences:

$$t(2^{\lambda}a_1b,c)+t(2^{\lambda}bc,a_1)-t(2^{\lambda}b,a_1c)\equiv 0 \pmod 2, \tag{10.41}$$

$$t(bc,2^{\lambda}a_1)-t(bca_1,2^{\lambda})-t(2^{\lambda}bc,a_1)\equiv 0 \pmod 2, \tag{10.42}$$

$$-t(b,2^{\lambda}a_1c)+t(bca_1,2^{\lambda})+t(2^{\lambda}b,a_1c)\equiv 0 \pmod 2, \tag{10.43}$$

where (10. 41) falls under Case I and (10. 42) and (10. 43) are covered by Case IIa. Adding (10. 41), (10. 42) and (10. 43) we get (10. 4). This concludes the proof of Theorem 20.

Finally, in order to dispose of the cases in which $\lambda=1$ and $\lambda=2$, we treat two cases of lesser generality, involving only two variables b and c.

THEOREM 21. *Let bc be odd, $3|bc$ and $(b,c)=1$. Then we have*

$$\left(s(4b,c)-\frac{b}{3c}\right)+\left(s(bc,4)-\frac{bc}{48}\right)-\left(s(b,4c)-\frac{b}{48c}\right) \equiv 1 \pmod 2. \tag{10.5}$$

Proof. This theorem resembles Case IIa of Theorem 20 with $\lambda=2$. Let us put

$$C=(48cs(4b,c)-16b)+(48cs(bc,4)-bc^2)-(48cs(b,4c)-b).$$

The proof of

$$C\equiv 0 \pmod{3c} \tag{10.51}$$

follows closely the proof of (10. 33) in Case IIa of the previous theorem. Moreover, since (10. 34) was generally valid for $\lambda\geqq 1$, we get from there

$$C\equiv -2^4b \pmod{2^5}$$

or, since $-b\equiv 3c \pmod 2$, we have

$$C\equiv 2^4\cdot 3c \pmod{2^5}. \tag{10.52}$$

But (10. 51) may also be written in the form

$$C\equiv 2^4\cdot 3c \pmod{3c}. \tag{10.53}$$

Combining (10.52) and (10.53) we infer that

$$(10.54) \qquad C \equiv 2^4 \cdot 3c \pmod{2^5 \cdot 3c}.$$

If we now divide both members of the last congruence by $48c$, Theorem 21 follows immediately.

We close this section with

THEOREM 22. *If bc is odd and $(b, c) = 1$, then*

$$s(2b, c) - s(b, 2c) + \frac{b(c^2-1)}{8c} \equiv 0 \pmod 2.$$

Proof.[9] We do not need to assume that $3 \mid bc$ this time. First we prove, along the lines established in the preceding proofs, that

$$(10.55) \qquad (24cs(2b, c) - 4b) + (24cs(bc, 2) - c^2b) - (24cs(b, 2c) - b) \equiv 0 \pmod{\theta c}.$$

But from the definition of $s(h, k)$ in (1.2) it is evident that

$$s(bc, 2) = 0$$

Hence (10.55) reduces to

$$24cs(2b, c) - 24cs(b, 2c) - 3b - c^2b \equiv 0 \pmod{\theta c}.$$

Furthermore, it follows from (10.34) that

$$24cs(2b, c) - 24cs(b, 2c) - 3b - c^2b \equiv -4b \equiv -4bc^2 \pmod{16},$$

and therefore that

$$24cs(2b, c) - 24cs(b, 2c) - 3b - c^2b \equiv -4bc^2 \pmod{16\theta c},$$

or

$$(10.56) \qquad 24cs(2b, c) - 24cs(b, 2c) + 3b(c^2-1) \equiv 0 \pmod{16\theta c}.$$

Now if $3 \nmid c$ we infer from (9.21) that 3 divides the left member of (10.56) anyway. Thus the congruence is established modulo $48c$. If we now divide by $24c$ we obtain the result announced in our theorem.

11. Factorization of the $A_k(n)$. The Dedekind sums appear, as we have mentioned in the introduction, in the theory of the modular transformations of $\log \eta(\tau)$. It is from this source that they appear as exponents in certain

[9] Theorem 22 may also be deduced from Theorem 10. The essential point in the argument is to show that if h and k are odd and $(h, k) = 1$, then $\sum_{\mu=1}^{(k-1)/2} [h\mu/k + 1/2] \equiv 0 \pmod 2$.

sums of roots of unity in the series for the number $p(n)$ of unrestricted partitions of n. These sums are defined by

(11.1) $$A_k(n) = \sum_{h \bmod k}' \exp\left(\pi i s(h, k) - 2\pi i \frac{hn}{k}\right),$$

where the dash $'$ beside the summation symbol indicates here and in the sequel that the letter of summation runs only through a reduced residue system with respect to the modulus. D. H. Lehmer **7** has studied these sums on the basis of a different expression for the roots of unity involved. He was able to reduce them to sums studied by H. D. Kloosterman and H. Salié. In the first place he factored the $A_k(n)$ according to the prime number powers contained in k. Secondly, by using Salié's formulae, he evaluated the $A_k(n)$ explicitly in the case in which k is a prime or a power of a prime. Both results together provide a method for calculating the $A_k(n)$. Our theorems of § 10 open a new approach to the factorization of the $A_k(n)$, whereas we have nothing new to offer with regard to the second of Lehmer's results.

In what follows, we shall derive three theorems [10] for expressing $A_k(n)$ as a product of two A's whose subscripts are coprime integers whose product is k.

THEOREM 23. *If* $k = k_1k_2$, $(k_1, k_2) = 1$, *and if furthermore,* $8 \mid k$ *in case* k *is even, then*

(11.2) $$A_k(n) = A_{k_1}(n_1) A_{k_2}(n_2),$$

where n_1 *and* n_2 *are determined by the congruences*

(11.21) $$k_2^2 v_2 w n_1 \equiv v_2 w n + \frac{k_2^2 - 1}{v_1} (\bmod k_1),$$

and

(11.22) $$k_1^2 v_1 w n_2 \equiv v_1 w n + \frac{k_1^2 - 1}{v_2} (\bmod k_2),$$

respectively, and where v_1, v_2, w *are defined by*

(11.23) $$24 = v_1 v_2 w,$$

and

(11.24) $$v_1 = (24, k_1), \qquad v_2 = (24, k_2).$$

Proof. Clearly $A_k(n)$ is periodic in n with respect to the modulus k.

[10] For the sake of clarity our theorems are stated somewhat more explicitly than those of Lehmer **7**. It should be observed that our theorems and Lehmer's theorems overlap to a certain extent. Lehmer's Theorem 1 may be deduced from our Theorem 23. His Theorem 2 follows from our Theorems 23 and 24. His theorem 4 is the same as our Theorem 25.

Let n_1 and n_2 be two integers to be determined explicitly later on. Then we have

$$A_{k_1}(n_1)A_{k_2}(n_2)$$
$$= \sum_{h_1 \bmod k_1}' \exp\left(\pi i s(h_1, k_1) - 2\pi i \frac{h_1 n_1}{k_1}\right) \cdot \sum_{h_2 \bmod k_2}' \exp\left(\pi i s(h_2, k_2) - 2\pi i \frac{h_2 n_2}{k_2}\right)$$
$$= \sum_{h_1 \bmod k_1}' \exp\left(\pi i s(k_2 h_1, k_1) - 2\pi i \frac{k_2 h_1 n_1}{k_1}\right) \cdot \sum_{h_2 \bmod k_2}' \exp\left(\pi i s(k_1 h_2, k_2) - 2\pi i \frac{k_1 h_2 n_2}{k_2}\right)$$

where we have replaced h_1 by $k_2 h_1$ and h_2 by $k_1 h_2$. (This is permissible since $(k_1, k_2) = 1$). Now for each pair of summation indices h_1, h_2 we define h by means of the pair of congruences

$$h_1 \equiv h \pmod{k_1},$$
$$h_2 \equiv h \pmod{k_2}.$$

It is clear that as h runs over a reduced residue system modulo $k_1 k_2$, then each pair h_1, h_2 (h_1 modulo k_1, k_2 modulo k_2) occurs once and only once. Hence

(11.3) $$A_{k_1}(n_1)A_{k_2}(n_2)$$
$$= \sum_{h \bmod k_1 k_2}' \exp\left(\pi i s(k_2 h, k_1) - 2\pi i \frac{k_2 h n_1}{k_1} + \pi i s(k_1 h, k_2) - 2\pi i \frac{k_1 h n_2}{k_2}\right).$$

We may moreover assume that $w \mid h$ since the assumptions of the theorem and the definitions (11.23) and (11.24) imply that $(w, k_1 k_2) = 1$. Therefore $24 \mid h k_1 k_2$. We are now in position to apply Theorem 20 which yields

$$\left(s(k_2 h, k_1) - \frac{k_2 h}{12 k_1}\right) + \left(s(k_1 h, k_2) - \frac{k_1 h}{12 k_2}\right) - \left(s(h, k_1 k_2) - \frac{h}{12 k_1 k_2}\right)$$
$$\equiv 0 \pmod 2.$$

A comparison of (11.1) and (11.3) on the other hand shows that our theorem requires

$$\left(s(k_2 h, k_1) - \frac{2 k_2 h n_1}{k_1}\right) + \left(s(k_1 h, k_2) - \frac{2 k_1 h n_2}{k_2}\right) - \left(s(h, k_1 k_2) - \frac{2hn}{k_1 k_2}\right)$$
$$\equiv 0 \pmod 2.$$

From the last two congruences we obtain by subtraction

(11.4) $$h(k_2^2(1 - 24 n_1) + k_1^2(1 - 24 n_2) - (1 - 24 n))$$
$$\equiv 0 \pmod{24 k_1 k_2}.$$

Consequently our object is to determine under what circumstances the congruence (11.4) holds. Now, since $w \mid h$ and $(h/w, v_1 v_2 k_1 k_2) = 1$, this congruence is equivalent to

$$k_2^2(1-24n_1)+k_1^2(1-24n_2)-(1-24n)\equiv 0 \pmod{v_1k_1v_2k_2},$$

which certainly possesses solutions in n_1 and n_2 if the two congruences

$$-24k_2^2n_1+(k_2^2-1)+24n\equiv 0 \pmod{v_1k_1},$$
$$-24k_1^2n_2+(k_1^2-1)+24n\equiv 0 \pmod{v_2k_2},$$

are separately solvable in n_1 and n_2 respectively. But these are the congruences (11.21) and (11.22). We have only to observe that $\dfrac{k_2^2-1}{v_1}$ is an integer since v_1 divides 24 and is prime to k_2; the analogous remark holds for $\dfrac{k_1^2-1}{v_2}$. Furthermore, the congruences (11.21) and (11.22) are solvable with unique solutions since $(k_2^2v_2w, k_1)=1$ and $(k_1^2v_1w, k_2)=1$. This completes the proof of the theorem.

The preceding theorem enables us to decompose $A_k(n)$ for composite k if k is odd or is divisible by 2^3. We now consider the cases in which k is even but is not divisible by 2^3.

THEOREM 24. *If $k=4k_1$, with k_1 odd, then*

(11.5) $$A_k(n)=-A_{k_1}(n_1)A_4(n_2),$$

where n_1 and n_2 are determined by

(11.51) $$128n_1\equiv 8n+5 \pmod{k_1},$$

and

(11.52) $$k_1^2n_2\equiv n-2-\frac{k_1^2-1}{8} \pmod{4},$$

respectively.

Proof. Exactly as in the proof of Theorem 23, we obtain

(11.6) $$A_{k_1}(n_1)A_4(n_2)$$
$$=\sum_{h \bmod 4k_1}{}' \exp\left(\pi i s(4h,k_1)-\frac{2\pi i 4hn_1}{k_1}+\pi i s(k_1h,4)-\frac{2\pi i k_1hn_2}{4}\right),$$

which is valid for any n_1 and n_2. If we now put $3=v\cdot u$, $v=(3,k_1)$, we have $(u,k_1)=1$ and may therefore assume that the summation index h in (11.6) is always divisible by u. Then $3\,|\,hk_1$ and Theorem 21 is applicable:

(11.7) $$\left(s(4h,k_1)-\frac{4h}{12k_1}\right)+\left(s(k_1h,4)-\frac{k_1h}{48}\right)-\left(s(h,4k_1)-\frac{h}{48k_1}\right)$$
$$\equiv 1 \pmod{2}.$$

Equations (11.1), (11.5) and (11.6), on the other hand, show that our theorem would be proved if we could establish

$$\left(s(4h, k_1) - \frac{8hn_1}{k_1}\right) + \left(s(k_1h, 4) - \frac{2k_1hn_2}{4}\right) - \left(s(h, 4k_1) - \frac{2hn}{4k_1}\right) \equiv 1 \pmod 2.$$

In view of (11.7) this congruence reduces to

$$h(16(1-24n_1) + k_1^2(1-24n_2) - (1-24n)) \equiv 0 \pmod{3.\,2^5k_1}$$

or

$$16(1-24n_1) + k_1^2(1-24n_2) - (1-24n) \equiv 0 \pmod{v.\,2^5k_1}.$$

The last congruence is, in turn, equivalent to the two congruences

(11.81) $$-3\cdot 2^7n_1 + 15 + 3\cdot 8n \equiv 0 \pmod{vk_1}$$

and

(11.82) $$-24k_1^2n_2 + 16 + 24n + k_1^2 - 1 \equiv 0 \pmod{2^5}.$$

If we divide both members of (11.81) by v we get

$$u(-2^7n_1 + 5 + 8n) \equiv 0 \pmod{k_1},$$

which is the same as (11.51) since $(u, k_1) = 1$. Furthermore, it is easy to see that (11.82) may be replaced by (11.52). This completes the proof. We conclude with

THEOREM 25. *If $k = 2k_1$, with k_1 odd, then*

(11.9) $$A_k(n) = A_{k_1}(n_1)A_2(n_2),$$

where n_1 and n_2 are solutions of the congruences

(11.91) $$32n_1 \equiv 8n + 1 \pmod{k_1},$$

(11.92) $$n_2 \equiv n - \frac{k_1^2 - 1}{8} \pmod 2.$$

Proof. Since the proof is very much like the proofs of the two preceding theorems, it will suffice to sketch it briefly. This time we make use of Theorem 22. We have merely to show that the congruences (11.91) and (11.92) are together equivalent to the single congruence

$$h\left(\frac{n}{k_1} - \frac{4n_1}{k_1} - n_2 - \frac{k_1^2 - 1}{8k_1}\right) \equiv 0 \pmod 2$$

or

$$n - 4n_1 - k_1n_2 - \frac{k_1^2 - 1}{8} \equiv 0 \pmod{2k_1},$$

which is clearly the case.

BIBLIOGRAPHY

1. R. Dedekind, "Erläuterungen zu den vorstehenden Fragmenten," Riemann's *Gesammelte Werke*, pp. 466-478. Dedekind's *Gesammelte Werke* (1930), Bd. I, pp. 159-172.

2. B. Riemann, "Fragmente über die Grenzfälle der elliptischen Modulfunktionen," *Gesammelte Werke* 2. Auflage (1892), pp. 461-465.

3. R. Dedekind, "Schreiben an Herrn Borchardt über die Theorie der elliptischen Modulfunktionen," *Journal für die reine und angewandte Mathematik*, vol. 83 (1877), pp. 265-292. Also Dedekind's *Gesammelte Werke*, Bd. I., pp. 174-201.

4. H. Rademacher, "Zur Theorie der Modulfunktionen," *Journal für die reine und angewandte Mathematik*, vol. 167 (1931), pp. 312-336.

5. H. Rademacher, "Eine arithmetische Summenformel," *Monatshefte für Mathematik und Physik*, vol. 34 (1932), pp. 221-228.

6. H. Rademacher, "Über eine Reziprozitätsformel aus der Theorie der Modulfunktionen," *Matematikai és Fizikai Lapok*, vol. 40 (1933), pp. 24-34.

7. D. H. Lehmer, "On the series for the partition function," *Transactions of the American Mathematical Society*, vol. 43 (1938), pp. 271-295.

UNIVERSITY OF PENNSYLVANIA,
HARVARD UNIVERSITY.

Notes to Paper 44

This paper, written jointly with A. L. Whiteman, has been reviewed in the JF, vol. 67 (1941), p. 291, in the Z, vol. 25 (1941-1942), p. 28, and in the MR, vol. 2 (1941), p. 249.

Page 377, equation (1.1).
This definitive and most satisfactory definition of the symbol $((x))$ occurs here for the first time (compare, e.g., with the corresponding one in paper 26, p. 298, equation (3), and in paper 32, equation (2), p. 24, of the present collection.)

Page 380, second display line.
Author's correction (read h instead of k).

Page 391, first display line.
Insert the factor 2 before the first Σ.

Page 393, tenth display line.
Add a parenthesis) after $s(h,2k)$.

Page 396, equation (9.5).
Author's correction (insert the coefficient 2).

Page 397, line 7 of text.
Read $\lambda \geqslant 0$ instead of $\lambda \equiv 0$.

Page 397, fifth display line.
In the upper limit read $(h-1)/2$ instead of $h-1/2$.

Page 397, footnote, third line from bottom.
Read $k/2 < r_\mu$ instead of $k/2 < k$.

Page 398, seventh display line.
Read $\equiv$ instead of $=$.

Page 399, second line from bottom.
Read (mod $3c$) instead of (mod 3)

Page 401, line 1.
Read ". . $a = 2^\lambda a_1, a_1, b, c$ be odd, . . ."

Page 407, footnote 4.
See paper 27.

Page 407, footnote 5.
See paper 28.

Page 407, footnote 6.
See paper 32.

THE RAMANUJAN IDENTITIES UNDER MODULAR SUBSTITUTIONS

BY

HANS RADEMACHER

1. Introduction. In connection with his discovery of certain divisibility properties of the partition function Ramanujan [1](1) stated the identities

$$\sum_{l=0}^{\infty} p(5l+4)x^l = 5\frac{\prod (1-x^{5m})^5}{\prod (1-x^m)^6}, \tag{1.1}$$

and

$$\sum_{l=0}^{\infty} p(7l+5)x^l = 7\frac{\prod (1-x^{7m})^3}{\prod (1-x^m)^4} + 49x\frac{\prod (1-x^{7m})^7}{\prod (1-x^m)^8}. \tag{1.2}$$

Here, as always in the sequel, the index m in the infinite products runs through all positive integers. If these identities, for which various proofs have been given, are expressed in terms of the Dedekind η-function

$$\eta(\tau) = e^{\pi i\tau/12}\prod (1-e^{2\pi i m\tau}), \qquad \mathfrak{I}(\tau) > 0, \tag{1.3}$$

they appear in a form which suggests certain group-theoretical considerations, similar to those employed by Hecke in his theory of modular forms. In this way we transform the identities into new ones which are noteworthy because of the occurrence of the Legendre symbol and which, by a simple further argument, lead also to a proof of (1.1) and (1.2). An analogous identity for the modulus 13, given by Zuckerman, can be treated in the same way.

G. N. Watson and H. S. Zuckerman have also derived identities for the moduli 5^2 and 7^2. These will lead us to certain modular equations, which in turn will shed some light on those identities.

PART I. IDENTITIES OF RAMANUJAN AND ZUCKERMAN

2. We have known since Euler

$$\sum_{n=0}^{\infty} p(n)x^n = \frac{1}{\prod (1-x^m)} \tag{2.1}$$

with $p(0)=1$, or

$$\frac{1}{\eta(\tau)} = e^{-\pi i\tau/12}\sum_{n=0}^{\infty} p(n)e^{2\pi i n\tau}. \tag{2.2}$$

Presented to the Society, February 22, 1941; received by the editors April 25, 1941.

(1) Numbers in square brackets refer to the bibliography at the end of this paper.

Hence we obtain

$$\sum_{\lambda=0}^{4} \frac{1}{\eta\left(\frac{\tau + 24\lambda}{5}\right)} = \sum_{\lambda=0}^{4} e^{-(\pi i\tau/60)-(2\pi i\lambda/5)} \sum_{n=0}^{\infty} p(n) e^{(2\pi i n\tau/5)+(48\pi i\lambda n/5)}$$

$$= e^{-\pi i\tau/60} \sum_{n=0}^{\infty} p(n) e^{2\pi i n\tau/5} \sum_{\lambda=0}^{4} e^{-(2\pi i/5)\lambda(1-24n)}$$

$$= 5e^{-\pi i\tau/60} \sum_{n\equiv 4 \pmod 5} p(n) e^{2\pi i n\tau/5},$$

and therefore

$$\sum_{l=0}^{\infty} p(5l+4) e^{2\pi i l\tau} = \frac{1}{5} e^{-19\pi i\tau/12} \sum_{\lambda=0}^{4} \frac{1}{\eta\left(\frac{\tau + 24\lambda}{5}\right)}.$$

If we also express the right-hand member of (1.1) in terms of $\eta(\tau)$, through its definition (1.3), we get

$$\sum_{\lambda=0}^{4} \eta\left(\frac{\tau + 24\lambda}{5}\right)^{-1} = 5^2 \frac{\eta(5\tau)^5}{\eta(\tau)^6} \tag{2.3}$$

as a restatement of (1.1).

In a similar way (1.2) can be rewritten as

$$\sum_{\lambda=0}^{6} \eta\left(\frac{\tau + 24\lambda}{7}\right)^{-1} = 7^2 \frac{\eta(7\tau)^3}{\eta(\tau)^4} + 7^3 \frac{\eta(7\tau)^7}{\eta(\tau)^8}. \tag{2.4}$$

3. We are now going to subject (2.3) and (2.4) to modular transformations, of which we need to test only the generators

$$S = \begin{pmatrix} 1 & 1 \\ 0 & 1 \end{pmatrix}, \qquad T = \begin{pmatrix} 0 & -1 \\ 1 & 0 \end{pmatrix}.$$

The definition (1.3) shows that

$$\eta(\tau + 24) = \eta(\tau).$$

Consequently the range of the summation in the left-hand member of (2.3) can be replaced by "modulo 5." Therefore S^{24} produces only a cyclical exchange of the terms of the sum and does not change the sum as a whole. It follows that S and S^{25} have the same effect on the left-hand member of (2.3), and since S^{25} means the replacement of $(\tau+24\lambda)/5$ by $((\tau+24\lambda)/5)+5$, this effect is clearly the appearance of a multiplier $e^{-5\pi i/12}$ in each summand. On the other side, the substitution S, that is, $\tau \to \tau+1$, provides the multiplier

$$e^{19\pi i/12} = e^{-5\pi i/12}$$

also on the right-hand side. Thus the equation (2.3) goes over into itself under the substitution S.

Similarly we see that S and S^{49} have the same effect on the left-hand member of (2.4), viz., multiplication of each summand by

$$e^{-7\pi i/12}.$$

This same factor is taken up also, as (1.3) shows, by each term of the right-hand member of (2.4) so that the equation (2.4) also remains invariant under the substitution S.

4. This is not so with T. Through the substitution

$$\tau \to -\tau^{-1}$$

equation (2.3) goes over into

$$\sum_{\lambda=0}^{4} \eta\left(\frac{-1+24\lambda\tau}{5\tau}\right)^{-1} = 5^2 \frac{\eta\left(-\frac{5}{\tau}\right)^5}{\eta\left(-\frac{1}{\tau}\right)^6}. \tag{4.1}$$

The left-hand member, which we designate by L_5, can be rewritten as

$$L_5 = \eta\left(\frac{-1}{5\tau}\right)^{-1} + \sum_{\lambda=1}^{4} \eta\left(\frac{24\lambda\dfrac{\tau+24\lambda'}{5}+b_\lambda}{5\dfrac{\tau+24\lambda'}{5}-24\lambda'}\right)^{-1} \tag{4.2}$$

with

$$\lambda\lambda' \equiv -1 \pmod{5}, \tag{4.21}$$

and

$$b_\lambda = -(24^2\lambda\lambda'+1)/5. \tag{4.22}$$

Now for a modular substitution

$$\begin{pmatrix} a & b \\ c & d \end{pmatrix}, \qquad c > 0,$$

we have [8],

$$\eta\left(\frac{a\tau+b}{c\tau+d}\right) = \exp\left(-\pi i\left(s(a, c) - \frac{a+d}{12c}\right)\right)(-i(c\tau+d))^{1/2}\eta(\tau), \tag{4.3}$$

where $s(a, c)$ is the "Dedekind sum"

$$s(a, c) = \sum_{\mu \bmod c}\left(\left(\frac{\mu}{c}\right)\right)\left(\left(\frac{a\mu}{c}\right)\right) \tag{4.31}$$

with

(4.32) $$((x)) = \begin{cases} x - [x] - \frac{1}{2} & x \text{ not an integer,} \\ 0 & x \text{ an integer.} \end{cases}$$

If we apply (4.3) on (4.2) and the right-hand member of (4.1) for the modular substitutions

$$\begin{pmatrix} 0 & -1 \\ 1 & 0 \end{pmatrix},$$

and

$$\begin{pmatrix} 24\lambda & b_\lambda \\ 5 & -24\lambda' \end{pmatrix},$$

we obtain after a few reductions

(4.4) $$\frac{1}{5^{1/2}\eta(5\tau)} + \sum_{\lambda=1}^{4} \frac{\exp\left(-\pi i(s(\lambda, 5) + \frac{2}{5}(\lambda - \lambda'))\right)}{\eta\left(\dfrac{\tau + 24\lambda'}{5}\right)} = \frac{1}{5^{1/2}} \frac{\eta\left(\dfrac{\tau}{5}\right)^5}{\eta(\tau)^6}.$$

The Dedekind sums enjoy the following properties for

$$(h, k) = 1, \; hh' \equiv -1 \pmod{k}:$$

(4.51) $$12ks(h, k) \equiv h - h' \pmod{k},$$

(4.52) $$12ks(h, k) \equiv 0 \pmod{3}, \qquad \text{for } 3 \nmid k,$$

and, for k odd,

(4.53) $$12ks(h, k) \equiv k + 1 - 2\left(\frac{h}{k}\right) \pmod{8},$$

with the Legendre-Jacobi symbol on the right-hand side of congruence (4.53) [9].

In our case $k = 5$, $(\lambda, 5) = 1$, we derive from these congruences

$$\frac{1}{2} s(\lambda, 5) + \frac{\lambda - \lambda'}{5} \equiv \frac{1}{4}\left(1 + \left(\frac{\lambda}{5}\right)\right) \pmod{1},$$

and therefore

$$\exp\left(-2\pi i\left(\frac{1}{2}s(\lambda, 5) + \frac{1}{5}(\lambda - \lambda')\right)\right) = -\left(\frac{\lambda}{5}\right).$$

We remark that $(\lambda'/5) = (\lambda/5)$ and obtain from (4.4)

$$(4.6)\qquad 5^{-1/2}\eta(5\tau)^{-1} - \sum_{\lambda=1}^{4}\left(\frac{\lambda}{5}\right)\eta\left(\frac{\tau+24\lambda}{5}\right)^{-1} = 5^{-1/2}\frac{\eta\left(\frac{\tau}{5}\right)^5}{\eta(\tau)^6}.$$

In virtue of (2.2) we can write for the sum on the left-hand side:

$$\begin{aligned}\sum_{\lambda=1}^{4} &= \sum_{\lambda=1}^{4}\left(\frac{\lambda}{5}\right)e^{-\pi i(\tau+24\lambda)/60}\sum_{n=0}^{\infty}p(n)e^{2\pi i n(\tau+24\lambda)/5}\\ &= e^{-\pi i\tau/60}\sum_{n=0}^{\infty}p(n)e^{2\pi i n\tau/5}\sum_{\lambda=1}^{4}\left(\frac{\lambda}{5}\right)e^{(2\pi i\lambda/5)(1-24n)}\\ &= 5^{1/2}e^{-\pi i\tau/60}\sum_{n=0}^{\infty}\left(\frac{n+1}{5}\right)p(n)e^{2\pi i n\tau/5},\end{aligned}$$

where we have evaluated a Gaussian sum, and where for $5\mid(n+1)$ the symbol $((n+1)/5)$ means 0, as customary. If we introduce this result into (4.6), apply (2.2) to its first term and finally replace $e^{2\pi i\tau/5}$ by x we obtain

$$(4.7)\qquad \sum_{n=0}^{\infty}p(n)x^{25n} - 5\sum_{n=1}^{\infty}\left(\frac{n}{5}\right)p(n-1)x^n = \frac{\prod(1-x^m)^5}{\prod(1-x^{5m})^6},$$

which is the new identity we wished to derive.

Incidentally, we can construct a formula which is free of infinite products. If we multiply (4.7) by (1.1) the right-hand side will appear as

$$5\frac{1}{\prod(1-x^m)}\cdot\frac{1}{\prod(1-x^{5m})}.$$

These infinite products can be replaced by series by means of (2.1), so that we get

$$(4.8)\qquad \left\{\sum_{n=0}^{\infty}p(n)x^{25n} - 5\sum_{n=1}^{\infty}\left(\frac{n}{5}\right)p(n-1)x^n\right\}\sum_{l=0}^{\infty}p(5l+4)x^l = 5\sum_{n=0}^{\infty}p(n)x^n\cdot\sum_{n=0}^{\infty}p(n)x^{5n}.$$

Comparison of coefficients would yield certain quadratic relations among the $p(n)$.

5. The identity (2.4) can be treated in the same manner. We give only a few highlights. Replacing τ by $-\tau^{-1}$ in (2.4) we have

$$(5.1)\qquad \sum_{\lambda=0}^{6}\eta\left(\frac{24\lambda\tau-1}{7\tau}\right)^{-1} = 7^2\frac{\eta\left(-\frac{7}{\tau}\right)^3}{\eta\left(-\frac{1}{\tau}\right)^4} + 7^3\frac{\eta\left(-\frac{7}{\tau}\right)^7}{\eta\left(-\frac{1}{\tau}\right)^8}.$$

Calling the left-hand member L_7 we rewrite it as

$$(5.2)\qquad L_7 = \eta\left(-\frac{1}{7\tau}\right)^{-1} + \sum_{\lambda=1}^{6} \eta\left(\frac{24\lambda\,.\,\dfrac{\tau+96\lambda'}{7} + b_\lambda}{7\dfrac{\tau+96\lambda'}{7} - 96\lambda'}\right)^{-1}$$

with

$$(5.21)\qquad \lambda\lambda' \equiv -1 \pmod 7,$$

and

$$(5.22)\qquad b_\lambda = -(24\cdot 96\lambda\lambda' + 1)/7.$$

By means of (4.3) the formula (5.2) goes over into

$$(5.3)\qquad L_7 = (-7i\tau)^{-1/2} + (-i\tau)^{-1/2}\sum_{\lambda=1}^{6} M_\lambda \eta\left(\frac{\tau+96\lambda'}{7}\right)^{-1},$$

where we have

$$M_\lambda = \exp\left\{\pi i\left((s(24\lambda, 7) - \frac{2\lambda - 8\lambda'}{7}\right)\right\}.$$

The congruences (4.51), (4.52), (4.53) yield

$$\frac{1}{2}s(24\lambda, 7) - \frac{\lambda - 4\lambda'}{7} \equiv \frac{1}{4}\left(\frac{\lambda}{7}\right) \pmod 1,$$

and therefore

$$(5.4)\qquad M_\lambda = \exp\left\{\frac{\pi i}{2}\left(\frac{\lambda}{7}\right)\right\} = .i\left(\frac{\lambda}{7}\right) = -i\left(\frac{\lambda'}{7}\right).$$

We introduce (5.2), (5.3), (5.4) into (5.1), carry out the modular transformation on the right-hand side, and get thereby

$$(5.5)\qquad 7^{-1/2}\eta(7\tau)^{-1} - i\sum_{\lambda'=1}^{6}\left(\frac{\lambda'}{7}\right)\eta\left(\frac{\tau+96\lambda'}{7}\right)^{-1} = 7^{1/2}\frac{\eta\left(\dfrac{\tau}{7}\right)^{3}}{\eta(\tau)^4} + 7^{-1/2}\frac{\eta\left(\dfrac{\tau}{7}\right)^{7}}{\eta(\tau)^8}.$$

The sum over λ' can be expanded into an infinite series by means of (2.2):

$$(5.6)\qquad \sum_{\lambda'=1}^{6}\left(\frac{\lambda'}{7}\right)\eta\left(\frac{\tau+96\lambda'}{7}\right)^{-1} = -i7^{1/2}e^{-7\pi i\tau/12}\sum_{n=2}^{\infty}\left(\frac{n}{7}\right)p(n-2)e^{2\pi i n\tau/7};$$

this derivation required the use of the Gaussian sum

$$\sum_{\lambda'=1}^{6}\left(\frac{\lambda'}{7}\right)e^{2\pi\lambda'(96n-4)/7} = -\,i7^{1/2}\left(\frac{n+2}{7}\right).$$

If we now insert (5.6) in (5.5), apply (2.2) and (1.3) in appropriate places, and finally change $e^{2\pi i\tau/7}$ into x we obtain

$$\sum_{n=0}^{\infty} p(n)x^{49n} - 7\sum_{n=2}^{\infty}\left(\frac{n}{7}\right)p(n-2)x^n = 7x\frac{\prod(1-x^m)^3}{\prod(1-x^{7m})^4} + \frac{\prod(1-x^m)^7}{\prod(1-x^{7m})^8}. \tag{5.7}$$

6. Up to this moment we have taken the Ramanujan identities (1.1) and (1.2) for granted and have inferred the identities (4.7) and (5.7) as direct consequences. From another point of view, however, we can take these new identities as bases for proofs of (1.1) and (1.2). For that purpose we consider the Ramanujan identities in the forms (2.3) and (2.4), which we prefer to write now as

$$\sum_{\lambda=0}^{4}\eta(5\tau)\eta\left(\frac{\tau+24\lambda}{5}\right)^{-1} = 25\left(\frac{\eta(5\tau)}{\eta(\tau)}\right)^6, \tag{6.1}$$

and

$$\sum_{\lambda=0}^{6}\eta(7\tau)\eta\left(\frac{\tau+24\lambda}{7}\right)^{-1} = 7^2\left(\frac{\eta(7\tau)}{\eta(\tau)}\right)^4 + 7^3\left(\frac{\eta(7\tau)}{\eta(\tau)}\right)^8. \tag{6.2}$$

We shall refer to these equations shortly in the abbreviations

$$L_5^*(\tau) = R_5^*(\tau); \qquad L_7^*(\tau) = R_7^*(\tau),$$

respectively.

We can show that $L_5^*(\tau)$ and $R_5^*(\tau)$ are both modular functions of "level 5" ("stufe 5" in Felix Klein's terminology), that is, belonging to a congruence subgroup modulo 5 of the modular group. The subgroup in question is $\Gamma_0(5)$, characterized by $c\equiv 0 \pmod 5$; it is of index 6 in the full modular group. As generators of $\Gamma_0(5)$ we can choose the substitutions (cf. [7, p. 147])

$$S = \begin{pmatrix}1 & 1\\ 0 & 1\end{pmatrix}, \quad V_2 = \begin{pmatrix}-2 & -1\\ 5 & 2\end{pmatrix}, \quad V_3 = \begin{pmatrix}-3 & -1\\ 10 & 3\end{pmatrix}$$

and need to test the invariance of $L_5^*(\tau)$ and $R_5^*(\tau)$ only with respect to these 3 substitutions. The discussion of S has already essentially been done in §3. The multiplier $e^{-5\pi i/12}$, which is mentioned there, is exactly absorbed by the factor $\eta(5\tau)$ by which (6.1) differs from (2.3).

As far as $R_5^*(\tau)$ is concerned we have, for V_2,

$$\eta\left(5\frac{-2\tau-1}{5\tau+2}\right)=\eta\left(\frac{-2\cdot 5\tau-5}{5\tau+2}\right)$$
$$=\exp\left\{-\pi i s(-2,1)\right\}(-i(5\tau+2))^{1/2}\eta(5\tau),$$

in virtue of (4.3), and also

$$\eta\left(\frac{-2\tau-1}{5\tau+2}\right)=\exp\left\{-\pi i s(-2,5)\right\}(-i(5\tau+2))^{1/2}\eta(\tau).$$

Since $s(-2, 1)=0$, as directly seen from (4.31) and (4.32), and

$$s(-2,5)=-s(2,5)=0$$

because of the property of the Dedekind sums

$$s(h,k)=0$$

for

$$h^2\equiv -1 \pmod{k},$$

we have

$$R_5^*(V_2\tau)=R_5^*(\tau).$$

Similarly we find

$$R_5^*(V_3\tau)=R_5^*(\tau).$$

As a matter of fact, not only $R_5^*(\tau)$ but already its sixth root $\eta(5\tau)/\eta(\tau)$ is invariant with respect to V_2 and V_3, but not with respect to S.

The expression $L_5^*(\tau)$ goes over under V_2 into

$$L_5^*(V_2\tau)=\sum_{\lambda=0}^{4}\eta\left(5\frac{-2\tau-1}{5\tau+2}\right)\eta\left(\frac{1}{5}\left(\frac{-2\tau-1}{5\tau+2}+24\lambda\right)\right)^{-1}.$$

In order to bring this back into the previous form we need modular substitutions

$$\begin{pmatrix}a & b\\ c & d\end{pmatrix}$$

and a summation variable μ such that

$$\frac{1}{5}\left(\frac{-2\tau-1}{5\tau+2}+24\lambda\right)=\frac{a\dfrac{\tau+24\mu}{5}+b}{c\dfrac{\tau+24\mu}{5}+d}.$$

A comparison of the coefficients of the linear functions of τ on both sides leads to

$$\begin{aligned} &\mu = 2 - \lambda, \\ (6.3)\qquad &a = 120\lambda - 2, \qquad b = -24^2\lambda(2 - \lambda) - 19, \\ &c = 25, \qquad d = -120(2 - \lambda) + 2. \end{aligned}$$

We have therefore:

$$L_5^*(V_2\tau) = \sum_{\lambda=0}^{4} \eta\left(\frac{-2\cdot 5\tau - 5}{5\tau + 2}\right)\eta\left(\frac{a\dfrac{\tau + 24(2-\lambda)}{5} + b}{c\dfrac{\tau + 24(2-\lambda)}{5} + d}\right)^{-1},$$

a, b, c, d being taken from (6.3). Application of (4.3) now shows that

$$L_5^*(V_2\tau) = \sum_{\lambda=0}^{4} M_\lambda \eta(5\tau)\eta\left(\frac{\tau + 24(2-\lambda)}{5}\right)^{-1}$$

with the multiplier

$$M_\lambda = \exp\left\{\pi i\left(s(120\lambda - 2, 25) - \frac{240\lambda - 240}{300}\right)\right\}.$$

Now we have

$$s(120\lambda - 2, 25) = -s(5\lambda + 2, 25),$$

and from (4.51), (4.52), (4.53)

$$12\cdot 25 s(5\lambda + 2, 25) \equiv \begin{cases} 5\lambda + 2 - (-5\lambda + 12) \pmod{25} \\ 0 \pmod{3} \\ 25 + 1 - 2 \pmod{8}, \end{cases}$$

from which we readily derive

$$s(5\lambda + 2, 25) \equiv -\tfrac{4}{5}(\lambda - 1) \pmod{2},$$

so that

$$M_\lambda = 1.$$

Therefore

$$L_5^*(V_2\tau) = L_5^*(\tau)$$

is proved. Reasonings of a similar kind verify the equation

$$L_5^*(V_3\tau) = L_5^*(\tau).$$

7. With $L_5^*(\tau)$ and $R_5^*(\tau)$ the difference

$$D_5(\tau) = L_5^*(\tau) - R_5^*(\tau)$$

belongs also to $\Gamma_0(5)$. If we now can show that $D_5(\tau)$ remains bounded in the whole fundamental region it must be a constant. Now in the interior of the upper τ-half-plane $\eta(\tau)$ is free of poles and zeros and $D_5(\tau)$ is therefore finite. The only parabolic points of the fundamental region of $\Gamma_0(5)$ are the points $\tau = i\infty$ and $\tau = 0$. Now for $\tau \to i\infty$ it is readily seen that $D_5(\tau) \to 0$ since L_5^* and R_5^* tend separately to 0, as their expansions in $e^{2\pi i\tau}$, which can be taken from (1.3), show directly.

In order to test $D_5(\tau)$ for τ near 0 we carry out the substitution $\tau \to -\tau^{-1}$ and study $D_5(-\tau^{-1})$ for τ near $i\infty$. This is now simple with (4.6), which in correspondence to (6.1) we shall have to write as

$$5^{-1/2}\eta\left(\frac{\tau}{5}\right)\eta(5\tau)^{-1} - \sum_{\lambda=1}^{4}\left(\frac{\lambda}{5}\right)\eta\left(\frac{\tau}{5}\right)\eta\left(\frac{\tau+24\lambda}{5}\right)^{-1}$$

$$= 5^{-1/2}\left(\frac{\eta\left(\frac{\tau}{5}\right)}{\eta(\tau)}\right)^{6} \tag{7.1}$$

Indeed, $D_5(-\tau^{-1})$ is the difference of the two members of equation (7.1). In the uniformizing variable $e^{2\pi i\tau/5}$ both members have a pole of the first order at $i\infty$. If therefore the two sides of (7.1) agree in their first term the difference $D_5(-\tau^{-1})$ remains bounded also at $i\infty$ or $D_5(\tau)$ at the second parabolic point $\tau = 0$. Instead, however, of comparing the coefficients of the first term of the members of (7.1), it is easier to do it with (4.7). This is equivalent since (7.1) is obtained from (4.7) through the multiplication by

$$5^{-1/2}x^{-1}\prod(1 - x^m).$$

Now indeed both sides in (4.7) begin with the term 1.

We have therefore proved that $D_5(\tau)$ is a constant, which can only be zero since $D_5(\tau) \to 0$ with $\tau \to i\infty$ as we have mentioned before. This proves (6.1) and therefore (2.3) and (1.1). Mordell in [2] also proves (6.1) by testing its two members at the parabolic points of $\Gamma_0(5)$. We used here for this purpose the independent theory of $\eta(\tau)$.

8. We can discuss (6.2) in the same manner. First we have to show that $L_7^*(\tau)$ and $R_7^*(\tau)$ are modular functions of level 7, belonging to $\Gamma_0(7)$ with $c \equiv 0 \pmod 7$. This step we could perform in analogy to the procedure in §6 by testing the generating substitutions of $\Gamma_0(7)$ which we can take as

$$S = \begin{pmatrix} 1 & 1 \\ 0 & 1 \end{pmatrix}, \qquad V_3 = \begin{pmatrix} -2 & -1 \\ 7 & 3 \end{pmatrix}, \qquad V_5 = \begin{pmatrix} -4 & -1 \\ 21 & 5 \end{pmatrix},$$

(cf. [8]). Such a procedure, however, would not only mean a repetition of previous arguments but would involve a good deal of numerical work, for which, by the way, the congruences (4.51)–(4.53) would not quite suffice as a basis[2]. We prefer therefore to discuss (6.2) on a more general ground, by taking recourse to the following theorems, in which p always designates a prime number greater than 3.

THEOREM 1. *The functions*

$$\Phi_{p,r}(\tau) = \left(\frac{\eta(p\tau)}{\eta(\tau)}\right)^r \tag{8.1}$$

with

$$r(p-1) \equiv 0 \pmod{24} \tag{8.2}$$

have the transformation equation

$$\Phi_{p,r}(V\tau) = \left(\frac{a}{p}\right)^r \Phi_{p,r}(\tau) \tag{8.3}$$

for the modular substitution

$$V\tau = \frac{a\tau + b}{c\tau + d}$$

of $\Gamma_0(p)$, (a/p) *being the Legendre symbol.*

THEOREM 2. *The function*

$$L_p^*(\tau) = \sum_{\lambda=0}^{p-1} \eta(p\tau)\eta\left(\frac{\tau + 24\lambda}{p}\right)^{-1} \tag{8.4}$$

is invariant under the modular substitutions of $\Gamma_0(p)$ *with* $c \equiv 0 \pmod{p}$.

In order not to interrupt the present line of thought we postpone the proof of these theorems to Part III of this paper.

For $p=7$ and $r=4$ and 8 the Theorems 1 and 2 show immediately that $D_7(\tau) = L_7^*(\tau) - R_7^*(\tau)$ as taken from (6.2) is an invariant of $\Gamma_0(7)$. We have now to show that $D_7(\tau)$ remains bounded in the fundamental region of this group. The only parabolic points of that region are again the points $\tau = i\infty$ and $\tau = 0$. For $\tau \to i\infty$ we have $D_7(\tau) \to 0$, since $L_7^*(\tau)$ and $R_7^*(\tau)$ tend separately to 0, in virtue of the factor $e^{\pi i\tau/12}$ before the infinite product in the definition (1.3) for $\eta(\tau)$.

Instead of investigating $D_7(\tau)$ directly for $\tau \to 0$ we carry out the substitution $T\tau = -\tau^{-1}$ and then let τ tend to $i\infty$. But this substitution has been studied in §5. We have therefore $D_7(-\tau^{-1})$ as the difference of the two members of the equation

---

(2) Cf. Lemmas 1 and 3, §13.

$$(8.5)\qquad \begin{aligned} 7^{-1/2}\eta\left(\frac{\tau}{7}\right)\eta(7\tau)^{-1} - i\sum_{\lambda'=1}^{6}\left(\frac{\lambda'}{7}\right)\eta\left(\frac{\tau}{7}\right)\eta\left(\frac{\tau+96\lambda'}{7}\right)^{-1} \\ = 7^{1/2}\left(\frac{\eta\left(\frac{\tau}{7}\right)}{\eta(\tau)}\right)^{4} + 7^{-1/2}\left(\frac{\eta\left(\frac{\tau}{7}\right)}{\eta(\tau)}\right)^{8}, \end{aligned}$$

which is obtained from (5.5) by multiplication with $\eta(\tau/7)$ and which is the result of the transformation of (6.2). The application of (1.3) shows that each member of (8.5) begins with terms in $e^{-4\pi i\tau/7}$, or, in other words has a pole of the second order in the uniformizing variable $e^{2\pi i\tau/7}$. If we can therefore verify that the two members of (8.5) have their pole terms, the first two terms, in common, then the difference $D_7(-\tau^{-1})$ remains bounded also at $\tau=i\infty$, and is bounded in the whole fundamental region. The comparison of the first two terms of each side of (8.5) is much easier to carry out in (5.7), which through multiplication by

$$7^{-1/2}x^{-2}\prod(1-x^m)$$

goes over into (8.5). Now the first two coefficients of both sides of (5.7) are indeed in agreement, they are 1 and 0 for both.

Since therefore $D_7(\tau)$ is bounded in the fundamental region it is a constant, and this constant is obviously 0, since $D_7(\tau)\to 0$ for $\tau\to i\infty$, as mentioned. But $D_7(\tau)=0$ means that the equation (6.2) must hold, and this is equivalent to a proof of (1.2) (cf. [2]).

9. All these reasonings apply also to an identity which Zuckerman [4] has given in the form

$$(9.1)\qquad \sum_{l=0}^{\infty} p(13l+6)x^l = \sum_{j=0}^{6} a_j x^j \frac{\prod(1-x^{13m})^{2j+1}}{\prod(1-x^m)^{2(j+1)}},$$

where the a_j are certain integers which are computed in Zuckerman's paper. The procedure which we applied to (1.1) and (1.2) in §§4 and 5 leads here to the transformed identity

$$(9.2)\qquad \begin{aligned} \sum_{n=0}^{\infty} p(n)x^{169n} - 13\sum_{n=7}^{\infty}\left(\frac{n}{13}\right)p(n-7)x^n \\ = \sum_{j=0}^{6} a_j 13^{1-j}x^{6-j}\frac{\prod(1-x^m)^{2j+1}}{\prod(1-x^{13m})^{2(j+1)}}. \end{aligned}$$

The numbers $13^{1-j}a_j$ are integers.

Our method yields now a direct proof of (9.1). We first express (9.1) and (9.2) in terms of $\eta(\tau)$. We have only to observe that $x=e^{2\pi i\tau}$ in (9.1) and $x=e^{2\pi i\tau/13}$ in (9.2). Moreover we multiply the resulting equations by $\eta(13\tau)$ and $\eta(\tau/13)$, respectively.

For the proof of (9.1) we have first to show that

$$\left(\frac{\eta(13\tau)}{\eta(\tau)}\right)^2,$$

and

$$\sum_{\lambda=0}^{12} \eta(13\tau)\eta\left(\frac{\tau+24\lambda}{13}\right)^{-1}$$

belong to the group $\Gamma_0(13)$ with $c \equiv 0 \pmod{13}$. This is at once inferred from the Theorems 1 and 2 for $p=13$ and $r=2$.

Secondly, we simply have to compare the first 7 coefficients of (9.2) since $\eta(\tau/13)\eta(13\tau)^{-1}$ as well as $\eta(\tau/13)^{14}\eta(\tau)^{-14}$ begin with $e^{-14\pi i\tau/13}$, that is, have a pole of 7th order in the uniformizing variable $e^{2\pi i\tau/13}$. Now the comparison of the first seven coefficients of (9.2) yields without too much effort the following seven equations, with $b_j = 13^{1-j}a_j$:

$$
\begin{aligned}
1 &= b_6,\\
0 &= -\ 13b_6 + b_5,\\
0 &= 65b_6 - 11b_5 + b_4,\\
0 &= -\ 130b_6 + 44b_5 - 9b_4 + b_3, \qquad (9.3)\\
0 &= -\ 65b_6 - 55b_5 + 9b_4 - 7b_3 + b_2,\\
0 &= 728b_6 - 110b_5 - 12b_4 + 14b_3 - 5b_2 + b_1,\\
0 &= -\ 871b_6 + 484b_5 - 90b_4 + 7b_3 + 5b_2 - 3b_1 + b_0.
\end{aligned}
$$

But these are exactly the equations by which Zuckerman (p. 104 of his paper) determines the coefficients which we here have called a_j. His derivation of these equations is based on an entirely different argument.

Part II. Identities of Watson and Zuckerman

10. G. N. Watson [3] and H. S. Zuckerman [4] have derived identities analogous to those of Ramanujan, but corresponding to powers of 5 and 7 as moduli:

$$\sum_{l=0}^{\infty} p(25l+24)x^l = \sum_{j=1}^{5} b_j x^{j-1} \frac{\prod (1-x^{5m})^{6j}}{\prod (1-x^m)^{6j+1}}, \tag{10.1}$$

and

$$\sum_{l=0}^{\infty} p(49l+47)x^l = \sum_{j=1}^{14} c_j x^{j-1} \frac{\prod (1-x^{7m})^{4j}}{\prod (1-x^m)^{4j+1}}. \tag{10.2}$$

The b_j and c_j are integers, which are computed in [4] and which incidentally, in accordance with Ramanujan's theorems about $p(25l+24)$ and $p(49l+47)$, have the properties $25 | b_j$ and $49 | c_j$. From our present point of view we can easily obtain a proof for (10.1) and (10.2). It may be sufficient to carry it out only for the first of these equations.

If in (6.1) we replace τ by $(\tau+24\mu)/5$ we get

$$\sum_{\lambda,\mu=0}^{4} \eta(\tau)\eta\left(\frac{\tau+24\mu+5\cdot 24\lambda}{25}\right)^{-1} = 5^2 \sum_{\mu=0}^{4}\left(\frac{\eta(\tau)}{\eta\left(\dfrac{\tau+24\mu}{5}\right)}\right)^6,$$

or

$$\text{(10.31)} \qquad \sum_{\lambda=0}^{24} \eta(\tau)\eta\left(\frac{\tau+24\lambda}{25}\right)^{-1} = 5^2\Psi_{5,6}(\tau)$$

with

$$\text{(10.32)} \qquad \Psi_{5,6}(\tau) = \sum_{\mu=0}^{4}\left(\frac{\eta(\tau)}{\eta\left(\dfrac{\tau+24\mu}{5}\right)}\right)^6.$$

Now $\Psi_{5,6}(\tau)$ belongs to the group $\Gamma_0(5)$ as we infer from the following theorem, whose proof we defer to Part III.

THEOREM 3. *The functions*

$$\text{(10.41)} \qquad \Psi_{p,r}(\tau) = \sum_{\lambda=0}^{p-1}\left(\frac{\eta(\tau)}{\eta\left(\dfrac{\tau+24\lambda}{p}\right)}\right)^r$$

with $r(p-1)\equiv 0 \pmod{24}$ *have the transformation equation*

$$\text{(10.42)} \qquad \Psi_{p,r}(V\tau) = \left(\frac{a}{p}\right)^r \Psi_{p,r}(\tau)$$

for the substitution

$$V\tau = \frac{a\tau+b}{c\tau+d}$$

with $c\equiv 0 \pmod{p}$, *p being a prime greater than* 3.

We can therefore try to construct $\Psi_{5,6}(\tau)$ as a polynomial in(³)

$$\Phi_{5,6}(\tau) = \left(\frac{\eta(5\tau)}{\eta(\tau)}\right)^6.$$

(³) The background of this possibility is, of course, the fact that (1) $\Phi_{5,6}$ is univalent in the fundamental region, having only a zero of order one at $\tau=i\infty$ and (2) $\Psi_{5,6}$ as well as $\Phi_{5,6}$ are regular in the interior of the fundamental region. However, we do not need this remark, since the following arguments are self-sufficient.

For this purpose we determine coefficients β_j so that

$$\Psi_{5,6}(\tau) = \sum_{j=1}^{N} \beta_j \Phi_{5,6j}(\tau). \tag{10.5}$$

We need to verify this equation only at the two parabolic points of the fundamental region of $\Gamma_0(5)$, namely, at $\tau = i\infty$ and $\tau = 0$. At the former point (10.5) is satisfied since both members tend to 0 as $\tau \to i\infty$. Instead of discussing (10.5) directly for $\tau = 0$ we subject it first to the transformation T, which yields, by the device employed in (4.2),

$$\eta\left(-\frac{1}{\tau}\right)^6 \eta\left(-\frac{1}{5\tau}\right)^{-6} + \sum_{\lambda=1}^{4} \eta\left(-\frac{1}{\tau}\right)^6 \eta\left(\frac{24\lambda\dfrac{\tau+24\lambda'}{5}+b_\lambda}{5\dfrac{\tau+24\lambda'}{5}-24\lambda'}\right)^{-6}$$

$$= \sum_{j=1}^{N} \beta_j \eta\left(-\frac{5}{\tau}\right)^{6j} \eta\left(-\frac{1}{\tau}\right)^{-6j}$$

or, in analogy to (4.6),

$$5^{-3}\left(\frac{\eta(\tau)}{\eta(5\tau)}\right)^6 + \sum_{\lambda=1}^{4}\left(\frac{\eta(\tau)}{\eta\left(\dfrac{\tau+24\lambda}{5}\right)}\right)^6 = \sum_{j=1}^{N} \beta_j 5^{-3j}\left(\frac{\eta\left(\dfrac{\tau}{5}\right)}{\eta(\tau)}\right)^{6j}. \tag{10.6}$$

Here both members show poles at $\tau = i\infty$: the left-hand member begins with a term in $e^{-2\pi i\tau}$, whereas the right-hand member begins with a term in $e^{-2\pi i\tau N/5}$. We have therefore $N = 5$, that is, a pole of order 5 in the uniformizing variable $e^{2\pi i\tau/5}$.

If now the coefficients β_j, $j = 1, \cdots, 5$, are determined in such a way that both members of (10.6) agree in their pole terms at $\tau = i\infty$, that is, in the 5 first terms, then the difference of the members of (10.5) remains bounded throughout the fundamental region and is therefore a constant, which in particular must be equal to 0. We rewrite (10.6) as

$$\eta(5\tau)^{-6} - 5^3\eta\left(\frac{\tau}{5}\right)^{-6} + 5^3\sum_{\lambda=0}^{4}\eta\left(\frac{\tau+24\lambda}{5}\right)^{-6} = \sum_{j=1}^{5}\beta_j 5^{3(1-j)}\frac{\eta\left(\dfrac{\tau}{5}\right)^{6j}}{\eta(\tau)^{6(j+1)}},$$

which, if we introduce $e^{2\pi i\tau/5} = x$, and

$$\frac{1}{\prod(1-x^m)^6} = \sum_{n=0}^{\infty} p_6(n)x^n, \qquad p_6(0) = 1,$$

leads to

$$\sum_{n=0}^{\infty} p_6(n)x^{25n} - 5^3x^6\sum_{n=0}^{\infty} p_6(n)x^n + 5^4x^{10}\sum_{l=0}^{\infty} p_6(5l+4)x^l$$
$$= \sum_{j=1}^{5} \beta_j 5^{-3j+3}x^{5-j}\frac{\prod (1-x^m)^{6j}}{\prod (1-x^{5m})^{6(j+1)}}.$$

We need to ensure only the agreement of the first 5 terms on each side, that is, the terms with x^0, x, $\cdots$, x^4. If we leave aside all unnecessary terms we have therefore to compute β_1, β_2, $\cdots$, β_5 from

$$\sum_{j=1}^{5} \beta_j 5^{-3j+3}x^{5-j}\prod_{m=1}^{4} (1-x^m)^{6j} = 1 + O(x^5).$$

This is equivalent to 5 linear equations for $\beta_j 5^{-3j+3}$, which are solved stepwise, beginning with $\beta_5 5^{-12}=1$. These, however, are exactly the equations which Zuckerman solves on pp. 100, 101 of his paper, and which we do not need to repeat here.

Therefore, the equation (10.5) is proved for $N=5$ and appropriate β_j, $j=1, \cdots, 5$. Comparing (10.5) with (10.31) we obtain

$$\sum_{\lambda=0}^{24} \eta\left(\frac{\tau+24\lambda}{25}\right)^{-1} = 5^2\sum_{j=1}^{5} \beta_j \frac{\eta(5\tau)^{6j}}{\eta(\tau)^{6j+1}}. \tag{10.7}$$

This in turn is equivalent to (10.1) for $b_j=\beta_j$, which therefore is proved.

We remark that for the construction of a similar identity for the modulus 5^3 the required step would be slightly different from that one described at the beginning of this paragraph. We should first have again to replace τ by $(\tau+24\mu)/5$ which would lead to

$$\sum_{\lambda=0}^{124} \eta\left(\frac{\tau+24\lambda}{125}\right)^{-1} = \sum_{j=1}^{5} \beta_j \sum_{\mu=0}^{4} \frac{\eta(\tau)^{6j}}{\eta\left(\frac{\tau+24\mu}{5}\right)^{6j+1}}.$$

In order to have modular functions belonging to $\Gamma_0(5)$ we should now have to multiply both sides by $\eta(5\tau)$ (and not by $\eta(\tau)$ as in (10.31)). As a matter of fact, we could prove by theorems analogous to Theorems 1 to 3 that the functions

$$\sum_{\mu=0}^{4} \frac{\eta(\tau)^{6j}\eta(5\tau)}{\eta\left(\frac{\tau+24\mu}{5}\right)^{6j+1}},$$

and hence, in virtue of the preceding equation,

$$\sum_{\lambda=0}^{124} \eta(5\tau)\eta\left(\frac{\tau+24\lambda}{125}\right)^{-1}$$

belong to $\Gamma_0(5)$ and therefore admit of a representation as a polynomial in $\Phi_{5,6}(\tau)$. Watson [3], indeed, observes that in an induction from 5^k to 5^{k+1} two different kinds of procedures are required according to the parity of k.

We could discuss and prove (10.2) in a manner completely similar to that applied to (10.1). We refrain from giving the details since no new ideas are involved. It is, moreover, clear that our method proves the existence of similar identities for any power of 13 as modulus.

11. In the preceding paragraph we have applied the substitution T to the right-hand side of (10.7) but not yet to its left-hand side. It is worth while to carry it out since it will lead to a modular equation. With $\tau' = -\tau^{-1}$ we have

$$\sum_{\lambda=0}^{24} \eta\left(\frac{\tau' + 24\lambda}{25}\right)^{-1} = \sum_{\lambda=0}^{24} \eta\left(\frac{24\lambda\tau - 1}{25\tau}\right)^{-1}$$

(11.1)

$$= \sum_{\mu=0}^{4} \eta\left(\frac{24\mu\cdot 5\tau - 1}{5\cdot 5\tau}\right)^{-1} + \sum_{\lambda \bmod 25,\,(\lambda,5)=1} \eta\left(\frac{24\lambda\dfrac{\tau + 24\lambda'}{25} + b_\lambda}{25\dfrac{\tau + 24\lambda'}{25} - 24\lambda'}\right)^{-1}$$

where $\lambda\lambda' \equiv -1 \pmod{25}$ and

$$b_\lambda = \frac{1}{25}(-24^2\lambda\lambda' + 1).$$

The first sum on the right side of (11.1) can be taken from (4.1), the second admits the application of (4.3), therefore

$$\sum_{\lambda=0}^{24} \eta\left(\frac{\tau' + 24\lambda}{25}\right)^{-1} = 5^2 \frac{\eta\left(-\dfrac{1}{\tau}\right)^5}{\eta\left(-\dfrac{1}{5\tau}\right)^6}$$

$$+ (-i\tau)^{-1/2} \sum_{\lambda \bmod 25,\,(\lambda,5)=1} M_\lambda \eta\left(\frac{\tau + 24\lambda}{25}\right)^{-1}$$

with

$$M_\lambda = \exp\left\{-\pi i\left(s(24\lambda, 25) - \frac{24\lambda - 24\lambda'}{12\cdot 25}\right)\right\} = 1$$

in virtue of the congruences (4.51), (4.52), (4.53). If we apply the substitution T also to the right-hand side of (10.7) we obtain

$$(11.2)\qquad \frac{1}{5}\,\frac{\eta(\tau)^5}{\eta(5\tau)} + \sum_{\lambda \bmod 25,\,(\lambda,5)=1} \eta\left(\frac{\tau + 24\lambda}{25}\right)^{-1} = 5^2 \sum_{j=1}^{5} \beta_j 5^{-3j} \frac{\eta\left(\dfrac{\tau}{5}\right)^{6j}}{\eta(\tau)^{6j+1}}.$$

We could now, following the procedure of §§4, 5, introduce here infinite series with $p(n)$ as coefficients and infinite products. Instead of doing that we write

$$\sum_{\lambda \bmod 25, (\lambda,5)=1} \eta\left(\frac{\tau+24\lambda}{25}\right)^{-1} = \sum_{\lambda=0}^{24} \eta\left(\frac{\tau+24\lambda}{25}\right)^{-1} - \sum_{\mu=0}^{4} \eta\left(\frac{\tau+24\cdot 5\mu}{25}\right)^{-1}.$$

Here we apply (10.7) and (2.2), the latter with $\tau/5$ instead of τ, so that (11.2) goes over into

$$\frac{1}{5}\frac{\eta(\tau)^5}{\eta(5\tau)^6} + 5^2\sum_{j=1}^{5}\beta_j\frac{\eta(5\tau)^{6j}}{\eta(\tau)^{6j+1}} - 5^2\frac{\eta(\tau)^5}{\eta\left(\frac{\tau}{5}\right)^6} = 5^2\sum_{j=1}^{5}\beta_j 5^{-3j}\frac{\eta\left(\frac{\tau}{5}\right)^{6j}}{\eta(\tau)^{6j+1}}.$$

If we multiply by $5\eta(\tau)$ and replace τ by 5τ we get

$$\Phi(5\tau)^{-1} - 5^3\Phi(\tau) + 5^3\sum_{j=1}^{5}\beta_j(\Phi(5\tau)^j - 5^{-3j}\Phi(\tau)^{-j}) = 0 \tag{11.3}$$

with

$$\Phi(\tau) = \Phi_{5,6}(\tau) = \left(\frac{\eta(5\tau)}{\eta(\tau)}\right)^6. \tag{11.31}$$

We have already found in §6, and again by means of Theorem 2, that $\Phi(\tau)$ is a modular function of level 5. Therefore (11.3) is a transformation equation of level 5 and order 5. In the form (11.3) it is reducible. We put

$$5^{-3}\Phi(\tau)^{-1} = X, \qquad \Phi(5\tau) = Y, \tag{11.4}$$

and have then, after multiplication of (11.3) by XY,

$$X - Y + 5^3 XY\sum_{j=1}^{5}\beta_j(Y^j - X^j) = 0,$$

and after the exclusion of the factor

$$5^3 XY\sum_{j=1}^{5}\beta_j\sum_{v=0}^{j-1}X^vY^{j-1-v} = 1. \tag{11.5}$$

This equation is of degree 5 in Y, as it has to be, because $\Phi(5\tau)$ belongs to the group $\Gamma_0(25)$ with $c\equiv 0 \pmod{25}$, which is of index 5 in the group $\Gamma_0(5)$ of $\Phi(\tau)$. Therefore (11.5) is irreducible.

Moreover the equation (11.5) is symmetric in X and Y. These two functions go over into each other by the substitution

$$\tau' = -(5\tau)^{-1}.$$

With these properties the equation (11.5) fulfills the definition of a modular equation (Klein-Fricke [5, vol. 2, pp. 56, 57]), it is a modular equation of level 5 and order 5.

There exists also an equation for

$$(11.6) \qquad \frac{\eta(5\tau)}{\eta(\tau)} = (\Phi(\tau))^{1/6}.$$

which in our notation is

$$(11.7) \quad 5^{5/2}X^{5/6}Y^{1/6}+5^3X^{4/6}Y^{2/6}+3\cdot 5^{5/2}X^{3/6}Y^{3/6}+5^3X^{2/6}Y^{4/6}+5^{5/2}X^{1/6}Y^{5/6}=1$$

(cf. [6, p. 395, formula (23)], and [3, p. 105, formula (3.2)]). The function (11.6) does not belong to $\Gamma_0(5)$, it is in fact a function of level 30. We can look upon (11.7) from our present point of view as on a modular equation of level 5 in "irrational form." By elementary algebraic processes the equation (11.5) can be regained from (11.7).

12. A similar treatment of (10.2) leads to an algebraically different situation. Instead of (11.3) we obtain this time

$$(12.11) \quad \begin{aligned} &7(\Phi(7\tau)^{-1} - 7^2\Phi(\tau)) + (\Phi(7\tau)^{-2} - 7^4\Phi^2(\tau)) \\ &\qquad + 7^3\sum_{j=1}^{14} c_j(\Phi(7\tau)^j - 7^{-2j}\Phi(\tau)^{-j}) = 0 \end{aligned}$$

with

$$(12.12) \qquad \Phi(\tau) = \Phi_{7,4}(\tau) = \left(\frac{\eta(7\tau)}{\eta(\tau)}\right)^4.$$

Theorem 2 shows that $\Phi(\tau)$ is a modular function of level 7, belonging to $\Gamma_0(7)$ and (12.11) is therefore a transformation equation of level 7 and order 7. Let us put

$$7^{-2}\Phi(\tau)^{-1} = X, \qquad \Phi(7\tau) = Y,$$

so that (12.11) can be written as

$$(12.2) \qquad (Y^{-2} - X^{-2}) + 7(Y^{-1} - X^{-1}) + 7^3\sum_{j=1}^{14} c_j(Y^j - X^j) = 0.$$

This equation can be freed of the factor $Y-X$. But then it would still remain of degree 15 in Y, whereas $\Phi(7\tau)$ belongs to $\Gamma_0(49)$ which is of index 7 in $\Gamma_0(7)$. Therefore even after division by $Y-X$ equation (12.2) cannot be irreducible. If we write it indeed in the form

$$(12.3) \quad Y^{-2} + 7Y^{-1} + 7^3\sum_{j=1}^{14} c_jY^j + C = X^{-2} + 7X^{-1} + 7^3\sum_{j=1}^{14} c_jX^j + C$$

it turns out by actual computation that for $C=(1/4)7^2\cdot 32145$ and the c_j which are numerically given in Zuckerman's paper [4] we have

$$X^{-2} + 7X^{-1} + C + 7^3\sum_{j=1}^{14} c_j X^j$$

$$= (X^{-1} + 7/2 + 82\cdot 7^4X + 176\cdot 7^6X^2 + 845\cdot 7^7X^3 + 272\cdot 7^9X^4 + 46\cdot 7^{11}X^5 + 4\cdot 7^{13}X^6 + 7^{14}X^7)^2, \tag{12.4}$$

a complete square. This permits the extraction of the square root in (12.3). The square root must be taken with the same sign on both sides so that the constant term disappears and a further division by $X - Y$ can reach the degree 7 of the equation which before that division is

$$\begin{aligned}(Y^{-1} - X^{-1}) &+ 82\cdot 7^4(Y - X) + 176\cdot 7^6(Y^2 - X^2) + 845\cdot 7^7(Y^3 - X^3)\\ &+ 272\cdot 7^9(Y^4 - X^4) + 46\cdot 7^{11}(Y^5 - X^5) + 4\cdot 7^{13}(Y^6 - X^6)\\ &+ 7^{14}(Y^7 - X^7) = 0.\end{aligned} \tag{12.5}$$

Multiplication by XY and then division by $Y-X$ establish an equation which is symmetric in X and Y and of degree 7 in Y and therefore irreducible, since 7 is the index of the group Y in that of X, as mentioned above. This equation is therefore a modular equation of level 7 and order 7. Again, an "irrational form" of it is known for

$$\frac{\eta(7\tau)}{\eta(\tau)} = (\Phi(\tau))^{1/4}. \tag{12.6}$$

It was given by Watson [3, p. 118, (5.2)] and appears in our notation as

$$\begin{aligned}&7^{7/2}X^{7/4}Y^{1/4} + 7^4X^{6/4}Y^{2/4} + 3\cdot 7^{7/2}X^{5/4}Y^{3/4}\\ &+ 7^4X^{4/4}Y^{4/4} + 3\cdot 7^{7/2}X^{3/4}Y^{5/4} + 7^4X^{2/4}Y^{6/4} + 7^{7/2}X^{1/4}Y^{7/4}\\ &+ 7^{5/2}X^{3/4}Y^{1/4} + 5\cdot 7^2X^{2/4}Y^{2/4} + 7^{5/2}X^{1/4}Y^{3/4} = 1.\end{aligned} \tag{12.7}$$

The equation (12.5) has as roots the 4th powers of the roots of (12.7) and can therefore (disregarding the factor $Y-X$ in (12.5)) be directly derived from
7 (12.6). A control by numerical calculation shows indeed complete agreement in the coefficients. The function (12.6), by the way, is of level 28.

PART III. PROOFS FOR THE THEOREMS 1, 2, AND 3

13. The proofs will be based on three lemmas, which occur as Theorems 17, 18, 19 in [9].

LEMMA 1. *Let $\Theta = \Theta_k$ denote 1 for $3\nmid k$ and 3 for $3 \mid k$ so that $\Theta\cdot k$ is either prime to 3 or divisible by 3^2. For $(h, k) = 1$ we have*

$$12hks(h, k) \equiv h^2 + 1 \pmod{\Theta\cdot k}. \tag{13.1}$$

Moreover

$$12ks(h, k) \equiv 0 \pmod 3, \qquad \text{if } 3\nmid k. \tag{13.2}$$

LEMMA 2. *For odd k we have*

$$12ks(h, k) \equiv k + 1 - 2\left(\frac{h}{k}\right) \pmod{8}, \tag{13.3}$$

where (h/k) denotes the Legendre-Jacobi symbol.

LEMMA 3. *If k is equal to $2^\lambda l$, $\lambda \geqq 0$ and l and h are odd integers, then*

$$12hks(h, k) \equiv h^2 + k^2 + 3k + 1 + 2k\left(\frac{k}{h}\right) \pmod{2^{\lambda+3}}. \tag{13.4}$$

We derive first two further lemmas about the "Dedekind sums" $s(h, k)$. In the sequel p will always be a prime number greater than 3, and r is an integer such that

$$r(p-1) \equiv 0 \pmod{24}. \tag{13.51}$$

The condition imposed on p will be used in the form

$$p^2 \equiv 1 \pmod{24}. \tag{13.52}$$

LEMMA 4. *Let a, b, c, d be integers with $ad - bc = 1$, $c > 0$, $p \mid c$. Put $c = p \cdot c_1$ and*

$$G = \left(s(a, c) - \frac{a+d}{12c}\right) - \left(s(a, c_1) - \frac{a+d}{12c_1}\right). \tag{13.61}$$

Then with r satisfying (13.51) *we have*

$$rG \equiv \frac{1}{2}\left\{1 - \left(\frac{a}{p}\right)^r\right\} \pmod{2}. \tag{13.62}$$

Proof. From (13.1) we obtain

$$12ac\left(s(a, c) - \frac{a+d}{12c}\right) \equiv a^2 + 1 - a(a+d) \equiv -bc \pmod{\Theta c},$$

where $\Theta = \Theta_c$ is defined as in Lemma 1. If we then apply (13.1) with $k = c/p = c_1$, we have, after multiplication by p,

$$12ac\left(s(a, c_1) - \frac{a+d}{12c_1}\right) \equiv pa^2 + p - pa(a+d) \equiv -pbc \pmod{\Theta c}.$$

It has to be remarked that Θ has the same value in both congruences, since c and c_1 have either both the factor 3 or not. With the definition (13.61), we obtain

$$12acr\, G \equiv r(p-1)\, bc \equiv 0 \pmod{\Theta c}.$$

For $3 \nmid c$ we have $12crG \equiv 0 \pmod 3$ as direct consequence from (13.2) and (13.51). Therefore

$$12crG \equiv 0 \pmod{3c}. \tag{13.71}$$

Now let c be odd. Then Lemma 2 is applicable and yields for $k=c$

$$12c\left(s(a,c) - \frac{a+d}{12c}\right) \equiv c + 1 - 2\left(\frac{a}{c}\right) - (a+d) \pmod 8,$$

and for $k = c/p = c_1$

$$12c\left(s(a,c_1) - \frac{a+d}{12c_1}\right) \equiv c + p - 2p\left(\frac{a}{c_1}\right) - p(a+d) \pmod 8.$$

We can replace (a/c_1) by $(a/c_1)(a/p^2) = (a/cp) = (a/c)(a/p)$, and get therefore

$$\begin{aligned}12crG &\equiv r(1-p) - 2r\left(\frac{a}{c}\right)\left\{1 - p\left(\frac{a}{p}\right)\right\} + r(p-1)(a+d) \\ &\equiv \pm\, 2r\left\{1 - p\left(\frac{a}{p}\right)\right\} \pmod 8,\end{aligned}$$

which implies

$$12crG \equiv 0 \pmod 8 \tag{13.72}$$

for r even. For r odd the condition (13.51) necessitates $p \equiv 1 \pmod 8$ and therefore

$$1 - p\left(\frac{a}{p}\right) \equiv 1 - \left(\frac{a}{p}\right) \pmod 4,$$

so that $(a/p)=1$ leads also to (13.72). For r odd together with $(a/p) = -1$ we find, however,

$$12crG \equiv 4 \pmod 8,$$

or

$$12crG \equiv 12c \pmod 8. \tag{13.73}$$

The congruences (13.71), (13.72), (13.73) prove (13.62) for odd c.

For even $c = 2^\lambda\gamma$ with γ odd we make use of Lemma 3 and obtain for $k=c$ and $k = c/p = c_1$

$$\begin{aligned}12ac\left(s(a,c) - \frac{a+d}{12c}\right) &\equiv a^2 + c^2 + 3c + 1 + 2c\left(\frac{c}{a}\right) - a(a+d) \\ &\equiv c^2 + 3c - bc + 2c\left(\frac{c}{a}\right) \pmod{2^{\lambda+3}},\end{aligned}$$

and

$$12ac\left(s(a, c_1) - \frac{a+d}{12c_1}\right) \equiv pa^2 + \frac{c^2}{p} + 3c + p + 2c\left(\frac{c_1}{a}\right) - pa(a+d)$$

$$\equiv \frac{c^2}{p} + 3c - pbc + 2c\left(\frac{c}{a}\right)\left(\frac{p}{a}\right) \pmod{2^{\lambda+3}},$$

and hence

$$12acrG \equiv r\frac{c^2}{p}(p-1) + r(p-1)bc + 2rc\left(\frac{c}{a}\right)\left\{1 - \left(\frac{p}{a}\right)\right\}$$

$$\equiv \pm 2^{\lambda+1}r\left\{1 - \left(\frac{p}{a}\right)\right\} \pmod{2^{\lambda+3}},$$

and therefore

(13.74) $$12crG \equiv 0 \pmod{2^{\lambda+3}}$$

for r even, as well as for $(p/a)=1$, which for the case of odd r and thus $p\equiv 1 \pmod 4$ is equivalent to $(a/p)=1$.

For r odd in conjunction with $(a/p)=-1$ we have

(13.75) $$12crG \equiv 2^{\lambda+2} \pmod{2^{\lambda+3}}.$$

The congruences (13.71), (13.74), (13.75) establish (13.62) also for the case of an even $c=2^{\lambda}\gamma$. This finishes the proof of Lemma 4.

LEMMA 5. *Let a, b, c, d be integers with $ad-bc=1$, $c>0$, $p^2 | c$. Then, with $c=p^2c_2$, we have*

(13.8) $$H = \left(s(a, c) - \frac{a+d}{12c}\right) - \left(s(a, c_2) - \frac{a+d}{12c_2}\right) \equiv 0 \pmod 2.$$

Proof. For $k=c$ Lemma 1 yields

$$12ac\left(s(a, c) - \frac{a+d}{12c}\right) \equiv a^2 + 1 - a(a+d) \equiv -bc \pmod{\Theta c},$$

and for $k=c/p^2=c_2$

$$12ac\left(s(a, c_2) - \frac{a+d}{c_2}\right) \equiv p^2(a^2+1) - p^2a(a+d) \equiv -p^2bc \pmod{\Theta c}.$$

Therefore

$$12cH \equiv 0 \pmod{\Theta c}.$$

This congruence holds, modulo $3c$, also if $3\nmid c$ since then according to (13.2)

$12c\,s(a,c)$ and $12c\,s(a,\ c_2)$ are separately divisible by 3, and so is $(p^2-1)a(a+d)$. Thus we obtain

$$12cH \equiv 0 \pmod{3c}. \tag{13.91}$$

Now suppose, first, c to be odd. Then we infer from Lemma 2

$$12c\left(s(a,c) - \frac{a+d}{12c}\right) \equiv c + 1 + 2\left(\frac{a}{c}\right) - (a+d) \pmod 8,$$

and

$$12c\left(s(a,c_2) - \frac{a+d}{12c_2}\right) \equiv c + p^2 + 2p^2\left(\frac{a}{c_2}\right) - p^2(a+d) \pmod 8.$$

But since $(a/c_2) = (a/c)$ and $p^2-1\equiv 0 \pmod 8$ we find by subtraction

$$12cH \equiv 0 \pmod 8, \tag{13.92}$$

which together with (13.91) proves (13.8) for odd c.

Secondly, in case we have $c=2^\lambda\gamma$, γ and a odd, we obtain from Lemma 3

$$\begin{aligned}12ac\left(s(a,c) - \frac{a+d}{12c}\right) &\equiv a^2 + c^2 + 3c + 1 + 2c\left(\frac{c}{a}\right) - a(a+d)\\ &\equiv c^2 + 3c - bc + 2c\left(\frac{c}{a}\right) \pmod{2^{\lambda+3}},\end{aligned}$$

and

$$\begin{aligned}12ac\left(s(a,c_2) - \frac{a+d}{12c_2}\right) &\equiv p^2a^2 + \frac{c^2}{p^2} + 3c + p^2 + 2c\left(\frac{c_2}{a}\right) - p^2a(a+d)\\ &\equiv \frac{c^2}{p^2} + 3c - p^2bc + 2c\left(\frac{c}{a}\right) \pmod{2^{\lambda+3}},\end{aligned}$$

nd therefore

$$12acH \equiv \frac{c^2}{p^2}(p^2-1) + (p^2-1)bc \equiv 0 \pmod{2^{\lambda+3}},$$

or

$$12cH \equiv 0 \pmod{2^{\lambda+3}}. \tag{13.93}$$

This together with (13.91) completes the proof of (13.8) for c even.

14. **Proof of Theorem** 1. Here and in the following proof we treat the modular substitutions with $c=0$ separately. Since all of these are iterations of $S\tau=\tau+1$, it suffices in this case to study only S. Now the definitions (8.1) and (1.3) show immediately

$$\Phi_{p,r}(\tau+1) = e^{\pi ir(p-1)/12}\Phi_{p,r}(\tau)$$

which because of (8.2) amounts to

$$\Phi_{p,r}(\tau + 1) = \Phi_{p,r}(\tau).$$

This can be subsumed under (8.3) for $a=1$. From now on we can assume that $c>0$, with $c\equiv 0 \pmod p$. The formula (4.3) yields here

$$\Phi_{p,r}(V\tau) = M_V \cdot \Phi_{p,r}(\tau)$$

with

$$M_V = \exp\left\{\pi i r\left(s(a, c) - \frac{a+d}{12c} - s(a, c_1) + \frac{a+d}{12c_1}\right)\right\}, \qquad c_1 = \frac{c}{p}.$$

Application of Lemma 4 gives

$$M_V = \exp\left\{\frac{\pi i}{2}\left(1 - \left(\frac{a}{p}\right)^r\right)\right\}.$$

This can be written briefly as

$$M_V = \left(\frac{a}{p}\right)^r,$$

which proves Theorem 1.

15. **Proof of Theorem** 2. We first consider the effect of S on $L_p^*(\tau)$. The definitions (8.4) and (1.3) show that in the sum defining $L_p^*(\tau)$ it is only essential that λ runs through a complete residue system modulo p. We can therefore write

$$L_p^*(\tau) = \sum_{\lambda=0}^{p-1} \eta(p\tau)\eta\left(\frac{\tau + 24N + 24\lambda}{p}\right)^{-1}$$

for any integer N. Hence we get

$$L_p^*(\tau + 1) = \sum_{\lambda=0}^{p-1} \eta(p(\tau + 1))\eta\left(\frac{\tau + 24N + 1 + 24\lambda}{p}\right)^{-1}.$$

If we choose here

$$24N + 1 = p^2,$$

we obtain

$$L_p^*(\tau + 1) = \sum_{\lambda=0}^{p-1} \eta(p\tau + p)\eta\left(\frac{\tau + 24\lambda}{p} + p\right)^{-1},$$

which is equal to $L_p^*(\tau)$ in virtue of (1.3). Having thus disposed of the case $c=0$ we can suppose from now on $c>0$ with $c\equiv 0 \pmod p$.

In

$$V = \begin{pmatrix} a & b \\ c & d \end{pmatrix}$$

we have therefore necessarily $a \not\equiv 0 \pmod{p}$. Since only a complete residue system of the index of summation was essential we can also write

$$L_p^*(\tau) = \sum_{\lambda=0}^{p-1} \eta(p\tau) \cdot \eta\left(\frac{\tau + (p^2 - 1)a\lambda}{p}\right)^{-1}.$$

We obtain therefore, with $c = pc_1$,

$$(15.1) \qquad L_p^*(V\tau) = \sum_{\lambda=0}^{p-1} \eta\left(\frac{ap\tau + pb}{c_1 p\tau + d}\right) \cdot \eta\left(\frac{1}{p}\left(\frac{a\tau + b}{c\tau + d} + (p^2 - 1)a\lambda\right)\right)^{-1}.$$

Proceeding here as we did in §6 for $p = 5$ we wish to construct modular substitutions

$$\begin{pmatrix} A & B \\ C & D \end{pmatrix}$$

such that

$$(15.2) \qquad \frac{1}{p}\left(\frac{a\tau + b}{c\tau + d} + (p^2 - 1)a\lambda\right) = \frac{A(\tau + (p^2 - 1)d\mu)/p + B}{C(\tau + (p^2 - 1)d\mu)/p + D},$$

where A, B, C, D and μ will depend on λ.

Comparison of coefficients shows that the equations

$$(15.3) \qquad \begin{aligned} a + c(p^2 - 1)a\lambda &= A, & b + d(p^2 - 1)a\lambda &= pB + A(p^2 - 1)d\mu, \\ pc &= C, & pd &= pD + C(p^2 - 1)d\mu \end{aligned}$$

are necessary and sufficient. It is clear that for any choice of λ and μ the numbers A, B, C, D are uniquely determined through (15.3) and satisfy

$$(15.4) \qquad \begin{vmatrix} A & B \\ C & D \end{vmatrix} = 1.$$

We can now tie μ to λ in such a way that B will become an integer, whereas A, C, D are obviously integers for any integers λ, μ. Since p divides c we have

$$a \equiv A \pmod{p},$$

and therefore

$$b - ad\lambda \equiv -ad\mu \pmod{p},$$

which implies already that B is an integer. Now

$$1 = ad - bc \equiv ad \pmod{p},$$

and therefore

$$\mu \equiv \lambda - b \pmod{p}.$$

The new summation index μ needs only to be determined modulo p and we can therefore, without loss of generality, put

$$\mu = \lambda - b. \tag{15.5}$$

This choice now completes the determination of A, B, C, D, which become

$$\begin{aligned} A &= a + c(p^2 - 1)a\lambda, & B &= padb - \frac{c}{p}b^2 - \frac{c}{p}(p^2-1)^2 ad\lambda(\lambda - b), \\ C &= pc, & D &= d - c(p^2-1)d(\lambda - b). \end{aligned} \tag{15.6}$$

From (15.1), (15.2), and (4.3) we derive now

$$L_p^*(V\tau) = \sum_{\lambda=0}^{p-1} M_\lambda \cdot \eta(p\tau) \cdot \eta\left(\frac{\tau + (p^2-1)d\mu}{p}\right)^{-1} \tag{15.7}$$

with

$$M_\lambda = \exp\left\{-\pi i\left(\left(s(a, c_1) - \frac{a+d}{12c_1}\right) - \left(s(A, C) - \frac{A+D}{12C}\right)\right)\right\}.$$

From (15.6) it follows that

$$a \equiv A \pmod{c}.$$

This permits us to write

$$s(a, c_1) = s(A, c_1) = s(A, C_2), \qquad C_2 = \frac{C}{p^2} = \frac{c}{p} = c_1.$$

Moreover the equations (15.6) show that $(a+d)/12c$, and $(A+D)/12c$ differ only by an even integer. We can therefore write

$$M_\lambda = \exp\left\{-\pi i\left(\left(s(A, C_2) - \frac{A+D}{12C_2}\right) - \left(s(A, C) - \frac{A+D}{12C}\right)\right)\right\},$$

and obtain then from Lemma 4

$$M_\lambda = 1.$$

If we observe that in (15.7) μ runs with λ through a complete residue system modulo p we have proved

$$L_p^*(V\tau) = L_p^*(\tau),$$

which finishes the proof of Theorem 2.

16. **Proof of Theorem** 3. The proof is closely similar to that of Theorem 2. The same auxiliary substitutions are used. It is then only necessary to apply Lemma 4 instead of Lemma 5 for the computation of the multiplier M_λ.

References

1. S. Ramanujan, *Some properties of $p(n)$, the number of partitions of n*, Collected Papers, pp. 210–213.

2. L. J. Mordell, *Note on certain modular relations considered by Messrs. Ramanujan, Darling, and Rogers*, Proceedings of the London Mathematical Society, (2), vol. 20 (1922), pp. 408–416.

3. G. N. Watson, *Ramanujans Vermutung über Zerfällungsanzahlen*, Journal für die reine und angewandte Mathematik, vol. 179 (1938), pp. 97–128.

4. H. S. Zuckerman, *Identities analogous to Ramanujan's identities involving the partition function*, Duke Mathematical Journal, vol. 5 (1939), pp. 88–110.

5. F. Klein und R. Fricke, *Vorlesungen über die Theorie der Elliptischen Modulfunktionen*, vol. 1, 1890, and vol. 2, 1892.

6. R. Fricke, *Die Elliptischen Funktionen und ihre Anwendungen*, vol. 1, 1916, and vol. 2, 1922.

7. H. Rademacher, *Über die Erzeugenden von Kongruenzuntergruppen der Modulgruppe*, Abhandlungen aus dem Mathematischen Seminar der Hamburgischen Universität, vol. 7 (1930), pp. 134–148.

8. ———, *Zur Theorie der Modulfunktionen*, Journal für die reine und angewandte Mathematik, vol. 167 (1931), pp. 312–336.

9. H. Rademacher and A. Whiteman, *On Dedekind sums*, American Journal of Mathematics, vol. 63 (1941), pp. 377–407.

University of Pennsylvania,
Philadelphia, Pa.

This paper has been reviewed in the Z, vol. 60 (1957), pp. 100-101, and in the MR, vol. 3 (1942), p. 271.

Page 609, equation (1.2).
Read $p(7l+5)x^l =$ instead of $p(7l+5)x$.

Page 618, second display line.
In the exponent read $-1/2$ instead of $-1j2$.

Page 618, second line from bottom.
The reference is to [7], rather than to [8].

Page 626, line 11 of text.
Read *factor* $Y - X$ instead of *factor*.

Page 628, marginal note.
Author's correction (read (12.7) for (12.6)).

Page 635, seventh line of text from bottom.
Read Lemma 5 instead of Lemma 4.

Page 636, footnote 7.
See paper 25 of the present collection.

Page 636, footnote 8.
See paper 27.

Page 636, footnote 9.
See paper 44.

TRENDS IN RESEARCH: THE ANALYTIC NUMBER THEORY

HANS RADEMACHER

Ladies and Gentlemen: I wish to express my gratitude for the invitation to address this meeting and to thank you for granting me this opportunity to contribute a little part to the celebration of the anniversary of our host, this renowned institution of higher learning, the University of Chicago. The research work emanating from here has in its mathematical branch strongly emphasized analysis as well as number theory. The name "analytic number theory," indicating a union of these two fields of mathematical endeavor, is about as old as this celebrating Institution. In 1894 a book was published under the title *Die Analytische Zahlentheorie*, by Paul Bachmann. The name "analytic number theory" itself was a program. It said more than "diverse applications of infinitesimal calculus to number theory" as Dirichlet, the real founder of our discipline, had modestly called one of his great memoirs (1839). The name "analytic number theory," implies, as I take it, a thorough fusion of analysis and arithmetic, in which as we shall see, analysis is not necessarily subordinate to arithmetic. Incidentally, "analysis" stands here always for function theory, especially the theory of analytic functions.

What has the theory of functions to do with the theory of integers? Let me, to gain a starting point, begin with a few examples of well established results of analytic number theory:

1. If $\pi(x)$ denotes as usual the number of prime numbers less than or equal to x, then

$$\lim_{x\to\infty} \frac{\pi(x)}{x/\log x} = 1 \tag{1}$$

as first proved by Hadamard and de la Vallée Poussin in the nineties of the last century.

2. The number of representations of a positive integer n as a sum of four squares is

$$r_4(n) = 8\left(\sum_{d|n} d - \sum_{4|d'|n} d'\right) \tag{2}$$

as Jacobi found as a corollary of his theory of elliptic functions.

An address delivered before the meeting of the Society in Chicago, Ill., on September 5, 1941, by invitation of the Program Committee; received by the editors January 6, 1942.

3. The number $h(D)$ of classes of positive primitive binary quadratic forms of determinant $D\equiv 7 \pmod 8$ is given by Dirichlet's formula

$$h(D) = \sum_a \left(\frac{a}{D}\right) \tag{3}$$

where $(a, D)=1$, and a runs through all positive integers less than $(1/2)D$.

Now, the first of these examples makes it quite obvious that analysis has to be applied, since the result does not only contain the continuous function $x/\log x$, but moreover involves the notion of limit, so that the two fundamental concepts of infinitesimal analysis are represented in (1).

This is not so in the two other examples. All expressions appearing in the equations (2) and (3) have an arithmetical meaning and are defined in elementary number theory without any reference to analysis. Nevertheless both results appear as the goal of analytic proofs in one of which appear ϑ-functions and in the other Dirichlet series. Both theorems can therefore be claimed by analytic number theory. However, there remains a big difference in these two cases: whereas the result (2) can also be proved by purely arithmetical methods, the equation (3) has defied all attempts of an elementary arithmetical proof.

This observation raises the question, which I will state in all its logical vagueness: what is the connecting link between arithmetic and analysis, or by what means can analysis take hold of arithmetical facts? In the steady flow of real numbers, called the continuum, which is the substratum of function theory, the integers are in no way distinguished from any other real numbers. The function e^z at first glance has no "affinity" to integers. We notice, however, the relation

$$e^{2\pi iz} = e^{2\pi i(z+n)}$$

for any integer n and all complex values z. The function has *the period* n where n can be any integer. The fundamental fact is that the periods of an analytic, or for that matter, a continuous function form a discrete module or lattice. Or, speaking a little more generally: the ideas connected with period, invariance, and substitution group are basic for the application of analysis to number theory, and, as we can remark already here, also for the inverse process. Let me give another example, which is a little more involved. For the modular invariant $J(\tau)$ we know that

$$J(\tau) = J(\tau') \tag{4a}$$

implies and is implied by

$$\text{(4b)} \qquad \tau' = \frac{a\tau + b}{c\tau + d}, \qquad ad - bc = 1,$$

a, b, c, d being integers. The function $J(\tau)$ appears thus as an embodiment of the discrete group of modular substitutions. This fact must have implications in both directions. Indeed, the group of modular transformations defines the classes of binary quadratic forms, and so we are led to the theory of singular moduli, the theory of complex multiplication, and finally to the Kronecker-Fueter theorem that any Abelian field over an imaginary quadratic field can be generated by the values of certain modular forms which they attain for special imaginary quadratic irrationalities. On the other side, equations (4a) and (4b) lead function-theoretically to the expansion (Petersson [**33**, **37**])

$$\text{(5a)} \qquad 12^3 J(\tau) = e^{-2\pi i \tau} + 744 + \sum_{n=1}^{\infty} c_n e^{2\pi i n \tau}$$

with

$$\text{(5b)} \quad c_n = \frac{2\pi}{n^{1/2}} \sum_{k=1}^{\infty} \frac{1}{k} I_1\left(\frac{4\pi n^{1/2}}{k}\right) \sum_{h \bmod k, (h,k)=1} e^{-(2\pi i/k)(nh+h')}, hh' \equiv -1(k),$$

where $I_1(z) = -iJ_1(ix)$, the Bessel function of the first order.

You will always find such a group-invariance or periodicity, sometimes conspicuous, sometimes concealed, as underlying the applications of analysis to number theory. Fourier series, Fourier integrals, and Poisson sums are, therefore, natural and frequently used tools. But also the configuration of a point lattice can be taken as to express periodicity (especially in Minkowski's work in the "geometry of numbers"), and even a power series is here often better understood as a series in $e^{2\pi i z}$, that is, a Fourier series with its obvious periodicity.

These analytic tools, developed in the 19th century, indicate sufficiently that our discipline must be a rather young one, so very much younger than elementary number theory, in whose history we meet the names of Euclid, Diophantos, Fermat. Indeed Lejeune Dirichlet can be regarded as the founder of our branch of research. Before him Euler, Gauss, and especially Jacobi, had found methods and results belonging to this field. After Dirichlet it was Riemann who with his epoch making memoir *Ueber die Anzahl der Primzahlen unter einer gegebenen Grösse* (1859) made the deepest mark in its history. In this brief survey I have also to mention the names of Hermite, Kronecker,

Hadamard, de la Vallée Poussin, and Minkowski, who all contributed works of lasting value.

The *present* standing of analytic number theory and its recognition as a fruitful branch of our science is, I think, entirely due to three mathematicians of this century: Landau, Hardy, and Littlewood. To Hardy and Littlewood we owe important progress in the theory of prime numbers and in our knowledge of the distribution of the complex zeros of the ζ-function. Their greatest contribution, which they share with Ramanujan, lies, however, in the development of a fundamentally new method in additive number theory, a method which yielded their celebrated theorems concerning partitions, and about additive representations of numbers as sums of powers (Waring's problem) and as sums of primes (Goldbach's problem).

Landau has exerted a wide influence through his untiring mathematical enthusiasm and through his forceful advocacy of utter precision. The substance of his work has proved to be less permanent; his results were soon surpassed by deeper and more comprehensive ones, and his papers have very often been only second redactions of discoveries of others. How much we owe him nevertheless becomes evident if we compare a publication on analytic number theory of the the 1890's (for example, the above mentioned book of Bachmann, but also such substantial papers like those of Hadamard and de la Vallée Poussin not excepted) with a contemporaneous paper. Landau developed the technique of analytic number theory and prepared the tools; he found certain theorems about power series, Dirichlet series, Fourier integrals, which now are generally used as lemmas; and he made us familiar with the usefulness of a number of devices and lemmas discovered by others, as for example, Abel's partial summation, Mellin's formula, and Carathéodory's lemma. Landau's handbook on prime numbers will remain a classical treatise. Not a small merit of it is that it shaped our notation: It brought the symbol "O" into general use (it had been invented by Pringsheim), and Landau introduced there the symbol "o" (a constant source of quarrel between him and Pringsheim, which the German mathematicians used to watch with amusement). Both these symbols, now so well known and generally adopted, are particularly fit to express asymptotic relations which so frequently occur in analytic number theory.

Landau is no longer with us; Hardy and Littlewood about 10 years ago turned to other fields of investigation. Hence I understand it as my main task today to give a survey of developments in analytic number theory which took place after these three outstanding authors and independent of them.

This development has proceeded chiefly in two directions. One of them can roughly be characterized by words like "asymptotic formula" and "error term;" inequalities play an important part in the technique to "estimate" the error. I think here mainly of problems like the prime number theorem, lattice points in the circle, asymptotic expressions for the number of representations of a number as a sum of one or the other sort, investigations in which Hardy, Littlewood, and Landau have distinguished themselves, so that this pursuit of research can be understood as a direct continuation of their work. Investigations of this sort have attracted general attention, so much so, indeed, that the impression may have prevailed that analytic number theory deals foremost with asymptotic expressions for arithmetical functions. This view, however, overlooks another side of analytic number theory, which I may indicate by the words "identities," "group-theoretical arguments," "structural considerations." This line of research is not yet so widely known; it may very well be that methods of its type will lead to the "deeper" results, will reveal the sources of some of the results of the first direction of approach. Remarks like those which I made about applicability of analysis to arithmetic are especially pertinent here. The names Artin, Hecke, Mordell and Siegel would have to be mentioned in this connection.

The first type of analytic number theory is at present carried on in the Russian school with Vinogradoff as the leading name. Vinogradoff has become widely known for his great contributions to Waring's problem and Goldbach's problem. In this work he depends entirely on Hardy's and Littlewood's precedency; he introduced essentially two improvements of their method. One is of minor importance and merely an affair of convenience. He observed that for the additive representation of the number n you obviously do not need summands surpassing n. This means that the power series which have been used by Hardy and Littlewood (and whose use in additive number theory really dates back to Euler) can be replaced by polynomials. This has certain advantages since it discards the difficulty of the natural boundary which the unit circle represents for those power series. Vinogradoff's second and decisive advance lies in his ingenious estimate of certain sums of roots of unity, as I shall now try to explain.

I begin with Waring's problem, in whose treatment Vinogradoff reaped his first success. For the discussion of

$$f(x) = \sum_{n=0}^{\infty} x^{n^k} \tag{6}$$

in the neighborhood of a root of unity $\xi = e^{2\pi i(h/q)}$ Hardy and Littlewood need expressions for the sums

$$S_q(M) = \sum_{n=0}^{M} e^{2\pi i (h/q) n^k} \tag{6a}$$

with the emphasis upon their dependence on q and M (for fixed given $k \geqq 3$; $k=2$ is in a different class since $f(x)$ is then essentially a ϑ function and $S_q(M)$ essentially a Gaussian sum, both of which are well known from other connections). Now for an M which is large compared with q an expression with a useful error term is easily found by partial summation. The error term, however, becomes unmanageable for q comparatively large. This case, which appears on the so-called "minor arcs" proved first to be a serious obstacle to Hardy's and Littlewood's efforts. It is known that this difficulty delayed their final success for two years at a time when the required device had already been found and published by H. Weyl [**58**] but due to the interruption of communications during the first World War had not yet come to Hardy's and Littlewood's notice.

Now Vinogradoff recognized that a certain k-fold iteration in Weyl's method could be dispensed with and replaced by one single step. Since that iteration was the source of a factor 2^k in the estimate for $G(k)$, the smallest number of kth powers sufficing to express every sufficiently large number additively, this change meant an extraordinary diminution of the estimated order of $G(k)$, a diminution as a matter of fact which was so considerable that it enabled Dickson to establish in many cases the true value $g(k)$ of kth powers which suffice to represent additively *all* positive integers, simply by bridging the finite gap not covered by $G(k)$ through a number of elementary computational steps [**6**, **7**].

Vinogradoff's lemma, which replaces Weyl's lemma, can be enunciated as follows (in the form which was given to it by Heilbronn [**21**]):

Let x run through X integers of an interval of length L, y through Y integers on an interval of length M, then

$$\left| \sum_{x,y} e^{2\pi i (xy/q)} \right|^2 \leqq \frac{XYML}{q}\left(1 + \frac{q}{M}\right)\left(1 + \frac{q \log q}{L}\right). \tag{7}$$

This lemma is not immediately applicable to (6a) and to (6). Vinogradoff employs the astonishing device of dropping certain Waring representations of the given number n, by selecting kth powers of a certain factorization. This diminishes the chance of a Waring representation. Nevertheless he finds that the number of representations is different from null for $s \geqq 6k \log k + (4 + \log 216)k$ summands and sufficiently large n [**50**].

Vinogradoff's most startling achievement was his proof in 1937 that every large odd number is the sum of three primes [52, 53], a result previously obtained by Hardy and Littlewood only under a generalized Riemann hypothesis. Vinogradoff's result is so far the closest approach to Goldbach's conjecture ever reached. The difficulties are here much greater than for the Waring problem, the distinction between major and minor arcs much subtler. The major arcs require for their treatment the classical theory of the distribution of primes founded on the ζ-function and the Dirichlet L-functions. In Vinogradoff's paper, moreover, a deep theorem of Siegel about the class-number of quadratic fields [46] was used, which can be expressed in the statement

$$|L(1, \chi)| > \frac{C}{k^{\epsilon}}$$

where χ is a real character modulo k, a theorem which widens Heilbronn's result [22] that the number of imaginary quadratic fields of class-number one is finite. Walfisz thereafter showed that Siegel's theorem can be dispensed with, at the cost of some loss of precision in the error term [54].

The sum to be appraised in Goldbach's problem is

$$T_q(M) = \sum_{p \leqq M} e^{2\pi i (h/q) p},$$

again in particular for the minor arcs, that is, for comparatively high denominator q. Now this sum does not fit into the scheme of Vinogradoff's lemma (7) since p alone and not a product appears in the exponent. Vinogradoff resorts here to the sieve of Eratosthenes [53] which had been so successfully used by Viggo Brun [3]. The argument runs like this:

$$T_q(M) = \sum_{\nu=2}^{M} e^{2\pi i (h/q)\nu} - \sum_{p_1 \leqq M^{1/2}} \sum_{2 \leqq m \leqq M/p_1} e^{2\pi i (h/q) m p_1} + \sum_{p_2 < p_1 \leqq M^{1/2}} \sum_{1 \leqq m \leqq M/p_1 p_2} e^{2\pi i (h/q) m p_1 p_2} + \cdots .$$

The first sum is a trivial geometric sum, and in the following sums we do find the desired products in the exponents. This opens the way to an application of (7). Still, the completion is not quite simple but is handled by Vinogradoff with superb technique. I have already quoted his final result regarding Goldbach's problem.

Before I drop this strand of my narration I wish to mention Tchudakoff's application of Vinogradoff's methods to the estimation of Riemann's ζ-function on $\Re(s) = 1$ [49]. Tchudakoff uses Vinogradoff's estimates **[51]** in the same way as Littlewood and Landau had used Weyl's method for the improvement of the error term in the prime number theorem to $O(x \exp(-a(\log x \log \log x)^{1/2}))$. Tchudakoff obtains the prime number theorem in the form

$$\pi(x) = \int_2^x \frac{dt}{\log t} + O(xe^{-a(\log x)^\alpha})$$

with a certain $\alpha > 1/2$. Until then the remainder term had for three decades shown the value $\alpha = 1/2$. Of course, any $\alpha < 1$ is still far off from the Riemann hypothesis which is equivalent to $\alpha = 1, a = (1/2) - \epsilon$.

Appraisals of sums of roots of unity have recently played an important role in several parts of analytic number theory. Mordell **[31]**, Davenport **[4]**, and Hasse **[5, 14]** discussed sums of the form

$$S_q = \sum_{n \bmod q} e^{(2\pi i/q)f(n)}$$

where $f(n)$ is a polynomial in n with integral coefficients. The aim is to give an estimate of S_q with respect to q. The problem is a generalization of the order of magnitude of Gaussian sums, that is, $f(n)$ of the second degree, in which case we have

$$S_q = O(q^{1/2}).$$

It is of interest that the exponent 1/2 which appears here in a number of cases has something to do with the abscissa $\mathfrak{R}(s) = 1/2$ in the Riemann hypothesis, not for the ordinary ζ-function, it is true, but for the ζ-function belonging to a field of algebraic functions over a finite field of characteristic p; these ζ-functions were introduced by E. Artin **[1]** in 1924 and later in a more general form defined by F. K. Schmidt **[44]**. The non-trivial zeros ρ of such a ζ-function turn out to be periodic and given by

$$\beta = p^\rho$$

where β are the roots of certain polynomials; clearly $|\beta| = p^{1/2}$ means

$$(*) \qquad \mathfrak{R}(\rho) = 1/2,$$

and the absolute value of β depends on the number of solutions of certain congruences modulo p, or, which is equivalent to this, on the value of certain exponential sums. Relation (*) has been proved for the Artin ζ-functions of the field of elliptic functions by Hasse **[12, 13]**, who also outlined a program for further research. Modern algebraic concepts are fundamental in this work. Quite recently, in this year, André Weil **[56, 57]** published two sketches of proofs for the

Riemann hypothesis in all algebraic function fields over a finite field. We are looking forward to a complete account of his proof. Here still a wide field is open to research, a field attractive at the same time by its large generality and its richness of specific details.

A particularly interesting case of sums of roots of unity are the so-called Kloosterman sums

$$S_q(u, v) = \sum_{h \bmod q, (h,q)=1} e^{(2\pi i/q)(uh+vh')}$$

with $hh' \equiv 1 \pmod{q}$. Since these sums are multiplicative with respect to q, they can be broken down to those with prime-number powers q. For $q = p^\alpha$, $\alpha \geqq 2$, the order $O(q^{1/2})$ has been established by Salié [43], and moreover an explicit formula involving Legendre symbols and trigonometric functions has been found. Surprisingly, however, the case of a prime number $q = p$ offers difficulties which so far have only permitted us to prove

$$S_p(u, v) = O(p^{2/3}).$$

Weil's above-mentioned results should establish here also $O(p^{(1/2)+\epsilon})$.

The Kloosterman sums appear in the theory of partitions and in the study of the coefficients of modular functions. Recent work connected with these problems can also be understood as an extension of the Hardy-Littlewood-Ramanujan work. Whereas Hardy and Ramanujan obtained only an asymptotic expression for the number $p(n)$ of partions of n, a refinement and at the same time a simplification of their method led to the equation [35, 36]

$$p(n) = \frac{1}{\pi 2^{1/2}} \sum_{q=1}^{\infty} A_q(n) \frac{d}{dn} \left\{ \frac{\sin h \dfrac{K}{q}\left(n - \dfrac{1}{24}\right)^{1/2}}{\left(n - \dfrac{1}{24}\right)^{1/2}} \right\},$$

$$A_q(n) = \sum_{h \bmod q, (h,q)=1} \omega_{h,q} e^{-2\pi(hn/q)}, \quad K = \pi(2/3)^{1/2},$$

where $\omega_{h,q}$ are certain roots of unity taken from the theory of transformation of the modular form $\eta(\tau)$, which have attracted the attention of quite a number of mathematicians since Hermite. The series in the equation just mentioned is absolutely convergent. It has been used successfully by D. H. Lehmer for the numerical computation of $p(n)$ for high values of n [26, 27]. Analogous results for other problems of partitions with certain congruence conditions imposed on the summands have been obtained by Zuckerman [59], Niven [32], Hua [23], Miss Haberzetle [9], and Lehner [28].

These investigations form a natural bridge between the two different ways of investigation I mentioned before. Technically they require estimates of error terms, but also some group-theoretical arguments. And they lead to identities, which in turn invite group-theoretical treatment and thus point in the second direction in our field.

Identities can be used as sources of other identities. In this way the famous Ramanujan identities, of which I quote here only

$$(**) \qquad \sum p(5l+4)x^l = 5\left[\prod (1-x^{5qm})^5 / \prod (1-x^m)^6\right],$$

can be proved by direct comparison of coefficients which are known in forms of series through the just mentioned methods. But the true source of the Ramanujan identities lies in the theory of transformation of modular forms, as it is seen at once if we rewrite the identity (**)

$$\sum_{\lambda=0}^{4} \frac{1}{\eta\left(\dfrac{\tau+24\lambda}{5}\right)} = 5^2 \frac{\eta(5\tau)^5}{\eta(\tau)^6}$$

[30, 39].

The formula (5) is an identity, which can also be derived by the Hardy-Littlewood-Ramanujan method. Here the problem arises to understand the invariance

$$J\left(\frac{a\tau+b}{c\tau+d}\right) = J(\tau)$$

as an outcome of the Fourier series on the right-hand member [38].

Hardy himself has contributed a remarkable paper to similar investigations [10], showing that for example,

$$\vartheta_3^8(0,\tau) = 1 + \frac{\pi^4}{3!} \sum_{n=1}^{\infty} n^3 S_8(n) e^{2\pi i n\tau},$$

$$(8a) \qquad S_8(n) = \sum_{q=1}^{\infty} \frac{1}{q^8} \sum_{h \bmod q,\,(h,q)=1} \left(\sum_{j=1}^{q} e^{2\pi i (hj^2/q)}\right)^8 e^{-2\pi i (hn/q)}.$$

Here $S_8(n)$ is the "singular series" of the problem of expressing a number as a sum of 8 squares. The proof of (8a) has to establish the fact that the Fourier series on the right of (8a) is a modular form of dimension -4.

The "singular series," first introduced into analytic number theory by Hardy and Littlewood, form objects of great mathematical interest of their own. They appear in problems of representation of integers in an additive way, for example, as $n = x_1^k + \cdots + x_r^k$. For the number of representations the Hardy-Littlewood method yields an asymptotic result of the form

$$N_r(n) \sim f_r(n) \cdot \mathfrak{S}_r(n),$$

where $f_r(n)$ is a continuous function of n, which can be interpreted as a density with respect to the r-dimensional point lattice. The other factor is the singular series $\mathfrak{S}_r(n)$, an arithmetical function of n. A remarkable feature of all singular series is that they are factorable according to prime numbers:

$$\mathfrak{S}_r(n) = \prod_p A_p^{(r)}(n). \tag{8b}$$

This fact finds its explanation in the meaning of $A_p^{(r)}(n)$. Whereas for the problem

$$n = x_1^k + \cdots + x_r^k$$

the function $f_r(n)$ means the derivative

$$f_r(n) = \frac{d}{dn} V_r(n)$$

of the volume

$$V_r(n) = \int \cdots \int_{x_1^k + \cdots + x_r^k \leqq n;\, 0 \leqq x_j} dx_1 \cdots dx_r$$

or, in other words, a measure of the area of the surface

$$n = x_1^k + \cdots + x_r^k, \qquad 0 \leqq x_j;\, j = 1, \cdots, r,$$

the $A_p^{(r)}(n)$ give the density of solutions of the congruence

$$n \equiv x_1^k + \cdots + x_r^k \pmod{p^l} \tag{9}$$

with respect to high powers of p as modulus, or explicitly:

$$A_p^{(r)}(n) = \lim_{l \to \infty} \frac{b^{(r)}(n;\, p^l)}{g^{(r)}(p^l)}$$

where $b^{(r)}(n; p^l)$ is the number of different solutions of (9) and $g^{(r)}(p^l)$ is the average of the number of solutions of (9), distributed over p^l different n's,

$$g^{(r)}(p^l) = \frac{p^{lr}}{p^l} = p^{l(r-1)}.$$

The multiplicativity of the $A_q^{(r)}(n)$ now becomes understandable, and the product (8b) means that $\mathfrak{S}_r(n)$ is the density of solutions of the congruence for, let us say, $m!$ as modulus with $m \to \infty$.

Recently Siegel has emphasized this interpretation and used it with great success in his analytic theory of quadratic forms [47]. In that connection the singular series had appeared before, in the works of Smith and Minkowski, as measure of genera of quadratic forms. Through this connection it becomes clear why the Hardy-Littlewood method leads to an exact formula for the number of representations of n as a sum of r squares for $5 \leqq r \leqq 8$, whereas for higher r it yields only an asymptotic result.

This interpretation of the singular series is possible in all those problems of additive number theory which preassign the number of summands or variables, as in Goldbach's problem, Waring's problem, or mixed próblems, as for example, the problem of representation as a sum of a square and two primes (Estermann [8]), or as a sum of r squares and s cubes. It fails, however, for problems with an undetermined number of variables, as in the case of unrestricted partitions, since there the transition from equation to congruence destroys the boundedness of the number of variables for each n, so that the number of solutions modulo p becomes meaningless. Indeed, the formula for $p(n)$, given above, is of quite a different structure.

Singular series appear as coefficients of Fourier series as in (8a). Siegel's analytic theory of quadratic forms can be regarded as a brilliant example of the study of such series. He shows that his series can be combined of ϑ-functions

$$\vartheta(\tau) = \sum_{m_1, \cdots, m_r} e^{2\pi i \tau Q(m_1, \cdots, m_r)}$$

where Q is a positive quadratic form with integral coefficients; such a combination includes ϑ-functions constructed with quadratic forms Q of all classes of the same genus. It is known that such ϑ-series are modular forms of dimension $-r/2$ and of a certain "level" (Stufe) depending essentially on the discriminant of the quadratic form $Q(m_1, \cdots, m_r)$ [45].

A general study of power series with singular series as coefficients has not yet been undertaken. It would be desirable to know more specific details about these functions than the general properties which can immediately be derived from Kac's, van Kampen's, and Wintner's remark [24, 25] that the singular series give rise to almost periodic arithmetical functions of n and from Bochner's and Bohnenblust's [2] theorems about power series with almost periodic coefficients.

The modular functions have from the beginning of our discipline played an important part in *additive* problems; I need in this respect only point to Euler's formula, fundamental for the whole theory of partitions

$$1 + \sum_{n=1}^{\infty} p(n)x^n = 1/\prod_{1}^{\infty} (1 - x^m)$$

in which the denominator is essentially the modular form

$$\eta(\tau) = e^{\pi i\tau/12}\prod_{1}^{\infty} (1 - e^{2\pi im\tau}),$$

or to equation (2), obtained by Jacobi from an expression for the 4th power of $\vartheta_3(0, \tau)$. Recently, however, they have also appeared in *multiplicative* number theory through the discovery of a surprising and most fruitful connection between modular functions and prime numbers. This connection is elaborated by Hecke in his beautiful theory, of which he published the first account in 1935 in the Royal Danish Academy [16]. As a matter of fact, a special result of Hecke's theory was anticipated by Ramanujan [41, 11] in a conjecture which was then proved by Mordell in 1917 [29]. This is again an instance of Ramanujan's gift of divination; not only that he foresaw or unveiled hitherto unknown facts, but most of his conjectures lead back to sources of great depth and abundant yield.

The "discriminant" in the theory of elliptic functions

$$\Delta(1, \tau) = g_2^3 - 27g_3^2 = (2\pi)^{12} x((1 - x)(1 - x^2) \cdots)^{24}$$

with $x = e^{2\pi i\tau}$ is a modular form of dimension -12, since for a modular substitution

$$\begin{pmatrix} a & b \\ c & d \end{pmatrix}$$

we have

$$\Delta\left(1, \frac{a\tau + b}{c\tau + d}\right) = (c\tau + d)^{12}\Delta(1, \tau).$$

The infinite product can be expanded into a power series

$$x\prod_{1}^{\infty} (1 - x^m)^{24} = \sum_{n=1}^{\infty} \tau(n)x^n.$$

If we now, following Ramanujan, associate with this power series the Dirichlet series

$$F(s) = \sum_{n=1}^{\infty} \frac{\tau(n)}{n^s},$$

it turns out that this series can be written as an infinite product

$$F(s) = \prod_p \frac{1}{1 - \tau(p)p^{-s} + p^{11-2s}}, \tag{10}$$

extended over all primes p. It is appropriate to call such a product an "Euler product" in view of the formula

$$\sum_{n=1}^{\infty} \frac{1}{n^s} = \prod_p \frac{1}{1 - p^{-s}},$$

first found and utilized by Euler. I shall have to make a remark about this latter product a little later. Not only the Euler product (10), but also a functional equation for $F(s)$: $(2\pi)^{-s}\Gamma(s)F(s) = (2\pi)^{s-12}\Gamma(12-s) \cdot F(12-s)$ reminds us strongly of the ζ-function and allied functions.

The coördination of a modular function on the one side and a Dirichlet series satisfying a functional equation on the other has now been systematized by Hecke [17, 18, 19]. Let us take an entire modular form $F(\tau)$ of dimension $-k$, k a positive even integer,

$$F\left(\frac{a\tau + b}{c\tau + d}\right) = (c\tau + d)^k F(\tau)$$

having the expansion

$$F(\tau) = c_0 + \sum_{n=1}^{\infty} c_n x^n, \qquad x = e^{2\pi i \tau},$$

which is everywhere convergent in the upper τ-half-plane. Then let us put

$$\phi(s) = \sum_{n=1}^{\infty} \frac{c_n}{n^s} \quad \text{(without } c_0).$$

Then the following statements are valid:

(1) The series for $\phi(s)$ is absolutely convergent in the half-plane $\Re(s) > k$.

(2) $\phi(s)$ is an entire function if $c_0 = 0$; if $c_0 \neq 0$ then $\phi(s)$ has a pole of the first order at $s = k$ with the residue $(-1)^{k/2}c_0(2\pi)^k/\Gamma(k)$ and $(s-k)\phi(s)$ is an entire function.

(3) $\phi(s)$ satisfies a functional equation:

$$(2\pi)^{-s}\Gamma(s)\phi(s) = (-1)^{k/2}(2\pi)^{s-k}\Gamma(k-s)\phi(k-s).$$

(4) $(s-k)\phi(s)$ is of finite order.

Of these statements (1) is implied by $c_n = O(n^{k-1+\epsilon})$, which in turn is a simple consequence of the dimension $-k$. For (2) and (3) we have the pattern of Riemann's classical proof for the continuation and the functional equation of the ζ-function by means of a ϑ-function. In its application here the proof runs like this:

We have from Euler's integral

$$\frac{\Gamma(s)}{(2\pi n)^s} = \int_0^\infty e^{-2\pi n y} y^{s-1} dy$$

and therefore for $\Re(s) > k$

$$(2\pi)^{-s}\Gamma(s)\phi(s) = \int_0^\infty (F(iy) - c_0) y^{s-1} dy$$

with

$$F(iy) - c_0 = \sum_{n=1}^\infty c_n e^{-2\pi n y}.$$

Since for the modular substitution

$$\begin{pmatrix} 0 & -1 \\ 1 & 0 \end{pmatrix}$$

we have $F(i/y) = (iy)^k F(iy)$ we can write

$$\begin{aligned}
&(2\pi)^{-s}\Gamma(s)\phi(s) \\
&\quad = \int_1^\infty (F(iy) - c_0) y^{s-1} dy + \int_0^1 ((iy)^{-k} F(i/y) - c_0) y^{s-1} dy \\
&\quad = \int_1^\infty (F(iy) - c_0) y^{s-1} dy + i^{-k} \int_1^\infty F(iw) w^{k-s-1} dw - \frac{c_0}{s} \\
&\quad = \int_1^\infty (F(iy) - c_0) y^{s-1} dy + (-1)^{k/2} \int_1^\infty (F(iw) - c_0) w^{k-s-1} dw \\
&\qquad + (-1)^{k/2} c_0 \int_1^\infty w^{k-s-1} dw - \frac{c_0}{s} \\
&\quad = \int_1^\infty (F(iy) - c_0)(y^s + (-1)^{k/2} y^{k-s}) \frac{dy}{y} \\
&\qquad - c_0 \left(\frac{1}{s} + (-1)^{k/2} \frac{1}{k-s} \right).
\end{aligned}$$

So far our reasoning required $\mathfrak{R}(s)>k$. Now we have arrived at an expression which remains meaningful for any complex s. We have, therefore, obtained an analytic continuation over the whole s-plane for $\phi(s)$. Moreover, the substitution $s\rightarrow k-s$ changes the right-hand member by the factor $(-1)^{k/2}$, which proves (3). At the same time we infer the validity of statement (2).

The argument runs also in reverse, from the Dirichlet series of properties (1) to (4) to a modular form of dimension $-k$. Instead of the Euler integral we have only to apply Mellin's formula.

Up to this point we have not encountered any trace of prime numbers. Hecke came to an Euler product by considering not a single modular form, but the whole set of all entire modular forms of dimension $-k$. They constitute a linear family of finite rank κ; the number κ of linearly independent modular forms in this set is a known function of k. A linear operator T will map the whole set on itself. If the system $F^{(\rho)}(\tau)$, $\rho=1, 2, \cdots, \kappa$ forms a basis then there exists a matrix $(\lambda_{\rho\sigma})$ such that

$$T(F^{(\rho)}(\tau)) = \sum_{\sigma=1}^{\kappa} \lambda_{\rho\sigma} F^{(\sigma)}(\tau).$$

As linear operator Hecke takes, for any natural number n,

$$T_n(F) = n^{k-1} \sum_{ad=n} \sum_{b \bmod d} F\left(\frac{a\tau + b}{d}\right) d^{-k},$$

for which it can be proved, by consideration of the subgroup

$$\begin{pmatrix} a & b \\ c & d \end{pmatrix} \equiv \begin{pmatrix} 1 & 0 \\ 0 & 1 \end{pmatrix} \pmod{n}$$

and its factor-group in the full modular group, that it furnishes again a modular form of dimension $-k$.

Hence we have

$$T_m(F^{(\rho)}(\tau)) = \sum_{\sigma=1}^{\kappa} \lambda_{\rho\sigma}(m) F^{(\sigma)}(\tau).$$

The application of the operator T_m on the Fourier expansion

$$F^{(\rho)}(\tau) = \sum_{n=0}^{\infty} a^{(\rho)}(n) e^{2\pi i n\tau}$$

yields by comparison of coefficients,

$$(11) \qquad \sum_{\sigma=1}^{\kappa} \lambda_{\rho\sigma}(m) a^{(\sigma)}(n) = \sum_{\delta/m, \delta/n} \delta^{k-1} a^{(\rho)}\left(\frac{mn}{\delta^2}\right), \qquad m \geqq 1,\ n \geqq 1.$$

It is decisive that here the right-hand member shows symmetry in m and n, which therefore must also prevail in the left-hand member:

$$\sum_{\sigma=1}^{\kappa} \lambda_{\rho\sigma}(m)a^{(\sigma)}(n) = \sum_{\sigma=1}^{\kappa} \lambda_{\rho\sigma}(n)a^{(\sigma)}(m).$$

The matrix $(\lambda_{\rho\sigma}(0))$, which appears here for the first time, can be uniquely determined as to fulfill also this relation. Now we obtain

$$\begin{aligned} T_m(F^{(\rho)}(\tau)) &= \sum_{\sigma=1}^{\kappa} \lambda_{\rho\sigma}(m) \sum_{n=0}^{\infty} a^{(\sigma)}(n)e^{2\pi i n\tau} \\ &= \sum_{\sigma=1}^{\kappa} a^{(\sigma)}(m) \sum_{n=0}^{\infty} \lambda_{\rho\sigma}(n)e^{2\pi i n\tau}. \end{aligned}$$

This leads to the introduction of the functions

$$f_{\rho\sigma}(\tau) = \sum_{n=0}^{\infty} \lambda_{\rho\sigma}(n)e^{2\pi i n\tau}$$

which span up the same linear set $\{F^{(1)}(\tau), \cdots, F^{(\kappa)}(\tau)\}$ and of which it is easily seen that they belong to the same set. With $\lambda(n) = (\lambda_{\rho\sigma}(n))$ we have in

$$B(\tau) = \sum_{n=0}^{\infty} \lambda(n)e^{2\pi i n\tau}$$

a matrix of modular forms. In connection with (11) it is found that

$$\lambda(m)\cdot\lambda(n) = \sum_{d|(m,n)} d^{k-1}\lambda\left(\frac{mn}{d^2}\right), \tag{12a}$$

which in particular for $(m, n) = 1$ furnishes

$$\lambda(m)\cdot\lambda(n) = \lambda(mn) \tag{12b}$$

and also shows that the different matrices $\lambda(n)$ are commutative.

Now we go over with Hecke to a set of associated Dirichlet series

$$\Phi(s) = (\phi_{\rho\sigma}(s)) = \sum_{n=1}^{\infty} \frac{\lambda(n)}{n^s},$$

a matrix which, of course, stands for

$$\phi_{\rho\sigma}(s) = \sum_{n=1}^{\infty} \frac{\lambda_{\rho\sigma}(n)}{n^s}, \qquad \rho, \sigma = 1, \cdots, \kappa.$$

Since each element of this matrix is associated to the modular form

$f_{\rho\sigma}(\tau)$ we infer the validity of statements (1) to (4) for each element and hence the functional equation

$$(2\pi)^{-s}\Gamma(s)\Phi(s) = (-1)^{k/2}(2\pi)^{s-k}\Gamma(k-s)\Phi(k-s)$$

in matrix notation. The relation (12b) makes it evident that we can write

$$\Phi(s) = \prod_p \sum_{l=0}^{\infty} \frac{\lambda(p^l)}{p^{ls}}$$

in which the summation of each infinite series can be effectuated by means of (12a) with the result:

$$\Phi(s) = \prod_p (\lambda(1) - \lambda(p)p^{-s} + \lambda(1)p^{k-1-2s})^{-1},$$

where $\lambda(1)$, according to definition, is the unit matrix of order κ. Here we have the analogue of the Euler product.

If we would start the whole process with a new basis, obtained from the former one by a linear substitution A, we would obtain the transformed matrices

$$\lambda^*(n) = A\cdot\lambda(n)\cdot A^{-1}$$

in our results. Such a transformation can be used to get matrices $\lambda(n)$ of an especially simple type. Since the $\lambda(n)$ are commutative they can simultaneously be transformed into matrices in which all elements beneath the principal diagonal are zeros. Beyond this H. Petersson [**34**] has shown that the λ's can be brought simultaneously into diagonal form, which means that we can start the process with κ linearly independent "eigen-functions" of all the linear operators

$$T_n(F^{(\rho)}(\tau)) = \overset{*}{\lambda}_{\rho\rho}(n)F^{(\rho)}(\tau), \qquad n = 1, 2, \cdots.$$

Petersson proves this in a very elegant manner by introducing a definition of orthogonality of modular forms based on integrals extended over the fundamental region. The $\lambda^*_{\rho\rho}$ are real algebraic numbers. We have now the Dirichlet series

$$\phi_{\rho\rho}(s) = \sum_{n=1}^{\infty} \frac{\overset{*}{\lambda}_{\rho\rho}(n)}{n^s} \tag{13a}$$

with an ordinary Euler product

$$\phi_{\rho\rho}(s) = \prod_p \frac{1}{1 - \lambda^*_{\rho\rho}(p)p^{-s} + p^{k-1-2s}}, \qquad \rho = 1, \cdots, \kappa, \tag{13b}$$

which is exactly of the form as it appears in Ramanujan's work on $\Delta(1, \tau)$. Summarizing, we can roughly say that the Riemann functional equation of a Dirichlet series already implies the Euler factorization with respect to prime numbers, certainly a result in which analytic and arithmetic elements are inseparably welded together.

Unfortunately, Riemann's original coordination of a modular form and a Dirichlet series, namely,

$$\vartheta(\tau) = \frac{1}{2} + \sum_{n=1}^{\infty} e^{\pi i n^2 \tau}, \qquad \zeta(2s) = \sum_{n=1}^{\infty} \frac{1}{n^{2s}}$$

with the Euler product

$$\prod_{1}^{\infty} (1 - p^{-2s})^{-1}$$

does not fit into the scheme; as already the comparison of this Euler product with (13b) shows; indeed, $\vartheta(\tau)$ has the dimension $-1/2$, and for fractional dimensions the complicated transformation formulae introduce multipliers which interfere with the definition of the operators T_n.

Let us go back for a moment to the functions $\phi_{\rho\rho}(s)$ in (13a) and (13b). We are here at once confronted with all the problems familiar from the theory of the Riemann ζ-function: What is the distribution of the zeros of $\phi_{\rho\rho}(s)$? The critical strip is $0 \leqq \Re(s) \leqq k$; do all non-trivial zeros of $\phi_{\rho\rho}(s)$ lie on the middle line $\Re(s) = k/2$? In case $\phi_{\rho\rho}(s)$ belongs to a modular form with $c_0 = 0$ ("cusp form") we know a few facts. The function $\phi_{\rho\rho}(s)$ has no pole, of course, its series is absolutely convergent for $\Re(s) > (k+1)/2$, $\phi_{\rho\rho}(s)$ has no zeros on $\Re(s) = (k+1)/2$ but infinitely many ones on $\Re(s) = k/2$, results all due to R. A. Rankin **[42]**. Ramanujan has for $\phi_{\rho\rho}(s) = \Delta(1, \tau)$ made the conjecture $|\tau(p)| \leqq 2p^{11/2}$ and Petersson **[34]** has generalized it to

$$|\lambda_{\rho\rho}(p)| \leqq 2p^{(k-1)/2}$$

or, which is equivalent, to the conjecture that the quadratic equation $z^2 - \lambda_{\rho\rho}(p)z + p^{k-1} = 0$ (derived from the denominator in (13b) with $p^s = z$) does not have two different real roots, which, since $\lambda_{\rho\rho}(p)$ is real would mean that the roots have to be of absolute value $p^{(k-1)/2}$. This conjecture has a close similarity to Hasse's theorem which confirms the Riemann hypothesis for elliptic function fields over finite fields.

Hecke has extended his theory also to modular forms of level (Stufe) Q. The operators T_n with $(n, Q) = 1$ have the same behavior as the operators for $Q=1$; the other ones, however, introduce complications on which I cannot dwell. Hecke applies this extension of his theory to a study of the ϑ-functions generated by quadratic forms of several variables [20].

So far I have spoken only of the theory of rational integers on one side and of functions of one variable on the other. I would overstep the limits which I had to set myself in this discussion if I were to enter upon the problems presented by the fields of algebraic numbers. In many respects a full analogy with the rational field can act as guide. However, some new features appear here, as, for example, the group of units. An algebraic field of degree n can be illustrated by an n-dimensional point lattice. Some problems, in particular additive ones, lead to analytic functions of n complex variables; some other problems, in particular the multiplicative ones, are handled by means of a set of ζ-functions with one complex variable s and $(n-1)$ integer parameters (Hecke's $\zeta(s, \lambda)$-functions) [15]. The multiplicative theory of algebraic fields is, through Hecke's fundamental investigations, about as far advanced as that for the rational field; the additive number theory in algebraic fields, however, is still in its beginnings.

The theory of quadratic forms offers another way to come to functions of several variables, as Siegel has recently shown in his introduction of modular functions of several variables [48].

This report could, of course, mention only selected topics of analytic number theory. I do not wish to imply that no recent progress has been made in subjects which I did not discuss. There are, for example, the important contributions to the estimates of lattice points in the circle and in other areas due to van der Corput and his successors, and the investigations of Siegel and Gelfond concerning the transcendentality of certain types of numbers. I have purposely chosen such portions of analytic number theory which seemed to me to exhibit most clearly the present tendency, the trend, in contemporary research work. Our discipline has grown beyond the stage of mere application of analysis to number theory; we can characterize the present tendency as a deep mutual penetration of analysis and arithmetic. It is still so that the theory of analytic functions helps to solve arithmetical problems which otherwise appear as inaccessible; but number theoretical problems are the sources of new functions, and number theoretical arguments lead to a deeper understanding of the structure of known functions. I think that analytic number theory is a branch of mathematics that fascinates through its harmony and beauty and appeals to the working mathematician through its variety and fruitfulness.

Bibliography

1. E. Artin, *Quadratische Körper im Gebiete der höheren Kongruenzen*, I, II, Mathematische Zeitschrift, vol. 19 (1924), pp. 153–206; 207–246.

2. S. Bochner and F. Bohnenblust, *Analytic functions with almost periodic coefficients*, Annals of Mathematics, (2), vol. 35 (1934), pp. 152–161.

3. Viggo Brun, *Le crible d'Eratosthène et le theorème de Goldbach*, Videnskapsselskapets Skrifter, Kristiania, 1920, no. 3.

4. H. Davenport, *On certain exponential sums*, Journal für die reine und angewandte Mathematik, vol. 169 (1933), pp. 158–176.

5. H. Davenport and H. Hasse, *Die Nullstellen der Kongruenzzetafunktionen in gewissen zyklischen Fällen*, Journal für die reine und angewandte Mathematik, vol. 172 (1935), pp. 151–182.

6. L. E. Dickson, *Universal Waring theorems*, Monatshefte für Mathematik und Physik, vol. 43 (1936), pp. 391–400.

7. ———, *The ideal Waring theorem for twelfth powers*, Duke Mathematical Journal, vol. 2 (1936), pp. 192–204.

8. T. Estermann, *Proof that every large integer is the sum of two primes and a square*, Proceedings of the London Mathematical Society, (2), vol. 42 (1937), pp. 501–516.

9. Mary Haberzetle, *On some partition functions*, American Journal of Mathematics, vol. 63 (1941), pp. 589–599.

10. G. H. Hardy, *On the representation of a number as the sum of any number of squares, and in particular of five*, Transactions of this Society, vol. 21 (1920), pp. 255–284.

11. ———, *Ramanujan, Twelve Lectures on Subjects Suggested by His Life and Work*, Cambridge, 1940, 236 pp.

12. H. Hasse, *Beweis des Analogons der Riemannschen Vermutung für die Artinschen und F. K. Schmidtschen Kongruenzzetafunktionen in gewissen elliptischen Fällen*, Göttinger Nachrichten, 1933, pp. 253–262.

13. ———, *Zur Theorie der abstrakten elliptischen Funktionenkörper*, I, II, III, Journal für die reine und angewandte Mathematik, vol. 175 (1936), pp. 55–62; 69–88; 193–208.

14. ———, *Über die Riemannsche Vermutung in Funktionenkörpern*, Comptes Rendus du Congrès International des Mathématiciens, Oslo, 1936, vol. 1, pp. 189–206.

15. E. Hecke, *Eine neue Art von Zetafunktionen und ihre Beziehungen zur Verteilung der Primzahlen*, I, II, Mathematische Zeitschrift, vol. 1 (1918), pp. 357–376; vol. 6 (1920), pp. 11–51.

16. ———, *Die Primzahlen in der Theorie der elliptischen Modulfunktionen*, K. Danske Videnskabernes Selskab, Mathematisk-fysiske Meddelelser, vol. 13 (1935), 16 pp.

17. ———, *Neuere Fortschritte in der Theorie der elliptischen Modulfunktionen*, Comptes Rendus du Congrès International des Mathématiciens, Oslo, 1936, vol. 1, pp. 140–156.

18. ———, *Über Modulfunktionen und die Dirichletschen Reihen mit Eulerscher Produktentwicklung*, I, II, Mathematische Annalen, vol. 114 (1937), pp. 1–28; 316–351.

19. ———, *Dirichlet series, modular functions and quadratic forms*, Lectures at the Institute for Advanced Study, Princeton, 1938, 48 pp.

20. ———, *Analytische Arithmetik der positiven quadratischen Formen*, K. Danske Videnskabernes Selskab, Mathematisk-fysiske Meddelelser, vol. 17 (1940), 134 pp.

21. Hans Heilbronn, *Über des Waringsche Problem*, Acta Arithmetica, vol. 1 (1936), pp. 212–221.

22. ———, *On the class-number of imaginary quadratic fields*, Quarterly Journal of Mathematics, Oxford, vol. 5 (1934), pp. 150–160.

23. L. K. Hua, *On the number of partitions of a number into unequal parts*, Transactions of this Society, vol. 51 (1942), pp. 194–201.

24. M. Kac, *Almost periodicity and the representation of integers as sums of squares*, American Journal of Mathematics, vol. 62 (1940), pp. 122–126.

25. M. Kac, E. R. van Kampen and Aurel Wintner, *Ramanujan sums and almost periodic functions*, American Journal of Mathematics, vol. 62 (1940), pp. 107–114.

26. D. H. Lehmer, *On a conjecture of Ramanujan*, Journal of the London Mathematical Society, vol. 11 (1936), pp. 114–118.

27. ———, *An application of Schläfli's modular equation to a conjecture of Ramanujan*, this Bulletin, vol. 44 (1938), pp. 84–90.

28. Joseph Lehner, *A partition function connected with the modulus five*, Duke Mathematical Journal, vol. 8 (1941), pp. 631–655.

29. L. J. Mordell, *On Mr. Ramanujan's empirical expansions of modular functions*, Proceedings of the Cambridge Philosophical Society, vol. 19 (1917), pp. 117–124.

30. ———, *Note on certain modular relations considered by Messrs. Ramanujan, Darling, and Rogers*, Proceedings of the London Mathematical Society, (2), vol. 20 (1922), pp. 408–416.

31. ———, *On a sum analogous to a Gauss sum*, Quarterly Journal of Mathematics, Oxford, vol. 3 (1932), pp. 161–167.

32. Ivan Niven, *On a certain partition function*, American Journal of Mathematics, vol. 62 (1940), pp. 353–364.

33. Hans Petersson, *Über die Entwicklungskoeffizienten der automorphen Formen* Acta Mathematica, vol. 58 (1932), pp. 169–215.

34. ———, *Konstruktion der sämtlichen Lösungen einer Riemannschen Funktionalgleichung durch Dirichlet-Reihen mit Eulerscher Produktentwicklung*, I, II, III, Mathematische Annalen, vol. 116 (1939), pp. 401–412; vol. 117 (1940), pp. 39–64; 277–300.

35. H. Rademacher, *A convergent series for the partition function $p(n)$*, Proceedings of the National Academy of Sciences, vol. 23 (1937), pp. 78–84.

36. ———, *On the partition function $p(n)$*, Proceedings of the London Mathematical Society, (2), vol. 43 (1937), pp. 241–254.

37. ———, *The Fourier coefficients of the modular invariant $J(\tau)$*, American Journal of Mathematics, vol. 60 (1938), pp. 501–512.

38. ———, *The Fourier series and the functional equation of the absolute modular invariant $J(\tau)$*, American Journal of Mathematics, vol. 61 (1939), pp. 237–248.

39. ———, *The Ramanujan identities under modular substitutions*, Transactions of this Society, vol. 51 (1942), pp. 609–636.

40. H. Rademacher and H. Zuckerman, *A new proof of two of Ramanujan's identities*, Annals of Mathematics, (2), vol. 40 (1939), pp. 473–489.

41. S. Ramanujan, *On certain arithmetical functions*, Transactions of the Cambridge Philosophical Society, vol. 22 (1916), pp. 159–184; also *Collected Papers*, pp. 136–162.

42. R. A. Rankin, *Contributions to the theory of Ramanujan's function $\tau(n)$ and similar arithmetical functions*, I, II, III, Proceedings of the Cambridge Philosophical Society, vol. 35 (1939), pp. 351–356; 357–372; vol. 36 (1940), pp. 150–151.

43. H. Salié, *Zur Abschätzung der Fourierkoeffizienten ganzer Modulformen*, Mathematische Zeitschrift, vol. 36 (1933), pp. 263–278.

44. F. K. Schmidt, *Analytische Zahlentheorie in Körpern der Charakteristik p*, Mathematische Zeitschrift, vol. 33 (1931), pp. 1–32.

45. B. Schoeneberg, *Das Verhalten von mehrfachen Thetareihen bei Modulsubstitutionen*, Mathematische Annalen, vol. 116 (1939), pp. 511–523.

46. C. L. Siegel, *Über die Classenzahl quadratische Zahlkörper*, Acta Arithmetica, vol. 1 (1936), pp. 83–86.

47. ———, *Über die analytische Theorie der quadratischen Formen*, I, II, III, Annals of Mathematics, (2), vol. 36 (1935), pp. 527–606; vol. 37 (1936), pp. 230–263; vol. 38 (1937), pp. 212–291.

48. ———, *Einführung in die Theorie der Modulfunktionen n-ten Grades*, Mathematische Annalen, vol. 116 (1939), pp. 617–657.

49. N. Tchudakoff, *On zeros of Dirichlet's L-functions*, Recueil Mathématique (Matematicheskii Sbornik), vol. 43 (1936), pp. 591–601.

50. I. Vinogradoff, *On Waring's problem*, Annals of Mathematics, (2), vol. 36 (1935), pp. 395–405.

51. ———, *On Weyl's sums*, Recueil Mathématique (Matematicheskii Sbornik), vol. 42 (1935), pp. 521–530.

52. ———, *Representation of an odd number as a sum of three primes*, Comptes Rendus de l'Académie des Sciences de l'URSS (Doklady), vol. 15 (1937), pp. 169–172.

53. ———, *Some theorems concerning the theory of primes*, Recueil Mathématique (Matematicheskii Sbornik), vol. 44 (1937), pp. 179–195.

54. Arnold Walfisz, *Zur additiven Zahlentheorie*, IV, Travaux de l'Institut Mathématique de Tblissi, vol. 3 (1938), pp. 121–192.

55. G. N. Watson, *Ramanujans Vermutung über Zerfällungsanzahlen*, Journal für die reine und angewandte Mathematik, vol. 179 (1938), pp. 97–128.

56. André Weil, *Sur les fonctions algébriques à corps de constantes fini*, Comptes Rendus de l'Académie des Sciences, Paris, vol. 210 (1940), pp. 592–594.

57. ———, *On the Riemann hypothesis in function-fields*, Proceedings of the National Academy of Sciences, vol. 27 (1941), pp. 345–347.

58. Hermann Weyl, *Über die Gleichverteilung von Zahlen mod. Eins.*, Mathetische Annalen, vol. 77 (1916), pp. 313–352.

59. H. S. Zuckerman, *On the coefficients of certain modular forms belonging to subgroups of the modular group*, Transactions of this Society, vol. 45 (1939), pp. 298–321.

60. ———, *Identities analogous to Ramanujan's identities involving the partition function*, Duke Mathematical Journal, vol. 5 (1938), pp. 88–110.

University of Pennsylvania

Notes to Paper 46

This paper has been reviewed in the MR, vol. 3(1942), p. 271.

Page 381, fourteenth line from bottom.
One may wish to suppress the word "as."

Page 382, line 20.
Suppress the last word "the" of this line.

Page 388, equation (**).
Read x^{5m} instead of x^{5qm}.

Page 400, reference 35.
See paper 37 of the present collection.

Page 400, reference 36.
See paper 38.

Page 400, reference 37.
See paper 39.

Page 400, reference 38.
See paper 41.

Page 400, reference 39.
See paper 45.

Page 400, reference 40.
See paper 42.

ON THE BLOCH-LANDAU CONSTANT

By Hans Rademacher.

1. In 1929 Landau, in a paper [1] on the Bloch constant, introduced besides the Bloch constant $\mathfrak{B}$ two other constants $\mathfrak{L}$ and $\mathfrak{A}$ of which only the first will interest us here. The definition of the Bloch constant is the following:

There exists a constant $\mathfrak{B}$ such that to every $\epsilon > 0$ and to every function $w = f(z)$ which is regular for $|z| < 1$ with $f'(0) = 1$ there exists a closed region $G = G_{f,\epsilon}$ in the interior of the unit-circle which is mapped by the function $w = f(z)$ in a one-to-one manner on a circle of radius $\mathfrak{B} - \epsilon$ in the w-plane; whereas there exist functions of the same specifications by which no subregion of the unit-circle is mapped in a one-to-one manner on a circle of radius $\mathfrak{B} + \epsilon$.

By dropping the condition of bi-uniqueness Landau defines another constant which we shall call the Bloch-Landau constant $\mathfrak{L}$: There exists a constant $\mathfrak{L}$ such that every function $w = f(z)$ which is regular in $|z| < 1$ with $f'(0) = 1$ assumes all values of a certain circle of radius $\mathfrak{L} - \epsilon$ in the w-plane, for any given positive ϵ; whereas to every $\epsilon > 0$ there exists a function of the same specifications whose values do not completely fill any circle of radius $\mathfrak{L} + \epsilon$.

Since every "Bloch circle" in the w-plane is a "Landau-circle," but not conversely, it is clear that we have $\mathfrak{B} \leqq \mathfrak{L}$. Landau, in the paper cited, proves also

$$0.39 < \mathfrak{B}, \qquad 0.43 \leqq \mathfrak{L} \leqq 2^{2/3}\pi\Gamma(\tfrac{1}{3})^3\Gamma(\tfrac{1}{4})^{-4},$$

where the latter upper bound is established by a certain example of $f(z)$ which does not permit a greater Landau circle. Landau gave no separate upper estimate for $\mathfrak{B}$. This was done some years ago by Ahlfors and Grunsky who proved [2] by a suitable example that

$$\mathfrak{B} \leqq (1 + 3^{1/2})^{1/2}\Gamma(\tfrac{1}{3})\Gamma(\tfrac{3}{4})\Gamma(\tfrac{1}{12})^{-1} = 0.4719 \cdots$$

I shall show that their method leads also to an improved upper estimate of $\mathfrak{L}$.

* Received March 21, 1942.

[1] E. Landau, "Über die Blochsche Konstante und zwei verwandte Weltkonstanten," *Mathematische Zeitschrift*, vol. 30 (1929), pp. 608-634.

[2] L. V. Ahlfors and H. Grunsky, "Über die Blochsche Konstante," *Mathematische Zeitschrift*, vol. 42 (1937), pp. 671-673.

2. The interior of the unit circle $|\zeta| < 1$ is mapped on an equilateral equiangular circular triangle of angles $\alpha\pi$ $(0 \leqq \alpha < 1)$ by means of the function

$$(1) \quad w = \phi_a(\zeta) = \zeta \cdot \frac{\int_0^1 t^{-1/2-a/2}(1-t)^{-1/6+a/2}(1-\zeta^3 t)^{-5/6+a/2}dt}{\int_0^1 t^{-1/2-a/2}(1-t)^{-5/6+a/2}(1-\zeta^3 t)^{-1/6+a/2}dt}$$

$$= \frac{\Gamma(2/3)\Gamma(5/6+\alpha/2)}{\Gamma(4/3)\Gamma(1/6+\alpha/2)} \cdot \zeta \cdot \frac{F(5/6-\alpha/2, 1/2-\alpha/2, 4/3; \zeta^3)}{F(1/6-\alpha/2, 1/2-\alpha/2, 2/3; \zeta^3)},$$

where $F(a, b, c; z)$ is that branch of the hypergeometric function which is, for $|z| < 1$, represented by the Gaussian hypergeometric series. In the mapping established by (1) the points $1, \rho, \rho^2$ $(\rho = e^{2\pi i/3})$ of the ζ-plane correspond to the vertices of the circular triangle in the w-plane.

The formula (1) is given by Ahlfors and Grunsky for $\alpha = 1/k$, $k > 0$. But it can be seen that the arguments which they set forth to establish its validity hold also for $\alpha = 0$. The case $\alpha = 1/3$ means the mapping of the unit circle on an equilateral straight triangle. We can, therefore, by combining two conformal mappings of the sort (1), map a zero-angled equilateral triangle on an equilateral straight triangle. If we introduce the inverse function

$$\zeta = \Phi_a(w)$$

of $w = \phi_a(\zeta)$ then

$$(2) \qquad w = f(z) = C_1\phi_{1/3}(\Phi_0(C_2 z))$$

will produce that conformal mapping; the constants C_1 and C_2 can be used to adjust the sizes of the triangles.

Now the conformal mapping (2) can be extended by reflection with respect to the sides of the triangles. The repeated images of the zero-angled triangle in the z-plane will just fill its orthogonal circle, which is its circumcircle, whereas the repeated images of the equilateral triangle in the w-plane will build up a Riemann surface in which the vertices of the triangle and of its reflections form a regular point lattice, and where all these vertices become branch-points of infinite order. The function $w = f(z)$ therefore maps, by this process of analytic continuation, the interior of a whole circle K on the described Riemann surface. We can normalize the mapping so that K is the unit circle $|z| < 1$ and so that $f'(0) = 1$.

Since, then, the image of any circle $|z| \leqq 1 - \epsilon$ will not cover any of the branch-points in the Riemann surface over the w-plane a Landau circle cannot be greater than the circumcircle of the equilateral straight triangle which was the element out of which the Riemann surface was constructed.

The normalization just mentioned requires

$$\Phi_0(C_2) = 1$$

or

$$C_2 = \phi_0(1) \tag{3}$$

and

$$f'(0) = C_1 \frac{\phi'_{1/3}(0)}{\phi'_0(0)} C_2 = 1. \tag{4}$$

Equation (3) has to be understood as

$$C_2 = \lim_{\xi \to 1} \phi_0(\xi) = \lim_{\xi \to 1} \frac{\Gamma(5/6)\Gamma(2/3)F(5/6, 1/2, 4/3; \xi)}{\Gamma(4/3)\Gamma(1/6)F(1/6, 1/2, 2/3; \xi)}.$$

Here we have hypergeometric functions $F(a, b, c; \xi)$ with $a + b = c$, in which case

$$F(a, b, c; \xi) \sim \frac{\Gamma(a+b)}{\Gamma(a)\Gamma(b)} \log \frac{1}{1-\xi}, \qquad \xi \to 1.$$

Hence

$$C_2 = 1,$$

and therefore formula (4) furnishes

$$C_1 = \frac{\phi'_0(0)}{\phi'_{1/3}(0)} = \frac{\Gamma(5/6)\Gamma(1/3)}{\Gamma(1/6)}.$$

The upper bound of the radii of all Landau circles, i. e. the circumradius of the equilateral triangle in the w-plane, is

$$f(1) = C_1\phi_{1/3}(1) = C_1 = \Gamma(5/6)\Gamma(1/3)\Gamma(1/6)^{-1} = 2\pi\Gamma(1/3)\Gamma(1/6)^{-2}.$$

We have therefore our result:

An upper bound of the Bloch-Landau constant $\mathfrak{L}$ *is given by*

$$\mathfrak{L} \leqq 2\pi\Gamma(1/3)\Gamma(1/6)^{-2}. \tag{5}$$

This result compares numerically with Landau's estimate quoted above as follows:

$$0.54325 < 2\pi\Gamma(1/3)\Gamma(1/6)^{-2} < 0.54326,$$ [3]

$$0.55488 < 2^{2/3}\pi\Gamma(1/3)^3\Gamma(1/4)^{-4} < 0.55489.$$

---

[3] In a letter of May 30, 1942, Professor R. M. Robinson informs me that he found the same upper bound for $\mathfrak{L}$ in 1937. His result is briefly mentioned in a footnote to Ahlfor's paper "An extension of Schwarz's Lemma," *Transactions of the American Mathematical Society,* vol. 43 (1938), p. 364 which I had overlooked: "In the other direction R. M. Robinson has proved $\mathfrak{L} < .544$. This result has not been published."

3. The configuration of the zero-angled circular triangle together with all its reflections is, of course, the well-known modular figure; and if it had not been for the convenience of the Ahlfors-Grunsky formula, I would have adhered to my first redaction of the present paper, which was based on the modular function $k^2(\tau) = \lambda(\tau)$ into which $\Phi_0(w)$ can be transformed. Landau in his paper quotes previous results of Carathéodory and Hartogs, which are also obtained from the theory of modular functions. An analysis of Landau's procedure shows that he has actually used the conformal mapping of the zero-angled triangle with the vertices 1, i, -1 on the equilateral triangle. It is through a gain in symmetry of our example over Landau's that we obtained the improved estimate (5).

Ahlfors and Grunsky in their paper make it plausible that their appraisal of Bloch's constant actually gives its true value. A similar reasoning would apply to the above example. The symmetry of the discussed conformal mapping suggests strongly that (5) may contain the best value of the Bloch-Landau constant $\mathfrak{L}$.

It is noteworthy that the Ahlfors-Grunsky example, as well as ours, exhibits functions which have the unit circle as natural boundary. This is in agreement with R. M. Robinson's statements about "Bloch functions," [4]

THE UNIVERSITY OF PENNSYLVANIA.

[4] Raphael M. Robinson, "Bloch functions," *Duke Mathematical Journal*, vol. 2 (1936), pp. 453-459.

Notes to Paper 47

This paper has been reviewed in the Z, vol. 61 (1961), p. 153, and in the MR, vol. 4 (1943), p. 270.

Page 389, marginal note.
Author's correction (suppress the first zero).

ON THE EXPANSION OF THE PARTITION FUNCTION IN A SERIES

By Hans Rademacher

(Received April 8. 1943)

1. A geometric property of the Farey series, discovered by L. R. Ford (**1**) is used in this note for the construction of a new path of integration to replace the circle carrying the Farey dissection, first introduced by Hardy and Ramanujan in their classical paper (**2**). This new path of integration will bring about an essential simplification in the treatment of the partition function and, in general, in the determination of the coefficients of modular functions of non-negative dimension. It seems to me that the new path exhibits more clearly than the Farey arcs do the different contributions of the approximation functions near the roots of unity. Moreover, only two estimations have to be performed, and they are direct consequences of the obvious statements (3.2) and (4.1) concerning the circle over the diameter 0 to 1.

Ford's theorem referred to above can be enunciated as follows:

If in a complex τ-plane we mark the points corresponding to the reduced fractions h/k and draw about the points

$$\tau_{h,k} = \frac{h}{k} + \frac{i}{2k^2}$$

the circles of radii $1/2k^2$, which touch the real axis at h/k, then these circles do not intersect. Two of them are tangent to each other if and only if their fractions h/k and l/m appear as neighbors in a Farey series of some order.

The proof is clear. Comparing the distance of the centers with the sum of the radii of two such circles we consider

$$|\tau_{h,k} - \tau_{l,m}|^2 - \left(\frac{1}{2k^2} + \frac{1}{2m^2}\right)^2 = \frac{(hm - lk)^2 - 1}{k^2m^2} \geqq 0.$$

The equality sign, indicating contact of the circles, is attained only for

$$hm - lk = \pm 1,$$

and this would mean that the fractions h/k and l/m are neighbors in some Farey series, e.g. that of order $N = k + m - 1$.

To each positive integer N we introduce now a path P_N which we shall later use for the complex integration. Let $c_{h,k}$ be the circle of Ford's theorem belonging to the (reduced) fraction h/k. We draw all circles $c_{h,k}$ for $k \leqq N$, $0 \leqq h/k < 1$, in other words all the circles belonging to the Farey series of order N. If $h_1/k_1 < h/k < h_2/k_2$ are three adjacent fractions of that series, then the circle $c_{h,k}$ has a point of contact with c_{h_1,k_1} as well as with c_{h_2,k_2}. These points of contact cut $c_{h,k}$ into two arcs, an upper one and a lower one. (The lower one touches the real axis.) As path P_N we take now the row of upper arcs $\gamma_{h,k}$, each traversed on its circle in the negative sense, on $c_{h,k}$ therefore from the point of contact with c_{h_1,k_1} to the point of contact with c_{h_2,k_2}. The figure shows P_N for $N = 3$. Because of the periodicity of the function to be integrated it does not matter that instead of the whole arc $\gamma_{0,1}$ we have drawn a part of $\gamma_{1,1}$ which is obtained from the omitted part of $\gamma_{0,1}$ through the translation $+1$.

For a later purpose we need the coordinates of the endpoints of the arc $\gamma_{h,k}$. They are, as simple geometric arguments show,

$$\frac{h}{k} + \left(-\frac{k_1}{k(k^2 + k_1^2)} + \frac{i}{k^2 + k_1^2}\right) = \frac{h}{k} + \zeta'_{h,k} \tag{1.1}$$

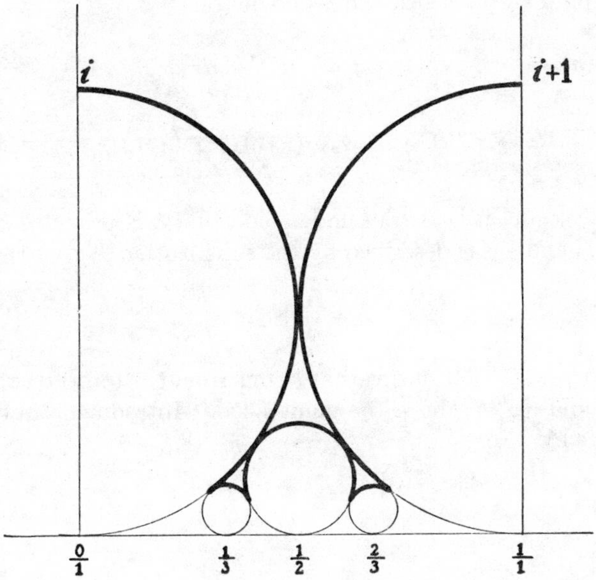

FIG. 1

and

$$\frac{h}{k} + \left(\frac{k_2}{k(k^2 + k_2^2)} + \frac{i}{k^2 + k_2^2}\right) = \frac{h}{k} + \zeta''_{h,k}. \tag{1.2}$$

Incidentally, the point $h/k + \zeta'_{h,k}$ lies on the semicircle over the diameter $(h_1/k_1, h/k)$; the whole path P_N lies above the row of semicircles connecting adjacent fractions of the Farey series of order N.[1]

[1] The imaginary part of $\zeta'_{h,k}$ is $1/(k^2 + k_1^2)$, and we have

$$\frac{1}{2N^2} \leq \frac{1}{k^2 + k_1^2} \leq \frac{2}{(k + k_1)^2} \leq \frac{2}{(N + 1)^2}.$$

Thus $\Im(\zeta'_{h,k})$ is of the order N^{-2}. This corresponds to the choice of a circle of radius $\exp(-2\pi N^{-2})$ as the path of integration associated with a Farey dissection of order N in my previous treatment of $p(n)$. (3).

2. In order to come to $p(n)$ we start with Euler's formula

$$(2.1)\qquad 1 + \sum_{n=1}^{\infty} p(n)x^n = \prod_{m=1}^{\infty} (1 - x^m)^{-1} = f(x), \qquad |x| < 1.$$

We have therefore

$$p(n) = \int_i^{i+1} f(e^{2\pi i\tau})e^{-2\pi i n\tau}\, d\tau$$

for any path of integration in the upper τ-halfplane connecting i and $i + 1$. We choose the path P_N described above and obtain

$$p(n) = \sum_{0 \leqq h < k \leqq N} \int_{\gamma_{h,k}} f(e^{2\pi i\tau})e^{-2\pi i n\tau}\, d\tau$$

$$= \sum_{0 \leqq h < k \leqq N} \int_{\zeta'_{h,k}}^{\zeta''_{h,k}} f\left(e^{2\pi i\left(\frac{h}{k}+\zeta\right)}\right)e^{-2\pi i n\left(\frac{h}{k}+\zeta\right)}\, d\zeta.$$

Here and in the sequel it is always understood that h is prime to k. The path of ζ between $\zeta'_{h,k}$ and $\zeta''_{h,k}$ is described by the substitution

$$\tau = \frac{h}{k} + \zeta,$$

where τ runs on $\gamma_{h,k}$. That means that ζ runs from $\zeta'_{h,k}$ to $\zeta''_{h,k}$ on the upper arc of a circle of radius $1/2k^2$ about the point $i/2k^2$. Introducing in each integral a new variable z by

$$\zeta = \frac{iz}{k^2}$$

we obtain

$$(2.2)\qquad p(n) = \sum_{0 \leqq h < k \leqq N} \frac{i}{k^2} e^{-\frac{2\pi i n h}{k}} \int_{z'_{h,k}}^{z''_{h,k}} f\left(e^{\frac{2\pi i h}{k} - \frac{2\pi z}{k^2}}\right)e^{\frac{2\pi n z}{k^2}}\, dz.$$

Here z runs in each integral on an arc of the circle K of radius $\frac{1}{2}$ about the point $\frac{1}{2}$ as center. The ends of the arc are

$$z'_{h,k} = -ik^2\zeta'_{h,k}, \qquad z''_{h,k} = -ik^2\zeta''_{h,k}$$

or, according to (1.1) and (1.2)

$$(2.3)\qquad \begin{aligned} z'_{h,k} &= \frac{k^2}{k^2 + k_1^2} + i\,\frac{kk_1}{k^2 + k_1^2}, \\ z''_{h,k} &= \frac{k^2}{k^2 + k_2^2} - i\,\frac{kk^2}{k^2 + k_2^2}. \end{aligned}$$

The points $z'_{h,k}$ and $z''_{h,k}$ divide the circle K into two arcs; that one is meant as path of integration which does not touch the imaginary axis. We have $\Re(z) > 0$.

On the integrands in (2.2) we apply now the transformation formula of the function $f(x)$, a formula, which stems from the theory of the elliptic modular functions (cf. (**2**), Lemma 4.32)

$$f(e^{\frac{2\pi ih}{k}-\frac{2\pi z}{k^2}}) = \omega_{h,k}\sqrt{\frac{z}{k}}\exp\left(\frac{\pi}{12z}-\frac{\pi z}{12k^2}\right)f(e^{\frac{2\pi ih'}{k}-\frac{2\pi}{z}}).$$

Here $\omega_{h,k}$ is a well-known root of unity, and h' is a solution of the congruence

$$hh' \equiv -1 \pmod{k};$$

for the square root the principal branch has to be taken. We get therefore from (2.2)

$$(2.4)\qquad p(n) = \sum_{0\leqq h<k\leqq N} ik^{-\frac{1}{2}}\omega_{h,k}e^{-\frac{2\pi ihn}{k}}\int_{z'_{h,k}}^{z''_{h,k}}\Psi_k(z)f(e^{\frac{2\pi ih'}{k}-\frac{2\pi}{z}})\,e^{\frac{2\pi nz}{k^2}}\,dz,$$

with the abbreviation

$$(2.5)\qquad \Psi_k(z) = z^{\frac{1}{2}}\exp\left(\frac{\pi}{12z}-\frac{\pi z}{12k^2}\right).$$

We rewrite (2.4) as

$$\begin{aligned}(2.6)\qquad p(n) &= \sum_{0\leqq h<k\leqq N} ik^{-\frac{5}{2}}\omega_{h,k}e^{-\frac{2\pi ihn}{k}}\int_{z'_{h,k}}^{z''_{h,k}}\Psi_k(z)e^{\frac{2\pi nz}{k^2}}\,dz\\ &+ \sum_{0\leqq h<k\leqq N} ik^{-\frac{5}{2}}\omega_{h,k}e^{-\frac{2\pi ihn}{k}}\int_{z'_{h,k}}^{z''_{h,k}}\Psi_k(z)\{f(e^{\frac{2\pi ih'}{k}-\frac{2\pi}{z}})-1\}\,e^{\frac{2\pi nz}{k^2}}\,dz\\ &= \sum_{0\leqq h<k\leqq N} ik^{-\frac{5}{2}}\omega_{h,k}e^{-\frac{2\pi ihn}{k}}I_{h,k} + \sum_{0\leqq h<k\leqq N} ik^{-\frac{5}{2}}\omega_{h,k}e^{-\frac{2\pi ihn}{k}}I^*_{h,k},\end{aligned}$$

where $I_{h,k}$ and $I^*_{h,k}$ are respectively abbreviations for the integrals.

3. We estimate first

$$(3.1)\qquad I^*_{h,k} = \int_{z'_{h,k}}^{z''_{h,k}}\Psi_k(z)\{f(e^{\frac{2\pi ih'}{k}-\frac{2\pi}{z}})-1\}\,e^{\frac{2\pi nz}{k^2}}\,dz.$$

The path of integration, which is an arc of the circle K, can here be replaced by the chord $s_{h,k}$ from $z'_{h,k}$ to $z''_{h,k}$. We have on and in the circle K

$$(3.2)\qquad 0 < \Re(z) \leqq 1, \qquad \Re\left(\frac{1}{z}\right) \geqq 1.$$

This yields the estimate

$$(3.3)\qquad \left|\Psi_k(z)\{f(e^{\frac{2\pi ih'}{k}-\frac{2\pi}{z}})-1\}\,e^{\frac{2\pi nz}{k^2}}\right| \leqq |z|^{\frac{1}{2}}e^{2\pi n}\sum_{m=1}^{\infty}p(m)e^{-2\pi\left(m-\frac{1}{24}\right)}.$$

On the chord $s_{h,k}$ we have $|z|$ less than or equal to the greater of the numbers $|z'_{h,k}|$ and $|z''_{h,k}|$. From (2.3) we derive readily

$$|z'_{h,k}| = \frac{k}{\sqrt{k^2 + k_1^2}}, \qquad |z''_{h,k}| = \frac{k}{\sqrt{k^2 + k_2^2}}.$$

Now

$$\sqrt{k^2 + k_1^2} \geqq 2^{-\frac{1}{2}}(k + k_1) \geqq 2^{-\frac{1}{2}}(N + 1)$$

since k and k_1 are the denominators of neighbor fractions in the Farey series of order N. Therefore we have on $s_{h,k}$

$$|z| \leqq 2^{\frac{1}{2}}k(N + 1)^{-1}. \tag{3.4}$$

The length of the chord $s_{h,k}$ is, according to (2.3),

$$\begin{aligned} & k\left\{\left(\frac{k}{k^2 + k_1^2} - \frac{k}{k^2 + k_2^2}\right)^2 + \left(\frac{k_1}{k^2 + k_1^2} + \frac{k_2}{k^2 + k_2^2}\right)^2\right\}^{\frac{1}{2}} \\ (3.5) \qquad & = \frac{k\,|k_1 - k_2|}{(k^2 + k_1^2)^{\frac{1}{2}}(k^2 + k_2^2)^{\frac{1}{2}}} \leqq \frac{2k\,|k_1 - k_2|}{(k + k_1)(k + k_2)} \\ & = 2k\left|\frac{1}{k + k_1} - \frac{1}{k + k_2}\right| < 2k(N + 1)^{-1}. \end{aligned}$$

From (3.1), (3.3), (3.4), (3.5) we obtain now

$$|I^*_{h,k}| < Ck^{\frac{3}{2}}N^{-\frac{3}{2}}$$

where the constant C contains n, which however we keep fixed. This estimate leads to

$$\left|\sum_{0 \leqq h < k \leqq N} ik^{-\frac{5}{2}}\omega_{h,k}e^{-\frac{2\pi ihn}{k}} I^*_{h,k}\right| < CN^{-\frac{3}{2}} \sum_{0 \leqq h < k \leqq N} k^{-1} < CN^{-\frac{1}{2}}.$$

We can thus replace (2.6) by

$$p(n) = \sum_{0 \leqq h < k \leqq N} ik^{-\frac{5}{2}}\omega_{h,k}e^{-\frac{2\pi ihn}{k}} I_{h,k} + O(N^{-\frac{1}{2}}), \tag{3.6}$$

$$I_{h,k} = \int_{z'_{h,k}}^{z''_{h,k}} \Psi_k(z)e^{\frac{2\pi nz}{k^2}}\,dz.$$

4. In $I_{h,k}$ we introduce now the whole circle K, from 0 around in the negative sense to 0, as path of integration:

$$I_{h,k} = \int_{K(-)} \Psi_k(z)e^{\frac{2\pi nz}{k^2}}\,dz - \int_0^{z'_{h,k}} \Psi_k(z)e^{\frac{2\pi nz}{k^2}}\,dz - \int_{z''_{h,k}}^{0} \Psi_k(z)e^{\frac{2\pi nz}{k^2}}\,dz.$$

We estimate the last two integrals. Since they are of the same type we need to consider only the first. On the circle K we have

$$\Re\left(\frac{1}{z}\right) = 1, \qquad 0 < \Re(z) \leqq 1. \tag{4.1}$$

The arc from 0 to $z'_{h,k}$ is less than

$$\frac{\pi}{2}\,|z'_{h,k}| < \pi 2^{-\frac{1}{2}} k N^{-1},$$

and (3.4) is also valid on that arc. Remembering the definition (2.5) we obtain therefore

$$\left| \int_0^{z'_{h,k}} \Psi_k(z) e^{\frac{2\pi n z}{k^2}}\, dz \right| < C e^{2\pi n} k^{\frac{3}{2}} N^{-\frac{3}{2}},$$

$$I_{h,k} = \int_{K^{(-)}} \Psi_k(z) e^{\frac{2\pi n z}{k^2}}\, dz + O(k^{\frac{3}{2}} N^{-\frac{3}{2}}).$$

Insertion of this into (3.6) yields

$$\begin{aligned} p(n) &= \sum_{0 \leqq h < k \leqq N} i k^{-\frac{5}{2}} \omega_{h,k} e^{-\frac{2\pi i h n}{k}} \int_{K^{(-)}} \Psi_k(z) e^{\frac{2\pi n z}{k^2}}\, dz \\ &\qquad + O(N^{-\frac{1}{2}}) + \sum_{0 \leqq h < k \leqq N} O(k^{-1} N^{-\frac{1}{2}}) \\ &= i \sum_{1 \leqq k \leqq N} A_k(n) k^{-\frac{5}{2}} \int_{K^{(-}} \Psi_k(z) e^{\frac{2\pi n z}{k^2}}\, dz + O(N^{-\frac{1}{2}}), \end{aligned}$$

with the abbreviation

(4.2) $$A_k(n) = \sum_{\substack{h \bmod k \\ (h,k)=1}} \omega_{h,k} e^{-\frac{2\pi i h n}{k}}.$$

If we now let N tend to infinity the error term goes to zero, and we obtain an infinite convergent series

(4.3) $$p(n) = i \sum_{k=1}^{\infty} A_k(n) k^{-\frac{5}{2}} \int_{K^{(-)}} z^{\frac{1}{2}} e^{\frac{\pi}{12 z} + \frac{2\pi z}{k^2}\left(n - \frac{1}{24}\right)}\, dz.$$

5. In order to carry out the integral we substitute

$$w = \frac{1}{z}.$$

The path of the integral in the w-plane is then the line parallel to the imaginary axis through the point 1. Therefore we have

$$p(n) = \frac{1}{i} \sum_{k=1}^{\infty} A_k(n) k^{-\frac{5}{2}} \int_{1-i\infty}^{1+i\infty} w^{-\frac{5}{2}} e^{\frac{\pi w}{12} + \frac{2\pi}{k^2}\left(n - \frac{1}{24}\right)\frac{1}{w}}\, dw.$$

The integral here is brought into a known form by the substitution

$$\frac{\pi}{12} w = t,$$

which yields

$$p(n) = \frac{1}{i}\left(\frac{\pi}{12}\right)^{\frac{3}{2}} \sum_{k=1}^{\infty} A_k(n)k^{-\frac{3}{2}} \int_{c-i\infty}^{c+i\infty} t^{-\frac{5}{2}} e^{t+\frac{\pi^2}{6k^2}\left(n-\frac{1}{24}\right)\frac{1}{t}}\, dt, \qquad c > 0.$$

We find[2] thus

$$p(n) = 2\pi(24n-1)^{-\frac{3}{4}} \sum_{k=1}^{\infty} \frac{A_k(n)}{k} I_{\frac{3}{2}}\left(\frac{\pi}{k}\sqrt{\tfrac{2}{3}(n-\tfrac{1}{24})}\right),$$

where $I_{\frac{3}{2}}$ is a "Bessel function of imaginary argument."

Finally, Bessel functions of half odd order can be reduced to elementary functions. In our case we apply the relation

$$I_{\frac{3}{2}}(z) = \sqrt{\frac{2z}{\pi}}\frac{d}{dz}\left(\frac{\sinh z}{z}\right).$$

Using the abbreviations

$$C = \pi\sqrt{\tfrac{2}{3}}, \qquad \lambda_n = \sqrt{n - \tfrac{1}{24}}$$

we obtain then the result (3)

$$p(n) = \frac{1}{\pi\sqrt{2}} \sum_{k=1}^{\infty} A_k(n)k^{\frac{1}{2}} \frac{d}{dn}\left(\frac{\sinh\left(\frac{C}{k}\lambda_n\right)}{\lambda_n}\right),$$

which we had set out to prove.

UNIVERSITY OF PENNSYLVANIA.

REFERENCES

(1) L. R. FORD, *Fractions*. American Mathematical Monthly, vol. 45, (1938), pp. 586–601.

(2) G. H. HARDY AND S. RAMANUJAN, *Asymptotic Formulae in Combinatory Analysis*. Proc. London Math. Soc. (2), vol. 17 (1918), pp. 75–115; also Ramanujan's Collected Papers (1927), pp. 276–309.

(3) H. RADEMACHER, *On the Partition Function p(n)*. Proc. London Math. Soc. (2) vol. 43 (1937), pp. 241–254.

[2] Watson, Bessel Functions, p. 181, formula (1), where, however, the path of integration is bent into a loop around the negative real axis; compare the remark to formula (8), p. 177.

Notes to Paper 48

This paper has been reviewed in the Z, vol. 60 (1957), p. 100, and in the MR, vol. 5 (1944), p. 35.

Page 418, marginal note.
Author's correction: the exponent should read $2\pi i n(h/k + \zeta)$.

Page 418, last display line.
The last fraction should read $kk_2/(k^2 + k_2^2)$.

Page 420, marginal note and equation (3.5).
Author's correction. The new inequality (3.5) is still not entirely correct. In fact, one has $|s_{hk}| \leqslant |z'_{hk}| + |z''_{hk}|$ so that, using the estimates at the top of the page,

$$|z'_{hk}| = \frac{k}{\sqrt{k^2 + k_1^2}} \leqslant \frac{\sqrt{2}\,k}{N+1}, \text{ and similarly, } |z''_{hk}| \leqslant \frac{\sqrt{2}\,k}{N+1},$$

leading to

$$|s_{hk}| \leqslant \frac{2\sqrt{2}\,k}{N+1} < \frac{2\sqrt{2}\,k}{N}.$$

The estimate for $|I^*_{hk}|$ remains correct.

Page 422, reference (3).
See paper 38 of the present collection.

AN ITERATION METHOD FOR CALCULATION WITH LAURENT SERIES*

BY

H. A. RADEMACHER AND I. J. SCHOENBERG

University of Pennsylvania and Ballistic Research Laboratories, Aberdeen Proving Ground

Introduction. The power series is a basic concept of Analysis which is of fundamental importance from the theoretical as well as from the computational point of view. The theoretical importance of power series springs from the fact that it represents any analytic function in the neighborhood of a regular point. The reason for its practical importance is the ease with which implicitly defined functions, by finite relations or differential equations, may be expanded in power series by the so-called method of undetermined coefficients, known and used since the dawn of mathematical analysis.

Laurent series play a definitely minor role as compared to power series. One reason is the more complicated nature of the connection between the sum of the series and its coefficients. Another reason, dependent on the first, is the difficulty of calculations with Laurent series.

The purpose of this paper is to describe a method whereby rational or algebraic operations with Laurent series may be performed with high accuracy at the expense of a reasonable amount of labor. A general approximation method to empirical data, developed by one of us,[1] required the very accurate reciprocation of certain Laurent series. This problem of reciprocation of Laurent series was the starting point of our investigation. Our method for solving this particular problem turned out to be identical with a method of reciprocation of finite matrices already investigated by H. Hotelling.[2] We finally point out that our method of computation with Laurent series extends to computations with trigonometric series provided these series converge absolutely.

1. Newton's algorithm and statement of the problem. Let

$$f(x) \equiv a_0x^m + a_1x^{m-1} + \cdots + a_m = 0, \qquad (a_0 \neq 0), \tag{1}$$

be an algebraic equation with numerical real or complex coefficients. If x is a simple root of this equation, then very close approximations to x may be readily computed by Newton's iterative algorithm represented by the recurrence relation

$$x_{n+1} = x_n - \frac{f(x_n)}{f'(x_n)}. \tag{2}$$

The reason for the fast convergence of x_n towards x is as follows: Expanding the right-

* Received Nov. 6, 1945.

[1] I. J. Schoenberg, *Contributions to the problem of approximation of equidistant data by analytic functions*, Part A, Quart. Appl. Math.

[2] H. Hotelling, *Some new methods in matrix calculation*, Ann. Math. Statist. **14**, 1–34 (1943), especially p. 14, and *Further points in matrix calculation and simultaneous equations*, Ann. Math. Statist. **14**, 440–441 (1943).

hand side in a power series in $x_n - x$, in the neighborhood of the simple root x, we find that (2) may be written as

$$x_{n+1} - x = c_2(x_n - x)^2 + c_3(x_n - x)^3 + \cdots, \tag{3}$$

with coefficients c_ν depending only on the root x. Because there is no linear term in $x_n - x$ on the right-hand side we find that from a certain point on the error $x_{n+1} - x$ is of the order of magnitude of the square of the previous error $x_n - x$. From this stage on, each step will approximately double the number of correct decimal places of the previous approximation x_n. This type of rapid convergence is sometimes referred to as "quadratic convergence."

Let us notice that the iterative process (2) requires a division, by $f'(x_n)$, at each step of the process. This is a serious handicap in computing with machines which do not perform the operation of division, as for example the standard punch-card machines. This division is likewise a handicap if we wish to extend the process to the realm of matrices where division is a difficult numerical operation.

We propose to modify Newton's algorithm (2) so as to require only the operations of addition, subtraction and multiplication in its performance. It will then be shown how the modified Newton algorithm allows us to carry out numerically rational as well as algebraic operations on Laurent series. The most general numerical problem whose solution is facilitated by our method may be formulated as follows.

PROBLEM. *Let*

$$f(w, z) \equiv a_0(z)w^m + a_1(z)w^{m-1} + \cdots + a_m(z) = 0 \tag{4}$$

be an equation with the following properties:

1. *The coefficients* $a_\nu(z)$ *are all regular and uniform functions of* z *in the ring*

$$R:\ r_1 < |z| < r_2. \tag{5}$$

2. *We have*

$$a_0(z) \neq 0 \quad \text{in} \quad R. \tag{6}$$

3. *The discriminant* $D(z)$ *of* (4) *satisfies*

$$D(z) \neq 0 \quad \text{in} \quad R \tag{7}$$

so that the equation (4) *has no critical point in* R. *Let now* $w = w(z)$ *be a branch of a solution of* (4) *which is necessarily regular in* R *but need not be uniform in* R. *Given the numerical values of the Laurent expansions of the coefficients* $a_\nu(z)$, *the problem is to find the values of the coefficients of the Laurent expansions of* $w(z)$.

Remarks. 1. The difficulty of this problem is due to its being concerned with Laurent series rather than ordinary power series. Indeed, if everything else is unchanged, we replace, in its formulation, the ring (5) by the circle $|z| < r_2$, then all Laurent series mentioned become power series, in which case the power series expansion of the branch $w = w(z)$ may be obtained by the method of undetermined coefficients (see the first paragraph of our Introduction).

2. We did not require the branch $w = w(z)$ to be uniform in R. However, we do not restrict our problem by assuming $w(z)$ to be uniform in R. Indeed, if $w(z)$ returns to its values after k turns in R, $k > 1$, we change variable by setting

$$z = \zeta^k.$$

Our equation (4) thereby becomes

$$f(w, \zeta^k) = 0$$

and the branch $w(z)$ becomes uniform in the corresponding ring in the ζ-plane. If we can determine its uniform Laurent series

$$w(z) = \sum_{-\infty}^{\infty} \omega_n \zeta^n$$

we also have its algebraic expansion

$$w(z) = \sum_{-\infty}^{\infty} \omega_n z^{n/k}.$$

3. Even the case $m=1$ is far from trivial. Thus

$$a_0(z)w - 1 = 0$$

amounts to the important problem of the reciprocation of a given Laurent series.

2. The modification of Newton's algorithm. We return in this section to the case of the ordinary algebraic equation (1). We now impose the restriction that

$$f(x) \text{ has only simple zeros.} \tag{8}$$

This condition implies that the polynomials $f(x)$, $f'(x)$ have no common divisors and that we can therefore determine uniquely, by rational operations alone, two polynomials $\phi(x)$ and $\psi(x)$ satisfying the identity

$$f(x)\phi(x) + f'(x)\psi(x) \equiv 1, \tag{9}$$

and such that the degrees of ϕ and ψ do not exceed $m-2$ and $m-1$, respectively. The coefficients of $\phi(x)$, $\psi(x)$ are rational functions of the coefficients a_ν. For later reference it is important to remark that the coefficients of $\psi(x)$ may be written as a quotient of polynomials in a_ν divided by the discriminant D of the polynomial $f(x)$. Indeed, the coefficients of ϕ and ψ are determined, in view of (9), by a system of linear equations whose determinant is precisely the discriminant D of $f(x)$. This procedure leads to explicit expressions of ψ and D in determinant form. Thus for $m=3$ we obtain

$$\psi(x) = \frac{1}{D}\begin{vmatrix} x^2 & x & 1 & 0 & 0 \\ 3a_0 & 0 & 0 & a_0 & 0 \\ 2a_1 & 3a_0 & 0 & a_1 & a_0 \\ a_2 & 2a_1 & 3a_0 & a_2 & a_1 \\ 0 & a_2 & 2a_1 & a_3 & a_2 \end{vmatrix}, \qquad D = \begin{vmatrix} 3a_0 & 0 & 0 & a_0 & 0 \\ 2a_1 & 3a_0 & 0 & a_1 & a_0 \\ a_2 & 2a_1 & 3a_0 & a_2 & a_1 \\ 0 & a_2 & 2a_1 & a_3 & a_2 \\ 0 & 0 & a_2 & 0 & a_3 \end{vmatrix}. \tag{10}$$

This expression, which generalizes to any value of m, indeed shows that the coefficients of $\psi(x)$ have the common denominator D if regarded as rational functions of the a_ν's.

Now we modify Newton's algorithm (2) to its new form[3]

[3] We learn from a note by J. S. Frame, *Remarks on a variation of Newton's method*, Amer. Math. Monthly, **52**, 212–214 (1945), that precisely the same modification of Newton's algorithm has already been used since 1942 by H. Schwerdtfeger, of the University of Adelaide, South Australia, for the numerical solution of ordinary algebraic and transcendental equations.

$$x_{n+1} = x_n - f(x_n)\psi(x_n). \tag{11}$$

Setting

$$F(x) \equiv x - f(x)\psi(x) \tag{12}$$

we may write (11) as

$$x_{n+1} = F(x_n). \tag{11'}$$

On comparing (2) and its modification (11) we see that the division required by (2), at each step of the process, is not present in (11). Now we want to show that the algorithm (11′) also enjoys the property of (2) of producing fast convergence towards the zeros of $f(x)$. Indeed, let x be a root of (1),

$$f(x) = 0, \tag{13}$$

and let us expand $F(x_n)$ about the point $x_n = x$. Writing for convenience $f^{(\nu)}(x) = f^{(\nu)}$, $\psi^{(\nu)}(x) = \psi^{(\nu)}$, we have by Taylor's formula

$$f(x_n) = f + f'(x_n - x) + \tfrac{1}{2}f''(x_n - x)^2 + \cdots,$$
$$\psi(x_n) = \psi + \psi'(x_n - x) + \tfrac{1}{2}\psi''(x_n - x)^2 + \cdots,$$

hence

$$f(x_n)\psi(x_n) = f\psi + (f\psi' + f'\psi)(x_n - x) + \tfrac{1}{2}(f\psi'' + 2f'\psi' + f''\psi)(x_n - x)^2 + \cdots.$$

By (9) and (13) we have $f = 0$, $f'\psi = 1$ and therefore

$$F(x_n) - x = -\tfrac{1}{2}(2f'\psi' + f''\psi)(x_n - x)^2 + \cdots.$$

This shows that *we may write our relation* (11) *in the form*

$$x_{r+1} - x = b_2(x)(x_n - x)^2 + b_3(x)(x_n - x)^3 + \cdots + b_{2m-1}(x)(x_n - x)^{2m-1}, \tag{14}$$

where the $b_\nu(x)$ are polynomials in x with coefficients which are polynomials in a_ν divided by the common denominator D.

Again the missing linear term in $x_n - x$, on the right-hand side of (14), shows that if x_0 is sufficiently close to x, then the algorithm (11) will insure that $x_n \to x$ with quadratic convergence.

An important special case of (1) is the equation

$$ax^m - 1 = 0, \qquad (a \neq 0). \tag{15}$$

The identity

$$(-1)(ax^m - 1) + \frac{x}{m}(max^{m-1}) = 1$$

shows that in this case

$$\psi(x) = \frac{1}{m}x.$$

The relation (11) now becomes

$$x_{n+1} = x_n + \frac{1}{m}x_n(1 - ax_n^m). \tag{16}$$

In particular if $m = 1$, (15) reduces to

$$ax - 1 = 0 \tag{17}$$

when (16) becomes

$$x_{n+1} = x_n + x_n(1 - ax_n). \tag{18}$$

3. The reciprocation of matrices. The advantage of the modified, or division-free, Newton algorithm (11) appears in connection with matrix calculations. In recent years H. Hotelling has recommended the following procedure of finding the reciprocal $X = A^{-1}$ of a given numerical non-singular matrix

$$A = \|\alpha_{ij}\| \qquad (i, j = 1, \cdots, m). \tag{19}$$

Obtain in some way, e.g. by the so-called Gauss, or Doolittle, process a good approximation X_0 to A^{-1}. Then improve this approximation by the recurrence relation

$$X_{n+1} = X_n + X_n(I - AX_n). \tag{20}$$

In the case of $m = 1$ this relation is identical with (18).

In studying the convergence of X_n towards $X = A^{-1}$ Hotelling metrizes the space of real $m \times m$ matrices by means of the absolute value or norm

$$N(A) = \sqrt{\sum_{i,j} \alpha_{ij}^2} \tag{21}$$

which enjoys the following properties

$$N(A + B) \leqq N(A) + N(B), \qquad N(AB) \leqq N(A)N(B). \tag{22}$$

By means of these inequalities Hotelling derived an estimate of $N(X_n - X)$ which was improved by A. T. Lonseth as follows:[4]

INEQUALITY OF HOTELLING AND LONSETH. *Let X_0 be an approximation to $X = A^{-1}$ such that*

$$N(I - AX_0) = k < 1. \tag{23}$$

Starting with X_0 we obtain the sequence X_n by (20). *Then*

$$N(X_n - X) \leqq N(X_0) \cdot k^{2^n} \cdot (1 - k)^{-1}. \tag{24}$$

This interesting result shows in particular that the inequality (23) is sufficient to insure the convergence of the process.

Our generalizations (16) and (11) of the recurrence relation (18) suggest similar iterative procedures for the solution of non-linear algebraic matrix equations. We prefer, however, to pass on to a discussion of calculations with Laurent series.

4. Calculations with Laurent series. Let

$$a(z) = \sum_{-\infty}^{\infty} \alpha_n z^n, \qquad (r_1 < |z| < r_2), \tag{25}$$

be a Laurent series converging in the ring (5). There is no inherent restriction of the generality of the Problem formulated in our Introduction if we assume that the ring R contains the unit-circle $|z| = 1$, i.e.

[4] See Hotelling's second note already mentioned.

$$r_1 < 1 < r_2. \tag{26}$$

An advantage of this normalization is that it implies that $\alpha_n \to 0$ exponentially as $n \to +\infty$ or $n \to -\infty$, insuring that the sequence $\{\alpha_n\}$ is "finite" to a fixed number of decimal places.

The relation (25) sets up a one-one correspondence

$$a(z) \sim \{\alpha_n\}$$

between functions $a(z)$ uniform and regular in R and sequences $\{\alpha_n\}$ for which the series (25) converges in R. To the function $a(z) \equiv 1$ corresponds the unit-sequence

$$I: \quad \alpha_0 = 1, \quad \alpha_n = 0 \quad \text{if} \quad n \neq 0.$$

This correspondence may be interpreted as an isomorphism concerning the operations of addition, subtraction, multiplication and multiplication by a scalar. Indeed, if

$$b(z) = \sum_{-\infty}^{\infty} \beta_n z^n, \qquad (r_1 < z < r_2), \tag{27}$$

is a second series then we find, on multiplying (25) and (27) that to the product

$$c(z) = a(z)b(z) \tag{28}$$

corresponds the series

$$c(z) = \sum_{-\infty}^{\infty} \gamma_n z^n \tag{29}$$

where

$$\gamma_n = \sum_{\nu=-\infty}^{\infty} \alpha_{n-\nu}\beta_\nu. \tag{30}$$

Thus to the operation (28) of multiplication of the functions a, b, corresponds the operation of *convolution* (30) of the two sequences $\{\alpha_n\}$, $\{\beta_n\}$, an operation which we write as

$$\gamma = \alpha\beta. \tag{31}$$

We mention incidentally a third interpretation of Laurent series isomorphic to the two already discussed. Indeed, consider the (4-way) infinite matrix

$$\|\alpha_{j-i}\| \tag{32}$$

in which α_{j-i} is the element in the ith row and jth column, both i and j assuming all integral values. Such matrices may be designated "striped" for the reason that all elements lying on a line, sloping down at a 45° angle, are identical. To every sequence $\{\alpha_n\}$ corresponds one such matrix, and conversely. The isomorphism between such matrices and sequences becomes evident if we remark that the multiplication of two striped matrices $\|\alpha_{j-i}\|$, $\|\beta_{j-i}\|$, is another striped matrix $\|\gamma_{j-i}\|$, where the sequence $\{\gamma_n\}$ is given by (30). This remark throws some light on the connection between Laurent series and the case of finite matrices of §3.

We finally define the *norm* of the function $a(z)$, or of the sequence $\{\alpha_n\}$, as the non-negative number

$$N(a) = N(\alpha) = \sum_{-\infty}^{\infty} | \alpha_n |. \tag{33}$$

This norm also enjoys the two properties (22) or

$$N(a + b) \leqq N(a) + N(b), \qquad N(ab) \leqq N(a)N(b). \tag{34}$$

Their verification is immediate in this case.

We are now able to attack the general Problem of section 1. However, it is essential to discuss first the important case $m=1$ of reciprocation.

4.1. The reciprocation of Laurent series. With the norm of a Laurent series as defined by (33), the result of Hotelling and Lonseth (section 3) applies to Laurent series without any change. Assuming that the sum $a(z)$ of the given Laurent series (25) does not vanish in R, we are to find the expansion

$$w(z) = \frac{1}{a(z)} = \sum_{-\infty}^{\infty} \omega_n z^n. \tag{35}$$

Let

$$w_0(z) = \sum_{-\infty}^{\infty} \omega_n^{(0)} z^n \tag{36}$$

be an approximation to (35) such that

$$N(1 - aw_0) = N(I - \alpha\omega^{(0)}) = k < 1. \tag{37}$$

The very important problem of how such approximations may be obtained will be discussed later (section 4.3). This starting sequence is now to be improved by the relation

$$\omega^{(n+1)} = \omega^{(n)} + \omega^{(n)}(I - \alpha\omega^{(n)}). \tag{38}$$

The rapid convergence is assured by the Hotelling-Lonseth inequality

$$N(\omega^{(n)} - \omega) \leqq N(\omega^{(0)}) \cdot k^{2^n} \cdot (1 - k)^{-1}. \tag{39}$$

Pending a discussion of procedures for obtaining the first approximation, we may therefore regard the numerical problem of reciprocation as solved. This implies that we may perform all *four* rational operations on Laurent series and that we may thus find the Laurent expansion of any *rational* function of Laurent series.

4.2. The general algebraic case. We turn now to the general case of the equation (4) with the two additional, and as we have seen, unessential restrictions that our ring R contains the unit-circle $|z| = 1$ and that the solution $w = w(z)$, of (4), be uniform in R. The problem is to find the numerical values of the coefficients of the expansion

$$w(z) = \sum_{-\infty}^{\infty} \omega_n z^n. \tag{40}$$

We return to our discussion (section 2) of the division-free Newton algorithm (11), especially in its expanded form (14). This discussion remains valid if applied to (4) rather than (1). The algorithm is in this case

$$w_{n+1} = w_n - f(w_n, z)\psi(w_n), \tag{41}$$

the expanded form of which is

$$w_{n+1} - w = b_2(w)(w_n - w)^2 + \cdots + b_{2m-1}(w)(w_n - w)^{2m-1}. \tag{42}$$

We have to remember, however, that $\psi(w_n)$ is a polynomial in w_n with coefficients which are polynomials in a_ν divided by $D = D(z)$. Since D is a polynomial in the a_ν, we may first derive its Laurent expansion by additions and multiplications from the given Laurent series of the coefficients $a_\nu(z)$ of (4). Secondly, since

$$D(z) \neq 0 \quad \text{in} \quad R,$$

we may also find by the method of section 4.1 the Laurent expansion of $1/D(z)$. In this way we arrive at the Laurent expansions of the coefficients of $\psi(w_n)$. These preliminary Laurent series operations allow to put the relation (41) in the form

$$w_{n+1} = w_n + (c_0(z) + c_1(z)w_n + \cdots + c_{2m-1}(z)w_n^{2m-1}), \tag{43}$$

where

$$c_\mu(z) = \sum_\nu \gamma_{\mu\nu} z^\nu \tag{44}$$

are numerically known Laurent expansions.

Now let

$$w_0(z) = \sum_{-\infty}^{\infty} \omega_\nu^{(0)} z^\nu \tag{45}$$

be an approximation to (40), (See section 4.3.) Starting from this approximation we obtain the successive series

$$w_n(z) = \sum_{-\infty}^{\infty} \omega_\nu^{(n)} z^\nu \tag{46}$$

by means of (43). This operation of deriving w_{n+1} from w_n is of course to be performed on the corresponding sequences of coefficients. By (43), (44), (46), the operation takes the form

$$\omega^{(n+1)} = \omega^{(n)} + (\gamma_0 + \gamma_1\omega^{(n)} + \cdots + \gamma_{2m-1}(\omega^{(n)})^{2m-1}). \tag{47}$$

Will the expansion (46) converge towards the expansion (40) of the solution? To answer this question we return to the form (42) of our relation. Taking the norms of both sides of (42) and using the properties (34) of the norm, we obtain

$$N(w_{n+1} - w) \leqq N(b_2(w))[N(w_n - w)]^2 + \cdots + N(b_{2m-1}(w))[N(w_n - w)]^{2m-1}. \tag{48}$$

This relation shows that if

$$N(w_0 - w) = \sum_{-\infty}^{\infty} | \omega_\nu^{(0)} - \omega_\nu | \tag{49}$$

is sufficiently small then (48) will indeed imply

$$\lim_{n\to\infty} N(w_n - w) = 0 \tag{50}$$

with quadratic convergence.

4.3. Derivation of an approximate Laurent expansion. The method for computation with Laurent series described in the previous sections will now become effective provided we can solve the following problem.

INITIAL APPROXIMATION PROBLEM. *Let*

$$F(z) = \sum_{-\infty}^{\infty} c_\nu z^\nu \tag{51}$$

be regular in the ring R containing the circle $|z| = 1$. *This function* $F(z)$, *whose Laurent coefficients* c_ν *are unknown, is defined by an algebraic equation which allows us to compute the value of* $F(z)$ *for any given z of R, in particular for any root of unity. We are to describe a practical method whereby, given* $\epsilon > 0$, *we may compute the coefficients* c_ν^* *of a Laurent series*

$$F^*(z) = \sum_{-\infty}^{\infty} c_\nu^* z^\nu \tag{52}$$

regular in R, such that

$$N(F - F^*) = \sum_{-\infty}^{\infty} | c_\nu - c_\nu^* | < \epsilon. \tag{53}$$

We shall now solve this problem by the method of trigonometric interpolation.[5] Let m be a positive integer and let

$$z_\mu = e^{2\pi i\mu/m}, \qquad (\mu = 0, 1, \cdots, m - 1), \tag{54}$$

be the mth roots of unity. These roots of unity satisfy the following orthogonality relations

$$\frac{1}{m} \sum_{\mu=0}^{m-1} z_\mu^\nu \bar{z}_\mu^s = \begin{cases} 1 & \text{if} \quad \nu \equiv s \pmod{m} \\ 0 & \text{if} \quad \nu \not\equiv s \pmod{m}. \end{cases} \tag{55}$$

If m is odd, $m = 2n + 1$, we consider the Laurent polynomial

$$F_m(z) = \sum_{\nu=-n}^{n} c_{m,\nu} z^\nu \tag{56}$$

having m arbitrary coefficients.

If m is even, $m = 2n$, we define our polynomial so as to contain again only m arbitrary coefficients as

$$F_m(z) = \sum_{\nu=-(n-1)}^{n-1} c_{m,\nu} z^\nu + \tfrac{1}{2} c_{m,n}(z^n + z^{-n}). \tag{57}$$

Whether m is even or odd we may always write

$$F_m(z) = {\sum_{\nu=-n}^{n}}' c_{m,\nu} z^\nu, \qquad \left(n = \left[\frac{m}{2}\right]\right), \tag{58}$$

[5] Concerning the subject of trigonometric interpolation we refer to the classical memoir by Ch. J. de la Vallée Poussin, *Sur la convergence des formules d'interpolation entre ordonées equidistantes*, Bulletin de l'Academie royale de Belgique, 319–410 (1908), and to Dunham Jackson, *The theory of approximation*, American Mathematical Society Colloquium Publications, vol. 11, New York, 1930, chap. IV.

where the summation symbol $\sum'$ is to indicate that if m is even, then

$$c_{m,-n} = c_{m,n}, \tag{59}$$

and that the terms of (58) for $\nu = \pm n$ are to be taken with half their value. The relation $n = [m/2]$ is to indicate that n is the greatest integer not exceeding $m/2$.

We shall now require the Laurent polynomial (58) to interpolate the function (51) in the points (54). This gives the m equations

$$\sum_{\nu=-n}^{n}{}' c_{m,\nu} z_\mu^\nu = F(z_\mu), \qquad (\mu = 0, 1, \cdots, m-1). \tag{60}$$

On multiplying (60) by $\bar{z}_\mu^s/m$ (s fixed, $-n \leqq s \leqq n$) we find in all cases, in view of (55) after summation by μ, that

$$c_{m,s} = \frac{1}{m} \sum_{\mu=0}^{m-1} F(z_\mu) \bar{z}_\mu^s, \qquad (-n \leqq s \leqq n). \tag{61}$$

The construction of our approximate Laurent expansion (58), i.e. (52), has now been completed. The following theorem will now show that the condition (53) may also be realized by the present method of construction.

THEOREM. *We assume the Laurent series*

$$F(z) = \sum_{\nu=-\infty}^{\infty} c_\nu z^\nu \tag{62}$$

to converge absolutely on the unit circle $|z| = 1$, *i.e.*

$$\sum_{-\infty}^{\infty} |c_\nu| < \infty.^{6} \tag{63}$$

Then our interpolating Laurent polynomial (58) *satisfies the condition*

$$\lim_{m \to \infty} N(F - F_m) = 0. \tag{64}$$

Remark. Notice that the regularity of $F(x)$ in a ring containing $|z| = 1$ implies our condition (63) but not conversely. This remark is of importance concerning calculations with absolutely convergent Fourier series. (See section 5.)

Proof. Let N be a positive integer. We shall restrict ourselves to values of $m \geqq 2N$, hence $n \geqq N$. We may then write

$$N(F - F_m) \leqq \sum_{\nu=-N+1}^{N-1} |c_\nu - c_{m,\nu}| + \sum_{N \leqq |s| \leqq n}{}' |c_{m,s}| + \sum_{|\nu| \geqq N} |c_\nu|. \tag{65}$$

We shall now estimate the three sums on the right-hand side.

[6] See Dunham Jackson, loc. cit., for other conditions insuring the convergence of trigonometric interpolation.

Let $\epsilon > 0$ be given. In view of our assumption (63) we may choose N such that

$$\sum_{|\nu| \geqq N} |c_\nu| < \epsilon. \tag{66}$$

An upper bound for the second sum of (65) may now be obtained as follows. By (61) and (62) we have

$$c_{m,s} = \frac{1}{m} \sum_{\mu=0}^{m-1} \bar{z}_\mu^{s} \sum_{\nu=-\infty}^{\infty} c_\nu z_\mu^{\nu} = \sum_{\nu=-\infty}^{\infty} c_\nu \frac{1}{m} \sum_{\mu=0}^{m-1} z_\mu^{\nu} \bar{z}_\mu^{s},$$

and finally by (55)

$$c_{m,s} = \sum_{\nu \equiv s \pmod{m}} c_\nu. \tag{67}$$

For all $m \geqq 2N$, whether m is even or odd, we now have

$$\sum_{N \leqq |s| \leqq n}' |c_{m,s}| \leqq \sum_{N \leqq |s| \leqq n}' \sum_{\nu \equiv s \pmod{m}} |c_\nu| \leqq \sum_{|\nu| \geqq N} |c_\nu| < \epsilon \tag{68}$$

by (66). We now return to (61). Since (62) converges uniformly on $|z| = 1$, $F(z)$ is continuous on $|z| = 1$ and (62) is its Fourier series. We therefore have the Fourier-Cauchy relations

$$c_\nu = \frac{1}{2\pi i} \int_{|z|=1} F(z) z^{-\nu-1} dz. \tag{69}$$

From the definition of this integral as a limit of Cauchy sums we may now write (with $z_m = 1$)

$$c_\nu = \lim_{m \to \infty} \frac{1}{2\pi i} \sum_{\mu=0}^{m-1} F(z_\mu) z_\mu^{-\nu-1}(z_{\mu+1} - z_\mu),$$

$$c_\nu = \lim_{m \to \infty} \frac{1}{2\pi i} \sum_{\mu=0}^{m-1} F(z_\mu) z_\mu^{-\nu}(e^{2\pi i/m} - 1),$$

and finally, by (61),

$$c_\nu = \lim_{m \to \infty} \frac{1}{m} \sum_{\mu=0}^{m-1} F(z_\mu) z_\mu^{-\nu} = \lim_{m \to \infty} c_{m\ \nu}. \tag{70}$$

We now have indeed

$$\sum_{\nu=-N+1}^{N-1} |c_\nu - c_{m,\nu}| < \epsilon, \quad \text{provided} \quad m > m_0(\epsilon). \tag{71}$$

By (65), (66), (68), and (71) we now have

$$N(F - F_m) < 3\epsilon, \quad \text{provided} \quad m > m_0(\epsilon), \tag{72}$$

and our theorem is established.

4.31. The 24-ordinate scheme of numerical harmonic analysis. The interpolation of our given function $F(z)$ in the 24th roots of unity will provide satisfactory approximation for most ordinary purposes. Let us assume for definiteness that $F(z)$ is *real* for real z. On the unit-circle $z=e^{i\theta}$, we then have

$$F(e^{i\theta}) = R(\theta) + iI(\theta), \tag{73}$$

where the real part $R(\theta)$ is an *even* function, while the imaginary part $I(\theta)$ is *odd*. Denote by

$$F_\mu = R_\mu + iI_\mu, \qquad (\mu = 0, 1, \cdots, 23), \tag{74}$$

the computed values of our function at 15°—intervals in θ, i.e. for the points (54) with $m=24$. We now interpolate the 13 ordinates R_μ $(\mu=0, 1, \cdots, 12)$ by a cosine polynomial

$$A_0 + A_1 \cos\theta + \cdots + A_{11}\cos 11\theta + A_{12}\cos 12\theta, \tag{75}$$

and the 13 ordinates $I_0=0, I_1, \cdots, I_{11}, I_{12}=0$ by a sine polynomial

$$B_1 \sin\theta + \cdots + B_{11}\sin 11\theta. \tag{76}$$

These polynomials are readily obtained by the 24-ordinate scheme as described in E. T. Whittaker and G. Robinson, *The calculus of observations*, ed. 3, 1940, section 137, pp. 273–278. The complex function (73) is now interpolated in the 24 points by the trigonometric polynomial

$$\begin{aligned} F_{24}(e^{i\theta}) = {} & A_0 + A_1\cos\theta + \cdots + A_{11}\cos 11\theta + A_{12}\cos 12\theta \\ & + iB_1\sin\theta + \cdots + iB_{11}\sin 11\theta. \end{aligned}$$

Setting

$$z = e^{i\theta}, \qquad \cos\nu\theta = \tfrac{1}{2}(z^\nu + z^{-\nu}), \qquad i\sin\nu\theta = \tfrac{1}{2}(z^\nu - z^{-\nu}), \tag{77}$$

we obtain the Laurent sum with real coefficients

$$\begin{aligned} F_{24}(z) = {} & A_0 + \sum_{\nu=1}^{11} \tfrac{1}{2}(A_\nu + B_\nu)z^\nu + \tfrac{1}{2}A_{12}z^{12} \\ & + \sum_{\nu=1}^{11} \tfrac{1}{2}(A_\nu - B_\nu)z^{-\nu} + \tfrac{1}{2}A_{12}z^{-12}. \end{aligned} \tag{78}$$

This initial approximate Laurent expansion will be used in section 6 in our example of reciprocation of a Laurent series.

5. Calculation with Fourier series. The method of calculation with Laurent series described in sections 4, 4.1, 4.2, 4.3 and 4.31, applies unchanged to the realm of absolutely convergent Fourier series written in the complex form

$$F(z) = \sum_{-\infty}^{\infty} c_\nu z^\nu, \quad \text{where} \quad z = e^{i\theta},$$

with the definition of the norm as

$$N(F) = \sum_{-\infty}^{\infty} |c_\nu|.$$

The general problem of section 1 may now be reformulated, replacing the ring R by the unit circle $|z|=1$. The coefficients $a_\nu(z)$ of the equation (4) are now defined by given absolutely convergent Fourier series. The conditions (6) and (7) remain unchanged. The important fact that a uniform, continuous solution $w=w(z)$ of (4) along the unit-circle admits of an *absolutely* convergent Fourier expansion is now assured by a general theorem of N. Wiener and P. Lévy.[7] The effectiveness of the interpolation method of section 4.3 for obtaining a satisfactory initial approximate Fourier series is secured by our theorem of section 4.3.

We finally mention briefly the special problem of the reciprocation of a non-vanishing absolutely convergent Fourier series

$$A(\theta) = \tfrac{1}{2}a_0 + \sum_{n=1}^{\infty} (a_n \cos n\theta + b_n \sin n\theta) \tag{79}$$

with *real* coefficients a_ν, b_ν. In applying our method, we have to pass to the complex variable z, by means of the relations (77), obtaining the series

$$A(\theta) = a(z) = \sum_{-\infty}^{\infty} \alpha_n z^n, \tag{80}$$

which is now to be reciprocated. The coefficients α_n being complex, it would appear that computations with complex numbers are unavoidable. This, however, is not the case since we may proceed as follows. Working with the real series

$$f(z) = \sum_{-\infty}^{\infty} R(\alpha_n)z^n, \qquad g(z) = \sum_{-\infty}^{\infty} I(\alpha_n)z^n, \tag{81}$$

we have by (80)

$$\frac{1}{a(z)} = \frac{1}{f+ig} = \frac{f}{f^2+g^2} - i\,\frac{g}{f^2+g^2} = \sum_{-\infty}^{\infty} \omega_n z^n. \tag{82}$$

Starting with (81) and operating with *real* series only, we now form the expansions of f^2, g^2 and then f^2+g^2. The real "Laurent" series of f^2+g^2 is now reciprocated by the method of section 4.1 and finally the series for

$$f/(f^2+g^2), \qquad g/(f^2+g^2)$$

obtained by multiplications. This furnishes the complex values of the ω_n of (82). Returning to the variable θ, by (77) we finally obtain the ordinary real Fourier expansion of

$$1/A(\theta)$$

6. An example of reciprocation of a Laurent series. Our numerical example will benefit by the following general remark concerning the modified Newton algorithm (11). For simplicity we limit ourselves to the case of the equation (15) or

$$ax^m - 1 = 0 \tag{83}$$

which is solved by the recurrent relation

[7] See Antoni Zygmund, *Trigonometrical series*, Warszawa-Lwów, 1935, pp. 140–142.

$$x_{n+1} = x_n + \frac{1}{m} x_n(1 - a x_n^m). \tag{84}$$

Let us assume that our first approximation x_0 is of such accuracy that x_2 will have all the accuracy we want, while x_1 does not quite do. More precisely we assume the "residual"

$$r = 1 - a x_0^m \tag{85}$$

so small that we may neglect r^3 everywhere in our calculations. We may use this fact in eliminating x_1 between the two equations

$$\begin{aligned} x_1 &= x_0 + \frac{1}{m} x_0(1 - x_0^m), \\ x_2 &= x_1 + \frac{1}{m} x_1(1 - x_1^m). \end{aligned} \tag{86}$$

Indeed, by (85), (86), we have

$$x_1 = x_0\left(1 + \frac{1}{m} r\right)$$

and neglecting r^3 we find

$$x_1^m = x_0^m\left(1 + r + \frac{m-1}{2m} r^2\right).$$

If we then compute x_2 in this way, i.e., neglecting r^3 wherever it appears, we easily find the following approximation to x_2:

$$x_2' = x_0 + \frac{1}{m} x_0\left(r + \frac{m+1}{2m} r^2\right). \tag{87}$$

We may interpret both equations (85), (87) as a recurrence relation furnishing x_2' in terms of the first approximation x_0. This process converges "cubically." Indeed, a simple calculation will show that we may write (87) as

$$x_2' - x = \frac{(m+1)(2m+1)}{6x^2}(x_0 - x)^3 + (\text{terms of order} > 3).$$

We note especially the following special case: To solve

$$ax - 1 = 0 \tag{88}$$

we set

$$r = 1 - a x_0 \tag{89}$$

and compute

$$x_2' = x_0 + x_0(r + r^2). \tag{90}$$

We turn now to our example which consists in expanding the reciprocal of the Bessel function

$$J_0(z) = 1 - \frac{z^2}{2^2} + \frac{z^4}{(2\cdot 4)^2} - \frac{z^6}{(2\cdot 4\cdot 6)^2} + \cdots \tag{91}$$

into a Laurent series between the first two positive roots of this function, which are approximately $\xi_1 = 2.4$, $\xi_2 = 5.5$. In order to avoid even exponents we consider

$$J_0(\sqrt{z}) = 1 - \frac{z}{2^2} + \frac{z^2}{(2\cdot 4)^2} - \cdots \tag{92}$$

whose reciprocal is to be expanded in Laurent series between its zeros

$$\xi_1^2 = 5.76 \quad \text{and} \quad \xi_2^2 = 30.25.$$

Let us notice that 13 is near the geometric mean of these numbers. In order to realize the condition (26) we replace in (92) z by $13z$, also changing the sign of the function for formal reasons. Thus let

$$a(z) = -J_0(\sqrt{13z}) = \sum_{n=0}^{\infty} \alpha_n z^n \tag{93}$$

be the entire function whose reciprocal

$$w(z) = -\frac{1}{J_0(\sqrt{13z})} = \sum_{-\infty}^{\infty} \omega_n z^n \tag{94}$$

we are to expand in a Laurent series convergent on and near the unit circle $|z| = 1$. Below are the 10-place values of the coefficients α_n of (93) as computed by

$$\alpha_n = (-1)^{n+1}(13)^n/(2^n \cdot n!)^2.$$

n	α_n
0	−1.00000 00000
1	3.25000 00000
2	−2.64062 50000
3	.95355 90278
4	−.19369 16775
5	.02517 99181
6	−.00227 31870
7	.00015 07726
8	−.00000 76564
9	.00000 03072
10	−.00000 00100
11	.00000 00003
$\sum$	= .39229 24951

μ	R_μ	I_μ	A_μ	B_μ
0	2.549 122	.000 000	.601 975	
1	2.262 721	−.257 032	1.063 727	−.361 583
2	1.655 481	−.395 409	.489 993	−.144 072
3	1.081 379	−.427 381	.219 762	−.062 309
4	.661 333	−.405 462	.097 293	−.028 189
5	.379 302	−.361 625	.042 831	−.012 990
6	.193 500	−.309 968	.018 815	−.006 018
7	.070 603	−.256 308	.008 261	−.002 786
8	−.011 124	−.202 973	.003 630	−.001 285
9	−.065 027	−.150 752	.001 604	−.000 588
10	−.099 093	−.099 722	.000 727	−.000 260
11	−.117 947	−.049 615	.000 368	−.000 099
12	−.123 985	.000 000	.000 136	

From these values, rounded to 6 places, we computed to 6 places the values of $a(z_\mu)$ at the 24th roots of unity

$$z_\mu = \cos(15\mu)^\circ + i \sin(15\mu)^\circ, \qquad (\mu = 0, 1, \cdots, 12),$$

and from these the values of the reciprocal

$$w(z_\mu) = 1/a(z_\mu) = R_\mu + iI_\mu, \qquad (\mu = 0, 1, \cdots, 12),$$

which are tabulated above. The coefficients A_μ and B_μ of the interpolating cosine and sine polynomials (75), (76) where then found by the 24-ordinate scheme. They are tabulated above.

From these values we computed the coefficients $\omega_n^{(0)}$ of the approximation

$$w_0(z) = \sum_{-12}^{12} \omega_n^{(0)} z^n$$

according to (78) by the formulae

$$\begin{aligned}
\omega_0^{(0)} &= A_0, \\
\omega_n^{(0)} &= \tfrac{1}{2}(A_n + B_n), \\
\omega_{-n}^{(0)} &= \tfrac{1}{2}(A_n - B_n), \\
\omega_{12}^{(0)} &= \omega_{-12}^{(0)} = \tfrac{1}{2}A_{12}, \qquad (n = 1, 2, \cdots, 11).
\end{aligned}$$

These values rounded to 5 places are in the first column of the following table which contains the complete computation according to the relations (89) and (90). The last column headed $\omega = \omega^{(0)} + \omega^{(0)}(r + r^2)$ gives the 9-place values of the coefficients ω_n of (94).

Remarks. 1. The basic numerical process in this computation is obviously the convolution of sequences. Thus the second column $\alpha\omega^{(0)}$ is obtained by the convolution of the column α with the column $\omega^{(0)}$. According to the formula (30) this is done very simply rewriting the column α, say, in reverse order, then matching it with the column $\omega^{(0)}$ such that the zero term of one column corresponds to the nth term of the other. The accumulated products of matching elements gives the nth term of the product column $\alpha\omega^{(0)}$. This operation is very familiar from the process of smoothing by means of a linear compound formula.

2. The operation of convolution of sequences implies an important check by means of their sums, for it is clear that the sum of the product column should equal the product of the sums of the factor sequence, except for the accumulated rounding error. At the very bottom of each column we wrote the actual sum of the sequence in that column. Directly below it we wrote (in parentheses) the value of this sum in terms of the sums of the columns which enter into its composition.

3. The column of final residuals $I - \alpha\omega$ was also computed (values not recorded here) and its terms were found to be so small that a further repetition of the process, with our 10-place values of the α_n, would not alter our 9-place values of the ω_n. As final checks we found by (93)

$$a(1)^{-1} = 2.54911\ 8356, \qquad a(-1)^{-1} = -.12398\ 5065,$$
$$a(\mathrm{i})^{-1} = .19349\ 9936 - .30996\ 7383\ \mathrm{i}.$$

The corresponding values of $w(z)$, computed by (94), were found to be

$$w(1) = 2.54911\ 8355, \qquad w(-1) = -.12398\ 5067,$$
$$w(\mathrm{i}) = .19349\ 9940 - .30996\ 7383\ \mathrm{i}.$$

n	$\omega^{(0)}$	$\alpha\omega^{(0)}$	$r=I-\alpha\omega^{(0)}$	r^2
−29				
−28				
−27				
−26				
−25				
−24				49
−23				4
−22				−102
−21				64
−20				18
−19				− 64
−18				85
−17				− 65
−16				9
−15				49
−14				− 80
−13				62
−12	.00007	−.00007 00000	.00007 00000	− 7
−11	.00023	−.00000 25000	.00000 25000	− 68
−10	.00049	.00007 26562	−.00007 26562	97
− 9	.00110	−.00004 80946	.00004 80946	− 42
− 8	.00246	.00002 68539	−.00002 68539	− 35
− 7	.00552	−.00000 52301	.00000 52301	61
− 6	.01242	−.00001 62992	.00001 62992	− 27
− 5	.02791	.00002 32702	−.00002 32702	− 17
− 4	.06274	−.00001 52800	.00001 52800	24
− 3	.14104	−.00000 13045	.00000 13045	− 14
− 2	.31703	.00001 89326	−.00001 89326	7
− 1	.71266	−.00002 53421	.00002 53421	− 13
0	.60198	1.00002 07580	−.00002 07580	−.00000 00056
1	.35107	.00000 39064	−.00000 39064	− 70
2	.17296	−.00001 57627	.00001 57627	237
3	.07873	.00000 43626	−.00000 43626	− 89
4	.03455	.00001 22969	−.00001 22969	−116
5	.01492	−.00001 61378	.00001 61378	174
6	.00640	.00000 57617	−.00000 57617	−145
7	.00274	.00000 04662	−.00000 04662	67
8	.00117	.00000 78363	−.00000 78363	30
9	.00051	−.00001 53488	.00001 53488	91
10	.00023	.00001 88876	−.00001 88876	91
11	.00013	−.00001 23781	.00001 23781	− 37
12	.00007	.00006 13042	−.00006 13042	− 29
13		.00002 88471	−.00002 88471	81
14		−.00009 48797	.00009 48797	− 75
15		.00004 63585	−.00004 63585	3
16		−.00001 07391	.00001 07391	53
17		.00000 14982	−.00000 14982	− 39
18		−.00000 01411	.00000 01411	− 13
19		.00000 00096	−.00000 00096	37
20		−.00000 00005	.00000 00005	− 11
21				− 11
22				0
23				32
24				− 21
25				80
26				−124
27				5
28				103
29				− 92
30				43
31				− 13
32				3
$\sum$	2.54913	1.00000 45679 (1.00000 45680)	−.00000 45679	.00000 00002 (.00000 00000)

n	$r+r^2$	$\omega^0(r+r^2)$	$\omega=\omega^{(0)}+\omega^{(0)}(r+r^2)$
−29		1	
−28		2	
−27		5^+	.00000 0001
−26		12	.00000 0001
−25		26	.00000 0003
−24	49	58	.00000 0006
−23	4	130	.00000 0013
−22	−102	292	.00000 0029
−21	64	657	.00000 0066
−20	18	1476	.00000 0148
−19	− 64	3318	.00000 0332
−18	85	7459	.00000 0746
−17	− 65	16767	.00000 1677
−16	9	37691	.00000 3769
−15	49	84725^-	.00000 8472
−14	− 80	1 90454	.00001 9045
−13	62	4 28120	.00004 2812
−12	6 99993	2 62369	.00009 6237
−11	24932	−1 36696	.00021 6330
−10	−7 26465	− 37118	.00048 6288
− 9	4 80904	− 68748	.00109 3125
− 8	−2 68574	− 27684	.00245 7232
− 7	52362	36008	.00552 3601
− 6	1 62965	− 35203	.01241 6480
− 5	−2 32719	9539	.02791 0954
− 4	1 52824	9186	.06274 0919
− 3	13031	− 49463	.14103 5054
− 2	−1 89319	21184	.31703 2118
− 1	2 53408	− 48016	.71265 5198
0	−.00002 07636	−.00000 53032	.60197 4697
1	− 39134	23489	.35107 2349
2	1 57864	7594	.17296 0759
3	− 43715	− 34341	.07872 6566
4	−1 23085	22715^-	.03455 2271
5	1 61552	4594	.01492 0459
6	− 57762	− 20946	.00639 7905
7	− 4595	− 47102	.00273 5290
8	− 78333	− 20304	.00116 7970
9	1 53397	−1 15311	.00049 8469
10	−1 88785	−1 73068	.00021 2693
11	1 23744	−3 92531	.00009 0747
12	−6 13071	−3 12836	.00003 8716
13	−2 88390	1 65178	.00001 6518
14	9 48722	70470	.00000 7047
15	−4 63582	30065^-	.00000 3006
16	1 07444	12827	.00000 1283
17	− 15021	5472	.00000 0547
18	1398	2335^-	.00000 0233
19	− 59	996	.00000 0100
20	− 6	425^+	.00000 0043
21	− 11	182	.00000 0018
22	0	78	.00000 0008
23	32	33	.00000 0003
24	− 21	14	.00000 0001
25	80	6	.00000 0001
26	−124	2	
27	5	1	
28	103	1	
29	− 92		
30	43		
31	− 13		
32	3		
$\sum$	−.00000 45677	−.00001 16443 (−.00001 14637)	2.54911 8355

Notes to Paper 49

This paper, written jointly with I. J. Schoenberg, has been reviewed in the Z, vol. 61 (1961), p. 267, and in the MR, vol. 8 (1947), p. 53.

Page 155, equation (86).
The parentheses should contain $1-ax_0^m$ and $1-ax_1^m$, respectively.

ON THE ACCUMULATION OF ERRORS IN PROCESSES OF INTEGRATION ON HIGH-SPEED CALCULATING MACHINES

HANS A. RADEMACHER
PROFESSOR OF MATHEMATICS
UNIVERSITY OF PENNSYLVANIA

Practically any known method of numerical integration can be adapted for use on digital computing machines. In all such methods integrals are replaced by sums, differential equations by difference equations. It is clear that the approximations given by these methods become better with increasing fineness of the integration steps. Appraisals of the errors can be given by means of well-known error terms, which show, for example, that the error in the "trapezoidal rule" is of order $(\Delta t)^2$, in Simpson's rule of the order $(\Delta t)^4$, where Δt is the integration step. Errors of this sort, pertaining to the special approximation formula, we call *truncation errors*.

Small steps of integration are therefore recommended. The number of steps required increases, of course, when the steps become smaller. A large number of steps is no serious obstacle when high-speed calculating machines with programmed sequencing are used. However, there appears the danger of an accumulation of the *rounding-off errors*. I shall try to analyze the influence of truncation and rounding-off errors on the result, in particular in the case of ordinary differential equations.

The astronomers seem to have been the first to discuss the rounding-off error to some extent. This was done by Frank Schlesinger for quadratures[1] and by Dirk Brouwer for differential equations.[2] Their arguments are somewhat qualitative in character and have to be made more specific.

Let us consider first a quadrature by means of the trapezoidal rule, which states

$$\int_a^b f(t)\,dt = \frac{\Delta t}{2}\left\{f(t_0) + 2\sum_{j=1}^{n-1} f(t_j) + f(t_n)\right\} + E, \tag{1}$$

with

$$E = -\frac{(\Delta t)^2}{12}\sum_{j=1}^{n} f''(\tau_j)\Delta t, \tag{1a}$$

where

$$\Delta t = \frac{b-a}{n},\ t_0 = a,\ t_n = b,\ t_j = a + j\Delta t, \tag{2}$$

[1] Frank Schlesinger, "On the Errors in the Sum of a Number of Tabular Quantities," *Astronomical Journal*, XXX (1917), 183–190.

[2] Dirk Brouwer, "On the Accumulation of Errors in Numerical Integration," *Astronomical Journal*, XLVI (1937), 149–153.

and the τ_j are certain mean values $t_{j-1} \leqq \tau_j \leqq t_j$. We have for the truncation error approximately

$$E \sim \frac{(\Delta t)^2}{12} \int_a^b f''(t)dt = -\frac{(\Delta t)^2}{12}(f'(b) - f'(a)). \tag{3}$$

Actually, the computing machine does not proceed exactly according to (1). Because of the limited fixed capacity of its counters it cannot handle the exact values of f(t) but only the rounded-off values. Let us assume that the machine can keep k digits after the decimal point. (Whether we use the decimal or perhaps the dyadic system is irrelevant for our argument.) Then we have for each summand

$$f(t_j) = \overline{f(t_j)} + \epsilon 10^{-k},$$

where the bar indicates the rounded-off value, and where

$$-0.5 \leqq \epsilon \leqq +0.5.$$

Instead of (1) we obtain therefore

$$\int_a^b f(t)dt = \frac{\Delta t}{2}\left\{\overline{f(t_0)} + 2\sum_{j=1}^{n-1} \overline{f(t_j)} + \overline{f(t_n)}\right\}$$
$$+ \frac{\Delta t}{2} 10^{-k}\left\{\epsilon_0 + 2\sum_{j=1}^{n-1} \epsilon_j + \epsilon_n\right\} + E. \tag{4}$$

Thus we have, besides the truncation error E, the rounding-off error

$$R = \frac{\Delta t}{2} 10^{-k}\left\{\epsilon_0 + 2\sum_{j=1}^{n-1} \epsilon_j + \epsilon_n\right\}, \tag{5}$$

which we can estimate as

$$|R| \leqq \frac{\Delta t}{2} 10^{-k}(2n)(0.5) = \frac{b-a}{2} 10^{-k}, \tag{6}$$

which is independent of n. Here, however, we have assumed that after the summation we introduce no new rounding-off error. If f is of the order of magnitude of one, then f fills $k+1$ digits. The sum fills then $K = k + 1 + \log n$ digits. This capacity may be provided for in the result counter, whereas the working counters have the capacity $k+1$ only. However, if K is limited with increasing n, then $n \times 10^k$ has to be kept in a constant order of magnitude 10^{k-1}; that is,

$$10^{-k} \sim \frac{n}{10^{K-1}}, \tag{7}$$

and then (6) would yield only

$$|R| \lesssim \frac{b-a}{2} \frac{n}{10^{K-1}},$$

showing an increase of the rounding-off error with n.

The rounding-off error (5) can also be treated statistically if the ϵ's are statistically independent. The dispersion of R is

$$\sigma^2(R) = \frac{(\Delta t)^2}{4} 10^{-2k}(4n-2) \int_{-1/2}^{1/2} \epsilon^2 d\epsilon \sim \frac{(\Delta t)^2}{12} 10^{-2k} n,$$

and the standard deviation

$$\sigma(R) \sim \frac{\Delta t}{2\sqrt{3}} 10^{-k} \sqrt{n} = \frac{b-a}{2\sqrt{3}} \frac{10^{-k}}{\sqrt{n}}.$$

Here the remarks about the capacity K of the result counter can also be made, and (7) could be applied if K is limited.

The statistical treatment however requires some precaution. If we had a column of values $f(t_j)$, like

3.4051538
3.4051539
3.4051537
3.4051535
3.4051532
3.4051529
- - - - -

and should abbreviate all these numbers to 3.40515, we should have a run of errors of about the same size 0.3×10^{-5}. Randomness of the errors would imply a change in the size and sign of the errors. For greater safety we may demand that also the last figures which are retained change from entry to entry, that is,

$$|f(t_j) - f(t_{j-1})| > 10^{-k}$$

or

$$|f'(\xi)| \Delta t > 10^{-k},$$

$$\Delta t > 10^{-k} |f'(\xi)|^{-1}, \tag{8}$$

which means that the rounding-off errors can safely be treated as random only if the integration steps are not too small.

Let us turn now to the discussion of the numerical integration of ordinary differential equations. It will suffice for our purpose to treat the system

$$\begin{cases} \dfrac{dx}{dt} = f(x, y) \\ \dfrac{dy}{dt} = g(x, y). \end{cases} \tag{9}$$

The two equations must be replaced by difference equations. For operation on the ENIAC, the Heun method was chosen (which is simpler but less accurate than the better known

178

Runge-Kutta method). Each integration step is a double step. Let x_0, y_0 be the initial values of $x(t)$, $y(t)$ for $t=t_0$. Then we compute first the auxiliary values

$$\begin{cases} x_1^* = x_0 + \Delta t\, f(x_0, y_0) \\ y_1^* = y_0 + \Delta t\, g(x_0, y_0), \end{cases} \tag{10a}$$

and then

$$\begin{cases} x_1 = x_0 + \dfrac{\Delta t}{2}\{f(x_0, y_0) + f(x_1^*, y_1^*)\} \\ y_1 = y_0 + \dfrac{\Delta t}{2}\{g(x_0, y_0) + g(x_1^*, y_1^*)\}. \end{cases} \tag{10b}$$

The values x_1, y_1 correspond to t_1. They are then used as initial values for the next step. In general the procedure is

$$\begin{cases} x_j^* = x_{j-1} + \Delta t\, f(x_{j-1}, y_{j-1}) \\ y_j^* = y_{j-1} + \Delta t\, g(x_{j-1}, y_{j-1}) \end{cases} \tag{10c}$$

$$\begin{cases} x_j = x_{j-1} + \dfrac{\Delta t}{2}\{f(x_{j-1}, y_{j-1}) + f(x_j^*, y_j^*)\} \\ y_j = y_{j-1} + \dfrac{\Delta t}{2}\{g(x_{j-1}, y_{j-1}) + g(x_j^*, y_j^*)\},\ j=1, 2, \cdots, n, \end{cases} \tag{10d}$$

where $t_0 \leq t \leq t_n = T$ is the range of integration and $T - t_0 = n\Delta t$.

If $x(t), y(t)$ is the solution of the system of differential equations (9) with the initial conditions $x(t_0) = x_0$, $y(t_0) = y_0$, then we can look upon x_j, y_j as approximations to $x(t_j)$, $y(t_j)$, and our first task is to estimate the truncation errors

$$u_j = x(t_j) - x_j,\quad v_j = y(t_j) - y_j. \tag{11}$$

Let $X(t)$, $Y(t)$ be another solution of (9), close to $x(t)$, $y(t)$, then we have, up to higher powers of $(X-x)$, $(Y-y)$,

$$\begin{aligned} \frac{d}{dt}(X-x) &= \frac{\partial f}{\partial x}(X-x) + \frac{\partial f}{\partial y}(Y-y) \\ \frac{d}{dt}(Y-y) &= \frac{\partial g}{\partial x}(X-x) + \frac{\partial g}{\partial y}(Y-y). \end{aligned} \tag{12}$$

Here it is supposed that $x(t)$, $y(t)$ are the variables in $\frac{\partial f}{\partial x}, \frac{\partial f}{\partial y}, \frac{\partial g}{\partial x}, \frac{\partial g}{\partial y}$. For these we form the "adjoint equations"

$$\begin{cases} \dfrac{d\lambda}{dt} = -\dfrac{\partial f}{\partial x}\lambda - \dfrac{\partial g}{\partial x}\mu \\ \dfrac{d\mu}{dt} = -\dfrac{\partial f}{\partial y}\lambda - \dfrac{\partial g}{\partial y}\mu \end{cases} \tag{13}$$

for the functions $\lambda(t)$, $\mu(t)$, which will presently be more completely specified.

From (12) and (13) we obtain

$$\lambda\frac{d}{dt}(X-x)+\mu\frac{d}{dt}(Y-y)=(\lambda\frac{\partial f}{\partial x}+\mu\frac{\partial g}{\partial x})(X-x)+(\lambda\frac{\partial f}{\partial y}+\mu\frac{\partial g}{\partial y})(Y-y)$$

$$=-\frac{d\lambda}{dt}(X-x)+\frac{d\mu}{dt}(Y-y)$$

or

$$\frac{d}{dt}\left\{\lambda(t)(X-x)+\mu(t)(Y-y)\right\}=0,$$

which we integrate in the interval from t_{j-1} to t_j;

$$\lambda(t)(X-x)+\mu(t)(Y-y)\Big|_{t_{j-1}}^{t_j}=0. \tag{14}$$

The functions $x(t)$, $y(t)$ have already been defined as solutions of (9) with the initial conditions

$$x(t_0)=x_0,\ y(t_0)=y_0,$$

that is, as the true solutions of our problem. As X and Y we take solutions $X_j(t)$, $Y_j(t)$ of (9) in the interval $t_{j-1}\leqq t\leqq t_j$ with the initial conditions

$$X_j(t_{j-1})=x_{j-1},\ Y_j(t_{j-1})=y_{j-1}. \tag{15}$$

The process (10c), (10d) will produce in the step from t_{j-1} to t_j a certain elementary truncation error R_j, S_j;

$$X_j(t_j)=x_j+R_j,\ Y_j(t_j)=y_j+S_j. \tag{16}$$

From (14) we get

$$\lambda(t)X_j(t)+\mu(t)Y_j(t)\Big|_{t_{j-1}}^{t_j}=\lambda(t)x(t)+\mu(t)y(t)\Big|_{t_{j-1}}^{t_j},$$

and, taking into account (11), (15), (16),

$$\lambda(t_j)R_j+\mu(t_j)S_j=[\lambda(t_j)u_j+\mu(t_j)v_j]$$
$$-[\lambda(t_{j-1})u_{j-1}+\mu(t_{j-1})v_{j-1}].$$

Summation over j from 1 to n yields

$$\sum_{j=1}^{n}[\lambda(t_j)R_j+\mu(t_j)S_j]=\lambda(T)u(T)+\mu(T)v(T), \tag{17}$$

where we have observed $u_0=v_0=0$ and have set $u_n=u(T)$, $v_n=v(T)$.

Heun's method, stated in equations (10), is devised in such a manner that it is correct up to terms of the second order. An application of Taylor's formula to (10c), (10d) leads to

$$R_j=\frac{1}{6}(\Delta t)^3\left\{x'''-\frac{3}{2}\Phi\right\}$$ (18)

$$S_j=\frac{1}{6}(\Delta t)^3\left\{y'''-\frac{3}{2}\Psi\right\},$$

if we neglect higher powers of Δt, and where

$$\Phi=\frac{\partial^2 f}{\partial x^2}x'^2+2\frac{\partial^2 f}{\partial x\partial y}x'y'+\frac{\partial^2 f}{\partial y^2}y'^2$$

$$\Psi=\frac{\partial^2 g}{\partial x^2}x'^2+2\frac{\partial^2 g}{\partial x\partial y}x'y'+\frac{\partial^2 g}{\partial y^2}y'^2,$$

the variables in $\partial^2 f/\partial x^2$ and so forth, to be taken for the values x_{j-1}, y_{j-1}.

Insertion of (18) into (17) leads to

$$\lambda(T)u(T)+\mu(T)v(T)=$$

$$\frac{1}{6}(\Delta t)^2\sum_{j=1}^{n}\left\{\lambda(t_j)\,(x'''-\frac{3}{2}\Phi)+\mu(t_j)\,(y'''-\frac{3}{2}\Psi)\right\}\Delta t.$$

This sum can be looked upon as a Riemann sum and can, by neglecting errors of higher order, be replaced by the integral

$$\int_{t_0}^{T}\left\{\lambda(t)\,(x'''-\frac{3}{2}\Phi)+\mu(t)\,(y'''-\frac{3}{2}\Psi)\right\}dt.$$

After an integration by parts we obtain finally

$$\begin{aligned}\lambda(T)u(T)&+\mu(T)v(T)\\ &\sim-\frac{(\Delta t)^2}{12}\Big[\lambda(t)x''(t)+\mu(t)y''(t)\Big]_{t_0}^{T}\\ &-\frac{(\Delta t)^2}{6}\int_{t_0}^{T}[\lambda'(t)x''(t)+\mu'(t)y''(t)]\,dt\end{aligned}$$ (19)

as the desired expression for the total *truncation errors* $u(T)$, $v(T)$ after the n steps of numerical integration (10).

In order to get $u(T)$ and $v(T)$ separately from (19) one has only to impose on $\lambda(t)$ and $\mu(t)$ the terminal conditions

$$\lambda(T)=1,\ \mu(T)=0,$$

or

$$\lambda(T)=0,\ \mu(T)=1,$$

each of which determines a solution of (13) uniquely.

The truncation error for the Heun method is of order $(\Delta t)^2$, as (19) shows. It could either be taken into account fully in order to improve the solution, or the formula (19) might be used for an appraisal of the expected truncation error. In the latter case a rough numerical approximation for the solution of (13) would suffice.

So far we have assumed that the numerical procedure would exactly follow the rules expressed in equations (10a) to (10d). This, however, is not so, since any process of computation can carry only a limited number of digits and has, therefore, to resort to rounding off. If we indicate rounded-off values by a bar, we have in the jth step instead of (10c) and (10d) actually equations like

$$\left\{\begin{aligned}\bar{x}_j^* &= \bar{x}_{j-1} + \Delta t\, f(\bar{x}_{j-1}, \bar{y}_{j-1}) + \sum_m \epsilon_{jm}\, r_{jm}^{(1)} \\ \bar{y}_j^* &= \bar{y}_{j-1} + \Delta t\, g(\bar{x}_{j-1}, \bar{y}_{j-1}) + \sum_m \epsilon_{jm}\, r_{jm}^{(2)}\end{aligned}\right. \tag{20c}$$

$$\left\{\begin{aligned}\bar{x}_j &= \bar{x}_{j-1} + \frac{\Delta t}{2}\left\{f(\bar{x}_{j-1}, \bar{y}_{j-1}) + f(\bar{x}_j^*, \bar{y}_j^*)\right\} + \sum_m \epsilon_{jm}\, r_{jm}^{(3)} \\ \bar{y}_j &= \bar{y}_{j-1} + \frac{\Delta t}{2}\left\{g(\bar{x}_{j-1}, \bar{y}_{j-1}) + g(\bar{x}_j^*, \bar{y}_j^*)\right\} + \sum_m \epsilon_{jm}\, r_{jm}^{(4)}\end{aligned}\right. \tag{20d}$$

where $|\epsilon_{jm}| = 0.5$ and where the $r_{jm}^{(k)}$ are certain coefficients, depending on the capacity of the counters of the machine and on the specific procedure to secure $f(\bar{x},\bar{y})$ from its arguments $\bar{x}$, $\bar{y}$ (that is, on the order and kind of operations, use of a function table in conjunction with interpolation and so on). In the first equation of (20c), for example, we have

$$\sum_m \epsilon_{jm}\, r_{jm}^{(1)} = \overline{\Delta t\, \overline{f(\bar{x}_{j-1}, \bar{y}_{j-1})}} - \Delta t\, f(\bar{x}_{j-1}, \bar{y}_{j-1}). \tag{21}$$

The determination of the $r_{jm}^{(k)}$ may be rather complicated and requires a detailed analysis of the programming of the calculation at hand. Since, however, we assume Δt to be a small quantity, we can single out one rather simple case. We assume that we can compute $f(\bar{x}_{j-1}, \bar{y}_{j-1})$ so accurately that the value $\overline{f(\bar{x}_{j-1}, \bar{y}_{j-1})}$ which the machine actually furnishes differs from it so little that the multiplication by Δt shifts the doubtful figures into those which are dropped in the last rounding-off step described in (21); in other words, that

$$\overline{\Delta t\, f(\bar{x}_{j-1}, \bar{y}_{j-1})} = \overline{\Delta t\, \overline{f(\bar{x}_{j-1}, \bar{y}_{j-1})}}. \tag{22}$$

In this case only one essential rounding-off remains to be taken into account and the first equation of (20c) becomes simply

$$\bar{x}_j^* = \bar{x}_{j-1} + \Delta t\, f(\bar{x}_{j-1}, \bar{y}_{j-1}) + \epsilon_{j1} 10^{-k},$$

if we keep k digits after the decimal point. Moreover, in the present discussion we are not interested in higher powers of Δt and we may thus replace the equations (20c) and (20d)

182

by the simpler difference equations

$$\begin{cases} \bar{x}_j = \bar{x}_{j-1} + 2\dfrac{\Delta t}{2} f(\bar{x}_{j-1}, \bar{y}_{j-1}) + 2\epsilon_{j1} 10^{-k} \\ \bar{y}_j = \bar{y}_{j-1} + 2\dfrac{\Delta t}{2} g(\bar{x}_{j-1}, \bar{y}_{j-1}) + 2\epsilon_{j2} 10^{-k}, \end{cases} \tag{23}$$

which we have to compare with

$$\begin{cases} x_j = x_{j-1} + \Delta t\, f(x_{j-1}, y_{j-1}) \\ y_j = y_{j-1} + \Delta t\, g(x_{j-1}, y_{j-1}), \end{cases} \tag{24}$$

which here may replace the more complicated Heun equations (10). (This simplification is nonessential and does not need to be made. The end results (28) and (29) are not affected by it.)

Since we are to compare the results of the difference equations (23) with those of (24) we consider

$$\bar{u}_j = x_j - \bar{x}_j,\ \bar{v}_j = y_j - \bar{y}_j.$$

From (23) and (24) follows

$$\begin{cases} \bar{u}_j = \bar{u}_{j-1} + \Delta t \left\{ f(x_{j-1}, y_{j-1}) - f(\bar{x}_{j-1}, \bar{y}_{j-1}) \right\} - 2\epsilon_{j1} 10^{-k} \\ \bar{v}_j = \bar{v}_{j-1} + \Delta t \left\{ g(x_{j-1}, y_{j-1}) - g(\bar{x}_{j-1}, \bar{y}_{j-1}) \right\} - 2\epsilon_{j2} 10^{-k} \end{cases}$$

or

$$\begin{cases} \bar{u}_j - \bar{u}_{j-1} = \Delta t \left\{ \dfrac{\partial f}{\partial x} \bar{u}_{j-1} + \dfrac{\partial f}{\partial y} \bar{v}_{j-1} \right\} - 2\epsilon_{j1} 10^{-k} \\ \bar{v}_j - \bar{v}_{j-1} = \Delta t \left\{ \dfrac{\partial g}{\partial x} \bar{u}_{j-1} + \dfrac{\partial g}{\partial y} \bar{v}_{j-1} \right\} - 2\epsilon_{j2} 10^{-k}. \end{cases} \tag{25}$$

We introduce, in analogy to (13), a system of adjoint difference equations

$$\begin{cases} \lambda_j - \lambda_{j-1} = -\Delta t \left\{ \dfrac{\partial f}{\partial x} \lambda_j + \dfrac{\partial g}{\partial x} \mu_j \right\} \\ \mu_j - \mu_{j-1} = -\Delta t \left\{ \dfrac{\partial f}{\partial y} \lambda_j + \dfrac{\partial g}{\partial y} \mu_j \right\} \end{cases} \tag{26}$$

whose solutions, for our present purpose, we may identify with $\lambda(t_j)$, $\mu(t_j)$ from (13). From (25) and (26) we infer

$$\begin{aligned} \lambda_j(\bar{u}_j - \bar{u}_{j-1}) + \mu_j(\bar{v}_j - \bar{v}_{j-1}) \\ = \Delta t \left\{ \frac{\partial f}{\partial x} \lambda_j + \frac{\partial g}{\partial x} \mu_j \right\} \bar{u}_{j-1} + \Delta t \left\{ \frac{\partial f}{\partial y} \lambda_j + \frac{\partial g}{\partial y} \mu_j \right\} \bar{v}_{j-1} \\ - 2\epsilon_{j1} 10^{-k} \lambda_j - 2\epsilon_{j2} 10^{-k} \mu_j \\ = -(\lambda_j - \lambda_{j-1}) \bar{u}_{j-1} - (\mu_j - \mu_{j-1}) \bar{v}_{j-1} \\ - 2\epsilon_{j1} 10^{-k} \lambda_j - 2\epsilon_{j2} 10^{-k} \mu_j, \end{aligned}$$

or

$$(\lambda_j\bar{u}_j - \lambda_{j-1}\bar{u}_{j-1}) + (\mu_j\bar{v}_j - \mu_{j-1}\bar{v}_{j-1})$$
$$= -2(\epsilon_{j1}\lambda_j + \epsilon_{j2}\mu_j)\,10^{-k}.$$

Summing the last equation over j from 1 to n we obtain

$$\lambda(T)\bar{u}(T) + \mu(T)\bar{v}(T) = -2\left\{\sum_{j=1}^{n}(\epsilon_{j1}\lambda_j + \epsilon_{j2}\mu_j)\right\}10^{-k}, \tag{27}$$

where we have observed $\bar{u}_0 = \bar{v}_0 = 0$ and have put $\bar{u}_n = \bar{u}(T)$, $\bar{v}_n = \bar{v}(T)$.

Equation (27) for the rounding-off errors $\bar{u}(T)$, $\bar{v}(T)$ corresponds to (17) for the truncation errors. Whereas in that case we had for the elementary errors the approximations (18) we have here only the knowledge

$$|\epsilon_{j1}| \leq 0.5, \; |\epsilon_{j2}| \leq 0.5.$$

From (27) we derive the following estimate for the maximum possible value of the rounding-off errors

$$|\lambda(T)\bar{u}(T) + \mu(T)\bar{v}(T)| \leq 10^{-k}\sum_{j=1}^{n}(|\lambda_j| + |\mu_j|) \tag{28}$$
$$\sim \frac{10^{-k}}{\Delta t}\int_{t_0}^{T}(|\lambda(t)| + |\mu(t)|)dt.$$

If, however, we can treat the ϵ_{j1} and ϵ_{j2} as random variables, we can write down the dispersion of (27)

$$\sigma^2(\lambda(T)\bar{u}(T) + \mu(T)\bar{v}(T)) = \left\{4\sum_{j=1}^{n}(\lambda_j^2 + \mu_j^2)\int_{-1/2}^{1/2}\epsilon^2 d\epsilon\right\}10^{-2k}$$
$$\sim \frac{10^{-2k}}{3\Delta t}\int_{t_0}^{T}[\lambda^2(t) + \mu^2(t)]\,dt,$$

and have, therefore, for the standard deviation of the rounding-off error

$$\sigma[\lambda(T)\bar{u}(T) + \mu(T)\bar{v}(T)]$$
$$\sim \frac{10^{-k}}{\sqrt{3}}(\Delta t)^{-1/2}\left[\int_{t_0}^{T}[\lambda^2(t) + \mu^2(t)]\,dt\right]^{1/2}. \tag{29}$$

Equations (28) and (29) show that the rounding-off errors grow with the number n of integration steps, as was to be expected. The statistical formula (29) which gives the better order of magnitude for the two formulas can, of course, be applied only if we take the ϵ_{j1} and ϵ_{j2} as statistically independent. When we look back to (23) for the origin of the rounding-off errors we can take statistical independence as assured if

$$\frac{\Delta t}{2}f(\bar{x}_{j-1}, \bar{y}_{j-1}) \text{ and } \frac{\Delta t}{2}f(\bar{x}_j, \bar{y}_j)$$

differ in the kth digit following the decimal point. This would mean

$$\frac{\Delta t}{2}\left|\frac{\partial f}{\partial x}(x_j - x_{j-1}) + \frac{\partial f}{\partial y}(y_j - y_{j-1})\right| > 10^{-k},$$

or

$$\frac{(\Delta t)^2}{2}\left|\frac{\partial f}{\partial x}x' + \frac{\partial f}{\partial y}y'\right| > 10^{-k},$$

or in view of (9)

$$(\Delta t)^2 > 2(10^{-k})|x''|^{-1}, \tag{30a}$$

and similarly

$$(\Delta t)^2 > 2(10^{-k})|y''|^{-1}, \tag{30b}$$

again a lower limitation for the size of the integration step.

For the programming of an integration of differential equation (9) on a high-speed numerical integrator, it will be necessary to decide first the size of the integration step Δt. We see from (19) that with smaller Δt, the truncation error will become smaller; however, according to (29), the probable rounding-off error will increase. An optimal Δt will be one for which we have about the same order of magnitude for the truncation and for the probable rounding-off errors or

$$(\Delta t)^{5/2} \sim 0.674D/|E|, \tag{31}$$

with

$$E = -\frac{1}{12}\Big[\lambda(t)x''(t) + \mu(t)y''(t)\Big]_{t_0}^{T} - \frac{1}{6}\int_{t_0}^{T}\Big[\lambda'(t)x''(t) + \mu'(t)y''(t)\Big]dt,$$

$$D = \frac{10^{-k}}{\sqrt{3}}\left[\int_{t_0}^{T}[\lambda^2(t) + \mu^2(t)]dt\right]^{1/2}.$$

RADEMACHER DISCUSSION

DR. STERNE: I should like to congratulate Dr. Rademacher for his very thorough and excellent analysis of the errors of all types that accumulate in the manner which he has so thoroughly studied in connection with the ENIAC. I may say that we realized some years ago at the Proving Ground that a study of this type was necessary.

You can always transform a system of equations into a system of first order equations. You can always form and solve to adequate accuracy the system adjoining the system of variational equations. You can always write down therefore the expressions for the accumulation both of round-off and of truncation errors.

We also carried on experiments to test the accuracy of the theory. You cannot work very well with a very accurate solution because the errors are so slow in building up. Therefore, we deliberately chose a very bad integration formula and introduced bad truncation errors. We found that repeated trials with very slightly different initial positions at which we started an integration, with the same type distribution of rounding-off errors, in the end gave the same distribution of the errors. I think that an investigation of this sort is essential for any extensive program of numerical integration of ordinary differential equations in order to carry out the work efficiently, and it appears that the type of equation has a marked effect on the outcome.

Some equations are of such a character that a small error introduced at any stage tends to be reduced in much the same way as the velocity or energy is diminished. The equations of ballistics for the motions of projectiles are of that type.

PROFESSOR RADEMACHER: May I make one remark. I think it is of interest to note that the rounding-off error is always of the order $(\Delta t)^{-\frac{1}{2}}$ irrespective of the order of the differential to be considered. Also with the choice method the truncation error is always of the same order. However, a coefficient which I wrote down explicitly will depend not only on the functions involved, but also on the order of the differential equation. They will look somewhat similar but slightly more involved. But the order is once more the same: $(\Delta t)^2$ for the truncation error, and $(\Delta t)^{-\frac{1}{2}}$ independent of the order or number of the ordinary differential equations.

DR. MAUCHLY: I just wanted to remark that this paper is probably one of the first to appear as a result of the interaction between computing machines and mathematics. I think many of us have been looking forward with certainty to the day when the advent of computing machines would influence the course of mathematics, and there are still many directions in which we can look forward to further such impacts. This reaction is not, of course, one way. Mathematics should also influence the computing machine.

I might say that historically, the role of mathematics was the original source of a particular problem; we had hoped to draw from the mathematicians some information as to the proper number of digits to be provided in the panels of the ENIAC. Accordingly, when we started, we observed that the adjoint equation treatment would be suitable for this problem and we asked Dr. Rademacher if he would be good enough to make a study that would enable us to set limits upon the size of the accumulators. Actually, we have obtained some experimental results, in a very crude way, which also have a bearing on this problem of rounding off and truncation error.

Long before the final ENIAC was completed, we had set up a few accumulators which were what you might call pilot models. About the only thing we could do with them was set up a second order differential equation with constant coefficients. We did this, and ran off sines and cosines which we observed visually (not having any printing mechanism associated with the ENIAC at that time).

One of the peculiarities that showed up was that the simplest method by which we could set up such a differential equation turned out to be the one in which the truncation error contained no exponential term — no term which differed more and more from the true solution. In other words, the truncation error was psychic.

A second development of probably greater interest was that we found the possibility (even with such simple apparatus) of controlling the calculations so as to stop on the occurrence of any particular digit in any particular column of the calculation. We did this more or less as a stunt and were very much surprised at the result which we obtained, which was that if we picked a column that presumably was almost entirely affected by rounding-off errors, one in which you might hope to find random numbers, and asked the machine to stop whenever a nine occurred in that column, the machine never stopped under the particular setup.

It is probable that all the other statistical theorems which you might want to use would continue to be valid. I cannot say for sure because we did not carry the experiments that far, but at least it was amusing that no nine ever occurred in this column which was on the borderline of the significant figures.

There is one other remark I might make about this adjoint equation treatment. It has always appeared to me that there ought to be some extension of this treatment, possibly to those fields which the economists have tried sometimes to treat using differential equations. Some years ago, I believe, someone wrote a paper on differential equations whose coefficients are subject to errors. Although I do not recall exactly the methods, it would appear that they were similar to those that Dr. Rademacher has used.

Note to Paper 50

This paper does not appear to have been reviewed in either the Z or the MR. A typewritten version of the paper was distributed by the National Bureau of Standards. The NBS version omitted the discussion (p. 185), but instead appended the following "Note on Rounding-Off Errors" by George E. Forsythe:

When ordinary (Gaussian) rounding off is used in the numerical integration of differential equations, the individual rounding-off errors ϵ are really not random variables—they are unknown constants. This fact is probably generally accepted in principle, but many persons have assumed that in a practical computation the distribution of values of the constants ϵ is close to that of an equal number of identical random variables independently and uniformly distributed on $(-.5, .5)$. If this assumption were correct, the ϵ's could be treated formally as independent random variables, and Professor Rademacher's estimates would apply to the ordinary rounding-off procedure.

In Dr. Huskey's integrations of the system $\dot{x} = y, \dot{y} = -x$ on the ENIAC,[1] the successive values of ϵ were found to be far from independent. In most intervals of the integration the ϵ's were so regularly distributed that the accumulated error was considerably less than that for independent random variables. In certain intervals, however, the ϵ's had a biased distribution which caused unexpectedly large accumulations of the rounding-off error. Following the lead of Professor Hartree,[1] the present writer (in a unpublished paper) has explained these phenomena and has shown how to predict the rounding-off errors for the system $\dot{x} = y, \dot{y} = -x$ (and to a certain extent for other systems). It seems clear that in the integration of smooth functions the ordinary rounding-off errors ϵ will seldom, if ever, be distributed like independent random variables.

To circumvent this difficulty the present writer[2] has proposed a *random rounding-off* procedure which makes ϵ a true random variable. Suppose, for example, that a real number u is to be rounded off to an integer. Let [u] be the greatest integer not exceeding u, and let $u - [u] = v$. In the proposed procedure u is "rounded up" to $[u] + 1$ with probability v, and "rounded down" to [u] with probability $1 - v$, the choice being made by some independent chance mechanism. The rounding-off error ϵ is thus a random variable with $E(\epsilon) = 0$ and $E(\epsilon^2) = v(1 - v)$. Since $v(1 - v) \leqslant 1/4$, one can give probabilistic bounds for the accumulated error which are independent of the distribution of v. The method is reasonably suited to machine computation: Instead of adding a 5 in the most significant position of the digits to be dropped (ordinary rounding off), one adds a random decimal digit to each of the digital positions to be dropped. As with ordinary rounding off, the addition carry-over determines whether the rounding off is "up" or "down."

Although Professor Rademacher's theory is generally inapplicable to ordinary rounding off, it is believed to be perfectly applicable to random rounding off. It is necessary only to replace the ordinary *approximation* $E(\epsilon^2) = 1/12$ by the *bound* $E(\epsilon^2) \leqslant 1/4$.

---

[1] Harry D. Huskey, "On the Precision of a Certain Procedure of Numerical Integration," *National Bureau of Standards Journal of Research,* vol. 42 (1949), pp. 57-62. Professor Hartree wrote an appendix explaining the worst rounding-off errors.

[2] Abstract: "Round-off errors in numerical integration on automatic machinery," *Bull. Amer. Math. Soc.,* vol. 56 (1950), p. 61.

COURANT
Anniversary Volume
1948

ON A THEOREM OF FROBENIUS

Hans Rademacher

Let γ be a set of commutative matrices of degree n with elements of an algebraically closed field Φ. The following theorem is due to Frobenius:[1)] If A,B,C,... are matrices of γ, then the eigen-values $\alpha^{(1)}, \alpha^{(2)}, \alpha^{(3)},\ldots$ of A, $\beta^{(1)}, \beta^{(2)}, \beta^{(3)},\ldots$ of B, $\gamma^{(1)}, \gamma^{(2)}, \gamma^{(3)},\ldots$ of C and so forth can be so arranged that for any rational function $f(x,y,z,\ldots)$ in Φ the eigen-values of $f(A,B,C,\ldots)$ are $f(\alpha^{(1)}, \beta^{(1)}, \gamma^{(1)},\ldots)$, $f(\alpha^{(2)}, \beta^{(2)}, \gamma^{(2)},\ldots)$, $f(\alpha^{(3)}, \beta^{(3)}, \gamma^{(3)},\ldots)$ and so forth.

The proof is usually based on a study of the characteristic equation of $xA + yB + zC + \ldots$, where x,y,z,... are variables. This provides also the special correspondence between the eigen-values of the different matrices which the theorem mentions.

A different proof which is short and simple can be obtained by taking into account the eigen-vectors (which appear only incidentally at the end of Frobenius' paper). The eigen-values of the same index belong to a common eigen-vector, as Frobenius had already observed.

Roman capital letters designate matrices of degree n, small German letters mean vectors, e.g. $\mathfrak{x} = (x_1, x_2, \ldots x_2)$. Small Greek letters stand for numbers of the field Φ.

Let α be an eigen-value of A. Let us assume first that $A-\alpha E$

---

1) G. Frobenius, Über vertauschbare Matrizen, Berliner Sitzungsberichte 1896, pp. 601-614.

is precisely of rank n-1. Then there exists a non-vanishing vector $\mathfrak{x}=(x_1,x_2,\dots x_n)$ such that

$$A\mathfrak{x}=\alpha\mathfrak{x} \tag{1}$$

and all other solutions of (1) are obtained from $\mathfrak{x}$ by multiplying it by a number.

Let B be any other matrix of γ. Then

$$BA\mathfrak{x} = B\alpha\mathfrak{x}$$

and because of the commutativity within the set γ

$$AB\mathfrak{x} = \alpha B\mathfrak{x},$$

which shows that $B\mathfrak{x}$ is also an eigen-vector of A to the eigen-value α and therefore in our case obtained from $\mathfrak{x}$ through multiplication by a certain number β

$$B\mathfrak{x} = \beta\mathfrak{x}$$

Thus $\mathfrak{x}$ is a common eigen-vector for all matrices of γ.

If, however, $A-\alpha E$ is of rank lower than n-1 then the non-trivial solutions $\mathfrak{x}$ of (1) form a linear set of higher dimension than 1. Let $\mathfrak{x}^{(1)},\dots,\mathfrak{x}^{(f)}$ be a basis of this set:

$$A\,\mathfrak{x}^{(j)} = \alpha\mathfrak{x}^{(j)}, \qquad j=1,\dots f. \tag{2}$$

As before $B\,\mathfrak{x}^{(j)}$ must belong to that linear set

$$B\,\mathfrak{x}^{(j)} = \sum_{k=1}^{f} b_{jk}\,\mathfrak{x}^{(k)} \qquad j = 1,\dots f.$$

Now let β be an eigen-value of the matrix $B^* = (b_{jk})$ of degree f, then there exists $\mathfrak{y}=(v_1,\dots,v_f)$ of dimension f such that

$$\sum_{j=1}^{f} b_{jk}\,v_j = \beta v_k \qquad k = 1,\dots f.$$

Then with

$$\mathfrak{y} = \sum_{j=1}^{k} v_j\,\mathfrak{x}^{(j)}$$

we have, firstly, from (2)

$$A\mathfrak{y} = \alpha\mathfrak{y}$$

and, secondly, we have

$$B\mathfrak{y} = B\sum_{j=1}^{f} v_j \mathfrak{c}^{(j)} = \sum_{j=1}^{f} v_j B\mathfrak{c}^{(j)}$$

$$= \sum_{j=1}^{f} v_j \sum_{k=1}^{f} b_{jk}\mathfrak{c}^{(k)} = \sum_{k=1}^{f} \mathfrak{c}^{(k)} \sum_{j=1}^{f} b_{jk}v_j$$

$$= \sum_{k=1}^{f} \mathfrak{c}^{(k)} \beta v_k = \beta\mathfrak{y}$$

Thus we have found an eigen-vector $\mathfrak{y}$ which A and B have in common, $\mathfrak{y}$ belonging, respectively, to the eigen-values α and β.

Now the set of all common eigen-vectors of A and B to the eigen-values α and β form again a linear set, and the process can be repeated for another matrix C of γ. Since there is only a finite number of linearly independent matrices in γ the process comes to an end after finitely many steps. We have the result:

There exists an eigen-vector $\mathfrak{z}$ of all matrices of γ. It produces a correspondence between the eigen-values $\alpha, \beta, \gamma, \ldots$ of A,B,C,... to which it belongs.

Since now the equations

$$A\mathfrak{z} = \alpha\mathfrak{z}, \qquad B\mathfrak{z} = \beta\mathfrak{z}$$

imply

$$(\lambda A + \mu B)\mathfrak{z} = (\lambda\alpha + \mu\beta)\mathfrak{z},$$

$$AB\mathfrak{z} = A\beta\mathfrak{z} = \alpha\beta\mathfrak{z}$$

and for $\alpha \neq 0$, i.e. for A non-singular,

$$A^{-1}\mathfrak{z} = \alpha^{-1}\mathfrak{z},$$

we infer immediately that if f(x,y,z,...) is a rational function of x,y,z... in the field Φ then the matrix f(A,B,C,..) has the eigen-value $f(\alpha, \beta, \gamma, \ldots)$ corresponding to the eigen- vector $\mathfrak{z}$.

We replace now the symbols $\mathfrak{z}, \alpha, \beta, \gamma, \ldots$ by $\mathfrak{z}^{(1)}, \alpha^{(1)}, \beta^{(1)}, \gamma^{(1)}, \ldots$ and look for other eigen-vectors.

If $A - \alpha^{(1)}E$ is of rank lower than $n-1$ then the linear set of eigen-vectors of A belonging to $\alpha^{(1)}$ is not exhausted by $\lambda\mathfrak{z}^{(1)}$. The eigen-vectors of A orthogonal to $\mathfrak{z}^{(1)}$ and belonging to $\alpha^{(1)}$ form again a linear set. We can then proceed as before and find among them a common eigen-vector $\mathfrak{z}^{(2)}$ of all matrices of Γ, belonging to certain eigen-values $\beta^{(2)}, \gamma^{(2)}, \ldots$. We put $\alpha^{(2)} = \alpha^{(1)}$ in this case and have a new string of eigen-values $\alpha^{(2)}, \beta^{(2)}, \gamma^{(2)}, \ldots$ for which again we state that $f(A,B,C,\ldots)$ has the eigen-value $f(\alpha^{(2)}, \beta^{(2)}, \gamma^{(2)}, \ldots)$.

If $\lambda\mathfrak{z}^{(1)} + \mu\mathfrak{z}^{(2)}$ does not yet exhaust all eigen-vectors of A to $\alpha^{(1)} = \alpha^{(2)}$ we discuss the linear set of those eigen-vectors of A belonging to $\alpha^{(1)} = \alpha^{(2)}$ and orthogonal to $\mathfrak{z}^{(1)}$ as well as $\mathfrak{z}^{(2)}$ and apply our method again.

After having exhausted the eigen-vectors of A belonging to $\alpha^{(1)}$ we go over to another eigen-value of A, provided there exists one.

When we have gone through all eigen-values and eigen-vectors of A, we must have also exhausted the bases of all linear sets of eigen-vectors of any other matrix M of Γ, because starting from M we could have found the eigen-vectors and eigen-values of A.

The whole procedure is of course finite, as we may infer from the characteristic equation. But we can leave the characteristic equation aside, since we can directly prove that the eigen-vectors which our procedure furnishes are linearly independent.

Suppose we had a linear relation

$$\sum_{j=1}^{k} \lambda^{(j)}\mathfrak{z}^{(j)} = 0$$

where we can take all α's different from zero, if we only possibly change the numeration of the eigen-vectors. As a result of

$$A\mathfrak{z}^{(j)} = \alpha^{(j)}\mathfrak{z}^{(j)}$$

we would get

$$0 = A\sum_{j=1}^{k} \lambda^{(j)}\mathfrak{z}^{(j)} = \sum_{j=1}^{k} \lambda^{(j)}\alpha^{(j)}\mathfrak{z}^{(j)}$$

and by iteration in general

$$(3) \qquad \sum_{j=1}^{k} \lambda^{(j)}\alpha^{(j)^{r}}\mathfrak{z}^{(j)} = 0, \quad r = 0,1,2,\ldots$$

Let $\alpha_1, \alpha_2, \ldots \alpha_\ell$ be the different ones among the eigen-values $\alpha^{(j)}$ If in (3) we bracket together those summands which contain equal eigen-values, we would have a system of equations, $r = 0,1,\ldots \ell-1$, with a non-vanishing Vandermonde determinant as its determinant. That would mean that some or all eigen-vectors belonging to, say, α_1 would be linearly dependent, which is impossible, since they were chosen as mutually orthogonal.

We have therefore found at most n strings of eigen-values of Γ, each string identified by a common eigen-vector. These strings of eigen-values $\alpha^{(1)}, \beta^{(1)}, \gamma^{(1)}, \ldots, \alpha^{(2)}, \beta^{(2)}, \gamma^{(2)}, \ldots, \alpha^{(m)}, \beta^{(m)}, \gamma^{(m)}, \ldots$, fulfill Frobenius' theorem.

Notes to Paper 51

This paper has been reviewed in the Z, vol. 33 (1950), p. 97, and in the MR, vol. 9 (1948), p. 264. The paper appeared in the *Courant Anniversary Volume* (Interscience, New York, 1948), and is reprinted by permission. The copy contained in Rademacher's reprint collection bore the penciled remark, "This paper contains an error. See review by Wielandt, Mathem. Zentralblatt 33 (1950), p. 97."

HELLY'S THEOREMS ON CONVEX DOMAINS AND TCHEBYCHEFF'S APPROXIMATION PROBLEM

HANS RADEMACHER AND I. J. SCHOENBERG

1. Introduction. Professor Dresden called to our attention the following theorem:[1]

If $S_1, S_2, \ldots, S_m$ are m line segments parallel to the y-axis, all of equal lengths, whose projections on the x-axis are equally spaced, and if we assume that a straight line can be made to intersect every set of three among these segments, then there exists a straight line intersecting all the segments.

This theorem was conjectured by M. Dresher; a first proof, unpublished, was communicated to us by T. E. Harris. Wide generalizations are possible. Dr. Harris noticed that we can dispense with the equidistance of the lines carrying our segments. We shall see in a moment that the equality of the lengths of the segments is likewise a superfluous assumption. A further generalization, also due to Dr. Harris, is as follows: The intersecting straight lines can be replaced by general parabolic curves

$$y = a_0x^n + a_1x^{n-1} + \ldots + a_n \qquad (n \leqslant m - 2); \tag{1}$$

again, if each set of $n + 2$ among our segments can be cut by such a parabola, then all may be simultaneously intersected by one such curve.

In this note we wish to point out the close connection of this problem, and of the more general problem of best approximation in the sense of Tchebycheff, with two remarkable theorems on convex domains, due to E. Helly, which may be stated as follows;[2]

THEOREM 1 (Helly). *If $C_1, C_2, \ldots, C_m$ is a finite collection of convex sets, which need not be closed or bounded, in the n-dimensional Euclidean space E_n $(m \geqslant n + 1)$, such that every $n + 1$ among the sets have a common point, then all m sets have a common point.*

THEOREM 2 (Helly). *Let $\{D\}$ be an infinite collection of closed and convex sets D, which need not be bounded, in E_n, such that every $n + 1$ among the sets have a common point. Then all the sets D have a common point, provided there exists a finite subcollection $D', D'', \ldots, D^{(k)}$, $(k \geqslant 1)$, of elements of $\{D\}$, such that their intersection $\Delta = D'D'', \ldots, D^{(k)}$ is non-void and bounded.*

Let us first see how very directly the Dresher-Harris theorem may be derived from Helly's Theorem 1. Let

$$S_\nu:\ x = x_\nu,\ b_\nu \leqslant y \leqslant c_\nu \qquad (\nu = 1, \ldots, m;\ \ x_1 < x_2 < \ldots < x_m,\ \ m \geqslant 3),$$

Received March 22, 1949.

[1] See [8], p. 4, where the theorem is stated without proof.

[2] See [4], [6], and [7].

be the m segments of the theorem. Consider the totality of lines $y = a_0x + a_1$ intersecting the νth segment S_ν; the requirement of intersection is expressed by the inequalities

$$b_\nu \leqslant a_0x_\nu + a_1 \leqslant c_\nu. \tag{2}$$

In the plane E_2 of the variables (a_0,a_1), the double inequality (2) defines a parallel strip of slope $-x_\nu$. This strip, which we denote by C_ν, is certainly a convex set in E_2. Let us now consider the collection of m convex sets $C_1, C_2, \ldots, C_m$, corresponding to the segments $S_1, S_2, \ldots, S_m$. From the assumption of the Dresher-Harris theorem we know that every three among these sets have a common point. Since all assumptions af Helly's Theorem 1 are satisfied in E_2, we may conclude that all m sets C_ν have a common point, hence all m segments S_ν are intersected by a line. This proof clearly extends to the case of the parabolic curves (1) by applying Helly's Theorem 1 in the space E_{n+1} of the variables $(a_0, a_1, \ldots, a_n)$.

As a second example of the versatility of Helly's ideas we shall again use Theorem 1 to give a new proof of the following separation theorem.[3]

THEOREM OF PAUL KIRCHBERGER. *Let $S = \{P\}$ and $S' = \{P'\}$ be two finite sets of points in E_n. We shall say that a hyperplane π separates strictly S from S', if all points of S are on one side of π, while all points of S' are on the other side, with none of the points lying on π. Such a strictly separating plane π exists if and only if the following condition is satisfied: For every set T of $n + 2$ points chosen arbitrarily from S and S', there should exist a hyperplane π_T which separates strictly the S-points of T from the S'-points of T.*

The necessity of the condition is obvious; to prove its sufficiency let us assume that it is satisfied and prove the existence of a strictly separating plane. We introduce in E_n a coordinate system $(x_1, \ldots, x_n)$. In the space E_{n+1} of the variables $(a_1, a_2, \ldots, a_{n+1})$, and corresponding to each point $P = (x_1, \ldots, x_n)$ of S, we define an open half-space H_P by the inequality

$$H_P:\ a_1x_1 + a_2x_2 + \ldots + a_nx_n + a_{n+1} > 0. \tag{3}$$

Likewise, corresponding to each point $P' = (x'_1, x'_2, \ldots, x'_n)$ of S', we define in E_{n+1} an open half-space $H_{P'}$ by the inequality

$$H_{P'}:\ a_1x'_1 + a_2x'_2 + \ldots + a_nx'_n + a_{n+1} < 0. \tag{4}$$

[3]See [5], where a proof of this theorem requires nearly 24 pages. The theorems of Kirchberger and Dresher-Harris are not unrelated. The following new generalization of the Dresher-Harris theorem indicates the connection: *Let S be a finite set of points $P_i = (x_i,y_i)$ in the plane and let S' be a second set of points $P'_j = (x'_j,y'_j)$. We say that a line $y + a_0x + a_1$ separates the sets S and S', if $y_i \geqq a_0x_i + a_1$ for all points of S, and $y'_j \leqq a_0x'_j + a_1$ for all points of S'. There exists a line $y = a_0x + a_1$ separating S from S' if and only if the following condition is satisfied: For every set T of three points chosen from $S + S'$ there should exist a line separating the S-points of T from the S'-points of T.* We obtain the Dresher-Harris theorem as a special case of this theorem if we take S to be the set of upper endpoints of the segments S_ν, while S' is the set of their lower endpoints. A proof of this generalization by means of Theorem 1 is obvious and so is its extension to parabolic curves (1).

In terms of the finite collection $\{H_P\} + \{H_{P'}\}$ of open half-spaces of E_{n+1}, Kirchberger's condition means that every $n + 2$ among these convex half-spaces have a common point. By Helly's Theorem 1 we conclude that there is a point $(a_1, \ldots, a_{n+1})$, with $\sum_1^n |a_\nu| > 0$, common to all of these half-spaces.[4] The corresponding plane $a_1x_1 + \ldots + a_nx_n + a_{n+1} = 0$ separates strictly S from S', and the proof is concluded. That Kirchberger's theorem becomes false if the number $n + 2$ is replaced by $n + 1$ is seen by the example of the set S being the set of $n + 1$ vertices of a simplex, while S' has only one element namely the centroid of the simplex. These two sets S, S' cannot be separated, though the points of S and S' occurring in any $(n + 1)$-tuple can be separated. We also wish to remark that the theorem becomes false if the sets S, S' are allowed to be infinite. Indeed, if in E_2 we take S to be the exponential curve $y = \exp x$ while S' is the x-axis, then clearly every $n + 2 = 4$ points of $S + S'$ can be strictly separated by a line, but not the sets S, S'.

Concerning Kirchberger's theorem the following remark is of interest. Let us replace the "strict separation" of the theorem by "separation" in the weaker sense that points of S or S' are also allowed to lie *on* the separating plane π. We may then state the following proposition;

Kirchberger's theorem in E_n remains true if in its statement "strict separation" is replaced by "separation" in the above wider sense, provided we replace in the theorem's condition the number $n + 2$ by $2n + 2$. Also no number smaller than $2n + 2$ will do. Moreover the sets S and S' may now also be infinite.

In order to prove this new result let us define in E_{n+1}, as we did above, the collection of *closed* half-spaces

$$\overline{H}_P:\quad a_1x_1 + \ldots + a_nx_n + a_{n+1} \geqslant 0, \text{ for } P = (x_1, \ldots, x_n) \in S, \tag{5}$$

$$\overline{H}_{P'}:\quad a_1x'_1 + \ldots + a_nx'_n + a_{n+1} \leqslant 0, \text{ for } P' = (x'_1, \ldots, x'_n) \in S'. \tag{6}$$

We wish to prove the existence of a point $(a_1, \ldots, a_{n+1})$, with $\sum_1^n |a_\nu| > 0$, which is common to *all* of these half-spaces. Helly's theorem does not help us here any more, but we can apply the following remarkable theorem of L. L. Dines and N. H. McCoy:[5]

A finite or infinite collection of closed half-spaces in E_{n+1}, each half-space having the origin 0 *on its boundary, do have a common point different from* 0, *if every set of $2n + 2$ among our half-spaces have a common point different from* 0.

This theorem assures the existence of a point $(a_1, \ldots, a_{n+1}) \neq (0, \ldots, 0)$ such that the inequalities (5) and (6) hold for all $P \in S$, and all $P' \in S'$,

[4]Actually we know only that $\sum_1^{n+1} |a_\nu| > 0$. However, all the points of sufficiently small spherical neighbourhood of the point $(a_0, \ldots, a_{n+1})$ likewise satisfy all conditions and among them we can certainly find one for which $\sum_1^n |a_\nu| > 0$.

[5]See [3], pp. 61-63; see also [2], pp. 962-963, where there are also references to a paper by C. V. Robinson.

respectively. This would mean that $a_1x_1 + \ldots + a_nx_n + a_{n+1} = 0$ is a separating hyperplane as soon as we know that $\sum_1^n |a_\nu| > 0$. This last point, however, is clear, for $a_1 = \ldots = a_n = 0$ and (5) and (6) would imply $a_{n+1} \geqslant 0$, $a_{n+1} \leqslant 0$, hence $a_{n+1} = 0$, which is impossible.

The following example shows that the number $2n + 2$ of the new version of Kirchberger's theorem may not be replaced by $2n + 1$: Let S consist of the $n + 1$ vertices $P_1, \ldots, P_{n+1}$ of simplex σ, and let S' consist of the same $n + 1$ points $P'_1, \ldots, P'_{n+1}$, with $P'_\nu = P_\nu$. Choosing $2n + 1$ points of $S + S'$ amounts to leaving out P_ν, or perhaps P'_ν. The remaining $2n + 1$ points are clearly separated by the $(n - 1)$-dimensional face of the simplex σ which is opposite to the vertex P_ν. Hence the conditions of the theorem are verified for every set of $2n + 1$ points, while there is no hyperplane π separating S from S'.

The connection of Helly's theorems with the idea of Tchebycheff approximation, i.e. the consideration of the minimum of a maximum, suggested to us a new proof[6] which we claim to be the first proof of Helly's theorems to be entirely geometric, in the sense that every single one of its steps has an intuitive geometric meaning. This proof is given in the first part of the paper. The second and last part is devoted to an application of Helly's theorems to Tchebycheff's approximation problem.

A New Proof of Helly's Theorems

2. On proximity points of convex domains. We shall see that the main point in proving Helly's theorems is to prove Theorem 1 for the special case when the convex sets $C_1, \ldots, C_m$ are also *closed* and *bounded*. A closed and bounded convex set in E_n will be referred to as a *convex domain*. Let $D_1, D_2, \ldots, D_m$ be such convex domains. If $Q \in E_n$, we denote by $d(Q,D_\nu)$ the distance from the point Q to the domain D_ν. The point function

$$f(Q) = \max_\nu d(Q,D_\nu)$$

is evidently non-negative and continuous throughout E_n. Since $f(Q) \to \infty$ as $Q \to \infty$, the function $f(Q)$ assumes somewhere its absolute minimum value.

Definition. *An absolute minimum point P, of $f(Q)$, will be called proximity point of our domains $D_1, \ldots, D_m$. It has the property*

$$\max_\nu d(P,D_\nu) = \min_Q \max_\nu d(Q,D_\nu).$$

[6]Three earlier proofs have come to our attention: By J. Radon [7], E. Helly [4], and D. König [6]. Radon's proof, which is the shortest, is analytic. The proofs by Helly and König, essentially equivalent to each other, are geometric. However, all three proofs use the method of mathematical induction, a fact which seems to obscure the intuitive background of the results. Our proof uses the *metric* of E_n and is therefore related to the ideas of Menger and Blumenthal (see [2]).

The minimal value $f(P) = \min f(Q)$ *will be called the proximity of these domains and denoted by the symbol* Prox $(D_1, \ldots, D_m)$.

Evidently we have Prox $(D_1, \ldots, D_m) = 0$ if and only if our m domains have a point in common and if this is the case, any point of their intersection is a proximity point.[7]

3. A characteristic property of proximity points.[8] The following theorem expresses a fundamental property of proximity points.

THEOREM 3. *Let* $D_1, \ldots, D_m$, $(m \geqslant 2)$, *be convex domains in* E_n *having no common point. Let P be a proximity point of these domains, the proximity* $\rho =$ Prox $(D_1, \ldots, D_m)$ *being necessarily positive. Let* $P_\nu \in D_\nu$ *be such that* $PP_\nu = d(P,D_\nu)$, *hence we have* $\rho = \max PP_\nu$. *Then there are s among the m normals* PP_ν *from P to our domains,* $PP_1, PP_2, \ldots, PP_s$, *say, such that*

(i) $2 \leqslant s \leqslant n+1$.

(ii) $PP_1 = PP_2 = \ldots = PP_s = \rho$.

(iii) *The points* $P_1, P_2, \ldots, P_s$ *are the vertices of a* $(s-1)$*-simplex* σ, *which simplex contains the point P in its* $(s-1)$*-dimensional interior.*

(iv) *The corresponding s domains* $D_1, D_2, \ldots, D_s$ *have no common point.*

The last conclusion is for us the important one. In fact we shall use this theorem only in the following abbreviated form:

COROLLARY. *If m convex domains, of* E_n, *have no common point, then some s among these domains have no common point, where* $2 \leqslant s \leqslant n+1$.

Proof of Theorem 3. Suppose that

$$PP_1 = PP_2 = \ldots = PP_h = \rho, \quad PP_{h+1} < \rho, \ldots, PP_m < \rho \qquad (h \leqslant m).$$

Clearly $h \geqslant 2$; for if $h = 1$, then P could not be a proximity point. Indeed, then $\max d(P,D_\nu)$ could be diminished below its present value ρ by moving P slightly along PP_1 towards P_1.

Consider now the convex hull of the points $P_1, \ldots, P_h$, which we denote by

$$K = K(P_1, \ldots, P_h).$$

We claim that $P \in K$, for otherwise let PP' be the shortest distance from P to K; we could then, again as before, diminish *all* distances $PP_\nu = d(P,D_\nu)$, $(\nu = 1, \ldots, h)$, by moving P slightly along PP' towards P'. Hence indeed $P \in K(P_1, \ldots, P_h)$.

We shall now use the following known result:[9] *If P is a point of the convex*

[7]As illustrations of the notion of a proximity point we mention the following two propositions of elementary geometry: *Let A, B, C, be the vertices of an acute-angled triangle in the plane. The proximity point of the three points A,B,C, is the circumcenter of the triangle. The proximity point of the three segments BC, CA, AB, is the incenter.*

[8]The properties (i), (ii), and (iii) of the point P, as described in Theorem 3 are indeed *characteristic* for a proximity point, a fact which we mention without proof because we do not use it.

[9]See [1], Satz IX on p. 607. This theorem is easily derived from a well known result of Caratheodory to the effect that every point of $K(P_1, \ldots, P_h)$ is a centroid with positive masses of at most $n+1$ points among the P_ν.

hull $K(P_1, \ldots, P_h)$, *then either P coincides with one of the points* P_ν, *or else we can find a simplex* σ, *of dimension* $s - 1$ *ranging from* 1 *to at most n, having as vertices only points from among the points* P_ν, *and such that P is in the* $(s-1)$-*dimensional interior of* σ. Returning to our proximity point P, we remark that P cannot possibly coincide with any of the points P_ν, since $PP_\nu = \rho > 0$, $(\nu = 1, \ldots, h)$. Therefore the above result assures us of the existence of a simplex of vertices $P_1, P_2, \ldots, P_s$, say, satisfying the conditions (i), (ii), and (iii), of Theorem 3.

There remains to prove the fourth and last statement of the theorem to the effect that $D_1 D_2, \ldots, D_s = \phi$. We consider the s unit vectors

$$\vec{a}_i = \overrightarrow{PP_i}/PP_i \qquad (i = 1, \ldots, s),$$

and the s half-spaces H_i defined by

$$H_i\colon\ \overrightarrow{PQ} \cdot \vec{a}_i \geqslant \rho \qquad (i = 1, \ldots, s). \tag{7}$$

Since $D_i \subset H_i$, it is sufficient to show that

$$H_1 H_2 \ldots H_s = \phi. \tag{8}$$

Suppose (8) were false and let $Q \in H_i$, $(i = 1, \ldots, s)$; then all inequalities (7) hold. However, since P is in the interior of σ, we have a vector relation of the form

$$\sum_1^s \kappa_i \vec{a}_i = \vec{0}, \text{ with all } \kappa_i > 0. \tag{9}$$

But then, on multiplying (9) scalarly by $\overrightarrow{PQ}$, in view of (7), we obtain

$$0 = \sum \kappa_i (\overrightarrow{PQ} \cdot \vec{a}_i) \geqslant \sum \kappa_i \rho = \rho \sum \kappa_i,$$

which clearly contradicts the positivity of ρ and κ_i. This completes our proof.

4. A proof of Helly's Theorem 1 concerning a finite collection of convex sets. We distinguish two cases.

First case: We assume that the m convex sets C_ν of the theorem are also *closed* and *bounded*, an assumption which we emphasize by writing $C_\nu = D_\nu$, $(\nu = 1, \ldots, m)$. This case is now immediately disposed of, for if we assume to the contrary that our m convex domains D_ν have no common point, then, by the Corollary of Theorem 3, some s among them $(2 \leqslant s \leqslant n + 1)$, have no point in common, a fact which contradicts the assumption of Theorem 1 to the effect that every $n + 1$ domains have a common point.

Second case. We assume that the C_ν are convex sets which need not be closed or bounded. By assumption every combination $C_{i_0}, C_{i_1}, \ldots, C_{i_n}$ of $n + 1$ distinct sets have a common point. Let such a point be $A_{i_0, i_1, \ldots, i_n}$ and let it be regarded as a symmetric function of its $n + 1$ distinct subscripts. Corresponding to each C_i we now define the convex domain

$$D_i = K(A_{i, j_1, \ldots, j_n})$$

which is defined as the convex hull of the $\binom{m-1}{n}$ points $A_{i,\ j_1,\ \ldots,\ j_n}$ where $j_1, \ldots, j_n$ runs over all combinations of n among the $m-1$ numbers $1, \ldots, i-1, i+1, \ldots, m$. Since $A_{i,j_1,\ldots,j_n} \in C_i$, we have

$$D_i \subset C_i, \tag{10}$$

because C_i is convex. Every set of $n+1$ among these domains, $D_{i_0}, D_{i_1}, \ldots, D_{i_n}$, say, have a point in common, namely the point $A_{i_0,\ i_1,\ \ldots,\ i_n}$. By the *first case* already established we conclude that all D_i have a common point. In view of (10) we now obtain the desired conclusion to the effect that the sets C_i have a point in common.

5. A proof of Helly's Theorem 2 concerning an infinite collection of closed convex sets. Let $\{D\}$ be the given infinite collection of closed convex sets. By Theorem 1 we know that the elements of every finite subcollection of $\{D\}$ have a common point. Consider the new collection $\{D^* = \Delta D\}$, where Δ is the non-void and bounded set defined in the statement of Theorem 2, while D ranges over the given collection $\{D\}$. The elements of $\{D^*\}$ have the following properties;

(i) They are closed, bounded and *non-void* convex sets.

(ii) The elements of every finite set of D^*'s have a common point. The desired conclusion to the effect that all the D^*'s, and therefore also all the D's, have a common point now follows from the following general Theorem:

THEOREM OF F. RIESZ:[10] *If a collection $\{A\}$ of bounded and closed sets in E_n has the property that the elements of every finite subcollection have a common point, then all A's have a common point.*

AN APPLICATION OF HELLY'S THEOREM TO TCHEBYCHEFF'S APPROXIMATION PROBLEM[11]

6. Approximations to discontinuous functions. We first derive somewhat differently a classical result concerning the following finite problem: Let there be given $n+1$ points

$$(x_\nu, y_\nu) \qquad (\nu = 0, 1, \ldots, n;\ \ x_0 < x_1 < \ldots < x_n); \tag{11}$$

we wish to determine the polynomial

$$P(x) = a_0 x^{n-1} + a_1 x^{n-2} + \ldots + a_{n-1} \tag{12}$$

which minimizes the expression

$$\max_\nu |y_\nu - P(x_\nu)|. \tag{13}$$

We need the following lemma: *If the real variables $(u_0, u_1, \ldots, u_n)$ are connected by the linear relation with real constant coefficients*

[10] For the first published proof of Riesz's theorem see [6], p. 210; it is an almost immediate consequence of the Heine-Borel theorem.

[11] An excellent reference to Tchebycheff's approximation problem is [9], Chapter VI.

(14) $$b_0 u_0 + b_1 u_1 + \dots + b_n u_n = c \qquad (b_0 b_1 \dots b_n \neq 0),$$

then the expression $\max_\nu |u_\nu|$ *has the minimal value*

(15) $$\rho = \frac{|c|}{|b_0| + |b_1| + \dots + |b_n|},$$

which is reached for just one set of values $u_\nu = u^*_\nu$ *given by*

(16) $$u^*_\nu = \rho \operatorname{sgn}(cb_\nu) \qquad (\nu = 0, \dots, n).$$

We lose no generality in assuming that $c > 0$, for if $c = 0$ the result is trivial and if $c < 0$ we may multiply both sides of (14) by -1. In view of (15), the relations (16) indeed define a solution of (14); (16) also imply that $|u^*_\nu| = \rho$, hence $\rho = \max |u^*_\nu|$. Let now (u_ν) be an arbitrary set satisfying the two relations

$$\sum b_\nu u_\nu = c \quad \text{and} \quad \max_\nu |u_\nu| \leqslant \rho.$$

This set (u_ν) must be of the form

$$u_\nu = \epsilon_\nu u^*_\nu = \epsilon_\nu \rho \operatorname{sgn}(b_\nu), \text{ where } -1 \leqslant \epsilon_\nu \leqslant 1, \qquad (\nu = 0, \dots, 1).$$

Now $c = \sum b_\nu u_\nu = \sum b_\nu \epsilon_\nu \rho \operatorname{sgn}(b_\nu) = \sum \epsilon_\nu \rho |b_\nu|$, or $\sum e_\nu \rho |b_\nu| = c$. In view of (15) or $\sum \rho |b_\nu| = c$, the last relation implies that $\epsilon_\nu = +1$ for all ν, and therefore $u_\nu = u^*_\nu$. This completes the proof of our lemma.

Returning to the problem of minimizing (13), let $u_\nu = y_\nu - P(x_\nu)$ be the discrepancies between the points (11) and the polynomial (12). These discrepancies are not independent variables, for they are obviously connected by the single linear relation

$$\begin{vmatrix} u_0 - y_0 & 1 & x_0 \dots x_0^{n-1} \\ u_1 - y_1 & 1 & x_1 \dots x_1^{n-1} \\ \dots & & \dots \\ u_n - y_n & 1 & x_n \dots x_n^{n-1} \end{vmatrix} = 0,$$

or

$$\begin{vmatrix} u_0 & 1 \dots x_0^{n-1} \\ u_1 & 1 \dots x_1^{n-1} \\ & \dots \\ u_n & 1 \dots x_n^{n-1} \end{vmatrix} = \begin{vmatrix} y_0 & 1 \dots x_0^{n-1} \\ y_1 & 1 \dots x_1^{n-1} \\ & \dots \\ y_n & 1 \dots x_n^{n-1} \end{vmatrix}.$$

Since the coefficients of (u_ν) on the left-hand side of this linear relation alternate in sign, we obtain by (15) for the minimal value of (13) the explicit expression

(17) $$\rho = \text{abs. val.} \begin{vmatrix} y_0 & 1 & x_0 \dots x_0^{n-1} \\ y_1 & 1 & x_1 \dots x_1^{n-1} \\ & \dots & \\ y_n & 1 & x_n \dots x_n^{n-1} \end{vmatrix} \div \begin{vmatrix} 1 & 1 & x_0 \dots x_0^{n-1} \\ -1 & 1 & x_1 \dots x_1^{n-1} \\ & \dots & \\ (-1)^n & 1 & x_n \dots x_n^{n-1} \end{vmatrix}.$$

By (16) we also know that this minimal value ρ is reached for *just one* polynomial $P(x)$ for which the discrepancies $u_\nu = y_\nu - P(x_\nu)$ are all equal in absolute value to ρ and alternate in sign.

Concerning the analytical problem of best approximation of functions we wish to prove the following

THEOREM 4. *Let $f(x)$ be a real function defined in $\alpha \leqslant x \leqslant \beta$ about which we only assume that it is bounded. Given n $(n \geqslant 1)$, there exists a real polynomial $P^*(x)$, of degree not exceeding $n-1$, which minimizes the expression*

$$\sup_{\alpha \leqq x \leqq \beta} |f(x) - P(x)|, \tag{18}$$

giving it its minimal value

$$\rho = \sup_{\alpha \leqq x \leqq \beta} |f(x) - P^*(x)|.$$

For this minimal value ρ we have the relation

$$\rho = \sup_{(x_\nu)} \rho(x_0, x_1, \ldots, x_n) \quad for \quad \alpha \leqslant x_0 < x_1 < \ldots < x_m \leqslant \beta, \tag{19}$$

where

$$\rho(x_0,x_1,\ldots,x_n) = \text{abs. val.} \begin{vmatrix} f(x_0) & 1 & x_0 & \ldots & x_0^{n-1} \\ f(x_1) & 1 & x_1 & \ldots & x_1^{n-1} \\ & & \ldots & & \\ f(x_n) & 1 & x_n & \ldots & x_n^{n-1} \end{vmatrix} \div \begin{vmatrix} 1 & 1 & x_0 & \ldots & x_0^{n-1} \\ -1 & 1 & x_1 & \ldots & x_1^{n-1} \\ & & \ldots & & \\ (-1)^n & 1 & x_n & \ldots & x_n^{n-1} \end{vmatrix}. \tag{20}$$

In words, (19) means that the best approximation ρ of our function is the supremum of its best approximation $\rho(x_0, \ldots, x_n)$ over sets of $n+1$ distinct points of the range $[\alpha,\beta]$.

Proof. Since $f(x)$ is bounded, so are the best approximations (20). Let

$$\rho_0 = \sup_{(x_\nu)} \rho(x_0,x_1, \ldots, x_n) \quad \text{for} \quad \alpha \leqslant x_0 < x_1 < \ldots < x_n \leqslant \beta. \tag{21}$$

In the space E_n of the variables $(a_0,a_1, \ldots, a_{n-1})$ and corresponding to each value of x in the range $[\alpha,\beta]$, we consider the parallel layer of space D_x defined by

$$D_x\colon |f(x) - a_0x^{n-1} - a_1x^{n-2} - \ldots - a_{n-1}| \leqslant \rho_0. \tag{22}$$

We claim that the collection $\{D_x\}$ of convex domains in E_n satisfies both assumptions of Helly's Theorem 2. Indeed, if $\alpha \leqslant \xi_1 < \xi_2 < \ldots < \xi_n \leqslant \beta$, then $\Delta = D_{\xi_1} D_{\xi_2}, \ldots, D_{\xi_n}$ is evidently non-void and bounded; in fact Δ is a proper parallelepiped, except in the case $\rho_0 = 0$ when Δ reduces to a point. Let us now consider $n+1$ distinct abscissae

$$\alpha \leqslant x_0 < x_1 < \ldots < x_n \leqslant \beta. \tag{23}$$

If $P(x)$ is the polynomial of best approximation to the points $(x_\nu, f(x_\nu))$ $(\nu = 0, \ldots, n)$, we have by (21)

$$|f(x_\nu) - a_0x_\nu^{n-1} - \ldots - a_{n-1}| = \rho(x_0,x_1, \ldots, x_n) \leqslant \rho_0 \qquad (\nu = 0, \ldots, n).$$

Geometrically this means that the $n+1$ convex domains $D_{x_0}, D_{x_1}, \ldots, D_{x_n}$ have the common point $(a_0, \ldots, a_{n-1})$. By Helly's theorem we conclude the existence of a point $(a^*_0, \ldots, a^*_{n-1})$ which is common to *all* the domains D_x, hence there exists a polynomial $P^*(x)$ satisfying the inequality (22) for *all* x. For this polynomial $P^*(x)$ we therefore have

$$\sup_{\alpha \leqq x \leqq \beta} |f(x) - P^*(x)| \leqslant \rho_0. \tag{24}$$

On the other hand, for an *arbitrary* polynomial $P(x)$ we have

$$\sup_{\alpha \leqq x \leqq \beta} |f(x) - P(x)| \geqslant \sup_{\nu} |f(x_\nu) - P(x_\nu)| \geqslant \rho(x_0, \ldots, x_n),$$

or

$$\sup_{\alpha \leqq x \leqq \beta} |f(x) - P(x)| \geqslant \rho(x_0, \ldots, x_n).$$

Taking the supremum of the right-hand side we find

$$\sup_{\alpha \leqq x \leqq \beta} |f(x) - P(x)| \geqslant \rho_0, \text{ for every } P(x), \tag{25}$$

in particular also

$$\sup_{\alpha \leqq x \leqq \beta} |f(x) - P^*(x)| \geqslant \rho_0. \tag{26}$$

Now by (24) and (26) we find that

$$\sup_{\alpha \leqq x \leqq \beta} |f(x) - P^*(x)| = \rho_0. \tag{27}$$

From (25) and (27) we see ρ_0 is the minimum value of (18), hence $\rho = \rho_0$, which is what we wanted to prove.

REMARKS. 1. The polynomial $P^*(x)$, whose existence has just been proved, need not be unique. Thus, if $n = 2$ and

$$f(x) = [x] \qquad (0 \leqslant x \leqslant 1), \tag{28}$$

then a graph will show that every polynomial of the family

$$P(x) = a_0(x - 1) + \tfrac{1}{2} \qquad (0 \leqslant a_0 \leqslant 1),$$

minimizes the expression (18), giving it its minimal value $\rho = \frac{1}{2}$.

2. The existence part of Theorem 4 is also easily established directly by familiar continuity arguments; however, a proof of the relation (19), which seems to be new at least for a discontinuous $f(x)$, would be difficult or at least involved without the use of Helly's theorem which bridges most naturally the gap between the finite (algebraic) best approximation problem for $n+1$ points and the analytical problem for an interval $[\alpha,\beta]$.

3. Theorem 4 immediately generalizes to the case when the interval $[\alpha,\beta]$ of definition of $f(x)$ is replaced by an arbitrary bounded point-set of the x-axis. A further possible extension of Theorem 3 is as follows: The inequality (22) means that the curve $y = P(x)$ intersects the family of vertical segments $f(x) - \rho_0 \leqslant y \leqslant f(x) + \rho_0$, which are all of the same length $2\rho_0$. As in the Dresher-Harris theorem, this length could be required to vary with x.

7. The classical case of continuous functions. We now add the important additional assumption that the function $f(x)$ is *continuous* in the range $[\alpha,\beta]$. Then it is clear that $\rho(x_0, \ldots, x_n)$, defined by (20), is a continuous function of $(x_0, \ldots, x_n)$ as long as the inequalities (23) hold. We now extend the definition of the function $\rho(x_0, \ldots, x_n)$ throughout the closed domain

$$\alpha \leqslant x_0 \leqslant x_1 \leqslant \ldots \leqslant x_n \leqslant \beta, \tag{29}$$

by the convention that

$$\rho(x_0, x_1, \ldots, x_n) = 0 \quad \text{if the } x_\nu \text{ are not all different.} \tag{30}$$

We claim that the extended function $\rho(x_0, \ldots, x_n)$ is continuous throughout the closed domain (29). Indeed, let $(x_0, \ldots, x_n)$ be a point of (29) for which at least two x_ν's coalesce, hence $\rho(x_0, \ldots, x_n) = 0$. Let this point $(x_0, \ldots, x_n)$ be the limit of a sequence of points $(x_0^{(k)}, \ldots, x_n^{(k)})$ with

$$\alpha \leqslant x_0^{(k)} < x_1^{(k)} < \ldots < x_n^{(k)} \leqslant \beta \qquad (k = 1,2, \ldots).$$

We have to show that

$$\lim_{k\to\infty} \rho(x_0^{(k)}, \ldots, x_n^{(k)}) = 0. \tag{31}$$

Clearly there exists a polynomial $P(x)$, of degree $n - 1$ or less, so that

$$f(x_\nu) - P(x_\nu) = 0 \qquad (\nu = 0, \ldots, n),$$

hence by the continuity of $f(x)$ and $P(x)$

$$\max_\nu |f(x_\nu^{(k)}) - P(x_\nu^{(k)})| \to 0, \text{ as } k \to \infty.$$

This, together with the inequality

$$\max |f(x_\nu^{(k)}) - P(x_\nu^{(k)})| \geqslant \rho(x_0^{(k)}, \ldots, x_n^{(k)}),$$

implies (31).

Let us now return to our Theorem 4 to note the effect of the continuity of $f(x)$. Let us assume that the best approximation ρ is positive. The continuous function $\rho(x_0, \ldots, x_n)$ *assumes* its maximum value ρ at a point $(x^*_0, \ldots, x^*_n)$ and because ρ is positive we must have

$$\alpha \leqslant x^*_0 < x^*_1 < \ldots < x^*_n \leqslant \beta.$$

We may now readily establish contact with the classical oscillation properties of the polynomial $P^*(x)$ of best approximation[12] ρ. In the first place $P^*(x)$ is now uniquely defined: Indeed, a polynomial $P(x)$ of best approximation $\rho = \rho(x^*_0, \ldots, x^*_n)$ must satisfy the inequalities

$$|f(x^*_\nu) - P(x^*_\nu)| \leqslant \rho = \rho(x^*_0, \ldots, x^*_n) \qquad (\nu = 0, \ldots, n), \tag{32}$$

while we know from the discussion of the case of $n + 1$ points that there is only one polynomial satisfying (32); since $P^*(x)$ does satisfy (32), $P^*(x)$ is uniquely defined. Secondly, we know that the sequence

$$u^*_\nu = f(x^*_\nu) - P^*(x^*_\nu) \qquad (\nu = 0, \ldots, n),$$

has all its elements of absolute value equal to ρ and that they alternate in sign. These are the classical oscillation properties referred to above. Our example (28) shows that no such properties, beyond the general relation (19) hold in the case of discontinuous functions.

APPENDIX (Added July 1, 1949). The authors are much indebted to the referee for the following two valuable references;

1. The theorem of Dines and McCoy of our Introduction is an immediate corollary of a theorem of E. Steinitz, *Bedingt konvergente Reihen und konvexe Systeme* II, Journal für Mathematik, vol. 144 (1941), pp. 1-40. On pp. 12-13

[12] See [9], pp. 76-78.

Steinitz defines a family of rays in E_n, with common initial point O, to be *all-sided* provided there are rays of the family on each side of every hyperplane through O. An all-sided family is *irreducible* if no proper sub-family is all-sided. Steinitz then proves the following

THEOREM. *In any all-sided family of rays, there is contained at least one irreducible sub-family; such a sub-family has at least $n+1$ and at most $2n$ rays.*

The Dines-McCoy theorem for the space E_n, rather than E_{n+1}, follows thus: Let $\{H_\nu\}$ be the collection of half-spaces of the theorem and let R_ν denote the interior ray through O normal to the hyperplane bounding H_ν. Suppose that these half-spaces have no common ray. Then for every ray ρ through O, for some ν we must have $\angle(R_\nu,\rho) > \pi/2$; applying this remark to ρ and $-\rho$, we see that $\{R_\nu\}$ is an all-sided family of rays. By Steinitz's theorem there is an all-sided sub-collection $R_1, R_2, \ldots, R_s$, say, with $n+1 \leqslant s \leqslant 2n$. But then the corresponding $H_1, H_2, \ldots, H_s$ have no ray in common, in contradiction to the assumption of the theorem.

2. The Dresher-Harris theorem of our Introduction was fully discussed (for $n = 1$) by L. A. Santaló, *Complemento a la nota: Un teorema sobre conjuntos de paralelepipedos de aristas paralelas*, Publicaciones del Instituto de Matematica de la Universidad Nacional del Litoral, vol. 3 (1942), pp. 203-210. Also its proof by means of Helly's theorem is found in footnote 4 on page 207 and attributed by Santaló to J. Rey Pastor.

REFERENCES

[1] P. Alexandroff and H. Hopf, *Topologie*, vol. 1, Berlin, 1935.

[2] L. M. Blumenthal, *Metric methods in linear inequalities*, Duke Math. J., vol. 15 (1948), 955-966.

[3] L. L. Dines and N. H. McCoy, *On linear inequalities*, Trans. Roy. Soc. Can., Third Series, Sec. III, vol. 27 (1933), 37-70.

[4] E. Helly, *Über Mengen konvexer Körper mit gemeinschaftlichen Punkten*, Jahresbericht der deutschen Mathematiker Vereinigung, vol. 32 (1923), 175-176.

[5] P. Kirchberger, *Über Tschebyschefsche Annäherungsmethoden*, Math. Ann., vol. 57 (1903), 509-540. The same paper appeared also in more elaborate form (96 pages) in 1902 as a doctoral dissertation written under Hilbert's guidance.

[6] D. König, *Über konvexe Körper*, Math. Zeit., vol. 14 (1922), 208-210.

[7] J. Radon, *Mengen konvexer Körper, die einen gemeinsamen Punkt enthalten*, Math. Ann., vol. 83 (1921), 113-115.

[8] A. Tarski, *A decision method for elementary algebra and geometry*, The Rand Corporation, 1948, 57 pages.

[9] Ch. J. de la Vallée Poussin, *Leçons sur l'approximation des fonctions d'une variable réelle*, Paris, 1919.

The University of Pennsylvania

Notes to Paper 52

This paper, written jointly with I. J. Schoenberg, has been reviewed in the Z, vol. 36 (1951), pp. 237-238, and in the MR, vol. 11 (1950), p. 681.

Page 252, line 9.
Read ϵ_ν instead of e_ν.

Die Reziprozitätsformel für Dedekindsche Summen.

Von H. RADEMACHER in Philadelphia.

Ich erinnere mich gern an einen Besuch in Budapest vor 20 Jahren, bei welcher Gelegenheit ich einen Vortrag vor einer Gruppe von FEJÉRS Freunden und Schülern halten durfte[1]). Die jetzige Gelegenheit gibt mir einen Anlaß, auf meinen damaligen Gegenstand, einen Reziprozitätssatz über gewisse Summen, die ich kurz Dedekindsche Summen nennen möchte, zurückzukommen.

1. Eine Dedekindsche Summe $s(h, k)$ ist folgendermaßen definiert. Es sei zur Abkürzung für reelle x

$$\text{(1a)} \qquad \begin{aligned} ((x)) &= x-[x]-1/2 && \text{für nicht-ganze } x, \\ ((x)) &= 0 && \text{für ganze } x \end{aligned}$$

gesetzt. Dann sei für zueinander teilerfremde h und k

$$\text{(1b)} \qquad s(h, k)=\sum_{\mu=1}^{k-1}\left(\left(\frac{\mu}{k}\right)\right)\left(\left(\frac{h\mu}{k}\right)\right).$$

Die erwähnte *Reziprozitätsformel* lautet nun

$$\text{(2)} \qquad s(h, k)+s(k, h)=-\frac{1}{4}+\frac{h}{12k}+\frac{k}{12h}+\frac{1}{12hk}$$

Die Dedekindschen Summen erscheinen zuerst in Dedekinds Kommentar zu Riemanns nachgelassenen Fragmenten über die Modulfunktionen[2]). Dort findet man auch Dedekinds Beweis für die Reziprozitätsformel, der auf der Gruppeneigenschaft der Modulsubstitutionen beruht. Die Dedekindschen Summen, die eine fundamentale Rolle in der Theorie der Theta- und Modulfunktionen spielen, treten auch in der von HARDY und RAMANUJAN geschaffenen Theorie der Partitionen auf. Die Reziprozitätsformel (2) ist jedoch eine rein zahlentheoretische Aussage. Ich habe dafür mehrere von der Theorie der Modulfunktionen unabhängige Beweise gegeben, zu denen ich hier einen besonders durchsichtigen hinzufügen möchte.

[1]) H. RADEMACHER, Egy reciprocitásképletről a modulfüggvények elméletéből, *Matematikai és Fizikai Lapok*, **40** (1933), S. 24–34.

[2]) Erläuterungen zu Riemanns Fragmenten über die Grenzfälle der elliptischen Modulfunktionen, *Riemanns Werke* (1876), S. 438–447, *Dedekinds Werke* (1930), Bd. 1, S. 159–172.

2. Wir benötigen zwei Lemmata:

Lemma I. *Sind m und n zwei teilerfremde positive ganze Zahlen, so ist, mit den Bezeichnungen* (1a),

(3) $$\int_0^1 ((mx))\,((nx))\,dx = \frac{1}{12\,mn}.$$ [3]

Lemma II. *Es seien* $f(x)$, $g(x)$, $h(x)$ *Funktionen von beschränkter Schwankung im Intervall* $a \leqq x \leqq b$, *die paarweise keine Unstetigkeit gemeinsam haben. Dann gilt die Gleichung*

(4) $$\int_a^b f(x)\,dg(x)h(x) = \int_a^b f(x)g(x)\,dh(x) + \int_a^b f(x)h(x)\,dg(x)$$

zwischen Stieltjes-Integralen.

Bemerkung. Für $f(x)$ identisch gleich 1 geht (4) über in die Formel für partielle Integration

(5) $$\Big[g(x)h(x)\Big]_a^b = \int_a^b g(x)\,dh(x) + \int_a^b h(x)\,dg(x)$$

unter den obigen Bedingungen für $g(x)$, $h(x)$.

Was den *Beweis* von Lemma II angeht, so bemerken wir zunächst, daß $g(x)h(x)$ von beschränkter Schwankung ist und daß dieses Produkt keine Unstetigkeit mit $f(x)$ gemeinsam hat.

Wir wissen ferner, daß

$$\int_a^b f(x)\,d\varphi(x)$$

existiert, wenn $f(x)$ und $\varphi(x)$ beide von beschränkter Schwankung sind und keine Unstetigkeit gemeinsam haben[4]). Die in (4) genannten Integrale existieren also alle.

Haben wir nun eine Einteilung des Intervalls (a, b)

$$a = x_0 < x_1 < x_2 < \ldots < x_{n-1} < x_n = b$$

vor uns und haben $x_{j-1} \leqq \xi_j \leqq x_j$ gewählt, so betrachten wir die Identität

(6) $$\begin{aligned}&\sum_{j=1}^{n} f(\xi_j)\{g(x_j)h(x_j) - g(x_{j-1})h(x_{j-1})\} = \\ &= \sum_{j=1}^{n} f(\xi_j)g(x_j)\{h(x_j) - h(x_{j-1})\} + \sum_{j=1}^{n} f(\xi_j)h(x_{j-1})\{g(x_j) - g(x_{j-1})\}.\end{aligned}$$

Hier verlangen nur die Summen auf der rechten Seite einige Überlegung. Wenn man $f(x)$, $g(x)$, $h(x)$ als Differenzen von positiven, monoton wachsen-

---

[3]) Einen Beweis findet man in LANDAU's „Vorlesungen über Zahlentheorie", Bd. 2, S. 171; siehe auch J. FRANEL, *Göttinger Nachrichten* 1924, S. 199 und E. LANDAU, *ibidem*, S. 203.

[4]) Vgl. z. B. D. V. WIDDER, *The Laplace Transform* (Princeton, 1941), p. 25, Theorem 14.

den Funktionen schreibt, kommt man auf Summen von der Form

(7a) $$\sum \varphi(\xi_j)\,\psi(x_j)\{\chi(x_j)-\chi(x_{j-1})\}$$

und

(7b) $$\sum \varphi(\xi_j)\,\psi(x_{j-1})\{\chi(x_j)-\chi(x_{j-1})\},$$

wo φ, ψ, χ positive und monoton wachsende Funktionen sind, die paarweise keine Unstetigkeiten gemeinsam haben. Diese Summen liegen nun offenbar zwischen

$$\sum \varphi(x_{j-1})\,\psi(x_{j-1})\{\chi(x_j)-\chi(x_{j-1})\}$$

und

$$\sum \varphi(x_j)\,\psi(x_j)\{\chi(x_j)-\chi(x_{j-1})\},$$

die beide nach dem oben zitierten Satz mit Verfeinerung der Teilung gegen

$$\int_a^b \varphi(x)\,\psi(x)\,d\chi(x)$$

konvergieren, gegen welchen Limes daher auch (7a) und (7b) streben. Somit folgt aus (6) die zu beweisende Gleichung (4).

3. Unser Beweis der Reziprozitätsformel (2) ist nun schnell geführt, indem wir das Stieltjes-Integral

(8) $$I_\varepsilon = \int_\varepsilon^{1-\varepsilon} ((x))\,d((hx))((kx)), \quad \varepsilon > 0$$

auf zwei verschiedene Weisen behandeln. Die Funktion $((x))$ ist stetig im Intervall $\varepsilon \leqq x \leqq 1-\varepsilon$, und die Funktionen $((hx))$ und $((kx))$ haben dort wegen $(h, k) = 1$ keine Unstetigkeiten gemeinsam.

Einerseits ist nun wegen (4)

$$I_\varepsilon = \int_\varepsilon^{1-\varepsilon} ((x))((hx))\,d((kx)) + \int_\varepsilon^{1-\varepsilon} ((x))((kx))\,d((hx)) =$$

$$= \int_\varepsilon^{1-\varepsilon} ((x))((hx))\,k\,dx - \sum_{\varepsilon < \frac{\mu}{k} < 1-\varepsilon} \left(\left(\frac{\mu}{k}\right)\right)\left(\left(\frac{h\mu}{k}\right)\right) + \int_\varepsilon^{1-\varepsilon} ((x))((kx))\,h\,dx - \sum_{\varepsilon < \frac{\nu}{h} < 1-\varepsilon} \left(\left(\frac{\nu}{h}\right)\right)\left(\left(\frac{k\nu}{h}\right)\right)$$

und daher

$$\lim_{\varepsilon \to +0} I_\varepsilon = k\int_0^1 ((x))((hx))\,dx - \sum_{\mu=1}^{k-1} \left(\left(\frac{\mu}{k}\right)\right)\left(\left(\frac{h\mu}{k}\right)\right) + h\int_0^1 ((x))((kx))\,dx - \sum_{\nu=1}^{h-1} \left(\left(\frac{\nu}{h}\right)\right)\left(\left(\frac{k\nu}{h}\right)\right).$$

Infolge von (3) und (1b) ergibt sich somit

(9) $$\lim_{\varepsilon \to +0} I_\varepsilon = \frac{k}{12h} - s(h, k) + \frac{h}{12k} - s(k, h).$$

Anderseits erhält man mittels der partiellen Integration (5) aus (8)

$$I_\varepsilon = \Big[((x))((hx))((kx))\Big]_\varepsilon^{1-\varepsilon} - \int_\varepsilon^{1-\varepsilon} ((hx))((kx))\,d((x)) =$$

$$= -2((\varepsilon))((h\varepsilon))((k\varepsilon)) - \int_\varepsilon^{1-\varepsilon} ((hx))((kx))\,dx$$

und somit

$$\lim_{\varepsilon\to+0} I_\varepsilon = -2\{\lim_{\varepsilon\to+0} ((\varepsilon))\}^3 - \int_0^1 ((hx))((kx))\,dx$$

oder, infolge von (1a) und (3),

$$(10) \qquad \lim_{\varepsilon\to+0} I_\varepsilon = \frac{1}{4} - \frac{1}{12hk}.$$

Die Gleichungen (9) und (10) ergeben nun zusammen die zu beweisende Formel (2).

(Eingegangen am 8. August 1949.)

This paper has been reviewed in the Z, vol. 37 (1951), p. 311, and in the MR, vol. 11 (1950), p. 642.

Page 57, footnote 1.
See paper 32 of the present collection.

ADDITIVE ALGEBRAIC NUMBER THEORY[1]

Hans Rademacher

Additive number theory in the rational field uses, since Euler (*De partitione numerorum* 1748) the tool of power series, based on the fact that

$$x^m \cdot x^n = x^{m+n}.$$

The power series

$$\sum a_n x^n$$

can first be understood in a purely *formal* way, as Euler himself did, an interpretation which led him to numerous results in additive theory and to such a fundamental identity as

$$\prod_{m=1}^{\infty} (1 - x^m) = \sum_{\lambda=-\infty}^{\infty} (-1)^\lambda x^{\lambda(3\lambda+1)/2}$$

The new method which Hardy, Littlewood, and Ramanujan introduced from 1917 on was based on the *function-theoretical* interpretation of the power series, which permitted in particular the application of Cauchy's integral theorem. The abundant results in Waring's problem, Goldbach's problem, and the theory of partitions, which ensued, are well-known.

The analytic additive theory in algebraic number fields could not be developed before we had the equivalent of power series for the algebraic fields. This was found by Erich Hecke, who in turn developed ideas of Hilbert and Blumenthal.

Let us for the sake of simplicity confine our attention to totally real fields, i.e., such which together with their conjugate fields are real. An algebraic number ν is called totally positive if it together with all its conjugates is positive; we write $\nu \succ 0$. Then

$$(1) \quad f(t) = f(t^{(1)}, t^{(2)}, \cdots, t^{(n)}) = \sum_{\nu \succ 0} a(\nu) \exp\left(-(\nu^{(1)}t^{(1)} + \cdots + \nu^{(n)}t^{(n)})\right)$$

is a "power series" in an algebraic field of degree n, where $\nu^{(1)}, \cdots, \nu^{(n)}$ are the conjugates of the integer ν and $t^{(1)}, t^{(2)}, \cdots, t^{(n)}$ are independent complex variables. Convergence depends on the coefficients $a(\nu)$, of course. For $a(\nu) = 1$ we obtain the "geometric series", for which convergence is easily established in the domain $\Re(t^{(j)}) > 0$, $j = 1, 2, \cdots, n$, which we, in an obvious extension of the meaning of the symbol for total positivity, may also indicate by $\Re(t) \succ 0$.

For the power series in one variable

$$f(t) = \sum_{n=0}^{\infty} a_n e^{-nt}, \qquad \Re(t) > c,$$

[1] This address was listed on the printed program under the title *Remarks on the theory of partitions*

the Cauchy formula can be written as

$$a_n = \int_0^1 f(z + 2\pi i\varphi)e^{n(z+2\pi i\varphi)}\, d\varphi, \qquad \Re(z) > c.$$

The analogue for a power series for algebraic fields (1) demands the introduction of a basis $(\rho_1, \rho_2, \cdots, \rho_n)$ of the ideal $\mathfrak{d}^{-1}$, where $\mathfrak{d}$ is the ramification ideal, with $N(\mathfrak{d}) = |d|$, d the discriminant of the algebraic field. In the space $(y^{(1)}, y^{(2)}, \cdots, y^{(n)})$ (real coordinates) the points of $\mathfrak{d}^{-1}$ form a point lattice, whose parallelepiped we may call E (it corresponds to the unit interval in the rational case). Then we get

$$(2)\quad a(\nu) = N(\mathfrak{d})^{1/2} \int \cdots \int_E f(z + 2\pi iy) \exp\left(\sum_q \nu^{(q)}(z^{(q)} + 2\pi iy^{(q)})\right) dy^{(1)} \cdots dy^{(n)},$$

since $\mathfrak{d}^{-1}$ has the property that for any number δ in $\mathfrak{d}^{-1}$ the trace $S(\delta)$ is a rational integer.

The extension of the Hardy-Littlewood method to algebraic number fields requires one more tool: the analogue of the Farey dissection of order N. This means here a suitable dissection of E, each Farey piece belonging to a fraction γ such that $\gamma\mathfrak{d}$ has a denominator $\mathfrak{a}$, with $N(\mathfrak{a}) \leqq N$. Siegel carried out the details by means of an approximation theorem of Minkowski.

The first example of a power series of type (1) was the ϑ-series

$$\vartheta(t) = \sum_\nu \exp\left(-(\nu^{(1)2}t^{(1)} + \cdots + \nu^{(n)2}t^{(n)})\right), \qquad \Re(t) > 0,$$

investigated by Hecke. Here ν runs over all integers of the field; in special cases ν may be restricted to numbers of a given ideal. Hecke used these ϑ-series for two important purposes, namely, to establish the existence of the Dedekind ϑ-functions in the whole plane, and secondly, for the discussion of Gaussian sums in algebraic fields, which in turn lead to the reciprocity laws of quadratic residues.

Siegel realized that these ϑ-functions could be used for the analytic theory of the representation of totally positive numbers as sums of squares. If $A_r(\nu)$ is the number of representations of ν as a sum of squares of integers, Siegel obtained for the real-quadratic field (and similarly in other algebraic number fields) the result, for $r \geqq 5$

$$(3)\qquad A_r(\nu) = \frac{\pi^r}{\Gamma^2(r/2)\, d^{(r-1)/2}} S_r(\nu) N(\nu)^{r/2-1} + o(N(\nu)^{r/2-1}).$$

Here S is the "singular series" which reflects the arithmetical properties of ν There exist two positive constants C'_r and C''_r independent of ν so that $0 < C'_r < S < C''_r$ for each $r \geqq 5$, unless $S = 0$, which takes place for those ν which cannot be represented as a sum of any number of squares. (That such numbers exist is seen in the example $R(2^{1/2})$, where a square has always the form $(a + b\, 2^{1/2})^2 = a^2 + 2b^2 + 2ab\, 2^{1/2}$ which shows that the integer $t + u\, 2^{1/2}$ cannot be a sum of squares if u is odd.)

The form of the result (3) shows complete analogy to the results in rational cases: the formula for $A_r(\nu)$ shows, besides coefficients depending solely on the field, two factors, one determining the order of magnitude depending on $N(\nu)$, the other, $S_r(\nu)$, which is a product over the prime ideals of the field, involving the arithmetical properties of ν.

In the case of the sum of squares, the power series, which are here ϑ-functions, are so well known that we obtain excellent approximations to them by elementary functions in each Farey piece of E. A distinction corresponding to that between major and minor arcs is therefore not necessary.

The same favorable situation prevailed in the Hardy-Littlewood treatment of Goldbach's problem. There the goodness of approximation was established by fiat, namely by the assumption of the generalized Riemann hypothesis concerning all $L(s, \chi)$-functions. For algebraic fields the situation was similar, no distinction between minor and major Farey pieces was necessary if a Riemann hypothesis was introduced covering the Dedekind ζ-function and all the Hecke $\zeta(s, \lambda\chi)$-functions of the field. In the case of a real quadratic field I obtained for the number $B_r(\nu)$ of representations of a totally positive integer ν as the sum of r totally positive prime numbers (i.e., numbers whose principal ideal is a prime ideal) the result

$$B_r(\nu) = C_r S_r(\nu) \frac{N(\nu)^{r-1}}{(\log N(\nu))^r} + o\left(\frac{N(\nu)^{r-1}}{(\log N(\nu))^r}\right), \qquad r \geqq 3,$$

where $S_r(\nu)$ is the singular series

$$S_r(\nu) = \prod_{\mathfrak{p} | \nu, N(\mathfrak{p}) > 2} \left(1 + (-1)^r \frac{N(\mathfrak{p})}{(N(\mathfrak{p}) - 1)^r - (-1)^r}\right),$$

and where C_r is a constant depending on the field and on the number r. The analogy to the Hardy-Littlewood formula in the rational case is complete.

Unfortunately this result shares with that of Hardy and Littlewood also the shortcoming that it is hypothetical, being based on a generalized Riemann hypothesis. For the rational field this objection has been met by Vinogradoff, Linnik, and Tchudakoff. Whereas the former had to distinguish between the major arcs, on which the Hardy-Littlewood method could be carried through unconditionally, and the minor arcs, on which he used his famous estimates of exponential sums, the latter two could remodel the Hardy-Littlewood method without using that distinction. They utilized the fact that if the $L(s, \chi)$-functions should have zeros close to the line $\sigma = 1$, these zeros are relatively few and could be accounted for in the estimations. The corresponding work has not yet been carried out in the algebraic case. It seems that we would need analogues for Page's theorems concerning the paucity of undesirable zeros of the L-functions and also some sort of estimation for the $\zeta(s, \lambda)$-functions as the "approximate functional equations" provide for $\zeta(s)$ and the $L(s, \chi)$-functions in the rational case. These seem to me worthy objects of function-theoretical research.

Recently Siegel has been able to use the analogues of the major and minor arcs together, in the treatment of Waring's problem. The problem is to represent a totally positive ν as a sum of kth powers of totally positive integral numbers

$$\nu = \lambda_1^k + \lambda_2^k + \cdots + \lambda_r^k .$$

The result is again as to be expected: the number of representations is

$$A_r(\nu) = C_r S_r(\nu) N(\nu)^{r/k-1} + o(N(\nu)^{r/k-1})$$

for $r > (2^{k-1} + n)k$, where n is the degree of the field K, C_r depends on K, k, r only, and $S_r(\nu)$ is the singular series, which again appears as an infinite product over prime ideals of K. The singular series vanishes if and only if ν does not belong to the ring J_k which consists of the integers obtainable as sums of any number of kth powers.

There remains one important problem of additive number theory to be investigated for algebraic number fields: the problem of partitions. Whereas so far we were guided by analogies to the rational cases, the situation here is completely different and I can make today only small contributions to its solution.

In the rational field we have for $p(n)$, the number of partitions of n, Euler's generating function

$$1 + \sum_{n=1}^{\infty} p(n)x^n = \prod_{m=1}^{\infty} (1 - x^m)^{-1}.$$

The same reasoning leads here for $P(\nu)$, the number of partitions of the totally positive integer ν into totally positive integers, to the identity

$$1 + \sum_{\nu \succ 0} P(\nu)e^{-S(\nu t)} = \prod_{\mu \succ 0} (1 - e^{-S(\mu t)})^{-1},$$

with

$$S(\nu t) = \nu^{(1)}t^{(1)} + \cdots + \nu^{(n)}t^{(n)},$$

where $t^{(1)}, \cdots, t^{(n)}$ are independent complex variables of positive real part. Let us consider, for simplicity's sake, only the real quadratic field, in which case we have

$$f(t, t') = 1 + \sum_{\nu \succ 0} P(\nu)e^{-\nu t - \nu' t'} = \prod_{\mu \succ 0} (1 - e^{-\mu t - \mu' t'})^{-1}. \tag{4}$$

That this is the right generating function and that we are dealing with the proper generalization of the ordinary $p(n)$ is shown also by the fact that a well-known device gives a statement analogous to an elementary theorem of Euler:

$$1 + \sum_{\nu \succ 0} Q(\nu)e^{-\nu t - \nu' t'} = \prod_{\mu \succ 0} (1 + e^{-\mu t - \mu' t'})$$

$$= \frac{\prod (1 - e^{-2\mu t - 2\mu' t'})}{\prod (1 - e^{-\mu t - \mu' t'})} = \prod_{\mu \succ 0, 2 \nmid \mu} (1 - e^{-\mu t - \mu' t'})^{-1},$$

in words: the number $Q(\nu)$ of partitions of ν into different parts is equal to the number of partitions with repetition into parts which have no divisor 2.

Let us now, before discussing $P(\nu)$ further, remember the formula for $p(n)$ in the rational field. We need only the asymptotic formula

$$p(n) \sim \frac{1}{4 \cdot 3^{1/2} n} \exp (\pi 2n/3))^{1/2}).$$

It is of interest to note that here the arithmetical properties of n are not used in the main term. They are put in evidence only in the subsequent terms of lower order of a convergent series which we do not need here.

We might now expect from our previous experience with analogies a hypothetical formula like

$$P(\nu) \sim \frac{C}{N(\nu)} \exp (c(N(\nu))^{1/2}),$$

but such a formula does certainly not exist. As a matter of fact, I can show

$$P(\nu) = O(\exp N(\nu)^{2/5})$$

and something better than that.

The discrepancy becomes clearer by the observation that

$$\prod_{m=1}^{\infty} (1 - e^{-mt})$$

is essentially a modular function, whereas

$$\prod_{\mu > 0} (1 - e^{-\mu t - \mu' t'})$$

is *not* the analogue of a modular function for the real quadratic field. Such analogues exist, but they are of a different structure. We can quickly see the difference by taking the logarithms:

With

$$f(t) = \prod_{m=1}^{\infty} (1 - e^{-mt})^{-1}$$

we have

$$\log f(t) = \sum_{n=1}^{\infty} \sum_{m=1}^{\infty} \frac{1}{n} e^{-mnt}.$$

and through Mellin's formula:

$$\log f(t) = \frac{1}{2\pi i} \sum_{n=1}^{\infty} \sum_{m=1}^{\infty} \int_{\sigma_0 - i\infty}^{\sigma_0 + i\infty} \frac{\Gamma(s)}{n(mnt)^s} ds$$

$$= \frac{1}{2\pi i} \int_{\sigma_0 - i\infty}^{\sigma_0 + i\infty} t^s \Gamma(s) \zeta(s) \zeta(1 + s) \, ds, \qquad \sigma_c > 1,$$

which in virtue of the functional equation for $\zeta(s)$ yields

$$\log f(t) = \frac{1}{4\pi i}\int_{\sigma_0-i\infty}^{\sigma_0+i\infty} (2\pi)^s t^{-s} \frac{\zeta(1-s)\zeta(1+s)}{\cos(\pi s/2)}\, ds. \tag{5}$$

Here we have symmetry s and $-s$, which sends $t/2\pi$ into $2\pi/t$. This shows that $f(t)$ is essentially a modular function.

On the other hand, for the function $f(t, t')$ in (4) we obtain

$$\log f(t, t') = \sum_{\mu>0}\sum_{n>0} \frac{1}{n} e^{-n(\mu t+\mu' t')}.$$

Here the Mellin formula is not directly applicable, but a formula due to Hecke furnishes

$$\log f(t, t') = \frac{1}{2\pi i \log \epsilon_1} \sum_{n=-\infty}^{\infty}{}' \int_{\sigma_0-i\infty}^{\sigma_0+i\infty} \frac{\Gamma(s+inc_1)\Gamma(s-inc_1)}{t^{s+inc_1}t'^{s-inc_1}} Z(s, \lambda^n)\zeta(1+2s)\, ds \tag{6}$$

where ϵ_1 is the totally positive fundamental unit,

$$c_1 = \frac{\pi}{\log \epsilon_1},$$

$$Z(s, \lambda^n) = \sum_{(\mu)_1, \mu>0} \frac{\lambda^n(\mu)}{N(\mu)^s}$$

with

$$\lambda(\mu) = \left|\frac{\mu'}{\mu}\right|^{ic_1}$$

and $(\mu)_1$ indicates that of a set of numbers μ which differ by powers of ϵ_1 as factors only one is taken. In order to leave aside complications which do not belong to the core of the matter let us now assume that

1) the fundamental unit ϵ is *not* totally positive, so that $\epsilon_1 = \epsilon^2$,

2) the class-number of the field is 1 (these two conditions are fulfilled, e.g. in the field $R(2^{1/2})$).

Then we obtain after Hecke and Siegel a functional equation

$$Z(s, \lambda^n)\Gamma(s+inc_1)\Gamma(s-inc_1) = \frac{d^{1/2}}{4} \frac{(2\pi/d^{1/2})^{2s}}{\cos(\pi/2)(s+n+inc_1)\cos(\pi/2)(s-n-inc_1)} Z(1-s, \lambda^{-n}),$$

through which (6) goes over into

$$\log f(t, t_1) = \frac{d^{1/2}}{8\pi i \log \epsilon_1} \sum_{n=-\infty}^{\infty} \int_{\sigma_0-i\infty}^{\sigma_0+i\infty} \frac{(2\pi/d^{1/2})^{2s}}{t^{s+inc1}t'^{s-inc1}} \cdot \frac{Z(1-s, \lambda^{-n})\zeta(1+2s)}{\cos(\pi/2)(s+n+inc_1)\cos(\pi/2)(s-n-inc_1)}\, ds. \tag{7}$$

Although this formula corresponds to (4), we find here no longer symmetry in s and $-s$.

Let us now return to $P(\nu)$. Applying (2) to (4) we obtain

$$(8) \qquad P(\nu) = d^{1/2} \iint_E f(z + 2\pi iy, z' + 2\pi iy')e^{\nu(z+2\pi iy, z'+2\pi iy')} \, dy \, dy'.$$

From (7) we can insert here $f(t, t')$. An estimate gives me then so far only

$$P(\nu) = O(\exp N(\nu)^{5/14+\epsilon}),$$

whereas, on the other hand, it is evident that for rational integers

$$P(n) \geqq p(n) \sim \frac{C}{n} \exp (\pi(2n/3)^{1/2})$$

and therefore, since $N(n) = n^2$

$$P(\nu) = \Omega(N(\nu)^{-1/2} \exp (\pi(2/3)^{1/2}N(\nu)^{1/4})).$$

I have not been able to evaluate the integral (8) for $P(\nu)$ asymptotically. A new feature, however, appears clearly in my calculations: $P(\nu)$ cannot be approximated by a function of $N(\nu)$ alone in combination with arithmetical properties of ν, but such a function must involve also the "angular character" $\lambda(\nu)$ of ν.

The $\lambda(\nu)$ enter into the discussion through $Z(s, \lambda^n)$. This they do also in the previously mentioned problems of additive algebraic number theory, but there their influence is restricted to the (additive) error terms, whereas they appear here in a multiplicative way. We can therefore not even expect a formula

$$P(\nu) \sim CN(\nu)^{\alpha} \exp (N(\nu)^{\beta})$$

with suitable α, β, but have to assume that $\lambda(\nu)$ will also appear in the presumptive asymptotic formula for the number of partitions in an algebraic number field.

UNIVERSITY OF PENNSYLVANIA,
PHILADELPHIA, PA., U. S. A.

This paper has been reviewed in the Z, vol. 49 (1959), p. 162, and in the MR, vol. 13 (1952), p. 326.

Page 357, line 19.
Read ζ instead of ϑ.

Page 357, tenth line from bottom.
Read "... representations of ν as a sum of r squares ..."

Page 358, line 15.
Read $\zeta(s, \lambda)$ instead of $\zeta(s, \lambda\chi)$.

Page 358, line 16.
See paper 17 of the present collection.

Page 360, first display line.
Read $\exp(\pi(2n/3)^{1/2})$ instead of $\exp(\pi\ 2n/3))^{1/2})$.

Page 361, equation (7),
Read $\log f(t,t')$ instead of $\log f(t,t_1)$.

Page 362, line 1.
Read (5) instead of (4).

ON THE CONDITION OF RIEMANN INTEGRABILITY

HANS RADEMACHER, University of Pennsylvania

Some time ago Professor M. J. Norris published in this journal a paper* in which he proved the Riemann integrability of continuous functions without taking recourse to the notion of uniform continuity. Of course, the proof requires somewhere a transition from local properties to overall properties, and the author used for this purpose the mean value theorem of differential calculus. Since this theorem is easier for the beginner to grasp than the notion of uniform continuity, Professor Norris' proof is of great didactic merit. It turns out, however, that exactly the same proof exists already in the literature. It was given by Gerhard Kowalewski on pp. 174–176 of his little book *Grundzüge der Differential- und Integralrechnung* (Leipzig 1909).

When I read the Kowalewski-Norris proof for the Riemann integrability of continuous functions again it occurred to me that it can easily be widened so that it gives the Jordan theorem of the necessary and sufficient condition for Riemann integrability, and in a rather simple way, avoiding the Heine-Borel theorem and the Lebesgue integral. This proof I shall give in the following lines.

1. I have first to recall the definition of one-sided upper and lower derivatives. Let $f(x)$ be given as a real function in $a \leqq x \leqq b$, then we define the right upper derivative

$$\overline{D}_r f(x) = \overline{\lim_{x_1 \to x+0}} \frac{f(x_1) - f(x)}{x_1 - x},$$

the right lower derivative

$$\underline{D}_r f(x) = \underline{\lim_{x_1 \to x+0}} \frac{f(x_1) - f(x)}{x_1 - x},$$

and the left upper and the left lower derivatives respectively as

$$\overline{D}_l f(x) = \overline{\lim_{x_1 \to x-0}} \frac{f(x_1) - f(x)}{x_1 - x},$$

$$\underline{D}_l f(x) = \underline{\lim_{x_1 \to x-0}} \frac{f(x_1) - f(x)}{x_1 - x},$$

where $x_1 \to x+0$ is the customary symbol for ($x_1 > x$ and $x_1 \to x$). At a only the right derivatives are defined, at b only the left ones. We have, of course,

$$\overline{D}_r f(x) \geqq \underline{D}_r f(x), \qquad \overline{D}_l f(x) \geqq \underline{D}_l f(x).$$

Otherwise our definition does not exclude that any of the four derivatives is $+\infty$ or $-\infty$. A function is called differentiable at x_0 in the usual sense, if all

* Integrability of continuous functions, this MONTHLY vol. 59, 1952, pp. 244–245.

derivatives at x_0 are equal and finite.

With this definition we state a generalization of the theorem of the mean, given first by W. H. and G. C. Young in 1909.† We start with the generalization of Rolle's theorem, which we enunciate as

LEMMA 1. *If* $f(x)$ *is a continuous function in* $a \leqq x \leqq b$, *and if* $f(a) = f(b)$, *then there exists either a* ξ_1, $a < \xi_1 < b$ *such that*

$$\underline{D}_l f(\xi_1) \geqq 0 \geqq \overline{D}_r f(\xi_1)$$

or there exists a ξ_2, $a < \xi_2 < b$, *such that*

$$\overline{D}_l f(\xi_2) \leqq 0 \leqq \underline{D}_r f(\xi_2).$$

Proof: If for all x, $a < x < b$, we have $f(x) = f(a) = f(b)$, we may take $\xi_1 = \frac{1}{2}(a+b)$ and have $a < \xi_1 < b$ and

$$\underline{D}_l f(\xi_1) = \overline{D}_r f(\xi_1) = f'(\xi_1) = 0.$$

Otherwise there will exist an x_0, $a < x_0 < b$, such that $f(x_0) \neq f(a)$. Let us assume $f(x_0) > f(a)$. Now $f(x)$ being continuous in a closed interval attains its maximum there, at ξ_1 say. Since $f(\xi_1) \geqq f(x_0) > f(a)$, $\xi_1 \neq a$ and $\xi_1 \neq b$, so that $a < \xi_1 < b$ follows. Now

$$\frac{f(x_1) - f(\xi_1)}{x_1 - \xi_1} \leqq 0 \qquad \text{for } x_1 > \xi_1$$

and thus

$$\overline{D}_r f(\xi_1) = \overline{\lim_{x_1 \to \xi_1 + 0}} \frac{f(x_1) - f(\xi_1)}{x_1 - \xi_1} \leqq 0,$$

and similarly

$$\underline{D}_l f(\xi_1) \geqq 0.$$

In the case $f(x_0) < f(a)$ we take the minimum $f(\xi_2)$ and obtain the other alternative.

If we now lift the condition $f(a) = f(b)$ and consider in the well-known way the function

$$f_1(x) = f(x) - \frac{f(b) - f(a)}{b - a} x,$$

we obtain from Lemma 1 the generalized mean value theorem which we state as

LEMMA 2. *If* $f(x)$ *is a continuous function in* $a \leqq x \leqq b$ *then there exists either a* ξ_1, $a < \xi_1 < b$, *such that*

---

† On derivatives and the theorem of the mean, Quarterly Journal of Math. vol. 40, 1909, pp. 1–26, esp. p. 10, Theorem 4.

$$\underline{D}_l f(\xi_1) \geqq \frac{f(b) - f(a)}{b - a} \geqq \overline{D}_r f(\xi_1),$$

or there exists a ξ_2, $a < \xi_2 < b$, *such that*

$$\overline{D}_l f(\xi_2) \leqq \frac{f(b) - f(a)}{b - a} \leqq \underline{D}_r f(\xi_2).$$

Corollary. If for the continuous function $f(x)$ *the lower derivatives* $\underline{D}_r f(\xi)$ *and* $\underline{D}_l f(\xi)$ *in all interior points* ξ *have the property*

$$\underline{D}_r f(\xi) \leqq B, \quad \underline{D}_l f(\xi) \leqq B$$

then

$$\frac{f(b) - f(a)}{b - a} \leqq B.$$

2. Whereas the two lemmas dealt with continuous functions only we consider now real functions which need only to be bounded. Let $g(x)$ be defined in the closed interval $[a, b]$. We introduce the following definitions

$$M_g(\alpha, \beta) = \sup_{\alpha \leqq x \leqq \beta} g(x), \tag{2.1}$$

$$m_g(\alpha, \beta) = \inf_{\alpha \leqq x \leqq \beta} g(x), \tag{2.2}$$

and

$$\sigma_g(\alpha, \beta) = M_g(\alpha, \beta) - m_g(\alpha, \beta), \tag{2.3}$$

the "supremum," the "infinum," and the "oscillation" of $g(x)$ in the closed interval $[\alpha, \beta]$. (It is to be understood that if $[\alpha, \beta]$ should partly exceed the interval $[a, b]$ in which $g(x)$ is defined, the instruction "$\alpha \leqq x \leqq \beta$" in the definitions is to be applied only where meaningful, *i.e.* in the common part of $[a, b]$ and $[\alpha, \beta]$.) Since we assumed boundedness of $g(x)$, the numbers defined in (2.1), (2.2), (2.3) are finite.

For $h > 0$ the function of h

$$M_g(x_0 - h, x_0 + h)$$

is monotone non-decreasing since a supremum can only grow if the sample of which it is taken is enlarged. Similarly

$$m_g(x_0 - h, x_0 + h)$$

is monotone non-increasing, and

$$\sigma_g(x_0 - h, x_0 + h)$$

monotone non-decreasing as functions of $h > 0$.

Since these functions are bounded, the limits

$$\lim_{h\to+0} M_g(x_0 - h, x_0 + h) = M_g(x_0),$$

$$\lim_{h\to+0} m_g(x_0 - h, x_0 + h) = m_g(x_0),$$

$$\lim_{h\to+0} \sigma_g(x_0 - h, x_0 + h) = \sigma_g(x_0),$$

the "local supremum," "local infinum," "local oscillation" of g at x_0 exist. We remark that a function is continuous at x_0 if its local oscillation at x_0 is 0.

LEMMA 3. *$M_g(x)$ and $\sigma_g(x)$ are upper semi-continuous and $m_g(x)$ is lower semi-continuous.*

Proof: Upper semi-continuity of $M_g(x)$ at x_0 means:

$$\overline{\lim_{x_1\to x_0}}\, M_g(x_1) \leqq M_g(x_0).$$

Let $|x_1-x_0|<\delta$, and $0<h<\delta-|x_1-x_0|$. Then the interval $[x_1-h, x_1+h]$ lies in the interior of the interval $[x_0-\delta, x_0+\delta]$, and therefore

$$M_g(x_1) \leqq M_g(x_1 - h, x_1 + h) \leqq M_g(x_0 - \delta, x_0 + \delta);$$

thus

$$\overline{\lim_{x_1\to x_0}}\, M_g(x_1) \leqq M_g(x_0 - \delta, x_0 + \delta).$$

This is true for any $\delta>0$, and thus, for $\delta\to+0$

$$\overline{\lim_{x_1\to x_0}}\, M_g(x_1) \leqq M_g(x_0),$$

as we had to prove. The other statements of the lemma follow in a similar manner.

3. We come now to the main object of our discussion.

THEOREM 1. *A necessary condition for the Riemann integrability of the function $f(x)$ bounded in the interval $a\leqq x\leqq b$ is that, for every $\eta>0$, the set S_η of all x for which $\sigma_f(x)\geqq\eta$ can be covered by finitely many intervals of total length at most η.*

Remark: If the condition of this theorem is fulfilled, then for a *fixed* $\eta>0$ the set S_η can be covered by finitely many intervals of any total length $\epsilon<\eta$. Indeed, if $0<\epsilon<\eta$ then by definition

$$S_\epsilon \supset S_\eta,$$

and thus since S_ϵ can be covered by a finite set of intervals of total length ϵ at most, S_η is in this way also covered.

Proof of THEOREM 1: Suppose the condition is not fulfilled. This would mean that there exists an exceptional $\eta_0>0$ such that any finite interval set covering S_{η_0} has a total length $\geqq\eta_0$. It is convenient to distinguish here two cases:

1) Suppose that S_{η_0} contains a whole interval I of length λ. We take now a partition $a=x_0<x_1<\cdots<x_{n-1}<x_n=b$ of the interval $[a, b]$ and form the difference of the upper and lower Darboux sums for that partition:

$$\overline{S}-\underline{S}=\sum_j M_f(x_{j-1}, x_j)(x_j-x_{j-1})-\sum_j m_f(x_{j-1}, x_j)(x_j-x_{j-1})$$

$$=\sum_j \sigma_f(x_{j-1}, x_j)(x_j-x_{j-1}).$$

Of this sum we consider only those summands which belong to intervals $[x_{j-1}, x_j]$ which have a point of I in the interior. Since such a point, say ξ, belongs to S_{η_0}, we can say for the interval $[x_{j-1}, x_j]$ surrounding it that

$$\sigma_f(x_{j-1}, x_f) \geqq \sigma_f(\xi) \geqq \eta_0.$$

On the other hand these intervals together cover I and therefore have a total length λ at least. Thus

$$\overline{S}-\underline{S} \geqq \eta_0\lambda$$

for Darboux sums belonging to any partition, hence the upper and lower Darboux integrals differ by at least $\eta_0\lambda$, and $f(x)$ cannot be integrable.

2) Suppose that S_{η_0} contains no interval, *i.e.*, does not possess an interior point. Assume that we have a partition of (a, b) of fineness $\frac{1}{2}\delta$, *i.e.*,

$$x_j - x_{j-1} \leqq \frac{\delta}{2}\cdot$$

We can see to it that no point of the partition (except possibly a and b) belongs to S_{η_0}. If x_j should belong to S_{η_0}, we can find in a distance less than $\frac{1}{4}\delta$ another point x_j' not belonging to S_{η_0}. The so corrected partition $(a<x_1'<\cdots<x_{n-1}'<b)$ has still the fineness δ. Then each point of S_{η_0} (except possibly a and b) is contained in the *interior* of an interval of the partition. We consider only such intervals $[x_{j-1}', x_j']$ which have a point ξ of S_{η_0} in their interior. For these we have again

$$\sigma_f(x_{j-1}', x_j') \geqq \sigma_f(\xi) \geqq \eta_0.$$

Since these intervals must cover S_{η_0} their total length adds up at least to η_0, and we have

$$\overline{S}-\underline{S}=\sum_j \sigma(x_{j-1}', x_j')(x_j'-x_{j-1}') \geqq \eta_0^2$$

which again excludes Riemann integrability.

4. We come now to that part of our argument which resembles that one given by Kowalewski and Norris, and which leads to the converse of Theorem I.

THEOREM II. *A sufficient condition for Riemann integrability in* $a \leqq x \leqq b$ *of* *the bounded function $f(x)$ is that for every $\eta>0$ the S_η, defined as in* THEOREM I, *can be covered by finitely many intervals of total length η at most.*

Proof: We consider the function

$$F(z) = \overline{\int_a^z} f(x)dx - \underline{\int_a^z} f(x)dx,$$

which certainly fulfills

$$F(z) \geqq 0.$$

We assume $|f(x)| \leqq M$, and have then as consequence of the well-known additivity of Darboux integrals for $h>0$

$$\begin{aligned} 0 \leqq F(z+h) - F(z) &= \overline{\int_a^{z+h}} f(x)dx - \overline{\int_a^z} f(x)dx - \underline{\int_a^{z+h}} f(x)dx + \underline{\int_a^z} f(x)dx \\ &= \overline{\int_z^{z+h}} f(x)dx - \underline{\int_z^{z+h}} f(x)dx \\ &\leqq Mh + Mh = 2Mh, \end{aligned}$$

and similarly for $h<0$

$$|F(z+h) - F(z)| \leqq 2M|h|,$$

which shows the continuity of $F(z)$.

As far as the derivatives are concerned we see first, for $h>0$.

$$\begin{aligned} \frac{F(z+h) - F(z)}{h} &= \frac{1}{h}\left\{\overline{\int_z^{z+h}} f(x)dx - \underline{\int_z^{z+h}} f(x)dx\right\} \\ &\leqq M_f(z, z+h) - m_f(z, z+h) = \sigma_f(z, z+h) \\ &\leqq \sigma_f(z-h, z+h) \end{aligned}$$

and therefore

$$\begin{aligned} \overline{D}_r F(z) &= \overline{\lim_{h\to+0}} \frac{1}{h}\{F(z+h) - F(z)\} \\ &\leqq \lim_{h\to+0} \sigma_f(z-h, z+h) = \sigma_f(z). \end{aligned}$$

In the same manner

$$\overline{D}_l F(z) \leqq \sigma_f(z)$$

follows. Altogether we have

$$\underline{D}_l F(z) \leqq \overline{D}_l F(z) \leqq \sigma_f(z), \tag{4.1}$$

(4.2) $$\underline{D}_r F(z) \leqq \overline{D}_r F(z) \leqq \sigma_f(z).$$

It will be convenient to introduce a function of an interval $I=(\alpha, \beta)$, for $\alpha<\beta$, as follows:

$$F(I) = F(\alpha, \beta) = F(\beta) - F(\alpha).$$

Now let S_η be covered by finitely many intervals $I_j=(\alpha_j, \beta_j)$ of total length η at most. The complementary intervals we call $\overline{I}_j'$. We have then

(4.3) $$F(b) = F(b) - F(a) = \sum_j F(I_j) + \sum_j F(I_j').$$

In $\overline{I}_j'$ we have throughout

$$\sigma_f(x) < \eta$$

whereas in general $|f(x)| \leqq M$ entails

$$\sigma_f(x) \leqq 2M.$$

The inequalities (4.1) and (4.2) together with the Corollary show therefore

$$F(I_j) \leqq 2M \cdot l(I_j)$$
$$F(I_j') \leqq \eta \cdot l(I_j')$$

where $l(I)$ stands for the length of I. We infer therefore from (4.3)

$$F(b) \leqq 2M \sum_j l(I_j) + \eta \sum_j l(I_j')$$
$$\leqq 2M\eta + \eta(b - a) = \eta(2M + b - a).$$

Since η is any positive number and $F(b) \geqq 0$ we obtain now

$$F(b) = 0,$$

or explicitly,

$$\overline{\int_a^b} f(x)dx - \underline{\int_a^b} f(x)dx = 0,$$

which proves our theorem.

5. So far we have remained in the realm of the elementary theory of Riemann-Darboux integrals. But we can now easily connect our two theorems with the Lebesgue theory and prove

THEOREM III. *Necessary and sufficient condition for the Riemann integrability of the bounded function $f(x)$ is that the set S_0 of those x where $\sigma_f(x)>0$ is of Lebesgue measure zero.*

Proof: It is clear firstly that

$$S_0 = S_1 \cup S_{1/2} \cup S_{1/3} \cup \cdots \cup S_{1/n} \cup \cdots . \tag{5.1}$$

Now the *necessary* condition of Theorem I implies, in virtue of the "Remark" following it, that all S_η are of Lebesgue measure 0, and then (5.1) shows that

$$m_L S_0 = 0 \tag{5.2}$$

(m_L meaning Lebesgue measure).

Assume now conversely that is true. Then, since

$$S_\eta \subset S_0$$

we have also

$$m_L S_\eta = 0 \tag{5.3}$$

for any $\eta > 0$. This means that S_η can be covered by *denumerably* many intervals of at most, let us say, total length η.

But now, as we shall see in a moment, S_η is closed, and can therefore, in virtue of the Heine-Borel theorem, be covered with only *finitely* many among those infinitely many covering intervals, and thus (5.3) implies that S_η can be covered with *finitely* many intervals of total length η at most, which shows, according to Theorem II, that (5.3) is also *sufficient* for the Riemann integrability of $f(x)$.

The closure of S_η, comprising those points x in which $\sigma_f(x) \geqq \eta$, follows immediately from Lemma 3.

Of course, the arguments of section 5, employing the Heine-Borel theorem, which is essentially equivalent to the theorem of uniform continuity of a continuous function in a closed interval, overstep the limits which this note had set itself. Its main purpose remains the proof given for Theorem II.

Note to Paper 55

This paper has been reviewed in the Z, vol. 55 (1955), p. 53, and in the MR, vol. 15 (1954), p. 783.

On Dedekind Sums and Lattice Points in a Tetrahedron

by

HANS RADEMACHER

The Dedekind sums are defined as

$$s(h,k) = \sum_{\mu=1}^{k} \frac{\mu}{k}\left(\left(\frac{h\mu}{k}\right)\right), \tag{1}$$

where $(h, k) = 1$ and

$$((\xi)) = \xi - [\xi] - \frac{1}{2}, \qquad \xi \text{ not an integer}$$
$$= 0, \qquad \xi \text{ an integer.}$$

Let a, b, c be 3 positive integers, which are pairwise coprime, and let $N_3(a, b, c)$ be the number of lattice points in the tetrahedron

$$0 \leqslant x < a, \qquad 0 \leqslant y < b, \qquad 0 \leqslant z < c, \qquad 0 < \frac{x}{a} + \frac{y}{b} + \frac{z}{c} < 1. \tag{2}$$

Mordell has recently proved [1] that

$$s(bc, a) + s(ca, b) + s(ab, c) + N_3(a, b, c)$$
$$= \frac{1}{6} abc + \frac{1}{4}(bc + ca + ab) + \frac{1}{4}(a + b + c) \tag{3}$$
$$+ \frac{1}{12}\left(\frac{bc}{a} + \frac{ca}{b} + \frac{ab}{c}\right) + \frac{1}{12\,abc} - 2.$$

If one compares this result with Theorem 20 of [2] one finds, by the use of the reciprocity formula for Dedekind sums,

$$N_3(a, b, c) \equiv \frac{1}{4}(a+1)(b+1) \pmod 2$$

under the assumption $8|c$, $3|abc$.

I shall now remove these restrictions and prove the

Theorem I: If a, b, c are pairwise coprime positive integers, then the number $N_3(a, b, c)$ of lattice points in the tetrahedron (2) satisfies the congruence

$$N_3(a, b, c) \equiv \frac{1}{4}(a+1)(b+1)(c+1) \pmod 2. \tag{4}$$

This formula shows that N_3 is mostly even, and is odd only if one of the numbers a, b, c is even and the other two are of the form $4n+1$.

In virtue of Mordell's formula (3) this theorem, as is easily seen, is equivalent with the

Theorem II: Under the above mentioned conditions for a, b, c we have

$$(5)\qquad \left(s(bc, a)-\frac{bc}{12a}\right)+\left(s(ca, b)-\frac{ca}{12b}\right)+\left(s(ab, c)-\frac{ab}{12c}\right)$$
$$\equiv -\frac{1}{4}-\frac{1}{12}abc+\frac{1}{12abc} \pmod 2$$

We apply the reciprocity formula for Dedekind sums

$$s(h, k)+s(k, h)=-\frac{1}{4}+\frac{h}{12k}+\frac{k}{12h}+\frac{1}{12hk}$$

(see e. g. [2], Theorem 3) to the third parenthesis of the left-hand member of (5) and find that Theorem II in turn is equivalent with the

Theorem III: Under the same conditions for a, b, c we have

$$(6)\qquad \left(s(bc, a)-\frac{bc}{12a}\right)+\left(s(ca, b)-\frac{ca}{12b}\right)-\left(s(c, ab)-\frac{c}{12ab}\right)$$
$$+\frac{abc}{12}\equiv 0 \pmod 2.$$

We shall prove this theorem under the assumption that ab is odd. Since among the numbers a, b, c at least two must be odd, we will then have proved Theorem 2 in its full generality. But this theorem is symmetric in a, b, c, and therefore, because of the equivalence, Theorem III is then proved without exception. After multiplication with $12ab$ the congruence (6) goes over into

$$12ab\,s(bc, a)+12ab\,s(ca, b)-12ab\,s(c, ab)+c(a^2-1)(b^2-1)\equiv 0 \pmod{24ab}.$$

Let us put

$$(7)\qquad D=12ab\,s(bc, a)+12ab\,s(ca, b)-12ab\,s(c, ab).$$

Since ab is assumed to be odd we have to prove

$$(8\text{ a})\qquad D\equiv -c(a^2-1)(b^2-1) \pmod{3ab}$$

and

$$(8\text{ b})\qquad D\equiv -c(a^2-1)(b^2-1) \pmod 8.$$

For (8 a) we require the

Lemma 1: For $(h, k) = 1$ we have

$$12\,h\,k\,s(h,k) \equiv (1-k^2)(1+h^2) \pmod{3k}, \tag{9 a}$$

$$12\,h\,k\,s(k,h) \equiv k^2(h^2-1) \pmod{3k}. \tag{9 b}$$

This lemma can be extracted from pp. 395, 396 of [2], in particular the formulae (9.1), (9.21), (9.4), and from

$$12\,k\,s(h,k) \equiv h(k^2-1) \pmod{3}$$

p. 396 of [2]. Now (9 a) yields

$$12\,a\,b\,c\,s(b\,c, a) \equiv (1-a^2)(1+b^2c^2) \pmod{3a}$$

and (9 b)

$$12\,a\,b\,c\,s(c\,a, b) \equiv c^2a^2(b^2-1) \pmod{3a},$$

moreover (9 a) again

$$12\,a\,b\,c\,s(c, a\,b) \equiv (1-a^2b^2)(1+c^2) \pmod{3a}.$$

We obtain thus

$$c\,D \equiv -a^2-c^2+b^2c^2-a^2c^2+a^2b^2+a^2b^2c^2 \pmod{3a} \tag{10 a}$$

and since D is symmetric in a and b, also the congruence

$$c\,D \equiv -b^2-c^2+a^2c^2-b^2c^2+a^2b^2+a^2b^2c^2 \pmod{3b} \tag{10 b}$$

follows in the same way. Now a little computation shows that the right-hand member of (10 a) is indeed congruent to

$$-c^2(a^2-1)(b^2-1) \pmod{3a},$$

the difference being congruent to

$$a^2(b^2-1)(1-c^2) \equiv 0 \pmod{3a}.$$

Thus we have proved that

$$c\,D \equiv -c^2(a^2-1)(b^2-1) \pmod{3a}$$

and this holds true, because of symmetry in a and b, also modulo $3\,b$. Hence we have now

$$c\,D \equiv -c^2(a^2-1)(b^2-1) \pmod{3\,a\,b}. \tag{11}$$

If c, which is prime to $a\,b$, is not divisible by 3, we derive (8 a) from (11) by cancelling c. If, however, $3|c$, we can from (11) only infer

$$D \equiv -c(a^2-1)(b^2-1) \pmod{a\,b}. \tag{11 a}$$

But (9 b) implies, if 3 is not a divisor of h,

$$12\,h\,s(k,h) \equiv 0 \pmod{3}.$$

If we apply this to the three members of D we see that $D \equiv 0 \pmod{3}$. Moreover the right-hand member of (11 a) is divisible by 3 since 3 is not a divisor of $a\,b$ in this case. Therefore (11 a) is also true modulo 3, and we have proved (8 a) also for $3|c$.

For the proof of (8 b) we need the

Lemma 2: For k odd we have

$$12\,k\,s\,(h, k) \equiv k + 1 - 2\left(\frac{h}{k}\right) \qquad (\text{mod } 8),$$

where (h/k) denotes the Jacobi symbol.

This lemma, which goes back to Dedekind, is to be found as Theorem 18 in [2].[1]

A direct application of Lemma 2 gives us

$$\begin{aligned} &D + c\,(a^2 - 1)\,(b^2 - 1) \\ &\equiv b\left\{a + 1 - 2\left(\frac{b\,c}{a}\right)\right\} + a\left\{b + 1 - 2\left(\frac{c\,a}{b}\right)\right\} - \left\{a\,b + 1 - 2\left(\frac{c}{a\,b}\right)\right\} \quad (\text{mod } 8), \end{aligned}$$

where we have noted $a^2 \equiv 1$ (mod 8). Thus,

$$\begin{aligned} D + c\,(a^2 - 1)\,(b^2 - 1) &\equiv a\,b + a + b - 1 + 2\left\{\left(\frac{c}{a}\right) - a\left(\frac{a}{b}\right)\right\}\left\{\left(\frac{c}{b}\right) - b\left(\frac{b}{a}\right)\right\} \\ &\quad - 2\,a\,b\left(\frac{a}{b}\right)\left(\frac{b}{a}\right) \\ &\equiv a\,b + a + b - 1 - 2\,a\,b\,(-1)^{\frac{1}{4}(a-1)(b-1)} \quad (\text{mod } 8), \end{aligned}$$

where we have used the reciprocity law of the Jacobi symbol and have observed that the expressions in the braces are even.

We finally get

$$D + c\,(a^2 - 1)\,(b^2 - 1) \equiv 2\,a\,b\left\{1 - (-1)^{\frac{1}{4}(a-1)(b-1)}\right\} - (a - 1)\,(b - 1) \quad (\text{mod } 8),$$

and since for any integer n

$$n \equiv \frac{1}{2}\,(1 - (-1)^n) \qquad (\text{mod } 2),$$

we have finally, with $n = 1/4\,(a - 1)\,(b - 1)$,

$$D + c\,(a^2 - 1)\,(b^2 - 1) \equiv 0 \qquad (\text{mod } 8),$$

which is (8 b).

Thus our theorems I, II, III are proved. A more direct proof of I might be desirable. Theorem III contains the Theorems 20, 21, 22 of [2] as special cases. The Theorems II and III can be applied through a process of iteration to combinations of Dedekind sums in which more than 3 parameters appear,

---

[1] Through a typographical error the factor 2 in front of the Jacobi symbol is missing in the statement of that theorem in [2].

which are pairwise coprime. Mordell gives at the end of his paper [1] an identity involving such a combination for four parameters. We obtain, less precisely, only a congruence.

Theorem IV: If a, b, c, d are positive integers, which are pairwise coprime, then

$$s(bcd, a) + s(cda, b) + s(dab, c) + s(abc, d)$$

$$\equiv -\frac{1}{4} + \frac{1}{12}\left(\frac{bcd}{a} + \frac{cda}{b} + \frac{dab}{c} + \frac{abc}{d}\right) - \frac{abcd}{6} + \frac{1}{12\,abcd} \pmod 2. \tag{12}$$

Indeed, keeping cd together, we obtain from Theorem II

$$\left(s(bcd, a) - \frac{bcd}{12a}\right) + \left(s(cda, b) - \frac{cda}{12b}\right) + \left(s(ab, cd) - \frac{ab}{12cd}\right) \tag{13}$$

$$\equiv -\frac{1}{4} - \frac{1}{12}abcd + \frac{1}{12\,abcd} \pmod 2,$$

and then from Theorem III, keeping ab together,

$$\left(s(abc, d) - \frac{abc}{12d}\right) + \left(s(dab, c) - \frac{dab}{12c}\right) \tag{14}$$

$$-\left(s(ab, dc) - \frac{ab}{12dc}\right) \equiv -\frac{abcd}{12} \pmod 2.$$

Addition of (13) and (14) yields (12).

Through repetition of this process formulae with any number of parameters can evidently be obtained.

UNIVERSITY OF PENNSYLVANIA

References

1. L. J. MORDELL, Lattice points in a tetrahedron and generalized Dedekind sums, *Journal Indian Mathem. Soc.* **15** (1951), 41–46.
2. H. RADEMACHER and A. WHITEMAN, Theorems on Dedekind Sums, *Amer. Math. Journal* **63** (1941), 377-407.

This paper has been reviewed in the Z, vol. 56 (1955), p. 274, and in the MR, vol. 16 (1955), p. 341.

Page 53, reference 2.
See paper 44 of the present collection.

$N_3(a,b,c) \equiv$

$\frac{1}{4}(a+1)(b+1)(c+1)$ modulo 2

$= \frac{1}{4}(a-1+2)(b-1+2)(c-1+2)$

$= \frac{1}{4}(a-1)(b-1)(c-1) + \frac{(a-1)(b-1)}{2} + \frac{(b-1)(c-1)}{2} + \frac{(c-1)(a-1)}{2}$

$+ (a-1) + (b-1) + (c-1)$

But:

$N_3(a,b,c) - \frac{(a-1)(b-1)}{2} - \frac{(b-1)(c-1)}{2} - \frac{(c-1)(a-1)}{2}$

$- (a-1) - (b-1) - (c-1)$

is exactly the number of lattice points in the ~~interior~~ of the tetrahedron (~~taken~~ the points in the faces and on the edges being taken away)

This therefore

$\equiv \frac{1}{4}(a-1)(b-1)(c-1)$ modulo 2

as to be found in

M.M. Artyuhov, On the number of lattice points in certain tetrahedra (Russian)

Mat. Sbornik (N.S.) 57 (99) (1962) pp 3-12

Mathem. Reviews 25 # 2039, pp 399, 400

Reviewed by J.W.S. Cassels.

Note to Paper 56a

These handwritten remarks by the author are not dated, but the quotation shows that they were written in 1963 at the earliest. If $N_3(a,b,c)$ is defined as in paper 56 and $M_3(a,b,c)$ stands for the number of *interior* lattice points of the tetrahedron defined by (2), then

$$M_3(a,b,c) = N_3(a,b,c) - \frac{1}{2}\{(a-1)(b-1) + (b-1)(c-1)+(c-1)(a-1)\}$$
$$-\{(a-1) + (b-1) + (c-1)\},$$

and it follows from (4) that

$$M_3(a,b,c) \equiv \frac{1}{4}(a-1)(b-1)(c-1) \pmod 2,$$

a result obtained by M. M. Artjuhov (*Mat. Sb.* (N.S.), vol. 57, no. 99 (1962), pp. 3-12; MR, vol. 25 (1963), #2039, pp. 359-400).

GENERALIZATION OF THE RECIPROCITY FORMULA FOR DEDEKIND SUMS

By Hans Rademacher

Recently Mordell [2] has proved a certain relation in which three Dedekind sums appear together and which includes the well-known reciprocity formula as a special case. However, Mordell's formula contains as a further summand the number of lattice points in a certain tetrahedron and therefore has to be looked upon rather as a determination of this number of lattice points by means of Dedekind sums. A method developed by Redei [4] and simplified by Carlitz [1] leads now to another relation between 3 Dedekind sums, a formula which contains otherwise only elementary expressions, and which can be looked upon as a natural generalization of the reciprocity formula, which it implies as a special case. The formula in question seems to go beyond those properties which express the group property of the modular substitution in its application to log $\eta(t)$. Since Redei's and Carlitz's notation differ from mine I shall briefly derive the basic formulae again.

1. With the notation

$$((x)) = \begin{cases} x - [x] - \frac{1}{2} & \text{for } x \text{ not integer} \\ 0 & \text{for } x \text{ integer} \end{cases}$$

the function $((\mu/a))$ is periodic in μ of period a, where $a > 0$ and μ are integers. With ξ a primitive a-th root of unity there exists therefore a "finite Fourier expansion"

$$\left(\left(\frac{\mu}{a}\right)\right) = \sum_{j=0}^{a-1} A_j \xi^{\mu j},$$

where

$$A_j = \frac{1}{a} \sum_{\mu=1}^{a} \left(\left(\frac{\mu}{a}\right)\right) \xi^{-j\mu}$$

and in particular

$$A_0 = \frac{1}{a} \sum_{\mu=1}^{a} \left(\left(\frac{\mu}{a}\right)\right) = 0,$$

and for $0 < j < a$

$$A_j = \frac{1}{a} \sum_{\mu=1}^{a-1} \left(\frac{\mu}{a} - \frac{1}{2}\right) \xi^{-j\mu} = \frac{1}{a} \frac{1}{\xi^{-j} - 1} + \frac{1}{2a},$$

Received June 23, 1953.

so that

$$\left(\left(\frac{\mu}{a}\right)\right) = \frac{1}{a}\sum_{j=1}^{a-1}\left(\frac{\xi^j}{1-\xi^j}+\frac{1}{2}\right)\xi^{\mu j}, \tag{1}$$

which is essentially a formula of Eisenstein's.

For the Dedekind sum

$$s(b, a) = \sum_{\mu \bmod a}\left(\left(\frac{\mu}{a}\right)\right)\left(\left(\frac{b\mu}{a}\right)\right) \tag{2}$$

with $(a, b) = 1$ we have therefore

$$\begin{aligned} s(b, a) &= \frac{1}{a^2}\sum_{j=1}^{a-1}\sum_{k=1}^{a-1}\left(\frac{\xi^j}{1-\xi^j}+\frac{1}{2}\right)\left(\frac{\xi^k}{1-\xi^k}+\frac{1}{2}\right)\sum_{\mu \bmod a}\xi^{\mu(j+bk)} \\ &= \frac{1}{a}\sum_{k=1}^{a-1}\left(\frac{\xi^{-bk}}{1-\xi^{-bk}}+\frac{1}{2}\right)\left(\frac{\xi^k}{1-\xi^k}+\frac{1}{2}\right), \end{aligned} \tag{3a}$$

a formula which in trigonometical form I have first used in [3]. However (2) can be rewritten, if one observes, as special case $\mu = 0$ in (1), that

$$0 = \frac{1}{a}\sum_{j=1}^{a-1}\frac{\xi^j}{1-\xi^j}+\frac{a-1}{2a}.$$

By means of this equality (2) goes over into

$$s(b, a) = \frac{1}{a}\sum_{k=1}^{a-1}\frac{\xi^k}{(\xi^{bk}-1)(1-\xi^k)}-\frac{a-1}{4a}. \tag{3b}$$

So far ξ has been a fixed primitive a-th root of unity. In the summation in (3) ξ^k runs through all a-th roots of unity except 1. We may write therefore

$$s(b, a) = \frac{1}{a}\sum_{\xi}{}'\frac{\xi}{(\xi^b-1)(1-\xi)}-\frac{a-1}{4a}, \tag{4}$$

where the prime indicates $\xi \neq 1$, and where ξ runs through all other a-th roots of unity. Since evidently $s(-b, a) = -s(b, a)$ we obtain from (4)

$$\begin{aligned} s(b, a) &= -\frac{1}{a}\sum_{\xi}{}'\frac{\xi}{(\xi^{-b}-1)(1-\xi)}+\frac{a-1}{4a} \\ &= -\frac{1}{a}\sum_{\xi}{}'\frac{\xi^{-1}}{(\xi^b-1)(1-\xi^{-1})}+\frac{a-1}{4a} \end{aligned}$$

because ξ and ξ^{-1} range over the same system of roots of unity. We have thus finally

$$s(b, a) = -\frac{1}{a}\sum_{\xi}{}'\frac{1}{(\xi^b-1)(\xi-1)}+\frac{a-1}{4a}. \tag{5}$$

2. Let now a, b, c be three positive integers, pairwise without common divisor. We put

$$F(x) = \frac{x^a - 1}{x - 1}, \qquad G(x) = \frac{x^b - 1}{x - 1}, \qquad H(x) = \frac{x^c - 1}{x - 1}. \tag{6}$$

Then $F(x)$, $G(x)$, $H(x)$ are three polynomials which are pairwise coprime. Therefore also $(G(x)H(x), H(x)F(x), F(x)G(x)) = 1$, and thus there exist three polynomials $\varphi(x)$, $\psi(x)$, $\chi(x)$ such that

$$G(x)H(x)\varphi(x) + H(x)F(x)\psi(x) + F(x)G(x)\chi(x) = 1. \tag{7}$$

It is easily seen that φ, ψ, χ can be so chosen that

$$\text{degree of } \varphi < \text{degree of } F = a - 1, \tag{8a}$$

$$\text{degree of } \psi < \text{degree of } G = b - 1, \tag{8b}$$

$$\text{degree of } \chi < \text{degree of } H = c - 1, \tag{8c}$$

and that these conditions determine φ, ψ, χ uniquely in (7). Indeed, let us consider first

$$G(x)H(x)\varphi(x) + F(x)\lambda(x) = 1. \tag{9}$$

Because $(G(x)H(x), F(x)) = 1$ the polynomials φ and λ are uniquely determined under the restrictions (8a) and

$$\text{degree of } \lambda < \text{degree of } GH. \tag{10}$$

In order to get from (9) to (7) we have then to solve $H(x)\varphi(x) + G(x)\,\chi(x) = \lambda(x)$, which because of (10) is possible and in only one way so as to fulfill also (8b) and (8c).

An exceptional case needs some clarification. We exclude the case $a = b = c = 1$ as of no interest from our consideration (although our final formula (30) will be trivially true also in this case). If one of the numbers a, b, c, let us say a, equals 1, i.e. $F(x) = 1$, then (8a) should be interpreted as $\varphi(x) = 0$. Similarly when also $b = 1$, i.e. $G(x) = 1$. It is clear that (7) can then also be fulfilled with the observation of (8a)-(8c).

Let now ξ be any a-th root of unity different from 1 (let us assume for the moment $a > 1$) then (6) and (7) show

$$\varphi(\xi) = \frac{1}{G(\xi)H(\xi)} = \frac{(\xi - 1)^2}{(\xi^b - 1)(\xi^c - 1)}. \tag{11}$$

Thus $\varphi(x)$ is known for the $(a - 1)$ possible values of ξ. Since it moreover is a polynomial whose degree does not exceed $a - 2$, $\varphi(x)$ is completely determined through the Lagrange interpolation formula

$$\varphi(x) = F(x) \sum_{\xi}' \frac{\varphi(\xi)}{(x - \xi)F'(\xi)}. \tag{12}$$

Now from $(x - 1)F(x) = x^a - 1$ follows $(x - 1)F'(x) + F(x) = ax^{a-1}$ and hence

$$F'(\xi) = \frac{a}{\xi(\xi - 1)}. \tag{13}$$

This with (11) and (12) yields

$$\varphi(x) = \frac{1}{a} F(x) \sum_{\xi}' \frac{\xi(\xi - 1)^3}{(\xi^b - 1)(\xi^c - 1)(x - \xi)}. \tag{14}$$

We have so far omitted $a = 1$. But in this case the sum in (14) is void, and the equation should simply be read as $\varphi(x) = 0$, which agrees with our remarks about the exceptional case.

If η means any b-th root of unity and ζ any c-th root of unity, both however different from 1, we can restate (7) by means of (14) as

$$\begin{aligned}\frac{1}{a} \sum_{\xi}' \frac{\xi(\xi - 1)^3}{(\xi^b - 1)(\xi^c - 1)(x - \xi)} + \frac{1}{b} \sum_{\eta}' \frac{\eta(\eta - 1)^3}{(\eta^c - 1)(\eta^a - 1)(x - \eta)} \\ + \frac{1}{c} \sum_{\zeta}' \frac{\zeta(\zeta - 1)^3}{(\zeta^a - 1)(\zeta^b - 1)(x - \zeta)} = \frac{1}{F(x)G(x)H(x)},\end{aligned} \tag{15}$$

where the prime refers to the omission of the value 1 for the roots of unity. We change the expression (15) slightly in order to avoid the single factors ξ, η, ζ in the numerators. Setting $x = 0$ we obtain

$$\begin{aligned}-\frac{1}{a} \sum_{\xi}' \frac{(\xi - 1)^3}{(\xi^b - 1)(\xi^c - 1)} - \frac{1}{b} \sum_{\eta}' \frac{(\eta - 1)^3}{(\eta^c - 1)(\eta^a - 1)} \\ - \frac{1}{c} \sum_{\zeta}' \frac{(\zeta - 1)^3}{(\zeta^a - 1)(\zeta^b - 1)} = 1.\end{aligned} \tag{16}$$

If we subtract (16) from (15) and divide by x we have the result

$$\begin{aligned}\frac{1}{a} \sum_{\xi}' \frac{(\xi - 1)^3}{(\xi^b - 1)(\xi^c - 1)(x - \xi)} + \frac{1}{b} \sum_{\eta}' \frac{(\eta - 1)^3}{(\eta^c - 1)(\eta^a - 1)(x - \eta)} \\ + \frac{1}{c} \sum_{\zeta}' \frac{(\zeta - 1)^3}{(\zeta^a - 1)(\zeta^b - 1)(x - \zeta)} = \frac{1 - F(x)G(x)H(x)}{xF(x)G(x)H(x)} = \Phi(x),\end{aligned} \tag{17}$$

say. In (17) we expand the rational functions of x in both members in power series about the point $x = 1$ which leads to

$$\begin{aligned}-\sum_{n=0}^{\infty} (x - 1)^n \left\{\frac{1}{a} \sum_{\xi}' \frac{(\xi - 1)^{2-n}}{(\xi^b - 1)(\xi^c - 1)} + \frac{1}{b} \sum_{\eta}' \frac{(\eta - 1)^{2-n}}{(\eta^c - 1)(\eta^a - 1)} \right. \\ \left. + \frac{1}{c} \sum_{\zeta}' \frac{(\zeta - 1)^{2-n}}{(\zeta^a - 1)(\zeta^b - 1)}\right\} = \sum_{n=0}^{\infty} \frac{\Phi^{(n)}(1)}{n!} (x - 1)^n.\end{aligned} \tag{18}$$

Comparison of the coefficients of $(x - 1)^2$ gives in particular

$$(19)\quad -\frac{1}{a}\sum_{\xi}{}' \frac{1}{(\xi^b - 1)(\xi^c - 1)} - \frac{1}{b}\sum_{\eta}{}' \frac{1}{(\eta^c - 1)(\eta^a - 1)} - \frac{1}{c}\sum_{\zeta}{}' \frac{1}{(\zeta^a - 1)(\zeta^b - 1)} = \frac{1}{2}\Phi''(1).$$

Incidentally, the relation (16) is not contained in (18), but is one of a sequence of relations obtained by expanding (17) near $x = \infty$.

3. The left-hand term in (19) can be expressed by Dedekind sums. For that purpose we define integers a', b', c' through the congruences

$$(20)\quad aa' \equiv 1 \pmod{bc}, \qquad bb' \equiv 1 \pmod{ca}, \qquad cc' \equiv 1 \pmod{ab}.$$

If ξ runs through all a-th roots of unity except 1 then $\xi^{c'}$ does the same, so that we derive from (5)

$$-\frac{1}{a}\sum_{\xi}{}' \frac{1}{(\xi^b - 1)(\xi^c - 1)} = -\frac{1}{a}\sum_{\xi}{}' \frac{1}{(\xi^{bc'} - 1)(\xi - 1)} = s(bc', a) - \frac{1}{4} + \frac{1}{4a}.$$

Thus (19) appears now as

$$(21)\quad s(bc', a) + s(ca', b) + s(ab', c) - \frac{3}{4} + \frac{1}{4}\left(\frac{1}{a} + \frac{1}{b} + \frac{1}{c}\right) = \frac{1}{2}\Phi''(1).$$

It may be remarked that because of $bc' \cdot b'c \equiv 1 \pmod{a}$ we have

$$(22)\quad s(bc', a) = s(b'c, a) = \sum_{\mu \bmod a} \left(\left(\frac{b\mu}{a}\right)\right)\left(\left(\frac{c\mu}{a}\right)\right),$$

which shows that the left-hand member of (21) is fully symmetric in a, b, c.

4. There remains the determination of the right-hand member of (21). From (17) we take

$$(23)\quad 1 - P(x) = xP(x)\Phi(x)$$

with

$$(24)\quad P(x) = F(x)G(x)H(x).$$

The definitions (6) show that $P(1) = abc$ so that

$$(25)\quad \Phi(1) = \frac{1}{abc} - 1.$$

Differentiation of (23) furnishes

$$\Phi'(1) = -\frac{1}{P(1)}\{P'(1) + (P'(1) + P(1))\Phi(1)\} \tag{26a}$$

$$\begin{aligned}\Phi''(1) = -\frac{1}{P(1)}\{P''(1) &+ (P''(1) + 2P'(1))\Phi(1) \\ &+ 2(P'(1) + P(1))\Phi'(1)\}.\end{aligned} \tag{26b}$$

Thus $P'(1)$ and $P''(1)$ have to be computed. Let us introduce, according to (6), $Q(x) = P(x)\,(x-1)^3 = (x^a - 1)\,(x^b - 1)\,(x^c - 1)$. The fourth and fifth derivatives of this equation give

$$24P'(1) = Q^{IV}(1), \tag{27a}$$

$$60P''(1) = Q^{V}(1). \tag{27b}$$

Now, since

$$Q(x) = x^{a+b+c} - x^{b+c} - x^{c+a} - x^{a+b} + x^a + x^b + x^c - 1$$

we have

$$\begin{aligned}\frac{Q^{IV}(1)}{4!} = \binom{a+b+c}{4} - \binom{b+c}{4} - \binom{c+a}{4} - \binom{a+b}{4} \\ + \binom{a}{4} + \binom{b}{4} + \binom{c}{4}\end{aligned} \tag{28a}$$

and

$$\begin{aligned}\frac{Q^{V}(1)}{5!} = \binom{a+b+c}{5} - \binom{b+c}{5} - \binom{c+a}{5} - \binom{c+b}{5} \\ + \binom{a}{5} + \binom{b}{5} + \binom{c}{5}.\end{aligned} \tag{28b}$$

These expressions are polynomials in a, b, c of the fourth and fifth degree respectively,

$$\frac{Q^{IV}(1)}{4!} = p_4(a, b, c), \qquad \frac{Q^{V}(1)}{5!} = p_5(a, b, c),$$

which are symmetric in their 3 variables. From (28a) and (28b) we get

$$p_4(0, b, c) = p_4(a, 0, c) = p_4(a, b, 0) = 0,$$

$$p_5(0, b, c) = p_5(a, 0, c) = p_5(a, b, 0) = 0,$$

and we infer that both polynomials must contain the factor abc. Observing the degrees of these polynomials we conclude that they must be of the forms

$$p_4(a, b, c) = abc\{A(a + b + c) + B\}$$
$$p_5(a, b, c) = abc\{C(a + b + c)^2 + D(a^2 + b^2 + c^2) + E(a + b + c) + F\}.$$

By testing these polynomials for some more special values by means of their definitions (28a), (28b) we determine the unknown coefficients A to F and obtain finally

$$\text{(29a)} \qquad \frac{Q^{IV}(1)}{4!} = p_4(a, b, c) = \frac{abc}{2}(a + b + c - 3)$$

and

$$\text{(29b)} \qquad \frac{Q^{V}(1)}{5!} = p_5(a, b, c) = \frac{abc}{24}\{3(a + b + c)^2 + (a^2 + b^2 + c^2) - 24(a + b + c) + 42\}.$$

Equations (29) and (27) determine $P'(1)$, $P''(1)$ which in turn together with (25) have to be used in (26). Some straightforward calculations yield

$$\Phi''(1) = \frac{1}{6}\left(\frac{a}{bc} + \frac{b}{ca} + \frac{c}{ab}\right) + \frac{1}{2}\left(\frac{1}{a} + \frac{1}{b} + \frac{1}{c}\right) - 2.$$

We combine this formula with (21) and formulate our main result in the

THEOREM. *If a, b, c are positive integers, pairwise without common divisor, and if the integers a', b', c' are defined by the congruences* (20) *then the Dedekind sums defined in* (2) *fulfill the relation*

$$\text{(30)} \qquad s(bc', a) + s(ca', b) + s(ab', c) = -\frac{1}{4} + \frac{1}{12}\left(\frac{a}{bc} + \frac{b}{ca} + \frac{c}{ab}\right).$$

The frequently treated reciprocity formula appears as a corollary by the specialization $c = c' = 1$.

REFERENCES

1. L. CARLITZ, *The reciprocity theorem for Dedekind sums*, Pacific Journal of Mathematics, vol. 3(1953), pp. 523–527.
2. L. J. MORDELL, *Lattice points in a tetrahedron and generalized Dedekind sums*, Journal of the Indian Mathematical Society (N. S.), vol. 15(1951), pp. 41–46.
3. H. RADEMACHER, *Über eine Reziprozitätsformel aus der Theorie der Modulfunktionen* (Hungarian with German resumé), Matematikai es Fizikai Lapok, vol. 40(1933), pp. 24–34.
4. L. REDEI, *Elementarer Beweis und Verallgemeinerung einer Reziprozitätsformel von Dedekind*, Acta Scientiarum Mathematicarum Szeged, vol. 12, part B (1950), pp. 236–239.

THE INSTITUTE FOR ADVANCED STUDY
AND
THE UNIVERSITY OF PENNSYLVANIA.

This paper has been reviewed in the Z, vol. 57 (1956), p. 38, and in the MR, vol. 17 (1956), p. 15.

Page 396, fifth display line from bottom.
Read $\binom{a+b}{5}$ instead of $\binom{c+b}{5}$.

Page 397, reference 3.
See paper 32 of the present collection.

ON THE TRANSFORMATION OF $\log \eta(\tau)$

By HANS RADEMACHER*

[Received January 7, 1955]

1. A particularly simple proof of the equation

$$\log \eta(-1/\tau) = \tfrac{1}{2}\log \tfrac{\tau}{i} + \log \eta(\tau)$$

has recently been given by C. L. Siegel [1]. In that proof contour integration and the calculus of residues are applied to a certain product of cotangents. In the following I widen this method so that it covers the general modular transformation, which I will write as

$$\tau = (h + iz)/k, \ \tau' = (h' + i z^{-1})/k$$

with $(h, k) = 1$, $k > 0$, $\mathcal{R}(z) > 0$, $hh' \equiv -1 \pmod{k}$.

Of special interest seems to me the way in which this time the Dedekind sum $s(h, k)$ appears.

We have to prove, with

$$\eta(\tau) = e^{\pi i\tau/12} \prod_{m=1}^{\infty} (1 - e^{2\pi i m\tau}), \ \mathrm{Im}(\tau) > 0,$$

$$\log \eta\left(\frac{h' + i z^{-1}}{k}\right) = \log \eta\left(\frac{h + i z}{k}\right) + \tfrac{1}{2}\log z + \frac{\pi i}{12k}(h' - h) + \\ + \pi i s(h, k), \qquad (1.1)$$

where

$$s(h,k) = \sum_{\mu=1}^{k-1} \left(\frac{\mu}{k} - \frac{1}{2}\right)\left(\frac{\mu h}{k} - \left[\frac{\mu h}{k}\right] - \frac{1}{2}\right). \qquad (1.2)$$

Now, after the definition of $\eta(\tau)$,

$$\log \eta\left(\frac{h + iz}{k}\right) = \frac{\pi i(h + iz)}{12k} + \sum_{m=1}^{\infty} \log(1 - e^{2\pi i h m/k} e^{-2\pi z m/k})$$

$$= \frac{\pi i(h + iz)}{12k} + \sum_{\mu=1}^{k} \sum_{q=0}^{\infty} \log(1 - e^{2\pi i h\mu/k} e^{-2\pi z(qk+\mu)/k}).$$

*Guggenheim Research Fellow 1954-1955.

$$= \frac{\pi i h}{12k} - \frac{\pi z}{12k} - \sum_{\mu=1}^{k} \sum_{q=0}^{\infty} \sum_{r=1}^{\infty} \frac{1}{r} e^{2\pi i h \mu r/k} e^{-2\pi z (qk+\mu) r/k},$$

$$= \frac{\pi i h}{12k} - \frac{\pi z}{12k} - \sum_{\mu=1}^{k} \sum_{r=1}^{\infty} \frac{1}{r} e^{2\pi i h \mu r/k} \frac{e^{-2\pi z \mu r/k}}{1 - e^{-2\pi z r}}.$$

The formula (1.1) reduces therefore to

$$\sum_{\nu=1}^{k} \sum_{r=1}^{\infty} \frac{1}{r} e^{+2\pi i h' \nu r/k} \frac{e^{-2\pi \nu r/kz}}{1 - e^{-2\pi r/z}} - \sum_{\mu=1}^{k} \sum_{r=1}^{\infty} \frac{1}{r} e^{2\pi i h \mu r/k} \frac{e^{-2\pi \mu z r/k}}{1 - e^{-2\pi z r}} +$$

$$+ \frac{\pi}{12k}\left(\frac{1}{z} - z\right) + \pi i s(h, k) = -\tfrac{1}{2} \log z. \tag{1.3}$$

The logarithm is here everywhere taken with its principal branch.

We try to obtain the left-hand side of (1.3) as a sum of residues of a suitable function. The second double sum in (1.3) suggests something like

$$\sum_{\mu=1}^{k} \frac{1}{x} \frac{e^{-2\pi \mu N x/k}}{1 - e^{2\pi N x}} \frac{e^{-2\pi i h \mu N x/kz}}{1 - e^{-2\pi i N x/z}},$$

however, for a reason of convergence the summand for $\mu = k$ has to be changed slightly, and secondly a sort of symmetry is needed between μ and $h\mu$. We introduce therefore

$$\mu^* \equiv h\,\mu \pmod{k}, \quad 1 \leqslant \mu^* \leqslant k - 1 \tag{1.4}$$

and consider the function

$$F_n(x) = -\frac{1}{4ix} \coth \pi N x \cot \frac{\pi N x}{z} +$$

$$+ \sum_{\mu=1}^{k-1} \frac{1}{x} \frac{e^{2\pi \mu N x/k}}{1 - e^{2\pi N x}} \cdot \frac{e^{-2\pi i \mu^* N x/kz}}{1 - e^{-2\pi i N x/z}}, \tag{1.5}$$

where $N = n + \frac{1}{2}$. We integrate $F_n(x)$ over the parallelogram C with the vertices z, i, $-z$, $-i$, collect the residues of the poles in C and compare with the value of the integral. Finally we let n go to ∞.

2. The function $F_n(x)$ has poles at

$$x = 0, \quad x = ir/N, \quad x = -zr/N, r = \pm 1, \pm 2, \ldots.$$

The function

$$-\frac{1}{4ix} \coth \pi Nx \cot \frac{\pi Nx}{z}$$

has the residue

$$-\frac{1}{12i}\left(z - \frac{1}{z}\right) \tag{2.1}$$

at $x = 0$, as a straight forward calculation shows.

The residue at $x = 0$ of

$$\frac{1}{x} \frac{e^{2\pi\mu Nx/k}}{1 - e^{2\pi Nx}} \frac{e^{-2\pi i\mu^* Nx/kz}}{1 - e^{-2\pi iNx/z}}$$

is

$$\left(\frac{1}{12} - \frac{\mu}{2k} + \frac{1}{2}\frac{\mu^2}{k^2}\right) zi + \left(\frac{\mu}{k} - \frac{1}{2}\right)\left(\frac{\mu^*}{k} - \frac{1}{2}\right) +$$
$$+ \left(\frac{1}{12} - \frac{\mu^*}{2k} + \frac{1}{2}\frac{\mu^{*2}}{k^2}\right)\frac{1}{iz}. \tag{2.2}$$

This has to be summed over μ from 1 to $k - 1$. It has to be observed that in view of (1.4) μ^* runs also from 1 to $k - 1$. The first and last terms in (2.2) offer no difficulty in summation. For the middle term in (2.2) we remark that (1.4) implies

$$\frac{\mu^*}{k} = \frac{h\mu}{k} - \left[\frac{h\mu}{k}\right],$$

so that

$$\sum_{\mu=1}^{k-1} \left(\frac{\mu}{k} - \frac{1}{2}\right)\left(\frac{\mu^*}{k} - \frac{1}{2}\right) = s(h, k)$$

according to definition (1.2). Summation of (2.2) over μ from 1 to $k - 1$ yields therefore

$$\left(-\frac{1}{12} + \frac{1}{12k}\right) zi + s(h, k) + \left(-\frac{1}{12} + \frac{1}{12k}\right)\frac{1}{zi}$$

and together with (2.1) the residue of $F_n(x)$ at $x=0$ appears as

$$\frac{1}{12k}\left(zi+\frac{1}{zi}\right)+s(h,k). \tag{2.3}$$

3. The residue of $F_n(x)$ at $x=\frac{ir}{N}$ is

$$\frac{1}{4\pi r}\cot\frac{\pi i r}{z}-\frac{1}{2\pi i}\sum_{\mu=1}^{k-1}\frac{1}{r}e^{2\pi i\mu r/k}\frac{e^{2\pi\mu^* r/kz}}{1-e^{2\pi r/z}}.$$

We observe that in virtue of (1.4)

$$h'\mu^*\equiv hh'\mu\equiv-\mu\ (\mathrm{mod}\,k)$$

so that we can write

$$\frac{1}{4\pi i r}\coth\frac{\pi r}{z}-\frac{1}{2\pi i}\sum_{\mu^*=1}^{k-1}\frac{1}{r}e^{-2\pi i h'\mu^* r/k}\frac{e^{2\pi\mu^* r/kz}}{1-e^{2\pi r/z}}. \tag{3.1}$$

The parallelogram C contains the poles $x=\frac{ir}{N}$ for $-n\leqslant r\leqslant -1$ and $1\leqslant r\leqslant n$. Accordingly we sum (3.1) over these poles and obtain

$$\frac{1}{2\pi i}\sum_{r=1}^{n}\frac{1}{r}\left(\frac{2e^{-2\pi r/z}}{1-e^{-2\pi r/z}}+1\right)+\frac{1}{2\pi i}\sum_{\mu^*=1}^{k-1}\sum_{r=1}^{n}\frac{1}{r}e^{2\pi i h'\mu^* r/k}\frac{e^{-2\pi\mu^* r/kz}}{1-e^{-2\pi r/z}}$$
$$-\frac{1}{2\pi i}\sum_{\mu^*=1}^{k-1}\sum_{r=1}^{n}\frac{1}{r}e^{2\pi i h'(k-\mu^*)r/k}\frac{e^{-2r(k-\mu^*)/kz}}{e^{-2\pi r/z}-1}.$$

In the last sum we replace $k-\mu^*$ by μ^* and combine it with the other sums, so that the sum of the residues of $F_n(z)$ at $x=\frac{ir}{N}$, $r=\pm 1,\pm 2,\dots\pm n$ appears now as

$$\frac{1}{2\pi i}\sum_{r=1}^{n}\frac{1}{r}+\frac{1}{\pi i}\sum_{\nu=1}^{k}\sum_{r=1}^{n}\frac{1}{r}e^{2\pi i h'\nu r/k}\frac{e^{-2\pi\nu r/kz}}{1-e^{-2\pi r/z}}, \tag{3.2}$$

where we have written ν instead of μ^*.

In a similar manner we find the sum of the residues of $F_n(x)$ at $x=-\frac{zr}{N}$, $r=\pm 1,\pm 2,\dots\pm n$ as

$$\frac{i}{2\pi}\sum_{r=1}^{n}\frac{1}{r}+\frac{i}{\pi}\sum_{\mu=1}^{k}\sum_{r=1}^{n}\frac{1}{r}e^{2\pi i h\mu r/k}\frac{e^{-2\pi\mu r z/k}}{1-e^{-2\pi r z}}. \tag{3.3}$$

The sum of all residues of $F_n(x)$ within C is therefore, from (2.3), (3.2) and (3.3)

$$\frac{1}{12\,ki}\left(\frac{1}{z}-z\right)+s\,(h,k)+\frac{1}{\pi i}\sum_{\nu=1}^{k}\sum_{r=1}^{n}\frac{1}{r}\,e^{2\pi i h'\nu r/k}\,\frac{e^{-2\pi\nu r/kz}}{1-e^{-2\pi r/z}}$$

$$-\frac{1}{\pi i}\sum_{\mu=1}^{k}\sum_{r=1}^{n}\frac{1}{r}\,e^{2\pi i h\mu r/k}\,\frac{e^{-2\pi\mu rz/k}}{1-e^{-2\pi rz}}. \tag{3.4}$$

This expression has for $n \rightarrow \infty$ a limit which, when multiplied by πi, will just give the left hand side of (1.3). In order to finish the proof of (1.3) we have therefore to show that

$$\lim_{n\rightarrow\infty}\int_C F_n(x)\,dx = -\log z.$$

4. Now on the four sides of C, with the exception of the vertices z, i, $-z$, $-i$, the part of $F_n(x)$ under the summation sign in (1.5) goes to zero with $N = n + \frac{1}{2} \rightarrow \infty$, whereas coth πNx.cot $\pi Nx/z$ goes to i on the sides i to $-z$ and $-i$ to z, and goes to $-i$ on the sides i to z and $-i$ to $-z$.

Therefore

$$\lim_{n\rightarrow\infty} x\,F_n(x) = \begin{cases} \frac{1}{4} \text{ on } i \text{ to } z \text{ and on } -i \text{ to } -z, \\ -\frac{1}{4} \text{ on } i \text{ to } -z \text{ and on } -i \text{ to } z, \end{cases}$$

always with the exception of the end points of the segments.

The convergence, of course, is not uniform, but is bounded since the denominators $1 - e^{2\pi Nx}$ and $1 - e^{-2\pi i Nx/z}$ stay away from zero because of $N = n + \frac{1}{2}$, n integer. We have therefore

$$\begin{aligned}\lim_{n\rightarrow\infty}\int_C F_n(x)\,dx &= \frac{1}{4}\left\{-\int_{-i}^{z}\frac{dx}{x}+\int_{z}^{i}\frac{dx}{x}-\int_{i}^{-z}\frac{dx}{x}+\int_{-z}^{-i}\frac{dx}{x}\right\}\\ &= \frac{1}{2}\left\{-\int_{-i}^{z}\frac{dx}{x}+\int_{z}^{i}\frac{dx}{x}\right\}\\ &= \frac{1}{2}\left\{-\left(\log z+\frac{\pi i}{2}\right)+\left(\frac{\pi i}{2}-\log z\right)\right\}\\ &= -\log z,\end{aligned}$$

where the principal branch of the logarithm is meant. This completes the proof of (1.3) and therefore of (1.1).

REFERENCE

1. C. L. Siegel: A simple proof of $\eta(-1/\tau) = \eta(\tau)\sqrt{\tau/i}$, *Mathematica*, 1 (1954), 4.

Tata Institute of Fundamental Research, Bombay
and
University of Pennsylvania, U.S.A.

Notes to Paper 58

This paper has been reviewed in the Z, vol. 64 (1956), p. 327-328, and in the MR, vol. 17 (1956), p. 15.

Page 25, first display line.
Author's correction (read τ/i for τ).

Page 25, second display line from bottom.
Author's correction (insertion of + in front of Σ).

Page 26, third display line.
Author's correction (suppress the negative sign of the first exponent).

Page 26, fifth display line.
The first exponential should read $e^{2\pi\mu N x/k}$ (no minus sign).

Page 30, reference.
Read Mathematika for Mathematica.

Zur Theorie der Dedekindschen Summen

I. Schur zum Gedächtnis

Von

Hans Rademacher [1])

1. Die Dedekindschen Summen $s(h, k)$, die in der Theorie der Dedekindschen Funktion $\log \eta(\tau)$ auftreten, sind definiert als

$$s(h,k) = \sum_{\mu \bmod k} \left(\left(\frac{\mu}{k}\right)\right)\left(\left(\frac{h\mu}{k}\right)\right) \qquad (h,k) = 1, \quad k > 0, \tag{1}$$

wo für reelles x

$$((x)) = \begin{cases} 0 \text{ für ganzes } x \\ x - [x] - \frac{1}{2} \text{ sonst} \end{cases} \tag{2}$$

gesetzt ist. Ihre bemerkenswerteste Eigenschaft wird durch die Reziprozitätsformel

$$s(h,k) + s(k,h) = -\frac{1}{4} + \frac{1}{12}\left(\frac{h}{k} + \frac{k}{h} + \frac{1}{hk}\right) \tag{3}$$

ausgedrückt, die einerseits aus dem Verhalten von $\log \eta(\tau)$ unter Modulsubstitutionen abzulesen ist, für die aber auch eine Anzahl direkter Beweise bekannt sind. Man kann (2) mit Hilfe eines Euklidischen Algorithmus, da $s(h_1, k) = s(h_2, k)$, für $h_1 \equiv h_2 \pmod k$ ist, zweckmäßigerweise zur numerischen Berechnung von $s(h, k)$ für größere Werte von k, wo die Summendefinition unbequem wäre, benutzen.

Es treten in (2) und in einigen Verallgemeinerungen, die in verschiedener Richtung von Mordell [*1*] und von mir [*2*] gegeben worden sind, jedoch mehrere $s(h, k)$ zugleich auf. Im folgenden will ich nun Eigenschaften studieren, die der einzelnen Summe $s(h, k)$ zukommen. Abschnitt I soll sich mit den Werten von $s(h, k)$ beschäftigen, Abschnitt II von einer Anwendung der Dedekindschen Summen auf das Problem der Transformationsklassen der Modulgruppe handeln, und Abschnitt III zieht quadratische Formen heran.

I

2. Hilfssatz 1. *Wenn h' durch*

$$hh' \equiv 1 \pmod k \tag{4}$$

bestimmt wird, ist

$$s(h', k) = s(h, k). \tag{4a}$$

Ferner ist

$$s(-h, k) = -s(h, k). \tag{4b}$$

---

[1]) John Simon Guggenheim Memorial Fellow 1954/55.

Beweis. Die zweite Aussage folgt sofort aus der Tatsache, daß $((x))$ eine ungerade Funktion von x ist.

Wenn ferner μ ein volles Restsystem modulo k durchläuft, so tut dies auch $h'\mu$. Wir haben also

$$s(h,k)=\sum_{\mu \bmod k}\left(\left(\frac{\mu}{k}\right)\right)\left(\left(\frac{h\mu}{k}\right)\right)=\sum_{\mu \bmod k}\left(\left(\frac{h'\mu}{k}\right)\right)\left(\left(\frac{hh'\mu}{k}\right)\right)=\sum_{\mu \bmod k}\left(\left(\frac{h'\mu}{k}\right)\right)\left(\left(\frac{\mu}{k}\right)\right)=s(h',k).$$

3. Da $((x))$ eine ungerade Funktion der Periode 1 ist, folgt sofort

$$\sum_{\lambda \bmod k}\left(\left(\frac{\lambda}{k}\right)\right)=0$$

und somit

$$s(h,k)=\sum_{\mu=1}^{k-1}\left(\frac{\mu}{k}-\frac{1}{2}\right)\left(\left(\frac{h\mu}{k}\right)\right)=\sum_{\mu=1}^{k-1}\frac{\mu}{k}\left(\left(\frac{h\mu}{k}\right)\right).$$

Explizit heißt dies

$$(5)\qquad \begin{cases} s(h,k) &= \displaystyle\sum_{\mu=1}^{k-1}\frac{\mu}{k}\left(\frac{h\mu}{k}-\left[\frac{h\mu}{k}\right]-\frac{1}{2}\right)\\ &= \displaystyle\frac{h}{k^2}\sum_{1}^{k-1}\mu^2-\frac{1}{k}\sum_{1}^{k-1}\mu\left[\frac{h\mu}{k}\right]-\frac{1}{2k}\sum_{1}^{k-1}\mu\\ &= \displaystyle\frac{h(k-1)(2k-1)}{6k}-\frac{1}{k}g-\frac{k-1}{4}, \end{cases}$$

wo g eine ganze Zahl ist. Man sieht daher, daß der Nenner von $s(h,k)$ höchstens $6k$ sein kann. Da ferner in (5) der Nenner des einzelnen Summanden höchstens k^2 für gerades k und $2k^2$ für ungerades k beträgt, so ist der Nenner von $s(h,k)$ höchstens

für k gerade: $\qquad (6k, k^2)=k(6,k)=2k\cdot(3,k)$,

für k ungerade: $\qquad (6k, 2k^2)=2k(3,k)$.

Damit haben wir erhalten den

Hilfssatz 2. *Der Nenner von $s(h,k)$ ist höchstens $2k\cdot(3,k)$.*

Dieser Wert des Nenners wird erreicht, z.B. ist $s(1,3)=\frac{1}{18}$. Der Nenner kann aber auch ein echter Teiler des angegebenen Wertes sein: $s(3,25)=\frac{2}{5}$.

Multiplikation von (3) mit $12hk$ ergibt

$$(6)\qquad 12hk\,s(h,k)+12hk\,s(k,h)=-3hk+h^2+k^2+1.$$

Im Hinblick auf Hilfssatz 2 lesen wir hieraus leicht ab den

Hilfssatz 3. *Es sei $\Theta=(3,k)$. Dann ist*

$$(7)\qquad 12hk\,s(h,k)\equiv h^2+1 \pmod{\Theta k}.$$

4. Wir beweisen nun den

Satz 1. *Der einzige ganzzahlige Wert, den $12s(h,k)$ annimmt, ist Null. Dies tritt dann und nur dann ein, wenn*

$$(8)\qquad h^2+1\equiv 0 \pmod{k}.$$

Beweis. Es sei zunächst (8) erfüllt. Dann ist, in der Bezeichnung Hilfssatz 1

$$h' = -h,$$

also

$$s(h, k) = s(-h, k) = -s(h, k),$$

woraus die zweite Hälfte des Satzes folgt.

Sei nun umgekehrt $12s(h, k)$ ganz. Dann ist nach (6) und Hilfssatz 2 die Kongruenz (8) erfüllt.

5. Satz 2. *Es ist* $s(h, k) < s(1, k)$ *für* $1 < h < k$.

Beweis. Wir haben nach (5)

$$k \cdot s(1, k) = \sum_{\mu=1}^{k-1} \mu\left(\frac{\mu}{k} - \frac{1}{2}\right),$$

$$k \cdot s(h, k) = \sum_{\mu=1}^{k-1} \mu\left(\frac{h\mu}{k} - \left[\frac{h\mu}{k}\right] - \frac{1}{2}\right).$$

Es genügt also zu zeigen, daß

$$\sum_{\mu=1}^{k-1} \mu^2 > \sum_{\mu=1}^{k-1} \mu\left(h\mu - k\left[\frac{h\mu}{k}\right]\right)$$

ist. Hier sind die Zahlen

$$a_\mu = h\mu - k\left[\frac{h\mu}{k}\right]$$

eine Permutation der Zahlen $1, 2, \ldots, k-1$. Für alle möglichen Permutationen $\{a_\mu\}$ der Zahlen 1 bis $k-1$ erreicht aber die Summe

$$\sum \mu\, a_\mu$$

ihr Maximum für die Anordnung $a_\mu = \mu$. In der Tat, wenn nicht durchgängig $\mu = a_\mu$ ist, dann gibt es in der Anordnung $a_1, a_2, \ldots, a_{k-1}$ ein größtes $\mu = \mu_0$, so daß $a_{\mu_0} \neq \mu_0$, während $a_\mu = \mu$ für $\mu > \mu_0$ (sofern ein solches μ vorrätig ist). Dann ist jedoch sogar $a_{\mu_0} < \mu_0$, denn die $a_\mu > \mu_0$ sind besetzt durch $a_\mu = \mu$. Es sei dann $\mu_0 = a_{\mu_1}$ mit $\mu_1 < \mu_0$. Wir vertauschen dann a_{μ_0} und a_{μ_1}, ersetzen also die beiden Summanden $\mu_1 a_{\mu_1} + \mu_0 a_{\mu_0}$ durch $\mu_1 a_{\mu_0} + \mu_0 a_{\mu_1}$, was eine Vergrößerung bedeutet:

$$\mu_1 a_{\mu_0} + \mu_0 a_{\mu_1} - (\mu_1 a_{\mu_1} + \mu_0 a_{\mu_0}) = (\mu_0 - \mu_1)(a_{\mu_1} - a_{\mu_0}) = (\mu_0 - \mu_1)(\mu_0 - a_{\mu_0}) > 0.$$

Satz 3. *Es ist* $s(h, k) > 0$ *für* $0 < h < \sqrt{k-1}$.

Beweis. Nach (3) und nach Satz 2 haben wir

$$s(h, k) = -\frac{1}{4} + \frac{1}{12}\left(\frac{h^2+1}{hk} + \frac{k}{h}\right) - s(k, h)$$

$$\geq -\frac{1}{4} + \frac{1}{12}\left(\frac{h^2+1}{hk} + \frac{k}{h}\right) - s(1, h).$$

Nun folgt aus (3), da $s(h, 1) = 0$ ist,

$$s(1, h) = -\frac{1}{4} + \frac{1}{12}\left(\frac{h^2+1}{h} + \frac{1}{h}\right),$$

so daß wir erhalten

$$s(h, k) \geq \frac{(k-1)(k-h^2-1)}{12hk},$$

woraus die Behauptung des Satzes folgt.

Da in dieser Überlegung das Ungleichheitszeichen nur bei der Ersetzung von $s(k, h)$ durch $s(1, h)$ auftritt, so haben wir das

Korollar. $s(h, k) = \dfrac{(k-1)\left(\dfrac{k-1}{h} - h\right)}{12k}$ *für* $k \equiv 1 \pmod{h}$.

5. Satz 4. *Es seien* $\frac{h_1}{k_1} < \frac{h_2}{k_2}$ *zwei benachbarte Brüche in einer Farey-Reihe. Dann ist*

$$s(h_1, k_1) - s(h_2, k_2) = \frac{1}{4} - \frac{1 + k_1^2 + k_2^2}{12 k_1 k_2}. \tag{9}$$

Beweis. Die kennzeichnende Eigenschaft für benachbarte Farey-Brüche $\frac{h_1}{k_1} < \frac{h_2}{k_2}$ ist

$$\begin{vmatrix} h_1 & h_2 \\ k_1 & k_2 \end{vmatrix} = -1,$$

so daß also

$$h_1 k_2 \equiv -1 \pmod{k_1},$$
$$h_2 k_1 \equiv 1 \pmod{k_2}.$$

Dann ist aber nach Hilfssatz 1

$$s(k_2, k_1) = -s(h_1, k_1),$$
$$s(k_1, k_2) = s(h_2, k_2),$$

womit (9) aus (3) folgt, wenn dort h, k durch k_1, k_2 ersetzt werden.

Satz 5. *Über* h_1/k_1 *und* h_2/k_2 *mögen die Voraussetzungen von Satz 4 gelten. Dann ist*

$$s(h_1 + h_2, k_1 + k_2) = \frac{1}{2}\{s(h_1, k_1) + s(h_2, k_2)\} + \frac{(k_2 - k_1)(1 + k_1^2 + k_1 k_2 + k_2^2)}{24 k_1 k_2 (k_1 + k_2)}. \tag{10}$$

Beweis. Die Anwendung von (9) auf $\frac{h_1}{k_1} < \frac{h_1+h_2}{k_1+k_2} < \frac{h_2}{k_2}$ ergibt

$$s(h_1, k_1) - s(h_1 + h_2, k_1 + k_2) = \frac{1}{4} - \frac{1 + k_1^2 + (k_1 + k_2)^2}{12 k_1 (k_1 + k_2)},$$

$$s(h_1 + h_2, k_1 + k_2) - s(h_2, k_2) = \frac{1}{4} - \frac{1 + (k_1 + k_2)^2 + k_2^2}{12 (k_1 + k_2) k_2}.$$

Durch Subtraktion dieser beiden Gleichungen voneinander ergibt sich (10).

Die Formel (10) eignet sich besonders gut zur schrittweisen tabellarischen Berechnung von $s(h, k)$, ausgehend von $s(0, 1) = s(1, 1) = 0$.

6. Die Sätze 4 und 5 legen nahe, $s(h/k)$ statt $s(h, k)$ zu schreiben. Dies kann man unbedenklich tun, wenn man vorübergehend die Bedingung $(h, k)=1$ fallen läßt. In der Tat hängt dann $s(h, k)$ nur von dem Verhältnis $h:k$ ab, wie man ersieht aus

Hilfssatz 4. *Für ganzes positives q ist*

$$s(qh, qk) = s(h, k).$$

Dies gilt nur für die Definition (1) von $s(h, k)$, nicht etwa für Formel (5). Dieser Hilfssatz ist Theorem 1 in [3] und dort bewiesen. [Alle Formeln setzen aber bisher $(h, k)=1$ voraus.] Damit ist eine Funktion $s(\varrho)$ für alle rationalen $\varrho=h/k$ definiert. Über den Verlauf dieser Funktion x macht der folgende Satz eine Aussage.

Satz 6. *Die für rationale $\varrho=h/k$ definierte Funktion $s(\varrho)$ ist in jedem Intervall nach oben und unten unbeschränkt.*

Beweis. In (9) möge h_1/k_1 festgehalten werden und h_2/k_2 eine Folge von benachbarten Brüchen durchlaufen die von oben gegen h_1/k_1 gehen (z.B. durch festgesetze Mediantenbildung). Mit k_1 fest, $k_2 \to +\infty$ geht die rechte Seite von (9) gegen $-\infty$, also $s(h_2/k_2) \to +\infty$. Ebenso, wenn man h_2/k_2 festhält und h_1/k_1 gegen h_2/k_2 gehen läßt, folgt $s(h_1/k_1) \to -\infty$.

II

7. Es sei Γ die Gruppe aller homogenen Modulsubstitutionen

$$M=\begin{pmatrix} a & b \\ c & d \end{pmatrix}, \qquad \begin{vmatrix} a & b \\ c & d \end{vmatrix} = 1.$$

Zwei Substitutionen M_1, M_2 heißen ähnlich, wenn ein $L \in \Gamma$ existiert, so daß

$$M_1 = L^{-1} M_2 L.$$

Die Gruppe Γ zerfällt durch die Ähnlichkeitsrelation in *Klassen*. Die Spur $a+d$ ist offenbar eine Invariante der Ähnlichkeitsklassen. In diesem Abschnitt behandeln wir eine weitere Invariante.

Satz 7. *Die Funktion*

$$(11)\quad \begin{cases} \Psi(M) = \Psi\begin{pmatrix} a & b \\ c & d \end{pmatrix} \\ = \begin{cases} \dfrac{b}{d} & \text{für } c=0, \\ \dfrac{a+d}{c} - 12 \operatorname{sign} c \cdot s(a, |c|) - 3 \operatorname{sign}\big(c(a+d)\big) & \text{für } c \neq 0 \end{cases} \end{cases}$$

ist eine Klasseninvariante. Außerdem ist

$$(12)\qquad \Psi(M) = \Psi(-M), \qquad \Psi(M^{-1}) = -\Psi(M),$$

und Ψ ist eine ganze Zahl.

Beweis. Die Aussagen (12) folgen aus der Definition (11), wenn man noch

$$M^{-1} = \begin{pmatrix} a & b \\ c & d \end{pmatrix}^{-1} = \begin{pmatrix} d & -b \\ -c & a \end{pmatrix}$$

beachtet. Ferner kann man aus (7) entnehmen

$$12\,|c|\,s(a,|c|) \equiv a + d \pmod{c},$$

also

$$\frac{a+d}{|c|} - 12\,s(a,|c|)$$

ganzzahlig, woraus die Ganzzahligkeit von Ψ in (11) folgt.

Für die Invarianz von $\Psi(M)$ brauchen wir nur die Gleichungen

(13) $$\Psi(S^{-1}MS) = \Psi(M),$$

(14) $$\Psi(T^{-1}MT) = \Psi(M),$$

mit

$$S = \begin{pmatrix} 1 & 1 \\ 0 & 1 \end{pmatrix}, \quad T = \begin{pmatrix} 0 & -1 \\ 1 & 0 \end{pmatrix}$$

zu beweisen.

Mit

$$M = \begin{pmatrix} a & b \\ c & d \end{pmatrix}$$

haben wir

(13a) $$S^{-1}MS = \begin{pmatrix} a-c & a-c+b-d \\ c & c+d \end{pmatrix},$$

(14a) $$T^{-1}MT = \begin{pmatrix} d & -c \\ -b & a \end{pmatrix}.$$

Dann ist (13) unmittelbar aus der Definition von Ψ abzulesen. Wir haben also nur noch (14) zu untersuchen. Zunächst bemerken wir, daß

(15) $$12\,s(1,k) = -3 + k + \frac{2}{k},$$

was aus dem Korollar zu Satz 3, oder auch direkt aus (3) folgt. Für den Beweis von (14) unterscheiden wir drei Fälle.

I. $c=0$, dann ist $a=d=1$. [$a=d=-1$ kann auf Grund von (12) beiseite gelassen werden.] Dann lautet (14), (14a):

(16) $$\Psi\begin{pmatrix} 1 & b \\ 0 & 1 \end{pmatrix} = \Psi\begin{pmatrix} 1 & 0 \\ -b & 1 \end{pmatrix}$$

oder, nach (11),

$$b = \begin{cases} 0 & \text{für } b = 0 \\ \dfrac{2}{-b} + 12\,\operatorname{sign} b \cdot s(1,|b|) + 3\,\operatorname{sign} b, \end{cases}$$

was in der Tat durch (15) für $k=|b|$ bestätigt wird.

II. $c \neq 0$, $b = 0$. Dann handelt es sich um

$$\Psi\begin{pmatrix}1 & 0\\ c & 1\end{pmatrix} = \Psi\begin{pmatrix}1 & -c\\ 0 & 1\end{pmatrix},$$

was aus (16) durch Vertauschung von c und $-b$ hervorgeht und somit unter I erledigt ist.

III. $b \neq 0$, $c \neq 0$. Hier ist also im Hinblick auf (11), (14), (14a) zu beweisen:

$$(17)\quad \left\{\begin{aligned} &12\,\mathrm{sign}\, c\cdot s(a, |c|) + 12\,\mathrm{sign}\, b\cdot s(d, |b|)\\ &= \frac{a+d}{c} + \frac{a+d}{b} - 3\,\mathrm{sign}\big(c(a+d)\big) - 3\,\mathrm{sign}\big(b(a+d)\big).\end{aligned}\right.$$

Wenn $d=0$ ist, dann muß $b=-c=\pm 1$ sein, und (17) ist dann trivialerweise erfüllt.

Wenn $d \neq 0$, dann kann wegen (12) $d>0$ angenommen werden. Dann ergibt (3)

$$12\,s(d, |c|) + 12\,s(|c|, d) = -3 + \frac{d}{|c|} + \frac{|c|}{d} + \frac{1}{d|c|},$$

$$12\,s(d, |b|) + 12\,s(|b|, d) = -3 + \frac{d}{|b|} + \frac{|b|}{d} + \frac{1}{d\,|b|}.$$

Die erste dieser Gleichungen multiplizieren wir mit sign c, die zweite mit sign b und beachten, daß zufolge (4b)

$$s(h, k) = \mathrm{sign}\, h\cdot s(|h|, k)$$

ist, so daß

$$(18)\quad \left\{\begin{aligned} &12\,\mathrm{sign}\, c\cdot s(d, |c|) + 12\,s(c, d) = -3\,\mathrm{sign}\, c + \frac{d}{c} + \frac{c}{d} + \frac{1}{dc},\\ &12\,\mathrm{sign}\, b\cdot s(d, |b|) + 12\,s(b, d) = -3\,\mathrm{sign}\, b + \frac{d}{b} + \frac{b}{d} + \frac{1}{db}\end{aligned}\right.$$

hervorgeht. Nach Hilfssatz 1 ist

$$s(a, |c|) = s(d, |c|),$$

$$s(c, d) = -\,s(b, d).$$

Addition der Gln. (18) ergibt also

$$(19)\quad \left\{\begin{aligned} &12\,\mathrm{sign}\, c\cdot s(a, |c|) + 12\,\mathrm{sign}\, b\cdot s(d, |b|)\\ &= -3\,\mathrm{sign}\, c - 3\,\mathrm{sign}\, b + \frac{d}{c} + \frac{c}{d} + \frac{ad-bc}{dc} + \frac{d}{b} + \frac{b}{d} + \frac{ad-bc}{db}.\end{aligned}\right.$$

Ein Vergleich von (17) und (19) zeigt, daß (17) bewiesen sein wird, sobald die Richtigkeit von

$$(20)\qquad \mathrm{sign}\big(c(a+d)\big) + \mathrm{sign}\big(b(a+d)\big) = \mathrm{sign}\, c + \mathrm{sign}\, b$$

gezeigt ist. Nun ist (20) klar für $a+d>0$. Für $a+d\leq 0$ und $d>0$ (was wir annehmen durften) folgt $ad<0$, und somit $bc=ad-1<0$, was entgegengesetzte Vorzeichen von b und c bedeutet. Dann sind beide Seiten von (20) gleich Null. Somit ist (17) bewiesen und damit der Beweis von Satz 7 erbracht.

8. Unter Einführung der Matrix $U=ST$ mit

$$T^2=U^3=1 \tag{21}$$

(wo Faktoren ± 1 vernachlässigt, d.h. die inhomogenen Modulsubstitutionen gemeint sind) kann jede Modulsubstitution *eindeutig* in kürzester Form dargestellt werden[2]) als

$$M=U^{\varepsilon_0}\,T\,U^{\varepsilon_1}\,T\ldots T\,U^{\varepsilon_{\nu+1}}, \tag{22}$$

wo

$$\varepsilon_j=\pm 1, \qquad j=1,\ldots,\nu$$

$$\varepsilon_0,\,\varepsilon_{\nu+1}=0 \quad \text{oder} \quad \pm 1.$$

Elemente M, die durch zyklische Vertauschung der Faktoren der Darstellung (22) auseinander hervorgehen, gehören zur selben Klasse. Wenn man das Produkt (22) zyklisch schließt, wird im allgemeinen eine Reduktion durch Anwendung von (21) eintreten. Dann gibt es also in einer Klasse immer eine Transformation von folgender Art

1. die elliptischen T, U, U^{-1}

oder

2. $T\,U^{\varepsilon_1}\,T\,U^{\varepsilon_2}\ldots T\,U^{\varepsilon_\nu}, \quad \varepsilon=\pm 1$.

Hilfssatz 4. *Ist*

$$M=\begin{pmatrix} a & b \\ c & d \end{pmatrix}=T\,U^{\varepsilon_1}\,T\,U^{\varepsilon_2}\ldots T\,U^{\varepsilon_\nu}, \quad \varepsilon_j=\pm 1, \tag{23}$$

dann gilt[3])

$$a<0, \quad d<0, \quad b\geq 0, \quad c\geq 0. \tag{24}$$

Beweis. Diese Behauptungen sind richtig für $\nu=1$, also für

$$TU=TST=\begin{pmatrix} -1 & 0 \\ 1 & -1 \end{pmatrix},$$

$$TU^{-1}=S^{-1}=\begin{pmatrix} -1 & 1 \\ 0 & -1 \end{pmatrix}.$$

Angenommen die Behauptungen (24) wären für M in der Form (23) gültig, dann haben wir zu zeigen, daß sie auch für

$$M\cdot T\,U^{\varepsilon_{\nu+1}}$$

[2]) Den Hinweis auf die Bedeutung der kürzesten Darstellung in diesem Zusammenhang verdanke ich Herrn Reidemeister; vgl. [*4*], S. 42, 43.

[3]) Möglicherweise zu erreichen durch Multiplikation der Matrix mit -1.

zutreffen. Nun ist

$$M\,T\,U=\begin{pmatrix}a & b\\ c & d\end{pmatrix}\begin{pmatrix}-1 & 0\\ 1 & -1\end{pmatrix}=\begin{pmatrix}-a+b & -b\\ -c+d & -d\end{pmatrix}=\begin{pmatrix}a-b & b\\ c-d & d\end{pmatrix}=\begin{pmatrix}a' & b'\\ c' & d'\end{pmatrix}$$

und

$$M\,T\,U^{-1}=\begin{pmatrix}a & b\\ c & d\end{pmatrix}\begin{pmatrix}-1 & 1\\ 0 & -1\end{pmatrix}=\begin{pmatrix}-a & a-b\\ -c & c-d\end{pmatrix}=\begin{pmatrix}a & b-a\\ c & d-c\end{pmatrix}=\begin{pmatrix}a'' & b''\\ c'' & d''\end{pmatrix}.$$

Aus (24) folgt dann dieselbe Behauptung auch für a', b', c', d', und a'', b'', c'', d'', womit der Hilfssatz bewiesen ist.

9. Nunmehr kann man der Klasseninvariante Ψ eine neue Bedeutung geben. Es gilt nämlich der

Satz 8. *Es sei M in seiner Klasse in der zyklisch reduzierten Form angenommen. Dann gilt:*

$$(25)\qquad \begin{cases}\Psi(T)=\Psi\begin{pmatrix}0 & -1\\ 1 & 0\end{pmatrix}=0,\\ \Psi(U)=\Psi\begin{pmatrix}1 & -1\\ 1 & 0\end{pmatrix}=-2,\quad \Psi(U^{-1})=\Psi\begin{pmatrix}0 & -1\\ 1 & -1\end{pmatrix}=2\end{cases}$$

und im Falle (23)

$$(26)\qquad \Psi(M)=\sum_{j=1}^{\nu}\varepsilon_j.$$

Beweis. Die Formeln (25) liest man direkt aus der Definition (11) von Ψ ab. Die Gl. (26) wird durch die für den Beweis von Hilfssatz 4 angewandte Induktion bewiesen.

Zunächst ist

$$\Psi(T\,U)=\Psi\begin{pmatrix}-1 & 0\\ 1 & -1\end{pmatrix}=1,$$

$$\Psi(T\,U^{-1})=\Psi\begin{pmatrix}-1 & 1\\ 0 & -1\end{pmatrix}=-1,$$

wie aus (11) folgt. Also stimmt (26) für $\nu=1$.

Es sei nun (26) für ein ν erfüllt, dann bleibt nur zu zeigen, daß

$$(27)\qquad \Psi(M\,T\,U^{\varepsilon_{\nu+1}})=\Psi(M)+\varepsilon_{\nu+1},\qquad \varepsilon_{\nu+1}=\pm 1$$

ist. Sei

$$M=\begin{pmatrix}a & b\\ c & d\end{pmatrix},$$

dann ist

$$M\,T\,U=\begin{pmatrix}a-b & b\\ c-d & d\end{pmatrix},$$

also

$$(28\text{a})\qquad \Psi(M) = \begin{cases} \frac{b}{d} & \text{für } c = 0 \\ \frac{a+d}{c} - 12s(d,c) + 3, & c > 0, \end{cases}$$

$$(28\text{b})\qquad \Psi(MTU) = \frac{a-b+d}{c-d} - 12s(d, c-d) + 3,$$

wo nach (24) $c-d=0$ nicht vorkommt, und wo, wieder nach (24), $a+d<0$ $a-b+d<0$ benutzt ist.

Sei zunächst $c=0$, also $a=d=-1$. Dann ist

$$(29)\qquad \begin{cases} \Psi(M) = \frac{b}{d} = -b, \qquad \Psi(MTU) = \frac{-1-b-1}{-d} + 3 \\ \qquad\qquad\qquad\qquad\qquad = -b+1, \end{cases}$$

in Übereinstimmung mit (27).

Sei zweitens $c>0$, also nach (28a), (28b)

$$\Psi(MTU) - \Psi(M) = \frac{a-b+d}{c-d} - \frac{a+d}{c} - 12s(d, c-d) + 12s(d,c).$$

Nun ist aber nach (3)

$$-12s(d, c-d) = 12s(c,d) + 3 - \frac{d^2 + (c-d)^2 + 1}{d(c-d)}$$

und somit

$$\Psi(MTU) - \Psi(M) = 12s(c,d) + 12s(d,c) + \\ + 3 - \frac{d^2+(c-d)^2+1}{d(c-d)} + \frac{a-b+d}{c-d} - \frac{a+d}{c}.$$

Eine nochmalige Anwendung von (3) ergibt dann

$$\Psi(MTU) - \Psi(M) = \frac{c^2+d^2+1}{cd} - \frac{a+d}{c} + \frac{b-c+d}{d} = 1.$$

Diese Gleichung und (29) bestätigen (27) für den Fall $\varepsilon_{\nu+1} = +1$.

Ferner ist

$$(30)\qquad \Psi(MTU^{-1}) = \Psi\begin{pmatrix} a & -a+b \\ c & -c+d \end{pmatrix} = \begin{cases} -\frac{a+b}{d} & c = 0 \\ \frac{a-c+d}{c} - 12s(d,c) + 3, & c > 0 \end{cases}$$

und somit für $c=0$, $a=d=-1$

$$\Psi(M) = -b, \qquad \Psi(MTU^{-1}) = -1-b,$$

während sich für $c>0$ aus (28a) und (30)

$$\Psi(MTU^{-1}) - \Psi(M) = -\frac{c}{c} = -1$$

ergibt. Damit ist (27) auch für den Fall $\varepsilon_{\nu+1} = -1$ bewiesen und der Induktionsbeweis von Satz 8 zu Ende geführt.

10. Im allgemeinen kann man nur aussagen, daß $\Psi(M_1 M_2)$ und $\Psi(M_1) + \Psi(M_2)$ sich nur um Vielfache von 3 unterscheiden können. Jedoch gilt der

Satz 9. *Wenn M keine elliptische Substitution ist, dann ist für ganzes k*

$$\Psi(M^k) = k\,\Psi(M).$$

Beweis. Dies folgt zunächst für $k > 0$ daraus, daß in dem zyklischen Zusammenschluß von M_k k mal die gleichen Reduktionen auftreten, die nur einmal bei dem zyklischen Zusammenschluß von M auftreten. Also besteht das zyklisch reduzierte M_k aus k Wiederholungen des zyklisch reduzierten M, was den Satz für $k > 0$ beweist. Für $k < 0$ folgt der Satz dann aus der zweiten Gl. (12), und für $k = 0$ aus $\Psi(E) = 0$, wo E die Einheitsmatrix ist.

In etwas anderer Richtung liegt der

Satz 10. *Wenn für*

(31) $$M_1 = \begin{pmatrix} a_1 & b_1 \\ c_1 & d_1 \end{pmatrix}, \quad M_2 = \begin{pmatrix} a_2 & b_2 \\ c_2 & d_2 \end{pmatrix},$$

die Ungleichungen

(32) $$\begin{cases} a_1 < 0, & d_1 < 0, \quad c_1 > 0, \quad b_1 > 0 \\ a_2 < 0, & d_2 < 0, \quad c_2 > 0, \quad b_2 > 0 \end{cases}$$

/ b_2

gelten, so ist

(33) $$\Psi(M_1 M_2) = \Psi(M_1) + \Psi(M_2).$$

Beweis. Die Ungleichungen (32) sorgen dafür, daß in der Normaldarstellung (22) M_1 so endet und M_2 so anfängt, daß keine gegenseitigen Reduktionen in dem Produkt $M_1 M_2$ stattfinden.

Um dies einzusehen, ändern wir die Bedingungen zu (22) dahin ab, daß wir nur nicht-negative Potenzen zulassen, was dadurch geschieht, daß jeder etwa auftretende Exponent -1 durch 2 ersetzt wird. Dann werde die Definition $U = ST$ benutzt, so daß (22) umgeformt wird in

(34) $$\begin{cases} M = (ST)^{\eta_0}\, T (ST)^{\eta_1}\, T \ldots (ST)^{\eta_\nu}\, T (ST)^{\eta_{\nu+1}}, \\ \eta_0, \eta_{\nu+1} = 0, 1, 2, \quad \eta_j = 1, 2, \quad j = 1, \ldots, \nu. \end{cases}$$

Dann und nur dann, wenn $\eta_0 = 0$, $\eta_{\nu+1} > 0$ wird M in (22) die spezielle Form (23) haben, d.h. mit T anfangen und mit einer Potenz von U enden. Wir wollen zeigen, daß aus

(35) $$a < 0, \quad d < 0, \quad c > 0$$

folgt $\eta_0 = 0$, $\eta_{\nu+1} > 0$.

In der Tat, (34) kann nach gehörigem Zusammenziehen auf Grund von $T^2 = 1$ geschrieben werden:

(36) $$M = S^{q_0}\, T S^{q_1} \ldots T S^{q_l}\, T S^{q_{l+1}}, \quad q_j \geq 1,\ j = 1, \ldots l, \quad q_0 \geq 0,\ q_{l+1} \geq 0,$$

wo $q_0 > 0$ aus $\eta_0 > 0$, und $q_{l+1} > 0$ aus $\eta_{\nu+1} = 0$ folgen würde. Man sieht ferner leicht, daß wegen (34) von den nicht-verschwindenden q_j nur das erste und letzte gleich 1 sein können, während für die übrigen sogar $q_j \geqq 2$ gilt. Nach (36) kann man

$$M(\tau) = \frac{a\tau + b}{c\tau + d}$$

als einen Kettenbruch

$$\frac{a\tau+b}{c\tau+d} = q_0 - \frac{1}{q_1 - \ddots - \cfrac{1}{q_l - \cfrac{1}{q_{l+1} + \tau}}}$$

schreiben, woraus durch $\frac{1}{\tau} \to 0$

$$\frac{a}{c} = q_0 - \frac{1}{q_1 - \ddots - \cfrac{1}{q}}$$

folgt, so daß $q_0 > 0$

$$\frac{a}{c} \geqq 0$$

nach sich ziehen würde, gegen die Ungleichungen (35). Ebenso beweist man aus

$$M^{-1} = S^{-q_{l+1}} T S^{-q} \ldots T S^{-q_1} T,$$

also aus

$$-\frac{d}{c} = -q_{l+1} - \frac{1}{-q_l - \ddots - \cfrac{1}{-q_1}},$$

daß $q_{l+1} > 0$ die Ungleichung

$$-\frac{d}{c} \leqq 0$$

zur Folge hätte, wieder gegen (35). Es ist also $q_0 = q_{l+1} = 0$ und folglich $\eta_0 = 0$, $\eta_{\nu+1} > 0$, und M hat die Form (23), wobei nur $\varepsilon_{\nu+1}$ statt ε_ν zu schreiben ist.

Daher haben infolge von (32) die Substitutionen M_1 und M_2 auch die Form (23). Dann aber tritt in dem Produkt $M_1 M_2$ keine Reduktion ein, da auf das letzte U^{ε_ν} von M, das erste T von M_2 folgt. Damit ist Satz 10 bewiesen.

Wenn man in (33) die Matrizen von (31) einträgt und die Definition (11) von Ψ aus Satz 7 heranzieht, so folgt nach einigen Rechnungen das

Korollar. *Unter den Bedingungen* (32) *ist*

$$s(d, |c|) + s(d_1, c_1) + s(d_2, c_2) = \frac{1}{4} - \frac{c_1^2 + c_2^2 + c^2}{12 c_1 c_2 |c|}$$

mit

$$c = c_1 a_2 + d_1 c_2, \qquad d = c_1 b_2 + d_1 d_2.$$

Diese Formel könnte auch analytisch aus der Theorie der Funktion log $\eta(\tau)$ gewonnen werden. Sie folgt übrigens auch aus der dreigliedrigen Formel (30) in [2].

III

11. Für die elliptischen Substitutionen, die mit T oder U oder U^{-1} äquivalent sind, also nur zu 3 Klassen gehören können, ist die Spur $a+d$ gleich 0 oder ± 1. Wenn man die Spur $a+d=m$ mit $|m|\geq 2$ vorgibt, wobei man $a+d$ negativ wählen kann, so kann man nach dem Hilfssatz 4 und den ihm vorausgehenden Erörterungen in derselben Klasse stets eine Substitution in der Form (23) finden, für die $a<0$, $d<0$ ist. Solche a, d gibt es aber bei festem $a+d=m$ nur endlich viele.

Ist $m\leq -3$, so ist $a\cdot d\neq 1$, also

$$bc = ad - 1 \neq 0,$$

und von den b, c (die beide nach Hilfssatz 4 positiv zu sein haben), gibt es daher wieder nur endlich viele. Also gibt es mit vorgegebener Spur $a+d=m\leq -3$ nur je endlich viele Klassen von Modulsubstitutionen.

Für $a+d=-2$, also $a=d=-1$, $bc=0$ gibt es die unendlich vielen Substitutionen

$$(TU)^c = \begin{pmatrix} -1 & 0 \\ c & -1 \end{pmatrix}, \qquad c>0,$$

$$(TU^{-1})^b = \begin{pmatrix} -1 & b \\ 0 & -1 \end{pmatrix}, \qquad b>0,$$

die verschiedene Normalformen (23) zeigen, und also zu verschiedenen Klassen gehören. Dies sind die parabolischen Substitutionen. Es ist übrigens

$$\Psi\big((TU)^c\big) = c, \qquad \Psi\big((TU^{-1})^b\big) = -b.$$

12. Einen anderen Einblick in die Verteilung auf die Klassen bei gegebenem $m=a+d$ gewinnt man durch eine Zuordnung von quadratischen Formen zu den Modulsubstitutionen.

Wir gehen dazu wieder zu den homogenen Modulsubstitutionen über, unterscheiden also die unimodularen Matrizen M und $-M$. Dann werde der Modulsubstitution

$$M = \begin{pmatrix} a & b \\ c & d \end{pmatrix}$$

die quadratische Form

$$Q(x, y) = c x^2 + (d-a)\, x y - b y^2 \tag{36}$$

mit der Diskriminante

$$\Delta = (d-a)^2 + 4bc = (a+d)^2 - 4 = m^2 - 4$$

zugeordnet. Es ist klar, daß zu

$$-M^{-1} = \begin{pmatrix} -d & b \\ c & -a \end{pmatrix}$$

dieselbe Form Q gehört wie zu M.

Umgekehrt, hat

$$Q(x, y) = A x^2 + B x y + C y^2 \tag{37}$$

die Diskriminante

$$\varDelta = B^2 - 4AC = m^2 - 4, \tag{38}$$

dann gehört sie zu

$$M = \begin{pmatrix} \frac{1}{2}(m - B) & -C \\ A & \frac{1}{2}(m + B) \end{pmatrix} \quad \text{und} \quad -M^{-1} = \begin{pmatrix} -\frac{1}{2}(m + B) & -C \\ A & \frac{1}{2}(m - B) \end{pmatrix}.$$

Die Elemente in M sind ganz, da aus (38)

$$B \equiv m \pmod 2$$

folgt.

Im allgemeinen liegen M und $-M^{-1}$ in verschiedenen Ähnlichkeitsklassen der homogenen Modulgruppe, da sie verschiedene Spuren $a + d$ und $-a - d$ haben. In derselben Klasse können sie nur für $a + d = 0$ liegen, und tun es auch wirklich, weil dann $M = -M^{-1}$.

Der Ähnlichkeitsklasseneinteilung der Modulgruppe entspricht nun die Einteilung der quadratischen Formen in eigentliche Äquivalenzklassen. Wenn die Koeffizienten der quadratischen Form (37) die Matrix

$$Q = \begin{pmatrix} A & \frac{B}{2} \\ \frac{B}{2} & C \end{pmatrix}$$

bilden, so handelt es sich also um den folgenden

Satz 11. *Wenn der quadratischen Form (oder ihrer Koeffizientenmatrix) Q die Modulsubstitution M (zusammen mit $-M^{-1}$) zugeordnet ist, dann gehört zu $L'QL$ die Matrix $L^{-1}ML$ (zusammen mit $-L^{-1}M^{-1}L$), wo L eine beliebige Modulsubstitution ist und L' die Transponierte von L bedeutet.*

Beweis. Der Satz braucht nur für $L = S$ und $L = T$ bewiesen zu werden.

Es sei

$$M = \begin{pmatrix} a & b \\ c & d \end{pmatrix},$$

dann ist [s. (13a)]

$$S^{-1} M S = \begin{pmatrix} a' & b' \\ c' & d' \end{pmatrix} = \begin{pmatrix} a - c & a - c + b - d \\ c & c + d \end{pmatrix}.$$

Zu M gehört nach (36) die Koeffizientenmatrix

$$Q = \begin{pmatrix} c & \frac{d - a}{2} \\ \frac{d - a}{2} & -b \end{pmatrix},$$

und in der Tat gilt dann

$$S'QS = \begin{pmatrix} c & c+\frac{d-a}{2} \\ c+\frac{d-a}{2} & c-b+d-a \end{pmatrix} = \begin{pmatrix} c' & \frac{d'-a'}{2} \\ \frac{d'-a'}{2} & -b' \end{pmatrix},$$

wie es sein muß. Ebenso ist [s. (14a)]

$$T^{-1}MT = \begin{pmatrix} a'' & b'' \\ c'' & d'' \end{pmatrix} = \begin{pmatrix} d & -e \\ -b & a \end{pmatrix}$$

und

$$T'QT = \begin{pmatrix} -b & \frac{a-d}{2} \\ \frac{a-d}{2} & c \end{pmatrix} = \begin{pmatrix} c'' & \frac{d''-a''}{2} \\ \frac{d''-a''}{2} & -b'' \end{pmatrix},$$

wie es der zu beweisende Satz verlangt.

(Bemerkung. LATIMER und MACDUFFEE [5] und O. TAUSSKY [6] haben eine Zuordnung von Matrizen n-ter Ordnung zu Idealen in algebraischen Körpern n-ten Grades studiert. Wir ziehen hier statt der Ideale in quadratischen Körpern die quadratischen Formen vor, die auch für $\Delta = 0$ brauchbar sind.)

13. Satz 7 zusammen mit Satz 11 ergeben nun den

Satz 12. *Für die quadratischen Formen $Q(x,y) = Ax^2 + Bxy + Cy^2$ mit der Diskriminante $\Delta = B^2 - 4AC = m^2 - 4$ bildet die Funktion*

$$\Psi\begin{pmatrix} \frac{1}{2}(m-B) & -c \\ A & \frac{1}{2}(m+B) \end{pmatrix}$$

eine Klasseninvariante.

Das folgende Beispiel ist lehrreich: Für $m = 45$, $\Delta = 45^2 - 4 = 43 \cdot 47 = 2021$ gehören die beiden Formen

$$x^2 + xy - 505y^2 \quad \text{und} \quad 17x^2 - 7xy - 29y^2$$

verschiedenen Klassen an, denn für die erstere findet man

$$\Psi\begin{pmatrix} 22 & 505 \\ 1 & 23 \end{pmatrix} = 42$$

und für die letztere

$$\Psi\begin{pmatrix} 19 & 29 \\ 17 & 26 \end{pmatrix} = -6.$$

Die beiden Formen gehören aber demselben Geschlecht an, weil

$$\left(\frac{A}{43}\right) = +1, \quad \left(\frac{A}{47}\right) = +1$$

für $A = 1$ und $A = 17$ ist[4]). Die Funktion Ψ ist also keine generische Invariante.

---

[4]) Ein anderes Beispiel dieser Art, für $\Delta = 25^2 - 4 = 621$ wurde mir von Herrn R. Ayoub mitgeteilt.

14. Man kann nun umgekehrt den Satz 12 benutzen, um in vielen Fällen $s(h, k)$ ohne eine auf Gl. (3) gegründete Rekursion zu finden, nämlich zunächst, wenn die Klassenzahl zu $\Delta = m^2 - 4$ gleich 1 ist. Dies trifft nur für $|m| = 0, 1, 3$, also für $\Delta = -4, -3, 5$ zu. Da für diese m alle unimodularen Substitutionen der Spur $a + d = m$ zur gleichen Klasse gehören müssen, so haben wir hier

$$\Psi\begin{pmatrix} a & b \\ c & m-a \end{pmatrix} = \Psi\begin{pmatrix} m & -1 \\ 1 & 0 \end{pmatrix}$$

mit $a(m-a) - bc = 1$, $m = 0, \pm 1, \pm 3$, wobei wir $c > 0$ nehmen können. Wenn man in diese Gleichung die Werte aus (11) einsetzt, so erhält man den

Satz 13. *Es ist für $c > 0$ und $a(m-a) \equiv 1 \pmod{c}$*

$$s(a, c) = \frac{(1-c)\,m}{12\,c}$$

für $m = 0, \pm 1, \pm 3$.

Dieser Satz besagt für $m = 0$, also $a^2 \equiv -1 \pmod{c}$ nur einen Teil von Satz 1. Für ungerades m aber ist $a(m-a)$ gerade, also c ungerade. Dann lassen sich die Aussagen für $m = \pm 1, \pm 3$ auch folgendermaßen schreiben:

Für ungerades $c > 0$ und $(2a \pm 1)^2 \equiv -3 \pmod{c}$ ist

$$s(a, c) = \pm \frac{c-1}{12c}; \tag{39}$$

und für $(2a \pm 3)^2 \equiv 5 \pmod{c}$ ist

$$s(a, c) = \pm \frac{c-1}{4c}. \tag{40}$$

Dieser Fall ist besonders bemerkenswert, da er zu $\Delta > 0$ gehört, wo $M(\tau)$ hyperbolisch ist und nur Fixpunkte auf der reellen Achse hat. Ein Beispiel für (40) ist

$$a = 16,\ c = 61,\ m = +3,\ (32 + 3)^2 = 1225 \equiv 5 \pmod{61},$$

also

$$s(16, 61) = \frac{60}{4 \cdot 61} = \frac{15}{61}.$$

Auch wenn die Klassenzahl von Δ nicht 1 aber klein ist, könnte man Ergebnisse erzielen. Zum Beispiel für $m = 4$, $\Delta = 12$ gibt es zwei Klassen quadratischer Formen, von denen

$$x^2 - 4xy + y^2 \quad \text{und} \quad -x^2 + 4xy - y^2$$

Repräsentanten sind. Diesen Formen sind beziehungsweise die unimodularen Matrizen

$$\begin{pmatrix} 4 & -1 \\ 1 & 0 \end{pmatrix} \quad \text{und} \quad \begin{pmatrix} 0 & 1 \\ -1 & 4 \end{pmatrix}$$

zugeordnet, für die man erhält:

$$\Psi\begin{pmatrix} 4 & -1 \\ 1 & 0 \end{pmatrix} = 1, \quad \Psi\begin{pmatrix} 0 & 1 \\ -1 & 4 \end{pmatrix} = -1.$$

Wir haben also mit $c>0$, $a(4-a)-bc=1$

$$\Psi\begin{pmatrix} a & b \\ c & 4-a \end{pmatrix} = \pm 1$$

oder

(41) $$s(a,c) = \frac{4-3c\mp c}{12c}.$$

Sei zunächst $3 \nmid c$. Dann wissen wir nach Hilfssatz 2, daß 3 kein Teiler des Nenners von $s(a, c)$ sein kann, so daß also folgt

(42) $$s(a,c) = \begin{cases} \dfrac{4-4c}{12c} = \dfrac{1-c}{3c} & c \equiv 1 \pmod 3, \\ \dfrac{4-2c}{12c} = \dfrac{2-c}{6c} & c \equiv 2 \pmod 3. \end{cases}$$

Ist ferner $3/c$, so zeigt die Bedingung $a(4-a) \equiv 1 \pmod c$ oder

(43) $$(a-2)^2 \equiv 3 \pmod c,$$

daß $9 \nmid c$, also nur $c \equiv \pm 3 \pmod 9$ in Frage kommen.

Nun folgt aus (41)

$$12c\,s(a,c) \equiv 4 \mp c \pmod{3c},$$

anderseits aus (7), Hilfssatz 3,

$$12ac\,s(a,c) \equiv a^2+1 \pmod{3c},$$

also

$$4a \mp ac \equiv a^2+1 \pmod 9.$$

Wegen (43) ist

$$a-2 \equiv 0 \pmod 3, \qquad (a-2)^2 \equiv 0 \pmod 9,$$

also

$$3 \mp ac \equiv 0 \pmod 9,$$

oder

$$3 \pm c \equiv 0 \pmod 9,$$

so daß, da die oberen und unteren Vorzeichen hier und in (41) sich entsprechen,

$$s(a,c) = \begin{cases} \dfrac{4-4c}{12c} & c \equiv -3 \pmod 9 \\ \dfrac{4-2c}{12c} & c \equiv 3 \pmod 9 \end{cases}$$

Zusammenfassend haben wir also

Satz 14. *Wenn* $c>0$, $(a-2)^2 \equiv 3 \pmod c$, *so ist*

$$s(a,c) = \begin{cases} \dfrac{1-c}{3c} & \textit{für}\quad c \equiv 1 \pmod 3 \quad \textit{und für} \quad c \equiv -3 \pmod 9, \\ \dfrac{2-c}{6c} & \textit{für}\quad c \equiv -1 \pmod 3 \quad \textit{und für} \quad c \equiv 3 \pmod 9. \end{cases}$$

Von dieser Art könnte man noch weitere Sätze für geeignetes $\Delta > 0$ aufstellen.

15. Es bleibt noch der parabolische Fall $\Delta=0$, $m=\pm 2$ zu diskutieren. Hier haben wir, wie schon in § 11 erwähnt ist, unendlich viele Klassen.

Sei zunächst

$$a+d=2, \tag{44}$$

also

$$(d-1)^2=-bc.$$

Zu

$$M=\begin{pmatrix} a & b \\ c & d \end{pmatrix},$$

gehört die quadratische Form

$$\begin{aligned} Q(x,y) &= cx^2+2(d+1)xy-by^2 \\ &= \frac{1}{c}(cx+(d-1)y)^2. \end{aligned}$$

Wenn

$$\lambda=(c, d-1), \tag{45}$$

$$c=\lambda\gamma, \quad d-1=\lambda\delta$$

gesetzt wird, erhalten wir

$$Q(x,y)=\frac{\lambda^2}{c}(\gamma x+\delta y)^2.$$

Hier ist einerseits λ^2/c ganz, denn

$$\lambda^2=\left(c^2,(d-1)^2\right)$$

zugleich mit

$$(d-1)^2\equiv 0 \pmod c. \tag{46}$$

Anderseits ist

$$(\gamma,\delta)=1,$$

also lassen sich α, β finden, so daß

$$\begin{aligned} x' &= \alpha x+\beta y \\ y' &= \gamma x+\delta y \end{aligned}$$

eine Modulsubstitution darstellt, und somit ist

$$Q(x,y)\sim\frac{\lambda^2}{c}y^2.$$

Daher gilt innerhalb der Modulgruppe die Äquivalenz

$$\begin{pmatrix} a & b \\ c & d \end{pmatrix}\sim\begin{pmatrix} 1 & -\frac{\lambda^2}{c} \\ 0 & 1 \end{pmatrix},$$

so daß also

$$\Psi\begin{pmatrix} a & b \\ c & d \end{pmatrix}=\Psi\begin{pmatrix} 1 & -\frac{\lambda^2}{c} \\ 0 & 1 \end{pmatrix}=-\frac{\lambda^2}{c}$$

ist. Ausführlich heißt dies, unter der Annahme $c > 0$ (sonst könnte man die Vorzeichen umkehren und $a + d = -2$ betrachten),

$$\frac{2}{c} - 12\,s(a, c) - 3 = -\frac{\lambda^2}{c}, \tag{47}$$

unter der Bedingung (46).

Wenn man endlich statt (44) die Bedingung

$$a + d = -2$$

nimmt, so werden (45), (46) durch

$$\lambda = (c, d + 1),$$
$$(d + 1)^2 \equiv 0 \pmod{c}$$

ersetzt, und statt (47) erscheint, $c > 0$ vorausgesetzt,

$$-\frac{2}{c} - 12\,s(a, c) + 3 = \frac{\lambda^2}{c}.$$

Schließlich beachten wir noch $s(a, c) = s(d, c)$ und erhalten somit den

Satz 15. *Ist* $c > 0$, $(d \pm 1)^2 \equiv 0 \pmod{c}$, *so ist*

$$s(d, c) = \mp \frac{\lambda^2 + 2 - 3c}{12c},$$

mit $\lambda = (c, d \pm 1)$, *wobei die unteren und die oberen Vorzeichen zusammengehören.*

Für quadratfreies c ist stets $\lambda = c$, und der Satz besagt dann nicht mehr als (15). Er enthält jedoch eine neue Aussage für c mit quadratischem Faktor. Wir erhalten z.B. für

$$c = 50, \quad d = 10j \pm 1, \quad j = 1, 2, 3, 4, \quad \text{also} \quad \lambda = 10$$
$$s(10j \pm 1, 50) = \pm \frac{100 + 2 - 150}{600} = \mp \frac{2}{25}.$$

Literatur

[1] Mordell, L. J.: Lattice points in a tetrahedron and generalized Dedekind sums. J. Indian Math. Soc. **15**, 41—46 (1951). — [2] Rademacher, H.: Generalization of the reciprocity formula for Dedekind sums. Duke Math. J. **21**, 391—397 (1954). — [3] Rademacher, H., and A. Whiteman: Theorems on Dedekind sums. Amer. Math. J. **63**, 377—407 (1941). — [4] Reidemeister, K.: Einführung in die kombinatorische Topologie. Braunschweig 1932. — [5] Latimer, C. G., and C. C. MacDuffee: A correspondence between classes of ideals and classes of matrices. Ann. of Math. (2) **34**, 313—316 (1933). — [6] Taussky, O.: On a theorem of Latimer and MacDuffee. Canad. J. Math. **1**, 300—302 (1949).

Prof. H. Rademacher, University of Pennsylvania, Philadelphia 4, Pa. (USA)

(Eingegangen am 24. August 1955)

This paper has been reviewed in the Z, vol. 71 (1958), p. 42, and in the MR, vol. 18 (1957), p. 114.

Page 445, lines 8 and 12.
Read (3) instead of (2).

Page 450, last display line.
Add "für $b \neq 0$."

Page 454, equation (30).
Read $(-a+b)/d$ instead of $-(a+b)/d$.

Page 455, lines 7 and 9.
Read M^k instead of M_k.

Page 455, equation (32).
Author's correction (read b_2 for d_2).

Page 456, last line of third display.
Read q_l instead of q.

Page 457, second display line.
The exponent should read c.

Page 459, second display line.
Read $-c$ instead of $-e$.

Page 459, fourth display line.
Read $-C$ instead of $-c$.

Page 462, fourth display line.
Read $(d-1)$ instead of $(d+1)$.

Page 463, reference [2].
See paper 57 of the present collection.

Page 463, reference [3].
See paper 44.

FOURIER ANALYSIS IN NUMBER THEORY

Hans Rademacher

The sequence of integers might be called the prototype of a periodic phenomenon. It seems to be obvious then that Fourier analysis should find applications in number theory. Hermann Weyl has expressed this thought the other way around with the words: "$e^{2\pi ix}$ ist die eigentliche analytische Invariante Modulo 1."

The scope of this subject is, of course, immense. I have therefore to make a selection. I shall only disouss the direct applicability of Fourier analysis to number theoretical problems, but I shall not investigate further the resulting trigonometric sums and exponential sums themselves (e.g., estimates of Weyl and Vinogradoff) nor their use in the treatment of analytic functions (e.g., thetafunctions, zeta-functions), which in turn are tools in number theory.

I shall demonstrate a number of typical examples, which I think will cover more or less the field of application of Fourier analysis so far explored in number theory.

I. Finite Fourier Series

1. These come into play when each period contains only finitely many elements. Let f(n) be an arithmetical function with the property

$$f(n) = f(n') \quad \text{if } n \equiv n' \pmod{k}$$

Then

$$f(n) = \sum_{j=1}^{k} a_j e^{\frac{2\pi ijn}{k}} \tag{1.1}$$

The a_j here are uniquely determined since they are the solutions of a system of k linear equations which has a non-vanishing Vandermonde determinant. In order to compute the a_j we use orthogonality:

$$(1.2)\qquad \sum_{n=1}^{k} f(n)e^{-\frac{2\pi ihn}{k}} = \sum_{j=1}^{k} a_j \sum_{n=1}^{k} e^{\frac{2\pi i(j-h)n}{k}} = ka_h.$$

A number of useful and important formulae belong to this class. Let

$$f(n) = ((\tfrac{n}{k}))$$

where, for real x we define

$$((x)) = \begin{cases} 0 & x \text{ integer} \\ x-[x]-\frac{1}{2} & \text{otherwise.} \end{cases}$$

A direct computation of (1.2) yields then for (1.1)

$$(1.3)\qquad ((\tfrac{n}{k})) = \frac{i}{2k} \sum_{j=1}^{k-1} \cot g \frac{\pi j}{k}\, e^{\frac{2\pi ijn}{k}}$$

If this is inserted in the definition for the Dedekind sum

$$s(h,k) = \sum_{\mu=1}^{k} ((\tfrac{\mu}{k}))((\tfrac{h\mu}{k})), \quad (h,k) = 1,$$

the formula

$$(1.4)\qquad s(h,k) = \frac{1}{4k} \sum_{\ell=1}^{k-1} \cot g \frac{\pi \ell}{k} \cot g \frac{\pi h \ell}{k}$$

emerges, which can be used for a simple proof of the reciprocity law of Dedekind sums.

2. The converse problem may arise, namely to sum explicitly a given finite Fourier series. The computation of the Gaussian sums is a case in point. Another example are the Ramanujan sums

$$c_k(n) = \sum_{h \bmod k}{}' e^{\frac{2\pi ihn}{k}}$$

where the dash indicates that h runs through numbers prime to k only. It is well-known that

$$c_k(n) = \sum_{d/(n,k)} d\,\mu\left(\frac{k}{d}\right)$$

II. Enumeration of lattice points.

2. The general lattice point problem for the plane can be formulated in the following way: Let G be the region in the μ - ν plane

$$a \leqq \mu \leqq b, \qquad 0 \leqq \nu \leqq f(\mu).$$

The problem is to give an estimate for A(G) the number of lattice points in G:

(2.1)
$$A(G) = \sum_{(\mu,\nu)\varepsilon G} 1.$$

We have evidently

$$A(G) = \sum_{n=a}^{b} ([f(n)]+1) = -\sum_{n=a}^{b} (f(n)-[f(n)]-\tfrac{1}{2}) + \sum_{n=a}^{b} (f(n)-\tfrac{1}{2}).$$

Here the second sum does in general not cause any difficulty, since the Euler-Maclaurin formula can be applied. The first sum is a special case of

(2.2)
$$\sum_{n=a}^{b} \psi(f(n))$$

where $\psi(t)$ is of period 1 in t, $\psi(t)$ monotone for $0 < t < 1$ and $\int_0^1 \psi(t)dt = 0$. The sum (2.2) can be treated by investigating the trigonometric sum

(2.3)
$$\sum_{n=a}^{b} e^{2\pi imf(n)}$$

for m = 1,2,3,..., and the result will depend on the estimations of these sums. Landau, Van der Corput, Weyl, Vinogradoff have devised different methods for the treatment of these sums.

Sierpinski's theorem

$$\sum_{u^2+v^2 \leq x} 1 = \pi x + O(x^{\theta})$$

with $\theta = \frac{1}{3}$ follows in this way, but by Weyl's method (which I have to mention later), one obtains $\theta < \frac{1}{3}$.

3. Sums extended over the weighted lattice points of the whole space can be treated by the Poisson method. Let f(x) be a function with suitable regularity conditions on the whole real axis.

In order to study

$$S = \sum_{n=-\infty}^{\infty} f(n)$$

we introduce

$$S(x) = \sum_{-\infty}^{\infty} f(n+x).$$

Since S(x) is periodic

$$S(x) = S(x+1)$$

we have a Fourier expansion

$$S(x) = \sum_{=-\infty}^{\infty} A_{\ell}\, e^{2\pi i \ell x}$$

with

$$A_{\ell} = \int_0^1 S(x) e^{-2\pi i \ell x}\, dx$$

$$= \int_0^1 \sum_{n=-\infty}^{\infty} f(n+x) e^{-2\pi i \ell (n+x)} dx$$

$$= \int_{-\infty}^{\infty} f(t) e^{-2\pi i \ell t} dt$$

so that

$$S = S(0) = \sum_{n=-\infty}^{\infty} f(n) = \sum_{\ell=-\infty}^{\infty} \int_{-\infty}^{\infty} f(t) e^{-2\pi i \ell t} dt.$$

This transformation can be carried over to any number of dimensions and applied in many variations. I give here an application by Siegel to Minkowski's theorem on convex bodies in a point lattice.

Let K be a bounded convex body, symmetric with respect to the origin in n-space. We write $x = (x_1, x_2, \ldots x_n)$ and simply dx for $dx_1, dx_2, \ldots, dx_n$. Let further $\phi(x)$ be a square-integrable function which vanishes everywhere outside the convex body K. We form

$$f(x) = \sum_g \phi(2x-2g), \tag{3.2}$$

where g runs over all lattice points. The sum f(x) actually contains only finitely many terms, is square-integrable and has the period 1 in each coordinate $x_1, x_2, \ldots, x_n$, since if e is any lattice point we have

$$f(x+e) = \sum_g \phi(2(x+e)-2g) = \sum_g \phi(2x-2(g-e)) = f(x),$$

which is true in particular if e has only one coordinate 1 and all others 0. We can therefore write down the n-dimensional Fourier expansion

$$f(x) \sim \sum_h a_h \, e^{2\pi i (hx)}$$

with

$$a_h = \int_Q f(x) e^{-2\pi i (hx)} dx,$$

where Q is the unit cube. Here we have

$$a_h = \int_Q \sum_g \phi(2(x-g))e^{-2\pi i(hx)}dx$$

$$= \int_Q \sum_g \phi(2(x-g))e^{-2\pi i(h(x-g))}dx$$

$$= \int_{E^n} \phi(2y)e^{-2\pi i(hy)}dy,$$

(3.2) $$a_h = 2^{-n}\int_{E^n} \phi(x)e^{-\pi i(hx)}dx$$

Parseval's formula is in our case

(3.3) $$\int_Q |f(x)|^2dx = \sum_h |a_h|^2,$$

where h runs over all lattice-points. The left-hand side can be rewritten:

$$\int_Q |f(x)|^2dx = \int_Q \sum_g \phi(2(x-g))\overline{f(x)}dx = \int_Q \sum_g \phi(2(x-g))\overline{f(x-g)}dx$$

$$= \int_{E^n} \phi(2y)\overline{f(y)}dy = \sum_g \int_{E^n} \phi(2y)\overline{\phi(2(y-g)}dy$$

(3.4) $$= 2^{-n}\sum_g \int_{E^n} \phi(x)\ \overline{\phi(x-2g)}dx.$$

This together with (3.2) and (3.3) gives

$$\sum_g \int_{E^n} \phi(x)\ \overline{\phi(x-2g)}dx = 2^{-n}\sum_h \left|\int_{E^n} \phi(x)e^{-\pi i(hx)}dx\right|^2.$$

If now $\phi(x)\overline{\phi(x-2g)} \neq 0$, then x and x-2g are both in K. Because of the symmetry of K, -x is a lattice point. Because of the convexity of K, $\frac{-x+(x-2g)}{2} = -g$, is a lattice point.

If therefore K contains no other lattice point besides 0 then the sum on the left-hand side consists only of the term g = 0 and we have then

$$\int_K |\phi(x)|^2 dx = 2^{-n} \sum_h |\int_K \phi(x) e^{-\pi i(hx)} dx|^2.$$

Putting here

$$\phi(x) = \begin{cases} 1 & \text{for } x\epsilon K \\ 0 & \text{otherwise} \end{cases}$$

we obtain, with V the volume of K

$$V = 2^{-n} V^2 + 2^{-n} \sum_{h\neq 0} |\int_K e^{-\pi i(hx)} dx|^2$$

or

$$2^n = V + V^{-1} \sum_{h\neq 0} |\int_K e^{-\pi i(hx)} dx|^2,$$

which implies Minkowski's famous theorem.

4. There are special lattices, arising in the theory of algebraic numbers, which besides their additive period exhibit also a multiplicative period. The resulting Fourier expansion has been utilized in particular by Hecke in his investigations of zeta-functions belonging to algebraic fields. Another example is the following.

Let us have a real-quadratic number field, with integers (μ,μ'). Let η be the totally positive fundamental unit, $\eta' = \eta^{-1}$ and $f(\mu,\mu')$ a function which has the same values for associate numbers:

(4.1) $$f(\mu,\mu') = f(\eta\mu, \eta'\mu').$$

We consider the sum

(4.2) $$N(x,x') = \sum_{\substack{0<\mu\leq x \\ 0<\mu'\leq x'}} f(\mu,\mu')$$

which evidently has the property

$$N(x,x') = N(\eta x, \eta' x').$$

If we introduce

(4.3) $$g(\nu) = N(\eta^{\nu} x, \eta^{-\nu} x')$$

then this function shows the periodicity

$$g(\nu+1) = g(\nu).$$

We can therefore write down the Fourier expansion:

$$g(\nu) = \sum_{n=-\infty}^{\infty} A_n e^{2\pi i n\nu}$$

with

$$A_n = \int_0^1 g(\nu) e^{-2\pi i n\nu} d\nu.$$

Use of (4.1), (4.2), (4.3) leads, after some calculations, to

(4.4) $$A_o = \frac{1}{\log \eta} \sum_{\substack{(\mu)_1 \\ \mu\gg 0 \\ N(\mu)\leq xx'}} f(\mu,\mu') \log \frac{xx'}{N(\mu)}$$

and for $n \neq 0$

(4.5) $$A_n = \frac{1}{2\pi i n} \sum_{\substack{(\mu)_1 \\ \mu\gg 0 \\ N(\mu)\leq ax'}} f(\mu,\mu') \left\{ \left(\frac{x}{\mu}\right)^{\frac{2\pi i n}{\log\eta}} - \left(\frac{\mu'}{x'}\right)^{\frac{2\pi i n}{\log\eta}} \right\}_1 ,$$

where $\mu\gg 0$ indicates that μ is totally positive, and $(\mu)_1$ means that of all μ which are associate in the narrowest sense only one number has to be taken. The further treatment of (4.4) and (4.5), which makes use of zetafunctions of the Hecke type, does not concern us here.

I may mention, however, one example for (4.2). Let

$$f(\mu,\mu') = \begin{cases} 1 & \text{if } (\mu) \text{ is a prime ideal} \\ 0 & \text{otherwise.} \end{cases}$$

Then

$$N(x,x') = \prod(x,x') = \sum_{\substack{0<\pi\leq x \\ 0<\pi'\leq x'}} 1 ,$$

where the π are prime numbers. The method yields finally

$$\prod(x,x') = \frac{1}{2h \log \varepsilon} \log xx' \int_2^{xx'} \frac{dt}{(\log t)^2} + O(xx' e^{-e\sqrt{\log xx'}})$$

by means of the theory of the Hecke zetafunctions. Here h means the class number and $\varepsilon > 1$ the fundamental unit, whereas $e > 0$ is a certain positive constant.

III. Diophantine approximations, densities and similar problems.

We start with Kronecker's theorem:

Let $\lambda_1, \lambda_2, \ldots, \lambda_N$ be linearly independent real numbers (with respect to rationals, let $\phi_1, \phi_2, \ldots \phi_N$ be arbitrary real numbers and moreover $\varepsilon > 0$. Then there exists a real number t such that

$$|t\lambda_k - \phi_k| < \varepsilon \quad (\text{mod } 1) \quad k = 1,2,\ldots N.$$

Explanation: Here

$$|\xi| < \varepsilon \quad (\text{mod } 1)$$

means: there exists an integer g such that

$$|\xi - g| < \varepsilon.$$

Bohr first puts this theorem into a form in which its connection with Fourier analysis becomes evident:

Theorem: With the λ_k and ϕ_k of Kronecker's theorem the function

$$(5.1)\ f(t) = 1 + e^{2\pi i(t\lambda_1 - \phi_1)} + e^{2\pi i(t\lambda_2 - \phi_2)} + \ldots + e^{2\pi i(t\lambda_N - \phi_N)}$$

has the supremum N+1.

This is clearly equivalent to Kronecker's theorem. Bohr has given at least 3 different proofs of the theorem in this form, the idea always being to demonstrate that the $t\lambda_k$ modulo 1 behave somehow like independent real variables. I choose Bohr's first proof. Besides f(t) he introduces the function

$$(5.2)\qquad G(x_1,x_2,\ldots,x_N) = 1 + e^{2\pi i x_1} + e^{2\pi i x_2} + \ldots + e^{2\pi i x_N}.$$

He compares now

$$(5.3)\quad f(t)^p = \{1 + e^{2\pi i(t\lambda_1-\phi_1)} + e^{2\pi i(t\lambda_2-\phi_2)} + \ldots + e^{2\pi i(t\lambda_N-\phi_N)}\}^p$$

with

$$(5.4)\quad G(x_1,x_2,\ldots,x_N)^p = \{1 + e^{2\pi i x_1} + e^{2\pi i x_2} + \ldots + e^{2\pi i x_N}\}^p.$$

Because of the linear independence of the λ_k the terms in both expressions must behave in complete analogy: to those terms in (5.4) which can be collected into one term correspond terms in (5.3) and vice versa. Both expressions have the same sum of the absolute values of the squares of their coefficients. These sums can be obtained as

$$(5.5)\qquad f_p = \lim_{T\to\infty}\frac{1}{2T}\int_{-T}^{T}|f(t)^p|^2dt$$

and

$$G_p = \int_0^1\ldots\int_0^1 |\,G(x_1,\ldots,x_N)^p|^2dx_1,\ldots,dx_N,$$

and we know

$$f_p = G_p.$$

Now it is clear that

$$\lim_{p\to\infty} G_p^{\frac{1}{2p}} = N+1.$$

(i.e., the maximum of G in the unit cube). We can therefore conclude

$$(5.6)\qquad \lim_{p\to\infty} f^{\frac{1}{2p}} = N+1.$$

If we put

$$L_f = \sup_t |f(t)|$$

then (5.5) shows that $f_p^{\frac{1}{2p}} \leq L_f$,

and from (5.6) follows thus

$$N+1 \leq L_f$$

which proves Bohr's statement of Kronecker's theorem.

In the above formulation of Kronecker's theorem t is a continuous variable. By a simple supplementary argument one can show that t can be restricted to integers if the numbers

$$1, \lambda_1, \lambda_2, \ldots, \lambda_N$$

are linearly independent.

Bohr has made most important applications of Kronecker's theorem in the theory of the Riemann ζ-function

$$\zeta(s) = \sum_{n=1}^{\infty} \frac{1}{n^s} = \prod_p (1 + e^{-s \log p} + e^{-2s \log p} + \ldots)$$

simply observing the fact that the numbers $\{\log p\}$ are linearly indepemdent with respect to rational coefficients, which is only another way of stating the fundamental theorem of number theory.

6. For studies in the theory of almost periodic functions Bogolyuboff used the following theorem of additive number theory, which however is remarkable for its own sake:

Theorem: If the positive integers

(6.11) $$n_1 < n_2 < n_3 < \ldots$$

form a sequence of positive density, i.e., if for a certain number $A \geqq 1$ we have always

(6.12) $$n_j \leqq A_j \quad j = 1,2,\ldots$$

and if

(6.2) $$q \geq (4A)^2$$

then there exist q real numbers $\lambda_1, \lambda_2, \ldots, \lambda_q$ with $\theta \leq \lambda_k \leq 1$ such that any solution n of the set of inequalities

(6.3) $$|\lambda_k n| < \frac{1}{4} \pmod 1 \qquad k = 1,2,\ldots,q$$

is of the form

$$n = n_p + n_r - n_s - n_t$$

with elements taken from the sequence (6.11).

Proof: We choose a q to be kept fixed satisfying (6.2).

We observe now that in view of Dirichlet's pigeon-hole principle the system (6.3) has certainly for any set of λ_k at least one solution even under the restriction $0 < n \leq 4^q + 1$. If we choose therefore m so that

(6.41) $$A\,m \geq 4^q + 1$$

and put

(6.42) $$M = [Am]$$

then (6.3) has a solution $0 < n \leq M$. From (6.11), (6.12), (6.42) we infer

(6.43) $$0 < n_j \leq M \qquad j = 1,2,\ldots,m.$$

We now define the function $f_M(n)$ of period 4M and with

(6.5) $$f_M(n) = \begin{cases} 1 & \text{for } n = n_j,\ j = 1,2,\ldots,n \\ 0 & \text{otherwise} \end{cases}$$

for $0 \leq n \leq 4M-1$. This function possesses a finite Fourier expansion

(6.61) $$f_M(n) = \sum_{\ell=0}^{4M-1} a_\ell\, e^{\frac{2\pi i \ell n}{4M}},$$

with

(6.62) $$a_0 = \frac{1}{4M} \sum_{\nu=0}^{4M-1} f_M(\nu) = \frac{m}{4M},$$

(6.63) $$\alpha_\ell = \frac{1}{4M} \sum_{\nu=0}^{4M-1} f_M(\nu) e^{-\frac{2\pi i \ell \nu}{4M}},$$

(6.64) $$\sum_{\ell=0}^{4M-1} |\alpha_\ell|^2 = \frac{1}{4M} \sum_{\nu=0}^{4M-1} |f_M(\nu)|^2 = \frac{m}{4M},$$

and evidently

$$|\alpha_\ell| \leq \alpha_0.$$

We form now the convolution

$$g_M(n) = \frac{1}{4M} \sum_{a=0}^{4M-1} f_M(n+a) f_M(a) = \sum_{\ell=0}^{4M-1} |\alpha_\ell|^2 e^{\frac{2\pi i \ell n}{4M}},$$

where in the brief calculation we have used the fact

$$\alpha_{4M-\ell} = \bar{\alpha}_\ell,$$

which follows from (6.63) since $f_M(n)$ is real.

Repeating this process we obtain

(6.7) $$h_M(n) = \frac{1}{4M} \sum_{b=0}^{4M-1} g_M(n+b) g_M(b)$$

$$= \sum_{\ell=0}^{4M-1} |\alpha_\ell|^4 e^{\frac{2\pi i \ell n}{4M}}$$

$$= \sum_{\ell=0}^{4M-1} |\alpha_\ell|^4 \cos \frac{2\pi \ell n}{4M},$$

the latter because $h_M(n)$ is real.

If we have $h_M(n) > 0$ for some n in the range $0 < n \leq M$ then the explicit definition

$$h_M(n) = \frac{1}{(4M)^3} \sum_{b=0}^{4M-1} \sum_{a=0}^{4M-1} f_M(n+b+a) f_M(a) \sum_{c=0}^{4M-1} f_M(b+c) f_M(c)$$

shows, in view of (6.5) that for some a,b,c the integers $n + b + a$, a, $b+c$, c must be congruent modulo 4M to certain n_j, $1 \leq j \leq m$, let us say to n_p, n_t, n_s, n_r respectively. This implies

(6.81) $$n \equiv n_p + n_r - n_s - n_t \pmod{4M}.$$

On the one hand after (6.43) we have

$$- 2M \leqq n_p + n_r - n_s - n_t \leqq 2M$$

on the other hand

$$0 < n \leqq M.$$

These inequalities turn the congruence (6.81) into the equation

$$n = n_p + n_r - n_s - n_t. \tag{6.82}$$

We now order the a_ℓ according to size

$$a_o \geqq |a_{\ell_1}| \geqq |a_{\ell_2}| \geqq \dots \geqq |a_{\ell_{4M-1}}|,$$

so that

$$k|a_{\ell_k}|^2 \leqq |a_o|^2 + |a_{\ell_1}|^2 + \dots + |a_{\ell_{k-1}}|^2 \leqq \frac{m}{4M} \tag{6.91}$$

according to (6.64).

Let us assume now that n is a solution of

$$|\frac{\ell_k}{4M} n| < \frac{1}{4} \pmod 1, \quad k = 1,2,\dots q \tag{6.92}$$

with $0 < n \leqq M$. Such a solution n exists according to the remarks at the beginning of this proof. The system (6.92) is equivalent to the system of inequalities

$$\cos \frac{2\pi\ell_k n}{4M} > 0 \quad k = 1,2,\dots q$$

By means of (6.7) we get then

$$h(n) \geq a_o^4 + \sum_{k=1}^{q} |a_{\ell_k}|^4 \cos \frac{2\pi \ell_k n}{4M} - \sum_{k=q+1}^{4M-1} |a_{\ell_k}|^4$$

$$\geq (\frac{m}{4M})^4 - \sum_{k=q+1}^{4M-1} (\frac{m}{4M})^2 \frac{1}{k^2}$$

$$> (\frac{m}{4M})^2 \{ (\frac{m}{4M})^2 - \sum_{k=q+1}^{\infty} \frac{1}{k^2} \}$$

$$> (\frac{m}{4M})^2 \{(\frac{m}{4M})^2 - \frac{1}{q}\}$$

$$\geq (\frac{m}{4M})^2 \{ \frac{1}{(4A)^2} - \frac{1}{q}\} \geq 0$$

in view of (6.42) and (6.2), so that for such an n

$$h(n) > 0$$

Remembering now the condition belonging to (6.82) we find that any solution n, $0 < n \leq M$ of (6.92) can be expressed as in (6.82).

Let us put

$$\lambda_k^{(M)} = \frac{k}{4M}, \quad k = 1,2,\ldots,q.$$

To each m, M satisfying (6.41), (6.42) we have such a set of $\lambda_k^{(M)}$ with

$$0 < \lambda_k^{(M)} < 1.$$

There exists then a certain sequence of m and M such that for this sequence

$$\lambda_k^{(M)} \longrightarrow \lambda_k \quad k = 1,2,\ldots,q,$$

with

$$0 \leq \lambda_k \leq 1$$

These are the λ_k in (6.3) of Bogolyuboff's theorem. Indeed, let n be a solution of the system

$$|\lambda_k n| < \frac{1}{4} \quad (\text{mod } 1), \quad k = 1,2,\ldots,q,$$

then for large enough $M > n$ we have also

$$|\lambda_k^{(M)} n| < \frac{1}{4} \quad (\text{mod } 1),$$

and this means according to our previous results that n can be written as

$$n = n_p + n_r - n_s - n_t$$

with $|n_j| < M$, q.e.d.

7. Kronecker's theorem states that any point $(\phi_1, \phi_2, \ldots, \phi_N)$ in N-space can be approximated modulo 1 by $(\lambda_1 n, \lambda_2 n, \ldots, \lambda_N n)$ if

$$1, \lambda_1, \lambda_2, \ldots, \lambda_N$$

are linearly independent with respect to rationals. Weyl investigated the uniformity of the distribution modulo 1. Let us first consider the one-dimensional case N=1.

Let

$$\alpha_1, \alpha_2, \ldots, \alpha_j, \ldots$$

be a sequence of real numbers, let I be an interval modulo 1 of length ℓ, N(I,n) the number of α_j with $j \leq n$ and

$$\alpha_j \varepsilon I \quad (\text{mod } 1).$$

Then the sequence is called equi-distributed ("gleichverteilt") modulo 1 if

$$(7.1) \qquad \lim_{n \to \infty} \frac{N(I,n)}{n} = \ell$$

for any interval I modulo 1. It follows that if f(x) is periodic of period 1 and Riemann integrable then for an equi-distributed sequence we have

$$(7.2) \qquad \lim_{n \to \infty} \frac{1}{n} \sum_{j=1}^{n} f(\alpha_j) = \int_0^1 f(x)\,dx.$$

This follows for step-functions directly from (7.1) and for a Riemann integrable function f(x) by inclusion between two step functions. Evidently (7.2) is not only necessary but also sufficient for the criterion (7.1) of equi-distribution.

If we take $f(x) = e^{2\pi imx}$, $m \neq 0$, we derive from (7.2) the necessary condition

$$(7.3) \qquad \lim_{n\to\infty} \frac{1}{n} \sum_{j=1}^{n} e^{2\pi im\alpha_j} = \int_0^1 e^{2\pi imx}dx = 0$$

for the equi-distribution of the sequence $\{\alpha_j\}$.

The fundamental rôle of the function $f(x) = e^{2\pi imx}$ in the theory of approximation modulo 1 is now exhibited by the fact that (7.3) is also sufficient:

If for each integer $m \neq 0$ (7.3) is fulfilled then the sequence $\{\alpha_j\}$ is equi-distributed modulo 1.

Indeed, for the function

$$f(x) = \frac{a_0}{2} + \sum_{\mu=1}^{m} (a_\mu \cos \mu x + b_\mu \sin \mu x)$$

the property (7.3) implies directly (7.2). Then the Weierstrass theorem shows that (7.2) follows from (7.3) also for any continuous periodic function f(x). And finally by approximation of step-functions through continuous functions of steep slopes we can obtain (7.2) also for step-functions, from which the validity of (7.2) then follows for all Riemann integrable functions.

The criterion (7.3) is of great applicability. Weyl shows immediately: if α is irrational (i.e., $1, \alpha$ linearly independent) then the sequence

$$\{j\alpha\}$$

is equi-distributed modulo 1. Indeed:

$$\sum_{j=1}^{n} e^{2\pi imj\alpha} = \frac{e^{2\pi im(n+1)\alpha} - e^{2\pi im\alpha}}{e^{2\pi im\alpha} - 1},$$

hence $$\left|\sum_{j=1}^{n} e^{2\pi imj\alpha}\right| \leq \frac{2}{|e^{\pi im\alpha} - e^{-\pi im\alpha}|} = \frac{1}{|\sin \pi m\alpha|}$$

Since α is irrational, certainly $\sin \pi m\alpha \neq 0$ so that

$$\sum_{j=1}^{n} e^{2\pi imj\alpha} = O(1),$$

where $o(n)$ would already suffice.

The transition to N-dimensional space for a sequence of vectors

$$(7.4) \qquad \{ \alpha_j^{(1)}, \alpha_j^{(2)}, \ldots, \alpha_j^{(N)} \} \quad j = 1,2,3,\ldots$$

is immediate: The sequence of vectors (7.4) is equi-distributed modulo 1 if

$$\sum_{j=1}^{n} \exp 2\pi i(m_1\alpha_j^{(1)} + m_2\alpha_j^{(2)} + \ldots + m_N\alpha_j^{(N)}) = o(n)$$

for any system of integers $(m_1, m_2, \ldots, m_N)$ which do not all vanish

Returning to the case N=1 Weyl shows also the equi-distributio modulo 1 of the sequence

$$\{\alpha_j k\} \quad j = 1,2,3,\ldots$$

for irrational α. For this purpose Weyl develops an original metho to treat sums of the form

$$(7.5) \qquad \sum_{j=1}^{n} e^{2\pi imf(j)}$$

where $f(x)$ can be a polynomial with at least one irrational coefficient. Weyl's estimation of (7.5) is based on an ingenious iteratio of the Cauchy-Schwarz inequality.

This was the starting point of many further investigations, in particular by van der Corput and Vinogradoff. The sums (7.5) are now called Weyl sums. The success of Hardy and Littlewood in their solution of Goldbach's problem depended on Weyl's method (applied on the "minor arcs"). As soon as Vinogradoff improved Weyl's estimate of (7.5) he could also improve the Hardy-Littlewood results concerning Waring's problem.

IV. Prime Numbers and the Zeros of the Riemann Zetafunction

8. Whereas the previous examples showed an immediate formulation of arithmetic problems in terms of Fourier analysis, in this section we shall start with a Fourier series arrived at by methods of analytic number theory, which do not concern us here, and try to understand some of its arithmetical implications.

If we put, as customary,

$$\psi(x) = \sum_{n\leq x} \Lambda(n)$$

(8.1)
$$\Lambda(n) = \begin{cases} \log p, & n = p^k \\ 0 & \text{otherwise,} \end{cases}$$

where p is a prime number, and if

$$\rho_\nu = \beta_\nu + i\gamma_\nu, \qquad 0 < \beta_\nu < 1$$

are the non-trivial zeros of the zetafunction then we have the Riemann-von Mangoldt formula

(8.2)
$$\psi^*(x) = x - \sum_{\rho_\nu} \frac{x^{\rho_\nu}}{\rho_\nu} - \frac{1}{2}\log\left(1-\frac{1}{x^2}\right) - \log 2\pi$$

where

$$\psi^*(x) = \frac{1}{2}\left(\psi(x+0) + \psi(x-0)\right).$$

The infinite sum in (8.2) is not absolutely convergent. Its terms are ordered with respect to increasing $|\gamma_\nu|$. We shall discuss this series which must produce the discontinuities of $\psi^*(x)$. We have

$$\sum_{\rho_\nu} \frac{x^\rho}{\rho_\nu} = \sum_{\gamma_\nu>0} \left(\frac{x^{\beta_\nu+i\gamma_\nu}}{\beta_\nu+i\gamma_\nu} + \frac{x^{\beta_\nu-i\gamma_\nu}}{\beta_\nu-i\gamma_\nu} \right)$$

$$= \sum_{\gamma_\nu>0} x^{\beta_\nu}\frac{2\beta_\nu}{\beta_\nu^2+\gamma_\nu^2} \cos(\gamma_\nu \log x)$$

$$+ \sum_{\gamma_\nu>0} x^{\beta_\nu}\frac{2\gamma_\nu}{\beta_\nu^2+\gamma_\nu^2} \sin(\gamma_\nu \log x) = f_1(x)+f_2(x).$$

Now because of

$$(8.3) \qquad \sum_{0<\gamma_\nu \leq T} 1 \sim \frac{1}{2\pi} T \log T$$

the first sum is absolutely and uniformly convergent in any finite interval of x, and $f_1(x)$ is therefore continuous. For the second sum we find

$$f_2(x) = -2 \sum_{\gamma_\nu > 0} x^{\beta_\nu} \frac{\beta_\nu^2}{(\beta_\nu^2+\gamma_\nu^2)} \sin (\gamma_\nu \log x)$$

$$+ 2 \sum_{\gamma_\nu > 0} x^{\beta_\nu} \frac{1}{\gamma_\nu} \sin (\gamma_\nu \log x) = f_3(x)+f_4(x).$$

Again because of absolute and uniform convergence, $f_3(x)$ is continuous, so that

$$(8.4) \qquad f_4(x) = 2 \sum_{\gamma_\nu > 0} x^{\beta_\nu} \frac{\sin(\gamma_\nu \log x)}{\gamma_\nu}$$

has the same discontinuities as $-\psi^*(x)$ for $x > 0$. In particular, since $\psi^*(x) = 0$ for $1 < x < 2$ we have

$$f_4(x) \sim \sum_{\rho_\nu} \frac{x^{\rho_\nu}}{\rho_\nu} \sim -\frac{1}{2} \log \left(\frac{x^2-1}{x^2}\right) \longrightarrow +\infty$$

as $x \to 1+0$.

We can go one step beyond (8.4) before we introduce the Riemann hypothesis, by separating the terms in (8.4) with $|\beta_\nu - \frac{1}{2}| > \delta$ from those with $|\beta_\nu - \frac{1}{2}| \leq \delta$, with $0 < \delta < \frac{1}{2}$.

Then

$$(8.5) \qquad f_4(x) = 2 \sum_{\substack{\gamma_\nu > 0 \\ |\beta_\nu - \frac{1}{2}| \leq \delta}} + 2 \sum_{\substack{\gamma_\nu > 0 \\ |\beta_\nu - \frac{1}{2}| > \delta}}$$

But since the number $N(\frac{1}{2}+\delta, T)$ of zeros with

$$0 < \gamma < T, \quad |\beta - \tfrac{1}{2}| < \delta$$

is $O(T^{1-4d^2})$ the second sum converges absolutely and uniformly in x and thus represents again a continuous function. The discontinuities of $f_4(x)$ are therefore only produced by the first sum, containing only terms with zeros close to the line $\sigma = \frac{1}{2}$.

9. We assume now in this section the Riemann hypothesis, i.e., that $\beta_\nu = \frac{1}{2}$ throughout. Then the second sum in (8.5) becomes empty and we have the trigonometric series

$$\frac{1}{2} x^{-\frac{1}{2}} f_4(x) = \sum_{\gamma_\nu > 0} \frac{\sin(\gamma_\nu \log x)}{\gamma_\nu}$$

with y = log x we conclude that

$$(9.1) \qquad S(y) = \sum_{\gamma_\nu > 0} \frac{\sin \gamma_\nu y}{\gamma_\nu}$$

has the same discontinuities as

$$(9.2) \qquad -\frac{1}{2} e^{-\frac{y}{2}} \psi^*(e^y).$$

These are jumps which occur at the points

$$(9.3) \qquad y = k \log p$$

and are of the size

$$(9.31) \qquad -\frac{1}{2} p^{-\frac{k}{2}} \log p.$$

A few heuristic remarks may be in order. The jumps at the points necessitate for the partial sums of (9.1) steeply descending slopes. Now Landau has proved a theorem which under the Riemann hypothesis gives

$$\sum_{0<\gamma_\nu<T} e^{i\gamma_\nu y} = \begin{cases} -T \frac{\log p}{2\pi} + O(\log T) & \text{for } y = k \log p \\ O(\log T) & \text{otherwise} \end{cases}$$

Since

$$N(T) = \sum_{0<\gamma_\nu<T} 1 \quad \frac{1}{2\pi} T \log T$$

we see that

$$(9.5)\qquad \lim_{T\to\infty} \frac{1}{N(T)} \sum_{0<\gamma_\nu<T} e^{i\gamma_\nu ym} = 0$$

for any y, that is the sequence

$$\frac{y\gamma_1}{2\pi}, \frac{y\gamma_2}{2\pi}, \frac{y\gamma_3}{2\pi}, \ldots$$

is <u>equi-distributed</u> modulo 1 for any y. However, a distinction arises if the <u>speed</u> of approaching 0 in (9.5) is observed. We have

$$\lim_{T\to\infty} \frac{1}{T} \sum_{0<\gamma_\nu<T} e^{i\gamma_\nu ym} = \begin{cases} -\frac{\log p}{2\pi} & \text{for } y = k \log p \\ 0 & \text{otherwise} \end{cases}$$

For y = k log p there must be therefore in the sequence a predominance of terms which fulfill

$$(9.6)\qquad \left| \frac{\gamma_\nu k \log p}{2\pi} - \frac{1}{2} \right| < \frac{1}{4} \mod 1.$$

We are approaching here the problem of the arithmetical connection between the primes and the zeros of the zetafunction, a problem for the first time very emphatically formulated by Landau (1909) in his prime number book:

"Die Tatsache, dass $\Sigma \frac{x^\rho}{\rho}$ gerade in der Nähe der Primzahlen und der höheren Primzahlpotenzer und sonst in der Nähe keiner Stelle > 1 ungleichmassig konvergiert, deutet auf eimen arithmetischen Zusammenhang zwischen den komplexen Wurzeln ρ der Zetafunktion und den Primzahlen p hin. Ich habe keine Ahnung, worin derselbe besteht."

The inequalities (9.6) remind strongly of inequalities which define the translation numbers of an almost periodic function by means of the exponents of its trigonometric series. Indeed, S(y) is almost periodic B_2, since $\Sigma \frac{1}{\gamma_\nu^2}$ is convergent. It has, however, to be observed that for translation numbers τ and exponents Λ_ν the inequalities in the theory of almost periodic functions are

$$\left| \frac{\Lambda_\nu \tau}{2\pi} \right| < \delta \quad (\mathrm{mod}\ 1)$$

i.e., $\frac{\Lambda_\nu \tau}{2\pi}$ is near integers, whereas (9.6) shows that $\frac{\gamma_\nu}{2\pi}$ k log p has to be frequently near half-integers. Nevertheless, it seems to be worth while to ask whether 2k log p in some generalized sense belongs as translation number to S(y).

As a possible guide for future research it may be desirable to compute a greater number of zeros of the zetafunction with high precision and then to classify these zeros for different k log p according to (9.6).

It is quite surprising that a graph of the partial sum of S(y) up to ν =2q gives already a clear indication of the jumps at the logarithms of prime powers below 20. (Unpublished computations carried out by S. Rosen and the author in 1948 at the Institute for Numerical Analysis, Los Angeles.)

The series (9.1) invites of course a comparison with

$$(9.7) \qquad \sum_{n=1}^{\infty} \frac{\sin ny}{n} .$$

Besides the fact that in the series (9.7) the jumps are positive, whereas for S(y) they are negative, there is also the other difference that all jumps of (9.7) at k·2π have the same size

π, whereas for $S(y)$ they decrease with increasing y, even for the multiples k log p for the same fixed prime number p.

It is possible to construct a trigonometric series

$$\sum_{\nu=1}^{\infty} a_\nu \sin \lambda_\nu y$$

which imitates this behavior for the multiples

(9.8) $k\omega = k \cdot 2\pi$, k positive integer

with preassigned jumps

$$\sigma_k = c^{-k}, \quad c > 1$$

The series converges moreover everywhere conditionally and in closed intervals free of (9.8) also uniformly. The series can further be so modified that it shows such properties also for finitely many linearly independent $\omega_1, \omega_2, \ldots, \omega_N$. All this, of course, falls far short from the intricacies of $S(y)$, a series which exhibits such decreasing jumps for the multiples of infinitely many linearly independent $\omega = \log p$, and moreover has the peculiar connection $a_\nu = \lambda_\nu^{-1}$, properties which the series I have in mind do not imitate. I shall give a detailed account of such a trigonometric series at another occasion.

I wish finally to draw the attention of computers and numerical analysts to the zetafunctions of algebraic fields. Let us for the sake of simplicity say to that of the Gaussian field $K = R(i)$. Here

$$\zeta_K(s) = \zeta(s) L(s),$$

$$L(s) = \sum_{n=1}^{\infty} \left(\frac{-1}{n}\right) n^{-s}$$

so that $\zeta_K(s)$ preserves all the zeros of the ordinary zetafunction $\zeta(s)$. (A similar situation prevails for all class fields.) We know here also a Riemann-von Mangoldt formula, in which the jumps occur at the multiples of N(p), the norms of the prime ideals p. Now N(p) is either a prime number $\equiv 1 \pmod 4$ or the square of a prime number $\equiv 3 \pmod 4$. The numbers k, log N(p) are thus to be found among the previous k log p. It would be of interest to compute some zeros $\rho^* = \frac{1}{2} + i\gamma^*$ of L(s) to great precision and to follow up

$$\sum_{\gamma^* > 0} \frac{\sin \gamma^* y}{\gamma^*}$$

numerically as far as possible. This sum should show <u>positive</u> jumps at y = log p with $p \equiv -1 \pmod 4$ and <u>negative</u> jumps at y = log p with $p \equiv 1 \pmod 4$.

Notes to Paper 60

This report does not seem to have been reviewed.

Page 3, line 5.
Read μ, ν—.

Page 3, second display line.
Replace v by ν.

Page 3, equation (2.1).
Replace v by ν under the summation sign.

Page 3, fourth display line.
Replace the last – by a +.

Page 5, fourth display line.
hx stands for the inner product $h_1 x_1 + \dots + h_n x_n$.

Page 9, line 9.
Read: (with respect to the rationals), let . . .

Page 11, line 11.
Read: independent.

Page 19, third display line.
Read: $\beta_\nu + i\gamma_\nu$

Page 19, sixth display line.
The second fraction should read $(x^{\beta_\nu - i\gamma_\nu})/(\beta_\nu - i\gamma_\nu)$.

Page 20, second display line from bottom.
Read: $f_4(x)$.

Page 21, last line.
Insert the symbol $\cong$ between 1 and $1/2\pi$.

Page 22, seventh line from bottom.
The word is: Primzahlpotenzen.

Page 22, sixth line from bottom.
The word is: einen

page 25, line 6.
Read $k \cdot \log N(p)$ instead of k, $\log N(p)$.

For further elaboration of the problems considered in Part IV, see paper 64 of the present collection. For a partial answer to the question raised see the paper by L. A. Rubel and E. G. Straus, Special trigonometric series and the Riemann hypothesis, *Math. Scand.*, vol. 18 (1966), pp. 35-44, and the abstract by the same authors in the Proceedings of the 1963 Number Theory Conference held at the University of Colorado, Boulder, Colorado, August 5-24, 1963.

ON THE SELBERG FORMULA FOR $A_k(n)$

By HANS RADEMACHER*

[Received November 17, 1954]

THE arithmetical constants $A_k(n)$ in the series for the partition function :

$$p(n) = \frac{1}{\pi\sqrt{2}} \sum_{k=1}^{\infty} A_k(n)\, k^{\frac{1}{2}} \frac{d}{dn}\left(\frac{\sinh(c/k)\lambda}{\lambda}\right),$$

with

$$c = \pi\sqrt{(2/3)},\ \lambda = \sqrt{(n - 1/24)},$$

are defined as

$$A_k(n) = \sum_{\substack{h \bmod k \\ (h,k)=1}} \omega_{hk}\, e^{-(2\pi i h n)/k}, \tag{0.1}$$

where ω_{hk} is an often discussed and complicated 24th root of unity, arising from the theory of the modular form $\eta(\tau)$. Now Selberg has discovered and Whiteman has discussed [1] the following much simpler formula for $A_k(n)$:

$$A_k(n) = \sqrt{(k/3)} \sum_{(3j^2+j)/2 \equiv -n (\bmod k)} (-1)^j \cos \frac{6j+1}{6k}\pi, \tag{0.2}$$

which contains no reference to the ω_{hk}.

In his discussion of (0.1) Whiteman, however, makes use of the explicit formula for ω_{hk} which he obtains through the evaluation of certain Gaussian sums, following a procedure of W. Fischer [2]. I wish to show here that any evaluation of Gaussian sums is unnecessary if it is only assumed that a transformation formula of the type

$$\eta\left(\frac{h+iz}{k}\right) = \frac{\epsilon}{\sqrt{z}}\, \eta\left(\frac{h'+i/z}{k}\right),\ hh'+1 \equiv 0 \ (\bmod k) \tag{0.3}$$

with a root of unity $\epsilon = \epsilon(h, k)$ exists, a fact which is easily ascertained through the study of $\vartheta'_1(0/\tau) = 2\pi\eta^3(\tau)$ or of $\Delta(\tau) = \eta^{24}(\tau)$.

*The author is a John Simon Guggenheim Fellow for 1954-1955.

Whiteman then uses the Selberg formula (0.2) in order to obtain Lehmer's factorization theorems [3].

We shall use a slightly different form of (0.2) namely

$$A_k(n) = \tfrac{1}{4}\sqrt{(k/3)} \sum_{\substack{l^2 \equiv \nu (\mathrm{mod}\, 24k) \\ (l,6)=1}} (-1)^{\{l/6\}} e^{(\pi i l)/6k}, \quad \nu = 1 - 24n, \tag{0.4}$$

which will make the computations more transparent. Moreover we do not insist on re-establishing exactly Lehmer's theorems, but shall reach the factorization theorem (§ 9), in which no case distinctions concerning the factor 2 appear any more.

Part I. The Selberg formula

1. The root of unity ω_{hk} in (0.1) is defined through

$$f\left(\exp 2\pi i \frac{h+iz}{k}\right) = \omega_{hk}\sqrt{z}\exp\left(\frac{\pi}{12k}\left(\frac{1}{z} - z\right)\right) \times \\ \times f\left(\exp 2\pi i \frac{h' + iz^{-1}}{k}\right) \tag{1.1}$$

with $\mathrm{Re}\, z > 0$, $(h, k) = 1$, $hh' + 1 \equiv 0 \pmod{k}$ and

$$f(\exp 2\pi i\tau) = \exp\left(\frac{\pi i\tau}{12}\right)\eta(\tau)^{-1}, \tag{1.2}$$

$$\eta(\tau) = e^{(\pi i\tau)/12}\prod_{n=1}^{\infty}(1 - e^{2\pi i n\tau}). \tag{1.3}$$

Now Euler's pentagonal number theorem yields immediately

$$\eta(\tau) = \sum_{\lambda=-\infty}^{\infty}(-1)^{\lambda} e^{3\pi i(\lambda - 1/6)^2\tau} \tag{1.4}$$

and thus, by means of $\lambda = 2kq + j$,

$$\eta\left(\frac{h+iz}{k}\right) = \sum_{q=-\infty}^{\infty}\sum_{j=0}^{2k-1}(-1)^j e^{3\pi i(2kq+j-1/6)^2(h+iz)/k}$$

$$= \sum_{j=1}^{2k-1}(-1)^j e^{(3\pi i h/k)(j-1/6)^2}\sum_{q=-\infty}^{\infty} e^{-(3\pi z/k)(2kq+j-1/6)^2}.$$

Applying in the inner sum Jacobi's transformation formula we obtain

$$\eta\left(\frac{h+iz}{k}\right)=\sum_{j=0}^{2k-1}(-1)^j e^{(3\pi ih/k)(j-1/6)^2}\frac{1}{2\sqrt{(3kz)}}\sum_{m=-\infty}^{\infty}e^{2\pi im(6j-1)/12k-\pi m^2/12kz}$$

$$=\frac{1}{\sqrt{(3kz)}}\sum_{m=-\infty}^{\infty}e^{-\pi m^2/12kz}\,T_m(k),$$

$$\eta\left(\frac{h+iz}{k}\right)=\frac{1}{\sqrt{(3kz)}}\left\{T_0(k)+\sum_{m=1}^{\infty}e^{-\pi m^2/12kz}\,(T_m(k)+T_{-m}(k))\right\}, \tag{1.5}$$

with

$$T_m(k)=\tfrac{1}{2}\sum_{j\bmod 2k}e^{\pi i(j+(3h/k)(j-1/6)^2+m(6j-1)/6k)}.$$

On the other hand we assume (0.3)

$$\eta\left(\frac{h+iz}{k}\right)=\frac{\epsilon}{\sqrt{z}}\,\eta\left(\frac{h'+iz^{-1}}{k}\right),$$

from which by means of (1.4) we obtain

$$\eta\left(\frac{h+iz}{k}\right)=\frac{\epsilon}{\sqrt{z}}\sum_{\lambda=-\infty}^{\infty}(-1)^j e^{3\pi i(\lambda-1/6)^2(h'+iz^{-1})/k}$$

$$=\frac{\epsilon}{\sqrt{z}}\sum_{\lambda}(-1)^\lambda e^{-\pi(6\lambda-1)^2/12kz}\,e^{\pi ih'(6\lambda-1)^2/12k}. \tag{1.6}$$

The symbol $\sqrt{z}$, by the way, means here and in the previous formulae always the principal branch of the square root. Leaving aside the factor $1/\sqrt{z}$ the equations show both a power series in $\exp(-\pi/12\,kz)$, which must be the same one in both cases. A comparison of coefficients yields then

$$(1/\sqrt{(3k)})\,(T_{6\lambda-1}(k)+T_{-6\lambda+1}(k))=\epsilon\,(-1)^\lambda e^{(\pi ih'(6\lambda-1)^2)/12k},$$

for any integer λ. But ϵ is the same in all these equations, and we can therefore take $\lambda=0$ so that

$$\epsilon = \frac{1}{\sqrt{(3k)}} e^{-\pi i h'/12k} (T_1(k) + T_{-1}(k))$$

$$= \frac{1}{2\sqrt{(3k)}} e^{\pi i (h-h')/12k} e^{-\pi i/6k} \sum_{j \bmod 2k} e^{\pi i (3hj^2 + j(k-h+1))/k} +$$

$$+ \frac{1}{2\sqrt{(3k)}} e^{\pi i (h-h')/12k} e^{\pi i/6k} \sum_{j \bmod 2k} e^{\pi i (3hj^2 + j(k-h-1))/k} \tag{1.7}$$

Now we have, after (1.2) and (0.3)

$$f\left(\exp 2\pi i \frac{h+iz}{k}\right) = \exp\left(\pi i \frac{h+iz}{12k}\right) \eta\left(\frac{h+iz}{k}\right)^{-1}$$

$$= \epsilon^{-1} \sqrt{z} \exp\left(\pi i \frac{h+iz}{12k}\right) \eta\left(\frac{h'+iz^{-1}}{k}\right)^{-1}$$

$$= \epsilon^{-1} \sqrt{z} \exp\left(\pi i \frac{h-h'}{12k}\right) \exp \frac{\pi i (z^{-1}-z)}{12k} \times$$

$$\times f\left(\exp 2\pi i \frac{h'+iz^{-1}}{k}\right).$$

This compared with (1.1) shows

$$\omega_{hk} = \epsilon^{-1} \exp \frac{\pi i (h-h')}{12k}. \tag{1.8}$$

Now ϵ is a root of unity and thus

$$\epsilon^{-1} = \bar{\epsilon}.$$

This together with (1.7) and (1.8) yields

$$\omega_{hk} = \frac{1}{2\sqrt{(3k)}} e^{\pi i/6k} \sum_{j \bmod 2k} e^{-(\pi i/k)(3hj^2 + j(k-h+1))} +$$

$$+ \frac{1}{2\sqrt{(3k)}} e^{-\pi i/6k} \sum_{j \bmod 2k} e^{-(\pi i/k)(3hj^2 + j(k-h-1))}. \tag{1.9}$$

2. So far $(h, k) = 1$ was assumed. We need now a special case of Whiteman's Lemma 1.

LEMMA. *For $(h, k) \neq 1$ we have*

$$S = \sum_{j \bmod 2k} e^{-(\pi i/k)(3hj^2 + j(k-h\pm 1))} = 0. \tag{2.1}$$

PROOF. We put

$$(h, k) = d > 1, \text{ and } h = h^* d,\ k = k^* d,\ j = r + 2lk^*$$

with $0 \leqslant l \leqslant d-1,\ 0 \leqslant r \leqslant 2k^* - 1$, so that

$$S = \sum_{l=0}^{d-1} \sum_{r=0}^{2k^*-1} e^{-(\pi i/dk^*)(3h^*d(r+2lk^*)^2 + (r+2lk^*)(d(k^*-h^*)\pm 1)}$$

$$= \sum_{\lambda=0}^{2k^*-1} e^{-(\pi i/k)(3hr^2 + r(k-h\pm 1))} \sum_{l=0}^{d-1} e^{\mp 2\pi i l/d} = 0$$

because the inner sum vanishes.

Equations (0.1) and (1.9) yield

$$A_k(n) = \frac{1}{2\sqrt{(3k)}} e^{\pi i/6k} \sum_{h \bmod k} e^{-2\pi i h n/k} \sum_{j \bmod 2k} e^{-(\pi i/k)(3hj^2 + j(k-h+1))} +$$

$$+ \frac{1}{2\sqrt{(3k)}} e^{-\pi i/6k} \sum_{h \bmod k} e^{-2\pi i h n/k} \sum_{j \bmod 2k} e^{-(\pi i/k)(3hj^2 + j(k-h-1))}$$

where, in virtue of the lemma, h can now run over a full residue system modulo k. We change the order of summation and obtain

$$A_k(n) = \frac{1}{2\sqrt{(3k)}} e^{\pi i/6k} \sum_{j \bmod 2k} (-1)^j e^{-\pi i j/k} \sum_{h \bmod k} e^{-(2\pi i/k)(n + j(3j-1)/2)h} +$$

$$+ \frac{1}{2\sqrt{(3k)}} e^{-\pi i/6k} \sum_{j \bmod 2k} (-1)^j e^{\pi i j/k} \sum_{h \bmod k} e^{-(2\pi i/k)(n + j(3j-1)/2)h},$$

and thus

$$A_k(n) = \frac{\sqrt{k}}{2\sqrt{(3k)}} \left(\sum_{j \bmod 2k} (-1)^j e^{\pi i(6j-1)/6k} + \sum_{j \bmod 2k} (-1)^j e^{-\pi i(6j-1)/6k} \right), \tag{2.2}$$

where in the sums respectively

$$\frac{j(3j-1)}{2} \equiv -n \ (\text{mod}) \quad \text{and} \quad \frac{j(3j-1)}{2} \equiv -n \ (\text{mod } k)$$

or

$$A_k(n) = \sqrt{(k/3)} \sum_{\substack{j \bmod 2k \\ j(3j-1)/2 \equiv -n (\text{mod } k)}} (-1)^j \cos \frac{\pi(6j-1)}{6k}, \tag{2.3}$$

The latter is the Selberg formula.

3. We prefer, however, for later applications, to treat (2.2) differently. The second of the two conditions of summation can be written

$$(6j-1)^2 \equiv \nu \pmod{24\,k}$$

where we have defined

$$\nu = 1 - 24\,n. \qquad (3.1($$

If we replace in the second sum of (2.2) j by $2k - j$ we obtain

$$A_k(n) = \frac{\sqrt{k}}{2\sqrt{(3)}}\Bigg(\sum_{\substack{j \bmod 2k \\ (6j-1)^2 \equiv \nu \pmod{24k}}} (-1)^j e^{\pi i(6j-1)/6k} + \sum_{\substack{j \bmod 2k \\ (6j+1)^2 \equiv \nu \pmod{24k}}} (-1)^j e^{\pi i(6j+1)/6k}\Bigg)$$

$$= \frac{\sqrt{k}}{2\sqrt{(3)}} \sum_{\substack{6j \pm 1 \bmod 12k \\ (6j\pm 1)^2 \equiv \nu \pmod{24)k}}} (-1)^j e^{\pi i(6j\pm 1)/6k}.$$

Let us put $6j \pm 1 = l$, then $(l, 6) = 1$ and $j = \{l/6\}$ where $\{x\}$ designates the nearest integer to x. Thus

$$A_k(n) = \tfrac{1}{4}\sqrt{(k/3)} \sum_{l^2 \equiv \nu \pmod{24k}} (-1)^{\{l/6\}} e^{\pi i l/6k}, \qquad (3.2)$$

where the outer factor $\frac{1}{4}$ appears since we admit here l modulo $24k$ instead of $12k$ as above. As a function of n, $A_k(n)$ has the period k. It will be more convenient to emphasize ν from (3.1) as variable, and for that purpose we write

$$A_k(n) = B_k(\nu) \qquad (3.3)$$

with

$$\nu \equiv 1 \pmod{24} \qquad (3.4)$$

always, which implies $(l, 6) = 1$. We shall use (3.2) instead of the Selberg formula.

PART II. THE EVALUATION OF $A_k(n)$

4. We notice, for $(l, 6) = 1$,

$$(-1)^{\{l/6\}} = \left(\frac{3}{l}\right) = \left(\frac{l}{3}\right)\left(\frac{-1}{l}\right)$$

and write instead of (3.2), (3.3),

$$A_k(n) = B_k(\nu) = \tfrac{1}{4}\sqrt{(k/3)} \sum_{l^2 \equiv \nu (\mathrm{mod}\, 24k)} \left(\frac{l}{3}\right)\left(\frac{-1}{l}\right) e^{\pi i l/6k}. \tag{4.1}$$

Now we define

$$d = (24, k^3),\ e = 24/d, \tag{4.2}$$

where d and e can have only the values $1, 3, 8, 24$, and

$$(e, k) = (e, d) = 1.$$

In (4.1) the condition of summation can be written

$$l^2 \equiv \nu\ (\mathrm{mod}\ dk) \tag{4.31}$$

$$l^2 \equiv \nu\ (\mathrm{mod}\ e), \tag{4.32}$$

of which the last one is fulfilled in any case if we add again the condition

$$(l, 6) = 1. \tag{4.4}$$

Now let r be a solution of

$$(er)^2 \equiv \nu\ (\mathrm{mod}\ dk). \tag{4.5}$$

Then

$$l = er + kdh \tag{4.6}$$

satisfies the conditions (4.31), (4.32), (4.4) if and only if h runs modulo e and $(h, e) = 1$. To different pairs r, h moduli dk and e respectively belong different l modulo $24k$. We obtain thus

$$B_k(\nu) = \tfrac{1}{4}\sqrt{(k/3)} \sum_{(er)^2 \equiv \nu (\mathrm{mod}\, dk)} e^{\pi i e r/6k}\, S_k(r) \tag{4.7}$$

with

$$S_k(r) = \sum_{\substack{h \,\mathrm{mod}\, e \\ (h,e)=1}} \left(\frac{er + kdh}{3}\right)\left(\frac{-1}{er + kdh}\right) e^{\pi i d h/6}\ . \tag{4.8}$$

5. We compute $S_k(r)$ in the four cases $d = 1, 3, 8, 24$.

I. $d = 1$, $e = 24$ (i.e. $(k, 6) = 1$).

$$S_k(r) = \sum_{\substack{h \,\mathrm{mod}\, 24 \\ (h,24)=1}} \left(\frac{kh}{3}\right)\left(\frac{-1}{kh}\right) e^{\pi i h/6}$$

$$= 2\left(\frac{3}{k}\right) \sum_{\substack{h \bmod 12 \\ (h,12)=1}} (-1)^{\{h/6\}} e^{\pi i h/6}$$

$$= 4\left(\frac{3}{k}\right)\left(\cos\frac{\pi}{6} - \cos\frac{5\pi}{6}\right) = 4\left(\frac{3}{k}\right)\sqrt{3},$$

so that in this case

$$B_k(\nu) = \left(\frac{3}{k}\right)\sqrt{k} \sum_{(24r)^2 \equiv \nu (\bmod k)} e^{4\pi i r/k}, \ (k, 6) = 1. \qquad (5.1)$$

II. $d = 3, e = 8$ (i.e. $(k, 6) = 3$).

$$S_k(r) = \sum_{\substack{h \bmod 8 \\ (h,2)=1}} \left(\frac{8r}{3}\right)\left(\frac{-1}{3kh}\right) e^{\pi i h/2}$$

$$= \left(\frac{r}{3}\right)\left(\frac{-1}{k}\right) \sum_{\substack{h \bmod 8 \\ h \text{ odd}}} \left(\frac{-1}{h}\right) i^h$$

$$= \left(\frac{r}{3}\right)\left(\frac{-1}{k}\right) 4i,$$

so that

$$B_k(\nu) = i\left(\frac{-1}{k}\right)\sqrt{\left(\frac{k}{3}\right)} \sum_{(8r)^2 \equiv \nu (\bmod 3k)} \left(\frac{r}{3}\right) e^{4\pi i r/3k}, \ (k, 6) = 3. \qquad (5.2)$$

III. $d = 8, e = 3$ (i.e. $(k, 6) = 2$).

$$S_k(r) = \sum_{\substack{h \bmod 3 \\ (h,3)=1}} \left(\frac{8kh}{3}\right)\left(\frac{-1}{3r}\right) e^{4\pi i h/3}$$

$$= \left(\frac{k}{3}\right)\left(\frac{-1}{r}\right) \sum_{h=1,2} \left(\frac{h}{3}\right) e^{4\pi i h/3}$$

$$= \left(\frac{k}{3}\right)\left(\frac{-1}{r}\right) 2i \sin\frac{4\pi}{3}$$

$$= -i\left(\frac{k}{3}\right)\left(\frac{-1}{r}\right)\sqrt{3},$$

so that

$$B_k(\nu) = \frac{1}{4i}\left(\frac{k}{3}\right)\sqrt{k} \sum_{(3r)^2 \equiv \nu (\bmod 8k)} \left(\frac{-1}{r}\right) e^{\pi i r/2k}, \ (k, 6) = 2. \qquad (5.3)$$

IV. $d = 24$, $e = 1$ (i.e. $(k, 6) = 6$).

$$S_k(r) = \left(\frac{r}{3}\right)\left(\frac{-1}{r}\right) = \left(\frac{3}{r}\right)$$

and thus

$$B_k(\nu) = \tfrac{1}{4}\sqrt{\left(\frac{k}{3}\right)} \sum_{r^2 \equiv \nu (\mathrm{mod}\, 24k)} \left(\frac{3}{r}\right) e^{\pi i r/6k}, \quad (k, 6) = 6, \tag{5.4}$$

which is already contained in (4.1).

The formulae (5.1), (5.2), (5.3), (5.4) correspond to Whiteman's Theorems 1 to 4.

6. The special cases in which k is a prime power follow now from (5.1), (5.2), (5.3) as in Whiteman's paper. We discuss them briefly for the sake of completeness.

Let $k = p^\lambda$ where $p > 3$ is a prime. Equation (5.1) has to be applied. The sum is void for $\left(\frac{\nu}{p}\right) = -1$ and thus

$$B_{p^\lambda}(\nu) = 0 \text{ for } \left(\frac{\nu}{p}\right) = -1. \tag{6.1}$$

If $\left(\frac{\nu}{p}\right) = +1$ the congruence

$$(24\, r)^2 \equiv \nu \,(\mathrm{mod}\, p^\lambda) \tag{6.2}$$

has two solutions and

$$B_{p^\lambda}(\nu) = 2\left(\frac{3}{p}\right)^\lambda p^{\lambda/2} \cos\frac{4\pi r}{p^\lambda}, \tag{6.3}$$

where r is a solution of (6.2).

If further $p \mid \nu$ we consider first $\lambda = 1$. Then (6.2) has only the solution $r = 0$ and

$$B_p(\nu) = \left(\frac{3}{p}\right) p^{1/2}, \quad p \mid \nu \tag{6.4}$$

follows.

For $\lambda > 1$, $p^\lambda \mid \nu$ we have the solutions

$$r = p^{[(\lambda+1)/2]}\, j,\ j = 1, \ldots, p^{\lambda - [(\lambda+1)/2]}$$

and therefore

$$B_{p^\lambda}(\nu) = \left(\frac{3}{p}\right)^\lambda p^{\lambda/2} \sum_{j=1}^{p^{\lambda-[(\lambda+1)/2]}} \exp \frac{4\pi i j}{p^{\lambda-[\lambda+1]/2}} = 0, \lambda > 1 \quad (6.5)$$

since here $\lambda - \left[\frac{\lambda+1}{2}\right] \geqslant 1$.

If finally $p \mid \nu$ and $p^\lambda \nmid \nu$, and if p^μ with $1 < \mu < \lambda$ is the highest power of p dividing ν then μ odd will make (6.2) unsolvable and leads therefore to $B_{p\lambda}(\nu) = 0$. If $\mu = 2\rho$ is even then the congruence becomes

$$(24\, r_1)^2 \equiv \nu_1 \pmod{p^{\lambda-2\rho}} \quad (6.6)$$

with $r = p^\rho r_1$ and $\nu = p^{2\rho} \nu_1$.

If now here $\left(\frac{\nu_1}{p}\right) = -1$ then $B_{p^\lambda}(\nu)$ vanishes again. But it vanishes also in the case $\left(\frac{\nu_1}{p}\right) = +1$. If namely r_1 is a solution of (6.6) then all solutions of (6.2) can be written as

$$r = \pm p^\rho (r_1 + hp^{\lambda-2\rho}) = \pm (p^\rho r_1 + hp^{\lambda-\rho})$$

with h modulo p^ρ, and we obtain

$$B_{p^\lambda}(\nu) = \left(\frac{3}{p}\right)^\lambda p^{\lambda/2} \sum_{+-} e^{\pm 4\pi i r_1/p^{\lambda-\rho}} \sum_{h \bmod p^\rho} e^{\pm 4\pi i h/p^\rho} = 0.$$

Collecting then the results we have the

THEOREM. *If $k = p^\lambda$, $\lambda \geqslant 1$, $p > 3$ then*

$$A_k(n) = B_k(\nu) = \begin{cases} 0, \textit{ for } \left(\frac{\nu}{p}\right) = -1 \\ 2\left(\frac{3}{p}\right)^\lambda p^{\lambda/2} \cos \frac{4\pi r}{p^\lambda}, \left(\frac{\nu}{p}\right) = +1 \\ \left(\frac{3}{p}\right) p^{\frac{1}{2}}, p \mid \nu, \lambda = 1 \\ 0, p \mid \nu, \lambda > 1, \end{cases}$$

where r is a solution of the congruence

$$(24\,r)^2 \equiv \nu \pmod{p^\lambda}.$$

This is Whiteman's Corollary 1.

Let now $k = 3^\lambda$. The congruence in (5.2)

$$(8\,r)^2 \equiv \nu \pmod{3^{\lambda+1}}$$

has two solutions, $\pm r$ let us say. Then we obtain from (5.2)

$$B_{3\lambda}(\nu) = (-1)^\lambda\, i\, 3^{\frac{1}{2}(\lambda-1)} \left(\frac{r}{3}\right) (e^{4\pi i r/3} - e^{-4\pi i r/3})$$

and thus the

THEOREM.

$$B_{3\lambda}(\nu) = 2\,(-1)^{\lambda+1}\, 3^{(\lambda-1)/2} \left(\frac{r}{3}\right) \sin (4\,\pi\, r/3^{\lambda+1}),$$

where r is a solution of the congruence

$$(8r)^2 \equiv \nu \pmod{3^{\lambda+1}}.$$

This is Whiteman's Corollary 2.

Finally for $k = 2^\lambda$ we have to consider equation (5.3). The congruence there

$$(3r)^2 \equiv \nu \pmod{2^{\lambda+3}}$$

has 4 solutions since $\nu \equiv 1 \pmod 8$. If r is one of them then all can be represented as

$$\pm\,(r + h\, 2^{\lambda+2}),\ h = 0, 1.$$

Then we derive from (5.3)

$$B_{2\lambda}(\nu) = \frac{1}{4\,i}\,(-1)^\lambda\, 2^{\lambda/2} \left(\frac{-1}{r}\right) \Big(e^{\pi i r/2^{\lambda+1}} - e^{-\pi i r/2^{\lambda+1}} + e^{\pi i (r+2^{\lambda+2})/2^{\lambda+1}} + e^{-\pi i (r+2^{\lambda+2})/2^{\lambda+1}} \Big)$$

and obtain thus the

THEOREM.

$$B_{2\lambda}(\nu) = (-1)^\lambda \left(\frac{-1}{r}\right) 2^{\lambda/2} \sin \frac{\pi r}{2^{\lambda+1}},$$

where r is a solution of

$$(3r)^2 \equiv \nu \pmod{2^{\lambda+3}}.$$

This theorem is the same as Whiteman's Corollary 3.

Part III. The Multiplication Theorem

7. For $k = k_1 k_2$, $(k_1, k_2) = 1$ we intend to establish a multiplication theorem of the form

$$B_{k_1}(\nu_1)\, B_{k_2}(\nu_2) = B_k(\nu)$$

for certain ν_1, ν_2, ν. Two different cases have to be considered: either at least one of the factors k_1, k_2 is prime to 6, or both have a divisor in common with 6, in which case we may assume $2 \mid k_1$, $3 \mid k_2$.

Let first, k_1 be prime to 6. Then $d_1 = 1$ and (5.1) can be applied for $B_{k_1}(\nu_1)$, whereas we use for $B_{k_2}(\nu)$ the general formula (4.1). We have

$$B_{k_1}(\nu_1)\, B_{k_2}(\nu_2)$$
$$= \left(\frac{3}{k_1}\right) \sqrt{k_1} \sum_{(24r)^2 \equiv \nu_1 (\mathrm{mod}\, k_1))} e^{4\pi i r/k_1} \times \tfrac{1}{4} \sqrt{\frac{k_2}{3}} \sum_{l^2 \equiv \nu_2 (\mathrm{mod}\, 24 k_2)} \left(\frac{l}{3}\right) \times$$
$$\times \left(\frac{-1}{l}\right) e^{\pi i l/6k_2},$$

$$B_{k_1}(\nu_1)\, B_{k_2}(\nu_2)$$
$$= \tfrac{1}{4} \left(\frac{3}{k_1}\right) \sqrt{(k_1 k_2/3)} \sum_{\substack{(24r)^2 \equiv \nu_1 (\mathrm{mod}\, k_1) \\ l^2 \equiv \nu_2 (\mathrm{mod}\, 24k_2)}} \sum \left(\frac{l}{3}\right) \left(\frac{-1}{l}\right) e^{\pi i (24 k_2 r + k_1 l)/6 k_1 k_2}. \quad (7.1)$$

We put

$$24\, k_2\, r + k_1\, l \equiv t \pmod{24\, k_1\, k_2} \quad (7.2)$$

and can then replace the conditions for r and l by

$$t^2 \equiv \nu \pmod{24\, k_1\, k_2}$$

if

$$t^2 \equiv (24\, k_2\, r + k_1 l)^2 \equiv (24\, k_2\, r)^2 \equiv \nu \pmod{k_1},$$

which implies

$$\nu \equiv k_2^2\, \nu_1 \pmod{k_1}, \quad (7.31)$$

and similarly

$$t^2 \equiv (24\,k_2\,r + k_1 l)^2 \equiv (k_1 l)^2 \equiv \nu \pmod{24\,k_2}$$

with

$$\nu \equiv k_1^2\,\nu_2 \pmod{24\,k_2}. \tag{7.32}$$

In (7.1) we have only to compare $\left(\frac{l}{3}\right)\left(\frac{-1}{l}\right)$ with $\left(\frac{t}{3}\right)\left(\frac{-1}{t}\right)$ in order to be able to rewrite the double sum as a simple sum in t defined by (7.2). Now

$$\left(\frac{t}{3}\right)\left(\frac{-1}{t}\right) = \left(\frac{24\,k_2\,r + k_1\,l}{3}\right)\left(\frac{-1}{24\,k_2\,r + k_1\,l}\right) = \left(\frac{k_1\,l}{3}\right)\left(\frac{-1}{k_1\,l}\right)$$

$$= \left(\frac{3}{k_1}\right)\left(\frac{l}{3}\right)\left(\frac{-1}{l}\right)$$

so that (7.1) can be replaced by

$$B_{k_1}(\nu_1)\,B_{k_2}(\nu_2) = \tfrac{1}{4}\sqrt{\frac{k_1\,k_2}{3}} \sum_{t^2 \equiv \nu \pmod{24 k_1 k_2}} \left(\frac{t}{3}\right)\left(\frac{-1}{t}\right) e^{\pi i t/6 k_1 k_2}$$

$$= B_k(\nu) \tag{7.4}$$

according to (4.1). Here $(k_1, 6) = 1$, $(k_1, k_2) = 1$, $k = k_1\,k_2$, and ν, ν_1, ν_2 are connected by (7.31), (7.32). Since $\nu_1 \equiv \nu_2 \equiv 1 \pmod{24}$ we have also $\nu \equiv 1 \pmod{24}$ from (7.32).

8. Let again $(k_1, k_2) = 1$ but let $2 \mid k_1$, $3 \mid k_2$. We have to apply (5.3) for B_{k_1} and (5.2) for B_{k_2}. This yields

$$B_{k_1}(\nu_1)\,B_{k_2}(\nu_2)$$

$$= \tfrac{1}{4}\left(\frac{k_1}{3}\right)\left(\frac{-1}{k_2}\right)\sqrt{(k_1 k_2/3)} \sum_{(3r)^2 \equiv \nu_1 \pmod{8k_1}} \left(\frac{-1}{r}\right) e^{\pi i r/2k_1} \times$$

$$\times \sum_{(8s)^2 \equiv \nu_2 \pmod{3k_2}} \left(\frac{s}{3}\right) e^{4\pi i r/3k_2}$$

$$= \tfrac{1}{4}\left(\frac{k_1}{3}\right)\left(\frac{-1}{k_2}\right)\sqrt{(k_1 k_2/3)} \sum\sum_{\substack{(3r)^2 \equiv \nu_1 \pmod{8k_1} \\ (8s)^2 \equiv \nu_2 \pmod{3k_2}}} \left(\frac{-1}{r}\right) \times$$

$$\times \left(\frac{s}{3}\right) e^{\pi i (8k_1 s + 3k_2 r)/6k_1 k_2}. \tag{8.1}$$

We put

$$8k_1s + 3k_2r \equiv t \pmod{24\,k_1\,k_2} \tag{8.2}$$

and can then replace the conditions for r and s by the single condition

$$t^2 \equiv \nu \pmod{24\,k_1\,k_2}$$

if we have

$$t^2 \equiv (8k_1s + 3k_2r)^2 \equiv (3k_2r)^2 \equiv \nu \pmod{8k_1}$$

with ν satisfying

$$k_2^2\nu_1 \equiv \nu \pmod{8k_1}. \tag{8.31}$$

Similarly

$$t^2 \equiv (8k_1s + 3k_2r)^2 \equiv (8k_1s)^2 \equiv \nu \pmod{3\,k_2}$$

with

$$k_1^2\nu_2 \equiv \nu \pmod{3k_2}. \tag{8.32}$$

We are going to apply (5.4) to (8.1) and have therefore to compare $\left(\frac{3}{t}\right)$ with $\left(\frac{-1}{r}\right)\left(\frac{s}{3}\right)$. We have from (8.2)

$$\left(\frac{3}{t}\right) = \left(\frac{t}{3}\right)\left(\frac{-1}{t}\right) = \left(\frac{8k_1s + 3k_2r}{3}\right)\left(\frac{-1}{8k_1s + 3k_2r}\right)$$

$$= \left(\frac{k_1s}{3}\right)\left(\frac{-1}{k_2r}\right) = \left(\frac{k_1}{3}\right)\left(\frac{-1}{k_2}\right)\left(\frac{s}{3}\right)\left(\frac{-1}{r}\right).$$

Therefore (8.1) can be rewritten as

$$B_{k_1}(\nu_1)\,B_{k_2}(\nu_2) = \tfrac{1}{4}\sqrt{\frac{k_1k_2}{3}} \sum_{t^2 \equiv \nu \pmod{24k_1k_2}} \left(\frac{3}{t}\right) e^{\pi i t/6k_1k_2}$$

$$= B_k(\nu) \tag{8.4}$$

according to (5.4), with ν, ν_1, ν_2 connected by (8.31) and (8.32).

9. The results can be summarized in the following

THEOREM. *Let* $k = k_1k_2$, $(k_1, k_2) = 1$ *and let* d, d_1, d_2 *be defined by*

$$d = (24, k^3),\ d_1 = (24, k_1^3),\ d_2 = (24, k_2^3)$$

Let moreover be

$$\nu \equiv \nu_1 \equiv \nu_2 \equiv 1 \pmod{24}.$$

Then

$$B_{k_1}(\nu_1).\, B_{k_2}(\nu) = B_k(\nu)$$

if

$$\nu \equiv k_2^2\, \nu_1 \pmod{d_1 k_1}$$

and

$$\nu \equiv k_1^2\, \nu_2 \pmod{d_2 k_2}.$$

In this form the multiplication theorem avoids any case distinctions with respect to the occurrence of prime factors 2 in k.

In order to return to $A_k(n)$ one has only to remember the definition $A_k(n) = B_k(\nu)$ with $\nu = 1 - 24n$.

REFERENCES

1. ALBERT L. WHITEMAN : A sum connected with the series for the partition function, *Pacific Jour. Math.* 6(1956), 159-176.

2. WILHELM FISCHER : On Dedekind's function $\eta(\tau)$, *Pacific Jour. Math.* 1 (1951), 83-95.

3. D. H. LEHMER : On the series for the partition function, *Trans. American Math. Soc.* 43 (1938), 271-295.

Tata Institute of Fundamental Research, Bombay
and
The University of Pennsylvania

Notes to Paper 61

This paper has been reviewed in the Z, vol. 85 (1961), p. 267, and in the MR, vol. 19 (1958), p. 1163.

Page 41, line 5.
It has been proven by T. Apostol and P. Hagis that ω_{hk} is actually a 12th root of unity (T. Apostol, Dissertation, Berkeley, 1948, unpublished; P. Hagis, *Proc. Amer. Math. Soc.*, vol. 26 (1970), pp. 579-581).

Page 42, equation (0.4).
Here $\{x\}$ stands for the integer nearest to x (see p. 46, lines 6-7 of text).

Page 42, equation (0.4) (index to summation sign).
Read ν instead of v.

Page 43, sixth display line.
Read $(-1)^\lambda$ instead of $(-1)^j$.

Page 45, third display line.
In the first summation read r instead of λ.

Page 45, equation (2.2).
Read k instead of $\sqrt{k}$ in the numerator.

Page 45, marginal note.
Author's correction; delete also the word "and."

Page 46, equation (3.1).
Read (3.1) for (3.1(.

Page 47, marginal note.
Author's correction; while rarely used, modulis is the correct form of the plural ablative of modulus.

Page 51, first marginal note.
Author's corrections (the denominator of the exponents is $3^{\lambda+1}$).

Page 51, second marginal note.
Author's correction (delete the symbol i).

Page 51, fifth line from bottom.
Replace $=$ by $\equiv$.

Page 52, fourth display line.
Read (mod k_1) instead of (mod k_1)) under the first summation sign.

Page 53, third display line from bottom.
The exponent should read $4\pi is/3k_2$.

The Phragmèn-Lindelof Theorem and Subharmonic Functions

-1-

1. In a paper on the Phragmen-Lindelöf method I have proved the theorem:

If in the strip $a \leq R(s) \leq b$ the regular analytic function $f(s)$ satisfies the conditions

$$f(s) = O(e^{|t|^c}) \, , \; c > 0$$

and

$$|f(a+it)| \leq A|Q+a+it|^{\alpha}$$

$$|f(b+it)| \leq B|Q+b+it|^{\beta}$$

with $\alpha \geq \beta$, $Q+a > 0$, then in the whole strip

$$(1.1) \qquad |f(s)| \leq \left(A|Q+s|^{\alpha}\right)^{\frac{b-\sigma}{b-a}} \left(B|Q+s|^{\beta}\right)^{\frac{\sigma-a}{b-a}}$$

The essential feature is here that no extraneous constant factors appear in (1.1).

The clue to the proof is the fact that

$$U(\sigma,t) = (\lambda\sigma+\mu) \log \left((Q+\sigma)^2+t^2\right)$$

is subharmonic in the strip $a \leq \sigma \leq b$ for $\lambda \geq 0$.

Some other subharmonic functions can be used in a similar way leading to further results. For this purpose we prepare first some lemmas.

<u>Lemma 1</u>. For $a < b$, $Q+a > 0$ the function

$$U(\sigma,t) = (\lambda\sigma+\mu)\left((Q+\sigma)^2 + t^2\right)^{\frac{p}{2}}$$

is subharmonic in the strip $a \leq \sigma \leq b$ if

(1.2) $$p > 0, \ \lambda > 0, \ Q+a \geq -p \frac{\lambda a+\mu}{2\lambda}$$

Proof: A simple computation shows

$$\Delta U(\sigma,t) = 2p(\ (Q+\sigma)^2+t^2)^{\frac{p}{2}-1} \ (\lambda\,(\sigma+Q)+\frac{p}{2}\ (\lambda\sigma+\mu),$$

so that

$$\Delta U\ (\sigma,t) \geq 0$$

under conditions (1.2).

<u>Lemma 2</u>. For $a < b$, $Q+a > 1$, the function

$$W(\sigma,t) = (\lambda\sigma+\mu)\ \log\log(\ (Q+\sigma)^2+t^2)$$

is subharmonic in the strip $a \leq \sigma \leq b$ if

(1.3) $$\lambda > 0, \quad (Q+a)^{Q+a} \geq e^{\frac{\mu+\lambda a}{2\lambda}}$$

Proof: We find by differentiation

$$\Delta W(\sigma,t) = 4\ \frac{\lambda\,(Q+\sigma)\log(\ (Q+\sigma)^2+t^2) - (\lambda\sigma+\mu)}{(\log(\ (Q+\sigma)^2+t^2))^2\ (\ (Q+\sigma)^2+t^2)}$$

which implies

$$\Delta W \geq 0 \text{ under conditions (1.3) .}$$

2. We can now prove the

<u>Theorem 1</u>. In the strip $a \leq R(s) \leq b$ there exists for $Q+a > 0$ a regular analytic function $\varphi(s)$ such that

(2.1) $$\varphi(s) = O(e^{|t|^c}), \quad |t| \to \infty$$

for a certain $c > 0$ and that

$$|\varphi(a+it)| = e^{\gamma}|Q+a+it|^p$$

(2.2)

$$|\varphi(b+it)| = e^{\delta}|Q+b+it|^p$$

with

(2.3) $$p > 0, \gamma < \delta, \quad Q+a \geqq -p\frac{(b-a)\gamma}{2(\delta-\gamma)},$$

and which has the property

(2.4) $$|\varphi(s)| \geqq e^{|Q+s|^p(\gamma\frac{b-\sigma}{b-a}+\delta\frac{\sigma-a}{b-a})}.$$

Proof: We solve the boundary value problem for a <u>harmonic</u> function $u(\sigma,t)$ with the boundary conditions

$$u(a,t) = \gamma|Q+a+it|^p = A(t)$$

(2.5)

$$u(b,t) = \delta|Q+b+it|^p = B(t).$$

If we put

(2.61) $$w(\sigma,t) = \frac{1}{2}\frac{\sin\pi\sigma}{\operatorname{Cosh}\pi t - \cos\pi\sigma}$$

then

$$u(\sigma,t) = \frac{1}{b-a}\int_{-\infty}^{\infty} w\left(\frac{\sigma-a}{b-a}, \frac{t-y}{b-a}\right)A(y)\,dy$$

(2.62)

$$+\frac{1}{b-a}\int_{-\infty}^{\infty} w\left(\frac{b-\sigma}{b-a}, \frac{t-y}{b-a}\right)B(y)\,dy$$

is a harmonic function which solves the boundary value problem (2.5). The formula (2.61), (2.62) is obtained from the Poisson formula for the unit circle by the substitution

$$z = \tan \frac{\pi}{2} \frac{s-m}{b-a}, \quad m = \frac{a+b}{2},$$

which maps the strip $a \leq R(s) \leq b$ conformally on the unit circle $|z| \leq 1$.(*)

We want to show now that $u(\sigma,t)$ under the conditions (2.5) satisfies

$$(2.7) \qquad |u(\sigma,t)| \leq \mathcal{C}(1+|t|)^p$$

for some $\mathcal{C}$. We need to discuss only one of the integrals in (2.62) and to prove

$$I = \int_{-\infty}^{\infty} w\left(\frac{\sigma-a}{b-a}, \frac{t-y}{b-a}\right) |Q+a+iy|^p dy = O(|t|^p)$$

uniformly in σ, $a \leq \sigma \leq b$. We take $t > 0$ and write

$$I = \int_{-\infty}^{-2t} + \int_{-2t}^{2t} + \int_{2t}^{\infty} = I_1 + I_2 + I_3 .$$

On I_2

$$|Q+a+iy|^p \leq |Q+a+2it|^p$$

and thus, because (2.62) furnishes $u(\sigma,t) = 1$ for $A(y) = B(y) \equiv 1$, and because the kernel $w(\sigma,t) \geq 0$, we conclude

---

(*) I did not find the formula (2.61), (2.62) explicitly in the literature. However, it is for a=0, b=1 a simplified version of a formula of I.I.Hirschman, Journal d'Analyse Mathématique, vol. 2 (1952), pp. 209,210.

(2.71) $$0 \leq I_2 \leq |Q+a+2it|^p .$$

We have

$$|I_3| \leq \frac{1}{4} \int_{2t}^{\infty} \frac{|Q+a+iy|^p}{\operatorname{Sinh}^2 \frac{\pi(y-t)}{b-a}} dy$$

$$\leq \frac{1}{4} \int_{2t}^{\infty} \frac{|Q+a+iy|^p}{\operatorname{Sinh}^2 \frac{\pi}{2} \frac{y}{b-a}} dy$$

$$= O\left(\int_{2t}^{\infty} y^p \, e^{-\frac{\pi y}{b-a}} dy\right)$$

(2.72) $$= O\left(e^{-\frac{\pi t}{b-a}}\right) = o(t^p) .$$

And finally we have

$$I_1 \leq \frac{1}{4} \int_{2t}^{\infty} \frac{|Q+a-iy|^p}{\operatorname{Sinh}^2 \frac{\pi(y+t)}{b-a}} dy$$

(2.73) $$\leq \frac{1}{4} \int_{2t}^{\infty} \frac{|Q+a+iy|^p}{\operatorname{Sinh}^2 \frac{\pi y}{b-a}} dy = O(t^p)$$

as before. The estimates (2.71), (2.72), (2.73) prove (2.7).

3. Now the function $U(\sigma,t)$ of Lemma 1 with

$$\lambda\sigma + \mu = \gamma \frac{b-\sigma}{b-a} + \delta \frac{\sigma-a}{b-a} ,$$

or

(3.1) $$\lambda = \frac{d-\gamma}{b-a}, \quad \mu = \frac{b\gamma - a\delta}{b-a}$$

shares with $u(\sigma,t)$ the boundary conditions (2.5). A simple computation shows that in view of (2.3) and (3.1) the conditions (1.2) are fulfilled. Therefore $U(\sigma,t)$ is subharmonic and we conclude

(3.2) $$U(\sigma,t) \leq u(\sigma,t).$$

Let now $v(\sigma,t)$ be the conjugate harmonic function to $u(\sigma,t)$.

We put

$$\varphi(s) = e^{u(\sigma,t) + iv(\sigma,t)}$$

Then $\varphi(s)$ is regular analytic in the strip $S(a,b)$, and

$$|\varphi(s)| = e^{u(\sigma,t)}$$

This implies that the boundary conditions (2.2) are fulfilled for $\varphi(s)$. Moreover we see that

$$|\varphi(s)| \geq e^{U(\sigma,t)}$$

which with the definition $U(\sigma,t)$ in Lemma 1 and with (3.2) proves (2.4). Finally (2.1) is also satisfies because of (2.7). This concludes the proof of Theorem 1.

4. We are now prepared to prove

Theorem 2: Let $f(s)$ be regular analytic in the strip $a \leq \sigma \leq b$ and there

(4.1) $$f(s) = O(e^{|t|^c})$$

for some $c > 0$. Let moreover $f(s)$ satisfy the conditions

(4.2) $$|f(a+it)| \leq e^{\alpha\ |Q+a+it|^p}$$
$$|f(b+it)| \leq e^{\beta\,|Q+b+it|^p}$$

with

(4.3) $$p > 0,\ \alpha > \beta.\ Q+a \geq \max\left(0, \frac{p(b-a)\alpha}{2(\alpha-\beta)}\right),$$

then for all s in the strip

(4.4) $$|f(s) \leq e^{|Q+s|^p \left(\frac{b-\sigma}{b-a}\alpha + \frac{\sigma-a}{b-a}\beta\right)}.$$

Proof: We take $\varphi(s)$ of Theorem 1 with the specifications $\gamma = -\alpha$, $\delta = -\beta$. Then (4.3) shows that the conditions (2.3) of Theorem 1 are fulfilled. That means we have

$$|\varphi(a+it)| = e^{-\alpha|Q+a+it|^p}$$

$$|\varphi(b+it)| = e^{-\beta|Q+b+it|^p}.$$

We form now the regular analytic function

$$F(s) = f(s) \cdot \varphi(s)$$

for which we have

$$|F(a+it)| \leq 1$$

$$|F(b+it)| \leq 1$$

and

$$F(s) = O(e^{|t|^c}).$$

Therefore, according to the Phragmén-Lindelöf argument, we have throughout the strip

$$|F(s)| \leq 1.$$

This implies

$$|f(s)| \leq |\varphi(s)|^{-1}$$

and thus (4.4) .

5. In a fully analogous manner we can derive from Lemma 2 the following

Theorem 3: In the strip $a \leq R(s) \leq b$, there exists a regular analytic function $\Psi(s)$ such that

$$\Psi(s) = O(e^{|t|^c})$$

and that

$$|\Psi(a+it)| = (\log|Q+a+it|)^{\gamma}$$

(5.1)

$$|\Psi(b+it)| = (\log|Q+b+it|)^{\delta}$$

with

$$\gamma \leq \delta, \quad Q+a > 1, \quad (Q+a)^{Q+a} \geq e^{\frac{(b-a)}{2(\delta-\gamma)}}$$

and which has the property

$$|\Psi(s)| \geq (\log|Q+s|)^{\frac{b-\sigma}{b-a}\gamma + \frac{\sigma-a}{b-a}\delta}$$

As far as the proof is concerned it may be observed that the proof for (2.7) amply implies the same result under the conditions (5.1).

From this theorem we obtain then again by the Phragmen-Lindelof argument the

Theorem 4 If $f(s)$ is regular analytic and satisfies

$$f(s) = O(e^{|t|^c})$$

in $a \leq R(s) \leq b$, and if

$$|f(a+it)| \leq (\log|Q+a+it|)^{a}$$

(5.2)

$$|f(b+it)| \leq (\log|Q+b+it|)^{\beta}$$

with

$$\alpha < \beta,\ 1 < Q+a,\ (Q+a)^{Q+a} \geq e^{\frac{(b-a)\,\alpha}{2(\beta-\alpha)}},$$

then for all s in the strip

$$|f(s)| \leq (\log|Q+s|)^{\frac{b-\sigma}{b-a}\alpha + \frac{\sigma-a}{b-a}\beta} \tag{5.3}$$

Finally, it is obvious that in the same manner theorems can be established which contain multiplicative combinations of the boundary conditions (4.2), (5.2) of Theorems 2 and 4 and of conditions (4.2) of my previous article.

Notes to Paper 62

This report contains material treated also in papers 62a and 62b of the present collection; it seems that the report has not been reviewed separately.

Page 1, line 1.
The paper referred to seems to be 62b, where the corresponding statement appears almost verbatim as Theorem 2. Paper 62a, while apparently written earlier, was presented and appeared only after the date of the present report.

ON THE PHRAGMÉN-LINDELÖF THEOREM AND SOME APPLICATIONS

Hans Rademacher

Included because Hadamard's Three Circle Theor[em] mentioned

The Phragmén-Lindelöf theorem plays an important role in the theory of the Riemann zeta function. It is used to obtain information about the growth of $\zeta(s)$ in a vertical parallel strip from the given growth on the boundaries of the strip. Mostly only a half strip $t > 1$ is considered. However, this is not precise enough in some other instances, where whole families of functions are considered and a uniform estimate for all these functions is needed. The purpose of this paper is to obtain a theorem with all its constants explicitly known.

The goal is reached in two steps; the first is a boundary value problem for harmonic functions, the second the application of the Phragmén-Lindelöf argument. We prove:

THEOREM 1. Let a, b, γ, δ, Q be real numbers,

$$-Q < a < b, \quad \gamma \leq \delta.$$

Then there exists an analytic function $\phi(s) = \phi(s; Q)$ regular in the strip

$$S(a, b)\colon a \leq \mathrm{Re}(s) \leq b,$$

and such that

$$(1)\quad \begin{aligned} |\phi(a+it)| &= |Q+a+it|^{\gamma} \\ |\phi(b+it)| &= |Q+b+it|^{\delta} \end{aligned}$$

and that in S(a, b)

$$(2)\quad |\phi(s)| \geq |Q+s|^{\gamma \frac{b-\sigma}{b-a} + \delta \frac{\sigma-a}{b-a}}$$

Moreover

$$(3)\quad \phi(s) = O(e^{|t|^c}), \qquad |t| \longrightarrow \infty$$

for a certain $c > 0$, $s = \sigma+it$.

It is essential that the equality sign is valid in (1). Moreover (2) is sharp in the sense that it does not contain any unknown constant and requires the equality sign on the boundaries.

The second step leads to the

THEOREM 2. Let $f(s)$ be regular analytic in the strip $S(a, b)$ including its boundaries and fulfill for certain positive c, C

$$|f(s)| < Ce^{|t|^c}. \tag{4}$$

Let moreover

$$\begin{aligned} |f(a+it)| &\le A|Q+a+it|^{\alpha}, \\ |f(b+it)| &\le B|Q+b+it|^{\beta}, \end{aligned} \tag{5}$$

with

$$Q+a > 0, \quad \alpha \ge \beta. \tag{6}$$

Then in the strip $S(a, b)$

$$|f(s)| \le (A|Q+s|^{\alpha})^{\frac{b-\sigma}{b-a}} (B|Q+s|^{\beta})^{\frac{\sigma-a}{b-a}}. \tag{7}$$

A special case of Theorem 2 is Hadamard's three circle theorem, which is obtained for $\alpha = \beta = 0$ and for a function $f(s)$ periodic in t.

As far as the proof of Theorem 1 is concerned, we solve for the harmonic function $u(\sigma, t)$ the boundary value problem

$$\begin{aligned} u(a,t) &= \gamma \log|Q+a+it| = \frac{\gamma}{2} \log[(Q+a)^2+t^2], \\ u(b,t) &= \delta \log|Q+b+it| = \frac{\delta}{2} \log[(Q+b)^2+t^2]. \end{aligned} \tag{8}$$

That this boundary value problem has a solution can be shown in the following way. Put

$$\omega(\sigma, t) = \frac{1}{2} \frac{\sin \pi\sigma}{\operatorname{Cosh} \pi t - \cos \pi\sigma}. \tag{9.1}$$

Then

$$\begin{aligned} u(\sigma, t) &= \frac{1}{b-a} \int_{-\infty}^{\infty} \omega\left(\frac{\sigma-a}{b-a}, \frac{t-y}{b-a}\right) A(y)\, dy \\ &+ \frac{1}{b-a} \int_{-\infty}^{\infty} \omega\left(\frac{b-\sigma}{b-a}, \frac{t-y}{b-a}\right) B(y)\, dy \end{aligned} \tag{9.2}$$

is a harmonic function in the strip $S(a, b)$ with the boundary values

$$u(a, t) = A(t), \quad u(b, t) = B(t).$$

The formula (9.1), (9.2) is obtained from the Poisson formula for the circle by the conformal mapping

$$t = \operatorname{tang} \frac{\pi}{2} \frac{s-m}{b-a}, \quad m = \frac{a+b}{2}.$$

The boundary values given in (8) ensure also

$$|u(\sigma, t)| < C|t|.$$

Let now $v(\sigma, t)$ be the harmonic conjugate to $u(\sigma, t)$. We put

(10) $$\phi(s) = e^{u(\sigma,t)+iv(\sigma,t)}$$

and have a regular analytic function $\phi(s)$ which satisfies (1) and (3).

Now the function

(11) $$U(\sigma,t) = (\lambda\sigma + \mu)\log|Q + \sigma + it|$$

is <u>subharmonic</u> in $S(a,b)$ for $\lambda \geq 0$, as

$$\Delta U = 4\lambda \frac{Q+\sigma}{(Q+\sigma)^2 + t^2} \geq 0$$

shows. This is seen immediately from

$$U = f\cdot g,$$

$$\Delta U = \Delta f\cdot g + 2(f_\sigma g_\sigma + f_t g_t) + f\cdot\Delta g.$$

If we determine λ, μ so that

$$\lambda a+\mu = \gamma$$

$$\lambda b+\mu = \delta,$$

i.e.

$$\lambda = \frac{\delta-\gamma}{b-a} \geq 0,$$

we see that $U(\sigma,t)$ has the same boundary values as $u(\sigma,t)$. Hence

$$U(\sigma, t) \leq u(\sigma, t)$$

and consequently

$$|\phi(s)| \geq e^{U(\sigma, t)},$$

which is (2)

For the proof of Theorem 2 we take the $\phi(s)$ of Theorem 1 with $\gamma = -\alpha$, $\delta = -\beta$ and put

$$F(s) = f(s) \cdot \phi(s) E^{-1} e^{-\nu s},$$

where E and ν are so determined that

$$(12) \qquad \begin{aligned} A &= E e^{\nu a} \\ B &= E e^{\nu b} \end{aligned}$$

Then this together with (1) and (5) shows

$$|F(a+it)| \leq 1$$
$$|F(b+it)| \leq 1$$

and, since

$$F(s) = O(e^{|t|^c}),$$

according to the Phragmén-Lindelöf argument, also

$$|F(s)| \leq 1$$

inside S(a, b). Then

$$|f(s)| \leq E e^{\nu\sigma} |\phi(s)|^{-1},$$

which with (2) and (12) gives the desired result (7).

Through the use of other suitable subharmonic functions more theorems like Theorem 2 can be obtained, e.g.

THEOREM 3. Let f(s) be regular in S(a, b) and on its boundaries and

$$f(s) = O(e^{|t|^c}).$$

Let moreover f(s) satisfy the conditions

$$|f(a+it)| \leq e^{\alpha|Q+a+it|^p}$$
$$|f(b+it)| \leq e^{\beta|Q+b+it|^p}$$

with

$$p > 0, \ \alpha > \beta, \ Q+a > \max(0, \frac{p(b-a)\alpha}{2(\alpha-\beta)}),$$

then for all s in S(a, b)

$$|f(s)| \leq e^{|Q+s|^{p}\left(\frac{b-\sigma}{b-a}\alpha + \frac{\sigma-a}{b-a}\beta\right)}$$

More important, however, are the applications of Theorem 2 to special functions. Already for the Γ-function the following estimates seem to be new:

THEOREM 4. For $Q \geq 0$, $-\frac{1}{2} \leq \sigma \leq \frac{1}{2}$ the inequalities

$$\left| \frac{\Gamma\left(\frac{Q}{2} + \frac{1-s}{2}\right)}{\Gamma\left(\frac{Q}{2} + \frac{s}{2}\right)} \right| \leq \left(\frac{1}{2}|Q+1+s|\right)^{\frac{1}{2}-\sigma}$$

$$\left| \frac{\Gamma(Q+1-s)}{\Gamma(Q+s)} \right| \leq |Q+1+s|^{1-2\sigma}$$

hold.

It may be remarked that here equality is true in both cases for $s = \frac{1}{2} + it$, $Q \geq 0$, and $s = -\frac{1}{2} + it$, $Q = 0$.

Further applications of Theorems 2 and 4 can be made in the theory of $L(s,\chi)$ and ζ_K-functions. We obtain the following theorems:

THEOREM 5. For $0 < \eta \leq \frac{1}{2}$, for all moduli $k > 1$ and all primitive characters χ modulo k the inequality

$$|L(s,\chi)| \leq \left(\frac{k|1+s|}{2\pi}\right)^{\frac{1+\eta-\sigma}{2}} \zeta(1+\eta)$$

holds in the strip $-\eta \leq \sigma \leq 1+\eta$.

THEOREM 6. Let $\zeta_K(s)$ be the Dedekind function of the algebraic number field K of degree n and discriminant d. Then

$$|\zeta_K(s)| < 3\left|\frac{1+s}{1-s}\right|\left[|d|\left(\frac{|1+s|}{2\pi}\right)^n\right]^\tau [\zeta(1+\eta)]^n$$

where $\tau = \frac{1+\eta-\sigma}{2}$ and

$$-\frac{1}{2} \leq \eta \leq \sigma \leq 1+\eta \leq \frac{3}{2}.$$

If K is Abelian over the rational field then

$$|\zeta_K(s)| \leq \left(|d|\left(\frac{|1+s|}{2\pi}\right)^{n-1}\right)^\tau \cdot |\zeta(s)|[\zeta(1+\eta)]^{n-1}$$

in the same strip.

University of Pennsylvania
Philadelphia, Pennsylvania

I-322

Notes to Paper 62a

This paper has been reviewed in the Z, vol. 94 (1962), p. 274.

Page I-314, marginal note (top of page).
Author's remark. Its purpose is to justify the inclusion of this paper in his collection, although it overlaps with two other papers (see papers 62 and 62b) of which one has the same title and much of the same text. Theorems 1 and 2 of the present paper are essentially the same as Theorems 1 and 2 in 62b; Theorem 4 corresponds to Lemmas 1 and 3 in 62b, Theorem 5 corresponds to Theorem 3 and Theorem 6 to Theorem 4 and (a weak form of) Theorem 4a of 62b. In general, 62b is a refined and improved version of the present paper.

Page I-315, equation (3).
This estimate should be compared with (2.7) of 62b. The stronger result is obtained there at the expense of the uniformity in Q.

Page I-316, equation (7).
Author's correction. The inequality should read

$$|f(s)| \leqslant (A|Q+s|^{\alpha})^{(b-\sigma)/(b-a)}\,(B|Q+s|^{\beta})^{(\sigma-a)/(b-a)}$$

exactly as in Theorem 2 of 62b.

Page I-316, line 5.
Hadamard's Theorem is obtained by the change of variable $z = e^s$.

Page I-318, third display line.
The numerical coefficient should be 2 rather than 4.

On the Phragmén-Lindelöf theorem and some applications

In memoriam LEON LICHTENSTEIN

By

HANS RADEMACHER

1. One of the most fruitful applications of the Phragmén-Lindelöf theorem occurs in analytic number theory in the discussion of the Riemann zetafunction. In its most frequently used form the theorem is applied to a half-strip and to a single function (cf. LANDAU, Vorlesungen über Zahlentheorie, vol. 2, Satz 404, 405, pp. 48—51). This, however, is not quite general enough for situations encountered e.g. in the theory of character sums and of primes in arithmetic progressions. We need here estimations which are uniform in several parameters for an infinite set of $L(s, \chi)$ or $\zeta(s, \lambda)$-functions, and the whole strip must be considered[1]). It is the purpose of the present paper to provide a suitable refinement of the Phragmén-Lindelöf theorem. The improvement is achieved through the use of certain subharmonic functions. The applications in Theorems 3, 4, 5 concern zetafunctions. But already the results about the Γ-function in Lemmas 1 to 3 are believed to be new.

2. Basic for our arguments is the following

THEOREM 1. *Let a, b, Q, γ, δ be real numbers,*

$$-Q<a<b, \qquad \gamma \leqq \delta. \tag{2.1}$$

Then there exists an analytic function $\varphi(s)=\varphi(s; Q)$, depending also on the parameters a, b, γ, δ, which is regular in the strip

$$S(a, b): \qquad a \leqq \Re(s) \leqq b$$

and such that

$$\begin{cases} |\varphi(a+it; Q)| = |Q+a+it|^{\gamma} \\ |\varphi(b+it; Q)| = |Q+b+it|^{\delta} \end{cases} \tag{2.2}$$

and that for $a \leqq \sigma \leqq b$

$$|\varphi(s; Q)| \geqq |Q+s|^{l(\sigma)} \tag{2.3}$$

where

$$l(\sigma) = \gamma \frac{b-\sigma}{b-a} + \delta \frac{\sigma-a}{b-a}. \tag{2.31}$$

---

1) Such a situation is found e.g. in the discussion of the Hecke functions in [5], Hilfssatz 15, pp. 363—365; where, however, a rather crude treatment if sufficient.

Moreover

(2.32) $$\varphi(s; Q) = O(|t|^c), \quad |t| \to \infty$$

for a certain $c > 0$.

REMARK. Because of (2.1) we have $|Q+s| > 0$ so that (2.3) implies

(2.33) $$\varphi(s; Q) \neq 0.$$

It is essential for our further arguments that we have *equality* in (2.2) and that (2.3) holds for all Q with $Q+a>0$.

PROOF. The case $\gamma=\delta$ is trivial since here $\varphi(s; Q)=(Q+s)^\gamma=(Q+s)^\delta$ furnishes the solution. Here, and in similar situations later, we take the principal branch of $(Q+s)^\gamma$ in $S(a, b)$, i.e. that one which is real for positive $Q+s$. We can now assume $\gamma<\delta$.

We construct a harmonic function $u(\sigma, t)$ in the strip $S(a, b)$ which fulfills the boundary conditions

(2.4) $$\begin{cases} u(a, t) = \gamma \log|Q+a+it| = \tfrac{1}{2}\gamma \log((Q+a)^2+t^2) \\ u(b, t) = \delta \log|Q+b+it| = \tfrac{1}{2}\delta \log((Q+b)^2+t^2). \end{cases}$$

The boundary problem for $\Delta u=0$ in a strip is solved by a transtormation of the Poission formula for the circle. Let us put

(2.51) $$\omega(\sigma, t) = \frac{1}{2}\,\frac{\sin \pi\sigma}{\cosh \pi t - \cos \pi\sigma}.$$

Then

(2.52) $$u(\sigma, t) = \frac{1}{b-a}\int_{-\infty}^{\infty} \omega\left(\frac{\sigma-a}{b-a}, \frac{t-y}{b-a}\right) A(y)\,dy + \frac{1}{b-a}\int_{-\infty}^{\infty} \omega\left(\frac{b-\sigma}{b-a}, \frac{t-y}{b-a}\right) B(y)\,dy$$

satisfies $\Delta(u)=0$ in the interior of $S(a, b)$ with the boundary conditions

(2.53) $$u(a, t) = A(t), \quad u(b, t) = B(t),$$

where $A(t)$, $B(t)$ are supposed to be continuous and $e^{-\frac{\pi|t|}{b-a}}|A(t)|$, $e^{-\frac{\pi|t|}{b-a}}|B(t)|$ integrable in the interval $(-\infty, \infty)$. (Cf [2], p. 550.)

Taking

(2.6) $$A(t) = \gamma \log|Q+a+it|, \qquad B(t) = \delta \log|Q+b+it|,$$

we have in (2.52) a solution of the boundary problem (2.4).

From (2.51), (2.52) (2.6) it follows by a simple estimation that

(2.7) $$u(\sigma, t) = O(\log(2+|t|))$$

in the strip $S(a, b)$ uniformly in σ, although not in Q.

Let now $v(\sigma, t)$ be a conjugate harmonic function to $u(\sigma, t)$. Since $u(\sigma, t)$ in view of the boundary conditions (2.4) is evidently symmetric with respect to the σ-axis, $v(\sigma, t)$ must be constant for $t=0$, and we can normalize $v(\sigma, t)$ through an additive constant so that $v(\sigma, 0)=0$. We put now

$$\varphi(s; Q) = \exp\left(u(\sigma, t) + i v(\sigma, t)\right) \tag{2.8}$$

which is a regular analytic function of $s=\sigma+it$ with the absolute value

$$|\varphi(s; Q)| = \exp u(\sigma, t). \tag{2.81}$$

The boundary conditions (2.4) show that (2.2) is satisfied. Also satisfied is (2.32) because of (2.7).

In order to finish the proof of Theorem 1 we consider the function

$$U(\sigma, t) = (\lambda\sigma + \mu) \log|Q + \sigma + it| = (\lambda\sigma + \mu)\log|Q + s|. \tag{2.9}$$

We have

$$\Delta U = 2\lambda \frac{Q+\sigma}{(Q+\sigma)^2 + t^2}, \tag{2.91}$$

which is easily seen, since $U(\sigma, t) = f\cdot g$ and $\Delta U = f\cdot\Delta g + 2(f_\sigma g_\sigma + f_t g_t) + g\cdot\Delta f$, where Δg, Δf and f_t obviously vanish. We choose now the constants λ and μ so that $U(\sigma, t)$ also satisfies the boundary conditions (2.4) i.e.

$$\lambda a + \mu = \gamma, \qquad \lambda b + \mu = \delta$$

or

$$\lambda = \frac{\delta - \gamma}{b - a}, \qquad \mu = \frac{\gamma b - \delta a}{b - a}. \tag{2.92}$$

The conditions (2.1) show then that

$$\lambda > 0.$$

(since we consider here only $\gamma < \delta$), and therefore in virtue of (2.91) and $Q + \sigma > 0$

$$\Delta U > 0.$$

In other words, $U(\sigma, t)$ is a *subharmonic* function in the strip $S(a, b)$. It has the same boundary values as the harmonic function $u(\sigma, t)$, and therefore we conclude

$$U(\sigma, t) \leqq u(\sigma, t)$$

in $S(a, b)$. This inequality for the infinite strip $S(a, b)$ is obtained by an application of the Phragmén-Lindelöf argument to the subharmonic function $D(\sigma, t) = U(\sigma, t) - u(\sigma, t)$ which, in view of (2.7) and (2.9), has the property $D(\sigma, t) = O(|t|^c)$, $|t| \to \infty$, in $S(a, b)$. But then it follows from (2.81) that

$$|\varphi(s; Q)| \geqq \exp U(\sigma, t).$$

If we take into account (2.9) and (2.92) we see that this inequality is exactly (2.3). This concludes the proof of Theorem 1.

3. We use this theorem now to prove the following one, which is of the Phragmén-Lindelöf type:

THEOREM 2. *Let $f(s)$ be regular analytic in the strip $S(a, b)$ and satisfy for certain positive constants c, C*

$$|f(s)| < C e^{|t|^c}. \tag{3.1}$$

Suppose moreover that

$$\begin{cases} |f(a+it)| \leqq A\,|Q+a+it|^{\alpha} \\ |f(b+it)| \leqq B\,|Q+b+it|^{\beta} \end{cases} \tag{3.2}$$

with

$$Q + a > 0, \tag{3.31}$$

$$\alpha \geqq \beta. \tag{3.32}$$

Then in the strip $S(a, b)$

$$|f(s)| \leqq (A\,|Q+s|^{\alpha})^{\frac{b-\sigma}{b-a}} (B\,|Q+s|^{\beta})^{\frac{\sigma-a}{b-a}}. \tag{3.4}$$

REMARK. It will be noticed that no constants enter into (3.4) which do not already appear in the conditions (3.2)

PROOF. We consider the function $\varphi(s; Q)$ of Theorem 1 with the parameter values

$$\gamma = -\alpha, \qquad \delta = -\beta, \tag{3.5}$$

which in view of (3.32) satisfy (2.1). Then we form the regular function

$$F(s) = f(s)\,\varphi(s; Q)\,E^{-1} e^{-\nu s}, \tag{3.6}$$

where E and ν are so determined that

$$A = E\,e^{\nu a}, \qquad B = e^{\nu b}. \tag{3.7}$$

We see that because of (3.1) and (2.32) an estimate

$$|F(s)| < C^{|t|^c} \tag{3.8}$$

holds with suitable positive c, C. Now the conditions (3.2), (3.5), (3.7) give together with (2.2) the boundary conditions for $F(s)$:

$$|F(a+it)| \leqq 1, \qquad |F(b+it)| \leqq 1.$$

The well-known Phragmén-Lindelöf argument, admissible in view of (3.8), applied on the rectangle $a \leqq \sigma \leqq b$, $|t| \leqq T$, $T \to \infty$, shows that

$$|F(s)| \leqq 1$$

throughout $S(a, b)$. The definition (3.6) of $F(s)$ shows that this means

$$|f(s)| \leqq E\,e^{\nu\sigma}\,|\varphi(s)|^{-1},$$

which because of (2.3), (2.31), (3.5) and (3.7) yields the assertion (3.4) of the theorem.

4. The case $\alpha<\beta$ can also be treated, but the resulting theorem is less concise, since a factor M depending on a, b, α, β, Q appears. We obtain here

THEOREM 2a. *Under the conditions of Theorem 2, except* (3.32) *to be replaced by*

$$\alpha<\beta, \tag{4.1}$$

we have in the strip $S(a, b)$

$$|f(s)| \leqq M_Q^2 (A|Q+s|^{\alpha})^{\frac{b-\sigma}{b-a}} (B|Q+s|^{\beta})^{\frac{\sigma-a}{b-a}}, \tag{4.2}$$

where

$$M_Q = \operatorname{Max}\left(\left(\frac{a+Q}{b+Q}\right)^{\alpha}, \left(\frac{b+Q}{a+Q}\right)^{\beta}\right). \tag{4.3}$$

The factor M_Q *is a monotone decreasing function of* Q *and*

$$\lim_{Q\to\infty} M_Q = 1. \tag{4.4}$$

PROOF. This theorem is derived from Theorem 2 by the substitutions

$$s' = a+b-s,$$

$$\alpha' = \beta, \qquad \beta' = \alpha, \qquad A' = B, \qquad B' = A.$$

We have again

$$a \leqq \sigma' = \Re(s') \leqq b,$$

and α', β' satisfy condition (3.32) of Theorem 2.

We put

$$g(s') = g(a+b-s) = f(s).$$

Then $g(s')$ satisfies the conditions

$$|g(a+it)| = |f(b-it)| \leqq A'|Q+b+it|^{\alpha'}$$

$$|g(b+it)| = |f(a-it)| \leqq B'|Q+a+it|^{\beta'},$$

which have not yet the form (3.2), a and b being exchanged on the right side of the inequalities. We have, however,

$$\begin{cases} |g(a+it)| \leqq A'|Q+a+it|^{\alpha'} \cdot M_Q \\ |g(b+it)| \leqq B'|Q+b+it|^{\beta'} \cdot M_Q \end{cases} \tag{4.5}$$

with

$$M_Q = \operatorname{Max}_t \left\{ \left|\frac{Q+b+it}{Q+a+it}\right|^{\beta}, \left|\frac{Q+a+it}{Q+b+it}\right|^{\alpha} \right\}.$$

Now

$$1 \leqq \left|\frac{Q+b+it}{Q+a+it}\right| \leqq \frac{Q+b}{Q+a},$$

so that M_Q takes the form (4.3). Application of Theorem 2 on $g(s')$ with conditions (4.5) now yields in $S(a, b)$

$$|g(s')| \leqq M_Q (A'|Q+s'|^{\alpha'})^{\frac{b-\sigma'}{b-a}} (B'|Q+s'|^{\beta'})^{\frac{\sigma'-a}{b-a}}$$

or

$$(4.6)\quad \begin{cases} |f(s)| \leqq M_Q (B|Q+a+b-s|^{\beta})^{\frac{\sigma-a}{b-a}} (A|Q+a+b-s|^{\alpha})^{\frac{b-\sigma}{b-a}} \\ \quad = M_Q (A|Q+s|^{\alpha})^{\frac{b-\sigma}{b-a}} (B|Q+s|^{\beta})^{\frac{\sigma-a}{b-a}} P_Q(s) \end{cases}$$

with

$$P_Q(s) = \left| \frac{Q+a+b-s}{Q+s} \right|^{\alpha\frac{b-\sigma}{b-a}+\beta\frac{\sigma-a}{b-a}}$$

But for $a \leqq \sigma \leqq b$, $-\infty < t < \infty$ we have

$$\left(\frac{Q+a}{Q+b}\right)^2 \leqq \frac{(Q+a)^2+t^2}{(Q+b)^2+t^2} \leqq \frac{(Q+a+b-\sigma)^2+t^2}{(Q+\sigma)^2+t^2} \leqq \frac{(Q+b)^2+t^2}{(Q+a)^2+t^2} \leqq \left(\frac{Q+b}{Q+a}\right)^2$$

and thus

$$(4.7)\quad \begin{cases} P_Q(s) \leqq \underset{a \leqq \sigma \leqq b}{\operatorname{Max}} \left\{ \left(\frac{Q+a}{Q+b}\right)^{\alpha\frac{b-\sigma}{b-a}+\beta\frac{\sigma-a}{b-a}}, \left(\frac{Q+b}{Q+a}\right)^{\alpha\frac{b-\sigma}{b-a}+\beta\frac{\sigma-a}{b-a}} \right\} \\ \quad \leqq \operatorname{Max} \left\{ \left(\frac{Q+a}{Q+b}\right)^{\alpha}, \left(\frac{Q+b}{Q+a}\right)^{\beta} \right\} = M_Q. \end{cases}$$

The inequalities (4.6) and (4.7) together prove (4.2) of Theorem 2a. The statement (4.4) es evident from the definition of M_Q.

5. It is fortunate that in the application of our theorems to the theory of Dirichlet series the situation $\alpha \geqq \beta$ of Theorem 2 prevails. We need, however, first some lemmas about Γ-quotients, which are simple consequences of Theorem 2.

LEMMA 1. *For* $Q \geqq 0$, $-\frac{1}{2} \leqq \sigma \leqq \frac{1}{2}$ *we have*

$$(5.1)\quad \left| \frac{\Gamma\left(\frac{Q}{2}+\frac{1-s}{2}\right)}{\Gamma\left(\frac{Q}{2}+\frac{s}{2}\right)} \right| \leqq \left(\frac{1}{2}|Q+1+s|\right)^{\frac{1}{2}-\sigma}.$$

REMARK. Equality holds here for $\sigma = \frac{1}{2}$, any $Q \geqq 0$, and for $\sigma = -\frac{1}{2}$ with $Q = 0$.

PROOF. Let us put $a = -\frac{1}{2}$, $b = \frac{1}{2}$ and

$$f(s) = \frac{\Gamma\left(\frac{Q}{2}+\frac{1-s}{2}\right)}{\Gamma\left(\frac{Q}{2}+\frac{s}{2}\right)},$$

which is regular in the strip $S(-\frac{1}{2}, \frac{1}{2})$. We have, because of $|\Gamma(s)| = |\Gamma(\bar{s})|$,

$$(5.2)\quad \begin{aligned} |f(a+it)| &= \left| \frac{\Gamma\left(\frac{Q}{2}+\frac{1}{2}+\frac{1}{4}-\frac{it}{2}\right)}{\Gamma\left(\frac{Q}{2}-\frac{1}{4}+\frac{it}{2}\right)} \right| = \left| \frac{Q}{2}-\frac{1}{4}+\frac{it}{2} \right| \\ &= \frac{1}{2}\left| Q-\frac{1}{2}+it \right| \leqq \frac{1}{2}\left| Q+\frac{1}{2}+it \right| = \frac{1}{2}\left| Q+1+a+it \right|, \end{aligned}$$

and similarly

$$|f(b+it)| = 1.$$

With $\alpha=1$, $\beta=0$ Theorem 2 is applicable with $Q+1$ instead of Q and yields (5.1).

For $Q\geqq\frac{1}{2}$ this lemma can be sharpened to

LEMMA 2. *For* $Q-\frac{1}{2}\geqq 0$, $-\frac{1}{2}\leqq\sigma\leqq\frac{1}{2}$ *the inequality*

$$\left|\frac{\Gamma\left(\frac{Q}{2}+\frac{1-s}{2}\right)}{\Gamma\left(\frac{Q}{2}+\frac{s}{2}\right)}\right|\leqq\left(\frac{1}{2}\,|\,Q+s\,|\right)^{\frac{1}{2}-\sigma} \tag{5.4}$$

holds.

REMARK. Equality in (5.3) on both boundaries of $S(-\frac{1}{2},\frac{1}{2})$ for all $Q\geqq\frac{1}{2}$.

PROOF. The proof is almost the same as the previous one. Only instead of (5.2) we write

$$|f(a+it)|=\tfrac{1}{2}\,|\,Q-\tfrac{1}{2}-it\,|=\tfrac{1}{2}\,|\,Q+a+it\,|,$$

which again permits the application of Theorem 2 for $Q+a=Q-\frac{1}{2}>0$. For $Q-\frac{1}{2}=0$ formula (5.3) follows by a passage to the limit.

LEMMA 3. *For* $Q\geqq 0$, $-\frac{1}{2}\leqq\sigma\leqq\frac{1}{2}$ *we have*

$$\left|\frac{\Gamma(Q+1-s)}{\Gamma(Q+s)}\right|\leqq|\,Q+1+s\,|^{1-2\sigma}. \tag{5.4}$$

REMARK. Equality holds here on the boundaries under the same circumstances as in Lemma 1.

PROOF. We have again $a=-\frac{1}{2}$, $b=\frac{1}{2}$. For

$$f(s)=\frac{\Gamma(Q+1-s)}{\Gamma(Q+s)}$$

we find easily

$$|f(a+it)|=|\,Q+\tfrac{1}{2}+it\,|\cdot|\,Q-\tfrac{1}{2}+it\,|\leqq|\,Q+\tfrac{1}{2}+it\,|^2=|\,Q+1+a+it\,|^2$$

and

$$|f(b+it)|=1.$$

Theorem 2 with $\alpha=2$, $\beta=0$ and $Q+1$ instead of Q yields immediately (5.4).

6. We apply our results to some classes of L- and ζ-functions. Let $k>1$ be an integer and $\chi(n)$ a primitive character modulo k. For $\sigma>1$ one defines

$$L(s,\chi)=\sum_{n=1}^{\infty}\frac{\chi(n)}{n^s}.$$

This function with its analytic continuation is regular in the whole s-plane. For $\eta>0$ we have

$$|L(1+\eta+it,\chi)|\leqq\sum_{n=1}^{\infty}\frac{1}{n^{1+\eta}}=\zeta(1+\eta). \tag{6.1}$$

On the other hand, $L(s,\chi)$ satisfies the functional equation

$$\left(\frac{\pi}{k}\right)^{-\frac{s}{2}}\Gamma\left(\frac{a+s}{2}\right)L(s,\chi)=\varepsilon(\chi)\left(\frac{\pi}{k}\right)^{-\frac{1-s}{2}}\Gamma\left(\frac{a+1-s}{2}\right)L(1-s,\bar{\chi}),$$

where $\varepsilon(\chi)$ is a certain root of unity depending on χ and $a=0$ or 1 so that

$$\chi(-1)=(-1)^a.$$

We have therefore

$$|L(s,\chi)|=\left(\frac{\pi}{k}\right)^{\sigma-\frac{1}{2}}\left|\frac{\Gamma\left(\frac{a}{2}+\frac{1-s}{2}\right)}{\Gamma\left(\frac{a}{2}+\frac{s}{2}\right)}\right||L(1-s,\bar{\chi})|.$$

For $-\frac{1}{2}\leqq\sigma\leqq\frac{1}{2}$ we apply on the Γ-quotient Lemma 1 or 2 according to the values 0 or 1 of a. This gives

$$|L(s,\chi)|\leqq\left(\frac{2\pi}{k}\right)^{\sigma-\frac{1}{2}}|1+s|^{\frac{1}{2}-\sigma}|L(1-s,\bar{\chi})|,$$

and in particular for $0<\eta\leqq\frac{1}{2}$

(6.2) $$|L(-\eta+it,\chi)|\leqq\left(\frac{k}{2\pi}\right)^{\eta+\frac{1}{2}}|1+s|^{\eta+\frac{1}{2}}\zeta(1+\eta).$$

From (6.1) and (6.2) we derive then through Theorem 2 the result:

THEOREM 3. *For $-\frac{1}{2}\leqq-\eta\leqq\sigma\leqq1+\eta\leqq\frac{3}{2}$, for all moduli $k>1$ and all primitive characters χ modulo k the inequality*

$$|L(s,\chi)|\leqq\left(\frac{k|1+s|}{2\pi}\right)^{\frac{1+\eta-\sigma}{2}}\zeta(1+\eta)$$

holds.

It will be noticed that k and t appear here in the same order of magnitude.

7. The Dedekind zetafunction $\zeta_K(s)$ for the algebraic number field K of degree n is defined for $\sigma>1$ as

$$\zeta_K(s)=\sum_{\mathfrak{a}}\frac{1}{N(\mathfrak{a})^s}=\prod_{\mathfrak{p}}\frac{1}{1-N(\mathfrak{p})^{-s}},$$

where $\mathfrak{a}$ runs through all ideals and $\mathfrak{p}$ through all prime ideals of K. Any rational prime number p has the prime ideal decomposition

$$(p)=\mathfrak{p}_1^{e_1}\cdots\mathfrak{p}_k^{e_k},$$

$$N(\mathfrak{p}_i)=p^{f_i},$$

so that

$$e_1f_1+\cdots+e_kf_k=n.$$

Therefore we can write

$$\zeta_K(s)=\prod_p\left\{\frac{1}{1-p^{-f_1s}}\cdots\frac{1}{1-p^{-f_ks}}\right\},$$

hence

$$|\zeta_K(s)|\leqq\prod_p\left\{\frac{1}{1-p^{-f_1\sigma}}\cdots\frac{1}{1-p^{-f_k\sigma}}\right\}$$

$$\leqq\prod_p\left\{\frac{1}{1-p^{-\sigma}}\right\}^k\leqq\prod_p\left\{\frac{1}{1-p^{-\sigma}}\right\}^n=\zeta(\sigma)^n$$

or, for $\eta > 0$.

(7.1) $$|\zeta_K(1+\eta+it)| \leqq \zeta(1+\eta)^n$$

(which is, of course, a rather crude estimate, since $\zeta_K(s)$ has only a pole of first order at $s=1$). The function $\zeta_K(s)$ satisfies after HECKE the functional equation ([4], p. 74)

(7.11) $$A^s \Gamma\left(\frac{s}{2}\right)^{r_1} \Gamma(s)^{r_2} \zeta_K(s) = A^{1-s} \Gamma\left(\frac{1-s}{2}\right)^{r_1} \Gamma(1-s)^{r_2} \zeta_K(1-s),$$

where

(7.12) $$A = 2^{-r_2} \pi^{-\frac{n}{2}} \sqrt{|d|},$$

and d is the discriminant of the field K, which has among its conjugates r_1 real and $2r_2$ complex fields, $r_1 + 2r_2 = n$.

We have thus

$$|\zeta_K(s)| = A^{1-2\sigma} \left| \frac{\Gamma\left(\frac{1-s}{2}\right)}{\Gamma\left(\frac{s}{2}\right)} \right|^{r_1} \left| \frac{\Gamma(1-s)}{\Gamma(s)} \right|^{r_2} |\zeta_K(1-s)|$$

and, from Lemmas 1 and 3, for $-\frac{1}{2} \leqq \sigma \leqq \frac{1}{2}$

$$|\zeta_K(s)| \leqq A^{1-2\sigma} (\tfrac{1}{2}|1+s|)^{r_1(\frac{1}{2}-\sigma)} |1+s|^{r_2(1-2\sigma)} |\zeta_K(1-s)|$$

and in particular for $0 < \eta \leqq \frac{1}{2}$, in view of (7.1)

(7.2) $$|\zeta_K(-\eta+it)| \leqq A^{1+2\eta} 2^{-r_1(\frac{1}{2}+\eta)} |1-\eta+it|^{n(\frac{1}{2}+\eta)} \zeta(1+\eta)^n.$$

Theorem 2 can be applied only to regular functions. We consider therefore $\zeta_K(s)\,(s-1)$. From (7.1) and (7.2) we obtain

(7.31) $$|\zeta_K(1+\eta+it)\,(\eta+it)| < \zeta(1+\eta)^n |1+(1+\eta+it)|$$

and

(7.32) $$\begin{cases} |\zeta_K(-\eta+it)\,(-\eta+it-1)| \\ \qquad \leqq \zeta(1+\eta)^n A^{1+2\eta} 2^{-r_1(\frac{1}{2}+\eta)} \dfrac{1+\eta}{1-\eta} |1-\eta+it|^{n(\frac{1}{2}+\eta)+1} \end{cases}$$

since

$$|-\eta+it-1| \leqq \frac{1+\eta}{1-\eta} |1-\eta+it|.$$

From (7.31), (7.32) we infer now by means of Theorem 2 that in the strip $-\eta \leqq \sigma \leqq 1+\eta$

$$|\zeta_K(s)\,(s-1)| \leqq \zeta(1+\eta)^n |1+s| \left\{ A^{1+2\eta} 2^{-r_1(\frac{1}{2}+\eta)} \frac{1+\eta}{1-\eta} |1+s|^{n(\frac{1}{2}+\eta)} \right\}^{\frac{1+\eta-\sigma}{1+2\eta}}.$$

With $1 < \frac{1+\eta}{1-\eta} \leqq 3$ for $0 < \eta \leqq \frac{1}{2}$, and with the meaning of A in (7.12) we obtain thus

THEOREM 4. *In the strip $-\eta \leqq \sigma \leqq 1+\eta$, $0 < \eta \leqq \frac{1}{2}$, the Dedekind function $\zeta_K(s)$ belonging to the algebraic number field K of degree n and discriminant d satisfies the inequality*

(7.4) $$|\zeta_K(s)| \leqq 3 \left| \frac{1+s}{1-s} \right| \left(|d| \left(\frac{|1+s|}{2\pi} \right)^n \right)^{\frac{1+\eta-\sigma}{2}} \zeta(1+\eta)^n.$$

We know through E. ARTIN and R. BRAUER [1] that RIEMANN'S $\zeta(s)$ divides DEDEKIND'S $\zeta_K(s)$ in case K is normal over the rational field. We can use this fact to obtain a sharper version of the previous theorem.

In R. BRAUER'S notation the quotient, which is an entire function, appears as

$$\frac{\zeta_K(s)}{\zeta(s)} = \prod_{\varrho} L(s;\omega_\varrho^*)^{c_\varrho} \tag{7.5}$$

where $c_\varrho > 0$ and the functions under the product sign are L-functions for certain congruence characters ω_ϱ^*. Let us put

$$\sum c_\varrho = q = q(K). \tag{7.51}$$

Included in (7.5) is, in view of the class field theory, the case of Abelian K, for which $c_\varrho = 1$, $q \leqq n-1$. Moreover, as Prof. RICHARD BRAUER pointed out to me in a letter of August 10, 1957, the inequality

$$q \leqq n-1 \tag{7.52}$$

holds also for all normal fields. From (7.5) and (6.1) follows then, for $\eta > 0$

$$\left|\frac{\zeta_K(1+\eta+it)}{\zeta(1+\eta+it)}\right| \leqq \zeta(1+\eta)^{n-1}. \tag{7.53}$$

On the other hand, the functional equation (7.11) of $\zeta_K(s)$ together with that of $\zeta(s)$ furnishes

$$\frac{\zeta_K(s)}{\zeta(s)} = A^{1-2s}\,\pi^{\frac{1}{2}-s}\left(\frac{\Gamma\left(\frac{1-s}{2}\right)}{\Gamma\left(\frac{s}{2}\right)}\right)^{r_1-1}\left(\frac{\Gamma(1-s)}{\Gamma(s)}\right)^{r_2}\frac{\zeta_K(1-s)}{\zeta(1-s)}. \tag{7.6}$$

If $r_1 \geqq 1$ we obtain from Lemmas 1 and 3, for $-\frac{1}{2} \leqq \sigma \leqq \frac{1}{2}$

$$\left|\frac{\zeta_K(s)}{\zeta(s)}\right| \leqq A^{1-2\sigma}\,\pi^{\frac{1}{2}-\sigma}\left(\frac{1}{2}\,|1+s|\right)^{(r_1-1)(\frac{1}{2}-\sigma)}|1+s|^{r_2(1-\sigma)}\left|\frac{\zeta_K(1-s)}{\zeta(1-s)}\right|$$

and in particular with $0 < \eta \leqq \frac{1}{2}$

$$\left|\frac{\zeta_K(-\eta+it)}{\zeta(-\eta+it)}\right| \leqq A^{1+2\eta}\,\pi^{\frac{1}{2}+\eta}\,2^{-(r_1-1)(\frac{1}{2}+\eta)}\,|1+s|^{(n-1)(\frac{1}{2}+\eta)}\,\zeta(1+\eta)^{n-1}. \tag{7.7}$$

From this and (7.53) we obtain just as before by means of Theorem 2 the estimation

$$\left|\frac{\zeta_K(s)}{\zeta(s)}\right| \leqq \left(|d|\left(\frac{|1+s|}{2\pi}\right)^{n-1}\right)^{\frac{1+\eta-\sigma}{2}}\zeta(1+\eta)^{n-1}, \quad -\eta \leqq \Re(s) \leqq 1+\eta. \tag{7.8}$$

If $r_1 = 0$ we write instead of (7.6)

$$\frac{\zeta_K(s)}{\zeta(s)} = A^{1-2s}\,\pi^{\frac{1}{2}-s}\left(\frac{\Gamma(1-s)}{\Gamma(s)}\right)^{r_2-1}\frac{\Gamma(1-s)}{\Gamma(s)}\,\frac{\Gamma\left(\frac{s}{2}\right)}{\Gamma\left(\frac{1-s}{2}\right)}\,\frac{\zeta_K(1-s)}{\zeta(1-s)}$$

$$= (2A)^{1-2s}\,\pi^{\frac{1}{2}-s}\left(\frac{\Gamma(1-s)}{\Gamma(s)}\right)^{r_2-1}\frac{\Gamma\left(\frac{1}{2}+\frac{1-s}{2}\right)}{\Gamma\left(\frac{1}{2}+\frac{s}{2}\right)}\,\frac{\zeta_K(1-s)}{\zeta(1-s)}.$$

The application of Lemmas 1 and 2 yields also here exactly (7.7), and (7.8) can then be established as before.

We have thus:

THEOREM 4a. *If K is a normal field over the rational field then in the strip $-\eta \leqq \sigma \leqq 1+\eta$, $0<\eta\leqq\frac{1}{2}$ the Dedekind function $\zeta_K(s)$ satisfies*

$$|\zeta_K(s)| \leqq |\zeta(s)| \left(|d| \left(\frac{|1+s|}{2\pi}\right)^{n-1}\right)^{\frac{1+\eta-\sigma}{2}} \zeta(1+\eta)^{n-1}.$$

For Abelian fields the Theorem 4a is a direct consequence of Theorem 3 because of $c_\varrho=1$ and because the discriminant d is the product of the conductors k_ϱ of the characters ω_ϱ^*.

8. The full strength of the uniformity with respect to the parameter Q which Theorem 2 and Lemma 3 show is established only in their application to HECKE's $\zeta(s, \lambda)$-functions.

Let K be any algebraic number field and $\mathfrak{f}$ an ideal in K. The Hecke functions are then defined as

$$\zeta(s, \lambda) = \sum_{\mathfrak{m}} \frac{\lambda(\mathfrak{m})}{N(\mathfrak{m})^s} = \sum_{(\hat{\mu})} \frac{\lambda(\hat{\mu})}{|N(\hat{\mu})|^s}, \qquad \Re(s) > 1, \tag{8.1}$$

where $\mathfrak{m}$ runs over all ideals of K, $\hat{\mu}$ over all corresponding non-associate ideal numbers, and $\lambda(\mathfrak{m}) = \lambda(\hat{\mu})$ is a "Größencharacter for ideals modulo $\mathfrak{f}$" defined as

$$\lambda(\hat{\mu}) = \prod_{q=1}^{r_1+r_2} |\hat{\mu}^{(q)}|^{-i v_q} \prod_{p=r_1+1}^{n} \left(\frac{\hat{\mu}^{(p)}}{|\hat{\mu}|^{(q)}}\right)^{a_p} \tag{8.2}$$

where the a_p ($p=r_1+1, \ldots, n$) are non-negative integers with the condition $a_p \cdot a_{p+r_2}=0$, and the v_q ($q=1, \ldots, r_1+r_2$) certain real numbers, whose definition does not concern us here ([*3*], p. 30 and [*5*], p. 343).

From (8.1) and (8.2) follows immediately

$$|\zeta(s, \lambda)| \leqq \zeta_K(\sigma), \qquad \sigma > 1. \tag{8.3}$$

We put with HECKE

$$\xi(s, \lambda) = \gamma(\lambda)\, \Gamma(s, \lambda)\, A^s \zeta(s, \lambda),$$

where

$$|\gamma(\lambda)| = 1,$$

$$A^2 = \frac{|d|\, N(\mathfrak{f})}{\pi^n\, 2^{2 r_2}},$$

$$\Gamma(s, \lambda) = \prod_{q=1}^{r_1} \Gamma\left(\frac{a_q + s + i v_q}{2}\right) \cdot \prod_{p=r_1+1}^{r_1+r_2} \Gamma\left(\frac{a_p + a_{p+r_2}}{2} + s + \frac{i v_p}{2}\right).$$

Here the a_q for $q=1, \ldots r_1$, which are not explicitly mentioned in (8.2), are 0 or 1 and are determined by the character χ belonging to the congruence group modulo $\mathfrak{f}$ in the narrowest sense and induced by the Größencharacter λ ([3], p. 21). The $\xi(s, \lambda)$-functions with primitive λ-character modulo $\mathfrak{f}$ fulfill a functional equation

$$\xi(s, \lambda) = W(\lambda)\, \xi(1-s, \bar{\lambda}),$$

where $W(\lambda)$ is a certain number of absolute value 1, so that

$$|\zeta(s, \lambda)| = A^{1-2\sigma} \prod_{q=1}^{r_1} \left| \frac{\Gamma\left(\frac{a_q + 1 - s - i v_q}{2}\right)}{\Gamma\left(\frac{a_q + s + i v_q}{2}\right)} \right| \times$$

$$\times \prod_{p=r_1+1}^{r_1+r_2} \left| \frac{\left(\Gamma \frac{a_p + a_{p+r_2}}{2} + 1 - s - \frac{i v_p}{2}\right)}{\Gamma\left(\frac{a_p + a_{p+r_2}}{2} + s + \frac{i v_p}{2}\right)} \right| \cdot |\zeta(1-s, \bar{\lambda})|.$$

If we apply on the first product, where $a_q = 0$ or 1, the Lemmas 1 and 2, and on the second product, where $(a_p + a_{p+r_2})$ are unbounded non-negative integers, the Lemma 3, we obtain for $-\frac{1}{2} \leqq \sigma \leqq \frac{1}{2}$

$$|\zeta(s, \lambda)| \leqq A^{1-2\sigma} \prod_{q=1}^{r_1} \left(\frac{1}{2} |1 + s + i v_q|\right)^{\frac{1}{2}-\sigma} \times$$

$$\times \prod_{p=r_1+1}^{r_1+r_2} \left| \frac{a_p + a_{p+r_2}}{2} + 1 + s + \frac{i v_p}{2} \right|^{1-2\sigma} \cdot |\zeta(1-s, \lambda)|$$

and in particular for $s = -\eta + it$, $0 \leqq \eta \leqq \frac{1}{2}$ in virtue of (8.3),

$$(8.4) \quad \begin{cases} |\zeta(-\eta + it, \lambda)| \leqq A^{1+2\eta} \prod_{q=1}^{r_1} \left(\frac{1}{2} |1 - \eta + i(t + v_q)|\right)^{\frac{1}{2}+\eta} \times \\ \quad \times \prod_{p=r_1+1}^{r_1+r_2} \left| \frac{a_p + a_{p+r_2}}{2} + 1 - \eta + i\left(t + \frac{v_p}{2}\right) \right|^{1+2\eta} \cdot \zeta_K(1+\eta). \end{cases}$$

We have not prepared a theorem which we could apply here directly as we could use Theorem 2 in the previous cases. We construct therefore a function $F(s)$, on which the Phragmén-Lindelöf argument can be applied, out of factors which are furnished by Theorem 1. This theorem yields functions according to the given boundary conditions, and we select, with $a = -\eta$, $b = 1 + \eta$ the two functions

$$\varphi_1(s; 1) \quad \text{with} \quad \gamma = -\tfrac{1}{2} - \eta, \quad \delta = 0$$

and

$$\varphi_2(s; Q) \quad \text{with} \quad \gamma = -1 - 2\eta, \quad \delta = 0$$

which show, in view of (2.3), the properties

$$(8.5) \quad \begin{cases} |\varphi_1(s; 1) \geqq |1 + s|^{-\frac{1}{2}(1+\eta-\sigma)} \\ |\varphi_2(s; Q)| \geqq |Q + s|^{-(1+\eta-\sigma)} \end{cases}$$

for $-\eta \leqq \sigma \leqq 1 + \eta$, $Q - \eta > 0$.

Then

$$F(s)=\zeta(s,\lambda)\cdot\prod_{q=1}^{r_1}\varphi_1(s+i v_q;1)\cdot\prod_{p=r_1+1}^{r_1+r_2}\varphi_2\left(s+\frac{i v_p}{2};1+\frac{a_p+a_{p+r_2}}{2}\right)\cdot E e^{\nu s},$$

where E and ν are so determined that

$$E e^{-\nu\eta}=\left(A\cdot 2^{-\frac{r_1}{2}}\right)^{-1-2\eta},$$
$$E e^{\nu(1+\eta)}=1,$$

has because of (8.3) and (8.4) the properties

$$|F(-\eta+it)|\leqq\zeta_K(1+\eta)$$
$$|F(1+\eta+it)|\leqq\zeta_K(1+\eta)$$

and therefore satisfies after PHRAGMÉN and LINDELÖF also

$$|F(s)|\leqq\zeta_K(1+\eta)$$

for $-\eta\leqq\Re(s)\leqq 1+\eta$.

From this we infer, valid in the same strip,

$$|\zeta(s,\lambda)|\leqq\zeta_K(1+\eta)\prod_{q=1}^{r_1}|\varphi_1(s+i v_q;1)|^{-1}\times$$
$$\times\prod_{p=r_1+1}^{r_1+r_2}\left|\varphi_2\left(s+\frac{i v_p}{2};1+\frac{a_p+a_{p+r_2}}{2}\right)\right|^{-1}\cdot E^{-1}e^{-\nu\sigma}.$$

If we now use (8.5) and take into account also (7.1) we obtain the

THEOREM 5. *Let for the field K of degree n, which is among r_1 real and $2r_2$ imaginary conjugate fields of discriminant d the Hecke function $\zeta(s,\lambda)$ be given by (8.1) with primitive Größencharacter $\lambda(\hat{\mu})$ modulo $\mathfrak{f}$ of (8.2). Then $\zeta(s,\lambda)$ satisfies in the strip $-\eta\leqq\sigma\leqq 1+\eta$, $0<\eta\leqq\frac{1}{2}$ the inequality*

$$|\zeta(s,\lambda)|\leqq\zeta(1+\eta)^n\left\{\frac{|d|\,N(\mathfrak{f})}{(2\pi)^n}\prod_{q=1}^{r_1}|1+s+i v_q|\times\right.$$
$$\left.\times\prod_{p=r_1+1}^{r_1+r_2}\left|1+\frac{a_p+a_{p+r_2}}{2}+s+\frac{i v_p}{2}\right|^2\right\}^{\frac{1+\eta-\sigma}{2}}$$

[1] BRAUER, R.: On the zeta-functions of algebraic numberfields. Amer. J. Math. **69**, 243–250 (1947). — [2] HARDY, G. H., A. E. INGHAM and G. POLYA: Theorems concerning mean values of analytic functions. Proc. Royal Soc. London (A) **113**, 542–569 (1927). — [3] HECKE, E.: Eine neue Art von Zetafunktionen und ihre Beziehungen zur Verteilung der Primzahlen. Math. Z. **6**, 11–51 (1920). — [4] LANDAU, E.: Einführung in die elementare und analytische Theorie der algebraischen Zahlen und der Ideale. Leipzig 1918. — [5] RADEMACHER, H.: Zur additiven Primzahltheorie algebraischer Zahlkörper. III. Math. Z. **27**, 321–426 (1928).

Dept. of Math., University of Pennsylvania, Philadelphia 4, Pa. (U.S.A.)

(Eingegangen am 7. November 1958)

Notes to Paper 62b

This paper has been reviewed in the Z, vol. 92 (1962), pp. 277-278, and in the MR, vol. 22 (1961), #7982.

The present paper is a refined version of the preceding paper 62a.

Page 195, equation (3.7).
Read $B = Ee^{\nu b}$.

Page 197, line 6 of text.
Read *is* for *es*.

Page 198, line 1.
One also has to set $A = 1/2$, $B = 1$ in Theorem 2.

Page 198, first display line.
Read (5.3) instead of (5.4).

Page 198, line 6.
Insert the word "holds" after Equality.

Page 198, sixth line of text from bottom.
Here one also needs $A = B = 1$.

Page 201, equation (7.6).
Insert a missing parenthesis in the numerator of the next to last factor.

Page 204, Reference [5].
See paper 20 of the present collection.

ON THE HURWITZ ZETAFUNCTION

by

Hans Rademacher

The Zetafunction of Hurwitz defined by

$$\zeta(s,a) = \sum_{n=0}^{\infty} \frac{1}{(n+a)^s} \qquad 0 < a \leqq 1, \quad R(s) > 1$$

can be continued over the whole s-plane as a meromorphic function with the only pole at $s = 1$ of residue 1. For rational $a = \frac{h}{k}$ it satisfies the functional relation

$$(1) \quad \zeta\left(s,\frac{h}{k}\right) = \frac{2\,\Gamma(1-s)}{(2\pi k)^{1-s}} \sum_{j=1}^{k} \sin\left(\frac{\pi s}{2} + 2\pi\frac{jh}{k}\right) \zeta\left(1-s, \frac{j}{k}\right)$$

$$h = 1, 2, \ldots k.$$

We put

$$(2) \qquad \varphi\left(s, \frac{h}{k}\right) = (k\pi)^{-s/2}\, \Gamma\left(\frac{s}{2}\right)\, \zeta\left(s, \frac{h}{k}\right)$$

and introduce the column vector

$$(3) \qquad v(s) = \begin{pmatrix} \varphi(s, \frac{1}{k}) \\ \varphi(s, \frac{2}{k}) \\ \vdots \\ \varphi(s, \frac{k}{k}) \end{pmatrix}$$

By means of the symmetric matrix

$$(4.1) \quad A(s) = \left(\left(A_{hj}\right)\right)$$

where

$$(4.2)\quad A_{hj} = \frac{\sin\left(\frac{\pi s}{2} + 2\pi\frac{jh}{k}\right)}{k^{1/2}\ \sin\frac{\pi s}{2}}$$

The k equations can be condensed into

$$(5)\quad v(s) = A(s)\cdot v(1-s)\ .$$

Substituting $1 - s$ for s we obtain

$$v(1-s) = A(1-s)\, v(s)$$

and thus

$$v(s) = A(s)\, A(1-s)\, v(s).$$

This equation suggests

$$(6)\quad A(s)\, A(1-s) = 1_n\ ,$$

the unit matrix. This is indeed the case as can be seen by direct computation from (4a) and (4b).

In more detail, we have

$$(7)\quad A(s) = \frac{1}{\sqrt{k}}\, B + \frac{1}{\sqrt{k}}\ \operatorname{cotg}\frac{\pi s}{2}\ C\ ,$$

where B and C are constant matrices:

$$B = \left(\left(B_{hj}\right)\right) = \left(\left(\cos 2\pi\frac{jh}{k}\right)\right)$$

and

$$C = \left(\left(C_{hj}\right)\right) = \left(\left(\sin\frac{2\pi jh}{k}\right)\right).$$

$$(8.1)\quad A(s)\, A(1-s) = \frac{1}{k}\ (B^2 + C^2) + \operatorname{cotg}\frac{\pi s}{2}\ CB + \operatorname{tang}\frac{\pi s}{2}\ BC$$

with

$$(8.3)\quad \begin{aligned} B^2 &= \frac{k}{2}\,(I + H) \\ C^2 &= \frac{k}{2}\,(I - H), \end{aligned}$$

(8.4) $BC = CB = 0$

where 0 is the zero matrix and

$$(9)\qquad H = \begin{pmatrix} 0 & 0 & \cdots & & 0 & 1 & 0 \\ & & & & & 1 & \\ & & & & 1 & & \\ & & & \iddots & & & \\ & & \iddots & & & & \\ 0 & 1 & \cdots & & & 0 & 0 \\ 1 & 0 & \cdots & & & 0 & 0 \\ 0 & 0 & \cdots & & & 0 & 1 \end{pmatrix}$$

Equations (8.1) to (8.4) imply (6).

So far we have only transcribed the system of equations (1). Let us now consider a special function. We choose

(10) $f(s) = v(s)^2 = v'(s)\, v(s)$

$$= (k\pi)^{-s}\, \Gamma\left(\frac{s}{2}\right)^2 \sum_{h=1}^{k} \zeta\left(s, \frac{h}{k}\right)^2 .$$

(The dash ' means here and in the sequel the transpose of a matrix).

From (5) we deduce

(11) $f(s) = v'(1 - s)\, A'(s)\, A(s)\, v(1 - s).$

By means of (7) and (8.1) to (8.4) we obtain

$A'(s)\, A(s) = A^2(s)$

$$\frac{1}{2\sin^2\frac{\pi s}{2}}\, I \;-\; \frac{\cos \pi s}{2\sin^2\frac{\pi s}{2}}\, H$$

so that (11) goes over into

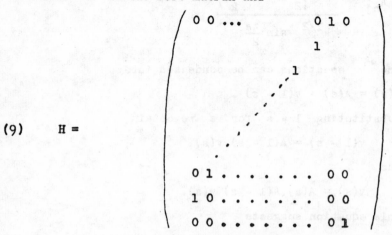

$$(12.1)\quad 2\sin^2 \frac{\pi s}{2}\ f(s) = f(1 - s) - \cos \pi s\ g(1 - s)$$

where

$$g(s) = v'(s)\ H\ v(s)$$

$$(12.2)\quad = \sum_{h=1}^{k-1} \zeta(s, \frac{k-h}{k})\ \zeta(s, \frac{h}{k}) + \zeta(s, \frac{k}{k})^2$$

by means of (9). Substitution of $1 - s$ for s in (12.1) yields

$$(12.3)\quad 2\cos^2 \frac{\pi s}{2}\ f(1 - s) = f(s) + \cos \pi s\ g(s).$$

From (12.1) and (12.3) we obtain by addition

$$(13)\quad f(s) + g(s) = f(1 - s) + g(1 - s).$$

Similarly subtraction and use of (13) leads to

$$(14)\quad \sin^2 \frac{\pi s}{2} \Big(f(s) - g(s)\Big) = \cos^2 \frac{\pi s}{2} \Big(f(1 - s) - g(1 - s)\Big).$$

Let us define

$$Z_1(s) = \sum_{h=1}^{k} \left\{ \sum_{\substack{n \equiv h(k) \\ n > 0}} \frac{1}{n^s} + \sum_{\substack{n \equiv -h\ (k) \\ n > 0}} \frac{1}{n^s} \right\}^2$$

and

$$Z_2(s) = \sum_{h=1}^{k-1} \left\{ \sum_{\substack{n \equiv h(k) \\ n > 0}} \frac{1}{n^s} - \sum_{\substack{n \equiv -h\ (k) \\ n > 0}} \frac{1}{n^s} \right\}^2$$

Then

$$\sum_{h=1}^{k} \zeta(s, \frac{h}{k})^2 + \sum_{h=1}^{k-1} \zeta(s, \frac{h}{k})\ \zeta(s, \frac{k-h}{k}) + \zeta(s, 1)^2$$

$$= 1/2\ k^{2s}\ Z_1(s)$$

and

$$\sum_{h=1}^{k-1} \zeta(s, \frac{h}{k})^2 - \sum_{h=1}^{k-1} \zeta(s, \frac{h}{k})\ \zeta(s, \frac{k-h}{k}) = 1/2k^{2s} Z_2(s).$$

The function $\mathcal{Z}_1(s)$ is meromorphic is the s-plane with a pole of second order at $s = 1$, whereas $\mathcal{Z}_2(s)$ is regular in the whole s-plane. With these definitions the equations (13) and (14) take the form

(15) $$\left(\frac{\pi}{k}\right)^{-s} \Gamma\left(\frac{s}{2}\right)^2 \mathcal{Z}_1(s) = \left(\frac{\pi}{k}\right)^{s-1} \Gamma\left(\frac{1-s}{2}\right)^2 \mathcal{Z}_1(1-s)$$

and

(16) $$\left(\frac{\pi}{k}\right)^{-s} \Gamma\left(\frac{1+s}{2}\right)^2 \mathcal{Z}_2(s) = \left(\frac{\pi}{k}\right)^{s-1} \Gamma\left(\frac{2-s}{2}\right)^2 \mathcal{Z}_2(1-s) .$$

Notes to Paper 63

This report does not seem to have been reviewed.

Page 74, equation (6).
Read I_n instead of 1_n.

Page 74, line 7.
(4a) and (4b) should read (4.1) and (4.2).

Page 74, equation (8.1).
Insert $\{$ after $1/k$, and $\}$ at the end of the line.

Page 74, second display line from bottom.
Number this equation with (8.2) and lower (8.3) to the last line.

Page 74, second display line from bottom.
Add an = sign after $A^2(s)$.

Page 76, second display line from bottom.
Read (1/2) instead of 1/2.

Page 76, last display line.
Read (1/2) instead of 1/2.

REMARKS CONCERNING THE RIEMANN-VON MANGOLDT FORMULA

by Hans Rademacher

This year sees the 100th anniversary of Riemann's hypothesis: Riemann's memoir "Über die Anzahl der Primzahlen unter einer gegebenen Grenze was submitted to the Prussian Academy in Berlin in November, 1859. Riemann's conjecture has withstood so far all efforts of generations of mathematicians, whereas the proofs of his theorems, partly only sketched in that memoir, have in the meantime been put on a solid basis. The important consequences of the Riemann hypothesis for our knowledge of the distribution of the prime numbers is well known.

My purpose today is to emphasize that even if the Riemann hypothesis should one day be proved, much would remain to be known about the zeroes of $\zeta(s)$ namely about the nature of the ordinates γ of the zeroes ρ.

This statement becomes evident if we look at the Riemann-von Mangoldt formula

$$(1) \qquad \psi^{*}(x) = 1/2\left(\psi(x+0) - \psi(x-0)\right)$$

$$\psi(x) = \sum_{n \leq x} \Lambda(n) = x - \sum_{\rho} \frac{x^{\rho}}{\rho} - 1/2 \log\left(1 - \frac{1}{x^2}\right)$$

$$- \log 2\pi$$

where

$$\Lambda(n) = \begin{cases} \log p \text{ if } & n = p^k \\ 0 \text{ otherwise} \end{cases}$$

and where $\sum_{\rho}$ which is not absolutely convergent, is arranged according to increasing ordinates $|\gamma_{\nu}|$.

The series has jumps at the powers of primes and therefore converges non-uniformly near these points. But Landau proved that its convergence is uniform in any closed interval not including a power of a prime.

In his now classical book on the distribution of primes Landau makes the following statement, which I quote in German, because of its felicitous phrasing:

"Die Tatsache, dass $\sum \frac{x^\rho}{\rho}$ gerade in der Nähe der Primzahlpotenzen und sonst in der Nähe keiner Stelle > 0 ungleichmässig konvergiert, deutet auf einen arithmetischen Zusammenhang zwischen den komplexen Wurzeln der Zetafunktion und den Primzahlen hin. Ich habe keine Ahnung, worin derselbe besteht." *

This statement of Landau gives rise to another anniversary: it was written in 1909 and is now just 50 years old. It will be more or less the theme of my remarks today.

Let us now consider the jumps of

$$f(x) = \sum \frac{x^\rho}{\rho}$$

Since the formula (1) shows that $f(x) + \psi^*(x)$ are continuous we have

$$\sigma(f(x)) = -\sigma(\psi^*(x)) ,$$

where we define

$$\sigma(f(x)) = f(x+0) - f(x-0)$$

Now

$$\sigma(\psi^*(p^k)) = \Lambda(p^k) = \log p$$

so that

$$\sigma(f(p^k)) = -\log p .$$

We put, for

$$x > 1$$
$$y = \log x,$$

(2)
$$f(x) = f(e^y) = \sum_\rho \frac{e^{\rho y}}{\rho} = F(y)$$

and have then

* Landau must himself have been aware of the felicity of his writing, since he repeated these words literally, only changing "Prime Number" into "prime ideal" in his later work "Elementare und Analytische Theorie der Algebraischen Zahlen "

(3) $\sigma F(k \log p) = -\log p, \quad k = 1, 2, 3 \ldots$

From now on we assume the Riemann hypothesis $\rho = 1/2 + i\gamma$

Then

(4) $F(y) = e^{\frac{y}{2}} \sum_{\gamma} \frac{e^{i\gamma y}}{1/2 + i\gamma}$

Simple computations show that the series

(5) $1/2 \sum_{\gamma} \frac{e^{i\gamma y}}{1/2 + i\gamma}$ and $\sum_{\gamma > 0} \frac{\sin \gamma y}{\gamma} = H(y)$

differ only by continuous functions, resulting from uniformly convergent series. (For these computations it is useful to observe that the ν th ordinate γ_{ν} can be estimated through

$$\gamma_{\nu} \sim \frac{2\pi\nu}{\log \nu},$$

The statements (4) and (5) show now that

$$\sigma\left(H(y)\right) = \sigma\left(1/2\, e^{-\frac{y}{2}} F(y)\right),$$

and in view of (3)

(6) $\sigma\left(H(k \log p)\right) = -1/2 \log p \cdot p^{-\frac{k}{2}}$

For fixed p the series H(y) has jumps repeated at distances log p. But the jumps are decreasing from step to step, whereas in the Fourier series $\sum_{n=1}^{\infty} \frac{\sin nx}{n}$

we have, because of periodicity, all jumps of equal size.

It seems to me to be worthwhile to construct a trigonometric series which will imitate the discontinuities of H(y) for $y > 0$. We realize immediately that the jumps of H(y) cannot characterize the single γ. Indeed, changing finitely many γ cannot have any influence on the jumps. And if we replace infinitely many γ by γ^* we have

$$\frac{\sin \gamma y}{\gamma} - \frac{\sin \gamma^* y}{\gamma^*} = (\gamma - \gamma^*)\left\{y \frac{\cos \tilde{\gamma} y}{\tilde{\gamma}} + \frac{\sin \tilde{\gamma} y}{\tilde{\gamma}^2}\right\}$$

where $\tilde{\gamma}$ is a mean value between γ and γ^*.

Choosing the γ^* so that

$$\sum \frac{|\gamma - \gamma^*|}{\tilde{\gamma}}$$

is convergent, the series

$$y \sum_{\gamma > 0} \frac{\gamma - \gamma^*}{\tilde{\gamma}} \cos \tilde{\gamma} y + \sum_{\gamma > 0} \frac{\gamma - \gamma^*}{\tilde{\gamma}} \sin \tilde{\gamma} y$$

will not furnish any discontinuities, and therefore

$$\sum \frac{\sin \gamma y}{\gamma} \quad \text{and} \quad \frac{\sin \gamma^* y}{\gamma^*}$$

will agree as far as discontinuities are concerned. It might, however, not be impossible in this way to deduce some asymptotic properties of the γ.

We construct first a trigonometric series which has prescribed jumps of πt_m at the points $x = 2\pi m$. We start with

$$\varphi(x) = \sum_{n=1}^{\infty} \frac{\sin nx}{n} = \frac{\pi - x}{2} \quad \text{for} \quad 0 < x < 2\pi$$

which has the jumps

$$\sigma(\alpha(2\pi m)) = \varphi(2\pi m + 0) - \varphi(2\pi m - 0) = \pi .$$

Then we form (assuming convergence for the time being)

(7) $$\Phi(x) = \sum_{k=1}^{\infty} a_k \varphi\left(\frac{x}{k}\right)$$

with certain coefficients a_k. Since $\varphi(\frac{x}{k})$ has jumps π at $2\pi km$ we find the jump of Φ at $2\pi m$ as

(8.1) $$\sigma(\Phi(2\pi m)) = \pi \sum_{k|m} a_k = \pi . t_m$$

If conversely the t_m are given, the usual application of the Moebius function gives

(8.2) $$a_m = \sum_{d|m} t_d \mu\left(\frac{m}{d}\right)$$

Equation (7) would be, explicitly

(7a) $$\Phi(x) = \sum_{k=1}^{\infty} a_k \sum_{n=1}^{\infty} \frac{\sin \frac{nx}{k}}{n}$$

We seek, eventually, to reorder such a series with respect to the frequencies $\frac{n}{k}$. But if we leave n, $k \geq 1$ unrestricted we obtain all positive rational numbers as frequencies. In order to avoid this accumulation of frequencies we change (7a) slightly and write

$$(9)\quad \Phi^*(x) = \sum_{k=1}^{\infty} a_k \sum_{n=\beta_k \cdot k} \frac{\sin \frac{nx}{k}}{n} = \sum_{k=1}^{\infty} a_k \varphi_k^* \left(\frac{x}{k}\right)$$

with suitably chosen β_k of the property

$$0 < \beta_1 < \beta_2 < \dots .$$

This will not change the jumps since each φ_k^* differs from $\varphi(\frac{x}{k})$ only by finitely many terms. But this device will prevent the accumulation of the frequencies. Indeed rational numbers $\frac{n}{j} < \beta_k$ can be admitted only if $j < k$, and

$$0 < \frac{n}{1}, \frac{n}{2} \dots \frac{n}{k-1} < \beta_k$$

admits only finitely many $\frac{n}{j}$ (some of which may already have been excluded through

$$\beta_1, \beta_2, \dots \beta_{k-1}).$$

Our plan is to rearrange $\Phi^*(x)$ according to increasing $r = \frac{n}{k}$. This would give

$$(10)\quad \Psi(x) = \sum_r \sin rx \sum_{\substack{n = kr \\ n \geq \beta_k k}} \frac{a_k}{n} = \sum_r \frac{\sin rx}{r} b_r$$

(10.1) where

$$b_r = \sum_{\substack{\beta_k \leq r \\ kr \text{ integer}}} \frac{a_k}{k} .$$

In the interval $0 < r < s$ there are only a finite number of $b_r \neq 0$ since there are only finitely many $\beta_k < s$, and to the finitely many k thus admitted belong only finitely many rational $r < s$ so that rk is an integer.

The β_k have now to be chosen in such a way that

(1) $\Phi^*(x)$ converges

(2) $\Phi^*(x) = \Psi(x)$

We need here only to consider $x \not\equiv 0 \pmod{2\pi}$

If we take

$$t_m = c^{-m} \quad \text{with } c > 1$$

then (8.2) shows that

(11) $$|a_n| < \frac{1}{c-1} = O(1)$$

For (9) we find, with the usual partial summation

$$\left| \varphi^*_k\left(\frac{x}{k}\right) \right| = \sum_{n=\beta_k \cdot k}^{\infty} \frac{\sin \frac{nx}{k}}{n} \leqslant \frac{2\pi}{x\beta_k}$$

for $k \geqslant \frac{x}{\pi}$, which is no loss of generality.

Hence, if we make

$$\sum_{k=1}^{\infty} \frac{1}{\beta k} < \infty$$

and have (11) then $\sum_{k=1}^{\infty} \left| a_k \varphi_k^* \left(\frac{x}{k}\right) \right|$

will converge, and uniformly in every bounded interval of x :

(12) $$\left| \Phi^*(x) - \sum_{k=1}^{K} a_k \sum_{n \geqslant \beta_k \cdot k} \frac{\sin \frac{nx}{k}}{n} \right| < C(x) \sum_{k=K+1}^{\infty} \frac{1}{\beta k}$$

The finitely many series $\varphi^*_k\left(\frac{x}{k}\right)$ can now be added and arranged according to increasing

$$r = \frac{n}{k} \quad \text{in (12)}$$

$$\sum_{k=1}^{K} a_k \sum_{n \geqslant \beta_k \cdot k} \frac{\sin \frac{nx}{k}}{n} = \sum_{0 < r} \frac{\sin rx}{r} \sum_{\substack{\beta_k \leq r,\ kr \text{ integer} \\ k \leq K}} \frac{a_k}{k}$$

Convergence in K still being uniform, we can let K go to infinity and have

(13) $$\Phi^*(x) = \Psi(x) = \sum_r \frac{\sin rx}{r} b_r .$$

This trigonometric series has the jumps

(14) $\quad \pi t_m = \pi c^{-m}$

at $x = 2\pi m$.

This we apply to our case with $c = p^{-\frac{k}{2}}$

and jumps equal to

$$-\frac{1}{2} \log p^{\frac{m}{2}}$$

at $y = m \log p$.

The series

(15) $$\psi_p(y) = -\sum_r \frac{\sin \frac{2\pi r}{\log p} y}{\frac{2\pi r}{\log p}} b_r(p)$$

with $b_r(p)$ obtained from (8.2) and (10.1) through $t_m = p^{\frac{m}{2}}$.

If we finally take the whole sequence of primes $p = 2, 3, \ldots$

we can put together the infinite sum $\sum_p \psi_p(y)$,

where we can enforce convergence by cutting off sufficiently many terms at the beginning of each ψ_p. This can be achieved by choosing each sequence $\beta_k = \beta_k(p)$ suitably.

The series (16) then will copy the convergence behavior of

$$H(y) = \sum_{\gamma > 0} \frac{\sin \gamma y}{\gamma}$$

even to the extent that it will assume the mean value at the points of discontinuity.

The series (16), however, possesses the undesirable coefficients $b_r(p)$ in comparison with $H(y)$. It would be desirable to construct a trigonometric series of the form $\sum_\lambda \frac{\sin \lambda y}{y}, 0 < \lambda_1 < \lambda_2 < \ldots$ which would imitate the convergence properties of $H(y)$ (always under the assumption of the Riemann hypothesis) .

Professors L. A. Rubel and E. G. Straus inform me that they have succeeded in doing this.

Notes to Paper 64

This report does not seem to have been reviewed. See also paper 60 of the present collection.

Page 33, line 6 of text.
Read ν instead of y.

Page 33, last line.
The last fraction has the denominator $\tilde{\gamma}^2$. Replace + by –.

Page 34, second display line.
Replace + by –; also $\tilde{\gamma}$ by $\tilde{\gamma}^2$ in the second denominator.

Page 37, seventh line from bottom.
(16) refers to the (not numbered) sum $\Sigma_p \psi_p (y)$ on page 37, line 8 of text.

Page 37, fifth line from bottom.
The denominator is λ.

A FUNDAMENTAL THEOREM IN THE THEORY OF MODULAR FUNCTIONS

by Hans Rademacher

A new and simple proof is given for the theorem: A modular function $\psi(\tau)$ belonging to a modular subgroup modulo N and which is regular and bounded in the half-plane $\operatorname{Re}\tau > 0$ is a constant.

The proof makes use of the following lemma : If $g(z)$ is regular and bounded in $|z|<1$, and if z_v, $v = 1, 2, 3, \ldots$, $z_v \neq 0$, are zeroes of $g(z)$ in the unit circle so that

$$\prod_{v=1}^{\infty} |z_v| = 0,$$

then $g(z)$ vanishes identically.

This lemma is a corollary of an inequality of Jensen.

For the proof of the theorem it suffices to consider the principal subgroup $\Gamma(N)$ modulo N

$$\begin{pmatrix} a & b \\ c & d \end{pmatrix} \equiv \begin{pmatrix} 1 & 0 \\ 0 & 1 \end{pmatrix} \pmod{N},$$

which is contained as subgroup in any congruence subgroup modulo N.

Since $\Gamma(N)$ contains $\begin{pmatrix} 1 & N \\ 0 & 1 \end{pmatrix}$ as element we have

$$\psi(\tau + N) = \psi(\tau)$$

and can thus restrict our discussions to the domain

$$(1) \qquad -\frac{N}{2} \leq \operatorname{Re}(\tau) < \frac{N}{2}, \qquad \operatorname{Im}\tau > 0 .$$

We introduce the variable z by

$$(2) \qquad \tau = \frac{N}{2\pi i} \log z$$

and obtain a uniquely defined function

$$g(z) = \psi(\tau), \quad |z| < 1,$$

The unit circle being the image of the domain (1) by means of (2).

The function $g(z)$ is bounded and regular in $|z| < 1$. The possible singularity at $z = 0$, stemming from $\tau = i\infty$, is a removable one since $g(z)$ is bounded.

We consider now the images $M(i)$ of $\tau = i$ under the substitutions M of $\Gamma(N)$. It can be shown that to each pair of natural numbers c, d with

(3) $c \equiv 0, \quad d \equiv 1 \pmod N, \quad (c, d) = 1,$

there exists exactly one pair a, b, which

$$M = \begin{pmatrix} a & b \\ c & d \end{pmatrix} \in \Gamma(N)$$

and
$$M(i) = \frac{ai + b}{ci + d} = \tau_{c,d}$$

lies in the domain (1). Here

$$\operatorname{Im}\left(M(i)\right) = \frac{1}{c^2 + d^2}.$$

In the z-plane we have correspondingly

$z_{c,d} = \exp \frac{2\pi i}{N} \tau_{c,d}$ and in particular

$|z_{c,d}| = \exp\left(-\frac{2\pi}{N} \cdot \frac{1}{c^2+d^2}\right)$. We want to prove

(4) $\prod |z_{a,d}| = 0$

where the multiplication is extended over all pairs of numbers (c,d) satisfying condition (3). A simple argument now shows that

$$\sum \frac{1}{c^2 + d^2}$$

for those (c,d) is divergent, which implies (4).

In view of the lemma we have therefore proved that $g(z) - \psi(i)$ is identically zero and thus that $\psi(\tau)$ is a constant.

A PROOF OF A THEOREM ON MODULAR FUNCTIONS.*

By Hans Rademacher.

The following theorem is fundamental in the theory of modular functions:

Theorem. *A modular function $\psi(\tau)$ belonging to a modular congruence subgroup modulo N and which is regular and bounded in the half-plane $\Re(\tau) > 0$ is a constant.*

This theorem is usually proved by a contour integration around the fundamental region of $\psi(\tau)$ and requires a study of the structure of this region, in particular of its vertices and cusps. The following proof avoids all this. All we need is Jensen's inequality:

Lemma. *If $f(z)$ is regular in the interior of the unit circle, $f(0) \neq 0$ and $|f(z)| < M$, then for any zeros $z_1, z_2, \cdots, z_n$ of $f(z)$ we have*

$$|z_1 z_2 \cdots z_n| \geqq |f(0)|/M.$$

This lemma implies the simple

Corollary. *If $g(z)$ is regular and bounded in $|z| < 1$, and if z_ν, $\nu = 1, 2, 3, \cdots, z_\nu \neq 0$, are zeros of $g(z)$ in the unit circle so that*

$$\prod_{\nu=1}^{\infty} |z_\nu| = 0,$$

then $g(z)$ vanishes identically.

For the proof of the theorem it suffices to consider the principal subgroup $\Gamma(N)$ modulo N

$$\begin{pmatrix} a & b \\ c & d \end{pmatrix} \equiv \begin{pmatrix} 1 & 0 \\ 0 & 1 \end{pmatrix} \pmod{N},$$

which is contained as subgroup in any congruence subgroup modulo N.

Now $\Gamma(N)$ has $\begin{pmatrix} 1 & N \\ 0 & 1 \end{pmatrix}$ as an element so that we can restrict our considerations to the domain

$$(1) \qquad -\tfrac{1}{2}N \leqq \Re(\tau) < \tfrac{1}{2}N, \qquad \Re(\tau) > 0.$$

* Received June 18, 1959.

Since $\psi(\tau+N)=\psi(\tau)$ we can introduce

$$\tau=(N/2\pi i)\log z \tag{2}$$

and obtain a uniquely defined function $g(z)=\psi(\tau)$ for $|z|<1$, the unit circle being the image of (1) by means of (2).

The function $g(z)$ is *bounded* and *regular* in $|z|<1$. The possible singularity at $z=0$, stemming from $\tau=i\infty$, is a removable one since $g(z)$ is bounded.

Let us consider now the images of $\tau=i$ under the group $\Gamma(N)$. For any pair of natural numbers $c\equiv 0$, $d\equiv 1 \pmod N$ and $(c,d)=1$ there exists exactly one pair a, b so that

$$M=\begin{pmatrix} a & b \\ c & d \end{pmatrix} \in \Gamma(N)$$

and that

$$M(i)=(ai+b)/(ci+d)=(i+(ac+bd))/(c^2+d^2)=\tau_{c,d}$$

has moreover the property

$$-\tfrac{1}{2}N \leqq \Re M(i)=(ac+bd)/(c^2+d^2)<\tfrac{1}{2}N. \tag{3}$$

Indeed, we have to solve the diophantine equation

$$ad-bc=1 \tag{4}$$

for a and b with the conditions

$$a\equiv 1, \qquad b\equiv 0 \pmod N.$$

Now (4) implies

$$ad\equiv 1 \pmod{Nc},$$

which is solvable since $(d,Nc)=1$. Let a_0 be a particular solution. All solutions are then of the form

$$a=a_0+mNc,$$

and the remaining b in (4) is given by

$$b=(a_0d-1)/c+mNd.$$

A short computation shows that (3) is expressed as

$$-\tfrac{1}{2}N \leqq a_0/c-d/c(c^2+d^2)+mN<\tfrac{1}{2}N,$$

which is satisfied by exactly one integer m, so that a and b and thus $\tau_{c,d}$ are now uniquely determined.

We have

$$\Im M(i)=\Im\tau_{c,d}=1/(c^2+d^2).$$

In the z-plane these values correspond to

$$|z_{c,d}| = \exp(-2\pi/N(c^2+d^2)).$$

We want to prove

$$\prod |z_{c,d}| = 0$$

where the multiplication is extended over all $c, d > 0$ with $c \equiv 0$, $d \equiv 1 \pmod N$, $(c,d) = 1$. Translated back into τ-coordinates our task is to prove that

$$\Sigma 1/(c^2+d^2) \tag{5}$$

is divergent.

But this is simple. Let us consider not the full series (5) but to any $d \equiv 1 \pmod N$ only those $c = \gamma N$ with $0 < \gamma \leqq d$ and, of course,

$$(c,d) = (\gamma, d) = 1.$$

These are $\phi(d)$ in number. For such a pair c, d we have $c^2 + d^2 \leqq (N^2+1)d^2$, so that the series (5) is estimated from below by

$$\Sigma 1/(c^2+d^2) > (N^2+1)^{-1} \sum_{\substack{d>0 \\ d\equiv 1 \pmod N}} \phi(d)/d^2.$$

But the latter series is indeed divergent. This can be seen in several ways, for instance by means of the trivial estimate

$$\phi(n)/n \geqq \log 2/\log 2n$$

(obtained in the following way:

$$\phi(n)/n = \prod_{p|n} (1-1/p) \geqq (1-\tfrac{1}{2})(1-\tfrac{1}{3}) \cdots (1-1/(r+1)) = 1/(r+1)$$

with

$$n \geqq \prod_{p|n} p = p_1 p_2 \cdots p_r \geqq 2^r,$$

$$r \leqq \log n/\log 2, \qquad r+1 \leqq \log 2n/\log 2).$$

And the series $\sum_{d>0} 1/d \log 2d$, $d \equiv 1 \pmod N$ is divergent.

This shows that the spots $z_{c,d}$ where $g(z) - \psi(i)$ vanishes make $\prod(z_{c,d})$ diverge to zero. Then, in view of the Corollary, $g(z)$ is identically equal to $\psi(i)$, and $\psi(\tau)$ is constant. Q. E. D.

THE UNIVERSITY OF PENNSYLVANIA.

Notes to Paper 65a

This paper has been reviewed in the Z, vol. 94 (1952), p. 57, and in the MR, vol. 22, #1670. A preliminary version of this paper had been presented to the Institute in the Theory of Numbers in Boulder (Colorado) during the summer of 1959; see paper 65 of the present collection.

Page 338, line 4.
Read $\mathfrak{I}$ instead of $\mathfrak{R}$.

Page 338, equation (1).
The second $\mathfrak{R}$ should be an $\mathfrak{I}$

Page 339, seventh line from bottom.
Read *which* instead of *whch.*

On a Theorem of Besicovitch

Hans Rademacher

Dedicated to George Pólya on his 75th birthday

Recently, Besicovitch's famous solution of Kakeya's problem [1] has again been discussed.* I submit another version of the proof in the hope that our friend to whom this volume is dedicated might find some pleasure in such elementary reasonings of non-trivial implications.

The basic idea of all the existing proofs of Besicovitch's theorem is to cut a given triangle into "slim" triangles with the same vertex and with bases on the old base line and then to shift them on the base line in such a manner that their union covers a small area. All these proofs, i.e., those by Besicovitch, Perron [2], and Schoenberg, prepare the figure by some *horizontal* divisions, the usefulness of which is difficult to foresee at the start of the proof. The following argument will dispense with such preparation.

We have to study only one figure: a triangle ABC with the median AD drawn.

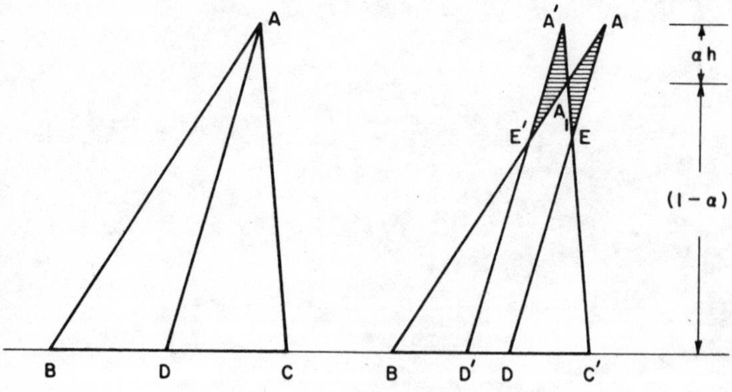

FIG. 1

* See I. J. Schoenberg, "On the Besicovitch-Perron Solution of the Kakeya Problem," in this volume.

The triangle is cut along AD, and then one of the two halves is shifted, with its basis remaining as the old base line, so that after the shift the two halves partly overlap.

Let the height of ABC be h. There appear now two "excess triangles" (shaded in the figure) AA_1E and $A'A_1E'$, and a "new" triangle A_1BC', which is similar to the old ABC. We choose α with $0 < \alpha < 1$, and shift DC to the new position $D'C'$ by such an amount that the height of A_1BC' is $(1-\alpha)h$ and its area thus $(1-\alpha)^2\Delta$, where Δ is the area of the original ABC. It is easily seen that each excess piece has $2\alpha^2$ times the area of that half-triangle from which it is cut, and the excess pieces together have therefore the area $2\alpha^2\Delta$.

Our proof will now consist only in an iteration of this process. Let a "big" triangle $A_0B_0C_0$ of area Δ be given. We divide the basis B_0C_0 into 2^n equal parts and connect each partition point with the vertex A_0. Then we obtain 2^n "slim" triangles. We form 2^{n-1} pairs of adjacent triangles of them, and on each pair we carry out the shifting process described above. The excess triangles, all of which have the same area, will have together an area $2\alpha^2\Delta$. The "new" triangles, each similar to an old pair, can now be pushed together, so that they fit together and form a triangle $A_1B_1C_1$, similar to $A_0B_0C_0$, of height $(1-\alpha)h$ and thus of area $(1-\alpha)^2\Delta$. This triangle carries now a division into 2^{n-1} slim triangles, stemming from the "new" triangles produced by the pairs. As the new triangles are pushed together to form $A_1B_1C_1$, the excess triangles may overlap to some extent, but so much the better, because that will only diminish the area of the union. We indicate this process by the self-explanatory symbols

$$\Delta \to (1-\alpha)^2\Delta + 2\alpha^2\Delta ,$$

where $2\alpha^2\Delta$ is actually an upper estimate of the area of the union of the excess triangles.

We repeat this process now with $A_1B_1C_1$, which is, as we observed, divided into 2^{n-1} slim triangles and indicate the result by

$$(1-\alpha)^2\Delta \to (1-\alpha)^4\Delta + 2\alpha^2(1-\alpha)^2\Delta ,$$

and so on:

$$(1-\alpha)^4\Delta \to (1-\alpha)^6\Delta + 2\alpha^2(1-\alpha)^4\Delta ,$$

and after n such steps,

$$(1-\alpha)^{2n-2}\Delta \to (1-\alpha)^{2n}\Delta + 2\alpha^2(1-\alpha)^{2n-2}\Delta .$$

We end with a triangle $A_nB_nC_n$, similar to the old $A_0B_0C_0$ and of area $(1-\alpha)^{2n}\Delta$ plus the union of all the excesses, so that the total figure of the union 2^n original "slim" triangles after these shifts is at most

$$\begin{aligned} U &\leqq \Delta\{(1-\alpha)^{2n} + 2\alpha^2[1 + (1-\alpha)^2 + \cdots + (1-\alpha)^{2n-2}]\} \\ &< \Delta\left[(1-\alpha)^{2n} + \frac{2\alpha^2}{1-(1-\alpha)^2}\right] \\ &= \Delta\left[(1-\alpha)^{2n} + \frac{\alpha}{1-(\alpha/2)}\right]. \end{aligned}$$

To obtain for a given $\varepsilon > 0$ the estimate $U < \varepsilon\Delta$ we choose α such that

$$\frac{\alpha}{1 - (\alpha/2)} < \frac{\varepsilon}{2}$$

and then choose n so large that $(1 - \alpha)^{2n} < \varepsilon/2$. Q.E.D.

Looking back at the whole process, we realize that we have in succession used triangles of the altitudes $h, (1 - \alpha)h, (1 - \alpha)^2 h, \cdots, (1 - \alpha)^n h$, forming a geometric progression. In Besicovitch's original proof the altitudes also form a geometric progression. However, there these altitudes are measured from the apex, whereas here they are measured from the basis. The altitudes lead thus to different "horizontal" divisions, as mentioned in the introduction. In Perron's and Schoenberg's proofs the altitudes form an arithmetic progression. The present proof thus employs a figure essentially different from those in the other known proofs.

Institute for Advanced Study
and
University of Pennsylvania

REFERENCES

[1] BESICOVITCH, A. S., On Kakeya's Problem and a Similar One, *Math. Z.*, **27** (1928), 312-20.

[2] PERRON, O., Über einen Satz von Besicovitch, *Math. Z.*, **28** (1928), 383-86.

Note to Paper 66

This paper has been reviewed in the Z, vol. 114 (1965), p. 384, and in the MR, vol. 26 (1963), #4217.

Some remarks on certain generalized Dedekind sums

by

H. Rademacher (Philadelphia, Pa.)

Dedicated to L. J. Mordell
on the occassion of his 75th birthday

1. In recent investigations concerning the functions

$$\sigma_{gh}(0, \tau) = \prod_{m=0}^{\infty} (1 - e^{2\pi i h/f} e^{2\pi i \tau (m+g/f)}) \prod_{n=1}^{\infty} (1 - e^{-2\pi i h/f} e^{2\pi i \tau (n-g/f)}) \tag{1.1}$$

Curt Meyer [1] and Ulrich Dieter [2] have introduced the following generalized Dedekind sums

$$s_{g,h}(a, c) = \sum_{\mu \bmod c} \left(\left(\frac{a\mu}{c} + \frac{ag+ch}{cf}\right)\right)\left(\left(\frac{\mu}{c} + \frac{g}{cf}\right)\right). \tag{1.2}$$

They derived a whole theory of these sums, including also a reciprocity theorem.

The functions (1.1) have been introduced by F. Klein [3], in the theory of "division" of σ-functions. C. Meyer uses these functions for investigations of class numbers of Abelian fields over quadratic ground fields, following the lead of Hecke. They appear, also, not recognized as Klein's σ_{gh} however, in papers by J. Lehner [4] and J. Livingood [5] where they play the role of conjugates of the generating function for certain partition numbers. In these papers one also finds certain generalizations of Dedekind sums, which turn out to be special cases of the sums (1.2).

I observe now that the rationality of g/f and h/f is completely irrelevant in the theory of the sums (1.2) and that the true generalization of the above kind is contained in the *definition*:

$$\begin{aligned} s(h, k; x, y) &= \sum_{\mu \bmod k} \left(\left(h\left(\frac{\mu+y}{k} + \frac{x}{h}\right)\right)\right)\left(\left(\frac{\mu+y}{k}\right)\right) \\ &= \sum_{\mu \bmod k} \left(\left(h\frac{\mu+y}{k} + x\right)\right)\left(\left(\frac{\mu+y}{k}\right)\right), \end{aligned} \tag{1.3}$$

where x and y are real numbers and h, k are coprime integers, $k > 0$. The symbol $((x))$ means here as usual the function

$$((x)) = \begin{cases} x-[x]-\frac{1}{2}, & x \text{ not integer}, \\ 0, & x \text{ integer}. \end{cases}$$

The definition (1.3) shows immediately that $s(h, k;\ x, y)$ has the period 1 in x as well as in y. We may therefore restrict our considerations to the range

$$0 \leqslant x < 1, \quad 0 \leqslant y < 1. \tag{1.4}$$

For $x = y = 0$ (and thus also for x, y both integers) the new sums are the classical Dedekind sums. The sums (1.2) are special cases of (1.3) with $h = a$, $k = c$, $x = h/f$, $y = g/f$.

2. We need further on the

THEOREM 1.

$$s(1, k; 0, 0) = s(1, k) = \frac{k}{12} + \frac{1}{6k} - \frac{1}{4}; \tag{2.1}$$

$$s(1, k;\ 0, y) = \frac{k}{12} + \frac{1}{k} B_2(y) \quad \textit{for} \quad 0 < y < 1, \tag{2.2}$$

where $B_2(y)$ is the second Bernoulli function, $B_2(y) = y^2 - y + 1/6$.

Proof. The case $y = 0$ is known from the theory of Dedekind sums and results from

$$s(1, k) = \sum_{\mu=1}^{k-1} \left(\left(\frac{\mu}{k}\right)\right)^2 = \sum_{\mu=1}^{k-1} \left(\frac{\mu}{k} - \frac{1}{2}\right)^2$$

by straightforward calculation.

The case $0 < y < 1$ requires

$$s(1, k;\ 0, y) = \sum_{\mu=0}^{k-1} \left(\left(\frac{\mu+y}{k}\right)\right)^2 = \sum_{\mu=0}^{k-1} \left(\frac{\mu+y}{k} - \frac{1}{2}\right)^2$$

from which (2.2) follows by simple calculation.

COROLLARY. *Because of the periodicity in x and y we have also*

$$s(1, k;\ 0, y) = \frac{k}{12} + \frac{1}{6k} - \frac{1}{4} \quad \textit{for } y \textit{ integer},$$

$$s(1, k;\ 0, y) = \frac{k}{12} + \frac{1}{k} \psi_2(y) \quad \textit{for } y \textit{ not integer},$$

where we use here and subsequently the abbreviation

$$\psi_2(y) = B_2(y - [y]). \tag{2.3}$$

For the establishment of the reciprocity formula we use a device by U. Dieter [2]. We need first the

LEMMA.

$$\sum_{\mu \bmod k} \left(\left(\frac{\mu+w}{k}\right)\right) = ((w)). \tag{2.4}$$

Proof. Let us investigate the difference

$$D(w) = \sum_{\mu \bmod k} \left(\left(\frac{\mu+w}{k}\right)\right) - ((w)).$$

This difference has obviously the period 1 in w, since both terms have it. We need thus only consider $0 \leqslant w < 1$. For $w = 0$ we have

$$D(0) = \sum_{\mu \bmod k} \left(\left(\frac{\mu}{k}\right)\right) = 0,$$

and for $0 < w < 1$

$$\begin{aligned} D(w) &= \sum_{\mu=0}^{k-1} \left(\left(\frac{\mu+w}{k}\right)\right) - ((w)) \\ &= \sum_{\mu=0}^{k-1} \left(\frac{\mu+w}{k} - \frac{1}{2}\right) - \left(w - \frac{1}{2}\right) \\ &= \frac{k-1}{2} + w - \frac{k}{2} - \left(w - \frac{1}{2}\right) = 0. \end{aligned}$$

Therefore $D(w) = 0$ for all w, which proves the lemma.

3. We study now, with $h > 0, k > 0, (h, k) = 1$,

$$S = s(h, k; x, y) + s(k, h; y, x) \tag{3.1}$$

$$= \sum_{\mu \bmod k} \left(\left(h\left(\frac{\mu+y}{k} + \frac{x}{h}\right)\right)\right)\left(\left(\frac{\mu+y}{k}\right)\right) + \sum_{\nu \bmod h} \left(\left(k\left(\frac{\nu+x}{h} + \frac{y}{k}\right)\right)\right)\left(\left(\frac{\nu+x}{h}\right)\right).$$

Applying now (2.4) to the two sums we obtain

$$\begin{aligned} S &= \sum_{\mu \bmod k} \sum_{\nu \bmod h} \left(\left(\frac{\nu}{h} + \frac{\mu+y}{k} + \frac{x}{h}\right)\right)\left(\left(\frac{\mu+y}{k}\right)\right) + \\ &\qquad + \sum_{\nu \bmod h} \sum_{\mu \bmod k} \left(\left(\frac{\mu}{k} + \frac{\nu+x}{h} + \frac{y}{k}\right)\right)\left(\left(\frac{\nu+x}{h}\right)\right) \\ &= \sum_{\mu=0}^{k-1} \sum_{\nu=0}^{h-1} \left(\left(\frac{\mu+y}{k} + \frac{\nu+x}{h}\right)\right)\left\{\left(\left(\frac{\mu+y}{k}\right)\right) + \left(\left(\frac{\nu+x}{h}\right)\right)\right\} \end{aligned} \tag{3.2}$$

We have to study only the range (1.4), with the exception of the case $x = 0$, $y = 0$, which is covered by the classical reciprocity formula for Dedekind sums. We have thus only

$$0 < x+y < 2.$$

It is preferable to postpone also the case $x = 0$ *or* $y = 0$ and to begin with

$$0 < x < 1, \quad 0 < y < 1.$$

Then we get

$$\text{(3.3)} \qquad S = \sum_{\mu=0}^{k-1} \sum_{\nu=0}^{h-1} \left(\left(\frac{\mu+y}{k} + \frac{\nu+x}{h}\right)\right)\left(\frac{\mu+y}{k} + \frac{\nu+x}{h} - 1\right).$$

Let us now look at the sum

$$\text{(3.4)} \qquad T = \sum_{\mu=0}^{k-1} \sum_{\nu=0}^{h-1} \left\{\left(\left(\frac{\mu+y}{k} + \frac{\nu+x}{h}\right)\right) - \left(\frac{\mu+y}{k} + \frac{\nu+x}{h} - 1\right)\right\}^2$$

$$= \sum_{\mu=0}^{k-1} \sum_{\nu=0}^{h-1} \left(\left(\frac{\mu+y}{k} + \frac{\nu+x}{h}\right)\right)^2 -$$

$$-2 \sum_{\mu=0}^{k-1} \sum_{\nu=0}^{h-1} \left(\left(\frac{\mu+y}{k} + \frac{\nu+x}{h}\right)\right)\left(\frac{\mu+y}{k} + \frac{\nu+x}{h} - 1\right) +$$

$$+ \sum_{\mu=0}^{k-1} \sum_{\nu=0}^{h-1} \left(\frac{\mu+y}{k} + \frac{\nu+x}{h} - 1\right)^2$$

or, say,

$$\text{(3.5)} \qquad T = S_1 - 2S_2 + S_3.$$

We discuss these sums separately. In S_1 the variables μ and ν need only to be taken modulo k and h, respectively. Then we have

$$S_1 = \sum_{\substack{\mu \bmod k \\ \nu \bmod h}} \left(\left(\frac{h\mu + k\nu}{hk} + \frac{hy + kx}{hk}\right)\right)^2$$

$$= \sum_{\varrho \bmod hk} \left(\left(\frac{\varrho + hy + kx}{hk}\right)\right)^2 = s(1, hk; 0, hy + kx).$$

4. Now we have to distinguish two cases:

I. $hy + kx$ *not integer*, II. $hy + kx$ *integer*.

We continue until further notice solely with Case I. Then we obtain from the Corollary to Theorem 1

$$\text{(4.1)} \qquad S_1 = \frac{hk}{12} + \frac{1}{hk}\,\psi_2(hy + kx).$$

Comparing (3.3), (3.4), (3.5) we recognize that

$$S_2 = S. \tag{4.2}$$

Furthermore, we have

$$S_3 = \sum_{\mu=0}^{k-1}\sum_{\nu=0}^{h-1}\left(\frac{\mu+y}{k}+\frac{\nu+x}{h}-1\right)^2.$$

This sum is *continuous* in x and y. We obtain first

$$S_3 = h\sum_{\mu=0}^{k-1}\left(\frac{\mu+y}{k}\right)^2 + k\sum_{\nu=0}^{h-1}\left(\frac{\nu+x}{h}\right)^2 + hk +$$

$$+2\sum_{\mu=0}^{k-1}\frac{\mu+y}{k}\sum_{\nu=0}^{h-1}\frac{\nu+x}{h} - 2h\sum_{\mu=0}^{k-1}\frac{\mu+y}{k} - 2k\sum_{\nu=0}^{h-1}\frac{\nu+x}{h}.$$

The computation yields finally

$$S_3 = \frac{h}{k}B_2(y) + \frac{k}{h}B_2(x) + 2\left(x-\frac{1}{2}\right)\left(y-\frac{1}{2}\right) + \frac{hk}{6}. \tag{4.3}$$

If on the other hand we use $((z)) - z = -[z] - 1/2$ in case z is not integer, then definition (3.4) for T gives

$$T = \sum_{\mu=0}^{k-1}\sum_{\nu=0}^{h-1}\left\{-\left[\frac{\mu+y}{k}+\frac{\nu+x}{h}\right]+\frac{1}{2}\right\}^2 \tag{4.4}$$

since in Case I none of the values

$$\frac{\mu+y}{k}+\frac{\nu+x}{h}, \qquad \mu = 0, \ldots, k-1, \; \nu = 0, \ldots, h-1,$$

can be an integer. We have thus

$$T = \sum_{\mu=0}^{k-1}\sum_{\nu=0}^{h-1}\left[\frac{\mu+y}{k}+\frac{\nu+x}{h}\right]\cdot\left(\left[\frac{\mu+y}{k}+\frac{\nu+x}{h}\right]-1\right)+\frac{hk}{4}.$$

Since with $0 < x < 1$, $0 < y < 1$ and within the range of μ and ν the bracket $\left[\frac{\mu+y}{k}+\frac{\nu+x}{h}\right]$ can only have the values 0 or 1, the double sum is 0 and we obtain thus

$$T = hk/4. \tag{4.5}$$

Putting together now formulae (3.5) and (4.1) to (4.5) we have

$$S = \tfrac{1}{2}(S_1 - T + S_3) \tag{4.6}$$

with

$$S_1 - T = \frac{1}{hk}\psi_2(hy+kx) - \frac{hk}{6} \tag{4.7}$$

and thus, finally,

$$S = ((x))((y)) + \frac{1}{2}\left\{\frac{k}{h}\psi_2(x) + \frac{1}{hk}\psi_2(hy+kx) + \frac{h}{k}\psi_2(y)\right\}, \tag{4.8}$$

where we have replaced $B_2(x)$, $B_2(y)$ by $\psi_2(x)$, $\psi_2(y)$, which is permissible in the range (4.1) of x and y.

5. We come now to Case II, $hy+kx$ integer, but still $0 < x < 1$, $0 < y < 1$. Formula (4.6) remains true here also, from the definitions. However, in this case a few modifications of our argument are necessary. We notice that now $(\mu+y)/k + (\nu+x)/h$ may be an integer for certain μ and ν. Indeed, this would require

$$\mu h + \nu k \equiv -(hy+kx) (\operatorname{mod} hk),$$

which has exactly one solution μ_0, ν_0 in the range of μ and ν since h and k are coprime. Equation (3.5) remains valid. But now we have, according to Theorem 1 and its Corollary,

(5.1)

$$S_1 = S(1, hk;\ 0, hy+kx) = \frac{hk}{12} + \frac{1}{6hk} - \frac{1}{4} = \frac{hk}{12} + \frac{1}{hk}\psi_2(hy+kx) - \frac{1}{4}.$$

The exiplicit result (4.5) for T will have to be revised since we have to remember the exceptional pair of values μ_0, ν_0. We have then from (3.4)

$$T = \sum_{\substack{\mu=0 \\ (\mu,\nu)\neq(\mu_0,\nu_0)}}^{k-1} \sum_{\nu=0}^{h-1} \left\{-\left[\frac{\mu+y}{k} + \frac{\nu+x}{h}\right] + \frac{1}{2}\right\}^2 + \left\{-\left[\frac{\mu_0+y}{k} + \frac{\nu_0+x}{h}\right] + 1\right\}^2$$

$$= \sum_{\mu=0}^{k-1} \sum_{\nu=0}^{h-1} \left\{-\left[\frac{\mu+y}{k} + \frac{\nu+x}{h}\right] + \frac{1}{2}\right\}^2 + 2\left\{-\left[\frac{\mu_0+y}{k} + \frac{\nu_0+x}{h}\right] + \frac{1}{2}\right\}\cdot\frac{1}{2} + \frac{1}{4}.$$

But because of $0 \leqslant \mu_0 < k$, $0 \leqslant \nu_0 < h$, $0 < x < 1$, $0 < y < 1$ the value of

$$\left[\frac{\mu_0+y}{k} + \frac{\nu_0+x}{h}\right]$$

can only be 1. We obtain therefore this time

$$T = \sum_{\mu=0}^{k-1}\sum_{\nu=0}^{h-1}\left\{-\left[\frac{\mu+y}{k}+\frac{\nu+x}{h}\right]+\frac{1}{2}\right\}^2-\frac{1}{4}=\frac{hk}{4}-\frac{1}{4}.$$

We see here and in (5.1) that in Case II S_1 and T are both diminished by $\frac{1}{4}$, compared to Case I. Hence S_1-T retains the value (4.7) it had for Case I. Since, moreover, S_3 is continuous in x and y, it is not effected in the distinction of Cases I and II. Therefore, since (4.6) remains true here also, the result (4.8) is correct also for Case II.

6. There remains now only the case $x = 0$, $0 < y < 1$ (or the symmetric one $0 < x < 1$, $y = 0$). We start anew from (3.2) with

$$S = \sum_{\mu=0}^{k-1}\sum_{\nu=0}^{h-1}\left(\left(\frac{\mu+y}{k}+\frac{\nu}{h}\right)\right)\left\{\left(\left(\frac{\mu+y}{k}\right)\right)+\left(\left(\frac{\nu}{h}\right)\right)\right\}.$$

Here the case $\nu = 0$ has to be specially treated.

$$S = \sum_{\mu=0}^{k-1}\sum_{\nu=1}^{h-1}\left(\left(\frac{\mu+y}{k}+\frac{\nu}{h}\right)\right)\left\{\frac{\mu+y}{k}+\frac{\nu}{h}-1\right\}+\sum_{\mu=0}^{k-1}\left(\left(\frac{\mu+y}{k}\right)\right)\left\{\frac{\mu+y}{k}-\frac{1}{2}\right\}$$

$$= \sum_{\mu=0}^{k-1}\sum_{\nu=0}^{h-1}\left(\left(\frac{\mu+y}{k}+\frac{\nu}{h}\right)\right)\left\{\frac{\mu+y}{k}+\frac{\nu}{h}-1\right\}+\frac{1}{2}\sum_{\mu=0}^{k-1}\left(\left(\frac{\mu+y}{k}\right)\right)$$

$$= S_2+\frac{1}{2}((y))$$

according to (3.4), (3.5), and (2.4). Thus (3.5) yields

$$2S = S_1-T+S_3+((y)). \tag{6.1}$$

We know that S_1-T does not depend on the arithmetical nature of $hy+k\cdot 0$ and is given by (4.7). Moreover, S_3 is continous. For $x = 0$, $0 < y < 1$ it can be written as

$$S_3 = \frac{h}{k}\varphi_2(y)+\frac{k}{h}\varphi_2(0)-((y))+\frac{hk}{6}.$$

If we put this and (4.7) into (6.1) we obtain

$$S = \frac{1}{2}\left\{\frac{k}{h}\varphi_2(0)+\frac{1}{hk}\varphi_2(hy)+\frac{h}{k}\varphi_2(y)\right\}.$$

Since $((0)) = 0$, this turns out to be the special case $x = 0$ of (4.8).

We can now lift the restriction of the range (4.1) on x and y. Our result is

THEOREM 2. *For x, y both integers the classical formula*

$$s(h,k;\ x,y)+s(k,h;\ y,x) = s(h,k)+s(k,h) = -\frac{1}{4}+\frac{1}{12}\left(\frac{h}{k}+\frac{1}{hk}+\frac{k}{h}\right)$$

remains in force.

If x, y are not both integers then the reciprocity formula is

$$s(h,k;\ x,y)+s(k,h;\ y,x)$$
$$= ((x))((y))+\frac{1}{2}\left\{\frac{h}{k}\psi_2(y)+\frac{1}{hk}\psi_2(hy+kx)+\frac{k}{h}\psi_2(x)\right\}.$$

7. The reciprocity theorem can be used in case of the ordinary Dedekind sums to compute $s(h, k)$ by means of a Euclidean algorithm. This can be done here also, however only with the rule given by the following theorem.

THEOREM 3. *Let m be an integer. Then*

$$s(h,k;\ x,y) = s(h-mk,k;\ x+my,y). \tag{7.1}$$

Proof. We have from the definition (1.3)

$$s(h-k,k;\ x+y,y) = \sum_{\mu \bmod k}\left(\left((h-k)\left(\frac{\mu+y}{k}\right)+x+y\right)\right)\left(\left(\frac{\mu+y}{k}\right)\right)$$
$$= \sum_{\mu \bmod k}\left(\left(h\left(\frac{\mu+y}{k}\right)-\mu+x\right)\right)\left(\left(\frac{\mu+y}{k}\right)\right)$$
$$= \sum_{\mu \bmod k}\left(\left(h\left(\frac{\mu+y}{k}+\frac{x}{h}\right)\right)\right)\left(\left(\frac{\mu+y}{k}\right)\right)$$
$$= s(h,k;\ x,y).$$

This settles the case $m = 1$. For any integer m the theorem follows now by iteration.

References

[1] C. Meyer, *Über einige Anwendungen Dedekindscher Summen*, Journal f. d. reine u. angewandte Mathem. 198 (1957), pp. 143-203.

[2] U. Dieter, *Zur Theorie der Dedekindschen Summen*, Inauguraldissertation, Kiel 1957 (mimeographed), in particular pp. 15, 52.

[3] F. Klein, *Über die elliptischen Normalkurven der n-ten Ordnung*, Abhandlungen d. Sächsischen Kgl. Ges. d Wiss. (Leipzig), 13 (1885) und Gesammelte Mathem. Abhandl. Berlin 1923, vol. 3, pp. 198-254.

[4] J. Lehner, *A partition function connected with the modulus five*, Duke Math. J. 8 (1941), pp. 631-655.

[5] J. Livingood, *A partition function with the prime modulus $p > 3$*, Amer. J. Math. 67 (1945), pp. 194-208.

NATIONAL BUREAU OF STANDARDS, WASHINGTON, D. C.
and UNIVERSITY OF PENNSYLVANIA, PHILADELPHIA, Pa.

Reçu par la Rédaction le 25. 7. 1963

Notes to Paper 67

This paper has been reviewed in the Z, vol. 128 (1967), p. 271, and in the MR, vol. 29 (1965), #1172.

Page 102, line 5.
Delete (4.1); meant is the range $0 < x < 1$, $0 < y < 1$.

ON THE NUMBER OF CERTAIN TYPES OF POLYHEDRA

BY

HANS RADEMACHER

1. Introduction. The discussions in the following pages are concerned with certain enumerations in the morphology of Eulerian polyhedra in 3-space. The "morphology" is actually the topology of the complexes consisting of the vertices, edges, and faces of polyhedra, with the restriction that these elements are linear. A class of polyhedra isomorphic to each other with respect to incidences is called a *type*. We impose also on this isomorphism the condition that it preserves the orientation. We shall call such a class a "type in the strict sense" in contradistinction to the usage in most classical papers in which the preservation of orientation was not required (Kirkman, Brückner). In Brückner's book [1] there is to be found the explicit statement "Verschiedene Typen ergeben sich nur, wenn . . ., denn weiterhin treten die Spiegelbilder der bisherigen Vielflache auf". That means that mirror-symmetric polyhedra belong in this sorting to the same type (in our terminology employed here "type in the wider sense"). Steinitz [9] in most parts of his book shares Brückner's point of view. However, on p. 86 he speaks of "direct isomorphy", which he defines as isomorphy under preservation of orientation. He formulates there his famous theorem on convex polyhedra, which is actually a homotopy theorem, stating that two convex polyhedra of the same type in the strict sense are homotopically equivalent, again with the "morphological" restriction that the complexes which constitute the continuous transition from one convex polyhedron to a directly isomorphic one remain always convex polyhedra of the same type in the strict sense.

It is clear that the number of types in the strict sense is at least as great as the number of types in the wider sense.

The number of types of polyhedra of a given number F of faces, whether the types are counted in the wider or the strict sense, is a problem mentioned by Euler, Steiner, Kirkman [4], Eberhardt [3], Brückner [1], and Steinitz [8]. Usually attention is only paid to trihedral polyhedra, i.e. those whose every vertex belongs to 3 faces and 3 edges. These polyhedra are considered as "general", whereas those with vertices of higher incidence are looked upon in such discussions as degenerate. We shall in this article also deal only with *trihedral polyhedra* and shall no longer mention this restriction.

For the types in the wider sense the enumeration has been carried out up to $F = 11$ by Brückner and recently, with the help of an electronic computer, by D. W. Grace, a student of G. Pólya. The number $\psi(F)$ of types (in the wider sense) increases rapidly with F, and no general formula has been found for it.

Received December 30, 1964.

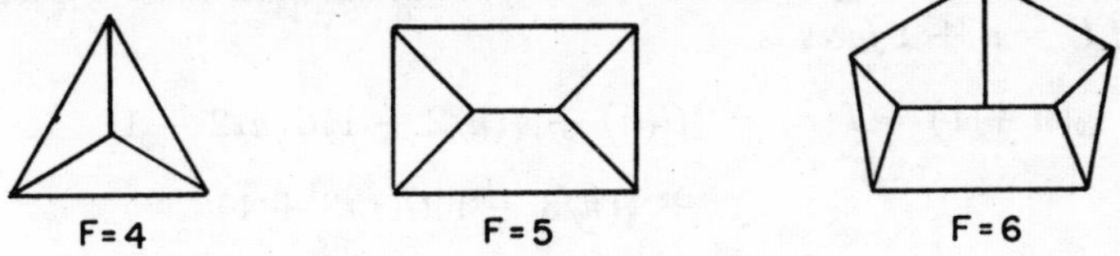

FIGURE 1

However, a special class of trihedral polyhedra has been singled out by Kirkman [4] namely those of F faces which possess a face of $F - 1$ sides, called the "base". Such polyhedra I shall call "based polyhedra"[1], and for these I shall derive explicit formulas for the number of types in the strict sense and in the wider sense. Examples of based polyhedra, seen in projection from above into the plane of the base, are given in Figure 1. Kirkman, in his paper [4] gives a recursive method to determine for each given F the number of types in the wider sense. However this method becomes soon unmanageable for increasing F. He does not arrive at an explicit formula nor an asymptotic expression for the number of types as a function of F. For the number $\chi_0(F)$ of types in the wider sense G. Pólya has proved that it satisfies the inequality

$$(1.1)\qquad \frac{1}{2(F-1)(F-2)}\binom{2F-6}{F-3} < \chi_0(F) \leqq \frac{6}{2(F-1)(F-2)}\binom{2F-6}{F-3}.$$

(Personal communication, in a letter dated 25th July 1964.)

Pólya's proof will be published in a dissertation of his student D. W. Grace. Pólya conjectured moreover that $\chi_0(F)$ is asymptotic to the left member of this inequality. I shall prove this conjecture and more, namely the following two theorems, which I enunciate with $F = n + 1$, n being the number of sides of the base of the based polyhedron.

THEOREM 1. *If $n \geqq 3$, the number of types in the strict sense of based polyhedra of $F = n + 1$ faces is*

$$(1.2)\qquad \psi_0(n+1) = g(n) = G(n) + \tfrac{1}{2}(n/2+1)G(n/2+1) + \tfrac{2}{3}(n/3+1)G(n/3+1),$$

where

$$(1.3)\qquad G(n) = \frac{(2n-4)!}{n!\,(n-2)!} \quad \text{for } n \text{ an integer} \geqq 2; \qquad = 0 \quad \text{otherwise.}$$

---

[1] Following a suggestion of M. Kac. G. Pólya in a correspondence, which was the start of these investigations, calls them "roofless".

THEOREM 2. *If $n \geqq 3$, the number of types in the wider sense of based polyhedra of $F = n + 1$ faces is*

$$\begin{aligned}\chi_0(n+1) = k(n) &= \tfrac{1}{2}G(n) + \tfrac{1}{4}(n/2+1)G(n/2+1) \\ &\quad + \tfrac{1}{3}(n/3+1)G(n/3+1) \\ &\quad + \tfrac{1}{2}([n/2]+1)G([n/2]+1).\end{aligned} \tag{1.4}$$

We shall devote Section I to Theorem 1 and Section II to Theorem 2. The short Section III is concerned with the special types of based polyhedra with only 2 triangles, which have been discussed and enumerated by O. Hermes.

I wish to thank Professor George Pólya for mentioning this problem to me and letting me know his inequality (1.1) together with his conjectures.

I. Number of types in the strict sense

2. Description of based polyhedra. In order to construct the based polyhedra for small n and to understand their generation for all n we have to analyze their structure. We can and shall always refer to the projection of the polyhedron into the plane of its base. The polyhedron is viewed so that it lies *above* the base. Such a projection forms a graph in the plane.

There are 3 sorts of edges:

(1) those of the base, n in number;
(2) those issuing from the base, one from each vertex, also n in number;
(3) the remaining ones, forming a complex which we call the "ridge".

Since we have only trihedral polyhedra with V vertices and E edges, so that

$$3V = 2E$$

and $F = n + 1$, we find from Euler's formula

$$E = 3n - 3. \tag{2.1}$$

The ridge consists thus of

$$r = E - 2n = n - 3 \tag{2.2}$$

edges. The whole graph contains

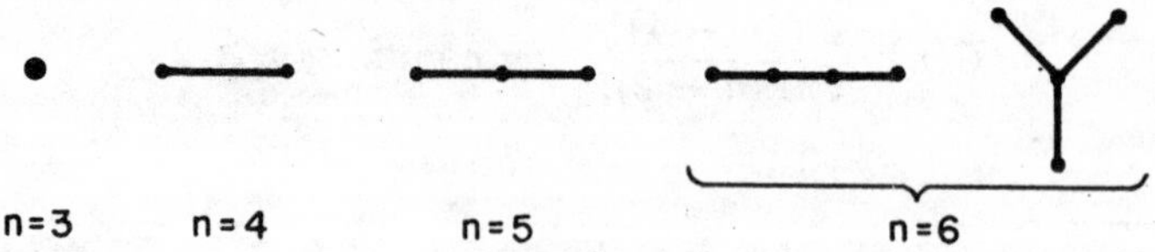

FIGURE 2

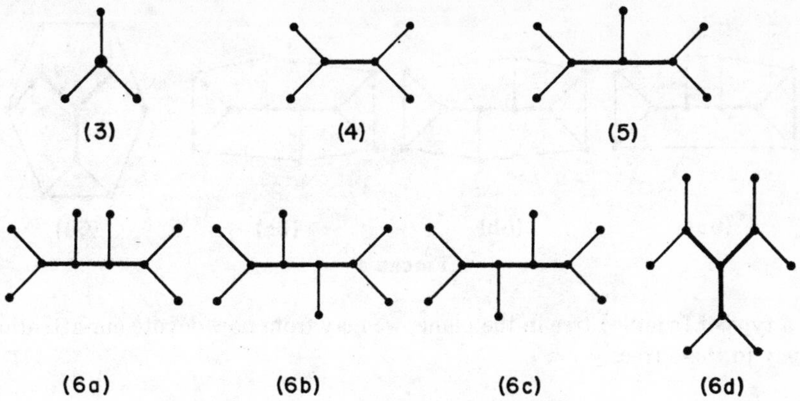

FIGURE 3

$$V = 2n - 2 \tag{2.3}$$

vertices. On the ridge are therefore $n - 2$ vertices.

Taking the n edges of the base away from the graph we obtain a graph T_n of $2n - 3$ edges and $2n - 2$ vertices. This operation does not destroy connectedness of the graph. Since the number of vertices exceeds the number of edges by 1 we see that T_n is a *tree*. It contains only nodes (vertices) of order 1 (the endpoints, i.e. vertices of the base) and of order 3 (the branchpoints or inner nodes). Such a tree we shall call a *simple tree*. The tree T_n has n endpoints and $n - 2$ branchpoints.

The ridge R_n is obtained by taking away from T_n all the ending edges (endsegments). No such amputation disturbs the connectivity. The ridge is therefore also a tree, and indeed we found its number of edges $r = n - 3$ by 1 smaller than its number of vertices $V - n = n - 2$.

The trees T_n and R_n have to be considered as imbedded in the oriented plane, and not only in a purely combinatorial manner as complexes of segments and nodes put together by incidence relations.[2] We construct the based polyhedra starting from the ridges. Examples of ridges R_n are given in Figure 2, from which we derive the trees T_n of Figure 3. Connecting now the endpoints in the order in which they appear in the oriented plane, we obtain the full projections of the based polyhedra; see Figure 1 for $n = 3, 4, 5$ and Figure 4 for $n = 6$. These are all possible based polyhedra to $F = 7$ or $n = 6$. We notice that $n = 3$ and (6d) have a rotational symmetry (R.S.) of order 3; $n = 4$ and (6b) and (6c) have a rotational symmetry of order 2. Moreover, (6c) is the mirror image of (6b).

Since each type of a based polyhedron stands in a one-to-one correspondence

---

[2] The Figures (6a), (6b), (6c) would represent the same tree combinatorially, but here they are distinguished as imbedded in the plane of the drawing.

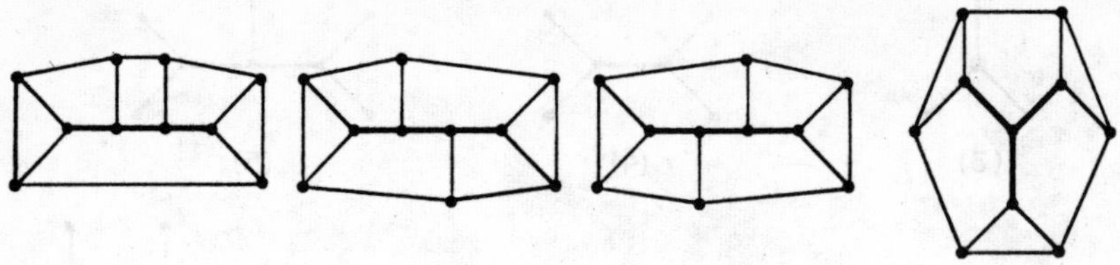

FIGURE 4

to a type of (simple) tree in the plane, we may from now devote our attention solely to these trees.

3. Rotational symmetry. There can be no other R.S.'s than those of order 2 and 3. Of course, it is necessary in these cases that $2|n$ and $3|n$ respectively.

That there are no other R.S.'s becomes clear if we look at more stringent necessary conditions in the cases of R.S. of orders 2 and 3. For R.S. of order 2 we remove pairs of segments which correspond to each other under the rotation. We first remove the sides of the base polygon in pairs, which shows that n is even. Then we continue with the inner segments, i.e. those of T_n. But the number of segments of T_n is $2n-3$, thus odd. One segment must remain unmatched, its central segment, and this has the R.S. of order 2.

Similarly, R.S. of order 3 first necessitates n divisible by 3. Then T_n has $2n-3$ segments, a number also divisible by 3. We continue to remove corresponding segments in triples. There remains in the end a branchpoint, the central node, which with its adjacent segments has the R.S. of order 3.

Since T_n is a simple tree no other outcome is possible for a rotational symmetry.

We notice further that each *endpoint* of the ridge R_n produces a triangle of the polyhedron. Since any R_n must have at least 2 endpoints we observe: A based polyhedron has at least 2 triangles. And conversely: if a based polyhedron has only two triangles then its ridge has only two ends. But that means that the ridge is a *chain* of $r = n - 3$ segments linked in succession to each other.

We remark finally that the indicated construction of based polyhedra leads to configurations which fulfill Steinitz's condition for K-polyhedra, which can all be realized as convex polyhedra in Euclidean 3-space, [9, p. 192, pp. 227 ff].

4. Distinctions according to rotational symmetry. Before continuing the investigation I wish to introduce an abbreviated notation. There is no need to distinguish between a based polyhedron and the type which it represents. We shall speak of a based polyhedron T_n or T''_n etc. as well as of the type T_n, T''_n, always understanding here the *type in the strict sense*. The set of all types T_n we shall write (T_n) and the cardinal number of this set as $|T_n|$.

We shall also speak of trees T_n, etc., corresponding to the type T_n. With $F = n + 1$ we put the number of types T_n

$$\psi_0(n + 1) = g(n)$$

and break this number down into the sum

$$g(n) = g_1(n) + g_2(n) + g_3(n), \tag{4.1}$$

where $g_1(n)$ is the number of types of polyhedra T_n' with base of n sides and without rotational symmetry (or R.S. of order 1), (see cases (5), (6a) in Figure 3), $g_2(n)$ the number of the types T_n'' of R.S. of order 2 (cases (4), (6b), (6c)) and $g_3(n)$ that of types T_n''' of R.S. of order 3 (cases (3), (6d)). It is clear that

$$\begin{aligned} g_2(n) &= 0 \quad \text{for} \quad 2 \nmid n \\ g_3(n) &= 0 \quad \text{for} \quad 3 \nmid n, \end{aligned} \tag{4.2}$$

and *only* in these cases, as we shall see.

Instead of taking up $g(n)$ directly we shall consider the modified number

$$G(n) = g_1(n) + \tfrac{1}{2}g_2(n) + \tfrac{1}{3}g_3(n). \tag{4.3}$$

The first values are

$$G(3) = \tfrac{1}{3}, \qquad G(4) = \tfrac{1}{2}, \qquad G(5) = 1, \tag{4.4}$$

as our examples in **2** show.

We can express $g_2(n)$ and $g_3(n)$ conversely by $G(n)$.

Indeed in a T_n'' (or R.S. of order 2) we cut its central segment into two segments and split thus T_n'' into two identical trees T_l with $l = n/2 + 1$ endpoints each. Now this T_l can be, according to its own rotational symmetry, of the sort T_l', T_l'', T_l'''. We can conversely use all T_l to construct all possible T_n'', choosing one of the l endpoints of T_l for attachment to an identical T_l (after rotation through 180°). We have only to observe that, whereas for a T_l' we have l choices of endpoints, for a T_l'' we have only $l/2$ choices in order to avoid repetition, and for a T_l''' only $l/3$ choices. This gives

$$\begin{aligned} g_2(n) &= lg_1(l) + (l/2)g_2(l) + (l/3)g_3(l) \\ &= lG(l) \end{aligned}$$

and thus

$$g_2(n) = (n/2 + 1)G(n/2 + 1), \qquad 2|n,\ n \geqq 4. \tag{4.5}$$

Splitting a T_n''' (of R.S. of order 3) at its central node into 3 identical T_k, $k = n/3 + 1$, we obtain similarly, distinguishing again the possibilities T_k', T_k'', T_k''' of T_k,

$$g_3 = (n/3 + 1)G(n/3 + 1), \qquad 3|n,\ n \geqq 4. \tag{4.6}$$

Our constructions show also that indeed (4.2) are the only cases for vanishing $g_2(n)$, $g_3(n)$ with $n \geqq 4$. From (4.1) and (4.3) we infer

$$g(n) = G(n) + \tfrac{1}{2}g_2(n) + \tfrac{2}{3}g_3(n)$$

and in view of (4.5), (4.6)

$$\begin{aligned}(4.7)\qquad g(n) &= G(n) + \tfrac{1}{2}(n/2 + 1)G(n/2 + 1) \qquad n \geqq 4,\\ &\qquad + \tfrac{2}{3}(n/3 + 1)G(n/3 + 1),\end{aligned}$$

where

$$G(x) = 0$$

for x not integer. It suffices therefore to find the function $G(x)$.

5. The recursion formula. We take now $n > 4$. Then the ridge R_n (which is a tree as we know) contains at least one segment.

We pick now a segment of R_n and split it into two segments, one going to a tree T_λ, the other to a tree T_μ. Since T_λ and T_μ each contain a new endpoint we have

$$\lambda + \mu = n + 2;$$

moreover

$$\lambda \geqq 3, \qquad \mu \geqq 3,$$

since the segment was chosen in $R_n \subset T_n$, and to each edge of R_n there are at least 2 further edges of T_n attached.

We have now to treat the T_n', T_n'', T_n''' separately. If we split a T_n' in any of its $r = n - 3$ ridge segments we obtain $2(n - 3)$ of *ordered* pairs of subtrees (T_λ, T_μ). These ordered pairs are all different since the identity of T_λ and T_μ (after suitable rotation) would mean a symmetry as it appears in T_n'' only. For a T_n''' we obtain only $2\cdot(n - 3)/3$ different ordered pairs of subtrees (T_λ, T_μ). For a T_n'' we have $2\cdot(n - 4)/2$ different ordered pairs (T_λ, T_μ) and from the dissection of the central segment *one* pair (T_l, T_l) with $l = n/2 + 1$. If we count now all T_λ, T_μ with $\lambda + \mu = n + 2$, $\lambda, \mu \geqq 3$, each T_n' occurs $2(n - 3)$ times, each T_n'' occurs $2\cdot(n - 4)/2 + 1 = n - 3$ times, and each T_n''' occurs $2\cdot(n - 3)/3$ times, so that altogether there are

$$2(n - 3)(g_1(n) + \tfrac{1}{2}g_2(n) + \tfrac{1}{3}g_3(n)) = 2(n - 3)G(n)$$

ordered pairs (T_λ, T_μ).

On the other hand, each pair T_λ, T_μ gives rise to several T_n, namely according to the endpoints which we choose for attachments. We have here to distinguish the cases T_λ', T_λ'', T_λ''' and also T_μ', T_μ'', T_μ'''. Each of the λ endsegments of a T_λ' can be joined to each of those of a T_μ'. Whereas, e.g. only $\lambda/3$ endsegments of a T_λ''' have to be joined to $\mu/2$ of a T_μ'' and so on. This gives

$$\sum_{\substack{\lambda+\mu=n+2\\ \lambda+\mu\geqq 3}} \left(\lambda g_1(\lambda) + \frac{\lambda}{2}\, g_3(\lambda) + \frac{\lambda}{3}\, g_3(\lambda)\right)\left(\mu g_1(\mu) + \frac{\mu}{2}\, g_2(\mu) + \frac{\mu}{3}\, g_3(\mu)\right)$$

$$= \sum_{\substack{\lambda+\mu=n+2\\ \lambda,\mu\geqq 3}} \lambda G(\lambda)\mu G(\mu)$$

recombinations of T_λ, T_μ into T_n. These recombinations (or conversely splittings) we have just counted starting from the T_n. We obtain thus the relation

$$(5.1)\qquad 2(n-3)G(n) = \sum\nolimits_{\lambda+\mu=n+2;\,\lambda,\mu\geqq 3} \lambda G(\lambda)\mu G(\mu), \qquad n \geqq 4.$$

Let us put

$$(5.2)\qquad nG(n) = h(n-2),$$

defined for $n \geqq 3$. We had in particular $G(3) = \frac{1}{3}$ and have thus

$$(5.3)\qquad h(1) = 1.$$

With $m = n - 2$ the formula (5.1) goes over into

$$(5.4)\qquad h(m) = \frac{m+2}{2(m-1)} \sum_{\substack{\rho+\sigma=m\\ \rho,\sigma\geqq 1}} h(\rho)h(\sigma), \qquad m \geqq 2.$$

Equation (5.3) is a recursion formula which gives the first few next values

$$h(2) = 2, \qquad h(3) = 5$$

in agreement with (4.4) and (5.2).

6. Solution of the recursion formula. We have now to find the solution of (5.4) with the initial condition (5.3). For this purpose we prove the

LEMMA. *The recursion formula*

$$(6.1)\qquad \omega(m) = \sum\nolimits_{\rho+\sigma=m;\,\rho,\sigma\geqq 1} \omega(\rho)\omega(\sigma), \qquad m \geqq 2,$$

with $\omega(1) = 1$ *is solved by*

$$(6.2)\qquad \omega(m) = \frac{1}{m}\binom{2m-2}{m-1}.$$

Proof. We consider the generating function

$$(6.3)\qquad \Psi(x) = \sum\nolimits_{m=1}^{\infty} \omega(m)x^m.$$

Then we have

$$\Psi(x) = x + \sum\nolimits_{m=2}^{\infty} x^m \sum\nolimits_{\rho+\sigma=m} \omega(\rho)\omega(\sigma) = x + \Psi^2(x)$$

so that

$$\Psi(x) = \tfrac{1}{2}(1 \pm \sqrt{1-4x}).$$

Since $\Psi(0) = 0$ only the minus sign can be valid:

$$\Psi(x) = \tfrac{1}{2}(1 - \sqrt{1-4x}) = -\tfrac{1}{2}\sum_{m=1} \binom{\frac{1}{2}}{m} (-4x)^m.$$

This can be rewritten as

$$\Psi(x) = \sum_{m=1}^{\infty} \frac{1}{m}\binom{2m-2}{m-1} x^m,$$

so that (6.2) is established through comparison with (6.3), and the Lemma is proved.

We have thus established the formula

$$\text{(6.4)} \qquad \frac{1}{m}\binom{2m-2}{m-1} = \sum_{\substack{\rho+\sigma=m \\ \rho,\sigma\geq 1}} \frac{1}{\rho}\binom{2\rho-2}{\rho-1}\frac{1}{\sigma}\binom{2\sigma-2}{\sigma-1}, \qquad m > 1.$$

Replacing here m by $m + 2$ we obtain

$$\text{(6.5)} \quad \frac{1}{m+2}\binom{2m+2}{m+1} = \sum_{\substack{\alpha+\beta=m \\ \alpha,\beta\geq 1}} \frac{1}{\alpha+1}\binom{2\alpha}{\alpha}\frac{1}{\beta+1}\binom{2\beta}{\beta} + \frac{2}{m+1}\binom{2m}{m}.$$

Now a simple computation shows that

$$\frac{1}{m+2}\binom{2m+2}{m+1} - \frac{2}{m+1}\binom{2m}{m} = \frac{2(m-1)}{(m+1)(m+2)}\binom{2m}{m},$$

so that (6.5) can be rewritten as

$$\text{(6.6)} \qquad \frac{1}{m+1}\binom{2m}{m} = \frac{m+2}{2(m-1)} \sum_{\substack{\alpha+\beta=m \\ \alpha,\beta\geq 1}} \frac{1}{\alpha+1}\binom{2\alpha}{\beta}\frac{1}{\beta+1}\binom{2\beta}{\beta}$$

We realize further that for $m = 1$

$$\frac{1}{m+1}\binom{2m}{m} = \frac{1}{2}\binom{2}{1} = 1.$$

If we compare this initial value and (6.6) with (5.3) and (5.4) we see that we have proved

$$\text{(6.7)} \qquad h(m) = \frac{1}{m+1}\binom{2m}{m}.$$

This implies after (5.2)

THEOREM 3. *If* $n \geq 3$,

$$\text{(6.8)} \qquad G(n) = \frac{1}{n\,(n-1)}\binom{2n-4}{n-4}.[3]$$

[3] The function $H(n) = nG(n)$ deserves some comments. It was already noticed by Euler (see e.g. [6, p. 102]) that it represents the number of different dissections of a convex polygon with n labelled vertices into triangles by diagonals. This has a connection with our problem, since the enumeration of based polyhedra is the dual to the counting of *topologically different* dissections of an unlabelled convex polygon of n sides into triangles by diagonals. This duality has already been noticed by Kirkman [5] and Steinitz [8]. On the other hand, Cayley [2, p. 114] investigated the number of rooted planar trees with m endpoints and a rootpoint, and interior nodes of order 3 only. He found $H(m + 1)$ as this number, which he, however, wrote "in the remarkably simple form"

$$\frac{1\cdot 3\cdot 5\cdot\ \cdots\ \cdot(2m-3)}{1\cdot 2\cdot 3\cdot\ \cdots\ \cdot m}\,2^{m-1}.$$

In view of (4.7) we have thus proved Theorem 1 for $n \geqq 4$. The case $n = 3$ of Theorem 1 is an easy verification. The formulas (1.2), (4.7) yield for some low values of F the following results:

F	4	5	6	7	8	9	10	11	12	13	14	15	16
$\psi_0(F)$	1	1	1	4	6	19	49	150	442	1424	4522	14924	49536

Remark. From its definition (4.3) it follows that the denominator of $G(n)$ is a divisor of 6. This can be verified by simple number theoretical arguments, also directly from the expression (6.8).

7. Another proof of Theorem 3. The determination of $G(n)$ can be achieved without the use of the Lemma by solving a differential equation for the generating function of $h(m)$.[4]

We define

$$\Phi(x) = \sum_{m=1}^{\infty} h(m)x^m. \tag{7.1}$$

Writing the recursion formula (5.4) in the form

$$2mh(m) - 2h(m) = m\sum_{\rho+\sigma=m;\,\rho,\sigma\geqq 1} h(\rho)h(\sigma) + 2\sum_{\rho+\sigma=m;\,\rho,\sigma\geqq 1} h(\rho)h(\sigma)$$

valid for $m \geqq 2$, we obtain

$$2x\frac{d}{dx}(\Phi(x) - x) - 2(\Phi(x) - x) = x\frac{d}{dx}(\Phi(x)^2) + 2\Phi(x)^2$$

or

$$x\Phi' - \Phi = x\Phi\Phi' + \Phi^2,$$

or

$$\left(\frac{1}{\Phi} - \frac{2}{1+\Phi}\right)\Phi' = \frac{1}{x}.$$

Integration on both sides yields

$$\log \Phi - 2\log(1+\Phi) = \log x + C$$

and thus

$$\Phi/(1+\Phi)^2 = Kx.$$

Now (7.1) shows that

$$\Phi(x)/x \to h(1) = 1 \quad \text{as} \quad x \to 0.$$

Cayley also observes that $H(m+1)$ moreover expresses the number of ways in which a product $A_1 \cdot A_2 \cdot \cdots \cdot A_m$ which does not obey the associative law can be understood through insertion of parentheses. Pólya, in his fundamental paper [6, footnote, p. 198] indicates an approach to the generating function of $H(n)$ through a functional equation. For comparison with our notation it should be noticed that he counts all nodes, except the rootpoint; his n is our $2n - 3$.

[4] I reproduce here a simplified version of my original proof, which I owe to a remark by Paul T. Bateman.

We conclude therefore $K = 1$ and have the quadratic equation for Φ

$$x(1 + \Phi)^2 = \Phi,$$

which has the solution

$$\Phi(x) = -1 + \frac{1}{2x} \pm \frac{1}{2x}\sqrt{1 - 4x}.$$

Here only the minus sign is acceptable since Φ has to be regular at $x = 0$. We obtain therefore

$$\Phi(x) = \frac{1}{2x}\left(-2x + 1 - \sum_{l=0}^{\infty}\binom{\frac{1}{2}}{l}\right)(-4x)^l$$

$$= \sum_{l=2}^{\infty} \frac{(2l-2)!}{l!\,(l-1)!} x^{l-1} = \sum_{m=1}^{\infty} \frac{1}{m+1}\binom{2m}{m} x^m.$$

From the comparison of this result with (7.1) we infer again the result (6.7).

8. Asymptotic estimates. Stirling's formula applied to (6.8) in the case of n integer yields immediately

$$G(n) \sim \frac{1}{16\sqrt{\pi}} n^{-5/2} 4^n$$

and thus

$$(n/2 + 1)G(n/2 + 1) = O(n^{-3/2} 4^{n/2}) = O(G(n)^{1/2}).$$

We have thus

THEOREM 4. *For $F = n + 1$ large*

(8.1) $$\psi_0(F) = \psi_0(n + 1) = g(n) = G(n) + O(G(n)^{1/2})$$

and less precisely

(8.2) $$\psi_0(F) \sim \frac{1}{64\sqrt{\pi}} F^{-5/2} 4^F.$$

II. Number of types in the wider sense

9. Reduction of types in the wider sense to those in the strict sense. As explained in the Introduction, two types in the strict sense which are mirror symmetric are counted as the same type in the wider sense. It stands to reason that the majority of types (in the strict sense) is not mirror symmetric, and we can thus expect that the number $k(n)$ of types in the wider sense is about $\frac{1}{2}$ of $g(n)$ the number of types in the strict sense.

We introduce some notations. A type (in the strict sense) which is *not* mirror symmetric we shall denote by $\tilde{T}_n$. A type with mirror symmetry we shall denote by T_n^*. If $\tilde{g}(n)$ and $g^*(n)$ are the numbers of types (in the strict sense) of $\tilde{T}_n$ and T_n^* respectively, the number of types in the wider sense will be

$$k(n) = \tfrac{1}{2}\tilde{g}(n) + g^*(n).$$

Since evidently the number of all types in the strict sense is

$$g(n) = \tilde{g}(n) + g^*(n)$$

we obtain

$$k(n) = \tfrac{1}{2}(g(n) + g^*(n)). \tag{9.1}$$

The function $g(n)$ is given in Theorem 1. There remains thus only the determination of $g^*(n)$.

10. Let us have a mirror-symmetric based polyhedron with n-sided base. We discuss its type T_n^* by considering again the projection which is a tree in the base plane. We call this tree, representing T_n^*, also T_n^* for short.

The number of segments of T_n^* is odd, viz. $(2n - 3)$. We pair off the segments corresponding to each other under the mirror symmetry. An odd number of segments must remain unmatched, which means that they are involutory under mirror symmetry. This involution can either keep the endpoints of a segment fixed or can exchange the endpoints: in the first case the segment lies in the symmetry axis (of the base plane), in the second orthogonal to it.

Actually there is exactly one involutory segment. Firstly there cannot be several involutory elements on the symmetry axis. Since the whole graph is connected, and the nodes are threefold, there must appear also some nodes outside the axis, and these, of course, pairwise. The connectivity would then imply one or more cycles, so that T_n^* could not be a tree. Also several involutory segments among which there is one orthogonal to the symmetry axis are impossible for the same reason. (See Figure 5.)

Taking the symmetry axis 'vertical" we say that T_n^* is an H_n if its involutory segment is orthogonal to the symmetry axis ("horizontal") and is a V_n if its involutory element lies in the symmetry axis (is "vertical").

A given T_n^* can, however, be an H_n as well as a V_n at the same time (see Fig. 6) after a rotation through 90°. These two symmetries, for group-theoretical

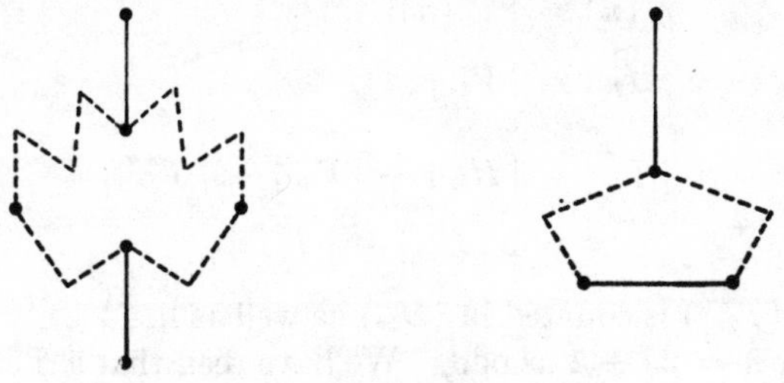

FIGURE 5

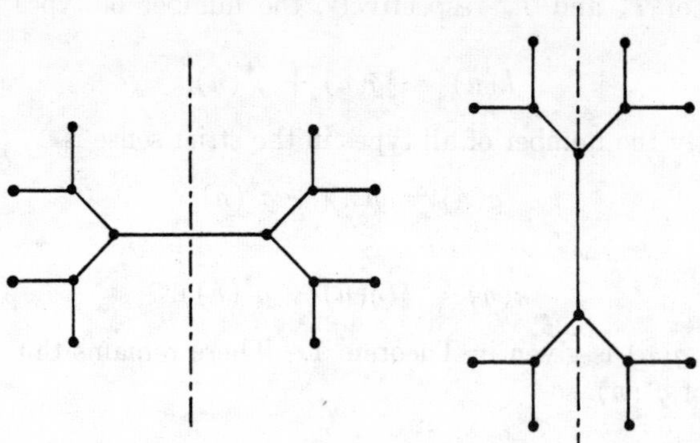

FIGURE 6

reasons, produce together a rotational symmetry of order 2, so that in such a case the T_n^* is also a T_n'', and conversely. We indicate this case by writing $T_n^{*''}$. In this case the number n of endpoints must clearly be divisible by 4.

With the pairing off of segments also the n endpoints are paired off. If n is odd there must be therefore an invariant endpoint under symmetry. This endpoint belongs to the involutory segment, and thus for n odd there exists only the class V_n.

11. Now let first $n = 2\nu$ be even. We shall establish among all the T_n^* a one-to-one correspondence between all the H_n and the V_n. We can change a T_n^* of class H_n into one (and only one) of class V_n by a process of "crossing-over". For this purpose we keep in a tree of class H_n the 4 nodes P_1, P_2, $\bar{P}_1$, $\bar{P}_2$ nearest to, but not on the involutory segment fixed. These 4 points are the endpoints of a subtree H_4. This subtree we replace by a subtree V_4 again with the endpoints P_1, $\bar{P}_1$, P_2, $\bar{P}_2$, as indicated in Figure 7. If and only if the tree subjected to the "crossing-over" process is of a type $T_n^{*''}$ then the two trees H_n and V_n related by the process will represent the same type (after rotation through 90°). Using the notation concerning the cardinal number of classes of types we can thus state

$$g^*(n) = |T^*(n)| \tag{11.1}$$
$$|H_{2\nu}| = |V_{2\nu}|,$$

$$|T_{2\nu}^*| = |H_{2\nu}| + |V_{2\nu}| - |T_{2\nu}^{*''}| \tag{11.2}$$
$$= 2\,|H_{2\nu}| - |T_{2\nu}^{*''}|,$$

since the class $(T_{2\nu}^{*''})$ is counted in $(H_{2\nu})$ as well as in $(V_{2\nu})$.

We take now $n = 2\nu + 1$ as odd. We have seen that a T_n^* for n odd must be of the sort V_n, with the involutory segment in the symmetry axis.

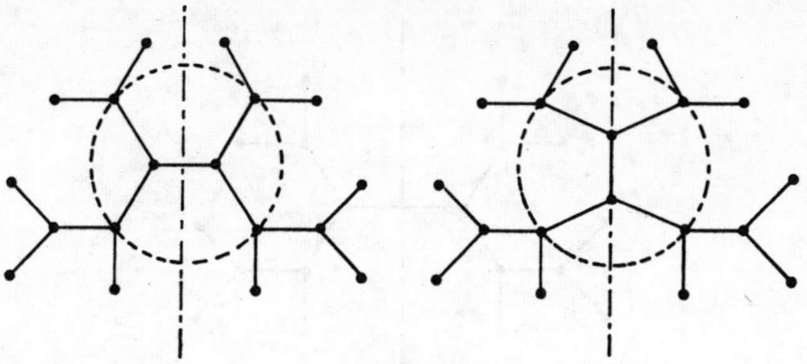

FIGURE 7

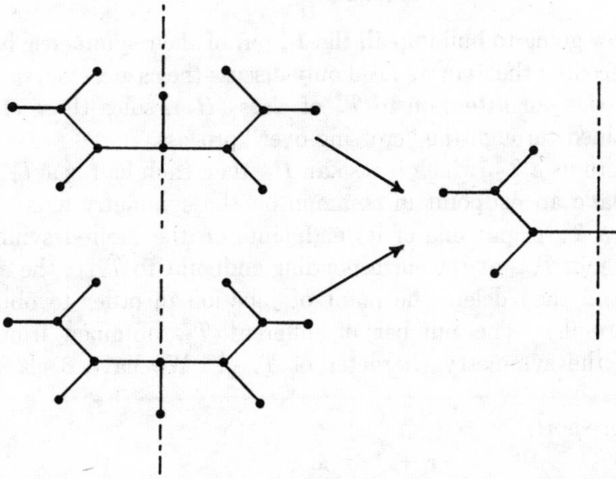

FIGURE 8

We take now this involutory segment away, with its two endpoints and retain a tree T_{n-1}^* of class H_{n-1}. (See Figure 8.) The same T_{n-1}^* is obtained whether the involutory element points "up" or "down". In general, therefore *two* trees T_n^* correspond to one T_{n-1}^*. However, if this T_{n-1}^* has also a second mirror symmetry, i.e. is of the class $(T_{n-1}^{*\prime\prime})$, then the two T_n^* just mentioned are identical, going over into each other by a rotation through 180°. The enumeration is thus

$$(11.3) \qquad | T_{2\nu+1}^{*} | = 2 | H_{2\nu} | - | T_{2\nu}^{*\prime\prime} |$$

since the $T_{n-1}^{*\prime\prime}$ are among the $H_{2\nu}$ and should be counted only once as we just stated.

Comparison of (11.2) and (11.3) leads to

THEOREM 5. *The number of mirror-symmetric types T_n^* is the same for $n = 2\nu$ and $n = 2\nu + 1$, that is,*

$$(11.4) \qquad g^*(2\nu) = g^*(2\nu + 1), \qquad \nu \geqq 2.$$

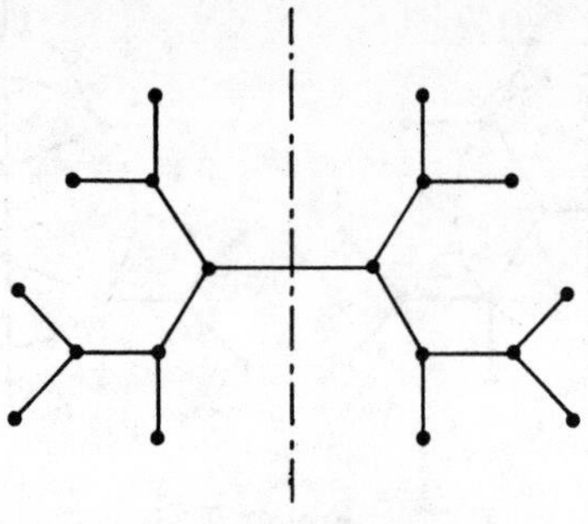

FIGURE 9

12. We are now going to build up all the T_n^* out of their symmetric halves. In view of the preceding theorem we need only discuss the case n even, $n = 2\nu$. Moreover we restrict our attention to T_n^* of class (H_n), since those of class (V_n) can be obtained through the "crossing-over" process.

If we have before us a $T_{2\nu}^*$ which is also an $H_{2\nu}$ then each half is a $T_{\nu+1}$ and the two halves have an endpoint in common on the symmetry axis. Conversely, we take a $T_{\nu+1}$, put one of its endpoints on the desired symmetry axis for $T_{2\nu}^*$, and join $T_{\nu+1}$ at the corresponding endpoint to $\bar{T}_{\nu+1}$, the mirror image of $T_{\nu+1}$, and then delete the point of junction in order to obtain a $T_{2\nu}^*$. (See Figure 9.) The number of different $T_{2\nu}^*$ obtained from one $T_{\nu+1}$ depends on the symmetry character of $T_{\nu+1}$. We have 6 classes of $T_{\nu+1}$ to consider.

Let us write, for short,

$$\nu + 1 = \mu.$$

A T_μ can either be a $\tilde{T}_\mu$ or T_μ^*, i.e. a tree without or with mirror symmetry. These trees can further be distinguished according to their rotational symmetry, so that we have to consider the six kinds $\tilde{T}'$, $\tilde{T}''$, $\tilde{T}'''$, $T^{*\prime}$, $T^{*\prime\prime}$, $T^{*\prime\prime\prime}$.

We firstly carry out the process of constructing a T_n^* out of a $\tilde{T}_\mu'$. Each of the μ ends of $\tilde{T}_\mu'$ can be joined to the corresponding end of $\bar{T}_\mu'$, the mirror image of $\tilde{T}_\mu'$. This will yield μ different T_n^*. However the $\bar{T}_\mu'$ (which is by definition different from $\tilde{T}_\mu'$) will also produce the same μ different T_n^*, which in our construction belong all to the class (H_n). We apply now the crossing-over process to each of the constructred H_n and obtain μ trees V_n so that the 2 trees $\tilde{T}_\mu'$, $\bar{T}_\mu'$ produce 2μ different T_n^*, $n = 2\nu = 2(\mu - 1)$.

Now let us take a $\tilde{T}_\mu''$. The rotational symmetry of order 2 in this case demands $2|\mu$. We can make use only of $\mu/2$ ends of $\tilde{T}_\mu''$ in order to obtain different T_n^*. Again $\bar{T}_\mu''$ will not produce new T_n^*, but only such which are obtained through rotation by 180° from the previous ones. The process of crossing-over will produce another set of T_n^*, of the sort V_n, so that we obtain altogether $2(\mu/2) = \mu$ different T_n^* from a pair $\tilde{T}_\mu^*$, $\bar{T}_\mu''$.

Similarly we obtain from a pair $\tilde{T}_\mu'''$, $\bar{T}_\mu'''$ by reflexion and crossing-over $2(\mu/3)$ different T_n^*.

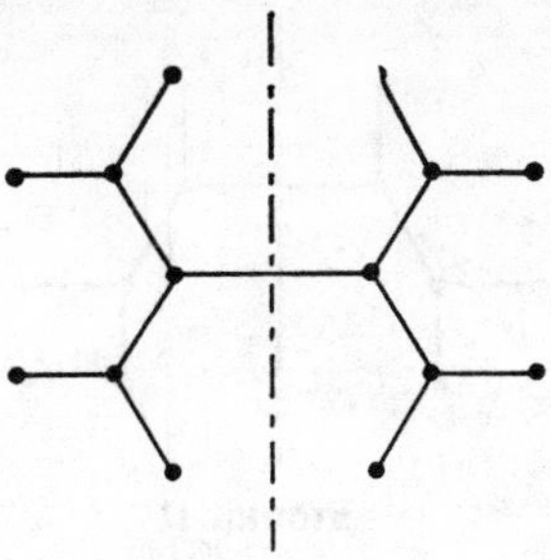

FIGURE 10

We have thus so far from all the $\tilde{T}_\mu$ constructed

$$\mu\cdot\{|\tilde{T}'_\mu| + \tfrac{1}{2}|\tilde{T}''_\mu| + \tfrac{1}{3}|\tilde{T}'''_\mu|\}$$

different T^*_n.

13. We come now to the class (T^*_μ) and its subclasses $(T^{*\prime}_\mu)$, $(T^{*\prime\prime}_\mu)$, $(T^{*\prime\prime\prime}_\mu)$.

In dealing with a $T^{*\prime}_\mu$ we have to distinguish μ even and odd. If, firstly, μ is even, then the endsegments and endpoints of $T^{*\prime}_\mu$ are paired off through the mirror symmetry. We take two copies of $T^{*\prime}_\mu$ and attach the partners of a pair of endsegments to each other, deleting the node in common. In this way we obtain from the use of each of the $\mu/2$ pairs of symmetric endsegments a symmetric T^*_n with $n = 2\mu - 2$ of the sort H_n. Using then the process of crossing-over we get again $\mu/2$ other T^*_n, (of the sort V_n), together thus μ different T^*_n.

If, however, μ is odd, then we have $(\mu - 1)/2$ pairs of corresponding endsegments and one unmatched (involutory) endsegment. Through the process just described we construct $(\mu - 1)/2$ T^*_n of the sort H_n and then also $(\mu - 1)/2$ further T^*_n of the sort V_n. The single unmatched segment of $T^{*\prime}_\mu$ lies on its symmetry axis. If we connect it to the corresponding $T^{*\prime}_\mu$ in mirrored position (see Figure 10) we obtain a T^*_n which is of the sort H_n, but also a V_n, by rotation through 90°. Again we have $2(\mu - 1)/2 + 1 = \mu$ different T^*_n gained from one $T^{*\prime}_\mu$.

If we have a $T^{*\prime\prime}_\mu$ to start with, then we know that $4|\mu$. Again we pair off the endsegments by mirror symmetry. Because of R.S. of order 2 each such pair goes over into another one by rotation through 180°. We use therefore only $\mu/4$ pairs for establishing a T^*_n by connecting in mirror fashion one end of $T^{*\prime\prime}_\mu$ with the corresponding end of another copy of it. (See Figure 11.) In this way we produce $\mu/4$ different T^*_n, all of the sort H_n. By means of crossing-over we get just as many different types T^*_n of sort V_n, thus $\mu/2$ altogether.

A $T^{*\prime\prime\prime}_\mu$ requires $3|\mu$ because of rotational symmetry of order 3. If now μ is *odd*, then the mirror symmetry expressed by the asterisk * requires that there

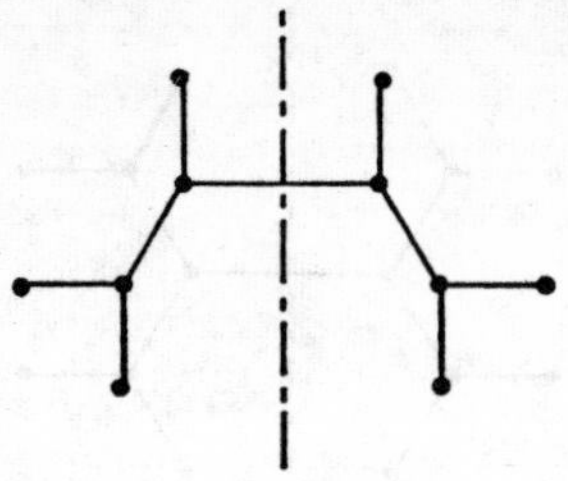

FIGURE 11

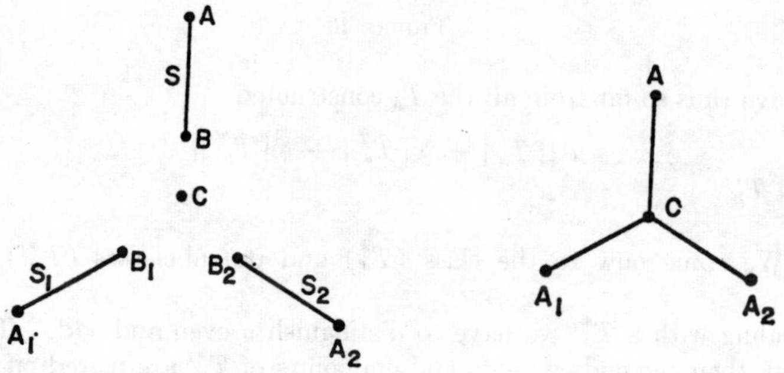

FIGURE 12

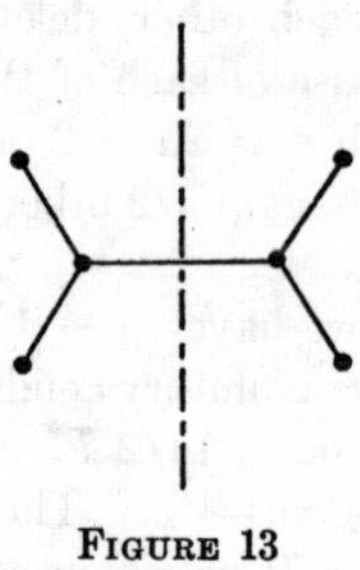

FIGURE 13

is an endsegment $s = AB$ in the symmetry axis with endpoints fixed under reflexion. Rotation around the central node C through 120° and 240° shows that then $T_\mu^{*\prime\prime\prime}$ possesses 2 other such endsegments $s = A_1 B_1$ and $s_2 = A_2 B_2$. (See Figure 12.) Since they are part of a connected graph, there must be a connection between s and s_1. The connection by segments BC and CB_1 would not be possible since it would create nodes of even order, which is excluded. Any other connection between s and s_1 would through rotation imply a connection between s_1 and s_2 and the corresponding one between s_2 and s and would thus mean the existence of a cycle in the tree $T_\mu^{*\prime\prime\prime}$ which is a contradiction. The only remaining possibility is that $B = B_1 = B_2 = C$ and that the whole $T_\mu^{*\prime\prime\prime}$ is $T_3^{*\prime\prime\prime}$, $\mu = 3$. This $T_3^{*\prime\prime\prime}$ produces only *one* T_4^* through reflexion, viz., the one shown in Figure 13. For $\mu > 3$ we can have a $T_\mu^{*\prime\prime\prime}$ only with $6|\mu$.

Through mirror symmetry we pair off the endsegments of $T_\mu^{*'''}$ into $\mu/2$ pairs. Through rotational symmetry each such pair goes over to 2 further equivalent pairs. We have thus $\mu/6$ pairs of endsegments which we use through the process of joining in reflexion to produce each a T_n^*. This gives $\mu/6$ different T^* of sort H_n. Crossing-over gives another set of $\mu/6$ different T_n^*, so that each $T_\mu^{*'''}$ produces $\mu/3$ types T_n^*.

If we collect the results of **12** and **13** we see that all $T_\mu = T_{\nu+1}$ produce the following number of T_n^*, $n = 2\nu$:

$$g^*(n) = \mu\{|T'_\mu| + \tfrac{1}{2}|T''_\mu| + \tfrac{1}{3}|T'''_\mu|\} + \mu\{|T^{*'}_\mu| + \tfrac{1}{2}|T^{*''}_\mu| + \tfrac{1}{3}|T^{*'''}_\mu|\}. \tag{13.1}$$

We have obtained all T_n^*, since the process is reversible and leads from a T_n^* to exactly one T_μ with one of its endsegments used for the process of reflexion and attachment.

Since the $\tilde{T}'_\mu, \cdots T^{*'''}_\mu$ are mutually exclusive we know that

$$g_1(\mu) = |\tilde{T}'_\mu| + |T^{*'}_\mu|$$

$$g_2(\mu) = |\tilde{T}''_\mu| + |T^{*''}_\mu|$$

$$g_3(\mu) = |\tilde{T}'''_\mu| + |T^{*'''}_\mu|.$$

Formula (13.1) implies thus, for $n = 2\nu$, $\mu = \nu + 1$:

$$g^*(2\nu) = (\nu + 1)\{g_1(\nu + 1) + \tfrac{1}{2}g_2(\nu + 1) + \tfrac{1}{3}g_3(\nu + 1)\}$$
$$= (\nu + 1)G(\nu + 1).$$

In view of Theorem 5 we have thus proved the important

THEOREM 6. *If $n \geqq 4$*

$$g^*(n) = ([n/2] + 1)G([n/2] + 1). \tag{13.2}$$

If we insert this result together with (1.2) in (9.1), we obtain for $n \geqq 4$ the statement of Theorem 2, which was our goal. For $n = 3$ the result of Theorem 2 is immediate. For a few low values of $F = n + 1$, formula (1.4) yields the following table:

F	4	5	6	7	8	9	10	11	12	13	14	15	16
$\chi_0(F)$	1	1	1	3	4	12	27	82	228	733	2282	7528	24834

for which the values up to $F = 11$ have already been found by Kirkman [4] and Brückner [1].[5] Theorem 2 implies Pólya's result (1.1) and proves also Pólya's conjecture

[5] Brückner [1a] gave also values up to $F = 16$, of which, however, those for 13, 15, and 16 are in error. In particular his value for $F = 16$ is too small by 522 and violates even Pólya's inequality (1.1).

$$\chi_0(F) \sim \frac{1}{2(F-1)(F-2)}\binom{2F-6}{F-3}.$$

III. On based polyhedra with only two triangles

14. The enumeration of certain very special types of based polyhedra, viz. of those with only two triangles has been carried out by O. Hermes (1894), see [1, p. 97] and [8, p. 55].

We can derive his formula easily from the fact that these polyhedra have a ridge consisting of a chain of $r = n - 3$ segments and $n - 2$ nodes, of which the two terminal ones are already occupied by the legs of the two triangles. We have thus only the choices to attach the missing $n - 4$ endsegments "up" or "down" at the $n - 4$ free nodes.

We first count the number $s(n)$ of different types in the strict sense. Let us take $n > 4$. We have 2^{n-4} possibilities to attach the endsegments. If n is odd all these choices are matched off in pairs through rotation by 180°, so that we obtain 2^{n-5} different polyhedra types (all belonging to T'_n).

If however n is even, there are among the 2^{n-4} graphs those in number $2^{(n-4)/2}$ which are not matched by another one through rotation by 180° but go over into themselves, belonging thus to T''_n. We have then

$$\tfrac{1}{2}(2^{n-4} - 2^{(n-4)/2}) + 2^{(n-4)/2} = 2^{n-5} + 2^{(n-6)/2}.$$

Since this last expression gives the correct value also when $n = 4$, we have proved

THEOREM 7. *If $n \geqq 4$, the number of types of based polyhedra with $F = n + 1$ faces and only 2 triangles, the types counted in the strict sense, is*

$$(14.1) \qquad \sigma(F) = \sigma(n+1) = s(n) = \begin{cases} 2^{n-5} & n \text{ odd} \\ 2^{n-5} + 2^{(n-6)/2} & n \text{ even.} \end{cases}$$

Let us now go over to the enumeration in the classical sense of Kirkman and Steinitz, where orientation does not have to be preserved.

We proceed as in the general case treated in Section II. Let $\bar{s}(n)$ be the number of types (in the strict sense) without mirror symmetry and $s^*(n)$ the number of those with mirror symmetry. Obviously we have

$$s(n) = \bar{s}(n) + s^*(n).$$

Steinitz now counts the number of types in the wider sense:

$$(14.2) \qquad t(n) = \tfrac{1}{2}\bar{s}(n) + s^*(n) = \tfrac{1}{2}(s(n) + s^*(n)).$$

We have thus still to determine $s^*(n)$. This time we take $n > 5$ (in order to avoid the possibility that the set of choices might be void in the following arguments).

Let first n be even. In view of the mirror symmetry we have only to make the choices "up" or "down" only for one half of the available free nodes

That gives $2^{(n-4)/2}$ possibilities. But if we interchange all choices "up" and "down" by the opposite ones we arrive at the same T_n^*, only turned through 180°. Therefore in this case

$$s^*(n) = 2^{(n-6/2}, \qquad n \text{ even.}$$

If n is odd we choose the middle endsegment as "up". Then there are $(n - 5)/2$ endsegments to be chosen as "up" or "down", which gives $2^{(n-5)/2}$ possibilities. This time rotation cannot be used. We have thus here

$$s^*(n) = 2^{(n-5)/2}, \qquad n \text{ odd.}$$

If we insert now these results and the statement (14.1) into (14.2) we have completed the proof of the theorem of O. Hermes:

THEOREM 8. *If $n \geqq 5$, the number of types in the wider sense of based polyhedra with $F = n + 1$ faces and only 2 triangles is*

$$t(F) = t(n + 1) = t(n) = 2^{n-6} + 2^{[(n-6)/2]}. \tag{14.3}$$

Of course the case $n = 5$ can be verified directly.

The formula found in [1] and [8] is expressed in F and distinguishes F even and odd, but is equivalent to (14.3).

BIBLIOGRAPHY

1. M. BRÜCKNER, *Vielecke und Vielflache*, Leipzig, 1900.
1a. ——— *Über die Anzahl $\psi(n)$ der allgemeinen Vielflache*, Atti del Congresso Internazionale dei Matematici, tomo IV, Bologna, 1928.
2. A. CAYLEY, *On the analytical forms called trees* (1859), Collected Mathematical Papers, vol. IV, pp. 112–115, Cambridge, 1891.
3. V. EBERHARD, *Zur Morphologie der Polyeder*, Leipzig, 1891.
4. TH. P. KIRKMAN, *On the enumeration of x-edra having triedral summits and an $(x - 1)$-gonal base*, Phil. Trans. Royal Soc. London, 1856, pp. 399–411.
5. ———, *On the partitions of the R-pyramid, being the first class of R-gonous X-edra*, Phil. Trans. Soc. London 1858, pp. 145–161.
6. G. PÓLYA, *Kombinatorische Anzahlbestimmungen für Gruppen, Graphen und chemische Verbindungen*, Acta Math. vol. 68 (1937), pp. 145–253.
7. ———, *Induction and analogy in mathematics*, Princeton, 1954, (vol. 1 of *Mathematics and plausible reasoning*).
8. E. STEINITZ, *Polyeder und Raumeinteilungen*, Enzyklopädie d. math. Wiss. vol, 3, part 1.2, B, 1–139, Leipzig 1922.
9. ———, *Vorlesungen über die Theorie der Polyeder*, edited by H. Rademacher, Berlin 1934.

THE ROCKEFELLER INSTITUTE
NEW YORK, NEW YORK

See also
J.W. Moon and L. Moser, Triangular dissections of N-gons,
Canad. Math. Bull, vol 6, no 2, May 1963
(received 9th November 1965)

This paper has been reviewed in the Z, vol. 151 (1968), p. 263, and in the MR, vol. 31 (1966), #3927.

Page 368, line 7.
Author's correction; read (5.4) instead of (5.3).

Page 369, marginal note.
Author's correction; the binomial coefficient should read $\binom{2n-4}{n-2}$.

Page 370, sixth display line.
Author's correction; read ϕ' instead of ϕ.

Page 371, third display line.
Read:

$$\phi(x) = \frac{1}{2x}\left(-2x+1 - \sum_{l=0}^{\infty} \binom{1/2}{l} (-4x)^l \right).$$

Page 373, second display line from bottom.
Author's correction; the left-hand member is $|T^*_{2\nu}|$.

Page 374, equation (11.3).
Author's correction; the left-hand member is $|T^*_{2\nu+1}|$.

Page 380, first display line.
Read $2^{(n-6)/2}$, instead of $2^{(n-6/2}$.

Page 380, marginal note (bottom of page).
Author's note.

A Packing Problem for Parallelepipeds

HANS RADEMACHER,* ROBERT DICKSON,† AND MORRIS PLOTKIN**

The Rockefeller University,
New York, New York

Communicated by S. Ulam

ABSTRACT

This article discusses finite sets of disjoint, closed, axis-parallel parallelepipeds ("boxes") in 3-space which have the property that their projections on the coordinate planes each fill a rectangle. Omitting the trivial case of just one box, then it is proved that 12 boxes are necessary and sufficient to form such a set. The minimal solution is described in detail.

1. Four axis-parallel, disjoint rectangles can be placed in such a manner (Figure 1) that their orthogonal projection on the x-axis as well as on the y-axis consists each of a single segment. As can be seen easily, four is the minimal number of rectangles fulfilling this requirement, if we leave out as uninteresting the case of one rectangle alone. We shall discuss the situation in more detail in § 2.

A number of years ago J. A. Clarkson proposed to the senior author the problem of the generalization of that configuration to 3-space: How many axis-parallel disjoint parallelepipeds (considered as closed sets, called boxes for short) are necessary and sufficient for a configuration P (of more than one box) which has on each coordinate plane a full rectangle as projection? Because of the availability of affine transfor-

* The Rockefeller University, New York, New York.
† The University of Vermont, Burlington, Vermont.
** The University of Pennsylvania, Philadelphia, Pennsylvania.

mations we can specify that these projections should be full squares. We shall say that such an arrangement P of boxes possesses the "shadow property."

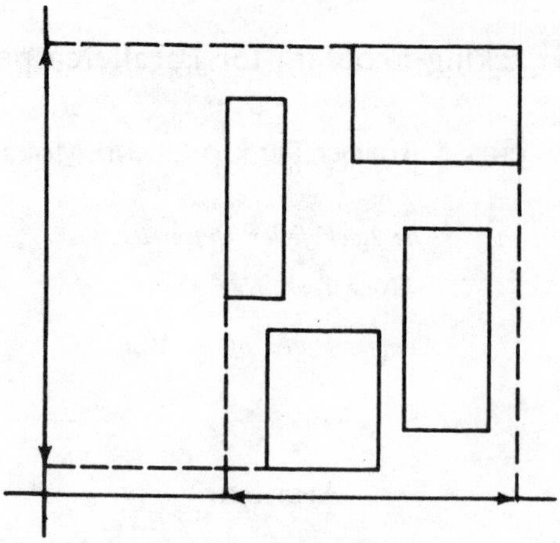

FIGURE 1

The problem seems to have come up first in a conversation between J. A. Clarkson and J. W. T. Youngs.

We prove here now the

THEOREM: *For an arrangement P of boxes in* 3-*space* 12 *boxes are necessary and sufficient.*

2. As a preparation for $n = 3$ we discuss in detail the simple case of dimension $n = 2$. Suppose we have an arrangement P of finitely many disjoint axis-parallel rectangles which has the shadow property, i.e., it has an orthogonal projection of a full interval on each axis. In view of a possible affine transformation we can, without loss of generality, take these shadows as of equal length. The arrangement P is then enclosed in a tightly fitting square Q.

We now enlarge each rectangle, wherever possible, so that it sits "flush" in the square Q, provided that such an enlargement does not lead to any interference of the rectangles, which have to remain disjoint. In Figure 2b the rectangles given in 2a are made "flush," the added areas being shown by hatching.

In this flush configuration P^*, which of course has the same number of rectangles as before, no rectangle can touch two opposite sides of Q, because otherwise a light beam could be shone along one side of it through the interior of Q, and this light beam would interrupt the shadow on the coordinate axis orthogonal to it.

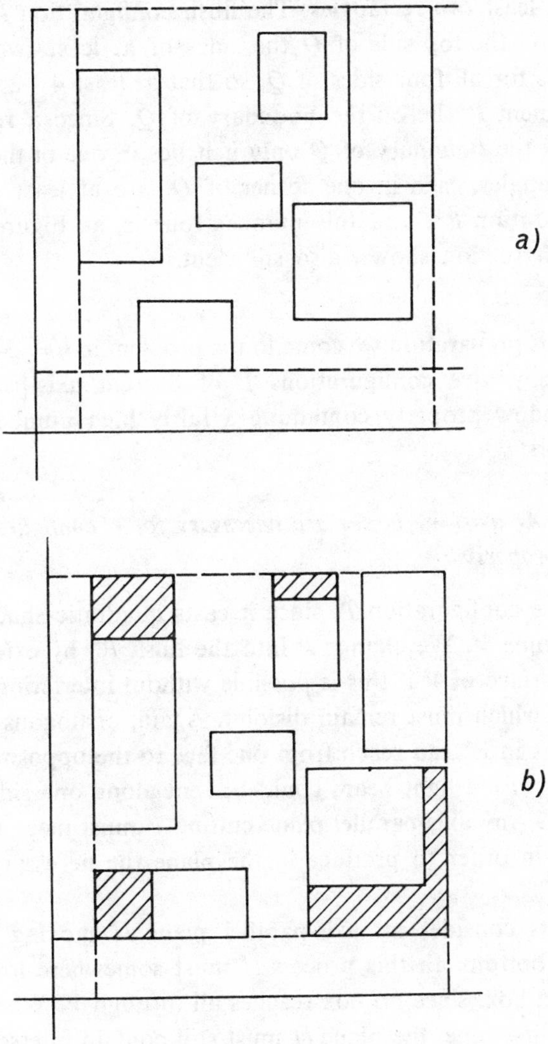

FIGURE 2

Now let us consider a line Λ parallel to the x-axis across Q. We lower Λ from the top of Q continuously to the bottom. It must always have an intersection with at least one rectangle. Now, since no rectangle can reach from top to bottom of Q, the line Λ in moving down must somewhere lose an intersection with one of the rectangles. When that happens for the first time, it must still intersect another rectangle (because of the shadow property). Therefore Λ started at the top with intersecting at least *two* rectangles. The flush configuration P^* must therefore have on the top side of Q the sides of at least two rectangles. But this goes for all four sides of Q, so that at least $4 \cdot 2 = 8$ sides of the arrangement P^* lie on the boundary of Q. Since a rectangle has two sides on the boundary of Q only if it lies in one of the corners of Q, four rectangles, each in one corner of Q, are at least needed in a flush configuration P^*, and this number four is, as Figure 1 even in the non-flush version shows, also sufficient.

3. After this preparation we come to the problem in $n = 3$ dimensions. One can easily give configurations P of disjoint axis-parallel boxes with the shadow property containing a fairly high number of boxes. We prove first

LEMMA 1. *At least* 11 *boxes are necessary for a configuration P with the shadow property.*

PROOF: The configuration P, since it casts quadratic shadows, is enclosed in a cube V. We change P into the flush P^* by extending each box to the surface of V if this is possible without interfering with other boxes, all of which must remain disjoint. Again, analogous to the case $n = 2$, no box in P^* can reach from one face to the opposite face of V, because otherwise a light beam could be sent along one side of such a box across V. Any axis-parallel plane cutting V must meet at least four boxes of P^* in order to produce in the plane the necessary condition for $n = 2$.

Now let us consider an axis-parallel plane Π moving through V from top to bottom. In this process Π must somewhere lose an intersection with a box, since no box reaches all through V. When this happens for the first time, the plane Π must still contain intersections with at least four boxes of P^*, so it must have begun with meeting at least five boxes. That means P^* must have at least five boxes touching the top face of V with one of its sides. But this goes for all six faces of V. Therefore P^* must have at least $6 \cdot 5 = 30$ sides on the surface of V.

A box has three sides on the surface of V if it lies in a corner of V, two sides if it has an edge in common with V, one side if it has a side in the interior of one side of V, and no side if it lies wholly in the interior of V. Let us call the respective numbers of boxes x_3, x_2, x_1, x_0. Then we have just seen that

$$3x_3 + 2x_2 + x_1 \geqq 30 \tag{1}$$

is a necessary condition for the shadow property. Clearly $x_3 \leqq 8$ since V has eight vertices. We are looking for a solution of the diophantine inequality for which

$$s = x_3 + x_2 + x_1 + x_0 = \text{minimum}.$$

This minimum condition demands the greatest possible value for x_3:

$$x_3 = 8. \tag{2a}$$

Then we retain

$$2x_2 + x_1 \geqq 6, \qquad x_2 + x_1 \text{ minimal},$$

which leaves only

$$x_2 = 3, \quad x_1 = 0, \quad x_0 = 0. \tag{2b}$$

and altogether

$$s = x_3 + x_2 + x_1 + x_0 = 11$$

as the minimal solution.

4. Now the solution (2a), (2b) held together with the condition that each face of V should show at least five sides of P^* leads to a configuration which is, up to rotations and reflections, unique, if we ignore the sizes of the participating boxes in P^*. Since the corners of V are occupied by eight boxes, the three boxes along the edges of V must contribute each one side to each face of V. That requires that the three occupied edges of V must be mutually orthogonal and must pairwise have no point in common. Figure 3 indicates their position with the names X, Y, Z, according to the direction of the edges which they occupy. The corners A, B, C, D on top and A', B', C', D' at the bottom are occupied by boxes

which are not shown in Figure 3, and to which we also give the names $A, B, ..., D'$.

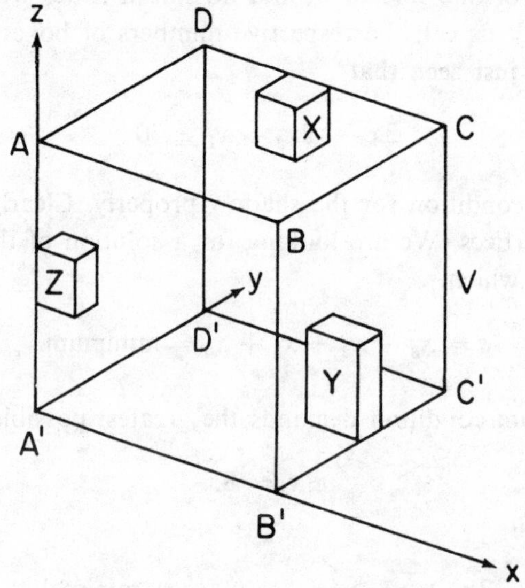

FIGURE 3

5. We prove now:

LEMMA 2. *A configuration* P^* *with* $s = 11$ *(described above) cannot possess the shadow property.*

PROOF: We can speak of an upper layer of boxes, consisting in counterclockwise order of A, B, C, X, D and of a lower layer, namely the boxes A', B', Y, C', D'. In between is box Z, which has no side in common with the top or with the bottom of V. Figures 4a-b show these arrangements, separately, seen from above. Now, Z is below A, but it cannot be below any of the other upper boxes B, C, X, D, because this would lead to a leak of light across V. Suppose, e.g., Z were partly below X, then a light beam could run in the y-direction above Z and below X unobstructed from the front to the rear, i.e., from face $A'B'BA$ to face $D'C'CD$. If Z were partly below B a light beam could run in the x-direction above Z and below B from right to left, i.e., from face $A'D'DA$ to face $B'C'CB$. And similarly in the other cases.

For the same reason Z, which is above A', cannot extend above any of the other boxes B', Y, C', D' of the lower layer. (It is this "inefficiency" of the intermediate boxes Z (and X, Y) which is the ultimate reason for the failure of the arrangement with $s = 11$.) Because Z does not overlap any of B', Y, C', D' seen from above there will be a gap between

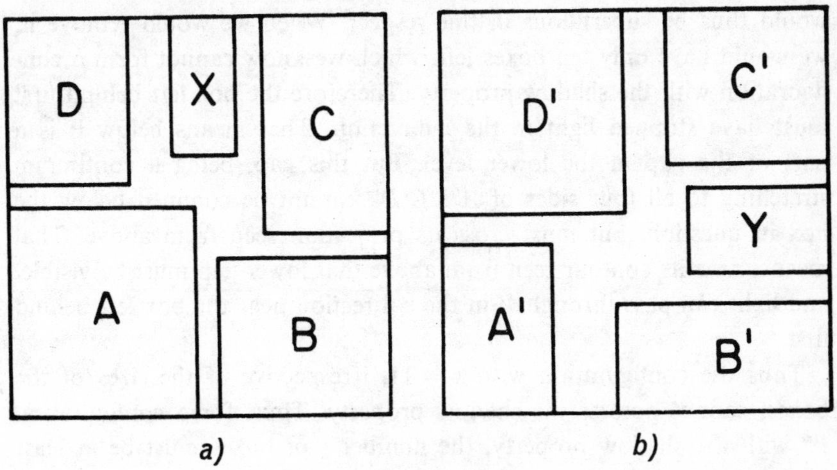

FIGURE 4

these projections which is a continuum and has points on all four sides of the square $A'B'C'D'$ (see Figure 5, where the gap is marked by hatching).

Now let a test plane Π parallel to the xy-plane move from the top of V to the bottom. It will in the beginning intersect the boxes A, B, C,

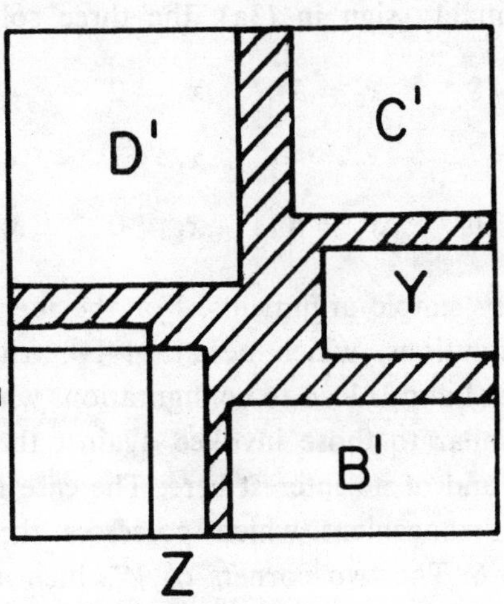

FIGURE 5

X, D. In moving down it will leave one of them behind. Let this have happened for the first time, and before Π meets any of the lower boxes. Then still no light can pass through V in the x- or the y-direction. The box left behind cannot have helped to stop light in those directions and would thus be superfluous in this respect. When we would remove it, we would have only ten boxes left, which we know cannot form a configuration with the shadow property. Therefore the box left behind first must have stopped light in the z-direction. That means below it is a part of the gap in the lower level. But this gap, being a continuum stretching to all four sides of $A'B'C'D'$, cannot be confined below the box in question, but must cross its projection seen from above. That means, near its contour seen from above that lower gap must be visible, and light can pass through V in the z-direction near the box left behind first.

Thus the configuration with $s = 11$, irrespective of the sizes of the boxes, cannot possess the shadow property. Thus, for a configuration P^* with the shadow property, the number s of boxes must be at least 'welve.

6. Now the diophantine conditions

$$3x_3 + 2x_2 + x_1 \geqq 30 \tag{3a}$$

$$s = x_3 + x_2 + x_1 + x_0 = 12, \quad x_3 \leqq 8 \tag{3b}$$

have, for the equality sign in (3a), the three solutions

I.	$x_3 = 8$	$x_2 = 2$	$x_1 = 2$	$x_0 = 0$
II.	$x_3 = 7$	$x_2 = 4$	$x_1 = 1$	$x_0 = 0$
III.	$x_3 = 6$	$x_2 = 6$	$x_1 = 0$	$x_0 = 0,$

as can be seen by simple arguments. For the inequality in (3a) there are four more solutions, which need not be written down.

The possibilities I and II lead to configurations which can be dismissed by arguments similar to those invoked against the case $s = 11$. They are cumbersome and of no interest here. The case III, however, can be realized by an arrangement which possesses the shadow property, shown in Figure 6. The two corners of V which are not occupied by boxes (x_3 being only 6) are the ends of a spatial diagonal, around which

P^* has a rotational symmetry of order 3. The arrangement P^* is moreover symmetric with respect to the center of V.

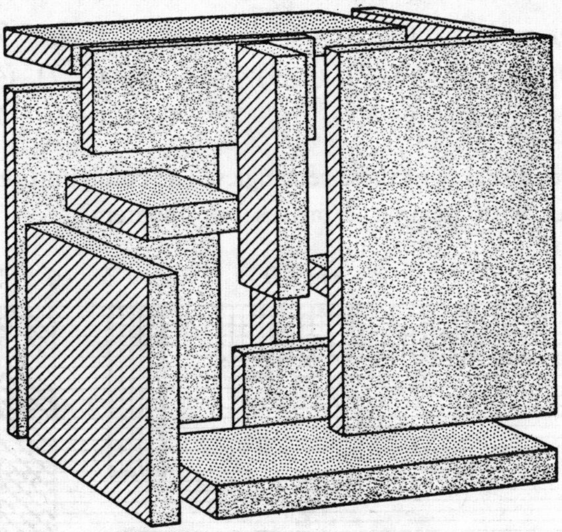

FIGURE 6

Since the boxes are axis-parallel it suffices to give for their determination only the coordinates of two opposite corners. Let V have the vertices $\pm 7, \pm 7, \pm 7$. We take as the distinguished spatial diagonal between unoccupied corners that one which runs from $-7, -7, -7$ to $+7, +7, +7$. The center is at 0. Since a rotation about the distinguished diagonal is expressed by a cyclic permutation of the xyz-coordinates, we obtain from one box two more boxes by cyclic permutation of the coordinates. Another set of three boxes is obtained by reflection at the point 0.

The whole configuration P consists now of the boxes of opposite corners

$$(7, -7, 7), \qquad (-5, 1, 6)$$

and

$$(7, -3, 7), \qquad (-1, 2, 4)$$

together with their cyclic permutations and their reflections at 0.

In order to check that these twelve boxes together have on each coordinate plane a full square as projection, it will suffice to suppress, e.g., the z-coordinate in the list of the twelve boxes. The two remaining coordinates determine opposite corners of their rectangular projections. A list of these rectangles follows:

(7, −7), (−5, 1);	(−7, 7), (1, 6);	(7, 7), (6, −5)
(−7, 7), (5, −1);	(7, −7), (−1, −6);	(−7, −7), (−6, 5)
(7, −3), (0, 2);	(3, 7), (2, 4);	(7, 7), (4, 0)
(−7, −3), (0, −2);	(−3, −7) (−2, −4);	(−7, −7), (−4, 0)

Figure 7 gives a survey of these rectangles. The configuration *P* described here together with Lemma 2 proves our theorem.

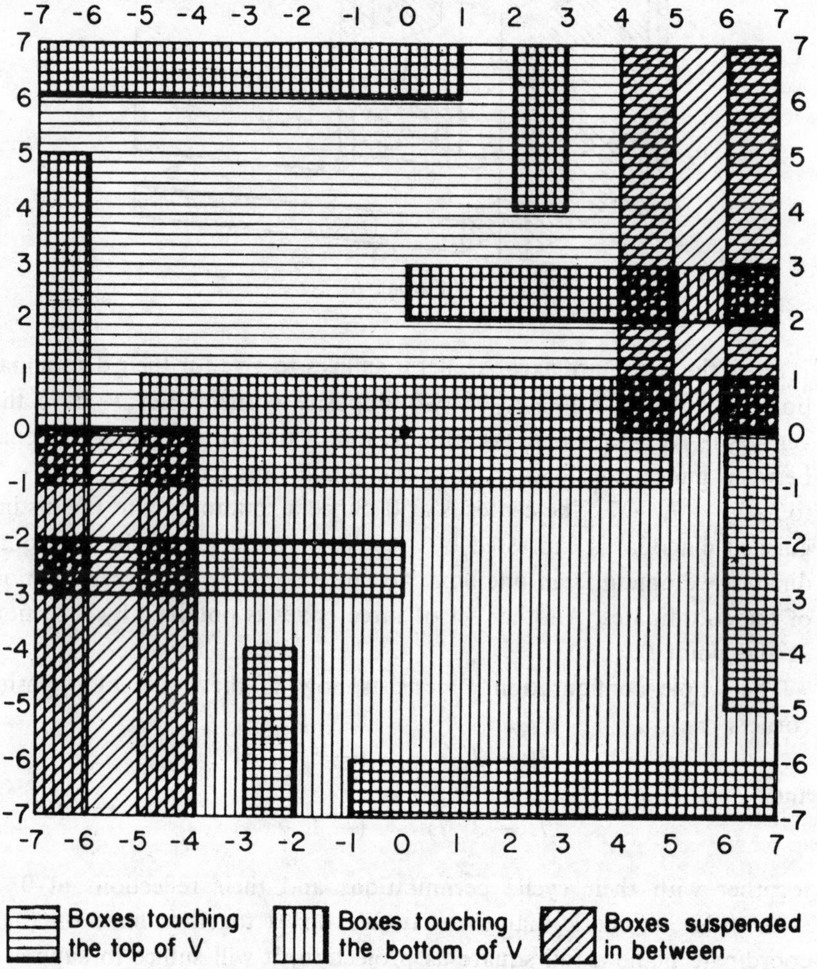

FIGURE 7

7. A few indications can be given about the situation in four and higher dimensions. In the case of $n = 4$ we enlarge the 4-dimensional boxes again until they are flush with the surface of the enclosing 4-dimensional cube W. A 3-dimensional hyperplane Σ orthogonal to one of the axes must always intersect at least twelve hyperboxes. But, since during the motion of Σ through W it must lose somewhere an intersection, it must have started with intersecting at least thirteen hyperboxes. The surface of W, consisting of eight 3-dimensional cubes, must therefore have contact with at least $8 \cdot 13 = 104$ sides of the hyperboxes. Let x_4 be the number of boxes having four sides in common with W, and correspondingly we define the numbers x_3, x_2, x_1, x_0. Thus we have the condition

$$4x_4 + 3x_3 + 2x_2 + x_1 \geqq 104, \tag{4a}$$

with

$$x_4 \leqq 16, \tag{4b}$$

the number of vertices of W. We look for the minimum of

$$t = x_4 + x_3 + x_2 + x_1 + x_0.$$

To reach this minimum we obviously have to make x_4 as large as possible:

$$x_4 = 16,$$

and are left with

$$3x_3 + 2x_2 + x_1 \geqq 40,$$

and the requirement to minimize

$$x_3 + x_2 + x_1 + x_0.$$

This leads to

$$x_3 = 14, \quad x_2 = x_1 = x_0 = 0,$$

$$x_3 = 13, \quad x_1 = 1, \quad x_2 = x_0 = 0,$$

$$x_3 = 12, \quad x_2 = 2, \quad x_1 = x_2 = 0$$

and thus to

$$x_3 + x_2 + x_1 + x_0 \geqq 14$$

and

$$t \geqq 30. \tag{5}$$

It has been conjectured that the true number of boxes needed is equal to the number of edges of the n-dimensional cube. In the case $n = 4$ we have thirty-two edges of the hypercube W. The conjecture is true for $n = 2$, 3, and in the case $n = 4$, as (5) shows, not implausible.

In the known cases $n = 2$, $n = 3$ it turns out that the boxes can be made arbitrarily thin, and thus each of them stops the light only in one direction. If one could show that this property characterizes the solutions for all n, it would follow that the number of boxes needed is at least the number of edges. For since no box can touch two parallel edges, at least one box would be needed to block rays parallel to and in the neighborhood of each edge.

This paper, written jointly with R. Dickson and M. Plotkin, has been reviewed in the Z, vol. 142 (1966), p. 202, and in the MR, vol. 33 (1967), #3192.

Page 12, third display line.
Read (7,3) instead of (7,–3).

Page 14, line 1.

A partial proof of this conjecture (namely that in any dimension n, $s = 2^{n-1}n$ (= number of edges of the n-dimensional cube)) was given by N. H. Bingham and J. M. Hammersley in "On a conjecture of Rademacher, Dickson and Plotkin," *Journal of Combinatorial Theory,* vol. 3 (1967), pp. 182-190. Specifically, it is proved there that in n dimensions,

$$2^{n-2}(n+4) - 2 \leq s \leq 2^{n-1}n,$$

and that the boxes may be taken arbitrarily thin. On account of the last paragraph of the present paper, it would follow that $s \geq 2^{n-1}n$, so that the present conjecture $s = 2^{n-1}n$ would follow. However, Bingham and Hammersley observe that the conclusion $s \geq 2^{n-1}n$ is warranted only if the boxes are of zero thickness and not necessarily if they are only known to be arbitrarily thin. It seems, therefore, that the present conjecture, while very likely true, is not yet proved.

Über die Transformation der Logarithmen der Thetafunktionen

CARL LUDWIG SIEGEL zu seinem siebzigsten Geburtstag gewidmet

HANS RADEMACHER

Es scheint mir wünschenswert zu sein, die Erläuterungen, die DEDEKIND zu einem Fragment von RIEMANN gegeben hat [1], noch in einer Richtung zu ergänzen. DEDEKIND hat Beweise für die Riemannschen Formeln nur insoweit gegeben, daß er die additiven Konstanten berechnet, die $\log\eta(\tau)$ unter Modulsubstitutionen aufnimmt, und dies auf die Logarithmen der Thetafunktionen angewandt, um RIEMANNs Resultate zu verifizieren (vgl. auch [2]). Weder RIEMANN noch DEDEKIND sind jedoch auf die Funktionen von τ eingegangen, in die sich z. B. $\log\vartheta_3(\tau)$ transformiert. Doch bemerkt DEDEKIND, daß sich hier ein besonderer Zugang zur Theorie der Transformation der Thetafunktionen eröffnet. Dies soll im folgenden etwas genauer verfolgt werden.

Wir haben mit $q=e^{\pi i\tau}$

(1) $$\eta(\tau)=q^{\frac{1}{12}}\prod_{n=}^{\infty}(1-q^{2n})$$

und

(2) $$\vartheta_3(\tau)=\Pi(1-q^{2n})(1+q^{2n-1})^2,$$

(3) $$\vartheta_4(\tau)=\Pi(1-q^{2n})(1-q^{2n-1})^2,$$

(4) $$\vartheta_2(\tau)=2q^{\frac{1}{4}}\Pi(1-q^{2n})(1+q^{2n})^2.$$

Die Eintragung von (1) in (2), (3), (4) liefert, nach Übergang zu den Logarithmen,

(2a) $$\log\vartheta_3(\tau)=5\log\eta(\tau)-2\log\eta\left(\frac{\tau}{2}\right)-2\log\eta(2\tau),$$

(3a) $$\log\vartheta_4(\tau)=\frac{\pi i}{12}+5\log\eta(\tau)-2\log\eta\left(\frac{\tau+1}{2}\right)-2\log\eta(2\tau),$$

(4a) $$\log\vartheta_2(\tau)=\log 2+\frac{\pi i}{12}+5\log\eta(\tau)-2\log\eta\left(\frac{\tau+1}{2}\right)-2\log\eta\left(\frac{\tau}{2}\right),$$

wo die Logarithmen mit ihrem Hauptwert genommen sind, so daß

$$\log\eta(\tau)=\frac{\pi i\tau}{12}+o(1),$$

wenn $\tau\to i\infty$ strebt.

Wir setzen

$$\tau = \frac{h+iz}{k}, \quad \tau' = \frac{h' + \dfrac{i}{z}}{k}$$

mit

$$\operatorname{Re} z > 0, \quad k > 0, \quad hh' \equiv -1 \,(\operatorname{mod} k).$$

Dann ist

$$\tau' = \frac{a\tau + b}{c\tau + d}$$

mit

$$(5) \qquad \begin{pmatrix} a & b \\ c & d \end{pmatrix} = \begin{pmatrix} h' & -\dfrac{hh'+1}{k} \\ k & -h \end{pmatrix}$$

Da $\vartheta_3(\tau)$ eine Modulform der Stufe 2 ist, haben wir die 6 Fälle

$$(6) \qquad \begin{pmatrix} a & b \\ c & d \end{pmatrix} \equiv \begin{pmatrix} 1 & 0 \\ 0 & 1 \end{pmatrix}, \begin{pmatrix} 1 & 1 \\ 0 & 1 \end{pmatrix}, \begin{pmatrix} 0 & 1 \\ 1 & 0 \end{pmatrix}, \begin{pmatrix} 1 & 1 \\ 1 & 0 \end{pmatrix}, \begin{pmatrix} 0 & 1 \\ 1 & 1 \end{pmatrix}, \begin{pmatrix} 1 & 0 \\ 1 & 1 \end{pmatrix} (\operatorname{mod} 2)$$

zu unterscheiden, die wir der Reihe nach als I, II, ..., VI bezeichnen wollen. Fall I charakterisiert die Hauptkongruenzuntergruppe modulo 2. Wir werden hier nur die Fälle I, IV, VI im einzelnen behandeln und für die übrigen, die analog verlaufen, nur die Resultate geben. Zugrunde liegt stets die Dedekindsche Transformationsformel für $\log \eta(\tau)$:

$$(7) \quad \log \eta\left(\frac{h+iz}{k}\right) = \log \eta\left(\frac{h' + \dfrac{i}{z}}{k}\right) - \frac{1}{2} \log z + \frac{\pi i}{12k}(h - h') - \pi i s(h, k),$$

wo

$$s(h, k) = \sum_{\mu \bmod k} \left(\left(\frac{\mu}{k}\right)\right) \left(\left(\frac{h\mu}{k}\right)\right)$$

die Dedekindsche Summe ist.

Fall I verlangt nach (6) und (5)

$$k \equiv 0, \quad h \equiv h' \equiv 1 \,(\operatorname{mod} 2), \quad hh' \equiv -1 \,(\operatorname{mod} 2k).$$

Wir erhalten dann nach (2a) und (7)

$$\log \vartheta_3\left(\frac{h+iz}{k}\right) = 5 \log \eta\left(\frac{h' + \dfrac{i}{z}}{k}\right) - 2 \log \eta\left(\frac{h' + \dfrac{i}{z}}{2k}\right) - 2 \log \eta\left(\frac{h' + \dfrac{i}{z}}{k/2}\right) -$$

$$- \frac{1}{2} \log z - \pi i \left(5 s(h, k) - 2 s(h, 2k) - 2 s\left(h, \frac{k}{2}\right)\right).$$

Ohne Einschränkung der Allgemeinheit können wir hier $h>0$ annehmen, da die rechte Seite offenbar nur von h modulo $2k$ abhängt. Dann haben wir

$$\log\vartheta_3\left(\frac{h+iz}{k}\right) = -\frac{1}{2}\log z + \log\vartheta_3\left(\frac{h'+\frac{i}{z}}{k}\right) - \pi i V(h,k)$$

mit

$$V(h,k) = 5s(h,k) - 2s(h,2k) - 2s\left(h,\frac{k}{2}\right),$$

was unter Anwendung der Reziprozitätsformel für die Dedekindschen Summen (siehe z. B. [2]) übergeht in

$$V(h,k) = -5s(k,h) + 2s(2k,h) + 2s\left(\frac{k}{2},h\right) - \frac{1}{4}. \tag{8a}$$

Wir definieren nun k' durch

$$kk' \equiv -1 \pmod{h}$$

und erhalten dann

$$\begin{aligned} V(h,k) &= 2(s(2k,h) - 2s(k,h)) - \left(s(k,h) - 2s\left(\frac{k}{2},h\right)\right) - \frac{1}{4} \\ &= -2S(k',h) + S(2k',h) - \frac{1}{4} \\ &= -S(k',h) + R(k',h) - \frac{1}{4} = -\frac{1}{2}d(k,h) - \frac{1}{4}, \end{aligned}$$

wo $d(k,h)$ den Überschuß der Anzahl der positiven über die Anzahl der negativen absolut kleinsten Reste von νk modulo h, $\nu = 1, \ldots, \frac{h-1}{2}$, bedeutet. (Siehe [2], Theorem 8; für den Vergleich ist zu bemerken, daß dort leider stets h' durch $hh' \equiv +1 \pmod{k}$ definiert ist, was an manchen Stellen eine Änderung des Vorzeichens für unseren jetzigen Zweck erfordert.)

Wir erhalten daher im Falle I

$$\log\vartheta_3\left(\frac{h+iz}{k}\right) = -\frac{1}{2}\log z + \log\vartheta_3\left(\frac{h'+\frac{i}{z}}{k}\right) + \frac{\pi i}{2}\left(d(k,h) + \frac{1}{2}\right). \tag{9}$$

Wenden wir uns nun dem Falle IV zu, wo

$$h \equiv 0, \quad k \equiv h' \equiv 1 \pmod{2}$$

ist. Dann ist überdies auch

$$2h \cdot \frac{h'+k}{2} \equiv -1 \pmod{k}.$$

Es folgt dann nach (2a) und (7)

$$\log\vartheta_3\left(\frac{h+iz}{k}\right)=5\log\eta\left(\frac{h+iz}{k}\right)-2\log\eta\left(\frac{\frac{h}{2}+\frac{iz}{2}}{k}\right)-2\log\eta\left(\frac{2h+2iz}{k}\right)$$

$$=5\log\eta\left(\frac{h'+\frac{i}{z}}{k}\right)-2\log\eta\left(\frac{2h'+\frac{2i}{z}}{k}\right)-$$

$$-2\log\eta\left(\frac{\frac{h'+k}{2}+\frac{i}{2z}}{k}\right)-\frac{1}{2}\log z+\frac{\pi i}{12}-\pi i V(h,k)$$

mit

$$V(h,k)=5s(h,k)-2s\left(\frac{h}{2},k\right)-2s(2h,k)$$

und somit nach (3a)

$$\log\vartheta_3\left(\frac{h+iz}{k}\right)=-\frac{1}{2}\log z+\log\vartheta_4\left(\frac{h'+\frac{i}{z}}{k}\right)-\pi i V(h,k)\,. \tag{10a}$$

Hier, in Analogie zu (8a), (8b), ist

$$V(h,k)=\frac{1}{2}d(h,k)\,, \tag{10b}$$

wo $d(h,k)$ den Überschuß der Anzahl der positiven über die Anzahl der negativen absolut kleinsten Reste von $\mu h \bmod k$, $\mu=1,2,\ldots,\frac{k-1}{2}$ angibt.

Die Fälle II und III zeigen keine weiteren Besonderheiten.

Wir betrachten endlich Fall VI, in dem die additive Konstante von etwas anderer Gestalt sein wird.

Wir haben hier

$$h\equiv k\equiv h'\equiv 1\ (\mathrm{mod}\,2)\,,$$

was

$$hh'\equiv -1\ (\mathrm{mod}\,2k)\,,\qquad 2h\cdot\frac{h'+k}{2}\equiv -1\ (\mathrm{mod}\,k)$$

zur Folge hat. Die Formeln (2a) und (7) ergeben hier

$$\log\vartheta_3\left(\frac{h+iz}{k}\right)=5\log\eta\left(\frac{h+iz}{k}\right)-2\log\eta\left(\frac{h+iz}{2k}\right)-2\log\eta\left(\frac{2h+2iz}{k}\right)$$

$$=5\log\eta\left(\frac{h'+\frac{i}{z}}{k}\right)-2\log\eta\left(\frac{h'+\frac{i}{z}}{2k}\right)- \tag{11}$$

$$-2\log\eta\left(\frac{\frac{h'+k}{2}+\frac{i}{2z}}{k}\right)-$$

$$-\frac{1}{2}\log z+\log 2-\frac{\pi i h'}{4k}+\frac{\pi i}{12}-\pi i V(h,k)$$

mit

$$\begin{aligned} V(h,k) &= 5s(h,k) - 2s(h,2k) - 2s(2h,k) \\ &= (s(2h,2k) - 2s(h,2k)) + 2(2s(h,k) - s(2h,k)) \\ &= -S(h',2k) + 2S(h',k) \\ &= -S(h',k) + R(h',k) + 2S(h',k) = S(2h',k) \end{aligned}$$

(siehe [2], (6.2), (5.64), (5.2)).

Nun ist

$$S(2h',k) = \sum_{\mu=1}^{\frac{k-1}{2}} \left(\left(\frac{2h'\mu}{k}\right)\right)$$

(12)
$$= \sum_{\mu=1}^{\frac{k-1}{2}} \left\{ \frac{2h'\mu}{k} - \left[\frac{2h'\mu}{k}\right] - \frac{1}{2} \right\}$$

$$= \frac{1}{4}(h'k - k + 1) - \frac{h'}{4k} - \sum_{\mu=1}^{\frac{k-1}{2}} \left[\frac{2h'\mu}{k}\right],$$

so daß im Hinblick auf (4a)

$$\log\vartheta_3\left(\frac{h+iz}{k}\right) = -\frac{1}{2}\log z + \log\vartheta_2\left(\frac{h' + \frac{i}{z}}{k}\right) - \\ - \frac{\pi i}{4}(h'k - k + 1) + \pi i \sum_{\mu=1}^{\frac{k-1}{2}} \left[\frac{2h'\mu}{k}\right].$$

Im Falle V gelangt man zu demselben Resultat, muß jedoch etwas anders schließen. Hier ist

$$h' \equiv 0, \qquad h \equiv k \equiv 1 \pmod 2$$

und somit

$$2h \cdot \frac{h'}{2} \equiv -1 \pmod k$$

$$h(h'+k) \equiv -1 \pmod{2k}.$$

Die additive Konstante ergibt sich dann als

$$\begin{aligned} V(h,k) &= 5s(h,k) - 2s(h,2k) - 2s(2h,k) \\ &= (s(2h,2k) - 2s(h,2k)) + 2(2s(h,k) - s(2h,k)) \\ &= -S(h'+k,2k) + 2S(h',k) \\ &= R(h',k) + S(h',k) + S(h',2k) = S(2h',k), \end{aligned}$$

da $S(h',2k) = 0$ wegen $h' \equiv 0 \pmod 2$, vgl. [2], (5.4), (6.2). Wir haben somit (12) wiedergewonnen.

Wenn wir noch bemerken, daß

$$\frac{1}{2}d(h,k) = \frac{k-1}{4} - N(h,k)$$

ist, wo $N(h,k)$ die Anzahl der negativen unter den absolut kleinsten Resten von $\mu h \bmod k$, $\mu = 1, 2, \ldots, \frac{k-1}{2}$ bedeutet, so erhalten wir die folgende Liste für die Fälle I bis VI von (6):

$$\text{I.}\qquad \log\vartheta_3\left(\frac{h+iz}{k}\right) = -\frac{1}{2}\log z + \log\vartheta_3\left(\frac{h'+\frac{i}{z}}{k}\right) + \frac{\pi i h}{4} - \pi i N(k,h)$$

$$= -\frac{1}{2}\log z + \log\vartheta_4\left(\frac{h'+\frac{i}{z}}{k}\right) + \frac{\pi i h}{4} - \pi i N(k,h)$$

$$\text{III.}\qquad = -\frac{1}{2}\log z + \log\vartheta_3\left(\frac{h'+\frac{i}{z}}{k}\right) - \frac{\pi i(k-1)}{4} + \pi i N(h,k)$$

$$\text{IV.}\qquad = -\frac{1}{2}\log z + \log\vartheta_4\left(\frac{h'+\frac{i}{z}}{k}\right) - \frac{\pi i(k-1)}{4} + \pi i N(h,k)$$

$$\text{V. u. VI.}\qquad = -\frac{1}{2}\log z + \log\vartheta_2\left(\frac{h'+\frac{i}{z}}{k}\right) - \frac{\pi i}{4}(h'k-k+1) +$$

$$+ \pi i \sum_{\mu=1}^{\frac{k-1}{2}} \left[\frac{2h'\mu}{k}\right].$$

Wenn wir von den Logarithmen zu den ϑ-Funktionen selbst zurückkehren, so erscheint in der Tat, wie zu erwarten, in den transformierten Ausdrücken eine achte Einheitswurzel. Diese Einheitswurzel enthält als Faktor ein Vorzeichen, das beziehungsweise als

$$(13)\qquad (-1)^{N(k,h)}, \quad (-1)^{N(h,k)}, \quad (-1)^{\Sigma\left[\frac{2h'\mu}{k}\right]}$$

auftritt. Da, wie leicht zu sehen,

$$\sum_{\mu=1}^{\frac{k-1}{2}} \left[\frac{2h'\mu}{k}\right] \equiv N(h',k) \pmod 2$$

ist, so sind die Vorzeichen (13) nach dem Gaußschen Lemma nichts anderes als die Legendre-Jacobischen Symbole

$$\left(\frac{k}{h}\right), \quad \left(\frac{h}{k}\right), \quad \left(\frac{h'}{k}\right).$$

10*

Im Gegensatz zu dieser Herleitung erscheint in der seit HERMITE [3] üblichen Bestimmung der Einheitswurzelfaktoren das quadratische Restsymbol als Konsequenz von Gaußschen Summen, während es sich hier aus dem Gaußschen Lemma ergibt. Dort aber geht man von den Summendarstellungen der Theta-Funktionen aus, während die vorliegende Untersuchung sich auf die Produktdarstellungen (2), (3), (4) stützt.

Literatur

[1] DEDEKIND, R.: Erläuterungen zu den vorstehenden Fragmenten. Riemanns Gesammelte Werke, pp. 466—478; Dedekinds Gesammelte Werke, **1**, 159—172 (1930).

[2] RADEMACHER, H., and A. WHITEMAN: Theorems on Dedekind Sums. Am. J. Math. **63**, 377—407 (1941).

[3] HERMITE, CH.: Oeuvres, tome **1** (1905), Sur quelques formules relatives à la transformation des fonctions elliptiques, pp. 482—496.

Professor Dr. HANS RADEMACHER
The Rockefeller University
New York, N.Y. 10021

(Eingegangen am 13. Januar 1966)

Notes to Paper 70

This paper has been reviewed in the Z, vol. 145 (1968), p. 319, and in the MR, vol. 34 (1967), #4218.

Page 144, fifth display line.
See Theorem 9 of [2] (paper 44 of the present collection).

Page 146, second display line from the bottom.
See Theorem 7 of [2] (paper 44).

Page 147, third display line.
Author's correction (insert II).

Page 148, reference [2].
See paper 44.

Eine Bemerkung über die Heckeschen Operatoren $T(n)$

Von Hans Rademacher

Die von Hecke [1] in die Theorie der Modulfunktionen eingeführten Operatoren $T(n)$ sind definiert durch die Transformation, die sie auf Modulformen $F(\tau)$ der Dimension $-k$ ausüben:

$$(1)\qquad F(\tau)\,|\,T(n) = n^{k-1} \sum_{\substack{AD=n\\ B \bmod D}} D^{-k} F\left(\frac{A\tau+B}{D}\right).$$

Sie stellen sich als assoziativ heraus und sind überdies multiplikativ für teilerfremde Parameter m, n:

$$T(m, n) = T(m)\,T(n), \qquad (m, n) = 1.$$

Die weitere Untersuchung dieser Operatoren kann sich daher auf Primzahlpotenzen $n = p^r$ beschränken. Hecke beweist, daß $T(p^r)$ ein Polynom $r$ten Grades in $T = T(p)$ ist, dessen expliziten Ausdruck er jedoch nicht angibt. Ich möchte nun zeigen, daß diese Polynome im wesentlichen Chebyshevsche Polynome zweiter Art sind.

Aus (1) folgt nach Hecke sofort die Rekursionsformel

$$(2)\qquad T(p)\,T(p^r) = T(p^{r+1}) + p^{k-1}\,T(p^{r-1}),$$

die insbesondere

$$T(p^2) = T(p)^2 - p^{k-1} I$$

ergibt, wo $I = T(p)^0$ die Identität ist. Formel (2) zeigt in der Tat durch Induktion, daß $T(p^r)$ ein Polynom in $T(p)$ ist. Dies hat wegen der Assoziativität und wegen

$$T(p)^r\,T(p)^s = T(p)^{r+s}$$

die Kommutativität der $T(p^r)$ zur Folge.

Wir setzen

$$(3)\qquad T(p^r) = P_r(T), \quad r \geqq 1,$$

wo zur Abkürzung $T = T(p)$ geschrieben ist.

Wir ersetzen nun T durch eine Unbestimmte x und haben dann die Polynome

$$P_1(x) = x,$$
$$P_2(x) = x^2 - p^{k-1}$$

und die Rekursionsformel

$$P_{r+1}(x) = x P_r(x) - p^{k-1} P_{r-1}(x), \quad r \geqq 2.$$

Durch die Substitutionen

$$z = \frac{x}{2} p^{\frac{1-k}{2}},$$
$$W_r(z) = p^{\frac{(1-k)r}{2}} P_r\left(2p^{\frac{k-1}{2}} z\right)$$

gehen diese Formeln über in

$$W_1(z) = 2z, \tag{4a}$$
$$W_2(z) = 4z^2 - 1, \tag{4b}$$
$$W_{r+1}(z) = 2z W_r(z) - W_{r-1}(z). \tag{4c}$$

Bekanntlich ist (4c) die Rekursionsformel für Chebyshevsche Polynome, und die Anfangsbedingungen (4a), (4b) zeigen, daß es sich um die Chebyshevschen Polynome zweiter Art

$$W_r(z) = U_r(z) = \frac{\sin(r+1)\vartheta}{\sin\vartheta}, \quad \cos\vartheta = z$$

handelt.

Damit haben wir explizite

$$T(p^r) = P_r(T) = p^{\frac{(k-1)r}{2}} U_r\left(\tfrac{1}{2} p^{\frac{1-k}{2}} T(p)\right), \tag{5}$$

was zu zeigen war.

Hecke beweist weiter

$$T(p^r) T(p^s) = \sum_{0 \leqq u \leqq \min(r,s)} p^{(k-1)u} T(p^{r+s-2u}). \tag{6}$$

Dies ist nun eine einfache Konsequenz unseres Ergebnisses (5). Im Hinblick auf die obigen Substitutionen sieht man nämlich, daß (6) hinauskommt auf

$$U_r(z) U_s(z) = \sum_{0 \leqq u \leqq \min(r,s)} U_{r+s-2u}(z).$$

Diese Identität für Chebyshevsche Polynome zweiter Art aber ist evident, wenn sie in trigonometrischer Form als

$$\sin(r+1)\vartheta \sin(s+1)\vartheta = \sum_{0 \leqq u \leqq \min(r,s)} \sin\vartheta \sin(r+s-2u+1)\vartheta$$

oder als

$$\cos(s-r)\vartheta - \cos(s+r+2)\vartheta = \sum_{0 \leqq u \leqq \min(r,s)} \{\cos(r+s-2u)\vartheta - \cos(r+s-2u+2)\vartheta\}$$

geschrieben wird.

Literatur

[1] Erich HECKE, Über Modulfunktionen und die Dirichletschen Reihen mit Eulerscher Produktentwicklung. I. Math. Annalen Bd. 114 (1937), 1—28, Mathematische Werke (Göttingen 1959) 644—671.

The Rockefeller University, New York, N. Y. 10021.

Eingegangen am 10. 1. 1966

Notes to Paper 71

This paper has been reviewed in the Z, vol. 159 (1969), p. 114, and in the MR, vol. 37 (1969), #2691.

Page 149, sixth line from bottom.
Read $T(p^0)$ instead of $T(p)^0$.

H. Rademacher in New York

COMMENTS ON EULER'S "DE MIRABILIBUS PROPRIETATIBUS NUMERORUM PENTAGONALIUM"

1. In the above mentioned article [1] Euler discusses consequences of his famous identity

$$\prod_{m=1}^{\infty} (1 - x^m) = \sum_{n=-\infty}^{+\infty} (-1)^n x^{\omega_n} = f(x), \tag{1.1}$$

where

$$\omega_n = \frac{n(3n-1)}{2}$$

are the so-called pentagonal numbers. Some of these consequences, in particular the recursion formula for $p(n)$, the number of unrestricted partitions of n, and the similar recursion formula for $\sigma(n)$, the sum of divisors of n, he had treated extensively in earlier papers [2], [3].

The main part of his memoir, however, is devoted to certain divergent series, connected with the pentagonal numbers. Euler makes the following statements

$$-1^\lambda - 2^\lambda + 5^\lambda + 7^\lambda - 12^\lambda - 15^\lambda + \cdots = 0 \tag{1.2}$$

or in our notation (1.2)

$$\sum_{n=-\infty}^{+\infty} (-1)^n \omega_n^\lambda = 0, \tag{1.3}$$

where λ is a positive integer. He states also the generalization

$$\sum_n (-1)^n \alpha^{\omega_n} \omega_n^\lambda = 0 \tag{1.4}$$

for α a root of unity.

In another set of equations the two "halves" $n > 0$ and $n < 0$ of the sequences of pentagonal numbers are involved. Euler states the equations

$$\begin{aligned} s_1 &= \sum_{n=1}^{\infty} (-1)^n \omega_n = \frac{1}{8}, \\ t_1 &= \sum_{n=1}^{\infty} (-1)^n \omega_{-n} = -\frac{1}{8}, \end{aligned} \tag{1.5}$$

17*

$$s_2 = \sum_{n=1}^{\infty} (-1)^n \omega_n^2 = -\frac{3}{16},$$

$$t_2 = \sum_{n=1}^{\infty} (-1)^n \omega_{-n}^2 = \frac{3}{16} \tag{1.6}$$

and he refers to the results

$$s_1 + t_1 = s_2 + t_2 = 0$$

as corroboration of (1.3). The formulae (1.5) and (1.6) are obtained by Euler through the process of summation named after him,

For (1.3) and (1.4) Euler gives a heuristic argument that treats the function (1.1) as a polynomial of a very high degree, which has multiple zeros at the roots of unity and must therefore vanish together with its derivatives at these roots of unity. The inadequacy of any argument of this sort is obvious, since the power series in (1.1) does not converge on the unit circle and has there a natural boundary, according to Fabry's theorem.

2. We have therefore, in order to make the statements (1.3) and (1.4) acceptable to look for a summation method and compare such a method with the Euler summation applied in (1.5) and (1.6).

Now the product (1.1) appears as the 24th root of the "discriminant" in Dedekind's function

$$\eta(\tau) = e^{\pi i\tau/12} \prod_{n=1}^{\infty} (1 - e^{2\pi i n\tau}), \quad \operatorname{Im} \tau > 0. \tag{2.1}$$

For this function the transformation formula

$$\eta\left(\frac{h + iz}{k}\right) = \varrho_{hk} \frac{1}{\sqrt{z}} \eta\left(\frac{h' + \frac{i}{z}}{k}\right) \tag{2.2}$$

is known [4], where $\frac{h}{k}$ is a reduced fraction, h' is chosen so that

$$hh' \equiv -1 \pmod{k},$$

and where ϱ_{hk} is a certain much discussed 24th root of unity, where finally

$$\operatorname{Re} z > 0, \quad |\arg \sqrt{z}| < \frac{\pi}{4}. \tag{2.3}$$

We have thus, after (1.1) and (2.1)

$$\eta(\tau) = e^{\pi i\tau/12} \sum_{-\infty}^{\infty} (-1)^n e^{2\pi i \omega_n \tau}$$

and therefore in view of (2.2)

$$\sum_n (-1)^n \alpha^{\omega_n} e^{-2\pi\omega_n z/k} \tag{2.4}$$

$$= \varrho_{hk} \left(\frac{\alpha_1}{\alpha}\right)^{1/24} \frac{1}{\sqrt{z}} e^{\pi z/12k} \sum_n (-1)^n \alpha_1^{\omega_n} e^{-2\pi(\omega_n+1/24)/kz}$$

where we have set

$$\alpha = e^{z\pi ih/k}, \quad \alpha_1 = e^{2\pi ih'/k}.$$

We define now by the Abel method of summation

$$S_0 = \sum_n (-1)^n \alpha^{\omega_n} = \lim_{z\to+0} \sum (-1)^n \alpha^{\omega_n} e^{-2\pi\omega_n z/k}$$

and have

$$S_0 = \varrho_{hk} \left(\frac{\alpha_1}{\alpha}\right)^{1/24} \lim_{z\to+0} \frac{1}{\sqrt{z}} c^{\pi z/12k} \sum_n (-1)^n \alpha_1^{\omega_n} e^{-2\pi(\omega_n+1/24)/kz}$$

$$= \lim_{z\to+0} O\left(\frac{1}{\sqrt{z}} e^{-(2\pi)/(kz)(1/24)}\right) = 0,$$

which agrees with (1.3) and (1.4) for the case $\lambda = 0$.

Similarly we define

$$S_\lambda = \sum_n (-1)^n \alpha^{\omega_n}\omega_n^\lambda = \lim_{z\to+0} \sum_n (-1)^n \alpha^{\omega_n}\omega_n^\lambda e^{-2\pi\omega_n z/k}$$

$$= \frac{(-k)^\lambda}{(2\pi)^\lambda} \lim_{z\to+0} \left(\frac{d}{dz}\right)^\lambda \sum_n (-1)^n \alpha^{\omega_n} e^{-2\pi\omega_n z/k}.$$

This gives us because of (2.4)

$$S_\lambda = \varrho_{hk} \left(\frac{\alpha_1}{\alpha}\right)^{1/24} \frac{(-k)^\lambda}{(2\pi)^\lambda} \lim_{z\to+0} \left(\frac{d}{dz}\right)^\lambda \left\{\frac{e^{\pi z/12k}}{z} \sum_n (-1)^n \alpha_1^{\omega_n} e^{-2\pi(\omega_n+1/24)/kz}\right\}$$

$$= \lim_{z\to+0} O\left(\frac{1}{z^{2\lambda+1/2}} e^{-2\pi/24kz}\right) = 0,$$

which gives (1.3) and (1.4) for $\lambda > 0$.

We have applied here the Abel summation by means of a lacunary power series in which only the powers x^{ω_n} appears. Such a method would be designated in Hardy's notation as $A(\omega_n)$.

3. On the other hand, for the functions

$$f_1(x) = \sum_{n=1}^{\infty} (-1)^n x^{\omega_n}, \tag{3.1}$$

$$f_2(x) = \sum_{n=1}^{\infty} (-1)^n x^{\omega_{-n}} \tag{3.2}$$

(the two "halves" of $f(x)$) we do not know anything about their behavior near the unit circle, which is again the natural boundary for them.

Euler, however, remarks that the sequences

$$\omega_n^{\lambda} \text{ and } \omega_{-n}^{\lambda}, \quad n = 1, 2, 3, \ldots$$

are arithmetical progressions of the order 2λ and as such very suitable to summation by his transformation (see e.g. [5], [6])

$$\begin{aligned}\sum_{n=0}^{\infty}(-1)^n a_n &= \sum_{0}^{\infty}\left(\frac{1}{2}\right)^{l+1} \Delta^l a_0 \\ &= \sum_{l=0}^{\infty}\left(\frac{1}{2}\right)^{l+1} \sum_{n=0}^{l} (-1)^n \binom{l}{n} a_n.\end{aligned} \tag{3.3}$$

In the cases (1.5) he has, for $n \geqq 0$,

ω_n:	0		1		5		12		22		35
$\Delta\omega_n$:		-1		-4		-7		-10		-13	
$\Delta^2\omega_n$:			3		3		3		3		
$\Delta^3\omega_n$:				0		0		0			

and obtains therefore

$$s_1 = \sum_{n=0}^{\infty} (-1)^n \omega_n = \sum_{l=0}^{2} \left(\frac{1}{2}\right)^{l+1} \Delta^l \omega_0 = -\frac{1}{4} + \frac{3}{8} = \frac{1}{8}.$$

For $-n \geqq 0$ the computation appears as

ω_{-n}:	0		2		7		15		26		40
$\Delta\omega_{-n}$:		-2		-5		-8		-11		-14	
$\Delta^2\omega_{-n}$:			3		3		3		3		
$\Delta^3\omega_{-n}$:				0		0		0			

and thus

$$t_1 = \sum_{n=0}^{\infty} (-1)^n \omega_{-n} = \sum_{l=0}^{2} \left(\frac{1}{2}\right)^{l+1} \Delta^l \omega_0 = -\frac{2}{4} + \frac{3}{8} = -\frac{1}{8},$$

both in agreement with (1.5). In the same way the values (1.6) are found. I may add to this list

$$s_3 = \sum_0^\infty (-1)^n \omega_n^3 = \frac{53}{64}.$$

$$t_3 = \sum_0^\infty (-1)^n \omega_{-n}^3 = -\frac{53}{64},$$

Euler finds in the results

$$s_1 + t_1 = s_2 + t_2 [= s_3 + t_3] = 0 \tag{3.4}$$

a corroboration of his statement (1.3), in which the summation is taken over the whole set of pentagonal numbers. It may be observed here that the Euler summation of s_λ and t_λ leads evidently always to rational numbers as limits.

Actually the statements (3.4) and (1.3) have nothing directly to do with each other. The Euler summation applied to the power series

$$g(z) = \sum_0^\infty (-1)^n a_n z^n \tag{3.5}$$

leads to

$$g(z) = \sum_{m=0}^{\infty} \left(\frac{1}{2}\right)^{m+1} \sum_{n=0}^{m} (-1)^n \binom{m}{n} a_n z^n, \tag{3.6}$$

and the domain of convergence of (3.5) induces that of (3.6) as is well-known. If in our case we put

$$a_n = \omega_n^\lambda \quad \text{or} \quad a_n = \omega_{-n}^\lambda$$

the function $g(z)$ becomes a rational function which has as its only singularity a pole of order $2\lambda + 1$ at $z = -1$. This follows from the fact that the ω_n^λ and ω_{-n}^λ form arithmetic sequences of order 2λ

It is known [5], [6] that then the series (3.6) for $g(z)$ converges in the circle

$$|z - 1| < 2 \tag{3.7}$$

which has the singularity $z = -1$ on its boundary. Clearly the point $z = +1$ lies in the interior of (3.7). Putting $z = 1$ in (3.5) and (3.6) we have (3.3). We have thus by (3.3) applied Abel's method to the power series (3.5), which is not lacunary and has $z = +1$ as a regular point on its circle of convergence. Euler's method thus has *not* summed the lacunary series (3.1) and (3.2) and their derivatives by Abel's summation near $x = 1$.

4. Nevertheless the results (3.4), generalized immediately to

$$s_\lambda = \sum_0^\infty (-1)^n \omega_n^\lambda, \quad t_\lambda = \sum_0^\infty (-1)^n \omega_{-n}^\lambda, \tag{4.1}$$

$$s_\lambda + t_\lambda = 0 \tag{4.2}$$

cannot be accidental.

Indeed we have

$$s_\lambda + t_\lambda = \sum_{n=0}^\infty (-1)^n (\omega_n^\lambda + \omega_{-n}^\lambda).$$

Now, for $\lambda > 0$,

$$\begin{aligned}\omega_n^\lambda + \omega_{-n}^\lambda &= 2^{-\lambda}\{(3n^2 + n)^\lambda + (3n^2 - n)^\lambda\} \\ &= A_\lambda n^{2\lambda} + B_\lambda n^{2(\lambda-1)} + \cdots + M_\lambda n^2,\end{aligned}$$

a polynomial in n^2 without constant term. Thus $s_\lambda + t_\lambda$ reduces to summation of

$$\sum_{n=1}^\infty (-1)^n n^{2k}.$$

But these sums have been discussed by Euler himself in another connection [7], where he summed them by a correct Abel summation to

$$(2^{2k+1} - 1)\zeta(-2k) = 0.$$

Landau has devoted a paper [8] to this investigation of Euler, in which by the way the Riemann functional equation for special values of the variable s appears for the first time. The statement (4.2) is thus a consequence of properties of $\zeta(s)$, and not of $\eta(\tau)$.

5. We wish, of course, for the sake of consistency, to sum s_λ and t_λ singly by an $A(\omega_n)$ and $A(\omega_{-n})$ method respectively. This can be done by theorems of G. H. Hardy and Miss M. L. Cartwright.

As a preparation we notice

$$24\omega_n + 1 = 36n^2 - 12n + 1 = (6n - 1)^2$$

so that

$$\omega_n = \frac{(6n-1)^2 - 1}{24}, \quad \omega_{-n} = \frac{(6n+1)^2 - 1}{24}, \quad n = 1, 2, 3, \ldots \tag{5.1}$$

We may restrict our treatment to ω_n; the treatment of ω_{-n} will be completely analogous.

The sum

$$s_\lambda(x) = \sum_{n=1}^\infty (-1)^n \omega_n^\lambda x^n$$

is thus a linear combination over the rational field of

$$\sum_{1}^{\infty} (-1)^n (6n-1)^{2q} x^n, \quad q = 0, 1, \ldots, \lambda$$

and we are interested in the limit of these power series for $x \to 1 - 0$. With a new variable we consider

$$\lim_{y \to +0} \sum_{n=1}^{\infty} (-1)^n (6n-1)^{2q} e^{-(6n-1)y} = l_q.$$

We know that through the Euler summation of

$$\sum_{1}^{\infty} (-1)^n (6n-1)^{2q},$$

applicable since the sequence $\{(6n-1)^{2k}\}$ forms an arithmetical progression of order $2k$, the limit l_k is a rational number.

The summation $A(\omega_n)$ would, however, require the limit

$$S_\lambda = \lim_{x \to 1-0} \sum (-1)^n \omega_n^\lambda x^{\omega_n}$$
$$= \lim_{y \to +0} \sum_{1}^{\infty} (-1)^n \omega_n^\lambda e^{\frac{-(6n-1)^2 - 1}{24} y}$$

or with a new variable y instead of $y/24$

$$\lim_{y \to 0} \sum_{1}^{\infty} (-1)^n \omega_n^\lambda e^{-(6n-1)^2 y},$$

which is again a linear combination over the rational field of

$$L_q = \lim_{y \to +0} \sum_{1}^{\infty} (-1)^n (6n-1)^{2q} e^{-(6n-1)^2 y}, \quad q = 0, 1, \ldots, \lambda.$$

We apply now theorems by Miss M. L. Cartwright [9] and G. H. Hardy [10]:

$$Z_1(s) = \sum_{n=1}^{\infty} (-1)^n (6n-1)^{-s}$$

is a Dirichlet series (actually the difference of two Hurwitz zetafunctions), which is regular at $s = 1$ and can be continued over the while s-plane as an entire function. If we write it as

$$\sum_{1}^{\infty} a_m m^{-s}$$

with

$$a_m = \begin{cases} 0 & \text{for } m \not\equiv -1 \pmod 6, \\ (-1)^{(m+1)/6} & \text{for } m \equiv -1 \pmod 6 \end{cases}$$

we see that it can be summed for any s by $A(m)$ as well as by $A(m^2)$ and to the same value. The conditions for the growth $Z_1(s)$ in the imaginary direction which Hardy's theorem requires [10] p. 180, are easily fulfilled.

Thus $l_k = L_k$, and by recombination we have proved

$$s_\lambda = \lim_{x \to 1-0} \sum_1^\infty (-1)^n \omega_n^\lambda x^n = \lim_{x \to 1-0} \sum_1^\infty (-1)^n \omega_n^\lambda x^{\omega_n} = S_\lambda.$$

In this way Euler's treatment of (1.3) (our section 2) can be reconciled with that of (1.5), (1.6) in section 3. The fact that

$$S_\lambda + T_\lambda = s_\lambda + t_\lambda = 0$$

is now in a new sense indeed a corroboration of Euler's results (1.3), (1.4).

6. We can add a little to Euler's statement and prove the

Theorem. *If α is a primitive k^{th} root of unity then*

$$\lim_{x \to 1-0} \sum_{n=1}^\infty (-1)^n \alpha^{\omega_n} \omega_n^\lambda x^{\omega_n} = S_\lambda(\alpha)$$

and

$$\lim_{x \to 1-0} \sum_{n=1}^\infty (-1)^n \alpha^{\omega_{-n}} \omega_{-n}^\lambda x^{\omega_{-n}} = T_\lambda(\alpha)$$

exist and are numbers of the field $R(\alpha)$. The limits exist in any Stolz angle $|\Theta| < \frac{\pi}{2}$.

We need the

Lemma.

$$\sum_{n=N+1}^{N+2k} (-1)^n \alpha^{\omega_n} = 0. \tag{6.1}$$

Proof. Two cases have to be distinguished, k odd and k even.

I. Let k be odd. Then

$$2\omega_{n+k} = (n+k)(3n+3k-1) = 2\omega_n + k(3k-1+6n) \equiv 2\omega_n \pmod{2k},$$

$$\omega_{n+k} \equiv \omega_n \pmod k$$

and therefore

$$(-1)^n \alpha^{\omega_n} + (-1)^{n+k} \alpha^{\omega_{n+k}} = 0, \tag{6.2}$$

which proves (6.1) for k odd.

II. Let k be even. Then we have

$$2\omega_{n+k} = 2\omega_n + k(3k + 6n) - k \equiv 2\omega_n + k \pmod{2k},$$

$$\omega_{n+k} \equiv \omega_n + \frac{k}{2} \pmod{k}$$

and

$$\alpha^{\omega_{n+k}} = \alpha^{\omega_n}\alpha^{k/2} = -\alpha^{\omega_n},$$

since α is a primitive k^{th} root of unity so that $\alpha^{k/2} \neq 1$. It follows again that

$$(-1)^n \alpha^{\omega_n} + (-1)^{n+k} \alpha^{\omega_{n+k}} = 0, \tag{6.2}$$

which proves (6.1) also in this case.

The lemma shows that

$$Z(s) = \sum_{n=1}^{\infty} (-1)^n \alpha^{\omega_n}(6n - 1)^{-s}$$

is convergent at $s = 1$ and can therefore be continued through the whole s-plane as an entire function.

We conclude again that

$$\lim_{x\to 1-0} \sum_{n=1}^{\infty} (-1)^n \alpha^{\omega_n}(6n - 1)^{2q} x^{6n-1} = l_q(\alpha) \tag{6.3}$$

exists and is equal to

$$\lim_{x\to 1-0} \sum_{n=1}^{\infty} (-1)^n \alpha^{\omega_n}(6n - 1)^{2q} x^{(6n-1)^2} = L_q(\alpha). \tag{6.4}$$

In view of (6.2) the series in (6.3) can be broken up into k series of the sort

$$\alpha^{\omega_l} \sum_{m=1}^{\infty} (-1)^m (6(l + km) - 1)^{2q} x^{6(l+mq)-1}, \tag{6.5}$$

in which the power series is a rational function and converges for $x \to 1$ to a rational number (obtainable, as above e.g. through the Euler transformation). Therefore $l_q(\alpha) = L_q(\alpha) \in R(\alpha)$. This proves the theorem in the beginning of this section.

The approach to the boundary in (6.3) and (6.4) does not have to be made on a radial path. Each component (6.5) converges to a limit for $x \to 1 - 0$ within a Stolz angle $-\frac{\pi}{2} + \varepsilon \leqq \Theta \leqq \frac{\pi}{2} - \varepsilon$. Therefore after Cartwright and Hardy (6.4) converges for $x \to 1 - 0$ within a Stolz angle $-\frac{\pi}{2} + 2\varepsilon \leqq \Theta \leqq \frac{\pi}{2} - 2\varepsilon$.

7. Summarizing we see that Euler was right in all his statements under proper interpretation. In modern notation we have shown that

$$\lim_{z\to 0}\left(\frac{d}{dz}\right)^{\lambda}\eta\left(\frac{h+iz}{k}\right)=0$$

in a Stolz angle $|\Theta|<\frac{\pi}{2}-\varepsilon$, a result which was known from the theory of the modular forms. If we introduce the two "halves" of $e^{-\pi i\tau/12}\eta(\tau)$, viz.

$$H_1(\tau)=\sum_{n=1}^{\infty}(-1)^n e^{2\pi i\omega_n\tau}, \tag{7.1}$$

$$H_2(\tau)=\sum_{n=-1}^{\infty}(-1)^n e^{2\pi i\omega_n\tau}, \tag{7.2}$$

we state the theorem of section 6 in the following way: the limits

$$\lim_{z\to +0}\left(\frac{d}{dz}\right)^{\lambda}H_1\left(\frac{h+iz}{k}\right)=\left(-\frac{2\pi}{k}\right)^{\lambda}S_\lambda, \tag{7.3}$$

$$\lim_{z\to +0}\left(\frac{d}{dz}\right)^{\lambda}H_2\left(\frac{h+iz}{k}\right)=\left(-\frac{2\pi}{k}\right)^{\lambda}T_\lambda \tag{7.4}$$

exist in any Stolz angle $|\Theta|<\frac{\pi}{2}$, where S_λ and T_λ are numbers of the field $R(e^{2\pi ih/k})$. The theory of modular forms would be of no assistance to the proofs of (7.3), (7.4).

References

[1] Leonard Euler, Opera Omnia (1), vol. 2, 480–496.
[2] L. Euler, De partitione numerorum, l.c., 254–294.
[3] L. Euler, Découverte d'une loi tout extraordinaire des nombres par rapport à la somme de leurs diviseurs, l.c. 241–253, and Demonstratio theorematis circa ordinem in summis divisorum observatum, l.c. 390–398.
[4] R. Dedekind, Erläuterungen zu den Fragmenten XXVIII in B. Riemann, Gesammelte Math. Werke, 2nd ed. (1892), 466–478.
[5] K. Knopp, Math. Z. **15** (1922), 226–253.
[6] H. Rademacher, Sitzungsber. Berliner Math. Ges. **21** (1922), 16–24.
[7] L. Euler, Remarques sur un beau rapport entre les séries des puissances tant directes que réciproques, Opera Omnia (1), vol. 15, 70–91.
[8] Edmund Landau, Euler und die Funktionalgleichung der Riemannschen Zetafunktion, Bibliotheca Math. (3) **7** (1906), 69–79.
[9] M. L. Cartwright, On the relation between the different types of Abel summation, Proc. London Math. Soc. (2) **31** (1930), 81–96.
[10] G. H. Hardy, The application of Abel's method of summation to Dirichlet's series, Quarterly J. of Math. **47** (1916), 176–192.

Notes to Paper 72

This paper has been reviewed in the MR, vol. 41 (1971), #6655, and in the Z, vol. 207 (1971), pp. 360-361.

Page 261, third display line.
Read 2 instead of z in the definition of α.

Page 261, fifth display line.
Read e instead of c.

Page 262, fourth display line.
Read $l = 0$ under the second summation symbol.

Page 265, fifth display line.
Replace -1 by $+1$ in the exponent.

Page 265, second line of text from bottom.
Read *whole* instead of *while*.

Page 266, seventh line from bottom.
Meant is the Stolz angle of approach to $x = 1$ from inside the unit circle.

Page 268, reference [6].
See paper 11 of the present collection.

DIVISION OF MATHEMATICS

ON THE ROUNDEST OVAL*

Hans Rademacher

Rockefeller University, New York, N.Y.

1. An oval is a closed convex curve. We shall consider here only ovals which are symmetric with respect to two orthogonal axes, for which the lengths 2a, 2b are given and kept fixed in the following discussion. For such ovals the example of the ellipse immediately comes to mind. Let $A_1 A_2 = 2a$ be the major axis, $B_1 B_2 = 2b$ the minor axis. The radius of curvature of the ellipse at A_1 and A_2 is $\rho_1 = \frac{b^2}{a}$ and at B_1 and B_2 is $\rho_0 = \frac{a^2}{b}$. This fairly large ratio

$$\frac{\rho_0}{\rho_1} = \left(\frac{a}{b}\right)^3$$

makes the ellipse appear to be rather sharply curved at A (from now on we suppress the subscripts) and rather flat at B. Some time ago, Mr. Joseph Diamond, a former student of mine and now an engineer in Denmark, asked me whether there exist ovals with the same semiaxes a and b, $a > b$, and which are less sharply curved at A and more strongly curved at B. The radius of curvature ρ is supposed to decrease monotonely and continuously from B to A. This leads to a problem of the calculus of variations, for which we shall show that there is no solution for the continuous radius of curvature ρ. The extremum is reached for a discontinuous function, as in many examples in the calculus of variation. Weierstrass was the first to stress the existence of such discontinuous solutions.

2. Let us look only at the fourth quadrant of the curve, which is sufficient because of the supposed symmetry. We take $\rho = \rho(u)$ as a function of the angle u which the tangent of the curve forms with the x-direction. We give the coordinates of the endpoints as $A = (a,0)$, $B = (0, -b)$. The angle u varies monotonely from 0 to $\frac{\pi}{2}$ on the curve BA.

We have on the curve $\Gamma = BA$, with ds as the arc element,

$$x = x\,(u) = \int_0^u \frac{dx}{du}\, du = \int_0^u \frac{dx}{ds}\frac{ds}{du}\, du = \int_0^u \cos u \;\rho(u)\, du$$

* This paper was presented at a meeting of the Division on April 6, 1967.

and thus

$$a = \int_0^{\frac{\pi}{2}} \rho(u) \cos u \, du$$

and similarly

$$b + y = b + y\,(u) = \int_0^u \frac{dy}{du}\, du = \int_0^u \frac{dy}{ds}\frac{ds}{du}\, du$$

$$= \int_0^u \sin u \;\; \rho(u)\, du$$

and therefore

$$b = \int_0^{\frac{\pi}{2}} \rho\,(u) \sin u \, du.$$

We obtain thus, in view of the assumed monotonicity of $\rho(u)$,

$$\rho\left(\frac{\pi}{2}\right) < \int_0^{\frac{\pi}{2}} \rho(u) \sin u \, du = b < a = \int_0^{2} \rho(u) \cos u \, du < \rho(0).$$

The inequality signs cannot be replaced by equality signs, as we have $\rho(u)$ actually decreasing, since for constant $\rho(u)$ we would get the circle against our assumption $b < a$.

Putting $\rho\frac{\pi}{2} = \rho_1$, $\rho(0) = \rho_0$, we have

$$\rho_1 < b < a < \rho_0. \tag{1}$$

Now I claim that any curve with continuous monotone $\rho(u)$ can be replaced by a better one, i.e., by a curve with greater ρ_1 and smaller ρ_0.

3. In order to see this we consider the evolute MN of the curve BA. The evolute is the locus of the centers of curvature of the curve Γ. Thus $MB = \rho_0$ and $NA = \rho_1$. Moreover, its length ℓ is the difference $\rho_0 - \rho_1$. We now deform the evolute leaving it convex and without changing its length. We push a small piece starting from N down towards the x-axis and push a small piece starting with M towards the y-axis. If the pieces are chosen

small enough, the rest of the curve can still be chosen as convex and tangent to the axes at the ends of the pieces. Then, since Γ is obtained by unwinding a string of length ρ_0 from the evolute, the effective radius of curvature at A is now longer than ρ_1, since it starts at a point to the left of N, and the radius of curvature at B is shorter than ρ_0, since it starts below M.

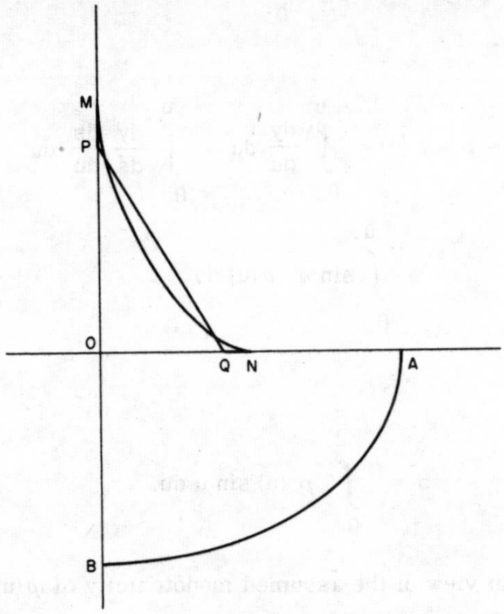

FIGURE 1.

The extreme deformation of the evolute in this manner would be obtained, if we replace the convex curve MN by the polygon MPQN, where PQ is now straight. Then we have the extreme radii of curvature QN = r at A and PB = R at B. Actually the curve Γ is now replaced by a curve Γ^* consisting of two circular arcs of radii r and R, as the unwinding of a string along MPQNA of length R demonstrates.

It can be easily shown, and would only necessitate unrewarding calculations, that the limit polygon MPQN of the evolute can be approximated as closely as we wish by a curve of class C^∞, which would then also lead to a curve as close to Γ^* as we wish and on which the curvature would change continuously and monotonely. Because of this fact, we consider from now on only Γ^*. The result is that any curve Γ = BA of an oval of continuously decreasing radius of curvature can be replaced by a "better" (in the sense given in section 1) curve.

4. We have now for Γ^* instead of (1) the inequality

$$r \leq b < a \leqq R. \tag{2}$$

In FIGURE 1 we find a right triangle with PQ = R-r, OP = R - b, OQ = a - r, so that

$$(R-b)^2 + (a-r)^2 = (R-r)^2.$$

A short computation reduces this to

$$2(R-a)(b-r) = (a-b)^2. \qquad (3)$$

It follows from this equation that if $r \to b$, then $R \to \infty$, and if $r \to 0$, then

$$R \to \frac{(a-b)^2}{2b} + a = \frac{a^2+b^2}{2b}$$

Both limit cases $r = 0$ and $r = b$ can be drawn. In the first case the oval consists of two circular arcs each going through B_1 and B_2 and intersecting at A_1 and A_2, where the oval ends in corners. In the second case, $R = \infty$ means that the oval is bounded by two lines parallel to the x-axis and closed at the ends A_1 and A_2 by two semicircles of radius b.

We have in both cases

$$\lim_{r \to b} \frac{R}{r} = \infty, \quad \lim_{r \to 0} \frac{R}{r} = \infty \qquad (4)$$

5. For given a and b we still have the choice of R and r, which have only to fulfill the inequality (2) and the equality (3). The problem outlined in section 1 is still rather vague: We have not stated whether we prefer a maximal ρ_1 or a minimal ρ_0. We now make it precise in asking for the "roundest oval," i.e., that oval which has the least variation of the radius of curvature; in other words, we ask for

$$\frac{R}{r} = \text{minimum}. \qquad (5)$$

We shall find only one extremum, and from (4) we infer that it can only be a minimum.

The method of the Lagrange multiplier demands, because of the side condition (3), the discussion of the function

$$F(R, r; \lambda) = \frac{R}{r} + \lambda(2(R-a)(b-r) - (a-b)^2)$$

and the vanishing of its partial derivatives

$$\frac{\partial F}{\partial R} = \frac{1}{r} + 2\lambda(b-r) = 0$$

$$\frac{\partial F}{\partial r} = -\frac{R}{r^2} - 2\lambda(R-a) = 0.$$

This can be written as

$$\frac{1}{r} = -2\lambda(b-r)$$

$$\frac{R}{r^2} = -2\lambda(R-a).$$

Division yields

$$\frac{R}{r} = \frac{R - a}{b - r}$$

and thus

$$bR - 2Rr + ar = 0 \tag{6}$$

We rewrite equation (3) as

$$2bR - 2Rr + 2ar = a^2 + b^2. \tag{7}$$

Subtraction of (6) from (7) yields

$$bR + ar = q^2, \tag{8}$$

where we have set

$$q^2 = a^2 + b^2. \tag{9}$$

Multiplying (6) by 2 and subtracting this from (7) leads to

$$2Rr = q^2. \tag{10}$$

We now define

$$ar = z_1, \ bR = z_2 \tag{11}$$

and read off from (2) immediately $ar \leq ab < Rb$ (the equalities can be dismissed because of (3)), so that

$$z_1 < z_2. \tag{12}$$

Equations (8) and (10) can now be replaced by

$$z_1 + z_2 = q^2, \ z_1 \cdot z_2 = \frac{1}{2} q^2 ab.$$

Therefore z_1 and z_2 are the roots of the quadratic equation

$$z^2 - q^2 z + \frac{1}{2} q^2 ab = 0,$$

which has the solutions

$$\left.\begin{matrix} z_1 \\ z_2 \end{matrix}\right\} = \frac{1}{2} q(q \mp d)$$

with

$$d = a - b. \tag{13}$$

Because of (12) the upper and lower signs correspond respectively to z_1 and z_2. Returning by (11) to r and R we obtain

$$r = \frac{q}{2a}(q - d), \quad R = \frac{q}{2b}(q + d) \tag{14}$$

as solution of the minimum problem (5).

6. Let us look at the quotient

$$V = \frac{\rho_0}{\rho_1}$$

which expresses the variation of curvature along the oval. For the ellipse it is $V = s^3$ where we have put $\frac{a}{b} = s > 1$. For the "roundest oval" we obtained for V its minimum

$$V_0 = \frac{R}{r} = s \cdot \frac{q + d}{q - d}.$$

A little computation leads to

$$V_0 = 1 - s + s^2 + (s - 1)\sqrt{1 + s^2}$$

and then to the consequences $V_0 < s^3$ for $s > 1$ and asymptotically

$$V_0 = 2s(s - 1) + \frac{3}{2} - \frac{\theta(s)}{s}$$

where

$$\frac{1}{2\sqrt{2}} < \theta(s) < \frac{5}{8}.$$

Since V_0 increases only with the second power of s, whereas V for the ellipse increases with the third power of s, the shape of the ellipse and that of the roundest oval will differ more and more for increasing s.

The following two figures show the ellipse and the roundest oval for $s = \frac{a}{b} = \frac{15}{8}$.

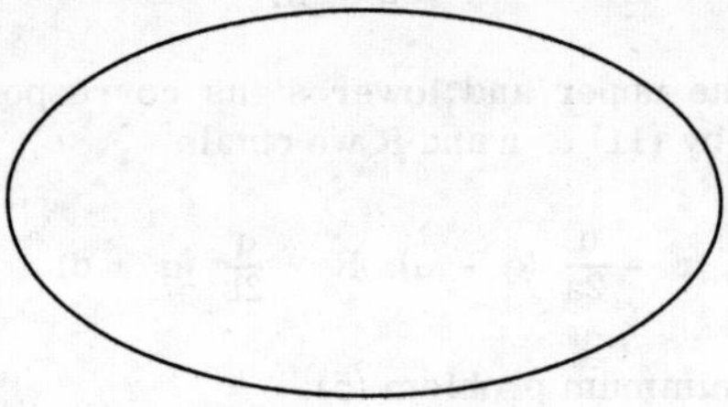

FIGURE 2.

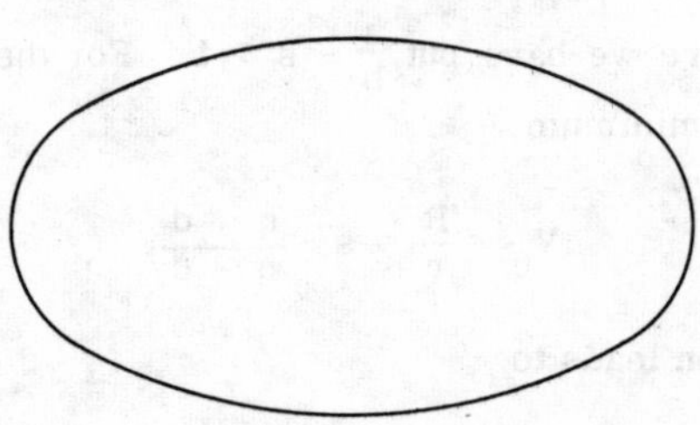

FIGURE 3.

Notes to Paper 73

This paper appeared posthumously; as of January 1, 1973, it had been reviewed only in the MR, vol. 44 (1972), #4640, p. 856.

Page 869, fifth display line.
The upper limit of the second integral is $\pi/2$ (not 2).

page 869, line 6.
Read $\rho(u)$ for ${}^{\rho}(u)$.

Page 869, line 8.
Read $\rho\left(\frac{\pi}{2}\right)$ for $\rho\frac{\pi}{2}$.

Page 870, line 8.
Read QA instead of QN.

Page 872, lien 8.
Read $ar < ab$ instead of $ar \leqslant ab$.

Abstracts

2. H. Rademacher: Über die Konvergenz von Reihen, die nach Orthogonalfunktionen fortschreiten.

Der Rieß-Fischersche Satz behauptet die Existenz einer Funktion $f(x)$, die eine gegebene Folge von reellen Konstanten $c_1, c_2, \ldots, c_\nu, \ldots$ mit konvergenter Quadratsumme $\sum c_\nu^2$ als Fourierkoeffizienten in bezug auf ein vorgegebenes Orthogonalsystem $\varphi_1(x), \varphi_2(x), \ldots \varphi_\nu(x), \ldots$ besitzt. Eine rechnerische Konstruktion der hiermit postulierten Funktion wird durch den folgenden Satz geliefert:

Ist $r_n = \sum_{\nu=n}^{\infty} c_\nu^2$, und Λ_ϱ die kleinste Zahl n, für welche $\frac{1}{\sqrt{r_n}} \geqq \varrho$ ist, so existiert fast überall der Limes

$$f(x) = \lim_{\varrho \to \infty} \sum_{\nu=1}^{\Lambda_\varrho} c_\nu \varphi_\nu(x).$$

Die Konvergenz der Reihe $\sum c_\nu \varphi_\nu(x)$ selbst kann man bisher nur unter einschränkenden Bedingungen über die Folge der c_ν beweisen. Es gilt hier der Satz:

Ist $\sum (c_\nu \log \nu)^2$ konvergent, so konvergiert $\sum c_\nu \varphi_\nu(x)$ fast überall im Orthogonalitätsintervall.

Dieses Ergebnis bedeutet eine Verschärfung eines Plancherelschen Satzes, in dem $\sum c_\nu^2 (\log \nu)^3$ als konvergent vorausgesetzt wird. Die Frage, ob sich der Exponent von $\log \nu$ auf 2 herunterdrücken ließe, war schon durch einen Satz von Hardy über Fouriersche Reihen nahegelegt. Aus unserem Satz folgt auch, daß bei Konvergenz von $\sum c_\nu^2$ fast überall $\sum_{\nu=1}^{n} c_\nu \varphi_\nu(x) = o(\log n)$ ist. Diesem Sachverhalte steht gleichfalls schon ein Satz von Hardy über Fouriersche Reihen gegenüber. Die Methode, welche den obigen Konvergenzsatz liefert, läßt sich auch anwenden zur Abschätzung von Ausdrücken der Form

$$\varrho_n(x) = \int_0^1 \left| \sum_{\nu=1}^{n} \varphi_\nu(x) \varphi_\nu(y) \right| dy,$$

die für den Fall der trigonometrischen Funktionen unter dem Namen der Lebesgueschen Konstanten bekannt sind. Man erhält das Ergebnis, daß fast überall $\varrho_n(x) = O\left(n^{\frac{1}{2}} \log n^{\frac{3}{2}+\varepsilon}\right)$ ist. Man kann ein Beispiel finden, in dem $\varrho_n \sim \sqrt{\frac{2n}{\pi}}$ ist, womit also der möglichen Verbesserung der eben angegebenen Abschätzung eine Grenze gezogen ist.

Der in dem Plancherelschen und in meinem verschärften Konvergenzsatz auftretende Logarithmus erscheint im Beweise durch den Umstand, daß die Minimalanzahl von ganzen Zahlen, aus denen sich alle ganzen Zahlen von 1 bis n additiv zusammensetzen lassen, asymptotisch gleich $\log n$ ist.

Die ausführliche Darstellung erscheint in den Math. Annalen.

A-2

57. Professor Hans Rademacher: *A new estimate of the number of prime numbers in a real quadratic field.*

By a detailed investigation of zero-free regions in the critical strip of Hecke's $\zeta(s, \lambda)$-functions, and by the application of a certain Fourier series, it is possible to obtain an estimate for the number $\Omega_m(Y, Y')$ of prime numbers satisfying the conditions $0<\omega\leqq Y$, $0<\omega'\leqq Y'$, $\omega\equiv\rho$ (mod $\mathfrak{n}$). The error term is of the form $O(x \exp(-c(\log x)^{1/2}))$. In the course of the proof we can state at the same time improvements of some previous results obtained by Hecke with the use of his $\zeta(s, \lambda)$-functions and with the application of Weyl's method of diophantine approximation. (Received November 23, 1934.)

A-3

366. Professor Hans Rademacher: *On prime numbers on real quadratic fields in rectangles.*

The existence of a totally positive fundamental unit gives rise to a periodicity in certain finite sums, of which the number of prime numbers in a rectangle is a special case. The periodicity is utilized for a Fourier expansion, which leads to an identity for those finite sums. In the case of prime numbers some estimates can be obtained, one of which has been already given by the author in Acta Arithmetica, vol. 1 (1935). (Received September 28, 1935.)

A-4

359. Professor Hans Rademacher: *Dedekind's ζ-function and Hecke's $\zeta(s, \lambda)$-functions in totally real algebraic fields.*

The sum $\Phi(a, s)=\sum_\mu N(a+i\mu)^{-s}$ for real $a_1, a_2, \cdots, a_n>0$, $R(s)>1$, where μ runs through all integers of a given totally real algebraic field, is treated in two different ways, one of which includes a Fourier development with respect to the point lattice of all integers of the field. Comparison of both results yields an identity which contains the functional equations of the Dedekind ζ-function and of all Hecke $\zeta(s, \lambda)$-functions simultaneously. (Received July 27, 1936.)

A-5

360. Professor Hans Rademacher: *On Waring's problem in algebraic fields.*

Hardy-Littlewood's method has already been applied to certain problems of additive number theory in algebraic fields. But Waring's problem in algebraic fields still offered a particular difficulty, arising from the "minor" Farey sections, since no analogue of Weyl's method of approximation was known. In the present paper this analogue is developed for totally real, in particular, real quadratic, fields. The generalized finite geometric series to which by Weyl's procedure a certain sum of roots of unity is reduced necessitates a Fourier expansion which can be avoided in the rational case. (Received July 27, 1936.)

A CERTAIN GENERALIZATION OF DEDEKIND SUMS

by Hans Rademacher

In recent investigations of C. Meyer on class numbers of Abelian fields over quadratic ground fields, but also in earlier work by Lehmer and others on partitions modulo p, there appear Dedekind sums in which the summation variable runs only over certain residue classes of a given modulus. These generalizations of Dedekind sums appear as special cases of sums defined as follows:

Let x, y be real numbers, and h, k coprime integers, then the generalized Dedekind sums are

$$S(h,k;x,y) = \sum_{\mu \bmod k} \left(\left(h\frac{\mu+y}{k} + x\right)\right)\left(\left(\frac{\mu+y}{k}\right)\right).$$

It is clear that $S(h,k;x,y)$ is periodic in x and y with period 1. The $S(h,k;x,y)$ satisfy a reciprocity relation. (An exposition of these sums has been submitted to the Acta Mathematica.)

Notes to Abstracts

A1. This abstract was expanded into paper 13 of the present collection.

A2. Abstract expanded into paper 34.

A3. Abstract expanded into paper 36. The second "on" of the abstract title should read "of."

A4 and A5. It seems that these two abstracts were never expanded into detailed papers.

A6. Abstract expanded into paper 67. In the last line "Mathematica" should read "Arithmetica."

Problems and Solutions

P-1

Lösungen.

Lösung der Aufgabe 26. (Dieser Jahresbericht Bd. 34, Heft 5/8, S. *98*.) „Für $0 \leqq \varphi \leqq \pi$ sei die integrierbare Funktion $g(\varphi)$ von folgender Beschaffenheit:

1. $0 \leqq g(\varphi) \leqq 1$,

2. $\int_0^\pi g(\varphi) \cos \varphi \, d\varphi = 0$,

3. es gebe ein α, $0 \leqq \alpha \leqq \frac{2\pi}{3}$, so daß $g(\varphi) = 0$ im Intervall $\alpha \leqq \varphi \leqq \alpha + \frac{\pi}{3}$.

Dann ist notwendig
$$\int_0^\pi g(\varphi) \sin \varphi \, d\varphi \leqq 1,$$
und zwar gilt das Gleichheitszeichen hier nur für die Funktion
$$g(\varphi) = \begin{cases} 0 \text{ für } \alpha \leqq \varphi \leqq \alpha + \frac{\pi}{3} \\ 0 \text{ für } \left|\alpha + \frac{\pi}{6} - \varphi\right| \geqq \frac{\pi}{2} \\ 1 \text{ sonst.} \end{cases}$$
und für die mit dieser bis auf eine Nullmenge übereinstimmenden Funktionen."

(H. Rademacher.)

Geometrisch bedeutet die Aufgabe folgendes: Durch
$$x = \int_0^\varphi g(\varphi) \cos \varphi \, d\varphi, \quad y = \int_0^\varphi g(\varphi) \sin \varphi \, d\varphi \quad (0 \leqq \varphi \leqq \pi;\ x(\pi) = 0)$$
ist ein konvexer Kurvenbogen definiert, falls $g(\varphi) \geqq 0$. φ ist der Neigungswinkel der Tangente und $g(\varphi)$ der Krümmungsradius im Punkt (x, y). Wenn der Krümmungsradius nirgends größer als 1 ist und der Bogen überdies noch eine Ecke besitzt, deren Ecktangenten mindestens den Winkel $\frac{\pi}{3}$ bilden, so soll gezeigt werden, daß der Abstand der Endpunkte des Bogens, also $y(\pi)$, höchstens 1 ist und daß dieser Extremwert nur angenommen wird für eine gewisse, aus zwei Kreisbogen vom Radius 1 zusammengesetzte Kurve.

Diese geometrische Formulierung führt leicht zu einer geometrischen Lösung, die, der Kürze halber in analytische Sprache übersetzt, hier folgt:

I. Wir nehmen an $\alpha \geqq \frac{\pi}{3}$, da der andere Fall durch die Substitution $\psi = \pi - \varphi$ auf diesen zurückgeführt werden kann. Funktionen, die sich nur in einer Menge vom Maß Null unterscheiden, betrachten wir im folgenden nicht als wesentlich verschieden.

II. Es sei $g(\varphi)$ irgendeine fest gegebene zulässige Funktion Aus den Voraussetzungen folgt dann

$$(1) \quad 0 \leqq \int_0^\alpha g(\varphi) \cos \varphi \, d\varphi = -\int_{\alpha+\frac{\pi}{3}}^\pi g(\varphi) \cos \varphi \, d\varphi \leqq -\int_{\alpha+\frac{\pi}{3}}^\pi \cos \varphi \, d\varphi = \sin\left(\alpha + \frac{\pi}{3}\right).$$

Es gibt nun sicher zulässige Funktionen $\bar{g}(\varphi)$, für die

(2) $$\bar{g}(\varphi) = 1 \quad \text{für} \quad \alpha + \frac{\pi}{3} < \varphi \leqq 1$$

(3) $$\int_0^\alpha \bar{g}(\varphi) \cos\varphi \, d\varphi = \sin\left(\alpha + \frac{\pi}{3}\right).$$

Insbesondere gibt es $\bar{g}$ für die

(4) $$\bar{g}(\varphi) \geqq g(\varphi) \quad \text{für} \quad 0 \leqq \varphi < \alpha.$$

Ein solches $\bar{g}(\varphi)$ erhält man z. B., wenn man für $0 \leqq \varphi \leqq \alpha$ setzt

$$\bar{g}(\varphi) = \frac{1 + \mu g(\varphi)}{1 + \mu} \qquad (\mu \geqq 0)$$

und μ so wählt, daß (3) erfüllt ist.

III. Für dieses letzte $\bar{g}(\varphi)$ ist

(5) $$\int_0^\pi \bar{g}(\varphi) \sin\varphi \, d\varphi \geqq \int_0^\pi g(\varphi) \sin\varphi \, d\varphi$$

und das Gleichheitszeichen kann nur stehen, wenn $g(\varphi)$ selbst ein $\bar{g}(\varphi)$ ist, d. h. (2) erfüllt. Wir suchen nun das $\bar{g}(\varphi)$ zu bestimmen, das in (5) die linke Seite möglichst groß macht. Wegen (2) ist dies dasselbe, das

(6) $$J(\bar{g}) = \int_0^\alpha \bar{g}(\varphi) \sin\varphi \, d\varphi$$

zum Maximum macht. Ist $\lambda < \frac{\pi}{2}$ eine Konstante, über die wir noch verfügen werden, so sind $J(\bar{g})$ und

(7) $$J^*(\bar{g}) = J(\bar{g}) \cos\lambda - \sin\left(\alpha + \frac{\pi}{3}\right) \sin\lambda = \int_0^\alpha \bar{g}(\varphi) \sin(\varphi - \lambda) \, d\varphi$$

gleichzeitig Maxima.

IV. Genügt $\bar{\bar{g}}$ den Bedingungen (2) und (3), ist ferner $\lambda < \alpha$ und

(8) $$\bar{\bar{g}} \geqq \bar{g} \quad \text{für} \quad \varphi > \lambda$$
$$\bar{\bar{g}} \leqq \bar{g} \quad \text{für} \quad \varphi \leqq \lambda,$$ so ist

(9) $$J^*(\bar{\bar{g}}) - J^*(\bar{g}) = \int_0^\alpha (\bar{\bar{g}} - \bar{g}) \sin(\varphi - \lambda) \, d\varphi \geqq 0,$$

da der Integrand nirgends negativ ist. Für jedes $\bar{g}$ ist (8) nur erfüllt, wenn man setzt:

(10) $$\bar{\bar{g}} = 0 \qquad \varphi \leqq \lambda$$
$$\bar{\bar{g}} = 1 \qquad \varphi > \lambda.$$

Dieses $\bar{\bar{g}}$ wird aber ein $\bar{g}$, also zulässig, erst, wenn (4) erfüllt ist. Wir verfügen über λ so, daß dies der Fall ist. Dann muß sein

(11) $$\lambda = \alpha - \frac{\pi}{3}.$$

4*

V. Da in (9) das Gleichheitszeichen nur gilt, wenn $\bar{g}$ von $\bar{\bar{g}}$ nicht wesentlich verschieden ist, und da dieses $\bar{\bar{g}}$ von der Ausgangsfunktion g gänzlich unabhängig ist, so folgt:

$\int_0^\pi g(\varphi)\sin\varphi\, d\varphi$ nimmt seinen größten Wert dann und nur dann an, wenn $g(\varphi)$ mit der Funktion $\bar{\bar{g}}$ zusammenfällt. Dabei ist

$$\begin{aligned} \bar{\bar{g}} &= 0 \quad \text{für} \quad 0 \leqq \varphi \leqq \alpha - \frac{\pi}{3} \quad \text{und} \quad \alpha \leqq \varphi \leqq \alpha + \frac{\pi}{3} \\ &= 1 \quad \text{für} \quad \alpha - \frac{\pi}{3} < \varphi < \alpha \quad \text{und} \quad \alpha + \frac{\pi}{3} < \varphi \leqq 1. \end{aligned}$$

Für dieses $\bar{\bar{g}}$ ist der Wert des Integrals gerade 1.

Unter Berücksichtigung von I. läßt sich das Ergebnis dann wie in der Aufgabe ausgesprochen formulieren.

Münster i. W., 23. Aug. 1925. M. Krafft.

(Eingegangen am 25. 8. 25.)

Diese Lösung war bereits gesetzt, als die nachstehende

zweite Lösung

einlief:

Es ist laut Voraussetzung 2.

$$\int_0^\pi g(\varphi)\cos\left(\alpha + \frac{\pi}{6} - \varphi\right) d\varphi = \sin\left(\alpha + \frac{\pi}{6}\right)\int_0^\pi g(\varphi)\sin\varphi\, d\varphi.$$

Bezeichnet $g_0(\varphi)$ die in der Aufgabe genannte „Extremalfunktion", dann ist, behaupte ich,

$$g(\varphi)\cos\left(\alpha + \frac{\pi}{6} - \varphi\right) \leqq g_0(\varphi)\cos\left(\alpha + \frac{\pi}{6} - \varphi\right).$$

Für $\alpha \leqq \varphi \leqq \alpha + \frac{\pi}{3}$ steht nämlich beiderseits 0; für $\left|\alpha + \frac{\pi}{6} - \varphi\right| \geqq \frac{\pi}{2}$ ist die linke Seite $\leqq 0$, die rechte Seite 0; für die sonstigen φ-Werte ist $\cos\left(\alpha + \frac{\pi}{6} - \varphi\right) > 0$, und die Ungleichung lautet: $g(\varphi) \leqq 1$, was wegen 1. richtig ist. Es gilt somit, da $\sin\left(\alpha + \frac{\pi}{6}\right) > 0$,

$$\int_0^\pi g(\varphi)\sin\varphi\, d\varphi \leqq \int_0^\pi g_0(\varphi)\sin\varphi\, d\varphi = 1.$$

Aus dem Beweis geht hervor, daß das Gleichheitszeichen nur dann eintreten kann, wenn bis auf eine Nullmenge $g(\varphi) = g_0(\varphi)$ gilt.

G. Szegö.

(Eingegangen am 31. 10. 25.)

Eine Lösung legte auch Herr Dr. S. C. van Veen in Dordrecht vor.

Lösung der Aufgabe 30. (Dieser Jahresbericht Bd. 34, S. *158*.) Die Aufgabe lautet:

Es seien M und n ganze positive Zahlen. Die Zahlen $r_1, r_2, \ldots, r_q$ $(q \geqq 2)$ sollen je ein verkürztes Restsystem mod M durchlaufen, doch mit der Einschränkung, daß stets

$$\sum_{j=1}^{q} r_j \equiv n \pmod{M}$$

ist. Wird dann

$$\Phi^{(q)}(M; n) = \sum_{r_1, r_2, \ldots, r_q} 1$$

gesetzt, so ist

$$\Phi^{(q)}(M; n) = M^{q-1} \prod_{p/(M,n)} \frac{(p-1)\left((p-1)^{q-1} - (-1)^{q-1}\right)}{p^q} \prod_{\substack{p/M \\ p \nmid n}} \frac{(p-1)^q - (-1)^q}{p^q},$$

worin p die Primzahlen durchläuft. Offenbar ist $\Phi^{(2)}(M; M)$ identisch mit der Eulerschen Funktion $\varphi(M)$.

RADEMACHER.

Lösung. Es seien $q \geqq 2$ und n fest, a und b zwei teilerfremde ganze Zahlen, dann wird behauptet, daß

$$\Phi^{(q)}(ab; n) = \Phi^{(q)}(a; n)\, \Phi^{(q)}(b; n)$$

ist. Sind nämlich $s_1, s_2, \ldots, s_q$ bzw. $t_1, t_2, \ldots, t_q$ zu a bzw. zu b teilerfremde Zahlen, die den Bedingungen

$$\sum_{j=1}^{q} s_j \equiv n \pmod{a}$$

bzw.

$$\sum_{j=1}^{q} t_j \equiv n \pmod{b}$$

genügen, so löse man für $j = 1, 2, \ldots, q$ die beiden simultanen Kongruenzen

$$r_j \equiv s_j \pmod{a}$$
$$r_j \equiv t_j \pmod{b}.$$

Für jedes j hat dieses System $(\operatorname{mod} ab)$ stets eine und nur eine Lösung. Die so gefundenen Werte $r_1, r_2, \ldots, r_q$ sind zu ab teilerfremd und genügen der Bedingung

$$\sum_{j=1}^{q} r_j \equiv n \pmod{ab}.$$

Umgekehrt liefert natürlich jedes System von Werten $r_1, r_2, \ldots, r_q$, das der letzten Kongruenz genügt und zu ab teilerfremd ist, je ein und nur ein System der s_j und der t_j. Daher genügt es, die Formel für Primzahlpotenzen $M = p^\alpha$ zu beweisen.

Es sei zunächst $q = 2$.

Für r_1 kommen $\varphi(p^\alpha)$ Werte in Betracht. Man muß zwei Fälle unterscheiden

1. $n \equiv 0 \pmod{p}$.

Dann läßt sich zu jedem der $\varphi(p^\alpha)$ Werte ein und nur ein Wert von r_2 angeben, derart, daß $(r_2, p) = 1$ und $r_1 + r_2 \equiv n \pmod{p^\alpha}$ ist. In diesem Fall wird also

$$\Phi^{(2)}(p^\alpha; n) = \varphi(p^\alpha) = p^{\alpha-1}(p-1) = p^{\alpha(2-1)} \frac{(p-1)\{(p-1)-(-1)\}}{p^2}.$$

2. $n \not\equiv 0 \pmod{p}$.

Von den $\varphi(p^\alpha)$ Werten von r_1 ergeben hier alle die keine Lösung von $r_1 + r_2 \equiv n \pmod{p^\alpha}$, für die $r_1 \equiv n \pmod{p}$ ist, da sonst $r_2 \equiv 0 \pmod{p}$ sein müßte; dies sind $\frac{\varphi(p^\alpha)}{p-1}$ Werte. Die übrigen Werte von r_1 ergeben wieder eine und nur eine Lösung. Daher wird in diesem Fall

$$\Phi^{(2)}(p^\alpha; n) = \varphi(p^\alpha)\left\{1 - \frac{1}{p-1}\right\} = p^{\alpha-1}(p-2) = p^\alpha \frac{(p-1)^2 - (-1)^2}{p^2}.$$

Für $q = 2$ ist die Formel also richtig; die Behauptung folgt nun durch Schluß von q auf $q+1$. Man nehme als bereits bewiesen an

$$\Phi^{(q)}(p^\alpha; n) = p^{\alpha(q-1)} \frac{(p-1)\{(p-1)^{q-1} - (-1)^{q-1}\}}{p^q} \quad \text{für} \quad n \equiv 0 \pmod{p},$$

$$\Phi^{(q)}(p^\alpha; n) = p^{\alpha(q-1)} \frac{(p-1)^q - (-1)^q}{p^q} \quad \text{für} \quad n \not\equiv 0 \pmod{p}.$$

Man erhält nun stets ein und nur ein brauchbares System $r_1, r_2, \ldots, r_q, r_{q+1}$, wenn $r_1 + r_2 + \cdots + r_q \not\equiv n \pmod{p}$ ist.

1. Es sei $n \equiv 0 \pmod{p}$. Dann ist

$$\Phi^{(q+1)}(p^\alpha; n) = \sum_{\substack{\nu=1 \\ (\nu,p)=1}}^{p^\alpha - 1} \Phi^{(q)}(p^\alpha; \nu) = \varphi(p^\alpha)\,\Phi^{(q)}(p^\alpha; 1)$$

$$= p^{\alpha-1}(p-1)\,p^{\alpha(q-1)} \frac{(p-1)^q - (-1)^q}{p^q} = p^{\alpha q} \frac{(p-1)\{(p-1)^q - (-1)^q\}}{p^{q+1}}.$$

2. Ist $n \not\equiv 0 \pmod{p}$, so wird

$$\Phi^{(q+1)}(p^\alpha; n) = \sum_{\substack{\nu=1 \\ \nu \not\equiv n \pmod{p}}}^{p^\alpha} \Phi^{(q)}(p^\alpha; \nu) = (p^\alpha - 2p^{\alpha-1})\,\Phi^{(q)}(p^\alpha; 1) + p^{\alpha-1}\Phi^{(q)}(p^\alpha; p)$$

$$= p^{\alpha-1}\left\{(p-2)p^{\alpha(q-1)} \frac{(p-1)^q - (-1)^q}{p^q} + p^{\alpha(q-1)} \frac{(p-1)\{(p-1)^{q-1} - (-1)^{q-1}\}}{p^q}\right\}$$

$$= p^{\alpha q} \frac{(p-1)^{q+1} - (-1)^{q+1}}{p^{q+1}}.$$

A. Brauer.

(Eingegangen am 21. 3. 26.)

P-3

Lösung der Aufgabe 31. (Dieser Jahresbericht Bd. 34, S. *158.*) Die Aufgabe lautet:

Es seien wieder M und n ganze positive Zahlen; $\mu(x)$ sei das Möbiussche Symbol. Dann gilt

$$\varphi(M) \sum_{\substack{d/M \\ (d,n)=1}} \frac{d}{\varphi(d)}\,\mu\left(\frac{M}{d}\right) = \mu(M) \sum_{\delta/(M,n)} \delta\,\mu\left(\frac{M}{\delta}\right),$$

worin über d bzw. δ zu summieren ist. RADEMACHER.

Lösung. $\frac{x}{\varphi(x)} = \prod_{p/x} \frac{1}{1-\frac{1}{p}}$ ist distributiv.[1]) Man setze für festes n

$$\psi(x) = \frac{x}{\varphi(x)} \quad \text{für} \quad (x, n) = 1,$$

$$\psi(x) = 0 \quad \text{für} \quad (x, n) \neq 1;$$

$$\chi(x) = x \quad \text{für} \quad (x, n) = x,$$

$$\chi(x) = 0 \quad \text{für} \quad (x, n) \neq x.$$

Nun sind $\psi(x)$ und $\chi(x)$ distributiv. Nach einem bekannten Satz[2]) sind dann die Summen auf beiden Seiten der zu beweisenden Formel als Funktionen von M distributiv. Daher genügt es, die Behauptung für Primzahlpotenzen $M = p^\alpha$ zu beweisen.

1. $n \not\equiv 0 \pmod p$

A. $\alpha \geqq 2$ $\quad \varphi(p^\alpha)\left\{\frac{p^\alpha}{\varphi(p^\alpha)} - \frac{p^{\alpha-1}}{\varphi(p^{\alpha-1})}\right\} = 0;$

B. $\alpha = 1$ $\quad (p-1)\left(\frac{p}{p-1} - 1\right) = (-1)(-1).$

2. $n \equiv 0 \pmod p$

A. $\alpha \geqq 2$ $\quad 0 = 0;$

B. $\alpha = 1$ $\quad (p-1)(-1) = (-1)(p-1).$

(Eingegangen am 21. 3. 26.) A. BRAUER.

1) Eine zahlentheoretische Funktion $f(n)$ heißt distributiv oder multiplikativ, wenn $f(a) \cdot f(b) = f(ab)$ für $(a, b) = 1$ ist.

2) Vgl. z. B. G. Pólya und G. Szegö, Aufgaben und Lehrsätze aus der Analysis II, Absch. VIII, Aufg. 59, S. 128 und 334 (Berlin 1925, Springer).

P-4

Lösung der Aufgabe 32. (Dieser Jahresbericht Bd. 34, S. *158.*) Die Aufgabe lautet:

Diese Formel läßt noch folgende Verallgemeinerung zu: Außer M und n seien noch der Teiler M_1 von M und eine zu M_1 teilerfremde Zahl a gegeben. Dann ist für $\left(\frac{M}{M_1}, M_1\right) = 1$

$$\varphi(M) \sum_{\substack{d/M \\ (d,n)=1 \\ (d, M_1)/n-a}} \frac{d}{\varphi(v)} \mu\left(\frac{M}{d}\right) = \mu\left(\frac{M}{M_1}\right) \sum_{\delta/\left(\frac{M}{M_1}, n\right)} \delta \mu\left(\frac{M}{M_1 \delta}\right) \sum_{\delta_1/(M_1, n-a)} \delta_1 \mu\left(\frac{M_1}{\delta_1}\right);$$

dagegen ist für $\left(\frac{M}{M_1}, M_1\right) > 1$

$$\varphi(M) \sum_{\substack{d/M \\ (d,n)=1 \\ (d, M_1)/n-a}} \frac{d}{\varphi(v)} \mu\left(\frac{M}{d}\right) = 0;$$

dabei soll in beiden Formeln die Variable v das kleinste gemeinsame Vielfache von d und M_1, also $v = \frac{d\,M_1}{(d, M_1)}$ bedeuten, und d bzw. δ und δ_1 sind die Summationsvariabeln. Für $M_1 = 1$ geht hieraus die Behauptung der Aufg. 31 hervor.

RADEMACHER.

Lösung. Ist $(M, M') = 1$, M_1'/M', so folgt aus d/M, d'/M' und $(d, M_1)/n - a$, $(d', M_1')/n - a$ auch $(dd', M_1 M_1')/n - a$. Setzt man $dd' = d^*$, $v' = \frac{d'\,M_1'}{(d', M_1')}$, $v^* = \frac{d\,d'\,M_1\,M_1'}{(d\,d', M_1\,M_1')}$, so ist $(v, v') = 1$, $v\,v' = v^*$ für jedes d und d'. Also ist

$$\varphi(M) \sum_{\substack{d/M \\ (d,n)=1 \\ (d,M_1)/n-a}} \frac{d}{\varphi(v)}\,\mu\left(\frac{M}{d}\right) \cdot \varphi(M') \sum_{\substack{d'/M' \\ (d',n)=1 \\ (d',M_1')/n-a}} \frac{d'}{\varphi(v')}\,\mu\left(\frac{M'}{d'}\right)$$

$$= \varphi(MM') \sum_{\substack{d/M \\ (d,n)=1 \\ (d,M_1)/n-a}} \sum_{\substack{d'/M' \\ (d',n)=1 \\ (d',M_1')/n-a}} \frac{dd'}{\varphi(vv')}\,\mu\left(\frac{MM'}{dd'}\right)$$

$$= \varphi(MM') \sum_{\substack{d^*/MM' \\ (d^*,n)=1 \\ (d^*,M_1M_1')/n-a}} \frac{d^*}{\varphi(v^*)}\,\mu\left(\frac{MM'}{d^*}\right).$$

Da nach Lösung 31 außerdem $\mu\left(\frac{M}{M_1}\right) \sum_{\delta/\left(\frac{M}{M_1}, n\right)} \delta\,\mu\left(\frac{M}{M_1\,\delta}\right)$ eine distributive Funktion von $\frac{M}{M_1}$ und $\sum_{\delta_1/(M_1, n-a)} \delta_1\,\mu\left(\frac{M_1}{\delta_1}\right)$ eine distributive Funktion von M_1, ferner M_1/M und $\frac{M}{M_1}\Big/M$ ist, genügt es, die Behauptung für Primzahlpotenzen $M = p^\alpha$ zu beweisen.

I. Ist $\left(\frac{M}{M_1}, M_1\right) = 1$, so muß entweder $M_1 = 1$ oder $M_1 = p^\alpha$ sein. Im ersten Fall erhält man die Aufgabe 31; im zweiten Fall ist

$$v = \frac{d\,p^\alpha}{(d, p^\alpha)} = p^\alpha = M.$$

Daher lautet die Behauptung

$$\sum_{\substack{d/p^\alpha \\ (d,n)=1 \\ (d,p^\alpha)/n-a}} d\,\mu\left(\frac{p^\alpha}{d}\right) = \sum_{\delta_1/(p^\alpha, n-a)} \delta_1\,\mu\left(\frac{p^\alpha}{\delta_1}\right),$$

da die übrigen Faktoren auf der rechten Seite gleich 1 werden. Es genügt daher, zu zeigen, daß d und δ_1 dieselben Werte durchlaufen. Für d kommen die und nur die Teiler von p^α in Betracht, für die $d/n - a$ ist; dann ist von selbst $(d, n) = 1$, da $(a, p^\alpha) = 1$ ist. Für δ_1 sind aber ebenfalls die und nur die Teiler von p^α zu nehmen, für die $\delta_1/n - a$ ist. Daher ist die Behauptung für $\left(\frac{M}{M_1}, M_1\right) = 1$ bewiesen.

II. Ist $\left(\frac{M}{M_1}, M_1\right) > 1$, so muß $M_1 \neq 1$ und $M_1 \neq p^\alpha$, ferner $\alpha > 1$ sein. In der zu betrachtenden Summe treten überhaupt nur Glieder auf, wenn $n \not\equiv 0 \pmod{p}$ und wenn $(p^{\alpha-1}, M_1)/n - a$ ist; dann ist aber auch $(p^\alpha, M_1)/n - a$, da $M_1 \neq p^\alpha$ ist. Man erhält

$$\varphi(p^\alpha)\left\{\frac{p^{\alpha-1}}{\varphi(p^{\alpha-1})}\mu(p)+\frac{p^\alpha}{\varphi(p^\alpha)}\mu(1)\right\}=0.$$

(Eingegangen am 21. 3. 26.) A. BRAUER.

P-5

98. H. Rademacher

Problem: Number of lattice points in a tetrahedron

Let $\{a_j\}$, $j = 1,\ldots m$ be positive coprime natural numbers and let $N_m(a_j)$ be the number of solutions of

$$0 < \sum_{j=1}^{m} \frac{x_j}{a_j} < 1, \qquad x_j \geqq 0$$

in non-negative integers x_j. Conjecture:

$$(*) \qquad N_m(a_j) \equiv \frac{1}{2^{m-1}} \Pi(a_j + 1) \pmod 2$$

For $m = 1,2$ N_m is explicitly known and trivial. For $m = 3$ there is a formula by Mordell (Journal of the Indian Math. Soc. ...) which brings N_3 in connection with the Dedekind sums, and (*) is correct in this case (Rademacher, Studies in Memory of R. v.Mises). Case $m = 4$ has also a connection with Dedekind sums, after Mordell loc. cit., but a proof of (*) would require a more precise knowledge of congruence properties of $s(h,k)$. If $a_2 = 1$ the case $m = 4$ reduces to $m = 3$. Also some numerical examples make the formula (*) plausible for $m = 4$.

P-6

99. H. Rademacher

Problem If we write $s(\frac{h}{k})$ instead of $s(h,k)$ then the Dedekind sum becomes a function of a rational variable r.

Question Is it true that the points $(r, s(r))$ lie everywhere dense in the r,s plane?

P-7

100. H. Rademacher

If k is a positive integer and h is an integer relatively prime to k, put

$$s(h,k) = \sum_{\mu \bmod k} ((\frac{\mu}{k}))((\frac{h\mu}{k})) ,$$

where

$$((x)) = x - [x] - \frac{1}{2} \quad \text{if } x \text{ is a non-integral real number,}$$

$$((x)) = 0 \quad \text{if } x \text{ is an integer.}$$

If h_1/k_1 and h_2/k_2 are adjacent fractions in the Farey series of some order and if $s(h_1,k_1)$ and $s(h_2,k_2)$ are both positive, is it necessarily true that

$$s(h_1 + h_2, k_1 + k_2) \geqq 0 \;?$$

(See Math. Zeit. 63(1956), 445-463.)

S-1

184. Es seien a und n beliebige ganze Zahlen, $n \geqq 1$. Dann ist die Kongruenz

$$x_1^n + x_2^n + \cdots + x_n^n \equiv a \pmod{p}$$

für alle Primzahlen p lösbar.

Uppsala. TRYGVE NAGELL.

(Eingegangen am 9. 1. 1934.)

Zweite Lösung.

Es läßt sich sogar etwas mehr beweisen, nämlich, daß schon durch $x_1^n + \cdots + x_d^n$ mit $d = (n, p-1)$ alle Restklassen mod p dargestellt werden. Bekanntlich ist die Anzahl der durch x_1^n mod p dargestellten Restklassen gleich $\frac{p-1}{d} + 1$. Für $d = 1$ ist also alles erledigt. Nun hat neuerdings Herr H. Davenport ("On the addition of residue classes", Journal of the London Math. Soc., Vol. 10, part 1, p. 30—32 [1935]) den folgenden Satz bewiesen: Sind $\alpha_1, \ldots \alpha_k$ k verschiedene Restklassen mod p und $\beta_1, \ldots \beta_l$ l unter sich verschiedene Restklassen mod p (die α_i brauchen von den β_j nicht verschieden zu sein), so gilt für die Anzahl m der durch $\alpha_i + \beta_j$ dargestellten Restklassen

$$m \geqq \mathrm{Min}\,(p, k + l - 1).$$

Sind für unsere Anwendung dieses Satzes die α die durch x_1^n und die β die durch x_2^n dargestellten Restklassen, so folgt, daß $x_1^n + x_2^n$ mindestens $2\frac{p-1}{d} + 1$ Restklassen durchläuft, womit auch der Fall $d = 2$ erledigt ist. Durch wiederholte Anwendung des Davenportschen Satzes folgt, daß $x_1^n + \cdots + x_d^n$ mindestens $d\frac{p-1}{d} + 1 = p$ verschiedene Restklassen mod p, also alle, durchläuft.

Philadelphia (USA.). HANS RADEMACHER.

(Eingegangen am 20. 6. 1935.)

Dieselbe Lösung sandte Herr E. Trost (Zürich) ein.

Wie Herr H. Hasse (Göttingen) bemerkt, ist die Behauptung der Aufgabe ein Spezialfall von Satz 300 in E. Landau, Vorlesungen über Zahlentheorie, Bd. 1, S. 289. Eine im wesentlichen auf den hier verwendeten Methoden beruhende Lösung sandte Herr A. Brauer (Berlin) ein. In diese Gruppe gehört auch die von dem Aufgabensteller selbst eingesandte Lösung.

Notes to Problems and Solutions

P1. This problem and its solutions were reviewed in the JF, vol. 52 (1926), p. 233.

P2, 3, 4. These problems and solutions have become rightly famous under the name of the "Rademacher-Brauer formula." The most frequent form in which the formula is quoted is that corresponding to problem P3. All three problems and solutions were reviewed in the JF, vol. 52 (1926), p. 139.

P5, 6, 7. It seems that these problems are still unsolved.

S1. The problem appeared in the *Jahresbericht der Deutschen Mathematiker Vereinigung,* vol. 45 (1935), p. 21. The problem and solution were reviewed in the JF, vol. 62 (1936), p. 131.

Appendix 1
Writings of Hans Rademacher That Are Not Included in the Present Collection

[1] *Von Zahlen und Figuren: Proben Mathematischen Denkens fur Liebhaber der Mathematik.* With O. Toeplitz. Julius Springer, Berlin, 1930. This book has been reviewed in the JF, vol. 56 (1930), pp. 62-64. A second, definitive edition came out in 1933 and has been reviewed in the JF, vol. 59 (1933), p. 69, and the Z, vol. 6 (1933), p. 146. A recent reprinting (1968) has been reviewed in the MR, vol. 40 (1970), and in the Z, vol. 169 (1969), p. 290.

An English translation by H. S. Zuckerman, based on the second German edition, and with two chapters added by H. S. Zuckerman, was published under the title *The Enjoyment of Mathematics* by the Princeton University Press in 1957. This translation has been reviewed in the Z, vol. 78 (1959), and in the MR, vol. 18 (1957), p. 454.

A translation into Russian, by V. I. Kontov, appeared under the editorship of I. M. Yaglom in 1962, under the title *Cisla i Figury.* It has been reviewed in the Z, vol. 111 (1965), p. 241.

A translation by Orhan S. Icen into Turkish, under the literal translation of the original German title, was published in 1965 as Publication No. 25 of the Turkish Mathematical Society. It has been reviewed in MR, vol. 37 (1969), #5065.

For a partial translation into Hebrew, see [3].

A Japanese translation (by Sabaro Yamasaki) was brought out in 1952, by Sogen-sha, Tokyo.

A Polish translation (by Abraham Goetz) was published by Pánstwowe Wydawnictwo Naukawe, Warsaw, 1956.

A French translation (from the English Translation) under the title *Plaisir des mathématiques* was published by Dunod, Paris, 1957.

A Bulgarian translation (by Maria Kr'steva) was brought out by the State Publishing House Nauka I Izkustoo, Sofia, in 1969.

A translation into Spanish (by Eva Rodriguez Halffter and Manuel Gregori Susa) was published by Alianza Editorial, Madrid, in 1970.

[2] *Vorlesungen über die Theorie der Polyeder unter Einschluss der Elemente der Topologie. Aus Ernst Steinitz' Nachlass herausgegeben und ergänzt von H. Rademacher.* Die Grundlehren der Mathematischen Wissenschaften in Einzeldarstellungen, vol. 41, Julius Springer, Berlin, 1934. This book has been reviewed in the Z, vol. 9 (1934), pp. 365-367.

[3] *The Necessity of the Compass in Elementary Geometric Constructions.* With O. Toeplitz. This is a translation into Hebrew (by Dov Jarden) of Chapter 26 of the book *Von Zahlen und Figuren* (see [1]) that was published in *Riveon Lematematika,* vol. 1 (1946), pp. 14-19. It was reviewed in the MR, vol. 8 (1947), p. 218.

[4] *Lecture Notes on elementary mathematics from an advanced viewpoint,* Eugene, Oregon, 1954.

[5] *Lecture Notes on analytic additive number theory,* Eugene, Oregon, 1954.

[6] *Lectures on Analytic Number Theory* (Notes by K. Balangangadharan and V. Venugopal Rao), Tata Institute for Fundamental Research, Bombay, 1954-55.

[7] *Lectures on Elementary Number Theory.* Blaisdell Publishing Company, New York, 1964. This book is based on the Philips Lectures delivered by the author at Haverford College in 1959-1960 and on a lecture delivered at Dartmouth College in 1960. The book has been reviewed in the MR, vol. 30 (1965), #1079.

Posthumous Manuscripts

[8] An almost complete posthumous manuscript of a book on analytic number theory exists. It is expected that this will be published, with a minimum of completion and editing, by three former students of the author (E. Grosswald, J. Lehner, and M. Newman).

[9] A detailed sketch for the Hedrik Lectures of 1963 on Dedekind sums forms the basis of a book (by Rademacher and Grosswald), which has appeared as Carus Monograph No. 16 (1972) of the Mathematical Association of America.

Appendix 2
Dissertations Directed by Hans Rademacher

[1] Theodor Estermann. Ueber Caratheodory's und Minkowski's Verallgemeinerung des Längenbegriffes (1925).

[2] Wolfgang Cramer. Die Reziprozitätsformel für Gauss'sche Summen in reell-quadratischen Zahlkörpern (1932).

[3] Otto Schulz. Ueber quaternäre Gruppen krystallographischer Bedeutung (1933).

[4] Käthe Silberberg. Ueber die Anzahl der Darstellungen ganzer, totalpositiver Zahlen eines beliebigen Zahlkörpers als Summen von Quadratzahlen und totalpositiven Primzahlen (1934).

[5] Albert L. Whiteman. Additive Prime Number Theory (1940).

[6] Joseph Lehner. A Partition Function connected with the Modulus Five (1942).

[7] Lowell Schoenfeld. A Transformation Formula in the Theory of Partitions (1944).

[8] Ruth E. Goodman. On the Bloch-Landau Constant for Schlicht Functions (1944).

[9] John Livingood. A Partition Function with the Prime Modulus $p > 3$ (1945).

[10] Paul T. Bateman. On the Representation of a Number as a Sum of 3 Squares (1946).

[11] Jean B. Walton. Theta Series in the Gaussian Field (1948).

[12] Emil Grosswald. On the Structure of some Subgroups of the Modular Group (1950).

[13] Saul Rosen. Modular Transformations of certain Series (1950).

[14] Leila A. Dragonette. Asymptotic Formulae for the Mock-Theta Series of Ramanujan (1951).

[15] Albert Schild. On a Problem in Conformal Mapping of Schlicht Functions (1951).

[16] Jean M. Calloway. On the Discriminant of Arbitrary Algebraic Number Fields (1952).

[17] Morris Newman. A Structure Theorem for certain Modular Subgroups, with Applications to the Construction of Modular Identities (1952).

[18] Frederick A. Homann. On some Integrals of Analytic Additive Number Theory (1959).

[19] William G. Spohn. Midpoint Regions and Simultaneous Diophantine Approximations (1962).

[20] George E. Andrews. On the theorems of Watson and Dragonette for Ramanujan's mock theta functions (1964).

Contents of Volume I

Papers

[28] Eine arithmetische Summenformel. *Monatshefte für Mathematik und Physik,* vol. 39 (1932), pp. 221-228 678